Statistical Mechanics of Liquids and Solutions

Intermolecular Forces, Structure and Surface Interactions

Volume 1

Statistical Mechanics of Liquids and Solutions

Intermolecular Forces, Structure and Surface Interactions

Volume 1

Roland Kjellander

CRC Press
Taylor & Francis Group
Boca Raton London New York

CRC Press is an imprint of the
Taylor & Francis Group, an **informa** business

CRC Press
Taylor & Francis Group
6000 Broken Sound Parkway NW, Suite 300
Boca Raton, FL 33487-2742

First issued in paperback 2020

© 2020 by Taylor & Francis Group, LLC
CRC Press is an imprint of Taylor & Francis Group, an Informa business

No claim to original U.S. Government works

ISBN-13: 978-1-4822-4401-4 (hbk)
ISBN-13: 978-0-367-47790-5 (pbk)

Visit the Taylor & Francis Web site at
http://www.taylorandfrancis.com

and the CRC Press Web site at
http://www.crcpress.com

Contents

PART II Fluid Structure and Interparticle Interactions

CHAPTER 5 ■ Interaction Potentials and Distribution Functions 165

Preface

THIS BOOK IS THE FIRST VOLUME of two on equilibrium statistical mechanics of liquids and solutions with focus on intermolecular forces, surface forces and the structure of fluids. A major theme is the intimate relationship between forces in a fluid and the fluid structure – a relationship that is paramount for the understanding of the subject of interactions in dense fluids. The book includes a comprehensive introduction to statistical mechanics, where the perspective is quantum mechanical, but the rest of the treatise deals with classical statistical mechanics of fluids. This first volume is limited to simple liquids in bulk, near surfaces and in confinement, but it includes interactions between particles of any size and shape immersed in such liquids, including surface interactions. In the second volume, the theory of liquids consisting of nonspherical particles like molecular liquids is a major topic.

The subject of interparticle interactions in liquids and solutions is immensely important in a large number of fields like solution chemistry, electrochemistry, soft matter physics, surface and colloid science, molecular biosciences and in many applications. Such interactions have traditionally been treated by using rather simple theories, but with the development of more sophisticated theoretical methods and, in particular, with the increasing power of computers, the realism of the theoretical treatments has been greatly improved during the last several decades.

A major tool in the theoretical investigation of liquids and solutions is computer simulations. There also exist various numerical methods based on approximations of considerable sophistication and accuracy. The use of such methods in parallel to simulations is complementary to the latter and their further development is important for future advances in the study of interactions in dense fluid media. This kind of methods differ from simulations in the respect that they rely more heavily on an understanding of fundamental statistical mechanics. Simulations are, however, more simple to perform and are more commonly used.

With the increased reliance on computer simulations for the study of molecular phenomena in liquids and solutions, there is, however, a risk of using such techniques as "black boxes" where an understanding of the statistical mechanical foundations is not prioritized. The results from simulations are numbers and without a sufficient theoretical framework, the relevance and reliability of these numbers can be hard to evaluate and may be overestimated. Furthermore, the links between various entities and their connection with the underlying phenomena may be out of sight. A deep statistical mechanical understanding of these matters can be of great advantage in the decisions of what calculations to do and how they can be interpreted. Thus, education in fundamental statistical mechanics for liquids and solutions is well motivated for all workers in the field.

For experimentalists, a knowledge of statistical mechanical theories of interparticle interactions beyond the traditional, simple ones is of great value for the design and interpretation of the increasingly more sophisticated methods available, like direct measurements of interparticle interactions and many other techniques where such interactions are paramount for the phenomena studied. In many systems with strong coupling, for example, concentrated electrolyte solutions and ionic liquids, and in a lot of other cases where many-body correlations are important, the traditional theories are clearly insufficient. The correct interpretation of the experimental results then must rely on more complete liquid state theory.

Insights into the underlying mechanisms for interactions in liquids constitute an essential ingredient in the analysis of many aspects of nature. Statistical mechanics of fluids should therefore be a part of the curriculum for a wider range of students than often is the case today. This is also important for the future development of the subject itself.

For these and many other reasons, a textbook like the present treatise has an important role to play in the education of chemists, physicists and students in neighboring fields. The book is designed in such a manner that it starts from the beginning of statistical mechanics, but it introduces the subject in a way that can be of substantial value also for students who have studied the subject before. The main part of the book deals with the theoretical treatment of fluids, where statistical mechanics of fluids is used to treat fluid structure and interactions in bulk phase and in inhomogeneous fluids like confined ones. A more comprehensive presentation of the contents of the book can be found in the following section "Overview of contents."

The inspiration to write the present treatise comes from my experiences from a course in statistical thermodynamics at the University of Gothenburg that I and Sture Nordholm have developed and that we ran for more than twenty years with several hundreds of participants from the entire country of Sweden and from abroad. The material that I wrote for this course, which includes basic statistical mechanics and an introduction to the theory of fluids, is to a large extent integrated in the current treatise. Together with Johan Bergenholtz, we also developed an advanced course in liquid state theory at the University of Gothenburg, and my material from that course is also integrated to a large extent in the two books. Major parts of the books are, however, newly written.

The feedback that I have received from the students of these courses has made me realize that my teaching approach to basic statistical mechanics and to the theory of fluids would be suitable for a book on the subject. The words from a colleague at another Swedish university have left a lasting impression on me in this regard. He said that when his PhD students had taken our course in statistical thermodynamics, they had virtually no remaining problems to understand and use statistical mechanics. This was certainly an exaggeration, but his words inspired me to continue in my effort to bridge and thereby overcome the conceptual problems that students have with statistical physics, including classical thermodynamics. A colleague at yet another university told me that some students thought that the course and its approach to the subject was one of the best experiences they had had as academic students.

It is indeed very rewarding as a teacher to inspire students. Their impressions communicated via course evaluations and by other means have made lasting marks on the teaching approach.

I am therefore very much indebted to our students for the feedback that has helped in the development of the contents of the current treatise. Both the questions and comments of the students during the lectures and after the courses, including ideas of what to improve and clarify, have been absolutely essential for me in the evolution of the material. Therefore, I would like to thank the students for their input in the process. I am also indebted to my present and former colleagues at the University of Gothenburg, in particular Kim Nygård for his feedback on parts of the present book. Finally, I want to express my deep gratitude to Gunnar Numeus. Without your support and understanding, this book would never have been written.

Overview of Contents

THIS BOOK STARTS FROM THE BEGINNING in *equilibrium statistical mechanics* and assumes only prior knowledge in basic classical thermodynamics. The subject is presented in a manner that can be of substantial value also for those who have studied statistical mechanics before. This is done by explaining the subject in a way that avoids, or at least minimizes, several of the common conceptual difficulties that many students encounter. A major theme of this introduction to statistical mechanics is to give, in a transparent manner, the link between the microscopic molecular world and macroscopic entities and phenomena, in particular entropy, temperature (which is not very simple!), free energy, reversible and irreversible changes. Simultaneously, the standard statistical mechanical machinery is introduced in a smooth manner. This machinery is then used throughout the book.

For conceptual reasons, the foundation of statistical mechanics is set up with a quantum mechanical perspective in **Chapters 1 and 2**, but the presentation leads over to statistical mechanics based on classical mechanics, given in **Chapter 3**, and it shows relationships between the quantum and the classical versions. The treatment of liquids and solutions in this treatise is entirely based on classical statistical mechanics, which is sufficient since we will only consider systems where quantum effects do not play any important role or at least can be incorporated in an approximate manner in the classical treatment.

The introductory part of the book terminates with some illustrative examples in **Chapter 4** from some *classical theories of gases and liquid mixtures*, including regular solution theory and the Flory-Huggins theory of polymer solutions and polymer mixtures. The reason for the inclusion of such theories in the book is that they are used to illustrate several important concepts for liquids and solutions in an explicit and direct fashion that is helpful for the understanding.

Thereafter, the treatment of *fluid structure and interparticle interactions* in fluids commences. In order to get a perspective on this, let us consider the interaction between, say, two particles in a liquid; they could be of the kind that constitutes the liquid or other kinds of particles immersed in the liquid. The forces that act on each of the two particles originate from the other particle in the pair and from all particles in the surroundings. The contribution from the latter depends on where they are located on average, that is, it depends on the structure of the surrounding liquid. This structure depends, in turn, on the locations of the two particles. There is accordingly a complex interplay between many particles that determine the forces that act on each pair of particles as a function of their separation. Therefore, structure and interactions have to be treated at the same time. Since interparticle interactions in fluids and solutions determine various properties of the system that are essential in both fundamental science and many applications, it is important to unravel the mechanisms in action.

In the present first volume of the book, the fluids are assumed to consist of spherically symmetric particles (so-called simple fluids), but the treatment includes interactions between particles of any size and shape immersed in such fluids, including forces between macroscopic surfaces. Statistical mechanics for fluids consisting of nonspherical particles is postponed to the second volume, which includes a much larger part of liquid state theory for both spherical and nonspherical particles than the present volume and uses more advanced statistical mechanical concepts[1] and techniques. One may say that the present volume has master-course contents, while the second volume lies on a higher level.

In **Chapter 5**, the central concepts of distribution functions and correlation functions of various orders, like density distribution profiles and pair distributions, are introduced. Distribution functions are first defined in an intuitive manner and then defined formally from probability functions. Both homogeneous bulk fluids and inhomogeneous fluids are included in the treatment. The relationships of these functions to interactional forces between the particles and to the chemical potential are outlined. This is done in a straightforward and transparent manner by analyzing force balances in the fluid and by linking such balances to equilibrium conditions described by free energy arguments. The links between distribution functions and basic thermodynamical entities are given, providing a basis for what follows. The microscopic meaning of the thermodynamical properties is thereby highlighted. Relationships between distribution functions of various orders are explained and simple mean field approximations are introduced. The chapter finishes with a section on the basics of computer simulations. In line with the main theme of this treatise, there is is a special emphasis on how to use simulations to determine distribution functions of various orders in both homogeneous and inhomogeneous fluids.

Chapter 6 gives a comprehensive treatment of simple bulk electrolytes. A study of electrolytes is important for many reasons, not least because of their presence in many applications, but also as they provide very suitable examples on how to use the statistical mechanical machinery for fluids. The chapter starts in Section 6.1 with the traditional mean field theory for electrolytes, the Poisson-Boltzmann (PB) approximation, which still is extensively used in various applications. The presentation includes the Debye-Hückel theory – the classical theory for bulk electrolyte solutions. In addition to these traditional aspects of the PB approximation, the treatment of this approximation includes the concept of effective charge, as defined in a strict manner, and recent findings regarding the decay behavior of the screened electrostatic potential from nonspherical particles immersed in simple electrolytes, including the electrostatic interactions between such particles. There is a fundamental difference between electrostatics in electrolytes, on the one hand, and in polar media or in vacuum, on the other, as regards the direction dependence of the potential from a nonspherical particle. This leads to the introduction of multipolar effective entities that are different from the usual multipolar moments of nonspherical charge distributions. Section 6.1 ends

---

[1] For instance, the introduction and use of direct correlation functions of various orders are placed in the second volume.

with a glimpse beyond the PB approximation and it is shown in a simple manner that if all ions in an electrolyte are treated on the same basis (which is not done in the PB approximation), there appears a whole range of phenomena like multiple decay lengths and oscillatory electrostatic interactions that cannot be treated by the PB theory.

In Section 6.2, it is shown how several of the findings in the PB theory, in particular regarding interparticle interactions, screening, effective charges and multipolar effective entities, can be generalized in a quite straightforward manner. Many of these findings are thereby found to be valid in the general case of exact statistical mechanics for bulk electrolytes under a wide variety of conditions. This makes it possible to understand quite complex properties of electrolytes in a manner that is similar to what was done in the PB approximation despite that the theory here is general. The nonlocal nature of electrostatics in electrolytes is explored and it is shown how both monotonically decaying electrostatic interactions with multiple decay lengths and oscillatory electrostatic interactions can be handled in one and the same formalism. Concrete explicit examples of such behaviors in ionic fluids of various densities, electrolyte solutions and molten salts are given and it is shown how these behaviors can be handled and analyzed.

Inhomogeneous and confined simple fluids, like hard sphere fluids, atomic fluids and simple electrolytes near or between surfaces, are treated in **Chapter 7**. Like for the homogeneous case, electrolyte systems provide very suitable examples on how the statistical mechanical machinery can be used for fluids: here inhomogeneous ones. The chapter therefore starts in Section 7.1 with a comprehensive treatment of electrolytes in contact with charged surfaces, the so-called diffuse electric double-layer systems. Electric double-layers occur, for instance, in clays, near colloidal particles, electrodes and surfaces of macroparticles in contact with electrolytes. The PB approximation for electric double-layers, also called the Gouy-Chapman theory, is included and apart from several traditional results of that theory, the presentation also contains a treatment of effective charges and newer findings regarding the decay of interactions between surfaces and nonspherical particles like macroions. Thereafter, the latter interactions are treated using the same general formalism as for bulk electrolytes, which is applicable in the general case of exact statistical mechanics under a wide variety of conditions. Thereby, both monotonically decaying and oscillatory decaying electrostatic interactions are included. Furthermore, the effects of ion-ion correlations, which are completely neglected in the PB approximation for double layers, are treated and surface polarization (image charge interactions) and dispersion interactions in electric double-layers are included. Several explicit, concrete examples are thereby given where such effects are important.

Section 7.2 is devoted to the structure of inhomogeneous fluids on the pair-correlation level and the relationship between this structure and surface-particle interaction forces, which determine the density distribution profiles near surfaces. Thereby, the interaction mechanisms behind the appearances of density profiles in inhomogeneous Lennard-Jones fluids, hard sphere fluids and simple electrolytes are treated. Explicit, graphical representations of the pair structure in the various inhomogeneous systems are given.

Surface forces are treated in **Chapter 8**. Such forces are fundamental for the understanding of, for example, colloidal systems, macromolecular solutions and lipid aggregates of various kinds. Also, in this case, electrolyte systems provide very good examples of the utilization of the statistical mechanical machinery. The chapter includes a comprehensive treatment of electric double-layer interactions between charged surfaces in electrolytes both in the PB approximation and the general case. The fundamental link between surface forces and pair correlations in inhomogeneous fluids is treated in some detail and examples include structural surface forces in simple fluids and electrostatic interactions in electrolytes. The intricate couplings between van der Waals forces, including dispersion forces, and the electrostatic interactions, including image charge interactions, are explained and several reasons for important deviations from classical theories of surface forces are explored.

Text marked with ★ is more advanced than the rest.

A *list of symbols* used in the book is included at the end.

Exercises are integrated in the text throughout the book and are marked with {**Exercise:** ... ...}.

Author

Roland Kjellander earned a master's degree in chemical engineering, a Ph.D. in physical chemistry, and the title of docent in physical chemistry from the Royal Institute of Technology, Stockholm, Sweden. He is currently a professor emeritus of physical chemistry in the Department of Chemistry and Molecular Biology at the University of Gothenburg, Sweden. His previous appointments include roles in various academic and research capacities at the University of Gothenburg, Sweden; Australian National University, Canberra; Royal Institute of Technology, Stockholm, Sweden; Massachusetts Institute of Technology, Cambridge, USA; and Harvard Medical School, Boston, USA. He was awarded the 2004 Pedagogical Prize from the University of Gothenburg, Sweden, and the 2007 Norblad-Ekstrand Medal from the Swedish Chemical Society. Professor Kjellander's field of research is statistical mechanics, in particular liquid state theory.

I

Basis of Equilibrium Statistical Mechanics

CHAPTER 1

Introduction

1.1 THE MICROSCOPIC DEFINITIONS OF ENTROPY AND TEMPERATURE

One of the main reasons why many persons experience conceptual difficulties when studying classical thermodynamics is probably that the concept of temperature and its main features are more or less taken for granted. Everybody has, of course, an everyday experience of temperature, but, as a consequence, when introducing the concept of entropy, S, through the second law of thermodynamics, entropy may become mysterious and quite hard to grasp. In actual fact, entropy is simpler and has a more direct connection to basic properties of matter than temperature. This connection is, however, outside the realm of classical thermodynamics; it can be established first in statistical thermodynamics.

To begin with, let us look at some basic properties of temperature. We shall make an operational definition of what we – from our everyday experience – mean by equal and different temperatures of two systems. A **system** is by definition the part of the universe that we are interested in and an **isolated system** cannot exchange either matter or energy with its surroundings. Let us take two isolated macroscopic systems A and B and bring them in thermal contact, as illustrated in Figure 1.1, so energy (denoted U) can flow from one to the other without any work being done and only heat can pass between A and B.[1] If no energy is spontaneously transferred between A and B, they have the *same temperature*; otherwise, the system that receives energy has a *lower temperature* than the other. This constitutes our operational definition of equal and different temperatures.

Why does energy flow between the systems in the latter case, that is, what property of the system describes which has the highest temperature? If the two systems were made of the same material (and have the same density), the answer is simple: the system with higher energy content per unit volume has the highest temperature. But how about systems made of different materials, for example, a block of metal and a container of some gas? Obviously, the gas can have a higher temperature than the metal and still have less energy content per

---

[1] We shall use the term "heat" for the part of the transferred energy that is not work. In this section, all transferred energy is only heat. The two concepts are defined more precisely in Section 2.1.

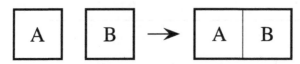

FIGURE 1.1 Two isolated systems A and B are brought into thermal contact with each other, but they remain isolated from the surroundings.

unit volume. Also, different materials of the same density can, in general, have different temperatures, even if they have the same energy per unit volume. Thus, what fundamental property of a system determines which temperature it has?

Before answering this question, let us remind ourselves how classical thermodynamics describes the process above (and we shall here assume that the machinery of this subject is already established). First, we shall assume that the systems have the same absolute temperature T (or rather that they have a small temperature difference ΔT so some energy can flow spontaneously and look at the systems in the limit $\Delta T \to 0$). We let the two systems to be in contact for a short time, so the amount of transferred energy dU is small. From the second law of thermodynamics, the entropy change dS when the amount of heat[2] $d\mathrm{q} \; (= dU$ in this case) is transferred equals

$$dS = \frac{d\mathrm{q}}{T} = \frac{dU}{T}$$

(a reversible change; we assume that dU is so small that T is not changed).

The amount of energy transferred from one system equals that received by the other $dU_A = -dU_B$. Thus,

$$dS_A = \frac{dU_A}{T} = -\frac{dU_B}{T} = -dS_B$$

since the temperature is equal. Thus, $dS_{tot} = dS_A + dS_B = 0$, that is, the total entropy is not changed.

Let us now assume that the systems have different temperatures, $T_A \neq T_B$, and let a very small amount of heat flow spontaneously between A and B so both systems remain in internal equilibrium at unchanged temperatures. The change in energy

$$dU_A = -dU_B > 0 \quad \text{if } T_A < T_B$$

and we have

$$dS_A = \frac{dU_A}{T_A} > 0 \quad \text{and} \quad dS_B = \frac{dU_B}{T_B} < 0.$$

Now, since $T_A < T_B$,

$$\frac{dU_A}{T_A} = -\frac{dU_B}{T_A} > -\frac{dU_B}{T_B}$$

---

[2] We use the symbol q to denote heat (to distinguish it from q, which will denote something else later). Since heat is not a state function, an infinitesimally small amount of heat, $d\mathrm{q}$, is not an exact differential in general, which is denoted by the symbol d.

and thus $dS_A > -dS_B$. We get $dS_{tot} = dS_A + dS_B > 0$, that is, the total entropy increases when the heat flow is spontaneous.

This is all fine but gives very little insight into the concepts of temperature and entropy. To achieve that, we need to look at the problem from a microscopic point of view, and here we need to use some quantum mechanical concepts. We will start with a simple example that illustrates the main points.

1.1.1 A Simple Illustrative Example

We shall first investigate a very simplified but illustrative example. Let system A consist of four "molecules," each of which has a set of energy levels that are equidistant in energy (i.e., the difference in energy between consecutive quantum states are equal).[3] We also assume that each level in a molecule consists of only one quantum state (the level is nondegenerate). The system can, for instance, consist of a set of four harmonic oscillators, which we assume interact so weakly that their energy levels are not perturbed by each other but nevertheless interact sufficiently strongly so they can exchange energy with each other. The seven lowest levels for the four molecules are illustrated in Figure 1.2.

Let us now investigate the system when it contains various amounts of energy; in particular, we shall find out in how many ways the energy can be distributed between the different quantum states of the molecules (the number of ways will be denoted Ω). The zero of the energy scale will be set at the ground state energy and the spacing between the levels will be used as the energy unit.

If the system has energy $U = 0$ (it is in its ground state), there is of course only one way to realize this (we will denote this symbolically as 0000) and we have $\Omega = 1$.

If $U = 1$, then there are four ways: any of the four molecules can be excited to its level 1 (we will denote this as 1000, 0100, 0010 and 0001) and we have $\Omega = 4$, that is, the first excited state of the whole system has a degeneracy of 4.

If $U = 2$, then it is possible for any one molecule to be excited to its level 2 and there are four ways to realize this (2000, 0200, 0020 and 0002). Alternatively, any two molecules can be excited to level 1 and there are 6 ways to do this (1100, 1010, 1001, 0110, 0101 and 0011). Totally, $\Omega = 10$ in this case, that is, there are 10 possibilities when $U = 2$.

FIGURE 1.2 System A.

---

[3] We here tacitly assume that the "molecules" are distinguishable.

Similarly, for $U = 3$, we have the possibilities
3000, 0300, 0030, 0003
2100, 2010, 2001, 1200, 0210, 0201, 1020, 0120, 0021, 1002, 0102, 0012
1110, 1101, 1011, 0111
We get $\Omega = 20$.

For $U = 4$, we have
4000, 0400, 0040, 0004
3100, 3010, 3001, 1300, 0310, 0301, 1030, 0130, 0031, 1003, 0103, 0013
2200, 2020, 2002, 0220, 0202, 0022
2110, 2101, 2011, 1210, 1201, 0211, 1120, 1021, 0121, 1102, 1012, 0112
1111
Here $\Omega = 35$.

For $U = 5$, we have
5000, 0500, 0050, 0005
4100, 4010, 4001, 1400, 0410, 0401, 1040, 0140, 0041, 1004, 0104, 0014
3200, 3020, 3002, 2300, 0320, 0302, 2030, 0230, 0032, 2003, 0203, 0023
3110, 3101, 3011, 1310, 1301, 0311, 1130, 1031, 0131, 1103, 1013, 0113
2210, 2201, 2120, 2021, 2102, 2012, 1220, 0221, 1202, 0212, 1022, 0122
2111, 1211, 1121, 1112
and $\Omega = 56$.

For $U = 6$, we have
6000, 0600, 0060, 0006
5100, 5010, 5001, 1500, 0510, 0501, 1050, 0150, 0051, 1005, 0105, 0015
4200, 4020, 4002, 2400, 0420, 0402, 2040, 0240, 0042, 2004, 0204, 0024
4110, 4101, 4011, 1410, 1401, 0411, 1140, 1041, 0141, 1104, 1014, 0114
3300, 3030, 3003, 0330, 0303, 0033
3210, 3201, 3120, 3021, 3102, 3012, 2310, 2301, 1320, 0321, 1302, 0312,
2130, 2031, 1230, 0231, 1032, 0132, 2103, 2013, 1203, 0213, 1023, 0123
3111, 1311, 1131, 1113
2220, 2202, 2022, 0222
2211, 2121, 2112, 1221, 1212, 1122
and $\Omega = 84$.

Each of these states is called a **microscopic state**, also called **microstate**, of the system (a quantum state of the whole system).

As system B, we shall select a slightly different one: levels 3 and 5 are missing in its "molecules," otherwise the lowest levels are the same as system A. The lowest energy levels for system B are shown in Figure 1.3. These molecules are different from those of A since their quantum states are different.

FIGURE 1.3   System B.

For this system, the energy can be distributed in the following ways:

For $U = 0$, we have
0000
and $\Omega = 1$.

For $U = 1$, we have
1000, 0100, 0010 and 0001
and $\Omega = 4$.

For $U = 2$, we have
2000, 0200, 0020 and 0002
1100, 1010, 1001, 0110, 0101 and 0011
and $\Omega = 10$.

For $U = 3$, system B starts to differ from system A. We have
2100, 2010, 2001, 1200, 0210, 0201, 1020, 0120, 0021, 1002, 0102, 0012
1110, 1101, 1011, 0111.
We get $\Omega = 16$.

For $U = 4$, we have
4000, 0400, 0040, 0004
2200, 2020, 2002, 0220, 0202, 0022
2110, 2101, 2011, 1210, 1201, 0211, 1120, 1021, 0121, 1102, 1012, 0112
1111
and $\Omega = 23$.

For $U = 5$, we have
4100, 4010, 4001, 1400, 0410, 0401, 1040, 0140, 0041, 1004, 0104, 0014
2210, 2201, 2120, 2021, 2102, 2012, 1220, 0221, 1202, 0212, 1022, 0122
2111, 1211, 1121, 1112
and $\Omega = 28$.

For $U = 6$, we have

6000, 0600, 0060, 0006

4200, 4020, 4002, 2400, 0420, 0402, 2040, 0240, 0042, 2004, 0204, 0024

4110, 4101, 4011, 1410, 1401, 0411, 1140, 1041, 0141, 1104, 1014, 0114

2220, 2202, 2022, 0222

2211, 2121, 2112, 1221, 1212, 1122

and $\Omega = 38$.

Table 1.1 summarizes our findings.

Let us now have systems A and B isolated from each other as depicted in Figure 1.4 and let them have an equal amount of energy $U_A = U_B = 3$. From the Table 1.1, we see that the degeneracies of the systems are $U_A = 20$ and $U_B = 16$. The total system composed of A and B has energy $U_{AB} = 6$. To find the degeneracy of the total system, we note that for each microscopic state of system A, system B can be in any of its Ω_B states (since they are independent). Thus, the degeneracy $\Omega_{AB} = \Omega_A \times \Omega_B = 320$.

A basic postulate in statistical mechanics is that all available microscopic states for an isolated system occur equally likely at equilibrium. This means that the probability that system A is in a particular microscopic state is $1/\Omega_A = 1/20 = 0.050$, and that system B is in a particular one of its states is $1/\Omega_B = 1/16 = 0.063$. The probability that system A has, say, any one of its molecules in energy level 3 is four times as large $4 \times 0.050 = 0.200$ since there are four such possibilities [3000, 0300, 0030 and 0003]. The probability to find the combined system AB in a specific microscopic state (i.e., in one particular A-state and one particular B-state simultaneously) is $1/\Omega_{AB} = 1/(\Omega_A \Omega_B) = 0.003$.

If the two systems are brought into thermal contact, Figure 1.5, we still have $U_{AB} = 6$, but the energy can now flow between A and B and it can be freely distributed among all microscopic states of the combined system that satisfy $U_A + U_B = 6$. There is no longer

TABLE 1.1

System A		System B	
U	$\Omega(U)$	U	$\Omega(U)$
0	1	0	1
1	4	1	4
2	10	2	10
3	20	3	16
4	35	4	23
5	56	5	28
6	84	6	38

FIGURE 1.4 Systems A and B are isolated from each other.

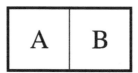

FIGURE 1.5 Systems A and B are in thermal contact with each other, but isolated from the surroundings.

TABLE 1.2

U_A	U_B	$\Omega_A \times \Omega_B$	=	Ω_{AB}
6	0	84×1	=	84
5	1	56×4	=	224
4	2	35×10	=	350
3	3	20×16	=	320
2	4	10×23	=	230
1	5	4×28	=	112
0	6	1×38	=	38
		Sum: Ω_{AB}^{tot}	=	1358

Note: $U_B = U_{AB} - U_A = 6 - U_A$ throughout.

any restriction that $U_A = U_B = 3$, but we can also have, for example, $U_A = 4$ and $U_B = 2$. Table 1.2 gives all possibilities.

From the basic probability postulate mentioned above, it follows that all of these 1358 microscopic states are equally likely. Since, for example, 84 of these 1358 occurrences have an energy $U_A = 6$, it means that the probability that A has energy $U_A = 6$ (and B hence 0) is $84/1358 = 0.062$. In general, the probability $\mathcal{P}_A$ that A has energy U_A is $\mathcal{P}_A(U_A) = \Omega_{AB}(U_A)/\Omega_{AB}^{tot}$. The average energy of A therefore becomes

$$\langle U_A \rangle = \sum_{U_A} U_A \mathcal{P}_A(U_A) = \sum_{U_A} U_A \frac{\Omega_{AB}(U_A)}{\Omega_{AB}^{tot}},$$

where the sum is over all possible values of U_A. Inserting the values from Table 1.2, we obtain $\langle U_A \rangle = 3.35$. Analogously, we get $\langle U_B \rangle = 2.65$, and we see that $\langle U_A \rangle + \langle U_B \rangle = 6$ as required.

{**EXERCISE 1.1**
Perform these calculations.}

Let us now see what these results really mean. We started by having two systems A and B that contained 3 units of energy each. When they were put in thermal contact, energy could be exchanged between the systems and, as a result, system A will contain on average 3.35 and system B 2.65 energy units. Thus, energy has spontaneously flowed from B to A, and from our operational characterization of temperature above, we conclude that B initially had a higher temperature than A. This is quite interesting since the only difference between the

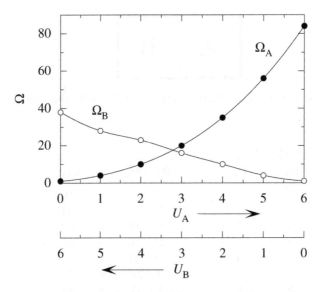

FIGURE 1.6 Ω for systems A and B as a function of their energy when $U_A + U_B = 6$.

two systems was that system B lacked a couple of energy levels that A contained. So we see that the temperature of a system reflects how the energy levels of the system are distributed as a function of energy. From Figure 1.6, where the data from Table 1.2 is plotted, we see that the Ω-curve of system A is much steeper than that of system B. We shall see later that a system with a steeper $\Omega(U)$ curve indeed has a lower temperature (it is actually the relative slope $\Omega'(U)/\Omega(U)$ that matters, where $\Omega'(U) = d\Omega(U)/dU$).

As we have seen, the probability for subsystem A to have energy U_A is equal to $\Omega_{AB}(U_A)/\Omega_{AB}^{tot}$. This means that the curve in Figure 1.7, which shows $\Omega_{AB}(U_A)$, is proportional to this probability (one simply divides its values by 1358 [$= \Omega_{AB}^{tot}$] to get the probability). We see that the probability is high near the average energy value $\langle U_A \rangle$, but

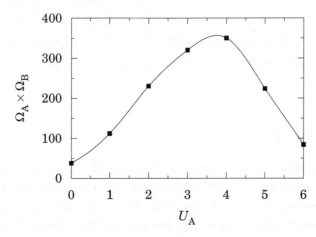

FIGURE 1.7 $\Omega_{AB} = \Omega_A \times \Omega_B$ for the combined system AB as a function of the energy of subsystem A.

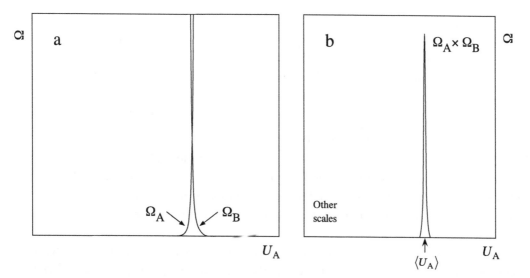

FIGURE 1.8 Sketches of $\Omega(U_A)$-functions for systems A and B (left) and the combined system AB (right) when they contain many molecules. In the right-hand frame, the average $\langle U_A \rangle$ virtually coincides with the maximum of $\Omega_{AB} = \Omega_A \Omega_B$. For a macroscopic system, the peak is enormously much more sharp than shown in the picture to the right.

that it is also quite high elsewhere. This means that the system will spend an appreciable fraction of time with a distribution of energy between A and B that is quite different from the average one, that is, we have large energy fluctuations in the system. This is characteristic of a small system.

For macroscopic systems with a huge number of molecules, the situation is different. Then the function $\Omega(U)$ is a rapidly increasing function[4] of U as illustrated in Figure 1.8a and $\Omega_{AB}(U_A)$ becomes sharply peaked (see frame b in the figure). The probability is very low that the system will have an energy distribution for which Ω_{AB} is small. A distribution of energy between A and B near the maximum in the Ω_{AB} curve is much more likely than the distributions away from the maximum. Therefore, the fluctuations in the combined system AB are very small and the average $\langle U_A \rangle$ virtually coincides with the U_A value for the maximum of Ω_{AB}. The quantity $\langle U_A \rangle$ is the *internal energy* of system A at equilibrium. The macroscopic (thermodynamic) state of system A is characterized by this energy and by the number of molecules it contains and its volume. Thus, *a thermodynamic state corresponds to the average of a large number of microscopic states and is dominated by the set of microscopic states that have the largest probability to occur.* (The corresponding statements apply to system B.)

---

[4] As we will see later, the function $\Omega(U)$ has typically an order of magnitude equal to a constant times U^{aN}, where N is the number of particles and a is a number of order 1 (for a monatomic ideal gas $a = 3/2$). For a macroscopic system with, say, N equal to 10^{20} or larger, $\Omega(U)$ hence grows extremely rapidly with U.

Note that when a system goes towards equilibrium, the value of Ω increases.[5] In our explicit example above for the combined system AB, we started with $\Omega_{AB} = 320$ when A and B could not exchange energy (when there were 3 units of energy in either subsystem). From this starting point, we let energy be freely exchanged between A and B, whereby we ended up with $\Omega_{AB}^{tot} = 1358$ at equilibrium.

1.1.2 Microscopic Definition of Entropy and Temperature for Isolated Systems

Bring two, initially isolated macroscopic systems A and B into thermal contact. Assume that A and B cannot do any work on each other and that they are weakly coupled, that is, that the interaction between them is so weak that the set of quantum states of one system is not affected by the presence of the other system. Assume also that the combined system AB is isolated.

The total energy is $U_{tot} = U_A + U_B$. For each particular distribution of energy between subsystems A and B, the degeneracy equals $\Omega_{AB} = \Omega_A(U_A)\Omega_B(U_{tot} - U_A)$. The most likely energy distribution is the one with maximal Ω_{AB}. Let us determine the conditions for this distribution. At the maximum, we have

$$\frac{\partial(\Omega_A\Omega_B)}{\partial U_A} = 0. \tag{1.1}$$

The derivative equals

$$\frac{\partial(\Omega_A\Omega_B)}{\partial U_A} = \frac{\partial\left[\Omega_A(U_A)\Omega_B(U_{tot} - U_A)\right]}{\partial U_A}$$

$$= \frac{\partial\Omega_A(U_A)}{\partial U_A}\Omega_B(U_{tot} - U_A) + \Omega_A(U_A)\frac{\partial\Omega_B(U_{tot} - U_A)}{\partial U_A}.$$

The last factor is

$$\frac{\partial\Omega_B(U_{tot} - U_A)}{\partial U_A} = \frac{\partial\Omega_B(U_{tot} - U_A)}{\partial(U_{tot} - U_A)} \times \frac{\partial(U_{tot} - U_A)}{\partial U_A}$$

$$= \frac{\partial\Omega_B(U_B)}{\partial U_B} \times (-1),$$

where we have used $U_{tot} - U_A = U_B$. At the maximum, we therefore have

$$0 = \frac{\partial(\Omega_A\Omega_B)}{\partial U_A} = \frac{\partial\Omega_A(U_A)}{\partial U_A}\Omega_B(U_B) - \Omega_A(U_A)\frac{\partial\Omega_B(U_B)}{\partial U_B}.$$

Thus,

$$\frac{\partial\Omega_A(U_A)}{\partial U_A}\Omega_B(U_B) = \Omega_A(U_A)\frac{\partial\Omega_B(U_B)}{\partial U_B}$$

---

[5] Strictly speaking, we are considering the passage between two equilibrium states: the initial state that we have before we let the energy transfer take place (when the two subsystems are isolated from each other) and the final state that ultimately is reached when energy exchange between the subsystems occurs freely.

which gives

$$\frac{\partial \Omega_A(U_A)/\partial U_A}{\Omega_A(U_A)} = \frac{\partial \Omega_B(U_B)/\partial U_B}{\Omega_B(U_B)}$$

and we conclude that

$$\frac{\partial \ln \Omega_A(U_A)}{\partial U_A} = \frac{\partial \ln \Omega_B(U_B)}{\partial U_B} \tag{1.2}$$

at the maximum, that is, for the most probable distribution of the total energy U_{tot} with U_A in system A and U_B in system B.[6]

Let us now investigate what happens if the initial energies of A and B (which we denote by U_A^0 and U_B^0) differ from the most probable distribution and we bring systems A and B in thermal contact so they can exchange energy as depicted in Figure 1.9. If U_A^0 is to the left of the maximum in Figure 1.8b, system A will increase its energy (at the expense of B) until it has reached $\langle U_A \rangle$ (= the most probable energy), that is, the temperature of A is lower than that of B. But to the left of the maximum,[7] we have $\partial(\Omega_A \Omega_B)/\partial U_A > 0$. By analogous arguments as those leading from Equations 1.1 to 1.2, we then can conclude that $\partial \ln \Omega_A(U_A)/\partial U_A > \partial \ln \Omega_B(U_B)/\partial U_B$, so it follows that:

If U_A^0 is to the left of max: $\frac{\partial \ln \Omega_A(U_A)}{\partial U_A} > \frac{\partial \ln \Omega_B(U_B)}{\partial U_B}$ and $\text{Temp}_A < \text{Temp}_B$.

Similarly:

If U_A^0 is to the right of max: $\frac{\partial \ln \Omega_A(U_A)}{\partial U_A} < \frac{\partial \ln \Omega_B(U_B)}{\partial U_B}$ and $\text{Temp}_A > \text{Temp}_B$.

If U_A^0 is at max: $\frac{\partial \ln \Omega_A(U_A)}{\partial U_A} = \frac{\partial \ln \Omega_B(U_B)}{\partial U_B}$ and $\text{Temp}_A = \text{Temp}_B$.

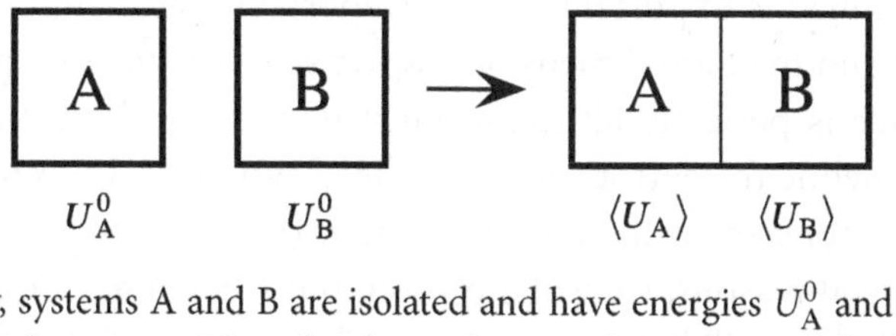

FIGURE 1.9 Initially, systems A and B are isolated and have energies U_A^0 and U_B^0, respectively. We bring them into thermal contact with each other so heat can be exchanged freely. When equilibrium is reached, A and B have the average energies $\langle U_A \rangle$ and $\langle U_B \rangle$, respectively.

---

[6] The condition that $\Omega_A \Omega_B$ has a maximum is that $\partial^2 \ln \Omega_\alpha/\partial U_\alpha^2 < 0$ where α = A or B. This follows from the fact that $\ln[\Omega_A \Omega_B]$ must also have a maximum and therefore $\partial^2 \ln[\Omega_A \Omega_B]/\partial U_A^2 = \partial^2 \ln \Omega_A/\partial U_A^2 + \partial^2 \ln \Omega_B/\partial U_B^2 < 0$. Once thermal equilibrium is established, the identities of A and B are not important and since the last equality must be true even in the special case when systems A and B are identical, the assertion follows. (Under rare circumstances, it is also possible that $\partial^2 \ln \Omega_\alpha/\partial U_\alpha^2 = 0$.)

[7] We assume that there is a single maximum.

Here, "Temp" stands for temperature. The number of microscopic states changes as follows:

$$\text{initially } \Omega_A(U_A^0)\Omega_B(U_B^0) \;\rightarrow\; \text{finally } \sum_{U_A} \Omega_A(U_A)\Omega_B(U_{tot} - U_A).$$

Obviously, the number of available states always increases when we remove the isolation of A and B from each other and let the energy flow freely between the subsystems (the initial term [to the left of the arrow] is one of the terms in the final sum [to the right] and all terms are positive). For a macroscopic system, the maximal term in the sum, $\Omega_A(\langle U_A\rangle)\Omega_B(U_{tot} - \langle U_A\rangle)$, dominates since $\Omega_A \times \Omega_B$ is extremely sharply peaked. Therefore, we have

$$\ln \sum_{U_A} \Omega_A(U_A)\Omega_B(U_{tot} - U_A) \approx \ln\left[\Omega_A(\langle U_A\rangle)\Omega_B(U_{tot} - \langle U_A\rangle)\right]$$

to a very good approximation.[8] Accordingly, the logarithm of the number of states changes as follows

$$\text{initially } \ln\left[\Omega_A(U_A^0)\Omega_B(U_B^0)\right] \;\rightarrow\; \text{finally } \ln\left[\Omega_A(\langle U_A\rangle)\Omega_B(U_{tot} - \langle U_A\rangle)\right].$$

We see that if $U_A^0 < \langle U_A\rangle$ or if $U_A^0 > \langle U_A\rangle$, the product $\Omega_A\Omega_B$ increases when the combined system goes to equilibrium (i.e., the most probable distribution of energy), compare with Figure 1.8. Thus, $\ln(\Omega_A\Omega_B)$ increases when the system goes from the initial state to the final equilibrium state. On the other hand, $\ln(\Omega_A\Omega_B)$ remains unchanged if $U_A^0 = \langle U_A\rangle$, that is, when we already are at equilibrium and the initial distribution equals the most probable one.

To summarize our findings (valid for the isolated system AB):

$\ln(\Omega_A\Omega_B)$ increases when the system goes to equilibrium and therefore $\ln(\Omega_A) + \ln(\Omega_B)$ increases too.

We have seen that $\frac{\partial \ln \Omega_A(U_A)}{\partial U_A}$ and $\frac{\partial \ln \Omega_B(U_B)}{\partial U_B}$ are related to temperature.

Obviously, the function $\ln \Omega$ is important. It is so important that it ought to have a name: *Entropy*. Since both Ω and $\ln \Omega$ are dimensionless, we see that the entropy, so defined, is also dimensionless. This is perfectly all right, but it is useful to have some unit for the entropy. Therefore, we define the entropy for an isolated system as $S = k \ln \Omega$, where k can be an arbitrary, positive constant with an arbitrary unit (its value can be defined by some convention). For historical reasons, we will, however, put k equal to Boltzmann's constant $k_B = 1.3807 \cdot 10^{-23}$ J K^{-1}, as will be discussed later. Thus, we have for an isolated system

$$S = k_B \ln \Omega$$

---

[8] A simplified example: If A contains 10^{20} molecules and each molecule has n states which contribute to $\Omega_A(\langle U_A\rangle)$, the latter has a magnitude of the order $n^{10^{20}}$. Likewise, say that $\Omega_B(\langle U_B\rangle)$ has the magnitude $m^{10^{20}}$ so the right-hand side (rhs) $\approx \ln\left[n^{10^{20}} \times m^{10^{20}}\right] = 10^{20} \ln[nm]$. Say that 10^{100} terms in the sum on the left-hand side (lhs) have a comparable magnitude to the maximal term, then lhs $\approx \ln\left[10^{100} \times n^{10^{20}} \times m^{10^{20}}\right] = 230 + 10^{20} \ln[nm] \approx$ rhs to a *very* good approximation.

and entropy has the unit J/K. Thereby, we have made the identification of entropy that originally was done by Ludwig Boltzmann in the 1870s. This identification agrees, in fact, in all practical aspects with the concept of entropy that had been introduced earlier in thermodynamics for systems at equilibrium.

The entropy is additive since

$$S_{AB} = k_B \ln \Omega_{AB} = k_B \ln(\Omega_A \Omega_B) = k_B \ln(\Omega_A) + k_B \ln(\Omega_B) = S_A + S_B.$$

Furthermore, S increases when the system goes from an initial state to the final equilibrium state, at which it attains its maximal value.

As we saw above, the derivative of the function $\ln \Omega(U)$ is also important: $\partial[k_B \ln \Omega(U)]/\partial U = \partial S/\partial U$ is larger for a system with a lower temperature than for one with a higher temperature. Therefore, we define *temperature* as[9]

$$T = \frac{1}{\partial S/\partial U} = \frac{1}{k_B(\partial \ln \Omega/\partial U)}. \tag{1.3}$$

It has the right properties as regards our operational characterization of temperature earlier.[10] Since U has the unit J, we see that T has the unit K, which is a reason for selecting the unit J/K for k_B. Furthermore, we see that from this definition follows

$$dS = \frac{dU}{T} = \frac{đq}{T}$$

since $dU = đq$ in our present case.[11] Thus, $1/T$ describes how rapidly the entropy changes when heat is transferred to the system.

If we let a system spontaneously attain the most probable energy distribution when a constraint is removed (e.g., an insulation between two subsystems as above), we have seen that the number of available microscopic states and hence the entropy increase. For our system AB we saw that S increased, $dS > 0$, despite that no energy was exchanged with its surroundings (AB was isolated). This increase in S originates from an *internal* exchange of energy (between subsystems A and B). The process is irreversible since there is a completely negligible probability that the system would spontaneously regain its initial energy distribution if we simply restored the constraint, that is, the system would not come back to its initial macroscopic state by doing so.

Thus, we see that the entropy of a system can be changed by energy (heat) exchange with the surroundings or by an internal entropy production due to some irreversible process. If we would let a system exchange some heat $đq$ with its surroundings while, at the same time,

---

[9] Expressed more accurately, we have $T = 1/(\partial S/\partial U)_{V,N}$ where we have specified that the volume V and the number of particles N are constant during the partial derivation.

[10] One can equally well use the dimensionless number $\beta = \partial \ln \Omega/\partial U$, which equals $1/(k_B T)$, as a measure of temperature, but then a low β-value ("temperature") would correspond to hot and high value to cold, which is not what we are used to. Therefore, we use $1/(k_B \beta)$ as a measure of temperature and call it T.

[11] We here monitor how much S varies $[= dS = (\partial S/\partial U)dU]$ when we change the energy of the system slightly $[= dU]$ (no work being done and the composition is unchanged) without leaving the condition of internal equilibrium.

some irreversible process took place internally, we would have $dS > đq/T$ because of the internal entropy production.

To summarize, we thus have

$$dS \geq \frac{đq}{T},$$

where the equality is valid in the absence of irreversible processes. We recognize this as the **second law of thermodynamics**. It can also be written $dS = đq/T + dS_{\text{irrev}}$, where $dS_{\text{irrev}} \geq 0$ is the entropy increase due to irreversible processes, if any. (Note that our discussion here does not constitute a proof of this law; we have, for instance, not described how energy transfer in terms of work enters into the picture. This will be done later.)

From the discussion above about spontaneous heat exchange and temperature, it is clear that if systems A and B are in thermal equilibrium ($T_A = T_B$) and if A is also in thermal equilibrium with system C ($T_A = T_C$), then B and C must also be in thermal equilibrium ($T_B = T_C$). This is the **zeroth law of thermodynamics**.

To be able to compare the temperatures of different systems empirically (which we do in our everyday life), we would need some system that changes the value of some macroscopic parameter (e.g., its volume or color) appreciably when its energy is changed by heat transfer, that is, a device called a thermometer. The thermometer should be much smaller than the system we measure so the energy change of the latter is negligible. From the values of the parameter at different temperatures, we can construct an empirical temperature scale for each thermometer. It follows from the zeroth law that if two bodies have the same temperature, according to this temperature scale they will be in thermal equilibrium when brought together: a very useful conclusion.

However, two different thermometers will not, in general, give the same value of empirical temperature for the same body.[12] Furthermore, if a body has a temperature, say, halfway between the temperatures of two other bodies as measured with the same thermometer, this conclusion is not necessarily true when measured with another kind of thermometer. These faults do not occur if we use the temperature scale defined by T. As we have seen, all bodies have the same value of T when they are in thermal equilibrium, so if this parameter is used for the construction of temperature scales for the different thermometers, they will all show the same temperature value for the same body. Furthermore, the thermometer then measures a fundamental property of the body, namely how rapidly the degeneracy of its energy levels changes with energy. To mark that this temperature scale is more fundamental than the other scales, the temperature T is called the **absolute temperature**.

Since the degeneracy of the energy levels always increases with increasing energy, we have $\partial \Omega / \partial U > 0$, and therefore

$$T = \Omega \left[k_B \frac{\partial \Omega}{\partial U} \right]^{-1} > 0.$$

---

[12] This is true unless we have calibrated them, that is, compared their values of the temperature and given them a common empirical scale.

The absolute temperature accordingly is always positive for a system at equilibrium. We also have $\partial T/\partial U \geq 0$,[13] so the temperature increases when we raise the energy (the equality applies at phase changes, for instance during melting, where T is constant while U increases).

The temperature can be decreased until we reach the ground state of the system, at which point $T = 0$. Let us denote the degeneracy of this state Ω_0. Thus, we have $S \to k_B \ln \Omega_0$ when $T \to 0$. The ground state for a pure crystalline substance is often nondegenerate, $\Omega_0 = 1$, (a perfect crystal) in which case we have $S \to 0$ when $T \to 0$. This is the **third law of thermodynamics**.

Finally, a word about the constant k_B. As we saw earlier, one can define entropy as $S = k \ln \Omega$ with an arbitrary k that has an arbitrary unit. One can even take k as a dimensionless number and we see from the definition of T that the temperature then would have dimension of energy. This is not the customary unit. For historical reasons – that is, the development of classical thermodynamics – we are used to measure (the absolute) temperature in Kelvin. Therefore, we put k equal to Boltzmann's constant $k_B = 1.3807 \cdot 10^{-23}$ J K^{-1}. Note that this is merely a convention; there is nothing fundamental about this value.

1.2 QUANTUM VS CLASSICAL MECHANICAL FORMULATIONS OF STATISTICAL MECHANICS: AN EXAMPLE

We will now investigate the number of microstates $\Omega(U)$ for a system at energy U in more detail. Let us assume that there is only one species of particles (molecules) in the system. Ω depends on the system's volume V and the number of particles N in the system, so in fact we have $\Omega = \Omega(U, V, N)$. When we hold V and N constant, we may, however, write $\Omega(U)$ in order to simplify the notation. As a concrete example, we will determine $\Omega(U, V, N)$ explicitly for a simple system, namely the monatomic ideal gas.

Firstly, in Section 1.2.1, we will determine the number of microstates for this gas in a similar manner as when we determined Ω for the system depicted in Figure 1.2 by distributing energy over the quantum states. For the monatomic ideal gas, the quantum states of the molecules are those of a well-known quantum mechanical model, the "particle in a box," which is an appropriate model in this case.

Secondly, in Section 1.2.2, we will treat the gas classically (i.e., in a classical mechanical manner) and determine an expression for the number of microstates, which turns out to be the same expression as that for the quantum case provided U is sufficiently large. Thereby, we will obtain some insights into how the two different manners to treat the gas, the quantum and the classical ones, differ conceptually.

The classical treatment is a preparation for the treatment of the liquid state in Part II of this treatise and the following. When dealing with liquids and solutions, classical statistical mechanics is much easier to use in practice than the quantum version. Since the classical theory is sufficient for the liquid systems we treat in this treatise, the main part of the treatise

---

[13] This follows from $\partial^2 \ln \Omega/\partial U^2 \leq 0$ (see footnote 6) and the definition of T, which implies that

$$k_B \frac{\partial[\partial \ln \Omega/\partial U]}{\partial U} = \frac{\partial[1/T]}{\partial U} = -\frac{1}{T^2}\frac{\partial T}{\partial U}$$

and hence $\partial T/\partial U \geq 0$ for constant V and N.

is based on classical statistical mechanics. The basis of statistical mechanics is, however, easier to understand conceptually when presented in a quantum mechanical manner and this motivates why we start by studying the quantum case.

When dealing with $\Omega(U)$, it is important to understand that the energy of a system cannot be determined with infinite precision in reality. It is therefore not relevant to talk about the number microstates $\Omega(U)$ with energy *exactly* equal to U.[14] It is more appropriate to consider the number of microstates with energies in the immediate vicinity of U, and, in fact, this is the relevant quantity that constitutes $\Omega(U)$. An appropriate question is how wide the "immediate vicinity" should be, that is, what energies should be included in the counting of microstates when we determine $\Omega(U)$. For macroscopic systems we will, however, see that $\ln \Omega$, which is the relevant quantity for S, is *extremely* insensitive to this width. Therefore, in practice, $\ln \Omega$ is a well-defined entity despite that the "immediate vicinity" is not precisely defined. Still, we will for simplicity continue talking about $\Omega(U)$ as "the number of quantum states at energy U," but this does not mean states with energy *exactly* equal to U.

1.2.1 The Monatomic Ideal Gas: Quantum Treatment[15]

Let us consider a container with a gas where the gas molecules interact very weakly with each other, so they do not affect each other's set of quantum states but can exchange energy. The set of quantum states of each molecule is thus the same as for a single molecule in the container. A gas with these features is, by definition, an **ideal gas**. Such a gas does, of course, not exist in reality, but the concept of an ideal gas is an excellent approximation for a real gas at very low densities.

For simplicity we restrict ourselves to a monatomic gas, whereby we for the moment ignore the electronic degrees of freedom and only consider the quantum states due to the translational degrees of freedom.[16] We furthermore assume that the container walls limit the space available for the molecules, but do not interact in any other manner with them. In order to make the discussion easier, we assume that the container is a cube with sides of length L. The sides of the box are aligned with the coordinate axes and the volume of

---

[14] Due to the presence of discrete energy levels, the function $\Omega(U)$ so defined would be a highly discontinuous function of U, although – for a macroscopic system – the levels lie so closely together that they nearly form a continuum. Furthermore, due to the Heisenberg uncertainty principle, it is not possible to specify the energy of a system precisely unless it has been observed for an infinitely long time. It is also impossible to isolate a system completely; there will always be some residual coupling with the rest of the universe. For these reasons, a better definition of Ω is the following:

> Ω is the number of states that lie between the energies $U - \delta U$ and U, where δU is a small but arbitrary energy interval.

It may appear that Ω so defined is quite arbitrary, but as discussed in Appendix 1A: one can show that $\ln \Omega$ and hence S is *extremely* insensitive to the magnitude of δU for a macroscopic system. The reason for this is that the number of quantum states that lie between $U - \delta U$ and U for reasonable values of δU is enormously much larger than the number below $U - \delta U$. Therefore, the change in the number of states is completely negligible when δU is increased. In fact, one can include *all* quantum states below U without changing the value of Ω noticeably, as shown in the Appendix. These facts are also demonstrated numerically in Section 1.2.1.

[15] A substantial part of the material in this section consists of a modification of Appendix E in the book *Thermodynamics Kept Simple – A Molecular Approach: What is the Driving Force in the World of Molecules?* by Roland Kjellander (CRC Press, 2015).

[16] We hence assume that the gas atoms are in their electronic ground states. This means that the temperature of the gas cannot be too high, but it is still sufficiently high so that the quantum numbers of the translational quantum states are high.

the gas is $V = L^3$. The fact that we choose a cubic box makes no difference for the macroscopic properties of a homogeneous gas since these properties do not depend on the shape of the box that the gas is contained in. The final results are therefore valid in general for the gas.

The appropriate quantum mechanical treatment of each gas particle is that of a "particle in a box," where the interaction potential between the particle and the walls, denoted by $v(\mathbf{r})$ where $\mathbf{r} = (x, y, z)$, is zero inside the box and infinitely large otherwise. The particle can thereby move freely in the box but cannot leave it. The quantum mechanics for a particle in a three-dimensional box is for completeness treated in the shaded text box below. The following description is, however, sufficient. Consider the momentum p_x the x direction for a particle. The momentum is quantized due to the fact that the standing wave (the solution to the Schrödinger equation) must have zero amplitude at the surfaces of the box walls. Using the de Broglie relationship between momentum and wavelength, $\lambda = h/p_x$, we can conclude that the available wavelengths of the standing wave satisfy the condition $j\lambda/2 = L$, where h is Planck's constant and j is any positive integer. We thus obtain $p_x = jh/2L$. The smallest possible positive integer is 1, so the minimum value of p_x is $p_{\min} = h/2L = h/(2V^{1/3})$. This means that $p_x = jp_{\min}$ for any integer $j \geq 1$. The same applies in the y and z directions, so we have $p_\alpha = j_\alpha h/2L = j_\alpha p_{\min}$ for $\alpha = x, y$ and z with integer $j_\alpha \geq 1$. The energy u of the particle, $\mathrm{u} = [p_x^2 + p_y^2 + p_z^2]/2m$, where m is the particle mass, is therefore also quantized.

The time-independent Schrödinger equation is $\mathscr{H}\Psi(\mathbf{r}) = \mathrm{u}\Psi(\mathbf{r})$, where $\Psi(\mathbf{r})$ is the wave function, $\mathscr{H} = -(h^2/8\pi^2 m)\nabla^2 + v(\mathbf{r})$ is the Hamiltonian operator, $\nabla^2 = \partial^2/\partial x^2 + \partial^2/\partial y^2 + \partial^2/\partial z^2$ is the Laplace operator and u is the energy. Inside the box, where $v(\mathbf{r})$ is zero in our case, the Schrödinger equation is

$$-\frac{h^2}{8\pi^2 m}\left[\frac{\partial^2\Psi(\mathbf{r})}{\partial x^2} + \frac{\partial^2\Psi(\mathbf{r})}{\partial y^2} + \frac{\partial^2\Psi(\mathbf{r})}{\partial z^2}\right] = \mathrm{u}\Psi(\mathbf{r})$$

with boundary condition $\Psi(\mathbf{r}) = 0$ at the boundary of the box, that is, whenever $x = 0$, $x = L, y = 0, y = L, z = 0$ or $z = L$. The general solution to the equation is a superposition of functions $\Psi(\mathbf{r}) = A\sin(k_x x + \vartheta_x)\sin(k_y y + \vartheta_y)\sin(k_z z + \vartheta_x)$, where A, ϑ_x, ϑ_y and ϑ_z are constants and $(k_x^2 + k_y^2 + k_z^2)h^2/(8\pi^2 m) = \mathrm{u}$. The quantity k_α is the wave number, which is related to the wave length λ_α as $k_\alpha = 2\pi/\lambda_\alpha$. Since the boundary condition requires that $\Psi(\mathbf{r}) = 0$ at the boundaries we have $\vartheta_\alpha = 0$ and $k_\alpha L = j_\alpha\pi$ for $\alpha = x, y$ and z, where j_α is a positive integer (in other words, L must be an integer multiple of $\lambda_\alpha/2$). Thus, solutions exist only when $[(j_x\pi/L)^2 + (j_y\pi/L)^2 + (j_z\pi/L)^2]h^2/(8\pi^2 m) = \mathrm{u}$, that is, when

$$\mathrm{u} = \frac{h^2}{L^2 8m}\left[j_x^2 + j_y^2 + j_z^2\right]. \tag{1.4}$$

This is the well-known quantization of the energy u for a particle in a box and j_α for $\alpha = x, y$ and z are quantum numbers of the quantum state.

Another manner to express the quantization is to consider the momentum p_α in each direction x, y and z, which equals $p_\alpha = h/\lambda_\alpha = hk_\alpha/2\pi$. The quantization for k_α obtained earlier, namely $k_\alpha L = j_\alpha \pi$, implies that the possible values of the momentum are $p_\alpha = j_\alpha h/2L = j_\alpha p_{min}$.

The possible momenta in each direction are integer multiples of p_{min} and the energy for the quantum states of the particle is

$$u = \frac{p_x^2 + p_y^2 + p_z^2}{2m} = \frac{p_{min}^2 \left[j_x^2 + j_y^2 + j_z^2 \right]}{2m}. \tag{1.5}$$

For instance, p_x can assume the values p_{min}, $2p_{min}$, $3p_{min}$ etcetera and the same thing applies for p_y and p_z. This can be illustrated as in Figure 1.10, which shows an example where $p_x = 4p_{min}$, $p_y = 5p_{min}$ and $p_z = 3p_{min}$. Another possibility is, for instance, that $p_x = 5p_{min}$, $p_y = 4p_{min}$ and $p_z = 3p_{min}$, where $4p_{min}$ and $5p_{min}$ have changed places. The number of possibilities is, as we can realize, very large. Each constitutes a possible quantum state of the monatomic molecule.

For a gas with N particles, there are a total of $3N$ momentum components (3 per particle). An example with three particles is given in Figure 1.11. Each component p_α, with $\alpha = x, y$ or z, gives the contribution $(p_\alpha)^2/2m$ to the energy since u for each particle is given by Equation 1.5. The total energy is the sum of u for all particles. Since the particles in an ideal gas interact very weakly, they can be treated independently of each other. If the energy is sufficiently high, it is very unlikely that two momentum components have the same value because the number of possible values is much larger than the number of particles. Then one can safely assume that the momentum components for all particles have different values. For the same reason, one can assume that all particles are in different quantum states when the energy is sufficiently high.

In a manner that corresponds to the arguments in the beginning of Section 1.1.1, we will now determine how many different energy distributions there are among the molecules

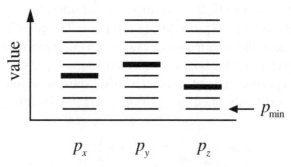

FIGURE 1.10 The possible values of the momenta p_x, p_y and p_z for a monatomic molecule in a box are represented by vertical bars (which continue infinitely upwards). Bold lines show each component's value for the molecule in the example.

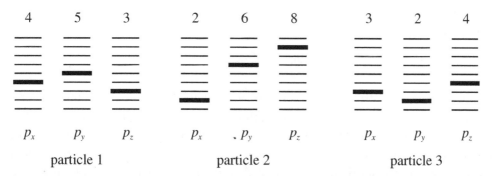

FIGURE 1.11 Illustration of three particles with examples of values of the momentum components. The numbers at the top show the value of each component (expressed as integer multiples of p_{min}). All possible distributions of values for the nine different components (three per particle) must be taken into account. The figure only shows one example of such a distribution.

when the total energy of the system is U, that is, the number of microstates $\Omega(U)$ for the entire system at energy U. Initially, we assume for simplicity that the molecules are distinguishable, which is not correct for identical particles in quantum mechanics. We will, however, correct for this later on. In order to find the number of distributions, we must take into account all possible distributions of energy between all $3N$ components for all molecules. It turns out to be easiest to first determine the number of possible energy distributions when the total energy is $\leq U$, which will be denoted by $\Phi(U)$, and then from this result determine $\Omega(U)$.

Let us first look at the case of a single momentum component p_α. The lowest energy is obtained when the component's absolute value is p_{min}, that is, the minimum energy is $u_{min} = (p_{min})^2/2m$. If we allow energies between u_{min} and U, the possible values of p_α are between p_{min} and $p_{max} = (2mU)^{1/2}$. All positive integer multiples of p_{min} are allowed, so the number of possible values of p_α is equal to $p_{max}/p_{min} = (U/u_{min})^{1/2}$ (or more specifically the integer closest below this number). Thus, the number of possible values is proportional to $\sqrt{U}$.

Next, we consider the number of different energy distributions for N molecules, that is, for $3N$ momentum components. We will show below that the number of possible energy distributions when the total energy is $\leq U$ is proportional to $(\sqrt{U})^{3N}$, provided that the energy is not very small. The number of distributions for $3N$ components is thus proportional to the corresponding number for a single momentum component to the power $3N$, that is, we have

$$\Phi(U) = \text{constant} \times U^{3N/2}, \tag{1.6}$$

where "constant" is independent of U.

To get a feeling for what this relationship implies, let us make a simple numerical exercise. For a macroscopic system, N is a very large number, say, 10^{20} or greater. $\Phi(U)$ thus increases faster than $U^{10^{20}}$, which is incredibly fast. Even if U only increases by, say, one billionth, the number of energy distributions with energies $\leq U$ increases by a factor that is greater than

$$(1.000000001)^{10^{20}} \approx 10^{4 \cdot 10^{10}}.$$

The number of energy distributions between $0.999999999\,U$ (that is $U/1.000000001$) and U is thus more than $10^{4 \cdot 10^{10}}$ times greater than those between 0 and $0.999999999\,U$ (!). The distributions for energies $\leq 0.999999999\,U$ are thus negligibly few compared to those with energies in the immediate vicinity of U. Obviously, this is also the case when the system contains far fewer molecules than in 10^{20} and when the margin is much smaller than one billionth. From this, we can make the important conclusion that for macroscopic systems the number of energy distributions with energies $\leq U$ is in practice *equal to the number in the immediate vicinity of the energy U*. As mentioned earlier, this latter number is the relevant quantity that constitutes $\Omega(U)$. It follows that in the definition of entropy, $S = k_B \ln \Omega$, one can replace $\ln \Omega$ by $\ln \Phi$ as an excellent approximation for a macroscopic system. These issues are discussed in more generality in Appendix 1A.

By applying these results, we conclude from Equation 1.6 that for a macroscopic system with N monatomic molecules, the following holds provided the energy is not very low

$$\ln \Omega(U) = \ln(\mathcal{K} U^{3N/2}), \tag{1.7}$$

where $\mathcal{K}$ is independent of U (the factor $\mathcal{K}$ depends only on N, the volume V and the mass m of a molecule and it is constant when only U varies). From this, we can immediately obtain a further interesting result. The definition (1.3) of the absolute temperature implies that we have

$$k_B T = \frac{1}{\partial \ln \Omega / \partial U} = \frac{1}{\partial \ln(\mathcal{K} U^{3N/2}) / \partial U}$$

$$= \frac{1}{\partial (\ln \mathcal{K} + (3N/2) \ln U) / \partial U} = \frac{U}{3N/2}.$$

This means that

$$U = \frac{3N k_B T}{2} \tag{1.8}$$

for a monatomic ideal gas when the electronic quantum states in the atoms are not considered, which is appropriate when the temperature is not too high (the zero of the energy is here set to the energy of the N molecules in their electronic ground state). The average energy per particle, which equals U/N, is accordingly equal to $3k_B T/2$. The energy that we have included here is the kinetic energy due to the translational motion of the particles. Thus, the kinetic energy of the gas is proportional to the absolute temperature.

Let us now demonstrate Equation 1.6, that is, find the number of different distributions of energy that exists for $3N$ momentum components for all possible total energies $\leq U$. Let us initially allow the value of each component to vary between p_{min} and $J p_{min}$, where J is an

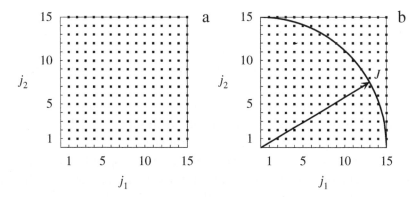

FIGURE 1.12 An example of the various possibilities for two momentum components, which can have values $j_1 p_{min}$ and $j_2 p_{min}$ where j_1 and j_2 are positive integers. Each possibility for the two components is indicated by a dot. (a) The two components can each adopt J different values independently of each other, $1 \leq j_1 \leq J$ and $1 \leq j_2 \leq J$. In this example, $J = 15$. (b) If one has the condition that the energy at most can be $U = J^2 \varepsilon_{min}$, only the points within the radius J are included.

arbitrarily large positive integer. How many ways are there to realize a system where each component may independently adopt J different values?

As an example, we take two components: the first component may assume the values $j_1 p_{min}$, $1 \leq j_1 \leq J$, and the second $j_2 p_{min}$, $1 \leq j_2 \leq J$, where j_1 and j_2 are integers, see Figure 1.12a. When component 1 has the value p_{min} ($j_1 = 1$), component 2 can have any of the values $p_{min}, 2p_{min}, \ldots, Jp_{min}$ ($1 \leq j_2 \leq J$). The same applies when component 1 has the value $2p_{min}$ ($j_1 = 2$). For each value of component 1, component 2 can thus assume J different values. In total, there are thus $J \times J = J^2$ possibilities, which is the number of integer points within the square in Figure 1.12a, that is, the area of the square.

This result can be easily generalized to more components. If one has three components, j_1 and j_2 can assume J^2 different values for each value that the third component adopts. There are thus $J^2 \times J = J^3$ possibilities. (This is the number of integer points within a cube with sides J and is equal to the volume of the cube.) For $3N$ components, there are in an analogous manner J^{3N} possibilities.

When we have a certain amount of energy to distribute, we are, however, not interested in all these possibilities. Let us therefore examine how many distributions of momentum there are when the total energy of the N molecules is lower than a certain value U. Each component p_α provides the contribution $(p_\alpha)^2/2m$ to energy. The smallest possible value of the energy contribution is $u_{min} = (p_{min})^2/2m$. When the component has the value $p_\alpha = jp_{min}$, where j is a positive integer, its contribution to the energy is $(jp_{min})^2/2m = j^2 u_{min}$.

We first examine the case of two components in Figure 1.12. The sum of the energy contributions cannot exceed U, that is

$$[(j_1)^2 + (j_2)^2]u_{min} \leq U = J^2 u_{min}, \tag{1.9}$$

where we have introduced

$$J = \left[\frac{U}{u_{min}} \right]^{1/2} \tag{1.10}$$

(J is not necessarily an integer; as we shall see, this does not matter). We can express the condition (1.9) as $(j_1)^2 + (j_2)^2 \leq J^2$, which means that the possible points (j_1, j_2) lie within the radius J (see Figure 1.12b), that is, within a quarter of a circle of radius J. The number of such points is proportional to the circle sector area (this is true at least as a very good approximation when J is a large number, that is, when the energy is large enough). For two momentum components, the number of different distributions of energy $\leq U$ is thus proportional to J^2. The number of possibilities is growing with increasing J in the same way in both Figure 1.12a and b (both the area of the square and of the circle sector grow proportionally to J^2).

If we have three components, we saw above for the case corresponding to Figure 1.12a that the number of possibilities is equal to J^3, which is the volume of a cube with the sides J. If we limit the energy $\leq U$ (which corresponds to the case of Figure 1.12b), we have $[(j_1)^2 + (j_2)^2 + (j_3)^2]u_{min} \leq U = J^2 u_{min}$, which implies that $(j_1)^2 + (j_2)^2 + (j_3)^2 \leq J^2$, where J is still given by Equation 1.10. Thereby, the possible points (j_1, j_2, j_3) lie within a sphere of radius J from the origin, as depicted in Figure 1.13 (the points lie within the part that has positive coordinates, that is, 1/8 sphere). The number of points grows with increasing J proportionally to the volume of the sphere, which means that the number grows proportionally to J^3. More precisely, the number of possible distributions of energy $\leq U$ (the number of points) is to an excellent approximation equal to the volume of 1/8 sphere of radius J, that is,

$$\Phi = \frac{1}{8} \mathscr{V}_s(J) = \frac{\pi J^3}{6} \quad \text{(three components)} \tag{1.11}$$

when U and therefore J are large. $\mathscr{V}_s(\mathcal{R}) = 4\pi\mathcal{R}^3/3$ is the volume of a sphere with radius $\mathcal{R}$.

We see that the number of possible energy distributions for both two and three components grows with increasing J as J^ℓ, where ℓ is the number of components, regardless

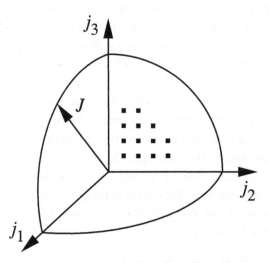

FIGURE 1.13 The various possibilities for three momentum components are given by the values $j_\alpha p_{min}$, where j_α, $\alpha = 1, 2, 3$, are positive integers. The integer points (j_1, j_2, j_3) are spread out in space in the figure; only a few of them are drawn. When we have the condition that the energy must be $\leq U = J^2 \varepsilon_{min}$, only the points within the radius J are counted.

of whether the energy is restricted to $\leq U = J^2 u_{min}$ or the values of the components are restricted to $\leq J p_{min}$ independently of each other. This applies generally. For N molecules, we have $3N$ components and we saw above that for the case corresponding to Figure 1.12a, the number of possibilities is equal to J^{3N}. When we instead restrict the energy to $\leq U$, we have the condition

$$[(j_1)^2 + (j_2)^2 + (j_3)^2 + \cdots + (j_{3N})^2] u_{min} \leq U, \tag{1.12}$$

which can be written

$$(j_1)^2 + (j_2)^2 + (j_3)^2 + \cdots + (j_{3N})^2 \leq J^2. \tag{1.13}$$

We cannot draw figures in $3N$ dimensions as illustrations, but also in this case the number of possibilities (microstates) is proportional to J^{3N} (the "volume" of a so-called hypercube with sides J and the "volume" of a hypersphere of radius J grow both as J^{3N} when J increases). Since J according to Equation 1.10 is proportional to $\sqrt{U}$, we can thus conclude that $\Phi(U)$ is growing like $U^{3N/2}$ when we increase U, so we obtain Equation 1.6.

In order to determine the proportionality constant in Equation 1.6, we need to know the volume of the hypersphere. In analogy to the case of 3 dimensions where the number of microstates is equal to the volume of 1/8 sphere, for $3N$ components this number is $1/2^{3N}$ times the volume $\mathscr{V}_s^{(3N)}$ of a hypersphere in $3N$ dimensions with radius J. For an M-dimensional hypersphere of radius $\mathcal{R}$, the volume is

$$\mathscr{V}_s^{(M)}(\mathcal{R}) = \frac{\pi^{M/2} \mathcal{R}^M}{\Gamma((M/2) + 1)} \approx \frac{(2\pi e/M)^{M/2} \mathcal{R}^M}{(\pi M)^{1/2}}, \tag{1.14}$$

where $\Gamma(x)$ is the gamma function.[17] In the last equality, we have used $\Gamma(x+1) \approx x^x e^{-x} (2\pi x)^{1/2}$, which is such an excellent approximation for large x that the right-hand side (rhs) of Equation 1.14 in practice equals $\mathscr{V}_s^{(M)}(\mathcal{R})$ when M is large.

Incidentally, the following remark is in order here: For a positive integer l we have $\Gamma(l+1) = l!$, so $l! \approx l^l e^{-l} (2\pi l)^{1/2}$, which is known as the **Stirling approximation** for the factorial. It is usually written in the form $\ln l! \approx l \ln l - l + (1/2) \ln(2\pi l)$, where the last term is negligible compared to the other terms for large l so we have in practice

$$\ln l! \approx l \ln l - l, \tag{1.15}$$

which is a very useful relationship.

We concluded earlier that the number of microstates is $1/2^{3N}$ times the volume $\mathscr{V}_s^{(3N)}$ for a hypersphere with radius J. Thus, we have $\Phi(U) = \mathscr{V}_s^{(3N)}(J)/2^{3N}$ with $J = [U/u_{min}]^{1/2}$. There is, however, one more thing to consider before we obtain the final result. We have assumed that the particles are distinguishable, which means that particle 1, particle 2,

---

[17] The definition of the gamma function is $\Gamma(x) = \int_0^\infty t^{x-1} e^{-t} dt$.

etcetera have a label with the number "attached" and when counting the number of possibilities we have taken into consideration whether it is particle 1 or particle 2 etcetera that had a certain value of its momentum component (like we did when we distributed energy for the quantum states of Figure 1.2 and like we have done in Figure 1.11). Since particles of the same species cannot be distinguished from each other, one cannot say which one is particle 1, which one is particle 2 etcetera. Therefore, we have obtained too many states. Since there are $N!$ possible permutations of the labels among the N molecules, Φ is a factor of $N!$ too large.[18] The correct value is therefore $\Phi(U) = \mathcal{V}_s^{(3N)}([U/u_{min}]^{1/2})/(2^{3N}N!)$. Since $u_{min} = (p_{min})^2/2m$ and $p_{min} = h/(2L) = h/(2V^{1/3})$, we obtain using Equation 1.14

$$\Phi(U, V, N) = \frac{1}{(3\pi N)^{1/2}} \left(\frac{4\pi me}{3h^2} \right)^{3N/2} \left(\frac{U}{N} \right)^{3N/2} \frac{V^N}{N!} \qquad (1.16)$$

which is valid for a macroscopic system provided U is not too small.

The entropy is $S = k_B \ln \Omega$ and, as we saw earlier, for macroscopic systems $\ln \Omega$ is in practice equal to $\ln \Phi$. Before proceeding, we make the following observation. In $\ln \Phi$, the first factor in the rhs of Equation 1.16 gives a contribution $-k_B T \ln(3\pi N)^{1/2}$ to the entropy, while, for example, the factor $(U/N)^{3N/2}$ gives a contribution $N \ln(U/N)^{3/2}$. For large N, the first contribution is completely negligible compared to the rest and can be excluded. The expression for S is therefore

$$S = k_B \ln \Omega(U, V, N) = k_B \ln \left[\left(\frac{4\pi me}{3h^2} \right)^{3N/2} \left(\frac{U}{N} \right)^{3N/2} \frac{V^N}{N!} \right].$$

By using the Stirling approximation (1.15) for $N!$ we can write this as

$$S = S(U, V, N) = Nk_B \ln \left[e^{5/2} \left(\frac{4\pi m}{3h^2} \right)^{3/2} \left(\frac{U}{N} \right)^{3/2} \frac{V}{N} \right], \qquad (1.17)$$

which is known as the **Sackur-Tetrode equation** for the entropy of the monatomic ideal gas. By inserting Equation 1.8, the equation can be written in terms of temperature instead of energy

$$S = Nk_B \ln \left[e^{5/2} \left(\frac{2\pi mk_B}{h^2} \right)^{3/2} T^{3/2} \frac{V}{N} \right]$$

$$= Nk_B \ln \left[\frac{e^{5/2}}{\Lambda^3} \frac{V}{N} \right], \qquad (1.18)$$

---

[18] Here we have tacitly assumed that the number of possible quantum states for each particle is much greater than the number of particles, so all particles are in different states. This is true provided the energy U is large, which we assume here. See Section 2.7 for a discussion of these matters.

where

$$\Lambda = \frac{h}{(2\pi m k_B T)^{1/2}} \tag{1.19}$$

is called the **thermal de Broglie wavelength**. These expressions for S are valid provided that the temperature and energy are not too low so the approximations we have made to obtain them are accurate. Note that if one changes the system size while holding the number density N/V and energy per particle U/N constant (and therefore the temperature is constant too), the entropy is proportional to N since the argument in the logarithm in the Sackur-Tetrode equation is constant. Hence, this equation shows that S is an extensive property of the system.

1.2.2 The Monatomic Ideal gas: Classical Treatment

Consider a system with N particles of mass m which move according to the laws of classical (Newtonian) mechanics. At time t the particles have positions $\mathbf{r}_1(t), \mathbf{r}_2(t), \ldots, \mathbf{r}_N(t)$ and velocities $\mathbf{v}_1(t), \mathbf{v}_2(t), \ldots, \mathbf{v}_N(t)$, where $\mathbf{r}_\nu = (x_\nu, y_\nu, z_\nu)$ [which we also will denote as $(r_{\nu,x}, r_{\nu,y}, r_{\nu,z})$] and $\mathbf{v}_\nu = (v_{\nu,x}, v_{\nu,y}, v_{\nu,z}) = d\mathbf{r}_\nu/dt$. It is customary (and suitable) to work with the momentum $\mathbf{p}_\nu = m\mathbf{v}_\nu$ rather than the velocity, so we will henceforth consider the momenta $\mathbf{p}_1(t), \mathbf{p}_2(t), \ldots, \mathbf{p}_N(t)$, where $\mathbf{p}_\nu = (p_{\nu,x}, p_{\nu,y}, p_{\nu,z})$.

A major difference between the quantum and the classical treatments is that the momenta and energies are not quantized, so if we consider the system when the maximal energy is U, all values of the total energy of the particles are possible up to U. Let us assume that the particles form an ideal gas, where we only deal with the translational energies. We assume that the particles are located in the same cubic box with side L as discussed in the previous section. The energy of particle ν is $u_\nu = |\mathbf{p}_\nu|^2/2m = (p_{\nu,x}^2 + p_{\nu,y}^2 + p_{\nu,z}^2)/2m$ and we have

$$\sum_{\nu=1}^{N} u_\nu = \sum_{\nu=1}^{N} \frac{|\mathbf{p}_\nu|^2}{2m} = \sum_{\nu=1}^{N} \frac{p_{\nu,x}^2 + p_{\nu,y}^2 + p_{\nu,z}^2}{2m} \leq U. \tag{1.20}$$

The spatial coordinates satisfy $0 \leq r_{\nu,\alpha} \leq L$ for $\alpha = x, y$ and z.

Let us introduce the vector $\mathbf{p}^N$ that consists of all momentum components of all particles, that is,

$$\mathbf{p}^N = \mathbf{p}_1, \mathbf{p}_2, \ldots, \mathbf{p}_N = (p_1, p_2, p_3, p_4, p_5, p_6, \ldots, p_{3N-2}, p_{3N-1}, p_{3N}),$$

where

$$(p_1, p_2, p_3) \equiv (p_{1,x}, p_{1,y}, p_{1,z}) = \mathbf{p}_1,$$

$$(p_4, p_5, p_6) \equiv (p_{2,x}, p_{2,y}, p_{2,z}) = \mathbf{p}_2,$$

$$\vdots$$

$$(p_{3N-2}, p_{3N-1}, p_{3N}) \equiv (p_{N,x}, p_{N,y}, p_{N,z}) = \mathbf{p}_N,$$

$\mathbf{p}^N$ is a vector in $3N$-dimensional space. The condition (1.20) implies

$$|\mathbf{p}^N|^2 = \sum_{l=1}^{3N} p_l^2 \leq 2mU,$$

so the vector $\mathbf{p}^N$ of all momentum components lies in the interior of a $3N$-dimensional hypersphere with radius $[2mU]^{1/2}$ and volume $\mathscr{V}_s^{(3N)}([2mU]^{1/2})$.

In the same fashion, we introduce the vector $\mathbf{r}^N$ that consists of all spatial coordinates of all particles, $\mathbf{r}^N = \mathbf{r}_1, \mathbf{r}_2, \ldots, \mathbf{r}_N$. This vector lies in the interior of a $3N$-dimensional hypercube with sides L, that is, $0 \leq r_l \leq L$ for $l = 1, \ldots, 3N$. The hypercube has volume $L^{3N} = V^N$.

It is practical to introduce a coordinate notation that combines the momentum and spatial coordinates, namely $\Gamma = (\mathbf{r}^N, \mathbf{p}^N)$, which is a vector in a $6N$-dimensional space. This space is called the **phase space**. A major reason for the introduction of phase space is that the positions and the momenta of all $3N$ particles are represented by a single point in this space. When time passes, the vectors $\mathbf{r}_1(t), \mathbf{r}_2(t), \ldots, \mathbf{r}_N(t)$ and $\mathbf{p}_1(t), \mathbf{p}_2(t), \ldots, \mathbf{p}_N(t)$ change, which means that the point $\Gamma(t)$ also changes. The motion of all particles is therefore represented by the motion of a single point in phase space, rather than N separate spatial points in three-dimensional space and likewise N separate momentum points. The trajectory of this single point when time passes thus describes the very complicated motions of all N particles.

Since $\mathbf{p}^N$ lies within the volume $\mathscr{V}_s^{(3N)}([2mU]^{1/2})$ and $\mathbf{r}^N$ within the volume V^N, the relevant volume for Γ is the product

$$\begin{aligned}
\mathscr{V}_\Gamma &= \mathscr{V}_s^{(3N)}([2mU]^{1/2})V^N \\
&= \frac{1}{(3\pi N)^{1/2}} \left(\frac{4\pi me}{3}\right)^{3N/2} \left(\frac{U}{N}\right)^{3N/2} V^N,
\end{aligned} \tag{1.21}$$

where we have used Equation 1.14. This can be compared with Φ in Equation 1.16 for the quantum case, which can be written as

$$\Phi = \frac{1}{(3\pi N)^{1/2}} \left(\frac{4\pi me}{3}\right)^{3N/2} \left(\frac{U}{N}\right)^{3N/2} V^N \frac{1}{N!h^{3N}}. \tag{1.22}$$

The only difference is the factor $1/(N!h^{3N})$. There is apparently a relationship between the number of microstates with energy $\leq U$ and the volume in phase space available for Γ, given a maximal energy U. Considering that classical mechanics is approached as a special case of quantum mechanics in the limit of large quantum numbers (large energies), it is perhaps not surprising that there is such a relationship [recall that our expression for Φ is valid in this limit, but not for small quantum numbers (small energies)]. The differing factor of $1/N!$ is readily understood from the fact that particles in classical mechanics are distinguishable. The factor $1/h^{3N}$ is more intriguing.

Due to the quantization in quantum mechanics, the number of microstates is well defined in that case, but in classical mechanics there is no such way to define a microstate and to calculate the number of such states. Since all values of the energy $\leq U$ are allowed in classical mechanics, the number of possible energy distributions between the particles is infinitely many and one cannot talk about the *number of* microstates. The same applies to the momenta that can assume an infinite number of values. Nevertheless, statistical mechanics was formulated in its classical version by Boltzmann and others long before quantum mechanics was developed, including the expression for entropy $S = k_B \ln \Omega$, which involves the number Ω. The definition of microstates and the counting of them is truly a conceptual problem in classical mechanics, but the visionary insights of Boltzmann led him to the right expression for entropy despite this fact.

One way out of this conceptual issue regarding microstates is to introduce a "graininess" of phase space by dividing it into small "volume cells," that is, a huge number of cells which subdivide the entire volume. Thereby, one can say that the system is in a certain microstate whenever the Γ point for the system is located within a certain cell; thereby, one has defined what constitutes a microstate in the classical case. If the size of all cells is the same and is equal to, say, $\delta \mathscr{V}_\Gamma$, the number of microstates with energies $\leq U$ is equal to $\mathscr{V}_\Gamma(U)/\delta \mathscr{V}_\Gamma$. In fact, quantum mechanics gives the value of the cell size equal to $\delta \mathscr{V}_\Gamma = h^{3N}$ as apparent from an identification of Equation 1.22, which contains the factor $1/h^{3N}$, and $\mathscr{V}_\Gamma(U)/\delta \mathscr{V}_\Gamma$ with $\mathscr{V}_\Gamma$ given by Equation 1.21. We see that in $\delta \mathscr{V}_\Gamma$ there is a factor of h for each translational degree of freedom for the particles (three per particle): in total h^{3N}. Thus, we can set in the classical case

$$\Phi(U, V, N) = \frac{\mathscr{V}_\Gamma(U, V, N)}{\delta \mathscr{V}_\Gamma N!} = \frac{\mathscr{V}_\Gamma(U, V, N)}{h^{3N} N!}. \tag{1.23}$$

The graininess of phase space is truly of quantum mechanical nature: the cell size contains Planck's constant, the ultimate quantum characteristic. Thus, quantum mechanics resolves this conceptual issue of classical statistical mechanics in a very elegant and fundamental manner. Note that the cells are not real; their only purpose is for bookkeeping of microstates. The motions of the particles, as governed by classical mechanics, are not influenced in any way by the cells.

This graininess of phase space is related to the Heisenberg uncertainty principle. For each degree of freedom of a particle, there is one position and one momentum component: r_α and p_α for $\alpha = x, y$ or z. The uncertainty principle says that these components can at most be resolved within a precision δr_α and δp_α, respectively, that satisfies $\delta r_\alpha \delta p_\alpha \gtrsim h$.[19] Thus, quantum mechanics says that it is not relevant to distinguish Γ points very close together in phase space, so there is indeed a natural "graininess."

Let us now turn to the case of constant energy U in the classical treatment. It is, of course, possible to have a system at a total energy exactly equal to U in classical mechanics, but from the previous discussion we see that this is not quite relevant. For the same reasons as in the quantum case, one can replace $\ln \Omega$ by $\ln \Phi$ as an excellent approximation in the definition

---

[19] More precisely, the Heisenberg uncertainty principle says that $\delta r_\alpha \delta p_\alpha \geq h/4\pi$, where δr_α and δp_α are the standard deviations in position and momenta, respectively.

of entropy, $S = k_B \ln \Omega$, for a macroscopic system.[20] One can therefore take $S = k_B \ln \Phi$ as an excellent approximation.

Since particles are distinguishable in classical mechanics, one may argue that the factor $1/N!$ should not be present in Φ and that one could take $\delta \mathscr{V}_\Gamma^* = v^{3N}$ for some value v (not necessarily equal to h). Let us therefore take Φ equal to $\Phi^* = \mathscr{V}_\Gamma(U)/\delta \mathscr{V}_\Gamma^*$ and take the entropy as $S^* = k_B \ln \Phi^* = k_B \ln(\mathscr{V}_\Gamma(U)/v^{3N})$. In entropy differences $S_2^* - S_1^* = k_B \ln(\Phi_2^*/\Phi_1^*)$, the value of $\delta \mathscr{V}_\Gamma^*$ and hence the value of v do not matter since $\delta \mathscr{V}_\Gamma^*$ is cancelled in the ratio Φ_1^*/Φ_2^*. For S^* itself, we obtain

$$S^* = N k_B \ln \left[\left(\frac{4\pi m e}{3v^2} \right)^{3/2} \left(\frac{U}{N} \right)^{3/2} V \right],$$

where we have neglected the contribution $-k_B T \ln(3\pi N)^{1/2}$ from the first factor in $\mathscr{V}_\Gamma(U)$ for the same reason as in the previous section. Now, if one changes the system size while holding the number density N/V and energy per particle U/N constant, the logarithm is not constant because of the factor V, so S^* is not proportional to N. Therefore, S^* is not an extensive quantity and cannot be identified with the thermodynamical entropy, which is extensive. This problem was recognized very early in the development of statistical mechanics and the correct solution, the inclusion of the factor $1/N!$, was applied. Quantum mechanics gives the correct reason for this factor, namely that particles of the same species are indistinguishable for fundamental reasons. Furthermore, it dictates that $v = h$ and hence $\delta \mathscr{V}_\Gamma = h^{3N}$.

To summarize, in classical statistical mechanics, $\Phi(U, V, N)$ for a monatomic ideal gas is given by Equation 1.23, where we have set $\delta \mathscr{V}_\Gamma = h^{3N}$ in order to get the same expression (1.22) as in the quantum case and we take $S(U, V, N) = k_B \ln \Phi(U, V, N)$. Thereby, the Sackur-Tetrode Equation 1.17 for the entropy is valid. Since the value of $\delta \mathscr{V}_\Gamma$ is cancelled in entropy differences, it is only in the value of S itself that the quantum mechanical nature of the granularity of phase space matters.

As we will see later, the inclusion of a granularity of phase space with a cell size $\delta \mathscr{V}_\Gamma = h^{3N}$ is relevant not only for the monatomic ideal gas. In fact, this constitutes the correct way in the general case to formulate a classical statistical mechanics, in agreement with quantum theory at large energies (large quantum numbers). The inclusion of the factor $1/N!$ is likewise necessary for the same reason.

{**EXERCISE 1.2**

　　a. A point particle of mass m is free to move in one dimension. Denote its position coordinate by x and its momentum p. Suppose that this particle is confined within a box so it is located between $x = 0$ and $x = L$ and suppose that its energy is known to lie

---

[20] The volume $\mathscr{V}_s^{(3N)}([2mU]^{1/2})$ where $\mathbf{p}^N$ is located for all energies $\leq U$ consists almost entirely of points corresponding to energies in the immediate vicinity of the energy U. In practice, the volume outside this vicinity is completely negligible for a macroscopic system. This follows from the fact that the volume $\mathscr{V}_s^{(3N)}(\mathcal{R})$ of a hypersphere with radius $\mathcal{R}$ is enormously much larger than one with radius $\mathcal{R} - \delta\mathcal{R}$, where $\delta\mathcal{R}$ is relatively small compared to $\mathcal{R}$. We have $(\mathcal{R} - \delta\mathcal{R})^{3N}/\mathcal{R}^{3N} = \exp(3N \ln[1 - \delta\mathcal{R}/\mathcal{R}]) \approx \exp(-3N\,\delta\mathcal{R}/\mathcal{R}) \lll 1$ provided that N is very large compared to $\mathcal{R}/\delta\mathcal{R}$.

between u and $u + \delta u$. Draw the classical phase space of this particle, indicating the regions of this space which are accessible to the particle.

b. Consider a system consisting of two weakly interacting point particles, each of mass m and free to move in one dimension. Denote the respective position coordinates of the particles by x_1 and x_2, their respective momenta by p_1 and p_2. The particles are confined within a box with end walls located at $x = 0$ and $x = L$. The total energy of the system is known to lie between U and $U + \delta U$. Since it is difficult to draw a four-dimensional phase space, draw separately the part of the phase space involving x_1 and x_2 and that involving p_1 and p_2. Indicate on these diagrams the regions of phase space accessible to the system.}

APPENDIX 1A: ALTERNATIVE EXPRESSIONS FOR THE ENTROPY OF AN ISOLATED SYSTEM

In this appendix, we consider an isolated macroscopic system containing particles of one species. In Section 1.2, we introduced $\Omega = \Omega(U, V, N)$ as the number of quantum states of the system at energy U when the volume is V and the system contains N particles. However, in footnote 14 we argued that it is not suitable (or even possible) to have Ω as the number of quantum states at *precisely* the energy U, and instead we took Ω as the number of quantum states with energies between $U - \delta U$ and U, where δU is a small but arbitrary energy interval. Here we shall look more closely at the properties of Ω so defined and hence the properties of $S = S(U, V, N) = k_B \ln \Omega(U, V, N)$. In particular, we shall see that S is well defined in practice despite that δU is arbitrary, which may seem at first sight to be a contradiction.

For a given V and N, the total number of quantum states with energies $\leq U$ for the system is denoted by $\Phi = \Phi(U, V, N)$. For a macroscopic system, the quantum states lie very densely on the energy axis, that is, the spacing between the energies for them is extremely small provided that U is well above the ground state energy for the system, which we assume here. Therefore, Φ can be considered to be a continuous function of U as a very good approximation. The interval δU mentioned above is assumed to be large compared to this spacing, but small compared to U.

We have

$$\Omega = \Phi(U, V, N) - \Phi(U - \delta U, V, N),$$

where the last term removes those states that are below $U - \delta U$ and hence are not in the interval between $U - \delta U$ and U. Obviously, Ω so defined will depend on δU but, as we shall see, for a macroscopic system the value of Ω is extremely insensitive of δU, so this dependence can be completely ignored. The reason for this is, as shown below, that Φ increases *extremely* fast as a function of U so that the number of quantum states with energies less than U', where $U' < U$, is *very* small compared to those between U' and U. Therefore, $\Phi(U, V, N)$ is dominated by the states at or just below the energy U on the macroscopic energy scale. (On the microscopic energy scale there is, however, a limit to how close U' can be to U for this to be true, as we will see below.)

For a macroscopic system, an order-of-magnitude estimate of the U dependence of Φ for large values of U is typically that $\Phi \sim \varphi(V, N)U^{\alpha N}$ where φ depends only on V and N, α is a number that is approximately proportional to the number of degrees of freedom per particle and U is counted from the ground state energy (i.e., $U = 0$ in the ground state). For example, for a monatomic ideal gas, where we only include the three translational degrees of freedom, it is shown in Section 1.2.1 that $\alpha = 3/2$. Since N is of the order of, say, 10^{20} or larger, we see that $U^{\alpha N}$ is an extremely rapidly increasing function of U.

We have $\Phi(U - \delta U, V, N) \sim \varphi(V, N)[U - \delta U]^{\alpha N}$ and hence

$$\frac{\Phi(U - \delta U, V, N)}{\Phi(U, V, N)} \sim \frac{[U - \delta U]^{\alpha N}}{U^{\alpha N}} = \left[1 - \frac{\delta U}{U}\right]^{\alpha N} = e^{\alpha N \ln[1 - (\delta U/U)]} \sim e^{-\alpha N \delta U/U}$$

since $\delta U/U$ is small. The average energy per particle is $\bar{u} = U/N$, so it follows that

$$\Phi(U - \delta U, V, N) \sim \Phi(U, V, N)e^{-\alpha \delta U/\bar{u}}. \tag{1.24}$$

Thus, the order of magnitude of Φ is a factor of $\exp(-\alpha \delta U/\bar{u})$ smaller at energy $U - \delta U$ than at energy U. Now, despite that δU is small on a macroscopic scale, it is typically large compared to the energy of a particle, so $\delta U/\bar{u}$ is a large number. This implies that $\exp(-\alpha \delta U/\bar{u})$ is a very small number, so $\Phi(U - \delta U, V, N)$ is negligible compared to $\Phi(U, V, N)$. This is the reason for the insensitivity of Ω with respect to the magnitude of δU.

This can be substantiated further for the entropy S. We have

$$\frac{S}{k_B} = \ln\left[\Phi(U, V, N) - \Phi(U - \delta U, V, N)\right]$$

$$= \ln \Phi(U, V, N) + \ln\left[1 - \frac{\Phi(U - \delta U, V, N)}{\Phi(U, V, N)}\right],$$

which implies, using the result in Equation 1.24, that

$$S = k_B \ln \Phi(U, V, N) + k_B \tau,$$

where $\tau \approx \ln[1 - \exp(-\alpha \delta U/\bar{u})]$. Now, $\ln \Phi \sim \ln(\varphi U^{\alpha N}) = \alpha N \ln U + \ln \varphi$ has a magnitude of the order N (one can show that $\ln \varphi$ also has this magnitude), while τ (which contains $\alpha \delta U/\bar{u}$) is independent of N for given δU and $\bar{u}$. When N is very large, the first term in S will therefore be much larger than the second one. Furthermore, when $\delta U/\bar{u}$ is a large number, we have seen that $\exp(-\alpha \delta U/\bar{u})$ is very small, so $\tau \approx -\exp(-\alpha \delta U/\bar{u})$, which hence is very small too. Thus, for macroscopic systems with large N, the last term in S is negligible compared to the first one. Thus, to a very good approximation

$$S = k_B \ln \Phi(U, V, N),$$

so $\ln \Omega$ can in practice be replaced by $\ln \Phi$ in the definition of S. Thus, the number of quantum states with energies between $U - \delta U$ and U can be replaced by the *total number of quantum states* below U in the definition of S. This is another way to see the consequences of the fact that $\Phi(U, V, N)$ is dominated by the number of quantum states at or just below U.

CHAPTER 2

Statistical Mechanics from a Quantum Perspective

2.1 POSTULATES AND SOME BASIC DEFINITIONS

An energy level consists of all microscopic (quantum) states of a system with the same energy. The energy levels (the energy values and degeneracies they have) of a system depend on the number of particles and the volume. For the time being, we restrict ourselves to systems containing only one kind of particles. Let us regard an isolated, macroscopic system as sketched in Figure 2.1. The number of (accessible) quantum states of the system at energy U when the volume is V and there are N number of particles present is

$$\Omega = \Omega(U, V, N).$$

We now (again) state the fundamental postulate of statistical mechanics:

> An isolated system in equilibrium is equally likely to be in any of its accessible microscopic (quantum) states.

This is called the postulate about **equal a priori probability**. An equilibrium situation is characterized by the fact that the probability to find the system in any one of its quantum states is independent of time, and we see that the probability for each particular state is $1/\Omega$. The validity of this postulate can be checked by making theoretical predictions based on it and by comparing these predictions with experimental results. A large amount of such calculations have been made, and in general they are in good agreement with observations for macroscopic systems.[1]

---

[1] However, one can make simple theoretical models that do not fulfill this postulate, for example a system in a pure quantum state (energy eigenstate) in which it would remain forever. The system would not make any transitions to other states. But, as we have said before, to be sure that a system has a precisely specified energy, one must observe the system for an infinitely long time, and, furthermore, it is not possible to isolate the system completely. Therefore, a real system will make transitions between the various accessible states.

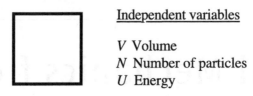

Independent variables

V Volume
N Number of particles
U Energy

FIGURE 2.1 An isolated macroscopic system consisting of one species of molecules.

Many macroscopic properties that we can observe for a system constitute *time averages*, whereby the average of a quantity X that depends on time t is $\langle X \rangle = \int_{t_i}^{t_f} dt\, X(t)/(t_f - t_i)$ in the limit $t_f \to \infty$, where t_i is the initial time and t_f is the final time. The time average is independent of t_i for equilibrium systems and one can in practice select a large but finite value of t_f. For example, the pressure of a gas equals the force per unit area that the gas exerts on its container walls when the gas molecules collide and interact with them. On a very short time scale the number of molecules that collide with a wall fluctuates wildly and so does the force they exert. However, the macroscopic pressure that we experience does not fluctuate in this manner. The walls have a large inertia (compared to the molecules) and do not respond to the impact of an individual molecule. The macroscopic pressure equals the *average* force per unit area calculated over a time interval that is sufficiently long to smooth out the fluctuations (this interval can, however, be rather short on a macroscopic scale).

Instead of calculating time averages from the time evolution of the system, which is an extremely difficult task due to the enormous number of molecules involved, Gibbs introduced an alternative procedure, the *ensemble method*. An ensemble is a (mental) collection of a very large number of macroscopic systems, all having the same values of the macroscopic parameters that define them (in the present case U, V and N). The members of the ensemble are thus *replicas* of the original system. All of these systems are accordingly macroscopically identical (they are all in the same thermodynamic state), but they are not equal on the microscopic level. In fact, there is an *extremely* large number of quantum states that are consistent with a given thermodynamic state (we are here talking about the quantum states of the entire system, not those of the individual molecules). Considering that we are dealing with, say, 10^{20} molecules or more, it is perfectly clear that it is not sufficient to specify three parameters (U, V and N) to define the detailed molecular state of the system.

At each instance in time, many different quantum states are represented in the various systems in the ensemble. For example, the instantaneous value of the pressure would generally be different for these different quantum states. Provided that the states are representative of the given thermodynamic state, the macroscopic value of a mechanical variable[2] can be calculated by taking the *ensemble average*, that is, by calculating the average value of the variable for all systems in the ensemble at a single instance of time giving equal weight to each of these systems. This procedure gives the correct value of the macroscopic variable if the following statement is true:

---

[2] Mechanical variables are those that can be defined in purely mechanical terms, for instance, pressure, energy and other properties that can be written as averages of functions that depend on the position coordinates and the momenta of the particles in the system. Volume and number of particles can also be considered as being mechanical variables. Entropy and temperature are examples of nonmechanical variables.

> The (long) time average of any mechanical variable is equal to the ensemble average of the same variable in the limit of infinite number of member systems in the ensemble.

This is taken as one of the basic postulates of statistical mechanics, and it is called the **ergodic hypothesis**. Note that it is *not* restricted to isolated systems. The only requirement is that the microscopic states of the systems in the ensemble are representative of the given thermodynamic state (this means that the probability of occurrence of the states has to be the appropriate one – more about this later). When applying the postulate, we will only be considering situations where all microscopic states are accessible when starting from any one of them. Again, the validity of this postulate can be checked by comparing its predictions with experimental observations, and so far, there is no experimental evidence available that falsifies it for the kind of systems we shall be concerned with.[3]

An ensemble that consists of isolated systems in the same thermodynamic state (same energy, volume and number of particles in our case) is called a *microcanonical ensemble*. An investigation of the properties of such ensembles is a good starting point in the systematic development of statistical thermodynamics, and we have actually already covered some ground in the first chapter. Later we will introduce other kinds of ensembles.

Let us first make some basic definitions for the general case. Some of the variables that define a macroscopic system, like the volume or the number of particles, affect the equations of motion of the system (i.e., they affect the Hamiltonian or the boundary conditions). They are called **external variables**.[4] The energies of the quantum states of the system, $\{\mathcal{U}_i\}$ where i is the state index, depend on these variables, for example in our case above we have $\mathcal{U}_i = \mathcal{U}_i(V, N)$. Consider two systems that interact with each other so they exchange energy. The interaction can be such that the external variables vary during the interaction or it can be such that these variables are constant. In the latter case, the possible energy levels of each system do not change and the energy transfer occurs entirely due to transitions between the levels. This is what we will call an entirely thermal interaction and the energy transferred is called heat (denoted q).

Two systems that cannot interact with each other when their external variables are fixed are **thermally insulated** from each other. It is, however, possible for such systems to interact by affecting each other's external variables; for instance, one system can increase its volume at the expense of the other system's volume. An interaction under these conditions is purely mechanical and we say that the process is **adiabatic**.[5] The transferred energy is called

---

[3] However, one can construct simple theoretical models that are non-ergodic, for example a system consisting of a collection of (perfect) harmonic oscillators. Each oscillator is uncoupled to all the other oscillators and the energy will not be redistributed among them. The system will therefore be stuck in one state. Any real oscillator is not perfectly harmonic, so the system will make transitions between the different states.

[4] Other examples of external variables are external fields – like electric or magnetic fields from a source outside the system – with which the molecules of the system interact.

[5] The term "adiabatic" is here used in the sense that is common in classical thermodynamics. This term may also mean that a process is done very slowly, but for the latter meaning we use the term "quasistatic" in this treatise, as will be described later.

work (denoted w).[6] In general, the amount of energy that is transferred depends on in what manner the external parameters are changed and how rapidly they are changed. Since the external parameters are varied, the energy levels of the systems change during the process. In addition, transitions between the various levels may be produced. The energy of each of the systems is affected by both.

In the most general case of interactions between two systems, the systems affect each other's external variables and they are not thermally insulated from each other. Then only a part of the energy change is due to the variation of the external parameters. This part is still called work and the remaining part is, by definition, **heat**. Together with the law of conservation of energy in quantum mechanics, this constitutes the **first law of thermodynamics** and we have[7]

$$dU = đ\mathbf{q} + đ\mathbf{w},$$

where $đ\mathbf{q}$ and $đ\mathbf{w}$ equal the heat *transferred to* and the work *done on* the system, respectively.[8]

In the last few paragraphs, we have considered quite general processes whereby two systems can interact with each other. A special, important case is a process that proceeds so slowly that at least one of the systems remains arbitrarily close to equilibrium at all times. The process is then said to be **quasistatic** for that particular system.[9] This simplifies the treatment considerably.

Consider a change in volume of a system, from V to $V + dV$ by moving one of the walls, as illustrated in Figure 2.2. If the system is in quantum state i, its energy equals $\mathcal{U}_i(V, N)$.

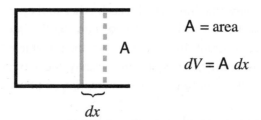

FIGURE 2.2   The volume of a system is changed by moving one of its walls a distance dx. The mobile wall, drawn in gray, has area A and acts like a frictionless piston. The motion of the wall is controlled by varying a force applied from the outside. At equilibrium, this force is equal to the force from the inside.

---

[6] Distinguish w from w, which will denote something else later on.

[7] Sometimes this expression is written as $dU = đ\mathbf{q} − đ\mathbf{w}$, where $đ\mathbf{w}$ is the work done *on the surroundings*. In this treatise, $đ\mathbf{w}$ and w always mean the work done *on the system*. The only difference between the two alternatives is a change in sign of $đ\mathbf{w}$ and w everywhere.

[8] The subdivision of dU into $đ\mathbf{q}$ and $đ\mathbf{w}$ depends on how the process is performed. This is the reason why we use $đ$ to indicate the small amount rather than d; the quantities $đ\mathbf{q}$ and $đ\mathbf{w}$ are not exact differentials.

[9] How slowly the process has to be carried out in order to be quasistatic depends on the system in question. Specifically, it depends on the time the system needs to attain equilibrium after being suddenly disturbed (the relaxation time). If the time scale of the process is, say, a few orders of magnitude larger than the relaxation time, then the process can be regarded as quasistatic to a good approximation.

During the volume change, the energy of this state changes according to

$$d\mathscr{U}_i = \left(\frac{\partial \mathscr{U}_i}{\partial x}\right)_N dx = -F_i dx,$$

where $(\partial \mathscr{U}_i/\partial x)_N = -F_i$ defines a force. The work done on the system while it remains in this state is by definition equal to $d\mathscr{U}_i$. Consider now an ensemble of macroscopically identical systems. In general, these systems are in different quantum states. Calculate the average force $F = \langle F_i \rangle$, where the average is taken over all systems of the ensemble and i indicates the quantum state that each particular system is in. Equivalently, in the limit of infinite number of member systems of the ensemble, the average can be taken over all quantum states, provided each state is weighted according to its probability. Thus, we have $F = \langle F_i \rangle_i$ where we have defined the average $\langle \cdot \rangle_i$ as

$$\langle F_i \rangle_i = \sum_i \mathfrak{p}_i F_i,$$

where the sum is over the quantum states and $\mathfrak{p}_i$ is the probability of state i. In the microcanonical ensemble for a system with energy U, we have $\mathfrak{p}_i = 1/\Omega$ when $\mathscr{U}_i = U$ and $\mathfrak{p}_i = 0$ otherwise.

If the volume change is done quasistatically, F has a well-defined value at any time. It can be calculated from the equilibrium ensemble that applies for the external parameter values at that time. The average work done on the systems is $\bar{d}w_{\text{qs}} = \langle d\mathscr{U}_i \rangle = -\langle F_i \rangle\, dx = -F dx$, that is, the macroscopic work done on the system that the ensemble represents. The subscript qs stands for quasistatic. During a quasistatic volume change, the system is in equilibrium at all times and the force on both sides of the wall is the same. Thus,

$$\bar{d}w_{\text{qs}} = -\frac{F}{A}A\,dx = -PdV,$$

where A is the area and P is the pressure (the average force per unit area). We can conclude that

$$P = -\frac{1}{A}\left\langle \left(\frac{\partial \mathscr{U}_i}{\partial x}\right)_N \right\rangle_i,$$

which we can write as

$$P = -\left\langle \left(\frac{\partial \mathscr{U}_i}{\partial V}\right)_N \right\rangle_i, \tag{2.1}$$

that is, the pressure is given by the average rate of change of the energy of the quantum states with varying volume. Furthermore, we have

$$\bar{d}w_{\text{qs}} = \left\langle \left(\frac{\partial \mathscr{U}_i}{\partial V}\right)_N \right\rangle_i dV \tag{2.2}$$

for a quasistatic volume change.

{**EXERCISE 2.1**

A particle in a box of dimensions $L_x \times L_y \times L_z$ has quantum states with energies

$$u_{j_x,j_y,j_z} = \frac{h^2}{8m}\left\{\left[\frac{j_x}{L_x}\right]^2 + \left[\frac{j_y}{L_y}\right]^2 + \left[\frac{j_z}{L_z}\right]^2\right\},$$

where j_x, j_y and j_z are positive integers. If $L_x = L_y = L_z = L$, determine the seven lowest energy levels of the particle and their degeneracies (Ω). What happens with the states in these energy levels if we decrease the volume of the box by decreasing L_x *very slightly* while $L_y = L_z = L$ are kept constant?}

2.2 ISOLATED SYSTEMS: THE MICROCANONICAL ENSEMBLE

As we have seen, a microcanonical ensemble describes an isolated system which has a fixed energy, volume and number of particles. The ensemble is, by definition, composed of a large number of replicas of the system, all having the same value of U, V and N.

As an illustration, let us return to our beloved system from Section 1.1 consisting of two isolated systems A and B shown in Figure 2.3. The quantities U, V and N are obviously additive: $U_{AB} = U_A + U_B$, $V_{AB} = V_A + V_B$ and $N_{AB} = N_A + N_B$. The number of microscopic states $\Omega = \Omega(U, V, N)$ of the combined system equals $\Omega_{AB} = \Omega_A\Omega_B$. We define the entropy $S = k_B \ln \Omega$, and we get $S_{AB} = k_B \ln(\Omega_A\Omega_B) = k_B \ln \Omega_A + k_B \ln \Omega_B = S_A + S_B$. Thus,

$$S = S(U, V, N)$$

is also additive.

Starting from the two isolated systems with initial energies U_A^0 and U_B^0, we now let A and B come in thermal contact with each other as shown in Figure 2.4 and let the combined system AB assume its equilibrium thermodynamical state. AB is isolated, but we have broken the isolation of subsystems A and B by allowing heat to pass between them. V and N remain unchanged for each of A and B. Still, U, V and N are obviously additive; for U, we now have $U_{AB} = \langle U_A\rangle + \langle U_B\rangle$. The logarithm of the total number of microscopic states is

$$\ln \Omega_{AB}(U_{AB}, V_{AB}, N_{AB}) = \ln \sum_{U_A} \Omega_A(U_A, V_A, N_A)\Omega_B(U_{AB} - U_A, V_B, N_B)$$

$$\approx \ln\left[\Omega_A(\langle U_A\rangle, V_A, N_A)\Omega_B(U_{AB} - \langle U_A\rangle, V_B, N_B)\right]$$

$$= \ln \Omega_A(\langle U_A\rangle, V_A, N_A) + \ln \Omega_B(\langle U_B\rangle, V_B, N_B)$$

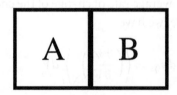

FIGURE 2.3   Two isolated systems A and B sitting next to each other.

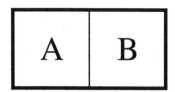

FIGURE 2.4 The systems A and B in Figure 2.3 are in thermal contact with each other, but are otherwise unchanged. They remain isolated from the surroundings.

to an extremely good approximation since the maximal term in the sum dominates for a macroscopic system ($\Omega_A \Omega_B$ in the sum is very sharply peaked). At equilibrium, we have the value $\langle U_A \rangle$ of the energy for subsystem A and the entropy is

$$S_A = k_B \ln \Omega_A(\langle U_A \rangle, V_A, N_A)$$

(and analogously for B), so the entropy remains additive $S_{AB} = S_A + S_B$ in practice. Note that if we would put back the insulating wall between system A and B when equilibrium is reached (without perturbing the system otherwise), then we would have the values of U and S for each isolated system equal to what they were (on average) in the combined system in the absence of the wall and the total entropy S_{AB} would not change.

Consider an isolated system that has some internal constraint such that only a subset $\Omega'(U, V, N)$ of all microscopic states of the system is available. From the postulate about equal a priori probability, it follows that the probability for each of the available states is $1/\Omega'$. If we remove the constraint (without changing U), more quantum states will become available. The number of states now available is $\Omega''(U, V, N)$ and we have $\Omega''(U, V, N) > \Omega'(U, V, N)$ because the original states constitute a subset of the new set of states (since U is the same). Thus, the probability to observe one of the original states *decreases*, $1/\Omega' > 1/\Omega''$. The probability to observe the system in any of the initial states becomes Ω'/Ω''. If Ω'' is *overwhelmingly* much larger than Ω', the system will be seen to spontaneously leave the constrained set of states since Ω'/Ω'' becomes vanishingly small.[10] The process is **irreversible** since, if the constraint would be put back again, the system is extremely much more likely to be in some other state than any of the original ones. Since $\Omega'' > \Omega'$, we see that $S'' > S'$, that is, the entropy has increased. This applies for all spontaneous (irreversible) processes in isolated systems for essentially the same reason. The constraint can be of a very general nature; it can even be an activation barrier which in practice (that is, for reasonably long times) restricts the number of available states to a small subset of all states. If the system is brought over the barrier (perhaps by external intervention), there is a vanishingly small probability that it will return to any of the initial states. In principle it could happen, but in practice is does not since the probability is so small.

---

[10] It is important here that $\Omega'' \gg \Omega'$. Otherwise, if, say, $\Omega'' = 2\Omega'$, half of all observations would show the system in one of the original states.

{**EXERCISE 2.2**

A monatomic ideal gas with N particles is enclosed in a container of volume V_1. We increase the container volume to $V_2 = 4V_1$ and let the gas fill this volume.

a. When equilibrium has been reached after the expansion, what is the probability that all gas particles simultaneously and spontaneously will be located in the original volume V_1? Calculate this probability explicitly for the two cases $N = 2$ and $N = 100$.

b. Consider the spontaneous process in a system of volume $4V_1$ where all particles initially are constrained to be in a volume V_1 and finally, after removal of the constraint, fill the volume $4V_1$. Thus, the gas has expanded from volume V_1 to $V_2 = 4V_1$. The system contains $N = 100$ molecules. Make use of the result in (a) to discuss this expansion in relation to the likelihood of the reverse process where all molecules gather in volume V_1.

c. Proceed to discuss the characteristics of a spontaneous, irreversible expansion of a macroscopic ideal gas (i.e., when there is a very large number of particles in the system). Is it possible for the reverse process to be spontaneous in the latter case? How can one motivate the use of the term "irreversible."}

Next, we define the temperature

$$T = \frac{1}{(\partial S/\partial U)_{V,N}}.$$

For a heat transfer $dU = d\!q$ (having V, N = constant) that is *quasistatic* (so T is well defined at all stages), we conclude that

$$dS = \frac{d\!q}{T}.$$

We have seen that it is natural to use the independent variables U, V, and N for the entropy: $S = S(U, V, N)$. Let us now invert this function and take S, V, and N as independent variables and U as the *dependent* one

$$U = U(S, V, N).$$

(This gives, of course, the same relation between S, U, V, and N as before; we are only asking the question: for which U does S have a given value when V and N are also given?) This set of independent variables for U is the normal choice in thermodynamics, and we can see that it is indeed quite natural. The temperature can be written

$$T = \frac{1}{(\partial S/\partial U)_{V,N}} = \left(\frac{\partial U}{\partial S}\right)_{V,N}.$$

Let us now investigate a quasistatic process where energy is changed as work. We let the systems A and B exchange energy by changing their volumes (N_A and N_B are kept constant and A and B are thermally insulated from each other). Each system shifts the energy of its quantum states when V is varied and, as we have seen, the average change of energy due to this effect equals

$$\text{\dj}w_{\text{qs}} = \langle d\mathcal{U}_i \rangle_i = \left\langle \left(\frac{\partial \mathcal{U}_i}{\partial V} \right)_{N} \right\rangle_i dV = -PdV$$

where i is the quantum state index.

Consider an isolated system with volume V and let us change the volume to $V + dV$, whereby the energy changes from U to $U + \text{\dj}w$. During the variation in volume, the system is not isolated since its energy changes, but after the volume change we let it be isolated again. Ω for the system changes from the initial value $\Omega(U, V, N)$ to the final value $\Omega(U + \text{\dj}w, V + dV, N)$. When the volume change is done quasistatically, we have $\text{\dj}w = \text{\dj}w_{\text{qs}}$. We can illustrate such a change as in Figure 2.5 (in a very simplified manner) and the reader is referred to the figure caption for a description of the notation used. Initially, the system is in an energy level[11] denoted as b which has $\Omega(U, V, N)$ quantum states, and finally, it is in level b' which has $\Omega(U + \text{\dj}w_{\text{qs}}, V + dV, N)$ states. It may appear from the figure as if the right set of energy levels simply equals the left set somewhat shifted upwards, but the situation is more complex than that in general. The different quantum states in, for example, level b are not shifted by the same amount when the volume is changed, and the same applies to all other quantum states.

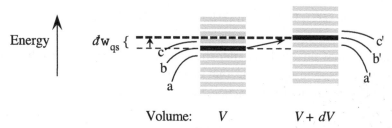

Volume: V $V + dV$

FIGURE 2.5 A very simplified sketch of energy levels of a system and what happens when the volume changes from V to $V + dV$. Each quantum state i is changed by $d\mathcal{U}_i = (\partial \mathcal{U}_i / \partial V)_N dV$ and therefore the energy levels of the system are also shifted. The left set of horizontal bands illustrates some of the energy levels when the volume assumes its initial value V; each band (at energy $\mathcal{U}$) contains $\Omega(\mathcal{U})$ quantum states; in reality, the energy levels are very close together on the energy axis. Three of the levels are marked a, b and c. The black one (b) indicates the initial energy level of the system. To the right, the energy levels are shown for the final value of the volume (the volume is decreased in this case, $dV < 0$, so the energy has increased). The black (b') indicates the energy level that differs in energy by $\text{\dj}w_{\text{qs}} = \langle d\mathcal{U}_i \rangle_i$ from that of b. Level b has $\Omega(U, V, N)$ states and level b', which is reached after a quasistatic change in volume, has $\Omega(U + \text{\dj}w_{\text{qs}}, V + dV, N)$ states.

---

[11] Here and in what follows, it would be more appropriate to talk about the energy levels within a band of (a quite arbitrary) width δU around the energy of the system (see footnote 14 of Chapter 1 on the definition of S). But to keep the argument simple, we shall continue to talk about single energy levels. A more complete argument is presented in Appendix 2A.

Mathematically, it follows formally that the change in $\ln \Omega (U, V, N)$ when N is constant is given by

$$d \ln \Omega = \left(\frac{\partial \ln \Omega}{\partial U} \right)_{V,N} dU + \left(\frac{\partial \ln \Omega}{\partial V} \right)_{U,N} dV, \tag{2.3}$$

so for a quasistatic volume change we have

$$d \ln \Omega = \ln \Omega (U + đw_{qs}, V + dV, N) - \ln \Omega (U, V, N)$$

$$= \left(\frac{\partial \ln \Omega}{\partial U} \right)_{V,N} đw_{qs} + \left(\frac{\partial \ln \Omega}{\partial V} \right)_{U,N} dV$$

$$= \left[\left(\frac{\partial \ln \Omega}{\partial U} \right)_{V,N} \left\langle \left(\frac{\partial \mathscr{U}_i}{\partial V} \right)_N \right\rangle_i + \left(\frac{\partial \ln \Omega}{\partial V} \right)_{U,N} \right] dV. \tag{2.4}$$

By investigating how much the number of quantum states changes on average due to a variation in V, it is shown in Appendix 2A (Equation 2.156) that

$$\left(\frac{\partial \ln \Omega}{\partial V} \right)_{U,N} = - \left(\frac{\partial \ln \Omega}{\partial U} \right)_{V,N} \left\langle \left(\frac{\partial \mathscr{U}_i}{\partial V} \right)_N \right\rangle_i \tag{2.5}$$

as an extremely good approximation for a macroscopic system. By inserting this in Equation 2.4, we see that the rhs is equal to zero, which implies that

$$\ln \Omega (U, V, N) = \ln \Omega (U + đw_{qs}, V + dV, N). \tag{2.6}$$

This means that each new energy level after the volume change in practice has the same value of $\ln \Omega$ as the corresponding old one, that is, level b′ has the same value of $\ln \Omega$ as level b, as illustrated in Figure 2.6. Thus, in practice, the levels are simply shifted by the amount

$$đw_{qs} = - \left\langle \left(\frac{\partial \mathscr{U}_i}{\partial V} \right)_N \right\rangle_i dV$$

while retaining their $\ln \Omega$ values. Since $\ln \Omega$ is retained, it means that $S = k_B \ln \Omega$ is unchanged too, so the *entropy of a system does not vary due to quasistatic work*. If we would reverse the volume change dV quasistatically, it also follows that if we are in level b′ we would come back to b, that is, the same level as we started from, so the process is **reversible**. This means that the entropy is not changed during reversible work, so this essential ingredient in the second law of thermodynamics is now explicitly demonstrated for the case of volume changes.

By inserting the definitions of S, T and P into Equation 2.5, we can write it as

$$\left(\frac{\partial S}{\partial V} \right)_{U,N} = \frac{P}{T}, \tag{2.7}$$

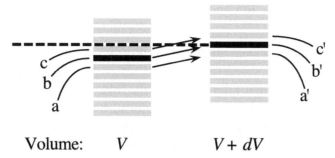

Volume: V $V + dV$

FIGURE 2.6 A sketch illustrating Equation 2.6, which can be interpreted in a simplified manner by saying that each energy level after a quasistatic volume change in practice has the same value of ln Ω as the corresponding old one, that is, that level b′ has the same value of ln Ω as level b, c′ the same as c, etcetera. (A more precise interpretation can be found in Appendix 2A.)

and by multiplying Equation 2.3 with k_B, we can hence write the latter as

$$dS = \frac{1}{T}dU + \frac{P}{T}dV, \tag{2.8}$$

which is well known from thermodynamics.

We are now ready to investigate processes with simultaneous, quasistatic heat and work transfer. Let us investigate the total change in energy during a quasistatic process when S and V may change while N is constant. We have $U = U(S, V, N)$, and hence

$$dU = \left(\frac{\partial U}{\partial S}\right)_{V,N} dS + \left(\frac{\partial U}{\partial V}\right)_{S,N} dV. \tag{2.9}$$

From our arguments above, it follows that the shift in energy due to the volume change equals $đw_{qs} = -PdV$, and we saw that this does not give rise to any change in S. Thus, we can identify

$$P = -\left(\frac{\partial U}{\partial V}\right)_{S,N}$$

(Thereby, we have at the same time given a microscopic interpretation of the differentiation condition S = constant in this well-known thermodynamic formula.) We have already seen that

$$T = \left(\frac{\partial U}{\partial S}\right)_{V,N}$$

so we obtain

$$dU = TdS - PdV,$$

which is the same as Equation 2.8. Now, since the work is $đw_{qs} = -PdV$, it follows that the remaining energy change equals $đq$ (this was our definition of heat, from which follows

$d\mathrm{q} = dU - d\mathrm{w}$), so we get $d\mathrm{q} = TdS$, and hence we have

$$dS = \frac{d\mathrm{q}}{T} \quad \text{(quasistatic process)}, \tag{2.10}$$

which means that this relationship is valid even when the volume of the system (or any other external parameter[12]) is varied quasistatically. Thus, for such a process, the entropy of the system can change only if heat is transferred to or from the system. The increase in entropy when a system is receiving heat occurs for the same reason as before: the system is excited to higher energy levels. On the other hand, during quasistatic, adiabatic work (thermally insulated systems) there are no excitations to other quantum states;[13] only the states themselves are changed. For an ensemble of systems we may express this as follows:

Heat transfer changes the population of the energy levels but leaves the levels themselves unchanged.

Quasistatic work does not change the population of the levels but changes their locations.

By reversing the heat transfer and the volume change (quasistatically), we will be able to return a system to the original thermodynamical state, so the process is reversible.

If we would do a fairly rapid volume change of a thermally insulated system, the energy levels will be changed in the same way as during a quasistatic volume change. However, due to the rapid change, the system will be excited to higher levels. During a volume *decrease* (compression), we add more energy to the system when we do it rapidly compared to when we do it slowly. Therefore, dU for the system is larger for a rapid change compared to a quasistatic process with the *same* volume change dV (in the example in Figures 2.5 and 2.6, the final energy level would be one of those above level b′). The final energy of the system is thus higher and since a higher energy means that Ω and therefore S is larger (Ω increases with energy), we conclude that S has increased after the rapid change in volume (for a quasistatic volume change S would have been unchanged).

A reversal of the volume change will not be sufficient to return the system to the initial thermodynamical state (in the example in the figures, the system would end up in one of the levels above level b; it is only a quasistatic reversal starting from b′ that will bring the system back to b). A rapid volume change is accordingly irreversible – the system cannot be brought back to the initial state by simply reversing the process (not even if the reversal is done quasistatically).

During a rapid *increase* in volume (expansion), we take out less energy from the system than during a slow increase, so we leave the system in a higher energy level compared to what

---

[12] This can be shown in a completely analogous manner as for the volume.

[13] The so-called adiabatic theorem in quantum mechanics states the following for an infinitely slow process that changes the Hamiltonian and therefore the quantum states: A system which initially is in one quantum state will remain in that state (or, more correctly expressed, in the state that derives from the initial state by continuity). In other words, there are no transitions between the states due to that kind of process.

would happen during a quasistatic process with the same volume change dV, and therefore the final S is larger after the rapid volume change.[14]

To summarize, we conclude that *the entropy increase due to irreversible work is a consequence of the fact that the final energy is higher than after reversible (quasistatic) work* for the same change dV in volume.

In the case with simultaneous work and heat exchange, the total change in entropy of the system is

$$dS = \left(\frac{\partial S}{\partial U}\right)_{V,N} dU + \left(\frac{\partial S}{\partial V}\right)_{U,N} dV,$$

$$= \frac{1}{T}(đq + đw) + PdV. \tag{2.11}$$

For a rapid volume decrease, we have $đw > -PdV > 0$ and for a rapid volume increase $-PdV < đw < 0$; the latter means that the work done on the surroundings, $-đw$, is less than the reversible work PdV that would be done on the surroundings during an equal but quasistatic volume increase (i.e., the maximal work). In both cases, we can write $đw = -PdV + dU^*$, where $dU^* > 0$, and by inserting this in Equation 2.11 we obtain

$$dS = \frac{đq}{T} + \frac{dU^*}{T} = \frac{đq}{T} + dS_{\text{irrev}}$$

where

$$dS_{\text{irrev}} = \frac{dU^*}{T} > 0$$

When the volume change is quasistatic, we have $dS_{\text{irrev}} = 0$ and Equation 2.10 applies even if the heat transfer is not quasistatic (T must, however, be defined during the transfer).

Note that it is not very important to distinguish between the various energy changes from excitations/deexcitations of the system like heat transfer, $đq$, and excitations due to irreversible work, dU^*. Together, they equal $đq + dU^*$ and give rise to an entropy change equal to $[đq + dU^*]/T$. If a part of dU^* would be classified as heat, it makes no difference to the outcome. This means, for example, that if some of the work performed on a piston during compression is "lost" due to friction but the "lost" energy is still transferred to the system, that part can be classified either as heat or as irreversible work without any problems.

---

[14] Let us consider a volume increase, where the mobile wall (for instance, a piston) is exposed to a force from the surroundings that is smaller than the force from the inside. In a classical mechanical picture of the process, the particles of the system transfer energy to the mobile wall when they set the latter in motion and increase its speed due to collisions with it. The wall does work on the surroundings when it moves. The faster the wall moves outwards, the less efficient is the transfer of energy from the particles of the system to the wall. Only particles that move towards the wall with a speed that is larger than the wall's speed can collide with it. In the extreme case when the wall is brought to its final position infinitely fast, no particles can collide with it. For an ideal gas, where the particles interact with the wall only at impact, no energy is then transferred from them. The gas does no work in this extreme case and its energy remains constant – a so-called *free expansion* of the gas.

In addition to work and heat exchange, there may also be an irreversible process going on internally in the system which gives an entropy contribution $dS_{irrev}^{int} > 0$ (e.g., from the release of a constraint). We then have

$$dS = \frac{d\mathchar'26 q}{T} + \frac{dU^*}{T} + dS_{irrev}^{int} = \frac{d\mathchar'26 q}{T} + dS_{irrev}$$

where in this case

$$dS_{irrev} = \frac{dU^*}{T} + dS_{irrev}^{int} > 0$$

is the total entropy contribution due to irreversible changes.

These results can be expressed either as

$$dS = \frac{d\mathchar'26 q}{T} + dS_{irrev} \quad \text{where } dS_{irrev} \geq 0, \tag{2.12}$$

or as

$$dS \geq \frac{d\mathchar'26 q}{T}, \tag{2.13}$$

where the equal signs apply when no irreversible changes (apart from heat transfer) take place. Both are a common manners to express the **second law of thermodynamics**.

Let us now regard the case where the number of particles N is changed. The development of the theory is parallel to that for volume changes – we only vary another external parameter this time. When a system is in quantum state i, its energy equals $\mathscr{U}_i(V, N)$. Let us assume that V is constant and write $\mathscr{U}_i = \mathscr{U}_i(N)$ for the time being. The ith state when the system has N particles and the ith state when the system has $N + dN$ particles differ in energy by

$$d\mathscr{U}_i = \mathscr{U}_i(N + dN) - \mathscr{U}_i(N).$$

We shall assume that dN is very small compared to N (of the order of a few molecules). We can define the "derivative" $\partial \mathscr{U}_i / \partial N$ from

$$\left(\frac{\partial \mathscr{U}_i}{\partial N} \right)_V = \lim_{dN \to 1} \frac{\mathscr{U}_i(N + dN) - \mathscr{U}_i(N)}{dN} = \mathscr{U}_i(N + 1) - \mathscr{U}_i(N), \tag{2.14}$$

that is, the increase in energy of quantum state i when we add one particle to the system. Normal differentiation rules will apply in practice for this "derivative" since N is very much larger than 1.

The average work done on the system when we add dN particles quasistatically[15] then equals

$$\bar{d}w_{qs} = \langle d\mathscr{U}_i \rangle_i = \left\langle \left(\frac{\partial \mathscr{U}_i}{\partial N} \right)_V \Big|_i \right\rangle dN$$

and we define the chemical potential

$$\mu = \left\langle \left(\frac{\partial \mathscr{U}_i}{\partial N} \right)_V \Big|_i \right\rangle. \tag{2.15}$$

In a similar manner as for volume changes we conclude that the degeneracy of the energy levels remain unchanged after the shift, so S does not vary during a quasistatic, adiabatic variation in N.

For a general quasistatic process (simultaneous heat transfer and work due to changes in both V and N), we conclude from $U = U(S, V, N)$ that

$$dU = \left(\frac{\partial U}{\partial S} \right)_{V,N} dS + \left(\frac{\partial U}{\partial V} \right)_{S,N} dV + \left(\frac{\partial U}{\partial N} \right)_{S,V} dN. \tag{2.16}$$

We can now identify

$$\mu = \left(\frac{\partial U}{\partial N} \right)_{S,V}.$$

By inserting T and P as before, we obtain

$$dU = TdS - PdV + \mu dN. \tag{2.17}$$

The sum of the last two terms equals $\bar{d}w$ and the first term in the rhs hence equals $\bar{d}q$. Thus, $\bar{d}q = TdS$ as before.

The differential expressions (2.16) and (2.17) are the appropriate ones for $U = U(S, V, N)$. When we consider $S = S(U, V, N)$, we have instead

$$dS = \left(\frac{\partial S}{\partial U} \right)_{V,N} dU + \left(\frac{\partial S}{\partial V} \right)_{U,N} dV + \left(\frac{\partial S}{\partial N} \right)_{U,V} dN$$

and we can write Equation 2.17 as

$$dS = \frac{1}{T} dU + \frac{P}{T} dV - \frac{\mu}{T} dN.$$

---

[15] A quasistatic addition of a particle to a system can be thought of as a gradual addition where the new particle initially does not interact with the other particles and finally behaves as any other particle. In theory one can imagine a process where one slowly brings the new particle into the system from a point at an infinite distance, where the particle does not interact with the other particles, and place it at a point inside the system. One then releases the particle slowly by letting it roam around in an increasing volume inside the system. The other particles move around at all times. The system (including the new particle) is allowed to relax to its equilibrium state at every stage of the slow, gradual process. Finally, the new particle is allowed to move around in the volume V like any other particle.

By identification of the coefficients in front of dU, dV and dN, respectively, in these two expressions, we see that $1/T = (\partial S/\partial U)_{V,N}$, which we recognize from our definition of temperature, $P/T = (\partial S/\partial V)_{U,N}$, which is the same as Equation 2.7, and

$$\frac{\mu}{T} = -\left(\frac{\partial S}{\partial N}\right)_{U,V}. \tag{2.18}$$

By inserting $S = k_B \ln \Omega(U, V, N)$, we can write these three relationships as

$$T^{-1} = k_B \left(\frac{\partial \ln \Omega}{\partial U}\right)_{V,N} \tag{2.19}$$

$$P = k_B T \left(\frac{\partial \ln \Omega}{\partial V}\right)_{U,N} \tag{2.20}$$

$$\mu = -k_B T \left(\frac{\partial \ln \Omega}{\partial N}\right)_{U,V}. \tag{2.21}$$

This implies that knowledge of the function $\Omega(U, V, N)$ is sufficient to be able to determine S, T, P and μ, and therefore all other thermodynamic quantities for an isolated system at equilibrium. $\Omega(U, V, N)$ is the degeneracy of the eigenstates of the Schrödinger equation for the whole system at energy U, so it is – at least in principle – possible to calculate Ω from the microscopic theory of matter. Thereby, *we have a recipe of how to pass from the microscopic (quantum mechanical) to the macroscopic (thermodynamical) description of matter!*

Example 2.1: The Ideal Monatomic Gas: Thermodynamic Quantities and the Ideal Gas Law

In Section 1.2.1, we determined $\ln \Omega(U, V, N)$ for the ideal monatomic gas and we found that [Equation 1.17]

$$\ln \Omega(U, V, N) = \frac{S(U, V, N)}{k_B} = N \ln \left[e^{5/2} \left(\frac{4\pi m}{3h^2}\right)^{3/2} \left(\frac{U}{N}\right)^{3/2} \frac{V}{N} \right]$$

$$= \frac{5N}{2} + \frac{3N}{2} \ln \left[\frac{4\pi m}{3h^2}\right] + \frac{3N}{2} \ln U + N \ln V - \frac{5N}{2} \ln N,$$

By taking the derivative of $\ln \Omega(U, V, N)$ with respect to U and using Equation 2.19, we obtain the inverse temperature

$$T^{-1} = k_B \left(\frac{\partial [\frac{3}{2} N \ln U]}{\partial U}\right)_{V,N} = \frac{3N k_B}{2U}$$

(the remaining terms give zero), which is the same as Equation 1.8. From the V derivative, we likewise obtain from Equation 2.20 the pressure of the gas $P = k_B T(\partial [N \ln V]/\partial V)_{U,N} = k_B T N/V$, which can be written as

$$PV = N k_B T, \tag{2.22}$$

that is, the **ideal gas law**. In molar units, it is written $PV = nRT$, where $n = N/N_{Av}$ is the number of moles, N_{Av} is Avogadro's constant, and $R = k_B N_{Av}$ is the **universal gas constant**.

The N derivative of $\ln \Omega(U, V, N)$ gives, according to Equation 2.21, the chemical potential

$$
\begin{aligned}
\mu &= -k_B T \left[\frac{5}{2} + \frac{3}{2} \ln \left(\frac{4\pi m}{3h^2} \right) + \frac{3}{2} \ln U + \ln V - \frac{5}{2} \ln N - \frac{5N}{2N} \right] \\
&= -k_B T \left[\frac{3}{2} \ln \left(\frac{4\pi m}{3h^2} \right) + \frac{3}{2} \ln \frac{U}{N} + \ln \frac{V}{N} \right] \\
&= -k_B T \left[\frac{3}{2} \ln \left(\frac{2\pi m k_B T}{h^2} \right) + \ln \frac{V}{N} \right],
\end{aligned}
$$

where we have used $U/N = 3k_B T/2$ to obtain the last equality. We can write this as

$$
\mu = k_B T \ln(\Lambda^3 N/V) = k_B T \ln(\Lambda^3 n), \tag{2.23}
$$

where Λ is the thermal de Broglie wavelength defined in Equation 1.19 and $n = N/V$ is the number density. By inserting $N/V = P/k_B T$ from the ideal gas law, we obtain

$$
\mu = k_B T \ln \frac{\Lambda^3 P}{k_B T}, \tag{2.24}
$$

which we can write as $\mu = k_B T \ln(\Lambda^3/k_B T) + k_B T \ln P$. In the latter expression, the argument of each logarithm has a unit, which should be avoided. One may therefore introduce an arbitrary pressure $P^{\ominus}$ and write $\mu = k_B T \ln(\Lambda^3 P^{\ominus}/k_B T) + k_B T \ln P/P^{\ominus}$. We can write this as

$$
\mu = \mu^{\ominus} + k_B T \ln \frac{P}{P^{\ominus}}, \tag{2.25}
$$

where

$$
\mu^{\ominus} = \mu^{\ominus}(T) = k_B T \ln \frac{\Lambda^3 P^{\ominus}}{k_B T}. \tag{2.26}
$$

The value of μ from these expressions is obviously not affected by the choice of $P^{\ominus}$. Since $P/P^{\ominus} = nk_B T/P^{\ominus}$, we can also write Equation 2.25 as

$$
\mu = \mu^{\ominus} + k_B T \ln \frac{n}{n^{\ominus}}, \tag{2.27}
$$

where $n^{\ominus} = P^{\ominus}/k_B T$ is the number density of the gas at pressure $P^{\ominus}$.

Likewise, by inserting $V/N = k_B T/P$ into Equation 1.18, we obtain

$$
S = N k_B \ln \left[\frac{e^{5/2} k_B T}{\Lambda^3 P} \right], \tag{2.28}
$$

which can be written as

$$S = N\left[s^{\ominus} - k_B \ln \frac{P}{P^{\ominus}}\right] = N\left[s^{\ominus} - k_B \ln \frac{n}{n^{\ominus}}\right], \tag{2.29}$$

where

$$s^{\ominus} = s^{\ominus}(T) = k_B \ln\left[\frac{e^{5/2}k_B T}{\Lambda^3 P^{\ominus}}\right] \tag{2.30}$$

is the entropy per molecule at pressure $P^{\ominus}$.

Equation 2.25 is often used for μ of an ideal gas in thermodynamics. Then $P^{\ominus}$ is the so-called **standard state pressure** and $\mu^{\ominus}$ is the **standard state chemical potential** of the gas. It is customary to select $P^{\ominus} = 1$ bar $= 10^5$ N m^{-2}.[16] Likewise, Equation 2.29 is used for the entropy of an ideal gas and $s^{\ominus}$ is the **standard entropy** of the gas ($s^{\ominus}$ is usually expressed per mole of substance rather than per molecule as we have done here). The concept of a **standard state** in thermodynamics is a practical notion used in calculations of, for example, μ and S at various pressures from Equations like (2.25) and (2.29). Then it is required that μ and S are known in one state, the standard state, where $\mu = \mu^{\ominus}$ and $S = S^{\ominus} = Ns^{\ominus}$, respectively (usually at a certain pressure). Extensive tables of experimental $\mu^{\ominus}$ and $s^{\ominus}$ values exist for various compounds. In statistical mechanics, the concept of a standard state is not equally important since the chemical potential and entropy can be calculated directly from the properties of the gas molecules from Equations 2.23 and 2.28, respectively. However, there are additional contributions to μ and S from the electronic quantum states that are not included in these expressions. They are important for high temperatures, where the molecules are excited to quantum states above the electronic ground state.

{EXERCISE 2.3

a. An ideal monatomic gas with N particles of mass m is enclosed in a cube with sides of length L so the volume is $V = L^3$. The gas is compressed adiabatically by varying L. If the process takes place quasistatically, there exists a simple relationship

$$U = BV^{-2/3} \tag{2.31}$$

between the volume and the energy of the gas. This relationship can be derived from the expression for the energy of the quantum states of each particle given by Equation 1.4, where j_x, j_y and j_z are the three quantum numbers (positive integers). Use Equation 1.4 to motivate why B is a constant independent of V.

b. Starting from Equation 2.31, determine how the entropy S, pressure P and temperature T vary with V. How much do these quantities change when the volume is halved?

---

[16] This is the current convention recommended by the International Union of Pure and Applied Chemistry (IUPAC). An earlier international convention was to set the standard state pressure to $P^{\ominus} = 1$ atm.

Comment on the results. Some useful formulas for the ideal gas that you may use for this task is the expression[17]

$$S = Nk_B \ln \left[\frac{(ek_B T)^{5/2} (2\pi m)^{3/2}}{h^3 P} \right],$$

the ideal gas law $PV = Nk_B T$ and the formula $U = 3Nk_B T/2$ for the energy.}

{**EXERCISE 2.4**

In the year 4004, scientists from Earth were negotiating with the blue-green, six-fingered and three-armed scientists from our nearest inhabited planetary system about common units for physical quantities. The scientists agreed that entropy for an isolated system should be defined from the natural logarithm of the number of possible quantum states, Ω, as $S = \mathbb{K} \ln \Omega$ where $\mathbb{K}$ is some constant. The most radical scientists argued that the value of $\mathbb{K}$ should be set to one and that entropy should be a dimensionless quantity. The more pragmatic scientists said that it would be better if S had some unit and that the values of S for macroscopic systems would be too large numbers to be practical if $\mathbb{K}$ would be set to one.

Earlier in the negotiations, they had agreed to put the new "Avogadro's constant" (but under a different name) equal to

$$\mathbb{N} = 2^{3^4} = 2^{81} = 2.41 \cdot 10^{24}.$$

The Earthians had tried very strongly to defend their preferred choice $2^{79} \approx 6.04 \cdot 10^{23}$ since it would change their ancient convention with less than one percent, but they lost. (The possible suggestion of the exact value $6 \cdot 10^{23}$ was not even tried since the alien scientists based their number system on 18 rather than 10.) The Earthians had to accept the change with a factor a 4.015 in the definition of the new mole concept, called nol, which was defined as $\mathbb{N}$ particles.

It had also been agreed to use a unit for energy called EU (= energy unit) and use the symbol € for it (this suggestion came from the Earthians who had some old myth that this symbol had been a very commonly used unit in the past). Converted to SI units $1€ = 4.336\,J$.

The final agreement was to put $\mathbb{K}$ numerically equal to $1/\mathbb{N}$ without any unit, thus making a compromise between the radical and the pragmatic scientists. Then the gas constant, $\mathbb{R} = \mathbb{K}\mathbb{N}$, became exactly 1 nol^{-1}, which was the reason why the radical ones could accept this definition.

a. What is the unit for absolute temperature in the new unit system? (No special unit was introduced for T, so it was expressed in terms of other units.) Motivate your answer from the definition of T from the entropy in the microcanonical ensemble.

b. How much is the SI temperature unit 1 K (Kelvin) in the new units? **Hint:** It is easiest to do the conversion in two steps, by first keeping the energy unit as J and then converting to the new energy unit €.}

---

[17] This follows from Equation 2.28 and the definition (1.19) of Λ.

2.3 THERMAL EQUILIBRIA AND THE CANONICAL ENSEMBLE

2.3.1 The Canonical Ensemble and Boltzmann's Distribution Law

As we have seen, the microcanonical ensemble deals with the case of systems at constant N, V and U. If we want to consider systems where U is not constant but where the temperature T is constant instead – a more common case – we have to proceed somewhat differently. The case of constant N, V and T is treated using the so-called **canonical ensemble** or the **NVT ensemble**. Such an ensemble consists, *by definition*, of a large number of replicas of the system, all having the same values of N, V and T.

Let us consider a system A in thermal contact with a reservoir R with constant temperature – a heat bath. Such a system is sketched in Figure 2.7. A and R are *weakly coupled* to each other; they can exchange heat through the wall that surrounds A, but they do not influence each other's sets of quantum states. Let the volume and the number of particles V_A, V_R, N_A and N_R be constant. Assume that R is *much* larger than A, so the change in the energy of R during the heat exchange between A and R is completely negligible. This ensures that R really acts as a heat bath at constant T; the energy exchanged with A does not alter the temperature of R.

Since R is so big, we may without loss of generality assume that the combined system AR is isolated from the environment. Therefore, $U_{tot} = U_{AR} = U_A + U_R = $ constant and the properties of AR can be obtained using the microcanonical ensemble. This means that all quantum states for the total system AR are equally probable; their probability is $1/\Omega_{tot}$, where $\Omega_{tot} = \Omega_{AR} = \Omega_{AR}(U_{AR})$ is the total number of quantum states for the whole system at energy U_{AR}. The canonical ensemble we will consider consists of subsystem A and its

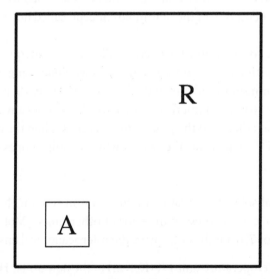

FIGURE 2.7 A macroscopic system A in contact with a heat reservoir R, which has constant temperature T and is much larger than A (larger than shown in the picture). Systems A and R are in thermal contact with each other so the temperature of A and R is the same. Subsystem A has constant volume and no particles can enter or leave it. The combined system AR is isolated from the environment.

replicas, all having the same values of T, V_A, and N_A. The reservoir is not included in the ensemble – its only role is to provide the constant temperature.

Since subsystem A can exchange energy with the reservoir R, it has many quantum states with different energies that are accessible. When A is in quantum state i with energy $\mathcal{U}_{A,i}$ the reservoir has energy $U_{tot} - \mathcal{U}_{A,i}$. Since R is so large compared to A, we can in practice assume that all quantum states of A are accessible. We shall now determine the probability that subsystem A is in quantum state i with the energy $\mathcal{U}_{A,i}$. We will denote this probability $\mathfrak{p}_A(\mathcal{U}_{A,i})$. Note that this is the probability for A to be in *one particular* quantum state.

When A is in quantum state i, the reservoir can be in any of its quantum states with energy $U_{tot} - \mathcal{U}_{A,i}$. There are $\Omega_R(U_{tot} - \mathcal{U}_{A,i})$ such states. The whole system AR hence has $1 \times \Omega_R(U_{tot} - \mathcal{U}_{A,i})$ states that are such that A is in quantum state i (factor 1 for A and factor $\Omega_R(U_{tot} - \mathcal{U}_{A,i})$ for R). Since AR is isolated, all these states for the whole system are equally probable. Each of them occur with the probability $1/\Omega_{tot}$. Thus, it follows that $\mathfrak{p}_A(\mathcal{U}_{A,i})$ is $\Omega_R(U_{tot} - \mathcal{U}_{A,i})$ times larger than $1/\Omega_{tot}$, so we have

$$\mathfrak{p}_A(\mathcal{U}_{A,i}) = \frac{\Omega_R(U_{tot} - \mathcal{U}_{A,i})}{\Omega_{tot}}. \tag{2.32}$$

The probability is accordingly proportional to the number of states of the reservoir with energy $U_{tot} - \mathcal{U}_{A,i}$.

If A instead would be in a different quantum state j with the *same value* of its energy, the number of states of the reservoir is the same. Thus, the probability $\mathfrak{p}_A(\mathcal{U}_{A,j})$ for state j is the same as $\mathfrak{p}_A(\mathcal{U}_{A,i})$ for state i when $\mathcal{U}_{A,j} = \mathcal{U}_{A,i}$. It follows that all quantum states of A with the same energy occur with the same probability. On the other hand, the probabilities for quantum states with different energies are not the same since Ω_R is different. From the formula for $\mathfrak{p}_A(\mathcal{U}_{A,i})$ above, we shall derive a very important expression for the probability called Boltzmann's distribution law. We shall do the derivation using a very simple argument. A more strict derivation is presented in Appendix 2B.

Let us focus our attention on the reservoir and consider the number of quantum states as a function of energy, $\Omega_R = \Omega_R(U_R)$. If system A were absent, R would be an isolated system and it follows from Equation 2.19 that

$$\left(\frac{\partial \ln \Omega_R(U_R)}{\partial U_R}\right)_{V_R, N_R} = \frac{1}{k_B T}. \tag{2.33}$$

Since A is so small compared to R, its presence does not influence the properties of R – in particular, the temperature of R is not affected by the presence of A. If we let U_R vary in the interval U_{tot} to $U_{tot} - \mathcal{U}_{A,i}$ (i.e., the energy of A varies between 0 and $\mathcal{U}_{A,i}$), the temperature remains constant since we always have $\mathcal{U}_{A,i} \ll U_R$. Constant temperature is guaranteed by the requirement that R is a large heat reservoir. Note that we are free to select R as large as we wish, so we can always make sure that this requirement is fulfilled.

From the definition of a derivative, it follows that

$$\frac{\partial \ln \Omega_R(U_R)}{\partial U_R} = \frac{\ln \Omega_R(U_{tot}) - \ln \Omega_R(U_{tot} - \mathcal{U}_{A,i})}{\mathcal{U}_{A,i}}$$

when $\mathscr{U}_{A,i}$ is sufficiently small (here this corresponds to $\mathscr{U}_{A,i} \ll U_{tot}$) and hence Equation 2.33 implies that

$$\frac{\ln \Omega_R(U_{tot}) - \ln \Omega_R(U_{tot} - \mathscr{U}_{A,i})}{\mathscr{U}_{A,i}} = \frac{1}{k_B T}.$$

(Alternatively, this result can be obtained by integrating Equation 2.33 between $U_{tot} - \mathscr{U}_{A,i}$ and U_{tot} at constant T.) We accordingly have

$$\ln \Omega_R(U_{tot} - \mathscr{U}_{A,i}) - \ln \Omega_R(U_{tot}) = -\frac{\mathscr{U}_{A,i}}{k_B T}$$

and hence

$$\Omega_R(U_{tot} - \mathscr{U}_{A,i}) = \Omega_R(U_{tot})e^{-\mathscr{U}_{A,i}/k_B T}.$$

Inserting this in our expression (2.32) for $\mathfrak{p}_A(\mathscr{U}_{A,i})$, we obtain

$$\mathfrak{p}_A(\mathscr{U}_{A,i}) = \mathcal{K}e^{-\mathscr{U}_{A,i}/k_B T}$$

where

$$\mathcal{K} = \frac{\Omega_R(U_{tot})}{\Omega_{tot}} \tag{2.34}$$

is a constant independent of $\mathscr{U}_{a,i}$. This constant is easy to determine. Since we must have $\Sigma_j \mathfrak{p}_A(\mathscr{U}_{A,j}) = 1$, where the sum is over all quantum states of subsystem A, we see that

$$\sum_j \mathcal{K}e^{-\mathscr{U}_{A,j}/k_B T} = 1$$

so

$$\mathcal{K} = \frac{1}{\sum_j e^{-\mathscr{U}_{A,j}/k_B T}}$$

and we obtain

$$\mathfrak{p}_A(\mathscr{U}_{A,i}) = \frac{e^{-\mathscr{U}_{A,i}/k_B T}}{\sum_j e^{-\mathscr{U}_{A,j}/k_B T}} = \frac{e^{-\mathscr{U}_{A,i}/k_B T}}{Q_A}, \tag{2.35}$$

where $Q_A = 1/\mathcal{K}$.

This result for $\mathfrak{p}_A(\mathscr{U}_{A,i})$ is the famous **Boltzmann's distribution law**. Note that the factor $\exp(-\mathscr{U}_{A,i}/k_B T)$ originates from the properties of the reservoir alone, as can be seen from the derivation above. One can say that $\exp(-U/k_B T)$ describes the "willingness" of a reservoir at temperature T to give away an amount of energy equal to U or, in other words, it describes the availability of the energy U at this temperature. The denominator, Q_A, is called the **canonical partition function** and equals

$$Q_A = \sum_j e^{-\mathscr{U}_{A,j}/k_B T}.$$

The entity Q_A is a property of subsystem A since it depends on the energies of all quantum states of A (subscript A means that Q_A is the partition function for A). From $Q_A = 1/\mathcal{K}$ and Equation 2.34 it follows that

$$Q_A = \frac{\Omega_{\text{tot}}}{\Omega_R(U_{\text{tot}})} = \frac{\Omega_{AR}}{\Omega_R(U_{AR})}, \tag{2.36}$$

which will be useful later.

Since the probability $\mathfrak{p}_A(\mathscr{U}_{A,i})$ only depends on the value of the energy of the quantum state, we can drop the explicit state index i and write that

$$\mathfrak{p}_A(\mathscr{U}_A) = \frac{e^{-\mathscr{U}_A/k_B T}}{Q_A} \tag{2.37}$$

is the probability for A to be in a particular quantum state with energy $\mathscr{U}_A$ at constant temperature T. This expresses that the probability for each quantum state with the same energy is the same, as observed earlier.

Let us now determine the probability that the *energy of subsystem* A is equal to U_A. We will denote this probability $\mathcal{P}_A(U_A)$. The energy of A is equal to U_A whenever A is in a quantum state with energy U_A, but not otherwise. Since there are $\Omega_A(U_A)$ quantum states with this energy and since each of these states occur with a probability equal to $\mathfrak{p}_A(U_A)$, it follows that $\mathcal{P}_A(U_A)$ must be $\Omega_A(U_A)$ times larger than $\mathfrak{p}_A(U_A)$, that is, we have

$$\mathcal{P}_A(U_A) = \Omega_A(U_A)\mathfrak{p}_A(U_A). \tag{2.38}$$

By inserting Equation 2.37, we obtain

$$\mathcal{P}_A(U_A) = \frac{\Omega_A(U_A)e^{-U_A/k_B T}}{Q_A}, \tag{2.39}$$

which is another form of Boltzmann's distribution law.

The *energy level* with energy U_A consists of the $\Omega_A(U_A)$ quantum states with the same energy. Hence, we can say that $\mathcal{P}_A(U_A)$ = the probability for A to have energy U_A = the probability that A is in the energy level with energy U_A = the probability for A to be in *any* of the quantum states with energy U_A. Distinguish this from $\mathfrak{p}_A(U_A)$ = the probability for A to be in *one particular* quantum state with energy U_A.

We can write the canonical partition function in two different ways

$$Q_A = \sum_j e^{-\mathscr{U}_{A,j}/k_B T} = \sum_{U_A} \Omega_A(U_A)e^{-U_A/k_B T},$$

where the first sum is taken over all *quantum states* of subsystem A and the second sum is taken over all *energy levels* of A. The second equality follows from the fact that the terms in the first sum that belong to the same energy level are equal (i.e., each set of $\Omega_A(U_A)$ terms with $\mathscr{U}_{A,j} = U_A$) and can be grouped together to form one term in the second sum.

The partition function Q_A depends only on the temperature and on the energies and degeneracies of the energy levels of A, that is, on the specific properties of A alone (the only role of R is to give a constant T). Since the quantum states of A depend on V_A and N_A, we have $Q_A = Q_A(T, V_A, N_A)$. Likewise, $\mathcal{P}_A$ and $\mathfrak{p}_A$ also depend on T, V_A, and N_A.

2.3.2 Calculations of Thermodynamical Quantities; the Connection with Partition Functions

2.3.2.1 The Helmholtz Free Energy

Let us investigate the properties of the canonical partition function a bit further. We saw in Equation 2.36 that it equals

$$Q_A = \frac{\Omega_{AR}(U_{AR})}{\Omega_R(U_{AR})}.$$

Therefore,

$$k_B \ln Q_A = k_B \ln \Omega_{AR}(U_{AR}) - k_B \ln \Omega_R(U_{AR}) = S_{AR}(U_{AR}) - S_R(U_{AR}).$$

Remember that $\Omega_{AR}(U_{AR}) = \Omega_{tot}$ is the total number of quantum states for the whole system aR at energy $U_{AR} = U_{tot}$ (= constant). S_{AR} is the entropy for the whole system, which is isolated. The second term $S_R(U_{AR})$ is the entropy that R would have if it contained all available energy and A was absent. (This term can in what follows be treated just as an additive constant without real relevance.)

Let us regard some spontaneous process in subsystem A. During the process, the only thing that happens to R is that it participates in energy transfer to or from A so the initial temperature of A (before the process) equals the final one. Since the whole system is isolated, we know that the total entropy increases when the process goes from the initial to the final state, $\Delta S_{AR} > 0$. Therefore, $\Delta(k_B \ln Q_A) = \Delta S_{AR} > 0$ (the term $S_R(U_{AR})$ stays constant and does not contribute to the difference). Obviously, the quantity $\Delta(k_B \ln Q_A)$ is important. Its sign tells the direction of a spontaneous process in A when the temperature is constant. We can also see that it has to be zero during a reversible process since the total entropy S_{AR} does not change. Thus, $k_B \ln Q_A$ plays the same role at constant temperature as S does for an isolated system and it ought to have a name. For historical reasons, however, we multiply $k_B \ln Q_A$ with a factor $-T$ and define the **Helmholtz free energy** (also called the **Helmholtz energy**), denoted by A, for system A as

$$A_A = -k_B T \ln Q_A.$$

The factor T in the definition of A does not change the properties described above since we deal with constant temperatures.[18] For a spontaneous process at constant T, V and N, *the*

---

[18] This factor and the minus sign occur in the definition of A because they are needed to obtain the quantity that corresponds to the Helmholtz free energy in classical thermodynamics.

free energy decreases and has a minimum at equilibrium. This follows from the fact that the entropy S_{AR} increases and has a maximum at equilibrium. Note that $k_B T$ has the unit of energy, so U and A have the same unit.

In general, we have

$$A = A(T, V, N) = -k_B T \ln Q(T, V, N) \tag{2.40}$$

and for processes at constant T, V and N

$$\Delta A = -T \Delta S_{tot} \leq 0 \tag{2.41}$$

where ΔS_{tot} is the entropy change in the system + its environment (this corresponds to ΔS_{AR} above, where A is the system and R the environment). Helmholtz's free energy of a system always decreases for a spontaneous process when the temperature T and volume V are constant and no particles are allowed to enter or leave the system – a closed system. (For a system with several species, the number of molecules inside the system may change due to chemical reactions between the species, but no molecules are allowed to pass the boundaries.)

Let us now regard two systems A and B in thermal contact with a heat reservoir R, as illustrated in Figure 2.8. At equilibrium, they have equal temperature = the reservoir temperature. When subsystem A is in a quantum state with energy U_A and B is in a state with energy U_B, the energy of the combined system AB equals $U_{AB} = U_A + U_B$ and the number of such states for AB is $\Omega_A(U_A) \times \Omega_B(U_B)$. The total number of quantum states for AB with

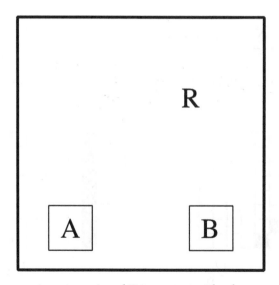

FIGURE 2.8 Two macroscopic systems A and B in contact with a heat reservoir R, which is much larger than both A and B and has constant temperature T. All three systems are in thermal contact with each other so the temperature of A, B and R is the same. V_A and V_B are constant and no particles can pass the boundaries of A and B. The combined system ABR is isolated from the environment.

energy U_{AB} is

$$\Omega_{AB}(U_{AB}) = \sum_{U_A} \Omega_A(U_A)\Omega_B(U_{AB} - U_A) = \sum_{\substack{U_A \\ (U_A+U_B=U_{AB})}} \Omega_A(U_A)\Omega_B(U_B).$$

The partition function for AB is

$$Q_{AB} = \sum_{U_{AB}} \Omega_{AB}(U_{AB})e^{-U_{AB}/k_BT}$$

$$= \sum_{U_{AB}} \left[\sum_{\substack{U_A \\ (U_A+U_B=U_{AB})}} \Omega_A(U_A)\Omega_B(U_B) \right] e^{-(U_A+U_B)/k_BT}.$$

In the second sum in the rhs, we have $U_B = U_{AB} - U_A$ with constant U_{AB}, and therefore one sums diagonally in, (U_A, U_B)-space, as illustrated in Figure 2.9a. Since the first sum runs over all possible values of U_{AB}, all possible values of U_A and U_B are thereby taken. Therefore, the two sums can be replaced as follows

$$\sum_{U_{AB}} \sum_{\substack{U_A \\ (U_A+U_B=U_{AB})}} = \sum_{U_B} \sum_{U_A},$$

where one holds U_B constant in the second sum in the rhs so the sum is done parallel to the U_A axis, as illustrated in Figure 2.9b.

Therefore, we obtain

$$Q_{AB} = \sum_{U_B} \sum_{U_A} \Omega_A(U_A)\Omega_B(U_B)e^{-(U_A+U_B)/k_BT}.$$

$$= \sum_{U_A} \Omega_A(U_A)e^{-U_A/k_BT} \sum_{U_B} \Omega_B(U_B)e^{-U_B/k_BT} = Q_AQ_B \qquad (2.42)$$

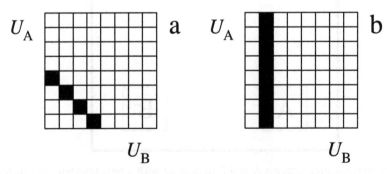

FIGURE 2.9 Illustration in (U_A, U_B)-space of U_A and U_B values with varying U_A when $U_A + U_B = U_{AB}$ = constant (left figure) and with varying U_A when U_B is constant (right figure).

and

$$A_{AB} = -k_B T \ln Q_{AB} = -k_B T \ln Q_A Q_B$$
$$= -k_B T (\ln Q_A + \ln Q_B) = A_A + A_B, \tag{2.43}$$

so we conclude that A is additive.

2.3.2.2 Thermodynamical Quantities as Averages

Consider an ensemble of macroscopically identical systems with specified T, V and N, that is, a canonical ensemble. All systems have the same set of quantum states of energies $\{\mathscr{U}_i\}$, where i is the state index (from now on we drop the subscript that denotes the system, like indices A and B, from the equations we obtained above). The average energy equals

$$\bar{U} \equiv \langle U \rangle = \sum_i \mathscr{U}_i \mathfrak{p}(\mathscr{U}_i) = \sum_U U \mathcal{P}(U), \tag{2.44}$$

where the first sum is over all quantum states of the system and the second sum is over all energy levels. The notation $\bar{U}$ for the average energy is an alternative to $\langle U \rangle$ that it is sometimes convenient to use; $\bar{U}$ and $\langle U \rangle$ mean exactly the same thing. We have

$$\mathfrak{p}(U) = \frac{e^{-U/k_B T}}{Q} \quad \text{and} \quad \mathcal{P}(U) = \frac{\Omega(U)e^{-U/k_B T}}{Q} \tag{2.45}$$

and the partition function Q equals

$$Q = Q(T, V, N) = \sum_i e^{-\mathscr{U}_i(V,N)/k_B T} = \sum_U \Omega(U, V, N)e^{-U/k_B T}. \tag{2.46}$$

Note that $\sum_i \mathfrak{p}(\mathscr{U}_i) = 1$ and $\sum_U \mathcal{P}(U) = 1$ as required.

The function $\mathcal{P}(U)$ is proportional to $\Omega(U) \times \exp(-U/k_B T)$, where we have suppressed V and N which are constant. $\Omega(U)$ is a very rapidly increasing function[19] of U and $\exp(-U/k_B T)$ is a very rapidly decreasing function of U. Therefore, their product and hence $\mathcal{P}(U)$ is expected to be a very sharply peaked function. This is verified in Section 2.6.1, where it is shown that the width of the probability $\mathcal{P}(U)$, that is, the magnitude of the fluctuations in energy, is in general extremely small for a macroscopic system. Thus, $\mathcal{P}(U)$ has a very sharp peak at the average volume $\langle U \rangle$ as illustrated in Figure 2.10 and $\langle U \rangle$ is in practice equal to the most probable energy. $\langle U \rangle$ is known in thermodynamics as the **internal energy** of the system, which hence equals the total average energy of all molecules in the system.

When we dealt with isolated systems, we concluded that all thermodynamic properties of a system at equilibrium could be calculated from the function $\Omega(U, V, N)$ or, equivalently, from $S(U, V, N) = k_B \ln \Omega(U, V, N)$. We are now going to show that all properties

---

[19] As mentioned in footnote 4 of Chapter 1, $\Omega(U)$ has typically an order of magnitude equal to a constant times $U^{\alpha N}$, where α is of the order 1.

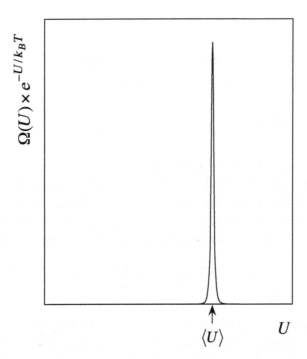

FIGURE 2.10 The product $\Omega(U) \times \exp(-U/k_BT)$, which is proportional to the probability $\mathcal{P}(U)$, is a very sharply peaked function of U as sketched in the figure. For a macroscopic system, the peak is enormously much more sharp than shown here. The average energy $\langle U \rangle$ coincides is practice with the most probable energy = the location of the maximum.

can be calculated from the partition function $Q(T, V, N)$ or, equivalently, from $A(T, V, N) = -k_BT \ln Q(T, V, N)$. Let us start from the expression

$$Q(T, V, N) = \sum_i e^{-\mathcal{U}_i(V,N)/k_BT} = \sum_i e^{-\beta \mathcal{U}_i(V,N)},$$

where we have introduced $\beta = 1/(k_BT)$. Since Q is a function of T it is, of course, alternatively a function of β, that is, $Q = Q(\beta, V, N)$. If we take the derivative of $\ln Q$ with respect to β at constant V and N, we obtain

$$\left(\frac{\partial \ln Q}{\partial \beta} \right)_{V,N} = \frac{(\partial Q/\partial \beta)_{V,N}}{Q} = \frac{-\sum_i \mathcal{U}_i e^{-\beta \mathcal{U}_i}}{Q}$$

$$= -\sum_i \mathcal{U}_i \mathfrak{p}(\mathcal{U}_i) = -\langle \mathcal{U}_i \rangle_i = -\bar{U}. \tag{2.47}$$

We can alternatively write this result as

$$\bar{U} = -k_B \left(\frac{\partial \ln Q}{\partial (1/T)} \right)_{V,N} = k_BT^2 \left(\frac{\partial \ln Q}{\partial T} \right)_{V,N} \tag{2.48}$$

or

$$\bar{U} = \left(\frac{\partial (A/T)}{\partial (1/T)} \right)_{V,N}. \tag{2.49}$$

The latter relationship is analogous to an equation known in thermodynamics as the Gibbs-Helmholtz equation (see Equation 2.89).

If we take the derivative of $\ln Q$ with respect to V at constant T and N (or constant β and N), we obtain

$$\left(\frac{\partial \ln Q}{\partial V} \right)_{T,N} = \frac{(\partial Q/\partial V)_{T,N}}{Q} = \frac{-\beta \sum_i (\partial \mathscr{U}_i/\partial V)_N \, e^{-\beta \mathscr{U}_i}}{Q}$$

$$= -\beta \sum_i \left(\frac{\partial \mathscr{U}_i}{\partial V} \right)_N \mathfrak{p}(\mathscr{U}_i) = -\beta \left\langle \left(\frac{\partial \mathscr{U}_i}{\partial V} \right)_N \right\rangle_i = \beta P,$$

where we have identified the average of $-(\partial \mathscr{U}_i/\partial V)_N$ as the pressure, in accordance with our general definition of pressure in Equation 2.1. In the present case, we take the canonical ensemble average, which is the appropriate one for the case of constant temperature. We can write this result as

$$P = k_B T \left(\frac{\partial \ln Q}{\partial V} \right)_{T,N} \tag{2.50}$$

or

$$P = - \left(\frac{\partial A}{\partial V} \right)_{T,N}, \tag{2.51}$$

which we can recognize from thermodynamics.

Let us consider two systems A and B that are in contact with each other via a mobile wall as illustrated in Figure 2.11. They are kept at the same constant temperature T by some heat reservoir. Let us assume that the pressures P_A and P_B are different, say, $P_A > P_B$, and that the wall between the two systems initially is locked at one position. Let us release the wall and let it move a small distance so the volumes of A and B change by dV_A and dV_B, respectively, and we have $dV_{AB} = dV_A + dV_B = 0$. A spontaneous process in the combined system AB is characterized by a decrease in free energy A_{AB} when T, N_{AB} and V_{AB} are constant. When the volume of A is changed by dV_A, its free energy changes by $dA_A = -P_A dV_A$ since the

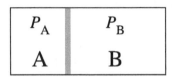

FIGURE 2.11 Two systems A and B in contact with each other via a wall (shown in gray) that can move without friction. No particles can penetrate the walls so N_A and N_B are constant. P_A and P_B, respectively, are the pressure in the two systems, and A and B are kept at the same constant temperature T by thermal contact with some heat reservoir (not shown in the figure).

temperature and number of particles are constant. Likewise, the change in B equals $dA_B = -P_B dV_B$. The total change in free energy is $dA_{AB} = dA_A + dA_B = -P_A dV_A - P_B dV_B = -(P_A - P_B)dV_A$ since $dV_B = -dV_A$. Since $P_A - P_B > 0$, we conclude that the spontaneous volume change is such that $dV_A > 0$ since $dA_{AB} < 0$. Thus, the wall moves to the right, which is the obvious direction: the force on the wall at the high pressure side is larger than the oppositely directed force at the low pressure side. If we let the wall move further, the volume changes will continue until $P_A = P_B$ and equilibrium is reached.

In a completely analogous fashion as in the derivation of Equations 2.50 and 2.51, if we take the derivative of $\ln Q$ with respect to N at constant T and V, we obtain

$$\mu = -k_B T \left(\frac{\partial \ln Q}{\partial N} \right)_{T,V} \tag{2.52}$$

and

$$\mu = \left(\frac{\partial A}{\partial N} \right)_{T,V}. \tag{2.53}$$

{**EXERCISE 2.5**
Derive Equation 2.52 by using the definition of chemical potential in Equation 2.15.}

To summarize the results in Equations 2.47, 2.50 and 2.52, we can write the differential of $\ln Q(\beta, V, N)$ as

$$d \ln Q = \left(\frac{\partial \ln Q}{\partial \beta} \right)_{V,N} d\beta + \left(\frac{\partial \ln Q}{\partial V} \right)_{\beta,N} dV + \left(\frac{\partial \ln Q}{\partial N} \right)_{\beta,V} dN$$
$$= -\bar{U} d\beta + \beta P dV - \beta \mu dN. \tag{2.54}$$

This will be useful in the next section.

Let us generalize our results to systems with several kinds of particles. The number of particles of species ν is equal to N_ν. The quantum state energies then depend on N_ν for all ν, so if we have three species we have $\mathscr{U}_i = \mathscr{U}_i(V, N_1, N_2, N_3)$ and the partition function becomes

$$Q(T, V, N_1, N_2, N_3) = \sum_i e^{-\beta \mathscr{U}_i(V, N_1, N_2, N_3)}.$$

The chemical potential for species ν then equals

$$\mu_\nu = -k_B T \left(\frac{\partial \ln Q}{\partial N_\nu} \right)_{T,V,\{N_{\nu'}\}_{\nu' \neq \nu}},$$

where $N_{\nu'}$ for all $\nu' \neq \nu$ are held constant, and we have

$$\mu_\nu = \left(\frac{\partial A}{\partial N_\nu} \right)_{T,V,\{N_{\nu'}\}_{\nu' \neq \nu}}. \tag{2.55}$$

Otherwise, our results above remain valid with N_1, N_2, N_3 inserted instead of N in the formulas. Equation 2.54 is replaced by

$$d \ln Q = -\bar{U} d\beta + \beta P dV - \beta \sum_{v} \mu_v dN_v,$$

where v runs over all species in the system.

Let us now treat transfer of particles between two systems. Consider two systems A and B that are in thermal contact with each other and are kept at the same constant temperature by some heat reservoir. If we open a connection between A and B for the passage of particles between them (V_A and V_B are assumed to remain constant), in which direction will the particles flow spontaneously? A spontaneous process in the combined system AB at constant T, V_{AB} and number of particles of each species is characterized by a decrease in free energy A_{AB}. When the number of particles of species v in A is changed by $dN_{v,A}$, the free energy of A changes by $dA_A = \mu_{v,A} dN_{v,A}$ since the temperature and volume are constant, and the change in B is $dA_B = \mu_{v,B} dN_{v,B}$. Since $dN_{v,B} = -dN_{v,A}$ we obtain $dA_{AB} = dA_A + dA_B = (\mu_{v,A} - \mu_{v,B}) dN_{v,A}$. If $\mu_{v,A} > \mu_{v,B}$ we conclude that $dN_{v,A} < 0$. Thus, a spontaneous flow of particles of species v occurs from the system with the highest chemical potential μ_v to the one with the lowest. This applies for each v and the direction of flow may be different for different species. If μ_v is equal in the two systems for all particle species, we have an equilibrium distribution of particles and no net particle flow.

2.3.2.3 Entropy in the Canonical Ensemble

Let us now consider a quasistatic process where β, V and N change so slowly that the system remains very close to equilibrium, so it is described by the canonical ensemble at each instant of time. The work done on the system when V is changed by dV and N is changed by dN equals $đw = -PdV + \mu dN$, so we can write Equation 2.54 as

$$d \ln Q = -\bar{U} d\beta - \beta \, đw.$$

We will now transform this relationship into something quite useful by adding $d(\beta \bar{U}) = \beta d\bar{U} + \bar{U} d\beta$ to either side and we obtain after simplification

$$d(\ln Q + \beta \bar{U}) = \beta(dU - đw).$$

From the definition of heat, it follows that $dU - đw = đq$ so we get after insertion of $\beta = 1/(k_B T)$ and multiplication by k_B

$$d\left(k_B \ln Q + \frac{\bar{U}}{T}\right) = \frac{đq}{T}.$$

The left-hand side (lhs) is clearly an exact differential (i.e., the differential of a function), so it follows that $đq/T$ is also an exact differential despite the fact that $đq$ is not. We term this exact differential dS and call the function S "entropy" in order to keep the same relationship $d\mathsf{q}/T = dS$ between S and q as in the microcanonical ensemble. (We shall soon see that this S is indeed nothing but our good old entropy.) Inserting the definition of the free energy A in the lhs, we obtain

$$d\left(\frac{\bar{U} - A}{T}\right) = dS,$$

so we can identify $(\bar{U} - A)/T = S + \mathcal{K}$, where $\mathcal{K}$ is an integration constant, which, as we will see, can be set equal to zero. Thus, we have

$$A = \bar{U} - TS$$

a well-known relationship in thermodynamics. S must be additive since A and $\bar{U}$ are additive.

To justify the identification of S above as entropy and the setting of $\mathcal{K}$ to zero, let us regard

$$\ln Q = \ln \sum_U \Omega(U)e^{-U/k_B T}.$$

As illustrated in Figure 2.10, $\Omega(U) \exp(U/k_B T)$ is a very sharply peaked function of U for a macroscopic system. Hence, we can approximate $\ln Q$ with the logarithm of the largest term in the sum (occurring at $U = \bar{U}$) and obtain the excellent approximation

$$\ln Q = \ln \left[\Omega(\bar{U})e^{-\bar{U}/k_B T}\right] = \ln \Omega(\bar{U}) - \frac{\bar{U}}{k_B T}.$$

By inserting the definition of A, Equation 2.40, we see that

$$k_B \ln \Omega(\bar{U}) = \frac{\bar{U} - A}{T}.$$

The rhs is equal to $S + \mathcal{K}$ according to our findings above, so if we set the constant $\mathcal{K} = 0$ we see that

$$k_B \ln \Omega(\bar{U}) = S, \tag{2.56}$$

which is the definition of entropy in the microcanonical ensemble with the average $\bar{U}$ inserted as the energy of the system. This justifies both our identification of S and the setting of $\mathcal{K}$ to zero. Equation 2.56 shows that the entropy in the canonical ensemble is for all practical purposes identical to that in the microcanonical provided $U_{\text{microcanonical}} = \bar{U}_{\text{canonical}}$ and that V and N are the same in the two cases. This observation is very important as it illustrates the fact that for macroscopic systems with extremely sharp probability distributions, *it does not matter which ensemble one uses to calculate the thermodynamical properties.* One can simply take the one that is most convenient. The canonical ensemble is often easier to treat mathematically than the microcanonical one since the condition of having constant total energy in the latter is tricky to implement in general.

For another way to see how these findings relate to the entropy definition $S = k_B \ln \Omega$ in the microcanonical ensemble, let us look at an isolated system with energy U. All $\Omega(U)$ quantum states have the same energy, $\mathcal{U}_i = U$. For this system, we therefore have

$$k_B T \ln \sum_i e^{-\mathcal{U}_i / k_B T} = k_B T \ln \left[\Omega(U) e^{-U/k_B T} \right]$$

$$= k_B T \ln \Omega(U) - U = TS - U$$

By defining the Helmholtz free energy in the usual manner as $A = U - TS$, we see that $A = -k_B T \ln \left[\sum_i \exp(-\mathcal{U}_i / k_B T) \right]$, which is the same expression as in the canonical ensemble. After this brief digression, let us return to the canonical ensemble.

The entropy can alternatively be written entirely in terms of the probability distributions in the canonical ensemble. To see this, we write

$$S = \frac{1}{T} \left[\bar{U} - A \right] = \frac{1}{T} \left[\sum_i \mathcal{U}_i \mathfrak{p}(\mathcal{U}_i) + k_B T \ln Q \right]$$

$$= -k_B \left[\sum_i \left\{ -\frac{\mathcal{U}_i}{k_B T} \right\} \mathfrak{p}(\mathcal{U}_i) - \left\{ \sum_i \mathfrak{p}(\mathcal{U}_i) \right\} \ln Q \right]$$

$$= -k_B \sum_i \mathfrak{p}(\mathcal{U}_i) \ln \frac{\exp(-\mathcal{U}_i / k_B T)}{Q}$$

$$= -k_B \sum_i \mathfrak{p}(\mathcal{U}_i) \ln \mathfrak{p}(\mathcal{U}_i),$$

where we in the second line have used the fact $\sum_i \mathfrak{p}(\mathcal{U}_i) = 1$ and written the first curly bracket in this line as $\ln \exp(-\mathcal{U}_i / k_B T)$ in the third line. If we write $\mathfrak{p}(\mathcal{U}_i) = \mathfrak{p}_i$ we can write this as

$$S = -k_B \sum_i \mathfrak{p}_i \ln \mathfrak{p}_i.$$

This is a very general form for S. It is sometimes used as a definition of the concept of entropy based on information theory (generally, $\sum \mathfrak{p} \ln \mathfrak{p}$ summed of all possibilities is a measure of the information contained in a probability distribution).

{EXERCISE 2.6

Show that S in the microcanonical ensemble can also be written in this general form.}

We have seen that the natural variables of Q and hence of the free energy are T, V and N. The differential of $A = A(T, V, N)$ equals

$$dA = \left(\frac{\partial A}{\partial T} \right)_{V,N} dT + \left(\frac{\partial A}{\partial V} \right)_{T,N} dV + \left(\frac{\partial A}{\partial N} \right)_{T,V} dN$$

and we have already determined the last two derivatives. Let us now find the first. From Equation 2.48, it follows that

$$\bar{U} = k_B T^2 \left(\frac{\partial \ln Q}{\partial T} \right)_{V,N} = -T^2 \left(\frac{\partial [A/T]}{\partial T} \right)_{V,N}$$

and by performing the differentiation of A/T, we can rewrite this as

$$\left(\frac{\partial A}{\partial T} \right)_{V,N} = \frac{A - \bar{U}}{T} = -S. \tag{2.57}$$

Thus, we have

$$dA = -SdT - PdV + \mu dN, \tag{2.58}$$

which we recognize from thermodynamics. In the case of several particle species, we have instead

$$dA = -SdT - PdV + \sum_{\nu} \mu_{\nu} dN_{\nu}, \tag{2.59}$$

where the sum is over all species in the system.

2.4 CONSTANT PRESSURE: THE ISOBARIC-ISOTHERMAL ENSEMBLE

We shall here treat the case with systems at constant temperature, pressure and number of particles. For such a system, the volume V is not an independent variable (exactly like U is not an independent variable in the canonical ensemble, for which N, V and T are the independent variables). Instead, the pressure P is an independent variable and one has to determine the average volume $\bar{V}$ for a given pressure P. To treat such systems we use the **isobaric-isothermal ensemble**, also called the **NPT ensemble**. It consists, by definition, of an ensemble of replicas of the macroscopic system, all replicas having the same N, P and T.

2.4.1 Probabilities and the Isobaric-Isothermal Partition Function

Let us regard a system A in contact with a reservoir R that has constant temperature T and pressure P, as illustrated in Figure 2.12. The two systems are in thermal contact with each other. The volume V_A of A is variable because one of its wall is mobile so V_A can change. Thus, energy can be transferred between systems A and R both as heat exchange and in the form of work due to volume changes (pressure-volume work). At equilibrium, the temperature of A and R is equal and, likewise, the pressure on both sides of the walls of A is the same. The average volume of A at equilibrium is denoted by $\bar{V}_A$ and the average energy by $\bar{U}_A$. The number of particles in each subsystem, N_A and N_R respectively, is constant. Without loss of generality, we can assume that the combined system AR is isolated, which means that $U_{tot} = U_{AR} = U_A + U_R = $ constant, $V_{tot} = V_{AR} = V_A + V_R = $ constant. The reservoir R is so large that exchange of heat between A and R and variations in the volumes do not change the temperature and pressure of R.

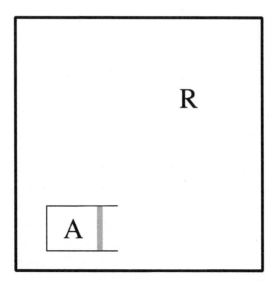

FIGURE 2.12 A macroscopic system A in contact with a reservoir R, which is much larger than A (larger than shown in the picture) and has constant temperature T and pressure P. Systems A and R are in thermal contact with each other. By means of a movable wall of A (a piston that can move without friction and that constitutes the right wall of A), the volume of A can change until the pressure on both sides of the wall is the same. Thus, at equilibrium, both the temperature and the pressure of A have the same values as in R. Subsystem A has a constant number of particles. The combined system AR is isolated from the environment.

In classical mechanics, the position of a particle can be determined with any precision, but in quantum theory the position of a particle and its momentum cannot be simultaneously determined by absolute accuracy due to the Heisenberg uncertainty principle. For a macroscopic solid body, the corresponding uncertainty in position and momentum is of very little consequence, but in principle it is still not possible to determine the position of a body with absolute precision (even if we disregard that it is made by atomic constituents), although the uncertainty is extremely small compared to its extent. Likewise, the volume V_A of our system A, which is determined by the positions of the movable wall and the stationary walls, can in principle not be determined with absolute precision. Therefore, we cannot distinguish values of the volume that differ by a very small amount. As a consequence, we will let V_A take on a set of discrete values rather than being a continuous variable. The difference between the discrete values is, say, of the order δV, where δV is very tiny but nonzero.

When we treat V_A in this manner, the total number of quantum states of the combined system AR for all possible values of volume and energy of A is given by

$$\Omega_{AR} = \Omega_{AR}(U_{AR}, V_{AR})$$
$$= \sum_{V_A} \sum_{U_A} \Omega_A(U_A, V_A)\Omega_R(U_{AR} - U_A, V_{AR} - V_A), \qquad (2.60)$$

where the sum over the discrete values of the volume is taken (for simplicity in notation we do not show the N_A and N_R dependences explicitly here and in several expressions below where N_A and N_R do not vary). If the volume instead is assumed to be a continuous variable like in classical mechanics, the number of possible states Ω_{AR} would be infinitely many since any volume then is possible (within some bounds). This would not make any sense since all bodies in AR are in principle governed by quantum mechanics, so the number of quantum states of AR must be finite. We will return to the question of discrete values of the volume later on.

Thus, Ω_{AR} is given by Equation 2.60. All Ω_{AR} quantum states are equally probable since AR is isolated. The isobaric-isothermal ensemble we consider consists of subsystem A and its replicas, all having the same values of T, P and N_A. The role of reservoir R is solely to provide the constant temperature and pressure. R is not included in the ensemble.

We shall determine the probability $\mathcal{P}_A(U_A, V_A)$ that system A has energy U_A and simultaneously has volume V_A. Thereby, we proceed similarly to the treatment of the canonical ensemble, so we start by determining the probability $\mathfrak{p}_A(\mathcal{U}_{A,i}, V_A)$ that subsystem A has volume V_A and is in quantum state i with the energy $\mathcal{U}_{A,i} = \mathcal{U}_{A,i}(V_A, N_A)$, that is, the ith quantum state for A when it has volume V_A (it contains N_A particles throughout). Since R is so huge compared to A, we can in practice assume that all quantum states of A are accessible for each value of the volume. As before, $\mathfrak{p}_A$ gives the probability for A to be in *one particular* quantum state.

When A has volume V_A and is in its quantum state i, the reservoir has volume $V_{AR} - V_A$ and energy $U_{AR} - \mathcal{U}_{A,i}$. R can thereby be in any of $\Omega_R(U_{AR} - \mathcal{U}_{A,i}, V_{AR} - V_A)$ quantum states and the whole system AR has $1 \times \Omega_R(U_{AR} - \mathcal{U}_{A,i}, V_{AR} - V_A)$ possible states [factor 1 for A and factor $\Omega_R(U_{AR} - \mathcal{U}_{A,i}, V_{AR} - V_A)$ for R]. Since AR is isolated, all these states for the whole system are equally probable. Each of them occurs with the probability $1/\Omega_{AR}$ and it follows that $\mathfrak{p}_A(\mathcal{U}_{A,i}, V_A)$ is $\Omega_R(U_{AR} - \mathcal{U}_{A,i}, V_{AR} - V_A)$ times larger than $1/\Omega_{AR}$, that is,

$$\mathfrak{p}_A(\mathcal{U}_{A,i}, V_A) = \frac{\Omega_R(U_{AR} - \mathcal{U}_{A,i}, V_{AR} - V_A)}{\Omega_{AR}}. \tag{2.61}$$

Hence, this probability is proportional to the number of states of the reservoir with energy $U_{AR} - \mathcal{U}_{A,i}$ when it has volume $V_{AR} - V_A$.

Like in the canonical ensemble, we note that if A instead would be in a different quantum state with the *same value* of its energy for the specified volume, the number of states of the reservoir is the same. It follows that for a given volume, all quantum states of A with the same energy occur with the same probability. On the other hand, the probabilities for quantum states with different energies are not the same since Ω_R is different.

We will first express $\mathfrak{p}_A$ as a Boltzmann distribution starting from formula (2.61) and then obtain the corresponding expression for $\mathcal{P}_A$. Like in the canonical case, we will use a quite simple argument (a more strict derivation is presented in Appendix 2B).

Consider the number of quantum states of the reservoir as a function of energy and volume, $\Omega_R = \Omega_R(U_R, V_R)$. If system A were absent, R would be an isolated system and we have

$$\left(\frac{\partial \ln \Omega_R(U_R, V_R)}{\partial U_R}\right)_{V_R, N_R} = \frac{1}{k_B T} \tag{2.62}$$

and

$$\left(\frac{\partial \ln \Omega_R(U_R, V_R)}{\partial V_R}\right)_{U_R, N_R} = \frac{P}{k_B T}, \tag{2.63}$$

as follows from Equations 2.19 and 2.20. Since A is so small compared to R, its presence does not influence the properties of R; in particular, the temperature and pressure of R are not affected by the presence of A.

Next, we investigate how much $\ln \Omega_R$ changes when the energy and volume changes by ΔU_R and ΔV_R, respectively. We will assume that the magnitudes of ΔU_R and ΔV_R are at most of the order U_A and V_A, that is, very small compared to U_R and V_R. Let the reservoir change its energy from an initial value U_R^0 to $U_R^0 + \Delta U_R$ and its volume from V_R^0 to $V_R^0 + \Delta V_R$. We have

$$\ln \Omega_R(U_R^0 + \Delta U_R, V_R^0 + \Delta V_R) - \ln \Omega_R(U_R^0, V_R^0)$$

$$= \int_{(U_R^0, V_R^0)}^{(U_R^0 + \Delta U_R, V_R^0 + \Delta V_R)} \frac{\partial \ln \Omega_R(U_R', V_R')}{\partial U_R'} dU_R' + \frac{\partial \ln \Omega_R(U_R', V_R')}{\partial V_R'} dV_R'$$

$$\tag{2.64}$$

$$= \int_{(U_R^0, V_R^0)}^{(U_R^0 + \Delta U_R, V_R^0 + \Delta V_R)} \frac{1}{k_B T} dU_R' + \frac{P}{k_B T} dV_R' = \frac{\Delta U_R}{k_B T} + \frac{P \Delta V_R}{k_B T},$$

where we have used the fact that the temperature and pressure remain constant throughout the integration since the magnitudes of ΔU_R and ΔV_R are at most of the order U_A and V_A (constant temperature and pressure are guaranteed by the requirement that R is very large compared to A). We can write this result as

$$\Omega_R(U_R^0 + \Delta U_R, V_R^0 + \Delta V_R) = \Omega_R(U_R^0, V_R^0) e^{(\Delta U_R + P \Delta V_R)/k_B T}. \tag{2.65}$$

Since $\Delta U_R = -\Delta U_A$ and $\Delta V_R = -\Delta V_A$, we see that the number of states of R changes by a Boltzmann-like factor $\exp[-(\Delta U_A + P \Delta V_A)/k_B T]$ when the energy and volume of A changes at constant T and P. One can say that this factor describes the "willingness" of a reservoir at a certain temperature and pressure to give away an amount of energy equal to ΔU_A and change its volume by ΔV_A (compare with the Boltzmann factor in the canonical ensemble).

Let us apply Equation 2.65 to Ω_R in Equation 2.61, that is, we set $U_R^0 = U_{AR}$, $\Delta U_R = -\mathscr{U}_{A,i}$, $V_R^0 = V_{AR}$ and $\Delta V_R = -V_A$, whereby we obtain

$$\Omega_R(U_{AR} - \mathscr{U}_{A,i}, V_{AR} - V_A) = \Omega_R(U_{AR}, V_{AR}) e^{-(\mathscr{U}_{A,i} + P V_A)/k_B T}$$

and Equation 2.61 yields

$$\mathfrak{p}_A(\mathscr{U}_{A,i}, V_A) = \mathcal{K}' e^{-(\mathscr{U}_{A,i}+PV_A)/k_BT},$$

where

$$\mathcal{K}' = \frac{\Omega_R(U_{AR}, V_{AR})}{\Omega_{AR}}. \tag{2.66}$$

Since $\Sigma_{V'_A}\Sigma_j\mathfrak{p}_A(\mathscr{U}_{A,j}, V'_A) = 1$, we have $\Sigma_{V'_A}\Sigma_j\mathcal{K}'\exp(-[\mathscr{U}_{A,j} + PV'_A]/k_BT) = 1$ and hence

$$\mathcal{K}' = \frac{1}{\sum_{V'_A}\sum_j e^{-(\mathscr{U}_{A,j}+PV'_A)/k_BT}}.$$

We obtain

$$\mathfrak{p}_A(\mathscr{U}_{A,i}, V_A) = \frac{e^{-(\mathscr{U}_{A,i}+PV_A)/k_BT}}{\sum_{V'_A}\sum_j e^{-(\mathscr{U}_{A,j}+PV'_A)/k_BT}} = \frac{e^{-(\mathscr{U}_{A,i}+PV_A)/k_BT}}{\Upsilon_A}, \tag{2.67}$$

where $\Upsilon_A = 1/\mathcal{K}'$ is called the **isobaric-isothermal partition function** and equals

$$\Upsilon_A = \sum_{V'_A}\sum_j e^{-(\mathscr{U}_{A,j}+PV'_A)/k_BT}. \tag{2.68}$$

Note the analogy with the corresponding results for the canonical ensemble where V_A was constant, the exponent in $\mathfrak{p}_A$ was $-\mathscr{U}_{A,i}/k_BT$ and the summation in the partition function was only over the quantum states. In the present case, the exponent contains $-PV_A/k_BT$ in addition to the energy term and in the partition function we sum over both quantum states and volume.

The partition function Υ_A in Equation 2.68 is a property of subsystem A and, as apparent from this equation, it depends on the values of the energies of all quantum states of A for all possible volumes of A. From Equation 2.66, we see that we can also express Υ_A as

$$\Upsilon_A = \frac{\Omega_{AR}(U_{AR}, V_{AR})}{\Omega_R(U_{AR}, V_{AR})}, \tag{2.69}$$

which will be used later.

The probability $\mathfrak{p}_A(\mathscr{U}_{A,i}, V_A)$ depends only on the *value* of the energy of the quantum state and therefore we can drop the explicit state index i. Thus,

$$\mathfrak{p}_A(\mathscr{U}_A, V_A) = \frac{e^{-(\mathscr{U}_A+PV_A)/k_BT}}{\Upsilon_A} \tag{2.70}$$

is the probability for A to have volume V_A, and, at the same time, be in a particular quantum state with energy $\mathscr{U}_A$ at constant T and P. This result can be compared to Boltzmann's distribution law in the canonical ensemble, Equation 2.37.

Let us now determine the probability $\mathcal{P}_A(U_A, V_A)$ that system A has volume V_A and simultaneously has energy U_A. There are $\Omega_A(U_A, V_A)$ quantum states with energy U_A when the volume of A is V_A. Since each of these states occurs with a probability equal to $\mathfrak{p}_A(U_A, V_A)$, we can conclude that $\mathcal{P}_A(U_A, V_A)$ must be $\Omega_A(U_A, V_A)$ times larger than $\mathfrak{p}_A(U_A, V_A)$, so we have

$$\mathcal{P}_A(U_A, V_A) = \Omega_A(U_A, V_A)\mathfrak{p}_A(U_A, V_A). \tag{2.71}$$

From Equation 2.67, it follows that

$$\mathcal{P}_A(U_A, V_A) = \frac{\Omega_A(U_A, V_A)e^{-(U_A+PV_A)/k_BT}}{\Upsilon_A}. \tag{2.72}$$

The isobaric-isothermal partition function can be written in two different manners

$$\Upsilon_A = \sum_{V'_A}\sum_j e^{-(\mathscr{U}_{A,j}+PV'_A)/k_BT} = \sum_{V'_A}\sum_{U'_A}\Omega_A(U'_A, V'_A)e^{-(U'_A+PV'_A)/k_BT}, \tag{2.73}$$

where – apart from the sums over volume – the sum in the first expression is taken over all *quantum states* of subsystem A and the sum in the second expression is taken over all *energy levels* of A.

So far, we have for simplicity in notation suppressed that all entities of A depend also on the number of particles N_A. We have $\mathscr{U}_{A,i} = \mathscr{U}_{A,i}(V_A, N_A)$ and $\Omega_A = \Omega_A(U_A, V_A, N_A)$, so from Equation 2.73 it follows that the partition function Υ_A depends on T, P and N_A, that is, the same quantities that defined the ensemble. So we have $\Upsilon_A = \Upsilon_A(T, P, N_A)$. Likewise, $\mathcal{P}_A$ and $\mathfrak{p}_A$ also depend on T, P and N_A.

2.4.2 Thermodynamical Quantities in the Isobaric-Isothermal Ensemble

2.4.2.1 The Gibbs Free Energy

From Equation 2.69, one can derive the relationship $\Delta(k_B \ln \Upsilon_A) = \Delta S_{AR} = \Delta S_{tot}$ in exactly the same way as we did for the partition function Q_A in the beginning of Section 2.3.2.1. The sign of $\Delta(k_B \ln \Upsilon_A)$, which is the same as that of ΔS_{tot}, tells whether a process at constant T, P and N_A is spontaneous or not. As we did in the canonical ensemble, we can obtain an appropriate free energy function from the partition function and we define $G_A = -k_B T \ln \Upsilon_A$, which is known as the **Gibbs free energy** for system A (also called the **Gibbs energy**).

From now on, we will drop the subscript A for the system (unless needed). In general, we have the Gibbs free energy

$$G = G(T, P, N) = -k_B T \ln \Upsilon(T, P, N) \tag{2.74}$$

and for processes at constant T, P and N

$$\Delta G = -T\Delta S_{tot} \le 0, \tag{2.75}$$

where ΔS_{tot} is the total entropy change of the system and its environment. For a spontaneous process in a system at constant T, P and N, the Gibbs free energy always decreases – this is analogous to Helmholtz free energy A, which decreases for a process at constant T, V and N.

Alternative expressions for the isobaric-isothermal partition function partition function are

$$\Upsilon = \sum_{V,U} \Omega(U, V, N) e^{-\beta[U+PV]} = \sum_{V,j} e^{-\beta[\mathscr{U}_j(V,N)+PV]}$$

$$= \sum_{V} \left[\sum_{j} e^{-\beta\mathscr{U}_j(V,N)} \right] e^{-\beta PV} = \sum_{V} Q(T, V, N) e^{-\beta PV}. \tag{2.76}$$

The last expression shows the connection to the canonical partition function.

In the beginning of the previous section (Section 2.4.1), we saw that the volume cannot be a continuous variable in the current ensemble. This is the reason why we have a sum over volumes in Equation 2.76, where V takes on values from a discrete set. If we would have treated the volume as a continuous variable and defined the partition function as, for example, $\int dV \sum_j \exp(-\beta[\mathscr{U}_j + PV])$, this function would have the unit of volume. When defining the Gibbs free energy, it would not be possible just to take the logarithm of this function since the argument of a logarithm should be dimensionless. Likewise, in the expressions for the probabilities,

$$\mathfrak{p}(\mathscr{U}, V) = \frac{e^{-(\mathscr{U}+PV)/k_B T}}{\Upsilon} \tag{2.77}$$

and

$$P(U, V) = \frac{\Omega(U, V) e^{-(U+PV)/k_B T}}{\Upsilon}, \tag{2.78}$$

the partition function Υ must be dimensionless. To obtain a reasonable Υ, we must therefore make it fulfill this requirement and we will now see how this could be done.

The only thing we have said so far about the set of V values is that the difference between the discrete values is of the order δV, where δV is very small but nonzero. We may say that space is approximately divided into "cells" of the size δV, so the possible V values differ by δV (as we will see, the precise value of δV does not have any practical consequences for a macroscopic system). A possibility is therefore to define the partition function in terms of a volume integral as

$$\Upsilon = \frac{1}{\delta V} \int dV \sum_{j} e^{-\beta[\mathscr{U}_j(V,N)+PV]} = \frac{1}{\delta V} \int dV \, Q(T, V, N) e^{-\beta PV}, \tag{2.79}$$

where we have replaced the sum over the volume by an integral divided by the "cell volume" δV. The factor $1/\delta V$ in front of the integral makes the partition function dimensionless, so the logarithm can be taken without problems.

In principle, the value of δV affects the value of G; we have

$$G = -k_B T \left[\ln \left(\int dV\, Q(T, V, N) e^{-\beta PV} \right) - \ln \delta V \right],$$

so if we would use two different values, say, δV_I and δV_{II} to evaluate G, we obtain different results G_I and G_{II} that differ by $G_{II} - G_I = k_B T \ln[\delta V_{II}/\delta V_I]$. Both δV_I and δV_{II} should be small compared to the macroscopic scale. While G_I and G_{II} are of the order $N k_B T$, their difference is of the order $k_B T$, which is completely negligible for a macroscopic system. So in practice the choice of the value of δV does not have any effect on G. Thus, it does not make any difference in practice if we take Υ from Equations 2.76 or 2.79.

Like the Helmholtz free energy, the Gibbs free energy is an additive function. This follows since Equation 2.76 implies that for two separate systems A and B at the same temperature and pressure we have

$$G_{AB} = -k_B T \ln \Upsilon_{AB} = -k_B T \ln \sum_{V_{AB}} Q_{AB}(T, V_{AB}, N_{AB}) e^{-\beta PV_{AB}}$$

$$= -k_B T \ln \sum_{V_A, V_B} Q_A(T, V_A, N_A) Q_B(T, V_B, N_B) e^{-\beta P[V_A + V_B]}$$

$$= -k_B T \ln \left[\sum_{V_A} Q_A e^{-\beta PV_A} \sum_{V_B} Q_B e^{-\beta PV_B} \right] = G_A + G_B,$$

where we have used the fact that $Q_{AB} = Q_A Q_B$ as we have demonstrated earlier. When two macroscopic bulk systems A and B at the same T and P consist of the same kind of particles and we merge them into one system AB by removing any wall that separates them (keeping T and P constant), we still have $G_{AB} = G_A + G_B$ to an excellent approximation. The effect of the presence or absence of a separating wall is negligible since a very small minority of the molecules are near this wall and thereby affected by it when it is present. Boundary effects are normally negligible[20] since the region near the boundary of the system constitutes an extremely small part compared to the bulk fluid (all three dimensions of the systems are here assumed to be macroscopically large). This means that G is an extensive quantity, so if we increase the number of particles by a factor α in a system at constant T and P, that is, from from N to αN, the Gibbs free energy changes from G to αG. Since $\alpha G/\alpha N = G/N$, the ratio G/N remains constant when N varies, that is, G is proportional to N.

2.4.2.2 Probabilities and Thermodynamical Quantities

At equilibrium, a system with constant pressure has a volume that fluctuates. It is of interest to determine the probability $\mathscr{P}(V)$ that a system at constant T, P and N has the volume V. Since $\mathfrak{p}(\mathscr{U}_i(V, N), V)$ is the probability that the system has volume V and, at the same time,

---

[20] A notable exception is pure polar liquids, where boundary effects can be important.

is in quantum state i, we can obtain $\mathscr{P}(V)$ by simply summing $\mathfrak{p}$ over all quantum states for a given V, that is,

$$\mathscr{P}(V) = \sum_i \mathfrak{p}(\mathscr{U}_i(V,N), V) = \frac{\left[\sum_i e^{-\beta \mathscr{U}_i(V,N)}\right] e^{-\beta PV}}{\Upsilon} = \frac{Q_V e^{-\beta PV}}{\Upsilon}, \tag{2.80}$$

where Q_V is the canonical ensemble partition function for a system with volume V [for the given T and N, that is, $Q_V = Q(T, V, N)$].

In fact, one can separately consider the probability for the system to have a certain volume and the probability to be in a certain state for each possible volume. To see this, we write

$$\mathfrak{p}(\mathscr{U}_i, V) = \frac{e^{-\beta \mathscr{U}_i} e^{-\beta PV}}{\Upsilon} = \frac{[\mathfrak{p}_V(\mathscr{U}_i)Q_V] e^{-\beta PV}}{\Upsilon} = \mathfrak{p}_V(\mathscr{U}_i)\mathscr{P}(V),$$

where $\mathfrak{p}_V(\mathscr{U}_i) = \exp(-\beta \mathscr{U}_i)/Q_V$ is the canonical ensemble probability to observe quantum state i in a system with volume V (for the given T and N). This implies that one can obtain the correct probability $\mathfrak{p}(\mathscr{U}_i, V)$ at constant pressure by multiplying the probability for the system to be in a certain quantum state at constant volume and the probability for the system to have that volume. Likewise, $\mathcal{P}(U, V) = \mathcal{P}_V(U)\mathscr{P}(V)$, where $\mathcal{P}_V(U)$ is the canonical ensemble probability to have energy U in a system with volume V.

The average volume equals

$$\bar{V} \equiv \langle V \rangle = \sum_{V,i} V \mathfrak{p}(\mathscr{U}_i, V) = \sum_V V \mathscr{P}(V) \tag{2.81}$$

and it can also be obtained from $\bar{V} = \sum_{V,U} V \mathcal{P}(U, V)$. The probability $\mathscr{P}(V)$ has a peak at around the average volume $\bar{V}$. The width of this peak, that is, the magnitude of the fluctuations in volume, is treated in detail in Section 2.6.3. There it is shown that this width is related to the system's compressibility, which gives the magnitude of the decrease in volume with increased pressure. A very compressible system has larger volume fluctuations than a less compressible one. For a macroscopic system the fluctuations are, however, extremely small in general, so $\mathscr{P}(V)$ has a very sharp peak at $V = \bar{V}$.

The average of a quantity $f_i(V, N)$, for example $f_i = \mathscr{U}_i(V, N)$ or $f_i = (\partial \mathscr{U}_i(V, N)/\partial N)_V$, equals

$$\bar{f} \equiv \langle f \rangle = \sum_{V,i} f_i \mathfrak{p}(\mathscr{U}_i, V) =$$

$$= \sum_V \left[\sum_i f_i \mathfrak{p}_V(\mathscr{U}_i) \right] \mathscr{P}(V) = \sum_V \bar{f}_V \mathscr{P}(V), \tag{2.82}$$

where $\bar{f}_V = \sum_i f_i(V, N)\mathfrak{p}_V(\mathscr{U}_i)$ is the canonical ensemble average of f in a system with volume V (for the given T and N).

When V is treated like a continuous variable, as in Equation 2.79, the averages are formed by integration over V instead of summation. This is explained in the following shaded text box.

The average in Equation 2.81 can be written as an integral

$$\bar{V} = \frac{1}{\delta V} \int dV \sum_i V \, \mathfrak{p}(\mathcal{U}_i, V) = \frac{1}{\delta V} \int dV \, V \, \mathscr{P}(V). \qquad (2.83)$$

Likewise, the average $\bar{f}$ can be written as

$$\bar{f} = \frac{1}{\delta V} \int dV \bar{f}_V \mathscr{P}(V). \qquad (2.84)$$

By inserting $\mathscr{P}(V)$ from Equation 2.80, we can write it in the following manner

$$\bar{f} = \frac{1}{\delta V} \int dV \bar{f}_V \frac{Q_V e^{-\beta PV}}{\Upsilon} = \frac{\int dV \bar{f}_V Q_V e^{-\beta PV}}{\int dV' \, Q_{V'} e^{-\beta PV'}},$$

where we have inserted Υ from Equation 2.79 to obtain the rhs; we see that the factor $1/\delta V$ is cancelled by the same factor in Υ, so the average is independent of δV. This is yet another demonstration of the fact that the value of δV does not have any significance in practice.

If we take the derivative of

$$\ln \Upsilon = \ln \sum_{V,i} e^{-\beta[\mathcal{U}_i(V,N)+PV]}$$

with respect to P, we obtain

$$\left(\frac{\partial \ln \Upsilon}{\partial P} \right)_{T,N} = \frac{(\partial \Upsilon / \partial P)_{T,N}}{\Upsilon} = \frac{-\beta \sum_{V,i} V e^{-\beta[\mathcal{U}_i(V,N)+PV]}}{\Upsilon}$$

$$= -\beta \sum_{V,i} V \mathfrak{p}(\mathcal{U}_i(V,N), V) = -\beta \bar{V}.$$

This implies that

$$\bar{V} = -k_B T \left(\frac{\partial \ln \Upsilon}{\partial P} \right)_{T,N} = \left(\frac{\partial G}{\partial P} \right)_{T,N}, \qquad (2.85)$$

where the rhs shows a result well-known in thermodynamics.

The derivative of $\ln \Upsilon$ with respect to N and the definition of chemical potential in Equation 2.15 yield in the same manner

$$\mu = -k_B T \left(\frac{\partial \ln \Upsilon}{\partial N} \right)_{T,P} = \left(\frac{\partial G}{\partial N} \right)_{T,P}, \tag{2.86}$$

also known from thermodynamics. Likewise, by taking the β derivatives of $\ln \Upsilon$, we have

$$\left(\frac{\partial \ln \Upsilon}{\partial \beta} \right)_{P,N} = -(\bar{U} + P\bar{V}). \tag{2.87}$$

{**EXERCISE 2.7**
Derive the results in Equations 2.86 and 2.87, whereby Equation 2.82 can be used.}

The combination $\bar{U} + P\bar{V}$ is, by definition, the **enthalpy** H

$$H = \bar{U} + P\bar{V},$$

so we have

$$H = -k_B \left(\frac{\partial \ln \Upsilon}{\partial (1/T)} \right)_{P,N}$$

$$= k_B T^2 \left(\frac{\partial \ln \Upsilon}{\partial T} \right)_{P,N} = -T^2 \left(\frac{\partial [G/T]}{\partial T} \right)_{P,N} \tag{2.88}$$

and

$$H = \left(\frac{\partial (G/T)}{\partial (1/T)} \right)_{P,N}. \tag{2.89}$$

The latter relationship is known as the **Gibbs-Helmholtz equation**, well-known from thermodynamics.

2.4.2.3 The Entropy in the Isobaric-Isothermal Ensemble

When we calculated the β derivative of $\ln \Upsilon$ earlier, we considered $\Upsilon = \Upsilon(\beta, P, N)$ rather than $\Upsilon = \Upsilon(T, P, N)$. From expressions for the derivatives of $\ln \Upsilon$ in the previous section, it follows that the differential of $\ln \Upsilon(\beta, P, N)$ is

$$d\ln \Upsilon = -(\bar{U} + P\bar{V})d\beta - \beta\bar{V}dP - \beta\mu dN.$$

If we add $d(\beta\bar{U} + \beta P\bar{V})$ to both sides, we obtain after simplification

$$d(\ln \Upsilon + \beta\bar{U} + \beta P\bar{V}) = \beta(d\bar{U} + Pd\bar{V} - \mu dN).$$

Since $Pd\bar{V} - \mu dN = -đw$, we have

$$dÜ + Pd\bar{V} - \mu dN = dÜ - đw = đq \tag{2.90}$$

and we obtain $d(\ln \Upsilon + \beta \bar{U} + \beta P \bar{V}) = \beta đq$, which we can write

$$d\left[k_B \ln \Upsilon + \frac{H}{T}\right] = \frac{đq}{T} = dS,$$

where we have identified the entropy S in analogy to the corresponding treatment in the canonical ensemble in Section 2.3.2.3. We integrate this expression, set the integration constant to zero and obtain

$$k_B \ln \Upsilon + \frac{H}{T} = S,$$

which implies the well-known thermodynamic relationship

$$G = H - TS. \tag{2.91}$$

Incidentally, we note that the relationship (2.90) above implies that when P and N are constant, we have $đq = dÜ + Pd\bar{V} = dH$. Thus, heat added to a system at constant pressure is equal to the increase in H provided that no other work than pressure-volume work is done. This is a characteristic property of the enthalpy.

By performing the differentiation of G/T in the rhs of Equation 2.88, we obtain after simplification

$$\left(\frac{\partial G}{\partial T}\right)_{P,N} = \frac{G - H}{T} = -S. \tag{2.92}$$

This result together with Equations 2.85 and 2.86 implies that the differential of $G(T, P, N)$ is

$$dG = -SdT + \bar{V}dP + \mu dN, \tag{2.93}$$

which we recognize from thermodynamics. In the case of several particle species, we have instead

$$dG = -SdT + \bar{V}dP + \sum_\nu \mu_\nu dN_\nu, \tag{2.94}$$

where the sum is over all species.

It is possible to write the entropy in the form

$$S = -k_B \sum_{V,i} \mathfrak{p}(\mathscr{U}_i(V), V) \ln \mathfrak{p}(\mathscr{U}_i(V), V) \tag{2.95}$$

(where we do not show the N dependence of $\mathscr{U}_i$). This formula implies that the general form for S that we mentioned in Section 2.3.2.3 applies also in the isothermal-isobaric ensemble.

{**EXERCISE 2.8**

Derive the expression (2.95) for the entropy.}

By inserting $H = \bar{U} + P\bar{V}$ into Equation 2.91, we obtain

$$G = \bar{U} + P\bar{V} - TS = A + P\bar{V}, \qquad (2.96)$$

where we have identified $\bar{U} - TS = A$ as the Helmholtz free energy, that is, the same relationship for A as in the canonical ensemble and in thermodynamics. This identification and the resulting expression $G = A + P\bar{V}$ can be justified in the current ensemble as follows. We have

$$G = -k_B T \ln \Upsilon = -k_B T \ln \left[\sum_V Q(T, V, N) e^{-\beta PV} \right]$$

$$\approx -k_B T \ln \left[Q(T, \bar{V}, N) e^{-\beta P\bar{V}} \right] = -k_B T \ln Q(T, \bar{V}, N) + P\bar{V},$$

where we have approximated the logarithm of the sum with the logarithm of its largest term. We have thereby used the fact that $Q(T, V, N) e^{-\beta PV}$ according to Equation 2.80 is proportional to $\mathscr{P}(V)$ and in general has a very sharp peak at $V = \bar{V}$ for a macroscopic system. Thus, to a very good approximation, we have $A = -k_B T \ln Q(T, \bar{V}, N)$, which is the Helmholtz free energy in the canonical ensemble evaluated at $V = \bar{V}$. From Equation 2.82 and the presence of the peak in $\mathscr{P}(V)$, it follows that the average of a quantity f in the isobaric-isothermal ensemble is equal to the corresponding average $\bar{f}_V$ in the canonical ensemble evaluated at $V = \bar{V}$ as a very good approximation.

As we have seen at the end of Section 2.4.2.1, for a single-component system G is proportional to N at constant T and P. The proportionality constant is $(\partial G/\partial N)_{T,P}$, so using Equation 2.86 we see that

$$G = \mu N. \qquad (2.97)$$

This also follows from Equation 2.93 by integration over N at constant T and P, that is, $G = \int dG = \mu \int dN = \mu N$ where the integration constant is zero since $G = 0$ when $N = 0$. Likewise, for a multicomponent system, we have

$$G = \sum_\nu \mu_\nu N_\nu, \qquad (2.98)$$

which follows from an integration of Equation 2.94 at constant T and P provided the composition of the system is unchanged when the number of particles is increased. (Note that the extensivity of G means that it changes to αG when the system size is changed by a factor α provided the composition is kept constant, that is, if the number of particles varies as αN_ν for all ν.) Finally, using Equations 2.96 and 2.98, we note that

$$\bar{U} + P\bar{V} - TS - \sum_\nu \mu_\nu N_\nu = 0. \qquad (2.99)$$

2.5 OPEN SYSTEMS: CHEMICAL POTENTIAL AND THE GRAND CANONICAL ENSEMBLE

Consider open systems where both energy and particles are allowed to be exchanged between them. Let us assume that the temperature of the systems is the same. As we saw at the end of Section 2.3.2.2, if the chemical potential μ of two systems are different, a spontaneous flow of particles occurs from the system with the highest μ to the one with the lowest. If μ is equal in the two systems, there is an equilibrium distribution of particles and the net flow of particle is zero.

For an open system at equilibrium with its surroundings, the variable N is not an independent variable. Instead, one takes the chemical potential μ of the system as an independent variable and then one has to evaluate the average number of particles $\bar{N}$ given the value of μ. We shall here treat the case of systems that are free to exchange particles and energy with their surroundings while keeping their volume constant. This will be done by using the **grand canonical ensemble,** also called the **μVT ensemble.** By definition, such an ensemble consists of replicas of the macroscopic system, all replicas having the same μ, V and T.

2.5.1 Probabilities and the Grand Canonical Partition Function

Let us regard a system A in contact with a reservoir R; the latter being so large that the exchange of energy and particles with A does not change the temperature and chemical potential of R (see Figure 2.13). Without loss of generality, we may assume that the combined system AR is isolated, which means that $U_{tot} = U_{AR} = U_A + U_R$ = constant, $N_{tot} = N_{AR} = N_A + N_R$ = constant and that all $\Omega_{AR}(U_{AR}, N_{AR})$ quantum states of AR are

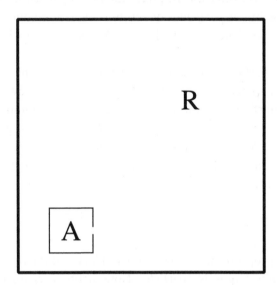

FIGURE 2.13 A macroscopic system A in contact with a reservoir R, which is much larger than A (larger than shown in the picture) and has constant temperature T and chemical potential μ. Systems A and R are in thermal contact with each other and can exchange particles via an opening between A and R, so both the temperature and the chemical potential of A and R are the same. R acts as an "energy and particle reservoir." Subsystem A has constant volume. The combined system AR is isolated from the environment.

equally probable, that is, all states of AB for all possible distributions of energies and particles between A and R

$$\Omega_{AR} = \sum_{N_A} \sum_{U_A} \Omega_A(U_A, N_A)\Omega_R(U_{AR} - U_A, N_{AR} - N_A). \tag{2.100}$$

Here we have distinguished between the quantum states of A and those of R, which is not entirely correct since A and R can exchange particles with each other via an opening in the wall surrounding A. However, both A and R are large macroscopic systems and the immediate neighborhood of the opening is a very small part of the system. Therefore, the influence of the hole on the properties of A and R at equilibrium is negligible in practice; the only role of the hole is to maintain the equilibrium by letting particles pass through it. Thus, $\Omega_A(U_A, N_A)$ is the number of states of A when it contains N_A particles and its energy is U_A. The corresponding considerations apply to R.

The grand canonical ensemble we consider consists of subsystem A and its replicas, all having the same values of T, V_A and μ. The role of the reservoir R is solely to provide the constant temperature and chemical potential and thereby to act as an "energy and particle reservoir." R is not included in the ensemble. We shall determine the probability $\mathcal{P}_A(U_A, N_A)$ that system A has energy U_A and simultaneously contains N_A particles. Thereby, we proceed similarly to the treatment of the other ensembles, so we start by determining the probability $\mathfrak{p}_A(\mathcal{U}_{A,i}, N_A)$ that subsystem A contains N_A particles and is in quantum state i with the energy $\mathcal{U}_{A,i} = \mathcal{U}_{A,i}(V_A, N_A)$, that is, the ith quantum state for A when it contains N_A particles in the volume V_A. Since R is so huge compared to A, we can in practice assume that all quantum states of A are accessible for the given number of particles. As before, $\mathfrak{p}_A$ gives the probability for A to be in *one particular* quantum state. For simplicity in notation, we do not show the volume dependencies explicitly in several expressions where V_A and V_R do not vary.

When A contains N_A particles and is in its quantum state i, the reservoir contains $N_{AR} - N_A$ particles and has energy $U_{AR} - \mathcal{U}_{A,i}$. R can thereby be in any of its $\Omega_R(U_{AR} - \mathcal{U}_{A,i}, N_{AR} - N_A)$ quantum states with this energy for the specified number of particles and the whole system AR has $1 \times \Omega_R(U_{AR} - \mathcal{U}_{A,i}, N_{AR} - N_A)$ possible states [factor 1 for A and factor $\Omega_R(U_{AR} - \mathcal{U}_{A,i}, N_{AR} - N_A)$ for R]. Since AR is isolated, all these states for the whole system are equally probable. Each of them occurs with the probability $1/\Omega_{AR}$ and it follows that $\mathfrak{p}_A(\mathcal{U}_{A,i}, N_A)$ is $\Omega_R(U_{AR} - \mathcal{U}_{A,i}, N_{AR} - N_A)$ times larger than $1/\Omega_{AR}$, that is,

$$\mathfrak{p}_A(\mathcal{U}_{A,i}, N_A) = \frac{\Omega_R(U_{AR} - \mathcal{U}_{A,i}, N_{AR} - N_A)}{\Omega_{AR}}. \tag{2.101}$$

Hence, this probability is proportional to the number of states of the reservoir with energy $U_{tot} - \mathcal{U}_{A,i}$ when it contains $N_{AR} - N_A$ particles. Note that this argument is analogous to the corresponding arguments in the canonical and isobaric-isothermal ensembles, and, in fact, the derivations of the various expressions for the probability functions in the three ensembles are very similar. Therefore, we will only give a shortened version of the derivation for the present case (like for the other ensembles, a more strict derivation is presented in Appendix 2B).

In the same manner as before, we can conclude that all quantum states of A with the same energy occur with the same probability (the number of states of the reservoir is the same when the energy of A is the same), but the probabilities for quantum states with different energies are not the same. Let us consider the number of quantum states of the reservoir as a function of energy and number of particles, $\Omega_R = \Omega_R(U_R, N_R)$, and we have (from Equations 2.19 and 2.21)

$$\left(\frac{\partial \ln \Omega_R(U_R, N_R)}{\partial U_R}\right)_{V_R, N_R} = \frac{1}{k_B T} \qquad (2.102)$$

and

$$\left(\frac{\partial \ln \Omega_R(U_R, N_R)}{\partial N_R}\right)_{U_R, V_R} = -\frac{\mu}{k_B T}. \qquad (2.103)$$

Since A is so small compared to R, its presence does not influence the properties of R – in particular, the temperature and chemical potential of R are not affected by the presence of A. If we let U_R vary in the interval U_{AR} to $U_{AR} - \mathcal{U}_{A,i}$ and N_R vary in the interval N_{AR} to $N_{AR} - N_A$ (i.e., the energy and number of particles of A vary from 0 to $\mathcal{U}_{A,i}$ and 0 to N_A, respectively), the temperature and chemical potential remain constant since we always have $\mathcal{U}_{A,i} \ll U_R$ and $N_A \ll N_R$ (constant temperature and chemical potential are guaranteed by the requirement that R is a large reservoir for energy and particles).

Since $\mathcal{U}_{A,i} \ll U_{AR}$ and $N_A \ll N_{AR}$ and since T and μ thereby are constant, we have

$$\ln \Omega_R(U_{AR}, N_{AR}) - \ln \Omega_R(U_{AR} - \mathcal{U}_{A,i}, N_{AR} - N_A)$$
$$= \frac{\partial \ln \Omega_R(U_{AR}, N_{AR})}{\partial U_R} \mathcal{U}_{A,i} + \frac{\partial \ln \Omega_R(U_{AR}, N_{AR})}{\partial N_R} N_A$$
$$= \frac{1}{k_B T} \mathcal{U}_{A,i} - \frac{\mu}{k_B T} N_A.$$

This can be written as

$$\Omega_R(U_{AR} - \mathcal{U}_{A,i}, N_{AR} - N_A) = \Omega_R(U_{AR}, N_{AR}) e^{-(\mathcal{U}_{A,i} - \mu N_A)/k_B T},$$

so Equation 2.101 yields

$$\mathfrak{p}_A(\mathcal{U}_{A,i}, N_A) = \mathcal{K}'' e^{-(\mathcal{U}_{A,i} - \mu N_A)/k_B T},$$

where

$$\mathcal{K}'' = \frac{\Omega_R(U_{AR}, N_{AR})}{\Omega_{AR}}. \qquad (2.104)$$

Since $\sum_{N_A'} \sum_j \mathfrak{p}_A(\mathcal{U}_{A,j}, N_A') = 1$, it follows that

$$\mathfrak{p}_A(\mathcal{U}_{A,i}, N_A) = \frac{e^{-(\mathcal{U}_{A,i} - \mu N_A)/k_B T}}{\sum_{N_A'=0}^{\infty} \sum_j e^{-(\mathcal{U}_{A,j} - \mu N_A')/k_B T}} = \frac{e^{-(\mathcal{U}_{A,i} - \mu N_A)/k_B T}}{\Xi_A}, \qquad (2.105)$$

where $\Xi_A = 1/\mathcal{K}''$ is called the **grand canonical partition function** and equals

$$\Xi_A = \sum_{N'_A=0}^{\infty} \sum_{j} e^{-(\mathcal{U}_{A,j}-\mu N'_A)/k_B T}.$$

Note the analogy with the corresponding results (2.35) and (2.67) for the canonical ensemble and isobaric-isothermal ensembles, respectively.

The sums over N'_A in the expressions for $\mathfrak{p}_A$ and Ξ_A go from zero to infinity. The latter is possible because in practice there is a limit to how many particles there can be in A at the same time. When the particles are densely packed in the volume V_A, it is difficult for more particles to enter due to the short-range interparticle repulsions. The probability to enter is low since the energy of A becomes very high when a further particle enters. This is taken care of automatically in the formulas. The terms with too many particles do not give any contribution in practice since the energy is very high and the exponential function is very small.

The partition function Ξ_A is a property of subsystem A and it depends on the values of the energies of all quantum states of A for all possible number of particles in A. From Equation 2.104, we see that we can also express Ξ_A as

$$\Xi_A = \frac{\Omega_{AR}}{\Omega_R(U_{AR}, N_{AR})}. \tag{2.106}$$

This will be useful later.

Let us now consider the probability $\mathcal{P}_A(U_A, N_A)$ that system A contains N_A particles and simultaneously has energy U_A. Analogous to the argument used in the other ensembles, we see that this probability must be $\Omega_A(U_A, N_A)$ times larger than $\mathfrak{p}_A(U_A, N_A)$ and we have

$$\mathcal{P}_A(U_A, N_A) = \Omega_A(U_A, N_A)\mathfrak{p}_A(U_A, N_A) \tag{2.107}$$

$$= \frac{\Omega_A(U_A, N_A)e^{-(U_A-\mu N_A)/k_B T}}{\Xi_A}. \tag{2.108}$$

The grand canonical partition function can be written as

$$\Xi_A = \sum_{N'_A=0}^{\infty} \sum_{j} e^{-(\mathcal{U}_{A,j}-\mu N'_A)/k_B T} = \sum_{N'_A=0}^{\infty} \sum_{U'_A} \Omega_A(U'_A, N'_A)e^{-(U'_A-\mu N'_A)/k_B T},$$

where – apart from the sums over number of particles – the sum in the first expression is taken over all quantum states of subsystem A and the sum in the last expression is taken over all energy levels of A.

All entities of A depend also on the volume V_A, like $\mathcal{U}_{A,i} = \mathcal{U}_{A,i}(V_A, N_A)$ and $\Omega_A = \Omega_A(U_A, V_A, N_A)$. Hence, the partition function Ξ_A depends on T, V_A, and μ, $\Xi_A = \Xi_A(T, V_A, \mu)$, that is, the same quantities that define the ensemble. Likewise, $\mathcal{P}_A$ and $\mathfrak{p}_A$ also depend on T, V_A, and μ.

2.5.2 Thermodynamical Quantities in the Grand Canonical Ensemble

From Equation 2.106, one can derive that $\Delta(k_B \ln \Xi_A) = \Delta S_{AR} = \Delta S_{tot}$ in exactly the same manner as for the partition functions Q_A and Υ_A. It follows that one can use the sign of this difference to decide whether a process at constant T, V_A and μ is spontaneous or not. The appropriate free energy function $\Theta_A = -k_B T \ln \Xi_A$ is called the **grand potential** for system A. It is also called the **grand free energy**.

Let us drop the subscript A that indicates the system. In the general case, we have the grand potential

$$\Theta = \Theta(T, V, \mu) = -k_B T \ln \Xi(T, V, \mu) \qquad (2.109)$$

and for processes at constant T, V and μ

$$\Delta\Theta = -T\Delta S_{tot} \leq 0 \qquad (2.110)$$

where ΔS_{tot} is the entropy change in the system + its environment. The grand potential of a system always decreases for a spontaneous process at constant temperature, volume and chemical potential. Θ is an additive function.

The grand canonical partition function can be written as

$$\Xi = \sum_{N,U} \Omega(U, V, N)e^{-\beta[U-\mu N]} = \sum_{N,j} e^{-\beta[\mathscr{U}_j(V,N)-\mu N]}$$

$$= \sum_N \left[\sum_j e^{-\beta \mathscr{U}_j(V,N)} \right] e^{\beta\mu N} = \sum_{N=0}^{\infty} Q(T, V, N)e^{\beta\mu N}, \qquad (2.111)$$

where the connection to the canonical partition function is shown in the last expression. Below, we will use the notation Q_N for the canonical ensemble partition function for a system with N particles (for the given T and V, that is, $Q_N = Q(T, V, N)$).

Let us now consider the probability $\mathscr{P}(N)$ that an open system for the given T, V and μ contains N particles. The probability $\mathscr{P}(N)$ is obtained by summing $\mathfrak{p}(\mathscr{U}_i(V, N), N)$ for each N over all possible microstates

$$\mathscr{P}(N) = \sum_i \mathfrak{p}(\mathscr{U}_i(V, N), N) = \frac{\left[\sum_i e^{-\beta \mathscr{U}_i(V,N)} \right] e^{\beta\mu N}}{\Xi} = \frac{Q_N e^{\beta\mu N}}{\Xi}. \qquad (2.112)$$

It is possible to factorize $\mathfrak{p}(\mathscr{U}_i, N)$ in two parts, namely $\mathscr{P}(N)$ times the probability that the system is in microstate i, given that its contains N particles. We have

$$\mathfrak{p}(\mathscr{U}_i, N) = \frac{e^{-\beta \mathscr{U}_i}e^{\beta\mu N}}{\Xi} = \frac{[\mathfrak{p}_N(\mathscr{U}_i)Q_N]e^{\beta\mu N}}{\Xi} = \mathfrak{p}_N(\mathscr{U}_i)\mathscr{P}(N),$$

where $\mathfrak{p}_N(\mathscr{U}_i) = \exp(-\beta \mathscr{U}_i)/Q_N$ is the canonical ensemble probability to observe quantum state i in a system with N particles (for the given T and V). Likewise, $\mathcal{P}(U, N) =$

$\mathcal{P}_N(U)\mathcal{P}(N)$, where $\mathcal{P}_N(U)$ is the canonical ensemble probability to have energy U in a system with N particles.

The average number of particles is given by

$$\bar{N} \equiv \langle N \rangle = \sum_{N,i} N \, \mathfrak{p}(\mathcal{U}_i, N) = \sum_N N \mathcal{P}(N).$$

It can also be obtained as $\bar{N} = \sum_{N,U} N P(U, N)$. The average of a quantity $f_i(V, N)$ equals

$$\bar{f} \equiv \langle f \rangle = \sum_{N,i} f_i \mathfrak{p}(\mathcal{U}_i, N) =$$

$$= \sum_N \left[\sum_i f_i \mathfrak{p}_N(\mathcal{U}_i) \right] \mathcal{P}(N) = \sum_N \bar{f}_N \mathcal{P}(N),$$

where $\bar{f}_N = \sum_i f_i(V, N) \mathfrak{p}_N(\mathcal{U}_i)$ is the canonical ensemble average of f in a system with N particles (for the given T and V). In Section 2.6.2, we will find that $\mathcal{P}(N)$ for a macroscopic system is in general sharply peaked at $N = \bar{N}$.

By taking the derivative of $\ln \Xi$ with respect to μ, we obtain

$$\left(\frac{\partial \ln \Xi}{\partial \mu} \right)_{T,V} = \frac{(\partial \Xi / \partial \mu)_{T,V}}{\Xi} = \frac{\sum_{N,i} \beta N e^{-\beta[\mathcal{U}_i(V,N) - \mu N]}}{\Xi}$$

$$= \beta \sum_{N,i} N \mathfrak{p}(\mathcal{U}_i(N), N) = \beta \bar{N},$$

so we have

$$\bar{N} = k_B T \left(\frac{\partial \ln \Xi}{\partial \mu} \right)_{T,V} = - \left(\frac{\partial \Theta}{\partial \mu} \right)_{T,V}. \tag{2.113}$$

Likewise, by taking the V and the β derivatives of $\ln \Xi$, we have

$$P = k_B T \left(\frac{\partial \ln \Xi}{\partial V} \right)_{T,N} = - \left(\frac{\partial \Theta}{\partial V} \right)_{T,N} \tag{2.114}$$

and

$$\left(\frac{\partial \ln \Xi}{\partial \beta} \right)_{V,\mu} = -\bar{U} + \mu \bar{N}, \tag{2.115}$$

which can be written

$$\bar{U} - \mu \bar{N} = k_B T^2 \left(\frac{\partial \ln \Xi}{\partial T} \right)_{V,\mu} = -T^2 \left(\frac{\partial [\Theta / T]}{\partial T} \right)_{V,\mu}. \tag{2.116}$$

Together with Equation 2.113, this gives a route to calculate the internal energy $\bar{U}$ of the system.

{**EXERCISE 2.9**
Derive the results in Equations 2.114–2.116.}

It follows that the differential of $\ln \Xi (\beta, V, \mu)$ is

$$d \ln \Xi = (-\bar{U} + \mu \bar{N}) d\beta + \beta P dV + \beta \bar{N} d\mu.$$

By adding $d(\beta \bar{U} - \beta \mu \bar{N})$ to both sides and doing similar manipulations as for the canonical and isobaric-isothermal ensembles, we obtain

$$d \left[k_B \ln \Xi + \frac{\bar{U}}{T} - \frac{\mu \bar{N}}{T} \right] = \frac{dq}{T} = dS, \tag{2.117}$$

where we have identified the entropy S in analogy to Sections 2.3.2.3 and 2.4.2.3.

{**EXERCISE 2.10**
Derive Equation 2.117.}

By integrating Equation 2.117 and setting the integration constant to zero, we obtain

$$k_B \ln \Xi + \frac{\bar{U}}{T} - \frac{\mu \bar{N}}{T} = S,$$

which implies that

$$\Theta = \bar{U} - TS - \mu \bar{N}. \tag{2.118}$$

If we perform the differentiation of Θ/T in the rhs of Equation 2.116, we can rewrite that equation as

$$\left(\frac{\partial \Theta}{\partial T} \right)_{V,\mu} = \frac{\Theta - \bar{U} + \mu \bar{N}}{T} = -S. \tag{2.119}$$

The entropy can be written in the form

$$S = -k_B \sum_{N,i} \mathfrak{p}(\mathcal{U}_i(N), N) \, \ln \mathfrak{p}(\mathcal{U}_i(N), N), \tag{2.120}$$

so the general form for S mentioned in Section 2.3.2.3 applies also in the grand canonical ensemble.

{**EXERCISE 2.11**
Derive Equation 2.120 for the entropy.}

Equations 2.113, 2.114 and 2.119 imply that the differential of $\Theta(T, V, \mu)$ is

$$d\Theta = -SdT - PdV - \bar{N}d\mu. \tag{2.121}$$

Let us keep μ and T constant and vary V. For an open system consisting of a homogeneous bulk fluid with volume V, the surroundings consist of the same fluid with the same density as inside the system. A variation of V simply means that more of the same fluid is located inside the system. Thus, P is not changed during such a variation, so if we vary the volume from V_1 to V_2, we have

$$\Delta\Theta = \int_{\Theta_1}^{\Theta_2} d\Theta = -P \int_{V_1}^{V_2} dV = -P\Delta V.$$

Applying this result for $V_1 = 0$ to $V_2 = V$ and using the obvious fact that $\Theta = 0$ when $V = 0$, we obtain

$$\Theta = -PV, \tag{2.122}$$

which means that the grand potential for a bulk fluid is a quite trivial quantity.[21] If we insert this in Equations 2.118 and 2.121, we conclude that

$$\bar{U} + PV - TS - \mu\bar{N} = 0 \tag{2.123}$$

and

$$-SdT + VdP - \bar{N}d\mu = 0. \tag{2.124}$$

The latter equation is known in thermodynamics as the **Gibbs-Duhem equation**. This equation expresses that for a system with three degrees of freedom (a one-component bulk system), it is impossible for three intensive variables (T, P and μ) to be independent of each other. To define the system, it is necessary to have at least one extensive variable as an independent variable (otherwise, the size of the system is not defined).

Incidentally, we note that if the Helmholtz and Gibbs free energies are defined in the grand canonical ensemble as $A = \bar{U} - TS$ and $G = \bar{U} + PV - TS$, we have from Equation 2.118

$$\Theta = A - \mu\bar{N}, \tag{2.125}$$

which is used in thermodynamics, and Equation 2.123 gives

$$G = \mu\bar{N},$$

which can be compared with Equation 2.97.

---

[21] For a system that contains an interface, the grand potential is far from trivial. For a system with a planar interface in its interior, $\Theta = \gamma^S A - PV$, where γ^S is the surface tension of the interface and A is the area of the latter.

For a multicomponent system with constant chemical potential μ_ν for each species ν, we have

$$\Xi = \sum_{U,N_1,N_2,...} \Omega(U, V, N_1, N_2, \ldots) e^{-\beta[U - \Sigma_\nu \mu_\nu N_\nu]}$$

$$= \sum_{i,N_1,N_2,...} e^{-\beta[\mathcal{U}_i(V, N_1, N_2, ...) - \Sigma_\nu \mu_\nu N_\nu]}$$

$$= \sum_{N_1,N_2,...} Q(T, V, N_1, N_2, \ldots) e^{\beta \Sigma_\nu \mu_\nu N_\nu} \tag{2.126}$$

and, for example,

$$\bar{N}_\nu = k_B T \left(\frac{\partial \ln \Xi}{\partial \mu_\nu} \right)_{T,V,\{\mu_{\nu'}\}_{\nu' \neq \nu}} = - \left(\frac{\partial \Theta}{\partial \mu_\nu} \right)_{T,V,\{\mu_{\nu'}\}_{\nu' \neq \nu}}.$$

The other results above are generalized in an analogous fashion. The relationship that corresponds to Equation 2.123 is

$$\bar{U} + PV - TS = \sum_\nu \mu_\nu \bar{N}_\nu$$

(compare with Equation 2.99) and the Gibbs-Duhem equation is

$$- SdT + VdP - \sum_\nu \bar{N}_\nu d\mu_\nu = 0. \tag{2.127}$$

Again, this set of intensive variables cannot be independent of each other.

{**EXERCISE 2.12**
Show that the grand canonical ensemble equations for S and P

$$S = k_B \left(\frac{\partial (T \ln \Xi)}{\partial T} \right)_{V,\mu} \quad \text{and} \quad P = k_B T \left(\frac{\partial \ln \Xi}{\partial V} \right)_{T,\mu}$$

go over into the canonical ensemble equations for the same variables

$$S = k_B \left(\frac{\partial (T \ln Q)}{\partial T} \right)_{V,N} \quad \text{and} \quad P = k_B T \left(\frac{\partial \ln Q}{\partial V} \right)_{T,N}$$

when $\ln \Xi$ is replaced by the logarithm of the largest term in the sum

$$\ln \Xi(T, V, \mu) = \ln \sum_{N=0}^{\infty} Q(T, V, N) e^{\beta \mu N}.$$

Note that different variables are kept constant in the two ensembles. The chain rule can be used to make the transition between the different sets of independent variables.}

2.6 FLUCTUATIONS IN THERMODYNAMICAL VARIABLES

In this section, we will derive certain relationships between fluctuations in thermodynamical variables and some other thermodynamical entities. In the various ensembles, there are different variables that fluctuate for example; in the grand canonical ensemble, N fluctuates and V is constant while, in the isobaric-isothermal ensemble V fluctuates and N is constant. The relationships that we derive are therefore not applicable in all ensembles.

2.6.1 Fluctuations in Energy in the Canonical Ensemble

Let us investigate how much the energy U of a system fluctuates around the mean value $\bar{U}$ in the canonical ensemble. As a measure of the deviation from $\bar{U}$, we will use the root mean-square deviation σ_U, which is defined from

$$\sigma_U^2 = \left\langle \left(U - \bar{U}\right)^2 \right\rangle = \left\langle U^2 + \bar{U}^2 - 2\bar{U}U \right\rangle$$
$$= \left\langle U^2 \right\rangle + \bar{U}^2 - 2\bar{U} \left\langle U \right\rangle = \left\langle U^2 \right\rangle - \left\langle U \right\rangle^2$$

(remember that $\bar{U}$ and $\langle U \rangle$ denote the same thing). From Equation 2.47, it follows that

$$\langle U \rangle = -\frac{1}{Q} \left(\frac{\partial Q}{\partial \beta} \right)_{V,N}$$

and, similarly, one can show that

$$\left\langle U^2 \right\rangle = \frac{1}{Q} \left(\frac{\partial^2 Q}{\partial \beta^2} \right)_{V,N}.$$

{**EXERCISE 2.13**
Show this expression for $\left\langle U^2 \right\rangle$.}

We have

$$\frac{\partial \langle U \rangle}{\partial \beta} = -\frac{\partial}{\partial \beta} \left[\frac{1}{Q} \frac{\partial Q}{\partial \beta} \right] = \frac{1}{Q^2} \left[\frac{\partial Q}{\partial \beta} \right]^2 - \frac{1}{Q} \frac{\partial^2 Q}{\partial \beta^2}$$
$$= \langle U \rangle^2 - \left\langle U^2 \right\rangle = -\sigma_U^2$$

and therefore

$$\sigma_U^2 = k_B T^2 \left(\frac{\partial \bar{U}}{\partial T} \right)_{V,N} = k_B T^2 C_V \tag{2.128}$$

where C_V is the heat capacity at constant volume. The size of the fluctuations in energy is accordingly directly related to the heat capacity of the system.

To estimate how large the fluctuations are, we use the fact that the average energy has a magnitude of the order $N k_B T$ and the heat capacity $N k_B$ (we have seen that this is true for ideal gases, but it is of more general validity). The relative size of the fluctuations, that is $\sigma_U/\bar{U} = (k_B T^2 C_V)^{1/2}/\bar{U}$, will therefore be of the same order of magnitude as $(k_B T^2 N k_B)^{1/2}/(N k_B T) = N^{-1/2}$. When N is of the order 10^{20}, the fluctuations in energy will be about 10^{-10} smaller than the energy itself, an extremely small quantity. All values of the energy other than $\bar{U}$ can therefore entirely be neglected.

2.6.2 Fluctuations in Number of Particles in the Grand Canonical Ensemble

Here we investigate how the number of particles N in a one-component open bulk system fluctuates around the average $\bar{N}$ in the grand canonical ensemble, that is, at constant T, V and μ. The treatment is similar to that for the energy fluctuations for the canonical ensemble in Section 2.6.1, so we give a shortened version here. The root mean-square deviation for the number of particles, σ_N, is given by

$$\sigma_N^2 = \left\langle (N - \bar{N})^2 \right\rangle = \langle N^2 \rangle - \langle N \rangle^2 .$$

and we have from Equation 2.113

$$\langle N \rangle = \frac{k_B T}{\Xi} \left(\frac{\partial \Xi}{\partial \mu} \right)_{T,V} .$$

One can show that {**EXERCISE 2.14:** Show the following two equations.}

$$\langle N^2 \rangle = \frac{(k_B T)^2}{\Xi} \left(\frac{\partial^2 \Xi}{\partial \mu^2} \right)_{T,V}$$

and

$$k_B T \left(\frac{\partial \langle N \rangle}{\partial \mu} \right)_{T,V} = \langle N^2 \rangle - \langle N \rangle^2 = \sigma_N^2.$$

This can be written

$$\sigma_N^2 = V k_B T \left(\frac{\partial n}{\partial \mu} \right)_{T,V} \tag{2.129}$$

where $n = \bar{N}/V$ is the number density.

Now, when T is constant, the Gibbs-Duhem Equation 2.124 becomes $V dP - \bar{N} d\mu = 0$, which can be written as $n d\mu = dP$. This implies $n(\partial \mu/\partial n)_T = (\partial P/\partial n)_T$ and hence, by inverting the derivatives, we have

$$\frac{1}{n} \left(\frac{\partial n}{\partial \mu} \right)_T = \left(\frac{\partial n}{\partial P} \right)_T = n \chi_T, \tag{2.130}$$

where

$$\chi_T = \frac{1}{n} \left(\frac{\partial n}{\partial P} \right)_T \tag{2.131}$$

is the **isothermal compressibility**. From Equations 2.129 and 2.130, we obtain $\sigma_N^2/n = Vk_BTn\chi_T$, which we can write as

$$\frac{\sigma_N^2}{\bar{N}} = \frac{\langle N^2 \rangle - \langle N \rangle^2}{\langle N \rangle} = k_BTn\chi_T. \tag{2.132}$$

Since $\langle N^2 \rangle > \langle N \rangle^2$, the compressibility cannot be negative.

The relative size of the fluctuations in number of particles is $\sigma_N/\bar{N} = [k_BTn\chi_T/\bar{N}]^{1/2}$. Since $k_BTn\chi_T$ is an intensive entity, $\sigma_N/\bar{N}$ goes to zero like $\bar{N}^{-1/2}$ when the system size increases at constant T and μ (an increase in V at constant density). Hence, the fluctuations will be very small for a large system. For an ideal gas, the ideal gas law $P = nk_BT$ implies that $(\partial n/\partial P)_T = 1/(k_BT)$ and hence $k_BTn\chi_T = 1$ so $\sigma_N/\bar{N} = \bar{N}^{-1/2}$, which means that when $\bar{N}$ is of the order 10^{20}, the fluctuations in N will be about 10^{-10} smaller than $\bar{N}$.

2.6.3 Fluctuations in the Isobaric-Isothermal Ensemble

In this section, we investigate the fluctuations in volume for a system at constant T, P and N. We will use similar arguments as in the previous two Sections 2.6.1 and 2.6.2, to which we refer. Here we consider the root mean-square deviation in volume σ_V defined from $\sigma_V^2 = \langle (V - \bar{V})^2 \rangle = \langle V^2 \rangle - \langle V \rangle^2$. In an analogous manner as the derivation of Equation 2.129, we can derive, using Equation 2.85,

$$\sigma_V^2 = -k_BT \left(\frac{\partial \bar{V}}{\partial P} \right)_{T,N} = -Nk_BT \left(\frac{\partial n^{-1}}{\partial P} \right)_{T,N} = \frac{\bar{V}k_BT}{n} \left(\frac{\partial n}{\partial P} \right)_{T,N} \tag{2.133}$$

where the number density $n = N/\bar{V}$ in this case.

{**EXERCISE 2.15**
Show the first equality in Equation 2.133.}

Using this result, we obtain

$$\frac{\sigma_V^2}{\bar{V}} = \frac{\langle V^2 \rangle - \langle V \rangle^2}{\langle V \rangle} = k_BT\chi_T, \tag{2.134}$$

so the fluctuations in V at constant T, P and N are related to the isothermal compressibility in a similar manner as fluctuations in N at constant T, V and μ (Equation 2.132). Incidentally, we note that the isothermal compressibility can be written as

$$\chi_T = \frac{\bar{V}}{N} \left(\frac{\partial (N/\bar{V})}{\partial P} \right)_{T,N} = \bar{V} \left(\frac{\partial \bar{V}^{-1}}{\partial P} \right)_{T,N} = -\frac{1}{\bar{V}} \left(\frac{\partial \bar{V}}{\partial P} \right)_{T,N}, \tag{2.135}$$

which is an alternative manner to define χ_T.

The relative size of the fluctuations in volume is $\sigma_V/\bar{V} = [k_B T \chi_T/\bar{V}]^{1/2}$ and since $k_B T \chi_T$ is an intensive entity, the relative fluctuation size goes to zero like $\bar{V}^{-1/2}$ when the system size increases at constant T and P (the number of particles is increased). For an ideal gas $k_B T \chi_T = 1/n$ so we have $\sigma_V/\bar{V} = (n\bar{V})^{-1/2} = N^{-1/2}$. Thus, the relative size of the fluctuations is 10^{-10} when $N = 10^{20}$ also in this case.

Exactly in the same manner as we derived Equation 2.128 for fluctuation in internal energy for the canonical ensemble, we can derive the following result for fluctuations in enthalpy. We define $\sigma_H^2 = \langle (U + PV - (\bar{U} + P\bar{V}))^2 \rangle$, and using Equations 2.76, and 2.87, we obtain

$$\sigma_H^2 = k_B T^2 \left(\frac{\partial H}{\partial T} \right)_{P,N} = k_B T^2 C_P, \tag{2.136}$$

where C_P is the heat capacity at constant pressure. Also, in this case, the relative fluctuation size goes to zero like $N^{-1/2}$ with the system size at constant T and P.

{**EXERCISE 2.16**
Derive Equation 2.136.}

2.7 INDEPENDENT SUBSYSTEMS

Let us assume that our system is composed of several subsystems a, b, c, ... which are very weakly coupled so the energy of a quantum state for the entire system is the sum of the energies of the quantum states $i, j, k, \ldots$ of each individual subsystem $\mathscr{U}_{a,i} + \mathscr{U}_{b,j} + \mathscr{U}_{c,k} + \cdots$. Likewise, the total internal energy is equal to the sum $\bar{U} = \bar{U}_a + \bar{U}_b + \bar{U}_c + \cdots$. Let us also assume that the system is in contact with a heat reservoir at constant temperature. Analogously to the proof of Equation 2.42, we can show that the canonical partition function factorizes like

$$Q = Q_a \times Q_b \times Q_c \times \cdots$$

and therefore the Helmholtz free energy of the system equals $A = A_a + A_b + A_c + \cdots$.

2.7.1 The Ideal Gas and Single-Particle Partition Functions

One case of particular interest is when each subsystem is composed of one single particle, for instance very weakly interacting particles adsorbed on a surface[22] or the particles in an ideal gas. If the particles are distinguishable, for example when each of them is adsorbed on an individual site on a surface (and remains there), the partition function factorizes as above

$$Q = \sum_{i,j,k,\ldots} e^{-\beta(u_{a,i} + u_{b,j} + u_{c,k} + \cdots)} = q_a \times q_b \times q_c \times \cdots, \tag{2.137}$$

---

[22] The particles interact weakly with one another but strongly with the surface.

where $u_{\alpha,i}$ is the energy of the single-particle quantum state i for particle α and where we have introduced the **single-particle partition function**

$$q_\alpha = \sum_i e^{-\beta u_{\alpha,i}}.$$

If the particles are not distinguishable – which is the normal case when they are of the same species – we have to proceed somewhat differently. We must, for instance, consider that the two possibilities

(particle a in state 1, particle b in state 2, ...) and
(particle a in state 2, particle b in state 1, ...)

then are not two separate states in the partition function Q for the whole system. Therefore, we would miscount the number of separate states for the system if we would calculate Q as in the previous case.

Let us assume that our system consists of N particles of the same species. In Equation 2.137, consider those terms $\exp[-\beta(u_{a,i} + u_{b,j} + u_{c,k} + \cdots)]$ in Q where all states $i, j, k, \ldots$ are different, that is, no two particles are in the same state. For each set of N distinct states $i, j, k, \ldots$, how many terms appear in Q? The answer is $N!$ because there are $N!$ possible permutations of the states among the N particles. Since the particles are indistinguishable, only one such term should be kept. Instead of removing all terms but one, we may utilize the fact that all $N!$ terms are equal and simply take

$$\frac{1}{N!} e^{-\beta(u_{a,i} + u_{b,j} + u_{c,k} + \cdots)}$$

in the sum.

Matters are not as simple when two of the states $i, j, k, \ldots$ are equal, for instance when $i = j$ in the exponent. If the particles are fermions, such a state is not possible due to the Pauli principle and should not be included in Q. For bosons, such states are allowed, but the number of permutations to be considered in the sum is not $N!$. However, these complications are only important at low temperatures. At higher T, the particles have an enormous number of states available, which means that we do not have to consider the possibility of more than one particle per state (the occupation number of most of the states is actually zero). Thus, we have to a very good approximation

$$Q = \frac{1}{N!} \sum_{i,j,k,\ldots} e^{-\beta(u_{a,i} + u_{b,j} + u_{c,k} + \cdots)} = \frac{1}{N!} q_a q_b q_c \times \cdots$$

provided that we avoid low temperatures. This approximation is called **Boltzmann statistics** and constitutes a limiting law for the behavior of fermions and bosons[23] at sufficiently high temperatures. In this text, we shall only consider Boltzmann statistics.

We can therefore factorize Q in the same manner and since we cannot distinguish between particles a, b, c, ... we drop that index and write

$$Q = \frac{1}{N!} q^N \quad \text{where} \quad q = \sum_i e^{-\beta u_i}. \tag{2.138}$$

For an ideal gas, the quantum state energy u_i only depends on V (apart from intra-particle parameters) so $q = q(T, V)$. Still, we have $Q = Q(T, V, N)$.

When the degrees of freedom of each particle can be divided up into, say, two groups that are independent of each other, the Hamiltonian can be written as a sum $\mathcal{H} = \mathcal{H}' + \mathcal{H}''$ where each part contains the degrees of freedom of each group. One says that the Hamiltonian is *separable*. Then the energy of each quantum state can be written as a sum $u = u' + u''$ where u' and u'' are the eigenvalues corresponding to $\mathcal{H}'$ and $\mathcal{H}''$, respectively. Furthermore, the labeling of the quantum states by the index i can be likewise separated so each quantum state can be labeled by a pair on indices i' and i''. This means that each value of i corresponds to one value of i' and one value of i'', for example, when $i = 1$ we may have $i' = 1$ and $i'' = 1$ while for $i = 2$ we have $i' = 1$ and $i'' = 2$, etc. until we have run through all pairs of values for i' and i''. We thereby obtain

$$q = \sum_i e^{-\beta u_i} = \sum_{i',i''} e^{-\beta(u'_{i'} + u''_{i''})}$$

$$= \sum_{i'} e^{-\beta u'_{i'}} \sum_{i''} e^{-\beta u''_{i''}} = q' q'', \tag{2.139}$$

so the single-particle partition function can be written as the product of the partition functions for each group of degrees of freedom.

For instance, for the particle in a cubic box in Section 1.2.1, the x, y and z coordinates are separable from each other and for each quantum state we have the quantum numbers j_x, j_y and j_z that are positive integers. From Equation 1.5, we have the energy $u_i = u_{j_x,j_y,j_z} = u_{\min} \left[j_x^2 + j_y^2 + j_z^2 \right]$ where $u_{\min} = (p_{\min})^2/2m$, so we obtain

$$q = \sum_i e^{-\beta u_i} = \sum_{j_x,j_y,j_z=1}^{\infty} e^{-\beta u_{j_x,j_y,j_z}} = q_x q_y q_z, \tag{2.140}$$

where $q_\alpha = \sum_{j_\alpha=1}^{\infty} \exp(-\beta u_{\min} j_\alpha^2)$ for $\alpha = x, y$ and z.

---

[23] The correct statistics for fermions – which considers the effects of the Pauli principle – is called Fermi-Dirac statistics. The correct statistics for bosons is called Bose-Einstein statistics. They both approach Boltzmann statistics at sufficiently high temperatures.

Example 2.2: The Ideal Gas Law: General Derivation

Let us investigate a grand canonical ensemble of a gas with very weakly coupled particles, that is, an ideal gas. The grand partition function from Equation 2.111 is

$$\Xi = \sum_{N=0}^{\infty} Q(T,V,N)e^{\beta\mu N} = \sum_{N=0}^{\infty} Q(T,V,N)\chi^N,$$

where $\chi = \exp(\beta\mu)$. We insert Equation 2.138 and obtain

$$\Xi = \sum_{N=0}^{\infty} \frac{1}{N!}(q\chi)^N = e^{q\chi}.$$

Thus, we have $\ln \Xi = q\chi = q\exp(\beta\mu)$. Using the formula (2.113) for the average number of particles, we obtain

$$\bar{N} = k_B T \left(\frac{\partial \ln \Xi}{\partial \mu} \right)_{T,V} = \beta^{-1} \left(\frac{\partial [q(T,V)\exp(\beta\mu)]}{\partial \mu} \right)_{T,V}$$

$$= \beta^{-1} q\,\beta \exp(\beta\mu) = q\chi$$

and we conclude that $\bar{N} = q\chi = \ln \Xi$. However, we know from Equation 2.122 that $k_B T \ln \Xi = -\Theta = PV$, so we obtain

$$PV = k_B TN$$

that is, the **ideal gas law** (we have here dropped the average notation for $\bar{N}$). Note the generality of this derivation. It applies for any $q(T,V)$ and is therefore independent of the internal properties of the particles.

Let us now generalize these results to a systems with several species. For distinguishable particles a, b, c, ... , r, s, t, ... we have seen that

$$Q = q_a \times q_b \times q_c \times \cdots q_r \times q_s \times q_t \times \cdots .$$

If we have two species, N_1 particles (a, b, c, ...) belong to species 1 and N_2 particles (r, s, t, ...) belong to species 2, we obviously have $Q = q_1^{N_1} q_2^{N_2}$, where q_v is the single-particle partition function for species v. If the particles of species 1 are indistinguishable and also those of species 2, we have under the same conditions as for the one-component case

$$Q = \frac{1}{N_1!}q_1^{N_1} \frac{1}{N_2!}q_2^{N_2}.$$

The generalization to more than two species is obvious.

In essentially the same manner as for one species, we can obtain the ideal gas law for two species {**EXERCISE 2.17:** Starting from Equation 2.126, show the following Equation 2.141.}

$$PV = k_B T(N_1 + N_2) \tag{2.141}$$

and we have the **partial pressure** for species ν

$$P_\nu = k_B T N_\nu / V = k_B T n_\nu, \tag{2.142}$$

where n_ν is the number density for this species.

2.7.2 Translational Single-Particle Partition Function

The quantum states for a monatomic particle in the cubic box of volume V were treated in Section 1.2.1, where we saw that the possible energies according to Equation 1.5 are $u = p_{min}^2 \left[j_x^2 + j_y^2 + j_z^2 \right] / 2m = u_{min} \left[j_x^2 + j_y^2 + j_z^2 \right]$, where j_α is the quantum number for $\alpha = x$, y and z (a positive integer), $p_{min} = h/(2V^{1/3})$ is the minimal value of the momentum and $u_{min} = h^2/(8mV^{2/3})$ is the minimal energy for each of x, y or z. The same applies for the translational motion for polyatomic molecules in an ideal gas, that is, the motion of the center of mass.[24] For monatomic particles, there is only the translational motion, while for polyatomic molecules there are also rotational and vibrational motions. Here we will, however, restrict ourselves to the monatomic case.

The single-particle partition function is $q = \sum_i \exp(-\beta u_i)$. We are here considering the situation at rather high temperatures so the average energy $\bar{u}$ per particle in the gas is large, much larger than u_{min}. The energy levels are very dense compared to $\bar{u}$; the difference in energy between them is of the order u_{min}. Therefore, the sum in the expression for q can be substituted by an integral and we obtain

$$q(T, V) = \int_0^\infty \omega(u, V) e^{-\beta u} du, \tag{2.143}$$

where $\omega(u, V)$ is the *density of states* on the energy axis, which means that $\omega(u, V)du$ equals the number of quantum states between u and $u + du$ when the volume is V and we have set the zero of the energy at the ground state $j_x = j_y = j_z = 1$ (do not confuse this ω with the orientational coordinates $\boldsymbol{\omega}$ used later). We can determine $\omega(u, V)$ from the quantity

---

[24] We here treat the translational degrees of freedom for the center of mass as separable from the other degrees of freedom, which at least is a very good approximation for the cases we are concerned with in this treatise.

$\Phi(u, V)$, which is the number of quantum states with energy less than u. Since $\omega(u, V)du = \Phi(u + du, V) - \Phi(u, V) \equiv d\Phi(u, V)$, we have $\omega = (\partial\Phi/\partial u)_V$.

For the current case of translational movement of one particle (three momentum components), Φ is given by Equation 1.11 with $J = [u/u_{min}]^{1/2}$ (cf. Equation 1.10), so we have

$$\omega^{tr}(u, V) = \frac{\partial}{\partial u}\left(\frac{\pi}{6}\left[\frac{u}{u_{min}}\right]^{3/2}\right) = \frac{\pi u^{1/2}}{4(u_{min})^{3/2}},$$

where superscript tr stands for "translational." Thus,

$$q^{tr}(T, V) = \frac{\pi}{4(u_{min})^{3/2}}\int_0^\infty u^{1/2}e^{-\beta u}du = \frac{\pi\beta^{-3/2}}{4(u_{min})^{3/2}}\int_0^\infty \tau^{1/2}e^{-\tau}d\tau,$$

where we have made the variable substitution $\tau = \beta u$. Since the integral in the rhs equals $\sqrt{\pi}/2$, we finally obtain $q^{tr}(T, V) = [\pi k_B T/(4u_{min})]^{3/2}$, which can be written as

$$q^{tr}(T, V) = \left[\frac{2\pi m k_B T}{h^2}\right]^{3/2}V = \frac{V}{\Lambda^3}, \tag{2.144}$$

where Λ is the thermal de Broglie wavelength defined in Equation 1.19.

Alternatively, this result can be obtained from Equation 2.140 by using the approximation $\sum_{\tau=1}^{\infty}\exp(-a\tau^2) \approx \int_0^\infty \exp(-a\tau^2)d\tau = (1/2)[\pi/a]^{1/2}$ in the calculations of q_x, q_y and q_z.

{**EXERCISE 2.18**
Show Equation 2.144 by using this approximation.}

Example 2.3: The Ideal Monatomic Gas Revisited

As an explicit example of how to calculate thermodynamical quantities in the canonical ensemble, let us (again) consider the ideal monatomic gas (cf. the derivations in Example 2.1 in the microcanonical ensemble). We have from Equations 2.138 and 2.144

$$Q(T, V, N) = \frac{1}{N!}[q^{tr}(T, V)]^N = \left[\frac{2\pi m k_B T}{h^2}\right]^{3N/2}\frac{V^N}{N!},$$

which implies that

$$\ln Q(T, V, N) = N\ln\left[e\left(\frac{2\pi m k_B T}{h^2}\right)^{3/2}\frac{V}{N}\right], \tag{2.145}$$

where we have used the Stirling approximation $N! = N \ln N - N = N \ln(N/e)$. The Helmholtz free energy is

$$A(T, V, N) = -k_B T \ln Q = N k_B T \ln \left[\frac{1}{e} \left(\frac{h^2}{2\pi m k_B T} \right)^{3/2} \frac{N}{V} \right]$$

$$= N k_B T \ln \left[\frac{\Lambda^3}{e} \cdot \frac{N}{V} \right]. \qquad (2.146)$$

The average energy can be obtained by using Equation 2.48 and the fact that we can write $\ln Q = N \ln T^{3/2} + N \ln(\text{remainder})$, where the last term contains the contributions in Equation 2.145 that do not depend on T. We obtain

$$\bar{U} = k_B T^2 \left(\frac{\partial \ln Q}{\partial T} \right)_{V,N} = k_B T^2 N \frac{\partial \ln T^{3/2}}{\partial T} = \frac{3 N k_B T}{2},$$

which agrees with Equation 1.8. The only kind of energy we have here is translational energy, so we have

$$\bar{U}^{\text{tr}} = \frac{3 N k_B T}{2}, \qquad (2.147)$$

which will be of more interest later.[25]
The entropy is

$$S = \frac{\bar{U} - A}{T} = \frac{3 N k_B}{2} - N k_B \ln \left[\frac{1}{e} \left(\frac{h^2}{2\pi m k_B T} \right)^{3/2} \frac{N}{V} \right]$$

$$= -N k_B \ln \left[\frac{1}{e^{5/2}} \left(\frac{h^2}{2\pi m k_B T} \right)^{3/2} \frac{N}{V} \right] = N k_B \ln \left[\frac{e^{5/2}}{\Lambda^3} \frac{V}{N} \right],$$

which is the same as Equation 1.18. The pressure is given by

$$P = -\left(\frac{\partial A}{\partial V} \right)_{T,N} = -N k_B T \frac{\partial \ln(1/V)}{\partial V} = \frac{N k_B T}{V},$$

that is, the ideal gas law, and the chemical potential is

$$\mu = \left(\frac{\partial A}{\partial N} \right)_{T,V} = k_B T \ln \left[\frac{\Lambda^3}{e} \cdot \frac{N}{V} \right] + N k_B T \frac{d \ln N}{dN},$$

so we have

$$\mu = k_B T \ln \left[\Lambda^3 \frac{N}{V} \right] = k_B T \ln(\Lambda^3 n),$$

which is the same as Equation 2.23.

---

[25] This is an example of the so-called equipartition of translational energy in classical statistical mechanics, which allocates $k_B T/2$ to each translational degree of freedom, that is, three per particle, as described in Section 3.2.

If we include the electronic degrees of freedom for the gas atoms, we have a single-particle partition function $q^{es}(T) = \sum_i \exp(-\beta u_i^{es})$ for them, where we sum over all electronic states (superscript es) of an atom. In most cases, only the electronic ground state contributes to q^{es} unless the temperature is very high, but let us consider the general case. The partition function for this case is

$$Q(T, V, N) = \frac{1}{N!}[q^{es}(T)q^{tr}(T, V)]^N$$

and the free energy becomes

$$A(T, V, N) = Nk_B T \ln\left[\frac{\Lambda^3}{eq^{es}} \cdot \frac{N}{V}\right] = Nk_B T \ln\left[\frac{1}{e\eta} \cdot \frac{N}{V}\right], \qquad (2.148)$$

where

$$\eta = \frac{q^{es}}{\Lambda^3}, \qquad (2.149)$$

which depends on temperature only, $\eta = \eta(T)$. The internal energy is

$$\bar{U} = \frac{3Nk_B T}{2} + N\langle u^{es}\rangle,$$

where $\langle u^{es}\rangle = \sum_i u_i^{es} \exp(-\beta u_i^{es})/q^{es}$ is the average electronic energy for a single atom. The chemical potential is

$$\mu = k_B T \ln\left[\frac{\Lambda^3}{q^{es}}\frac{N}{V}\right] = k_B T \ln\left[\frac{n}{\eta}\right],$$

which we can write as in Equations 2.25 and 2.27, that is,

$$\mu = \mu^\ominus + k_B T \ln \frac{n}{n^\ominus} = \mu^\ominus + k_B T \ln \frac{P}{P^\ominus}, \qquad (2.150)$$

where $\mu^\ominus = \mu^\ominus(T) = k_B T \ln[n^\ominus/\eta] = k_B T \ln[\beta P^\ominus/\eta]$ is the chemical potential for the standard state of density $n^\ominus$ and pressure $P^\ominus$.

In classical thermodynamics, one can derive formula (2.150) for the chemical potential by using the Gibbs-Duhem equation $dP = nd\mu$ at constant T and the ideal gas law. We have $d\mu = dP/n = k_B T dP/P = k_B T d\ln P$, which yields the rhs of Equation 2.150 upon integration from $P^\ominus$ to P. Then the standard chemical potential $\mu^\ominus$ is simply an integration constant, while from the statistical mechanical derivation we obtain it is expressed in molecular quantities.

{**EXERCISE 2.19**

Assume that we have two containers with ideal gases; one with N_1 particles of species 1 in a volume V_1 and one with N_2 particles of species 2 in a volume V_2. Combine the two containers and let the two gases mix in the volume $V = V_1 + V_2$. Show that the change in entropy equals

$$\Delta_{mix}S = -k_B N_1 \ln \frac{V_1}{V_1 + V_2} - -k_B N_2 \ln \frac{V_2}{V_1 + V_2}. \tag{2.151}$$

If the initial number densities of the two gases are equal, that is, $N_1/V_1 = N_2/V_2$, show that

$$\Delta_{mix}S = -k_B N_1 \ln \frac{N_1}{N_1 + N_2} - -k_B N_2 \ln \frac{N_2}{N_1 + N_2}.$$

Per particle in the mixture, we thus have in this case

$$\Delta_{mix}\widehat{S} \equiv \frac{\Delta_{mix}S}{N_1 + N_2} = -k_B [x_1 \ln x_1 + x_2 \ln x_2], \tag{2.152}$$

where $x_i = N_i/(N_1 + N_2)$ is the mole fraction of species i.}

{**EXERCISE 2.20**

a. For an ideal gas of independent particles $Q = q(T, V)^N/N!$. Use the relationship [Equation 2.50]

$$P = k_B T \left(\frac{\partial \ln Q}{\partial V} \right)_{T,N}$$

and the ideal gas law $PV = Nk_B T$ to show that $q = Vf(T)$, where f is a function of T only. This and the following tasks should be solved *for the general case* of an ideal gas of independent particles, so you may not use arguments that are only valid for the monatomic gas.

b. Utilize the result in (a) to show that the chemical potential can be written in the form $\mu = \mu^{\ominus}(T) + k_B T \ln[n/n^{\ominus}]$, where $\mu^{\ominus}$ is a standard chemical potential that, as always, depends on T only.

c. Using Equation 2.143, where the energy of the ground state is set to $u = 0$, show also that $q = Vf(T)$ implies that the number $\Phi(u)$ of single-particle quantum states with energy less than u is proportional to V.}

APPENDIX 2A: THE VOLUME DEPENDENCE OF S AND QUASISTATIC WORK

In Appendix 1A, we saw that for a macroscopic system the number of quantum states with energies immediately below U (in a narrow band $\leq U$) is *enormously* much larger than the number of states with lower energies. If Φ is the total number of quantum states with energies $\leq U$ and we define Ω in accordance with footnote 14 of Section 1.2 as

$$\Omega = \Phi(U, V, N) - \Phi(U - \delta U, V, N),$$

we found that for reasonable selections of δU, the first term in the rhs is enormously much larger than the second term. Therefore, we found that we can take

$$S = k_B \ln \Phi(U, V, N)$$

as an extremely good approximation, that is, $\ln \Omega$ can in practice be replaced by $\ln \Phi$ in the definition of S.

In this appendix, we will investigate the V dependence of Φ and S, in particular for cases with a quasistatic volume change for thermally insulated systems. When we change V by dV while keeping the number of particles N constant, the energy $\mathscr{U}_i = \mathscr{U}_i(V, N)$ of quantum state i is changed by $(\partial \mathscr{U}_i / \partial V)_N dV$. The various quantum states changes by different amounts, that is, their derivatives are different. For a quasistatic volume variation, the average change in energy U of the system is $dw_{qs} = \langle (\partial \mathscr{U}_i / \partial V)_N \rangle_i \, dV$ (see Equation 2.2), where subscript qs stands for quasistatic.

We shall show below that the change of $\Phi(U, V, N)$ when V varies quasistatically is given by

$$\left(\frac{\partial \Phi}{\partial V} \right)_{U,N} dV = -\left(\frac{\partial \Phi}{\partial U} \right)_{V,N} \left\langle \left(\frac{\partial \mathscr{U}_i}{\partial V} \right)_N \right\rangle dV$$

$$= -\left(\frac{\partial \Phi}{\partial U} \right)_{V,N} dw_{qs}, \qquad (2.153)$$

so when V is increased by dV, the number of quantum states below U is decreased by the product of (i) the derivative of Φ with respect to U and (ii) the average shift in location of the quantum states on the U axis, $dU_{qs} = dw_{qs}$, when the work is performed quasistatically.[26] The rhs of Equation 2.153 can be written $-\omega(U, V, N)dU_{qs}$, where

$$\omega(U, V, N) = \left(\frac{\partial \Phi}{\partial U} \right)_{V,N} \qquad (2.154)$$

is the *density of states* on the energy axis (do not confuse this ω with the orientational coordinates ω used later).

Before we prove Equation 2.153, let us investigate some of its consequences. In general, when V is changed by dV at the same time as U is changed by dU while N is kept constant, we have

$$d\Phi = \Phi(U + dU, V + dV, N) - \Phi(U, V, N)$$

$$= \left(\frac{\partial \Phi}{\partial U} \right)_{V,N} dU + \left(\frac{\partial \Phi}{\partial V} \right)_{U,N} dV.$$

---

[26] Most quantum states are actually moved downwards in energy when V is increased, $(\partial \mathscr{U}_i / \partial V)_N < 0$, so $dU_{qs} < 0$ and the number of states below U is increased. For an ideal gas, all quantum states are moved downwards, but for other systems they may move in either direction.

By inserting Equations 2.153 and 2.154, we can write this as

$$\Phi(U + dU, V + dV, N) = \Phi(U, V, N) + \omega(U, V, N)(dU - \dddot{w}_{qs}). \qquad (2.155)$$

For the particular case when the volume variation is done quasistatically and hence $dU = \dddot{w}_{qs}$, it follows that

$$\Phi(U + \dddot{w}_{qs}, V + dV, N) = \Phi(U, V, N).$$

Since $\Phi(U, V, N)$ is dominated by the number of quantum states in a narrow band $\leq U$ (immediately below U), this result says that after a quasistatic volume change the corresponding narrow band $\leq U + \dddot{w}_{qs}$ has the same number of states inside it. It is this fact that is illustrated in Figures 2.5 and 2.6, rather than the corresponding statements for individual energy levels as expressed in the simplified description given in the figure captions and the text. In practice, the number of states in each narrow band equals Ω.

During a rapid *compression*, $dV < 0$, the system is excited to some higher energy level and $dU > \dddot{w}_{qs}$. If $dU^* = dU - \dddot{w}_{qs}$ is the extra energy added to the system due to the excitation, the number of states is increased by ωdU^*, as follows from Equation 2.155. During a rapid *expansion*, $dV > 0$, the system is left behind in a higher energy level than what it would have been after a quasistatic expansion. Then $dU^* = dU - \dddot{w}_{qs}$ is the excess of energy left behind and the number of states is increased by ωdU^*.

By multiplying both sides of Equation 2.153 by $1/\Phi$, we can write the first equality as

$$\left(\frac{\partial \ln \Phi}{\partial V}\right)_{U,N} dV = -\left(\frac{\partial \ln \Phi}{\partial U}\right)_{V,N} \left\langle \left(\frac{\partial \mathscr{U}_i}{\partial V}\right)_N \right\rangle_i dV.$$

For the reasons explained above and in Appendix 1A, for a macroscopic system $\ln \Phi$ and $\ln \Omega$ can in practice replace each other and we have

$$\left(\frac{\partial \ln \Omega}{\partial V}\right)_{U,N} = -\left(\frac{\partial \ln \Omega}{\partial U}\right)_{V,N} \left\langle \left(\frac{\partial \mathscr{U}_i}{\partial V}\right)_N \right\rangle_i \qquad (2.156)$$

as an extremely good approximation. This is the result we use in Section 2.2, Equation 2.5.

Let us finally prove Equation 2.153. Consider the quantum states with energy below U, that is, $\mathscr{U}_i < U$. When V is changed by dV, the energy of a quantum state that has a positive derivative $(\partial \mathscr{U}_i/\partial V)_N$ will end up above U provided $\mathscr{U}_i + (\partial \mathscr{U}_i/\partial V)_N dV > U$, which means $\mathscr{U}_i > U - (\partial \mathscr{U}_i/\partial V)_N dV$, so the state must initially lie in the interval between $U - (\partial \mathscr{U}_i/\partial V)_N dV$ and U. Likewise, a quantum state that initially lies above U will end up below U provided $\mathscr{U}_i + (\partial \mathscr{U}_i/\partial V)_N dV < U$, where the derivative must be negative in this case. Thus, the state must initially lie in the interval between U and $U - (\partial \mathscr{U}_i/\partial V)_N dV$.

Say that there are $\Phi_D(U', V, N)$ quantum states with energy below U' that have a derivative with the value $D = (\partial \mathscr{U}_i/\partial V)_N$. Hence, we use subscript D to indicate the value of the derivative. We first consider states with positive values of D. For each such D, the number of states that lies between $U - DdV$ and U is equal to $\Phi_D(U, V, N) - \Phi_D(U - DdV, V, N)$. All of them end up above U, so if we consider all states with positive D, the function Φ evaluated at U must decrease by $\sum_{D>0}[\Phi_D(U, V, N) - \Phi_D(U - DdV, V, N)]$, where the sum is over all possible values of positive D. Likewise, due to the states with negative D values, Φ evaluated at U must increase by $\sum_{D<0}[\Phi_D(U - DdV, V, N) - \Phi_D(U, V, N)]$. In total, we accordingly have

$$\Phi(U, V + dV, N) - \Phi(U, V, N) = -\sum_D [\Phi_D(U, V, N) - \Phi_D(U - DdV, V, N)],$$

(2.157)

where the sum is over all D values. Now, we can write the left-hand side of Equation 2.157 as

$$\Phi(U, V + dV, N) - \Phi(U, V, N) = \left(\frac{\partial \Phi(U, V, N)}{\partial V}\right)_{U,N} dV$$

and in the right-hand side we have

$$\Phi_D(U, V, N) - \Phi_D(U - DdV, V, N) = \left(\frac{\partial \Phi_D(U, V, N)}{\partial U}\right)_{V,N} DdV$$

$$= \omega_D(U, V, N)DdV,$$

where $\omega_D(U, V, N)$ is the density of states for the quantum states with derivative D. Thus, we can write Equation 2.157 as

$$\left(\frac{\partial \Phi(U, V, N)}{\partial V}\right)_{U,N} dV = -\sum_D \omega_D(U, V, N)DdV.$$ (2.158)

Obviously, $\sum_D \omega_D(U, V, N) = \omega(U, V, N)$, that is, the total density of states at energy U. If we multiply and divide the right-hand side of Equation 2.158 by $\omega(U, V, N)$, we obtain

$$\left(\frac{\partial \Phi(U, V, N)}{\partial V}\right)_{U,N} dV = -\omega(U, V, N)\left[\sum_D \frac{\omega_D(U, V, N)}{\omega(U, V, N)}D\right] dV,$$ (2.159)

where the ratio $\omega_D(U, V, N)/\omega(U, V, N)$ gives the fraction of states at energy U that have the derivative D. Since all states at energy U are equally probable, this ratio also gives the probability for occurrence of the states with derivative D, which thus equals $\mathcal{P}_D = \omega_D(U, V, N)/\omega(U, V, N)$. The bracket in the right-hand side of Equation 2.159 is accordingly $\sum_D \mathcal{P}_D D = \langle D \rangle$, that is, the average of $D = (\partial \mathcal{U}_i/\partial V)_N$ for the states at energy U. Now, since ω is the derivative of Φ with respect to U, Equation 2.154, we finally obtain from Equation 2.159

$$\left(\frac{\partial \Phi}{\partial V}\right)_{U,N} dV = -\left(\frac{\partial \Phi}{\partial U}\right)_{V,N} \left\langle \left(\frac{\partial \mathcal{U}_i}{\partial V}\right)_N \right\rangle dV,$$

that is, the result that we anticipated above in Equation 2.153.

APPENDIX 2B: STRICTER DERIVATIONS OF PROBABILITY EXPRESSIONS

In this Appendix, we shall present more strict derivations of the probability expressions in the canonical, isobaric-isothermal and grand canonical ensembles. We start with the first one.

2B.1 The Boltzmann Distribution in the Canonical Ensemble

We start the derivation of Boltzmann's distribution law (2.39) from Equation 2.38

$$\mathcal{P}_A(U_A) = \Omega_A(U_A)\mathfrak{p}_A(U_A)$$

and Equation 2.32, which we can write as

$$\mathfrak{p}_a(U_A) = \frac{\Omega_R(U_{AR} - U_A)}{\Omega_{AR}},$$

where we have inserted $U_{tot} = U_{AR}$ and $\Omega_{tot} = \Omega_{AR}$. Since

$$\Omega_{AR} = \sum_{U_A'} \Omega_A(U_A')\Omega_R(U_{AR} - U_A'),$$

where the sum is over all energy levels of A, we have

$$\mathcal{P}_A(U_A) = \frac{\Omega_A(U_A)\Omega_R(U_{AR} - U_A)}{\sum_{U_A'} \Omega_A(U_A')\Omega_R(U_{AR} - U_A')}.$$

Note that $\sum_{U_A} \mathcal{P}_A(U_A) = 1$ as required. We can write $\mathcal{P}_A$ in a simpler form by making the following manipulations. Divide both the numerator and the denominator by $\Omega_R(U_{AR})$ and obtain

$$\mathcal{P}_A(U_A) = \frac{\Omega_A(U_A)[\Omega_R(U_{AR} - U_A)/\Omega_R(U_{AR})]}{Q_A}, \qquad (2.160)$$

where

$$Q_A = \sum_{U_A'} \Omega_A(U_A')\frac{\Omega_R(U_{AR} - U_A')}{\Omega_R(U_{AR})} = \frac{\Omega_{AR}}{\Omega_R(U_{AR})},$$

is the same quantity Q_A as we introduced in Equation 2.36.

Let us investigate $\Omega_R(U_{AR} - U_A)/\Omega_R(U_{AR})$ which occurs in both the numerator and the denominator (Q_A) of $\mathcal{P}_A(U_A)$. Since the reservoir R is so large, the presence or absence of subsystem A is of virtually no consequence for the properties of R. In the absence of A, the system R constitutes an isolated system and we can utilize the relationship $k_B \ln \Omega_R = S_R$, which can be written $\Omega_R = \exp(S_R/k_B)$. Thus,

$$\frac{\Omega_R(U_{AR} - U_A)}{\Omega_R(U_{AR})} = \frac{\exp(S_R(U_{AR} - U_A)/k_B)}{\exp(S_R(U_{AR})/k_B)} = e^{-[S_R(U_{AR}) - S_R(U_{AR} - U_A)]/k_B}.$$

Now, $U_A \ll U_{AR}$ since A is much smaller than R. Therefore, we can expand $S_R(U_{AR} - U_A)$ in a Taylor series around U_{AR} and we obtain

$$S_R(U_{AR} - U_A) = S_R(U_{AR}) - U_A \left.\frac{\partial S_R}{\partial U_R}\right|_{U_R = U_{AR}}$$

$$+ \frac{U_A^2}{2} \left.\frac{\partial^2 S_R}{\partial U_R^2}\right|_{U_R = U_{AR}} + \dots, \qquad (2.161)$$

where the partial derivatives are taken with V and N constant. But $(\partial S_R/\partial U_R)_{V,N} = 1/T_R$, the inverse temperature of R, and therefore

$$\frac{\partial^2 S_R}{\partial U_R^2} = \frac{\partial}{\partial U_R}\left[\frac{1}{T_R}\right] = -\frac{1}{T_R^2}\frac{\partial T_R}{\partial U_R}.$$

The product $U_A(\partial T_R/\partial U_R)$ gives an estimate of how much the temperature of R varies when the energy of R is changed by U_A, but according to our condition that R is a heat reservoir, this variation is vanishingly small. Note that $(\partial T_R/\partial U_R)_{V,N} = 1/(\partial U_R/\partial T_R)_{V,N} = 1/C_{V,R}$, where $C_{V,R}$, the heat capacity for the reservoir, is very large. If the variation in temperature is not small enough, we just make R even larger and thereby $C_{V,R}$ larger until this variation is negligibly small. Therefore, the last term in the expansion (2.161) is vanishingly small compared to the lower order terms (the same is true for the higher order terms), and we obtain the excellent approximation

$$S_R(U_{AR} - U_A) = S_R(U_{AR}) - \frac{U_A}{T_R}.$$

We conclude that

$$\frac{\Omega_R(U_{AR} - U_A)}{\Omega_R(U_{AR})} = e^{-U_A/k_B T_R},$$

which means that the reservoir only enters via its temperature in this ratio. Inserting this result in Equation 2.160, we obtain Boltzmann's distribution law

$$\mathcal{P}_A(U_A) = \frac{\Omega_A(U_A)e^{-U_A/k_B T}}{Q_A} = \frac{\Omega_A(U_A)e^{-U_A/k_B T}}{\sum_{U_A'} \Omega_A(U_A')e^{-U_A'/k_B T}},$$

where we have written T instead of T_R, indicating that this is the equilibrium temperature of the whole system including A.

2B.2 The Probability Distribution in the Isobaric-Isothermal Ensemble

In this Appendix, we shall derive the probability distribution (2.72) from Equations 2.71

$$\mathcal{P}_A(U_A, V_A) = \Omega_A(U_A, V_A)\mathfrak{p}_A(U_A, V_A)$$

and 2.61, which we can write as

$$p_A(U_A, V_A) = \frac{\Omega_R(U_{AR} - U_A, V_{AR} - V_A)}{\Omega_{AR}}.$$

These expressions yield

$$\mathcal{P}_A(U_A, V_A) = \frac{\Omega_A(U_A, V_A)\Omega_R(U_{AR} - U_A, V_{AR} - V_A)}{\Omega_{AR}}.$$

By dividing both the numerator and the denominator by $\Omega_R(U_{AR}, V_{AR})$, we can write

$$\mathcal{P}_A(U_A, V_A) = \frac{\Omega_A(U_A, V_A)[\Omega_R(U_{AR} - U_A, V_{AR} - V_A)/\Omega_R(U_{AR}, V_{AR})]}{\Upsilon_A}, \qquad (2.162)$$

where we in accordance with Equation 2.69 have defined

$$\Upsilon_A = \frac{\Omega_{AR}}{\Omega_R(U_{AR}, V_{AR})} = \sum_{N_A, U_A} \Omega_A(U_A, V_A)\frac{\Omega_R(U_{AR} - U_A, V_{AR} - V_A)}{\Omega_R(U_{AR}, V_{AR})}, \qquad (2.163)$$

with Equation 2.60 inserted.

Next, we investigate the ratio $\Omega_R(U_{AR} - U_A, V_{AR} - V_A)/\Omega_R(U_{AR}, V_{AR})$. Like in the previous section, we utilize the fact that the reservoir R is so large that the presence or absence of subsystem A is of virtually no consequence for the properties of R. In the absence of A, the system R constitutes an isolated system and we can write $\Omega_R = \exp(S_R/k_B)$. Thus,

$$\frac{\Omega_R(U_{AR} - U_A, V_{AR} - V_A)}{\Omega_R(U_{AR}, V_{AR})} = \frac{\exp(S_R(U_{AR} - U_A, V_{AR} - V_A)/k_B)}{\exp(S_R(U_{AR}, V_{AR})/k_B)}$$

$$= e^{-[S_R(U_{AR}, V_{AR}) - S_R(U_{AR} - U_A, V_{AR} - V_A)]/k_B}. \qquad (2.164)$$

Since A is much smaller than R, we have $U_A \ll U_{AR}$ and $V_A \ll V_{AR}$ and we can expand $S_R(U_{AR} - U_A, V_{AR} - V_A)$ in a two-dimensional Taylor series around (U_{AR}, V_{AR}). We obtain

$$S_R(U_{AR} - U_A, V_{AR} - V_A)$$

$$= S_R(U_{AR}, V_{AR}) - U_A \left.\frac{\partial S_R}{\partial U_R}\right|_{U_{AR}, V_{AR}} - V_A \left.\frac{\partial S_R}{\partial V_R}\right|_{U_{AR}, V_{AR}}$$

$$+ \frac{U_A^2}{2} \left.\frac{\partial^2 S_R}{\partial U_R^2}\right|_{U_{AR}, V_{AR}} + U_A V_A \left.\frac{\partial^2 S_R}{\partial V_R \partial U_R}\right|_{U_{AR}, V_{AR}} + \frac{V_A^2}{2} \left.\frac{\partial^2 S_R}{\partial V_R^2}\right|_{U_{AR}, V_{AR}} + \cdots,$$

where the derivatives are evaluated at $U_R = U_{AR}$ and $V_R = V_{AR}$. Now, $\partial S_R/\partial U_R = 1/T_R$ and $\partial S_R/\partial V_R = P_R/T_R$ and therefore

$$\frac{\partial^2 S_R}{\partial U_R^2} = \frac{\partial}{\partial U_R}\left[\frac{1}{T_R}\right] = -\frac{1}{T_R^2}\frac{\partial T_R}{\partial U_R},$$

$$\frac{\partial^2 S_R}{\partial V_R \partial U_R} = \frac{\partial}{\partial V_R}\left[\frac{1}{T_R}\right] = -\frac{1}{T_R^2}\frac{\partial T_R}{\partial V_R},$$

and

$$\frac{\partial^2 S_R}{\partial V_R^2} = \frac{\partial}{\partial V_R}\left[\frac{P_R}{T_R}\right] = \frac{1}{T_R}\frac{\partial P_R}{\partial V_R} - \frac{P_R}{T_R^2}\frac{\partial T_R}{\partial V_R}.$$

The products $U_A(\partial T_R/\partial U_R)$, $V_A(\partial T_R/\partial V_R)$ and $V_A(\partial P_R/\partial V_R)$ give estimates of how much the temperature and the pressure of R vary when R changes its energy by U_A and its volume by V_A. According to our conditions, R is so large that these variations are vanishingly small. The terms of second order in the expansion are therefore completely negligible compared to the first order terms. The same applies to the higher order terms. Thus, we have to a very good approximation

$$S_R(U_{AR}, V_{AR}) - S_R(U_{AR} - U_A, V_{AR} - V_A) = \frac{1}{T_R}U_A + \frac{P_R}{T_R}V_A$$

and we obtain from Equation 2.164

$$\frac{\Omega_R(U_{AR} - U_A, V_{AR} - V_A)}{\Omega_R(U_{AR}, V_{AR})} = e^{-(U_A + P_R V_A)/k_B T_R}.$$

The reservoir only enters via its temperature and chemical potential in this ratio. Inserting this in Equations 2.162 and 2.163, we obtain

$$\mathcal{P}_A(U_A, V_A) = \frac{\Omega_A(U_A, V_A)e^{-(U_A + PV_A)/k_B T}}{\Upsilon_A}$$

$$= \frac{\Omega_A(U_A, V_A)e^{-(U_A + PV_A)/k_B T}}{\sum_{V'_A, U'_A} \Omega_A(U'_A, V'_A)e^{-(U'_A + PV'_A)/k_B T}},$$

where we have written T and P instead of T_R and P_R since they are the equilibrium values for the whole system.

2B.3 The Probability Distribution in the Grand Canonical Ensemble

Here we shall derive the probability distribution (2.108) from Equation 2.107 and 2.101. The derivation is analogous to that in the isobaric-isothermal case, so we give a shortened version. We have

$$\mathcal{P}_A(U_A, N_A) = \frac{\Omega_A(U_A, N_A)\Omega_R(U_{AR} - U_A, N_{AR} - N_A)}{\Omega_{AR}}$$

and by dividing both the numerator and the denominator by $\Omega_R(U_{AR}, N_{AR})$, we obtain

$$P_A(U_A, N_A) = \frac{\Omega_A(U_A, N_A)[\Omega_R(U_{AR} - U_A, N_{AR} - N_A)/\Omega_R(U_{AR}, N_{AR})]}{\Xi_A}, \quad (2.165)$$

since

$$\Xi_A \equiv \frac{\Omega_{AR}}{\Omega_R(U_{AR}, N_{AR})}$$

$$= \sum_{N_A, U_A} \Omega_A(U_A, N_A) \frac{\Omega_R(U_{AR} - U_A, N_{AR} - N_A)}{\Omega_R(U_{AR}, N_{AR})}. \quad (2.166)$$

Furthermore, we can write,

$$\frac{\Omega_R(U_{AR} - U_A, N_{AR} - N_A)}{\Omega_R(U_{AR}, N_{AR})} = e^{-[S_R(U_{AR}, N_{AR}) - S_R(U_{AR} - U_A, N_{AR} - N_A)]/k_B}$$

and since $U_A \ll U_{AR}$ and $N_A \ll N_{AR}$, we can expand $S_R(U_{AR} - U_A, N_{AR} - N_A)$ in a Taylor series around (U_{AR}, N_{AR}). We obtain

$$S_R(U_{AR} - U_A, N_{AR} - N_A)$$

$$= S_R(U_{AR}, N_{AR}) - U_A \left.\frac{\partial S_R}{\partial U_R}\right|_{U_{AR}, N_{AR}} - N_A \left.\frac{\partial S_R}{\partial N_R}\right|_{U_{AR}, N_{AR}}$$

$$+ \frac{U_A^2}{2} \left.\frac{\partial^2 S_R}{\partial U_R^2}\right|_{U_{AR}, N_{AR}} + U_A N_A \left.\frac{\partial^2 S_R}{\partial N_R \partial U_R}\right|_{U_{AR}, N_{AR}} + \frac{N_A^2}{2} \left.\frac{\partial^2 S_R}{\partial N_R^2}\right|_{U_{AR}, N_{AR}} + \cdots.$$

From $\partial S_R / \partial U_R = 1/T_R$ and $\partial S_R / \partial N_R = -\mu_R/T_R$, it follows that

$$\frac{\partial^2 S_R}{\partial U_R^2} = \frac{\partial}{\partial U_R}\left[\frac{1}{T_R}\right] = -\frac{1}{T_R^2}\frac{\partial T_R}{\partial U_R}$$

$$\frac{\partial^2 S_R}{\partial N_R \partial U_R} = \frac{\partial}{\partial N_R}\left[\frac{1}{T_R}\right] = -\frac{1}{T_R^2}\frac{\partial T_R}{\partial N_R}$$

and

$$\frac{\partial^2 S_R}{\partial N_R^2} = -\frac{\partial}{\partial N_R}\left[\frac{\mu_R}{T_R}\right] = -\frac{1}{T_R}\frac{\partial \mu_R}{\partial N_R} + \frac{\mu_R}{T_R^2}\frac{\partial T_R}{\partial N_R}.$$

Since the reservoir is very large, the temperature and the chemical potential of R vary negligibly when R changes its energy by U_A and number of particles by N_A. We can conclude that the second order term (and the higher ones) in the expansion are completely negligible compared to the first order terms and we obtain to a very good approximation

$$S_R(U_{AR}, N_{AR}) - S_R(U_{AR} - U_A, N_{AR} - N_A) = \frac{1}{T_R}U_A - \frac{\mu_R}{T_R}N_A,$$

so we have

$$\frac{\Omega_R(U_{AR} - U_A, N_{AR} - N_A)}{\Omega_R(U_{AR}, N_{AR})} = e^{-(U_A - \mu_R N_A)/k_B T_R},$$

where the reservoir only enters via its temperature and chemical potential. By inserting this in Equations 2.165 and 2.166, we obtain

$$\mathcal{P}_A(U_A, N_A) = \frac{\Omega_A(U_A, N_A)e^{-(U_A - \mu N_A)/k_B T}}{\Xi_A}$$

$$= \frac{\Omega_A(U_A, N_A)e^{-(U_A - \mu N_A)/k_B T}}{\sum_{N'_A, U'_A} \Omega_A(U'_A, N'_A)e^{-(U'_A - \mu N'_A)/k_B T}},$$

where we have written T and μ instead of T_R and μ_R since they are the equilibrium values for the whole system.

CHAPTER 3

Classical Statistical Mechanics

T HE BASIS OF STATISTICAL mechanics has been explored in the previous chapters from a quantum perspective and we will now switch to classical statistical mechanics, where the particles are governed by the laws of classical mechanics. We have already covered some ground in Section 1.2.2. In that Section, we saw that the classical theory has conceptual difficulties that are resolved by quantum mechanics. Since the classical theory from a fundamental point of view is a special case of the quantum theory at large quantum numbers (large energies), it is very natural that there remains traces of the quantum treatment in the classical case as demonstrated in Section 1.2.2. The kind of systems we will deal with in this treatise is such that quantum effects do not play any important role or at least can be incorporated in an approximate manner in the classical treatment.

We will consider systems of particles that interact with each other and possibly with an external field that originates from a source that is not part of the system. The particles can, for example, be atoms, molecules, ions and/or macroparticles. Our primary objective here is fluids consisting of spherical particles, so-called **simple fluids**, so in the present treatment we have spherical particles in the system. In the second volume of this treatise, we will generalize the theory to nonspherical particles, but some aspects of such particles will be included in Chapters 6 and 7 of the present volume.

In classical mechanics, the Hamiltonian of a system is equal to the total energy, that is, the sum of the kinetic and potential energy of all particles in the system. By stating an expression for the Hamiltonian, one has given the physical assumptions for the system one deals with. We will include both exact and approximate statistical mechanical theories based on classical mechanics in this treatise. An exact theory gives the statistical mechanical consequences for a system that is governed by a given set of interaction potentials between the constituent particles – or, expressed more precisely, one has a given Hamiltonian for the system and one does not introduce any approximations in the mathematical treatment of its consequences. The Hamiltonian does not necessarily represent a real system in all aspects, so one may not be able to make successful comparisons with experimental data. But if there are any differences between experimental data and the results of an exact theory, the discrepancy depends on an incorrect Hamiltonian and not any mathematical approximations in the statistical mechanical treatment. If one compares the predictions of an approximate theory with experiments, one does not know whether a discrepancy is a consequence of the approximations done in the approximate theory or due to an inappropriate underlying Hamiltonian. Therefore, one does not know if the physical basis (the Hamiltonian) is inaccurate and needs to be improved or if simply some mathematical approximations need to be eliminated. It is important to note, regarding the predictions of an exact theory, that certain conditions may have to be fulfilled for the predictions to be valid and one has to keep that in mind when using the latter. Exact results can sometimes be derived in certain limits, such as sufficiently high temperatures, low densities or, for dense systems, large separations between two interacting particles immersed in a fluid medium.

3.1 SYSTEMS WITH N SPHERICAL PARTICLES

Let us treat a system containing N spherical particles that interact with each other and, in general, also with an external field. In Section 1.2.2, we saw that the motion of N particles is conveniently described in phase space: the $6N$-dimensional space of position and momentum coordinates of all particles. The trajectory of the vector $\mathbf{\Gamma} = (\mathbf{r}_1, \mathbf{r}_2, \ldots, \mathbf{r}_N, \mathbf{p}_1, \mathbf{p}_2, \ldots, \mathbf{p}_N)$ as a function of time in phase space represents the time evolution of the positions and momenta of all particles. Remember that $\mathbf{r}_\nu$ is the location of the center of particle ν.

To obtain the mean values of various properties of the system, one may follow the motion of the particles in time [either $(\mathbf{r}_\nu(t), \mathbf{p}_\nu(t))$ for each particle ν or the phase space point $\mathbf{\Gamma}(t)$] as governed by classical mechanics and form the time average. Thereby, $\mathbf{r}_\nu$ and $\mathbf{p}_\nu$ for $1 \leq \nu \leq N$ as functions of t can be obtained (at least in principle) by solving Newton's equations of motion. Alternatively, one can calculate the ensemble average of the same property, which, according to the ergodic hypothesis, gives the same result as the time averaging (cf. Section 2.1). To calculate the ensemble average, one needs in principle the probability to observe each possible value of $\mathbf{\Gamma}$. In practice, it suffices with a finite but large number of values that give a sufficiently good representation of the system.

The instantaneous potential energy of the system depends on the coordinates of the particles, $\check{U}_N^{\text{pot}} = \check{U}_N^{\text{pot}}(\mathbf{r}_1, \mathbf{r}_2, \mathbf{r}_3, \ldots, \mathbf{r}_N)$, where subscript N means that we have N particles in the system.[1] In general, we will treat both homogeneous bulk fluids and inhomogeneous fluids. The latter are exposed to an external field that interacts with the particles and makes the fluid inhomogeneous. Such a field arises from a source that is not part of the system, for example the walls of a container that encloses the fluid or a body that is immersed in the fluid but is not counted as being part of the system. As always, it is up to us to decide what the system consists of and what is outside it, but as we will see it is often convenient to classify some bodies as external ones that give rise to an external field that the system is

---

[1] The symbol "˘" over a quantity, like in $\check{U}$, is used throughout this treatise to indicate that the value is the instantaneous one.

exposed to. This field is described by a potential $v(\mathbf{r})$, a so-called **external potential**, such that a fluid particle at position $\mathbf{r}$ has the potential energy $v(\mathbf{r})$ in the field.

The interaction between the fluid particles themselves is, in the simplest case, given by a so-called **pair potential** between each pair of particles. For two particles located at $\mathbf{r}_1$ and $\mathbf{r}_2$, this potential is given by $u(\mathbf{r}_1, \mathbf{r}_2)$. The interaction energy between N particles is then equal to $\sum_{\nu=1}^{N} \sum_{\nu'=\nu+1}^{N} u(\mathbf{r}_\nu, \mathbf{r}_{\nu'})$, where the sums go over all distinct pairs of particles. This is called **pairwise additivity** of the interparticle interactions. In the present case of spherical particles, a pair potential is spherically symmetric, $u(\mathbf{r}_1, \mathbf{r}_2) = u(r_{12})$, where $r_{12} = |\mathbf{r}_{12}|$ is the distance between the sphere centers and $\mathbf{r}_{12} = \mathbf{r}_2 - \mathbf{r}_1$ is the vector that connects them. In the presence of an external potential, the total potential energy $\check{U}_N^{\text{pot}}$ also contains the contribution $\sum_{\nu=1}^{N} v(\mathbf{r}_\nu)$.

However, the interaction energy between the particles cannot, in general, be written as a sum of pair potentials. The interaction energy between, for example, three particles can be a complicated function of all three positions since the interaction between two of the particles can depend on the position of the third. An example occurs for electrostatic interactions when the third particle polarizes particles one and two and thereby changes the electrostatic interaction between the latter two. In the present section, we will not assume that the interparticle interaction can be written as a sum of pair potentials, so $\check{U}_N^{\text{pot}}(\mathbf{r}_1, \mathbf{r}_2, \mathbf{r}_3, \ldots, \mathbf{r}_N)$ can be a more complicated function of the coordinates. Later, we will, in several cases, restrict ourselves to pairwise additive interactions between the fluid particles.

The total energy is given by $\check{U}_N^{\text{tot}} = \check{U}_N^{\text{pot}} + \check{U}_N^{\text{kin}}$, where $\check{U}_N^{\text{kin}}$ is the total kinetic energy. In the current case, the kinetic energy is equal to the translational energy $\check{U}_N^{\text{kin}} = \check{U}_N^{\text{tr}}$ given by

$$\check{U}_N^{\text{tr}}(\mathbf{p}_1, \mathbf{p}_2, \ldots, \mathbf{p}_N) = \sum_\nu u_\nu^{\text{tr}} = \sum_\nu \frac{p_\nu^2}{2m_\nu}, \tag{3.1}$$

where u_ν^{tr} is the translational energy of particle ν and m_ν is its mass, $p_\nu = |\mathbf{p}_\nu|$ and $\mathbf{p}_\nu$ is its momentum. The sum is over all particles in the system.

For an isolated system, the total energy U_N of the system is constant and we have $\check{U}_N^{\text{tot}}(\boldsymbol{\Gamma}(t)) = U_N$. This means that the movement of the phase space point $\boldsymbol{\Gamma}$ in time is restricted to the part of this space with energy U_N, that is, on a hypersurface of dimension $6N - 1$ in the $6N$-dimensional space. In the microcanonical ensemble, all points $\boldsymbol{\Gamma}$ available for the system lie on this hypersurface. The postulate of equal a priori probability means that the point $\boldsymbol{\Gamma}$ that represents the system has an equal probability to be located anywhere on this hypersurface (that is, on any accessible part of the hypersurface that can be reached by $\boldsymbol{\Gamma}$ given sufficient time). For example, in the case of an ideal monatomic gas, where $\check{U}_N^{\text{pot}} = 0$ inside the vessel that contains the gas and $\check{U}_N^{\text{pot}} = \infty$ otherwise, the hypersurface accessible to $\boldsymbol{\Gamma}$ is equal to the $3N - 1$ dimensional surface of the hypersphere given by $\sum_\nu p_\nu^2/(2m_\nu) = U_N$ (a hypersurface where $\mathbf{p}_1, \mathbf{p}_2, \ldots, \mathbf{p}_N$ are located) combined with the $3N$-dimensional volume available for $\mathbf{r}_1, \mathbf{r}_2, \ldots, \mathbf{r}_N$. When the particles interact with each other, the hypersurface is much more complicated.

In other ensembles, where the energy is not constant, the point $\boldsymbol{\Gamma}$ that represents the system is not restricted to such a hypersurface, but the probability for $\boldsymbol{\Gamma}$ is equal anywhere on each hypersurface with constant energy. The probability for Γ is, however, different on different hypersurfaces depending on the value of the energy.

3.2 THE CANONICAL ENSEMBLE

Let us consider the canonical (NVT) ensemble, in which the number of particles N, volume V, and absolute temperature T are constant. As in Section 1.2.2, it is convenient to use a compact notation for the variables. The spatial coordinates for the N particles are denoted by $\mathbf{r}^N \equiv \mathbf{r}_1, \mathbf{r}_2, \ldots, \mathbf{r}_N$ and the momenta are denoted by $\mathbf{p}^N \equiv \mathbf{p}_1, \mathbf{p}_2, \ldots, \mathbf{p}_N$. We now introduce the **probability density** $\mathcal{P}_N^{(N)\text{tot}}(\mathbf{r}^N, \mathbf{p}^N)$ to observe the N particles at specific positions and with specific momenta. The notation (N) in the superscript indicates that $\mathcal{P}_N^{(N)\text{tot}}$ is a function of N positions $\mathbf{r}_\nu$ and N momenta $\mathbf{p}_\nu$, while subscript N indicates, as before, that there are N particles in the system. $\mathcal{P}_N^{(N)\text{tot}}$ is given by a simple Boltzmann distribution

$$\mathcal{P}_N^{(N)\text{tot}}(\mathbf{r}^N, \mathbf{p}^N) = C_{\text{tot}} e^{-\beta \breve{U}_N^{\text{tot}}} \tag{3.2}$$

$$= C_{\text{tot}} e^{-\beta \left[\breve{U}_N^{\text{pot}}(\mathbf{r}^N) + \breve{U}_N^{\text{kin}}(\mathbf{p}^N) \right]}$$

where C_{tot} is a normalization constant

$$C_{\text{tot}}^{-1} = \int d\mathbf{r}^N d\mathbf{p}^N e^{-\beta \left[\breve{U}_N^{\text{pot}} + \breve{U}_N^{\text{kin}} \right]}, \tag{3.3}$$

which ensures that the integral of $\mathcal{P}_N^{(N)\text{tot}}$ over the whole available phase space is equal to one. Here we have the differential $d\mathbf{r}^N \equiv d\mathbf{r}_1 d\mathbf{r}_2 \ldots d\mathbf{r}_N$ and likewise for $d\mathbf{p}^N$.

Whenever an integral appears without any explicit limits like in Equation 3.3, it will be assumed throughout this treatise that the integration is taken over all possible values of the integration variables (like over the whole available space). Likewise, whenever a sum is written without explicit limits, one sums over all relevant possibilities (like over all particles in the system as in Equation 3.1).

The product of the probability density and an appropriate "volume element" is equal to a probability. The interpretation of $\mathcal{P}_N^{(N)\text{tot}}$ is that the product $\mathcal{P}_N^{(N)\text{tot}} d\mathbf{r}^N d\mathbf{p}^N$ is equal to the probability to *simultaneously* observe

> particle 1 with its center in volume element $d\mathbf{r}_1$ located at position $\mathbf{r}_1$ (cf. Figure 3.1),
> particle 2 with its center in volume element $d\mathbf{r}_2$ at position $\mathbf{r}_2$,
> . . .,
> and particle N with its center in volume element $d\mathbf{r}_N$ at position $\mathbf{r}_N$ **when** (at the same time as)
> particle ν, for $\nu = 1, 2, \ldots, N$, has a momentum within $d\mathbf{p}_\nu$ from $\mathbf{p}_\nu$.

FIGURE 3.1   A volume element $d\mathbf{r} = dx\,dy\,dz$ located at position $\mathbf{r}$.

In reality, particles of the same species are not distinguishable, so one cannot say whether it is particle ν that is located at $\mathbf{r}_\nu$ and has momentum $\mathbf{p}_\nu$ or if it is some other particle of the same kind. Let us assume that all N particles are of the same species. In order to acquire the physically relevant probability, one has to take into account that any of the N particles can be located at $\mathbf{r}_1$ and have momentum $\mathbf{p}_1$, any of the remaining $N - 1$ particles be at $\mathbf{r}_2$ with momentum $\mathbf{p}_2$, ... , any of two particles at $\mathbf{r}_{N-1}$ with $\mathbf{p}_{N-1}$ and the only remaining one at $\mathbf{r}_N$ with $\mathbf{p}_N$. Therefore, one should multiply $\mathcal{P}_N^{(N)\text{tot}}$ by $N(N - 1) \times \cdots \times 2 \times 1 = N!$ to make it relevant.

Furthermore, it is not physically relevant to distinguish positions x and momenta p_x with higher precision than δx and δp_x, respectively, where the product $\delta x\,\delta p_x$ is of the order Planck's constant h. The same applies in the y and z directions. As we saw in Section 1.2.2, in order to have compatibility with quantum theory one must consider that the classical phase space has a "graininess." For an ideal gas, we saw that the graininess corresponds to "volume cells" of size $\delta \mathscr{V}_\Gamma$ obtained by multiplying a factor of h for each translational degree of freedom for the particles (three per particle); in total, $\delta \mathscr{V}_\Gamma = h^{3N}$. This is, in fact, true not only for ideal gases, but holds in general. Therefore, an arbitrarily small volume element $d\mathbf{r}_\nu d\mathbf{p}_\nu$ is not relevant for the probability at coordinates $\mathbf{r}_\nu$ and $\mathbf{p}_\nu$, but rather a volume element h^3. For these reasons, $\mathcal{P}_N^{(N)\text{tot}}h^{3N}N!$ is the physically adequate probability rather than $\mathcal{P}_N^{(N)\text{tot}}d\mathbf{r}^N d\mathbf{p}^N$.

Let us now compare the probability

$$\mathcal{P}_N^{(N)\text{tot}}(\mathbf{r}^N, \mathbf{p}^N)h^{3N}N! = \frac{e^{-\beta \breve{U}_N^{\text{tot}}(\mathbf{r}^N, \mathbf{p}^N)}}{\int d\mathbf{r}'^N d\mathbf{p}'^N e^{-\beta \breve{U}_N^{\text{tot}}(\mathbf{r}'^N, \mathbf{p}'^N)}} \times h^{3N}N! \tag{3.4}$$

with its quantum counterpart

$$\mathfrak{p}_N(\mathscr{U}_i) = \frac{e^{-\beta \mathscr{U}_i}}{\sum_j e^{-\beta \mathscr{U}_j}} = \frac{e^{-\beta \mathscr{U}_i}}{Q_N},$$

where subscript N indicates the number of particles. Since classical mechanics is a special case of quantum theory, we see that in classical statistical mechanics Q_N should be identified with the integral in the denominator together with the factor $h^{3N}N!$, that is,[2]

$$Q_N = \frac{1}{h^{3N}N!} \int d\mathbf{r}'^N d\mathbf{p}'^N e^{-\beta \breve{U}_N^{tot}(\mathbf{r}'^N, \mathbf{p}'^N)}. \tag{3.5}$$

Note that the canonical partition function Q_N differs from the inverse normalization factor C_{tot}^{-1} in Equation 3.3 by the prefactor $(h^{3N}N!)^{-1}$ in the rhs.

Since $\breve{U}_N^{pot}$ and $\breve{U}_N^{kin}$ are additive, $\mathcal{P}_N^{(N)tot}$ (given by Equation 3.2) factorizes into a potential and a kinetic part $\mathcal{P}_N^{(N)tot} = \mathcal{P}_N^{(N)pot}\mathcal{P}_N^{(N)kin}$ with

$$\mathcal{P}_N^{(N)pot}(\mathbf{r}^N) = C_{pot}e^{-\beta \breve{U}_N^{pot}(\mathbf{r}^N)} \tag{3.6}$$

$$\mathcal{P}_N^{(N)kin}(\mathbf{p}^N) = C_{kin}e^{-\beta \breve{U}_N^{kin}(\mathbf{p}^N)},$$

where C_{pot} and C_{kin} are appropriate normalization constants and $C_{pot}C_{kin} = C_{tot}$. Thus, the probability to observe a certain kinetic state of the system is completely *independent of the interactions*; it is the same for strongly interacting molecules in a liquid as it is for a thin gas of the same molecules at the same temperature. Furthermore, this probability does not depend on where the molecules are located. (Incidentally, we note that this is true in classical mechanics, but not in the quantum case where the Hamiltonian cannot be subdivided in this manner.)

$\mathcal{P}_N^{(N)kin}$ can be factorized in the same manner since the kinetic energy consists of the sum of the translational energy for each particle. We have

$$\mathcal{P}_N^{(N)kin} = \prod_{\nu=1}^{N} \mathcal{P}_\nu^{(1)tr}(\mathbf{p}_\nu),$$

where

$$\mathcal{P}_\nu^{(1)tr}(\mathbf{p}_\nu) = C_{\nu,tr}e^{-\beta u_{tr,\nu}} = C_{\nu,tr}e^{-\beta p_\nu^2/2m_\nu}$$

and

$$C_{\nu,tr}^{-1} = \int d\mathbf{p}_\nu e^{-\beta p_\nu^2/2m_\nu}$$

$$= \int_{-\infty}^{\infty} dp_{\nu,x}e^{-\beta p_{\nu,x}^2/2m_\nu} \int_{-\infty}^{\infty} dp_{\nu,y}e^{-\beta p_{\nu,y}^2/2m_\nu} \int_{-\infty}^{\infty} dp_{\nu,z}e^{-\beta p_{\nu,z}^2/2m_\nu}$$

$$= \left[\sqrt{2\pi m k_B T}\right]^3,$$

---

[2] This result, which we inferred here by a rather intuitive argument, can be derived by making a proper limit of quantum statistical mechanics to its classical version [see, for example, Section 22–6 in the book *An introduction to statistical mechanics* by T. L. Hill (Addison-Wesley, 1960)].

where we have used $\int_{-\infty}^{\infty} dx \exp(-x^2) = \sqrt{\pi}$. We have thereby obtained the **Maxwell-Boltzmann distribution law** for translational motion of a molecule

$$\mathcal{P}_\nu^{(1)\text{tr}}(\mathbf{p}_\nu) = \frac{e^{-(p_{\nu,x}^2 + p_{\nu,y}^2 + p_{\nu,z}^2)/(2m_\nu k_B T)}}{(2\pi m k_B T)^{3/2}}. \tag{3.7}$$

This holds for molecules in a gas or in a liquid, *irrespective of the strength of the intermolecular interactions* (in classical statistical mechanics). Totally, we have the **kinetic probability density**

$$\mathcal{P}_N^{(N)\text{kin}}(\mathbf{p}^N) = \frac{1}{(2\pi m k_B T)^{3N/2}} \prod_{\nu=1}^{N} e^{-\beta p_\nu^2 / 2m_\nu} \tag{3.8}$$

for the whole system.

In general, if we have an additive contribution $\varepsilon(s)$ to the total energy (the Hamiltonian) from one of the degrees of freedom, here described by the variable s (for example, $s = p_{1,x}$), the probability density to observe a certain value of it is $\mathfrak{p}(s) = \exp[-\beta\varepsilon(s)]/\int ds' \exp[-\beta\varepsilon(s')]$, irrespective of the other degrees of freedom (provided $\varepsilon(s)$ is the sole contribution that contains s). The mean energy due to this contribution is

$$\langle \varepsilon \rangle = \int ds\, \varepsilon(s) \mathfrak{p}(s) = \int ds\, \varepsilon(s) \frac{e^{-\beta\varepsilon(s)}}{\int ds'\, e^{-\beta\varepsilon(s')}} = -\frac{\partial}{\partial\beta} \ln \int ds\, e^{-\beta\varepsilon(s)}.$$

A particularly important case is when $\varepsilon(s)$ is a square function $\varepsilon(s) = bs^2$, where b is a constant, like the kinetic energy contribution in the x direction $p_x^2/2m$ above. Then

$$\langle \varepsilon \rangle = -\frac{\partial}{\partial\beta} \ln \int_{-\infty}^{\infty} ds\, e^{-\beta bs^2} = -\frac{\partial}{\partial\beta} \ln \left[\int_{-\infty}^{\infty} d\tau\, e^{-b\tau^2} / \sqrt{\beta} \right]$$

$$= -\frac{\partial}{\partial\beta} \left[\ln \int_{-\infty}^{\infty} d\tau\, e^{-b\tau^2} - \ln \sqrt{\beta} \right] = \frac{1}{2} \frac{\partial \ln \beta}{\partial\beta} = \frac{1}{2\beta} = \frac{k_B T}{2}$$

where we have made the substitution $\tau = s\sqrt{\beta}$. This is the **equipartition theorem**:

> *For each degree of freedom that contributes with a square term in the Hamiltonian, there is a contribution of $k_B T/2$ to the mean energy.*

Thus, the three translational degrees of freedom for each particle give the contribution $3k_B T/2$ to the mean energy and the total mean kinetic energy from translational motions of the N particles is

$$\langle U_N^{\text{tr}} \rangle = \frac{3N k_B T}{2}$$

in classical statistical mechanics. This is the same result as we found for a monatomic ideal gas in Equation 2.147. It is valid not only for ideal gases, but also for strongly interacting particles of any shape and size, whereby we consider the translational motion of the center of mass of the particle.

Our main concern in what follows is the probability density for the positions of N spherical particles, $\mathcal{P}_N^{(N)\text{pot}}(\mathbf{r}^N)$. Since the kinetic probability density is explicitly known, it can be included if needed in the following by forming the product $\mathcal{P}_N^{(N)\text{tot}}(\mathbf{r}^N, \mathbf{p}^N) = \mathcal{P}_N^{(N)\text{pot}}(\mathbf{r}^N)\mathcal{P}_N^{(N)\text{kin}}(\mathbf{p}^N)$. Therefore, we shall drop "pot" in the superscript for the positional part and henceforth write $\mathcal{P}_N^{(N)}(\mathbf{r}^N) \equiv \mathcal{P}_N^{(N)\text{pot}}(\mathbf{r}^N)$. The set of positions $\mathbf{r}^N$ will be denoted as the **configuration** of the particles in the system and $\mathcal{P}_N^{(N)}(\mathbf{r}^N)$ is called the **configurational probability density**, which hence is equal to

$$\mathcal{P}_N^{(N)}(\mathbf{r}^N) = \frac{1}{Z_N} e^{-\beta \check{U}_N^{\text{pot}}(\mathbf{r}^N)}, \tag{3.9}$$

where

$$Z_N = C_{\text{pot}}^{-1} = \int d\mathbf{r}^N e^{-\beta \check{U}_N^{\text{pot}}(\mathbf{r}^N)}. \tag{3.10}$$

$Z_N = Z_N(T, V)$ is called the **configurational partition function**. $\mathcal{P}_N^{(N)}(\mathbf{r}^N)$ thus gives the probability density to observe the particles in a particular configuration $\mathbf{r}^N$.

Example 3.1: Ideal Gas in an External Field

Let us consider an ideal gas where the particles do not interact with each other. Assume that the gas is exposed to an external potential $v(\mathbf{r})$ that interacts with the particles; in the macroscopic domain this may, for example, be the gravitational field. In the microscopic domain, it may be the interaction with a body that is external to the system. The total potential energy is

$$\check{U}_N^{\text{pot}}(\mathbf{r}^N) = \sum_{\nu=1}^N v(\mathbf{r}_\nu).$$

Since the energy is a sum over the energies of the individual particles, the configurational partition function factorizes

$$Z_N = \int d\mathbf{r}^N \prod_{\nu=1}^N e^{-\beta v(\mathbf{r}_\nu)} = \prod_{\nu=1}^N \int d\mathbf{r}_\nu e^{-\beta v(\mathbf{r}_\nu)}$$

$$= \left[\int d\mathbf{r} e^{-\beta v(\mathbf{r})} \right]^N = (Z_1)^N,$$

where the second last equality follows from the fact that all integrals have the same values. Likewise, the probability density (3.9) factorizes

$$\mathcal{P}_N^{(N)}(\mathbf{r}^N) = \prod_{\nu=1}^{N} \frac{e^{-\beta v(\mathbf{r}_\nu)}}{Z_1},$$

so each particle has a probability density

$$\mathcal{P}_1(\mathbf{r}_\nu) = \frac{e^{-\beta v(\mathbf{r}_\nu)}}{Z_1} \quad \text{(ideal gas)} \tag{3.11}$$

that is independent of the positions of the other particles (this is, of course, a consequence of the fact that the particles do not interact with each other).

The probability to find a particular particle in the volume element $d\mathbf{r}$ at position $\mathbf{r}$ is accordingly equal to $\mathcal{P}_1(\mathbf{r})d\mathbf{r}$. Since there are N particles in the system, the probability to find *any* particle in this volume element is equal to $N\mathcal{P}_1(\mathbf{r})d\mathbf{r}$, which is the only relevant physical quantity since the particles are indistinguishable from each other. This probability is proportional[3] to the average density $n(\mathbf{r})$ of the gas at position $\mathbf{r}$. We have

$$n(\mathbf{r}) = N\mathcal{P}_1(\mathbf{r}) = \frac{N}{Z_1} e^{-\beta v(\mathbf{r})} \quad \text{(ideal gas)}, \tag{3.12}$$

which satisfies $\int d\mathbf{r}\, n(\mathbf{r}) = N$ as required (hence, the proportionality factor has been selected correctly). If $n(\mathbf{r}_0)$ is the density at another point $\mathbf{r}_0$, it follows that $n(\mathbf{r})/n(\mathbf{r}_0) = \exp[-\beta v(\mathbf{r})]/\exp[-\beta v(\mathbf{r}_0)]$ and hence

$$n(\mathbf{r}) = n(\mathbf{r}_0)e^{-\beta[v(\mathbf{r})-v(\mathbf{r}_0)]} \quad \text{(ideal gas)}. \tag{3.13}$$

This relationship is called the **"barometric formula"** since it relates the local density and thereby the local pressure $P(\mathbf{r}) = n(\mathbf{r})k_B T$ of an ideal gas to the external potential, for instance the gravitational potential in the atmosphere at various heights.

Let us consider the average of a quantity X that depends on particle coordinates and has the instantaneous value $\check{X}_N = \check{X}_N(\mathbf{r}^N)$ for a system with N particles. This average is calculated from

$$\bar{X}_N \equiv \langle X_N \rangle = \int d\mathbf{r}^N \check{X}_N(\mathbf{r}^N)\mathcal{P}_N^{(N)}(\mathbf{r}^N) \tag{3.14}$$

in the canonical ensemble. For example, the average potential energy is

$$\left\langle U_N^{\text{pot}} \right\rangle = \int d\mathbf{r}^N \check{U}_N^{\text{pot}}(\mathbf{r}^N)\mathcal{P}_N^{(N)}(\mathbf{r}^N).$$

---

[3] The proportionality can be understood from the fact that an increase in density with a factor α implies an α-fold increase in probability to find a particle there.

The sum of $\langle U_N^{\text{pot}} \rangle$ and the average kinetic energy $\langle U_N^{\text{kin}} \rangle$ is, by definition, equal to the **internal energy**

$$\bar{U}_N \equiv \langle U_N \rangle = \langle U_N^{\text{pot}} \rangle + \langle U_N^{\text{kin}} \rangle \tag{3.15}$$

of the system with N particles.

Helmholtz free energy A_N is defined as before from

$$A_N = -k_B T \ln Q_N$$

with the canonical partition function from Equation 3.5

$$Q_N = \frac{1}{h^{3N} N!} \int d\mathbf{r}^N d\mathbf{p}^N e^{-\beta \left[\breve{U}_N^{\text{pot}} + \breve{U}_N^{\text{kin}} \right]}.$$

The $\mathbf{p}$ integral in Q_N can be evaluated analytically (compare with the derivation of Equation 3.7) and after this evaluation we can express Q_N as

$$Q_N = \frac{1}{N! \Lambda^{3N}} \int d\mathbf{r}^N e^{-\beta \breve{U}_N^{\text{pot}}} = \frac{Z_N}{N! \Lambda^{3N}}, \tag{3.16}$$

where Λ is the thermal de Broglie wavelength defined in Equation 1.19.

We can generalize this result to particles with internal degrees of freedom. For ideal gases, the electronic states of a monatomic fluid will then appear as an extra factor $(q^{\text{es}})^N$ in the numerator of Q_N, where q^{es} is the single-particle partition function for the electronic states (cf. Example 2.3 in Section 2.7). For interacting particles, the electronic states contribute to the intermolecular interactions and thereby to $\breve{U}_N^{\text{pot}}$. We still take q^{es} as the single-particle partition function for a particle in the absence of other particles and let $\breve{U}_N^{\text{pot}}$ include the effects of deviations therefrom due to interactions with other particles.[4] We will therefore write

$$Q_N = \frac{Z_N \eta^N}{N!}, \tag{3.17}$$

where (cf. Equation 2.149)

$$\eta = \begin{cases} q^{\text{es}}/\Lambda^3, & \text{electronic states are considered} \\ 1/\Lambda^3, & \text{no internal degrees of freedom are present.} \end{cases} \tag{3.18}$$

Note that η depends on T only, $\eta = \eta(T)$.

---

[4] In this treatise, the interactions included in $\breve{U}_N^{\text{pot}}(\mathbf{r}^N)$ are mostly from model potentials that may include quantum effects in terms of effective interaction potentials, for example dispersion interactions between particles. In any other case that we will be concerned with, the only significant contribution to $\breve{U}_N^{\text{pot}}(\mathbf{r}^N)$ originates from the electronic ground state of the system for given positions $\mathbf{r}^N$ of the particles, which has energy $\mathscr{U}_{\text{ground}}^{\text{es}}(\mathbf{r}^N)$. Then the contribution to the potential energy is approximated as $\mathscr{U}_{\text{ground}}^{\text{es}}(\mathbf{r}^N) - N u_{\text{ground}}^{\text{es}}$, where $u_{\text{ground}}^{\text{es}}$ is the ground state energy of a single particle.

When the electronic states are included, the average energy of the system also contains single-particle contributions $\langle U_N^{es} \rangle = N \langle u^{es} \rangle$ from them

$$\bar{U}_N = \left\langle U_N^{pot} \right\rangle + \left\langle U_N^{kin} \right\rangle + N \langle u^{es} \rangle,$$

where $\langle u^{es} \rangle$ is the average energy per molecule of the electronic states. All contributions that depend on spatial coordinates are included in U_N^{pot}. In what follows, we will in most cases set $\eta = 1/\Lambda^3$ for spherical particles and $\langle U_N^{es} \rangle = 0$.

The free energy of the system is

$$A_N = -k_B T \ln Q_N = k_B T \ln \frac{N!}{Z_N \eta^N}. \tag{3.19}$$

It is a function of T, V and N. For a monatomic ideal gas in the absence of an external potential, we have $\breve{U}_N^{pot} = 0$ and $Z_N = \int d\mathbf{r}^N = (\int d\mathbf{r})^N = V^N$. We thereby obtain $A_N = A_N^{ideal}$, where

$$A_N^{ideal} \equiv k_B T \ln \frac{N!}{V^N \eta^N} = N k_B T \ln \left[\frac{1}{e\eta} \cdot \frac{N}{V} \right] \tag{3.20}$$

and we have used Stirling's approximation $\ln N! = N \ln N - N$ to obtain the last equality. This is the same result for A as in Equation 2.148 for the ideal gas in the quantum case.

Example 3.2: Free Energy for an Inhomogeneous Ideal Gas

Let us investigate the Helmholtz free energy of the gas in Example 3.1, that is, an ideal gas exposed to an external potential $v(\mathbf{r})$. Since we do not consider the internal degrees of freedom of the particles, we have $\eta = 1/\Lambda^3$. For this system, we have seen that $Z_N = (Z_1)^N$, so the free energy from Equation 3.19 becomes

$$A_N = k_B T \left[\ln N! + N \ln \frac{\Lambda^3}{Z_1} \right] = k_B T N \left[\ln \frac{\Lambda^3 N}{Z_1} - 1 \right],$$

where we have used Stirling's approximation. Since $N = \int d\mathbf{r}\, n(\mathbf{r})$, we can write this as

$$A_N = k_B T \int d\mathbf{r}\, n(\mathbf{r}) \left[\ln \frac{\Lambda^3 N}{Z_1} - 1 \right]$$

and by inserting $N/Z_1 = n(\mathbf{r}) \exp(\beta v(\mathbf{r}))$ from Equation 3.12 we obtain

$$A_N = k_B T \int d\mathbf{r}\, n(\mathbf{r}) \left[\ln(\Lambda^3 n(\mathbf{r})) - 1 \right] + \int d\mathbf{r}\, n(\mathbf{r}) v(\mathbf{r}) \quad \text{(ideal gas)}. \tag{3.21}$$

The last term is the mean interaction energy of the gas with the external field, which can be understood from the fact that the average number of particles in the volume element $d\mathbf{r}$ at position $\mathbf{r}$ is equal to $n(\mathbf{r})d\mathbf{r}$ and each of them contributes with the interaction energy $v(\mathbf{r})$. The first term is the **ideal part of the free energy** (the non-interactional part) for the inhomogeneous gas.

In the general case, in the presence of interactions, we can split A_N in Equation 3.19 in two parts

$$A_N = k_B T \ln \frac{N!}{Z_N \eta^N} = k_B T \ln \frac{N!}{V^N \eta^N} + k_B T \ln \frac{V^N}{Z_N}, \tag{3.22}$$

where the first term in the rhs is the ideal contribution A_N^{ideal} (the non-interactional contribution) given by Equation 3.20 and the second is the **excess free energy** A_N^{ex}, which arises from the interactions

$$A_N^{\text{ex}} = k_B T \ln \frac{V^N}{Z_N} = -k_B T \ln \left[\frac{1}{V^N} \int d\mathbf{r}^N e^{-\beta \breve{U}_N^{\text{pot}}(\mathbf{r}^N)} \right]. \tag{3.23}$$

Thus, we have $A_N = A_N^{\text{ideal}} + A_N^{\text{ex}}$.

The **chemical potential** can be obtained from $\mu = A_N - A_{N-1}$, that is, the change in free energy when the number of particles is changed by one (compare with the original definition of μ in Equation 2.14). This is in agreement with $\mu = (\partial A / \partial N)_{T,V}$ (Equation 2.53) since we have $\Delta A = (\partial A / \partial N)_{T,V} \Delta N = \mu \Delta N$ for small ΔN and can apply this formula to the special case $\Delta N = 1$.

From $\mu = A_N - A_{N-1}$ and Equation 3.19, we obtain

$$\mu = k_B T \left[\ln \frac{N!}{Z_N \eta^N} - \ln \frac{(N-1)!}{Z_{N-1} \eta^{N-1}} \right]$$

$$= k_B T \ln \frac{N Z_{N-1}}{Z_N \eta}. \tag{3.24}$$

In the absence of interactions, $\breve{U}_N^{\text{pot}} = 0$, we have $Z_N = V^N$ and the system consists of an ideal gas with uniform number density $n = N/V$. We then have

$$\mu = k_B T \ln \left(\frac{N}{V \eta} \right) = k_B T \ln \left(\frac{n}{\eta} \right) \quad \text{(uniform ideal gas)}. \tag{3.25}$$

In the general case (in the presence of interactions), the **activity** ζ of a fluid is defined from

$$\mu = k_B T \ln \left(\frac{\zeta}{\eta} \right), \tag{3.26}$$

which can be written as

$$\zeta = \eta e^{\beta \mu}. \tag{3.27}$$

Thus, for a fluid with chemical potential μ, its activity ζ is equal to *the density that a uniform ideal gas must have in order to have the same* μ (the uniform ideal gas is an imagined system that consists of the same particles in the absence of all interactions). Explicitly, we have

$$\zeta = \frac{NZ_{N-1}}{Z_N}, \tag{3.28}$$

as obtained from Equation 3.24. Note that Z_{N-1}/Z_N has the unit of inverse volume, m^{-3}.

A uniform gas at very low density n is nearly ideal, so its activity is approximately equal to the density and when $n \to 0$ we have $\zeta/n \to 1$. At higher densities, where the particles interact more strongly with each other, the activity of the fluid deviates from n. The **activity coefficient** defined as $\gamma = \zeta/n$ is given by $\gamma = [NZ_{N-1}/Z_N]/n = VZ_{N-1}/Z_N$. It gives the deviation from ideality and we have

$$\mu = k_BT \ln\left(\frac{\gamma n}{\eta}\right) = k_BT \ln\left(\frac{n}{\eta}\right) + k_BT \ln\gamma \quad \text{(uniform fluid)}, \tag{3.29}$$

where the first term on the rhs is called the **ideal part of the chemical potential**, μ^{ideal} (the non-interactional part) often simply called the **ideal chemical potential** and the last term is called the **excess chemical potential**,

$$\mu^{\text{ex}} = k_BT \ln\gamma = k_BT \ln\left(\frac{VZ_{N-1}}{Z_N}\right) \tag{3.30}$$

(the interactional part), which also can be expressed as

$$\mu^{\text{ex}} = A_N^{\text{ex}} - A_{N-1}^{\text{ex}}. \tag{3.31}$$

We have

$$\mu = \mu^{\text{ideal}} + \mu^{\text{ex}} = k_BT \ln\left(\frac{n}{\eta}\right) + k_BT \ln\gamma. \tag{3.32}$$

The first term is the chemical potential of an ideal gas, Equation 3.25. When $n \to 0$, the activity coefficient $\gamma \to 1$.

We can write Equation 3.29 as

$$\mu = \mu^{\ominus} + k_BT \ln\left(\frac{n}{n^{\ominus}}\right) + k_BT \ln\gamma, \tag{3.33}$$

where $\mu^{\ominus} = \mu^{\ominus}(T) = k_BT \ln[n^{\ominus}/\eta]$ is a standard state chemical potential (cf. Equations 2.27 and 2.150). Here the standard state is an ideal gas at density $n^{\ominus}$ and pressure $P^{\ominus} = k_BTn^{\ominus}$.

When $P^{\ominus} = 1$ bar, an actual gas is very close to being ideal provided that the temperature is not very low, so we may say that the actual gas with density $n^{\ominus}$ constitutes the standard state. If $n^{\ominus}$ is selected appreciably larger so the actual gas deviates significantly from an ideal gas, the standard state is instead an imagined ideal gas at density $n^{\ominus}$ consisting of the same particles in the absence of interactions.

{**EXERCISE 3.1**

Consider an ideal gas in the presence of an external field that interacts with the particles as in Example 3.1 (we have $\eta = 1/\Lambda^3$). For this inhomogeneous gas, show that $\mu = k_B T \ln(N\Lambda^3/Z_1)$ and use this to show that

$$\mu = k_B T \ln(\Lambda^3 n(\mathbf{r})) + v(\mathbf{r}) \quad \text{(ideal gas)} \tag{3.34}$$

for all positions $\mathbf{r}$. The first term is the ideal part of the chemical potential at position $\mathbf{r}$ and the expression says that the total chemical potential is equal everywhere at equilibrium – a condition that holds in general as we have seen.}

3.3 THE GRAND CANONICAL ENSEMBLE

Let us now turn to open systems and the grand canonical (μVT) ensemble, where the chemical potential, volume and temperature of the system are constant while the number of particles can vary. The probability $\mathscr{P}(N)$ that the system contains N particles is given by (Equation 2.112)

$$\mathscr{P}(N) = \frac{Q_N e^{\beta \mu N}}{\Xi},$$

where $Q_N \equiv Q(T, V, N)$ is the canonical partition function and Ξ is the grand canonical partition function (Equation 2.111)

$$\Xi(T, V, \mu) = \sum_{N=0}^{\infty} Q(T, V, N) e^{\beta \mu N}. \tag{3.35}$$

Note that $\mathscr{P}(N)$ depends on T, V and μ, but this is not explicitly shown here. The average number of particles in the system is

$$\bar{N} = \sum_{N=0}^{\infty} N \mathscr{P}(N) = \frac{1}{\Xi} \sum_{N=0}^{\infty} N Q_N e^{\beta \mu N} = \left(\frac{\partial \ln \Xi}{\partial [\beta \mu]} \right)_{T,V} \tag{3.36}$$

as follows directly from a differentiation of $\ln \Xi$ and the definition of Ξ above.

Using Equation 3.17 and $\zeta = \eta e^{\beta \mu}$ (Equation 3.27), we obtain $Q_N e^{\beta \mu N} = \zeta^N Z_N/N!$, so we can write

$$\Xi = \sum_{N=0}^{\infty} Q_N e^{\beta \mu N} = \sum_{N=0}^{\infty} \frac{\zeta^N Z_N}{N!} = \sum_{N=0}^{\infty} \frac{\zeta^N}{N!} \int d\mathbf{r}^N e^{-\beta \breve{U}_N^{\text{pot}}(\mathbf{r}^N)}, \tag{3.37}$$

and

$$\mathscr{P}(N) = \frac{\zeta^N Z_N}{N! \, \Xi}, \tag{3.38}$$

where the factor $N!$ in the denominator assures that the particles are indistinguishable despite that they are treated as being distinguishable in Z_N. In Equation 3.37, the $N = 0$ term in all sums is equal to 1 since $Q_0 = 1$ (in the rhs, the integral is replaced by 1 in the $N = 0$ term).

The **grand potential** is defined like before

$$\Theta(T, V, \mu) = -k_B T \ln \Xi(T, V, \mu). \tag{3.39}$$

As we saw in Section 2.5.2, various properties of the system can be obtained by differentiation of Θ with respect to these variables, for instance $\bar{N} = -(\partial\Theta/\partial\mu)_{T,V}$ (Equation 2.113), which also follows from Equation 3.36.

For a fluid in a container, we assume that the interactions with the container walls are included in the external potential $v(\mathbf{r})$ and a natural choice of the system volume V may be the space inside the walls. Since the system is open, it is, however, possible for the particles to move between the system and a particle reservoir with chemical potential μ so the chemical potential in the system is equal to that of the reservoir. Another possibility is to select some other volume V, for instance a smaller volume than the whole container, and define the system as what is inside this volume.

In the grand canonical ensemble, the average of a quantity X, which depends on the number of particles and their spatial coordinates, $\check{X}_N(\mathbf{r}^N)$, is obtained from

$$\bar{X} \equiv \langle X \rangle = \sum_{N=0}^{\infty} \mathscr{P}(N) \int d\mathbf{r}^N \check{X}_N(\mathbf{r}^N) \mathcal{P}_N^{(N)}(\mathbf{r}^N)$$

$$= \sum_{N=0}^{\infty} \mathscr{P}(N) \bar{X}_N, \tag{3.40}$$

with $\mathcal{P}_N^{(N)} = \exp(-\beta \check{U}_N^{\text{pot}})/Z_N$ (Equation 3.9) and $\bar{X}_N$ given by Equation 3.14. The spatial integrals are over the volume V for each $\mathbf{r}_\nu$ in $\mathbf{r}^N$. The external potential that confines the fluid is contained in the potential energy $\check{U}_N^{\text{pot}}$ and if the confining potential is taken to be infinitely large outside the container, $\mathcal{P}_N^{(N)}$ and hence the integrand are equal to zero there. So if V is equal to or greater than the container volume, the spatial integrals can be extended to a larger region.

By introducing

$$\mathcal{P}^{(N)}(\mathbf{r}^N) \equiv \mathscr{P}(N)\mathcal{P}_N^{(N)}(\mathbf{r}^N) = \frac{\zeta^N}{N!\,\Xi} e^{-\beta \check{U}_N^{\text{pot}}(\mathbf{r}^N)}, \tag{3.41}$$

where we have used Equations 3.9 and 3.38, we can write

$$\bar{X} \equiv \langle X \rangle = \sum_{N=0}^{\infty} \int d\mathbf{r}^N \check{X}_N(\mathbf{r}^N) \mathcal{P}^{(N)}(\mathbf{r}^N) \tag{3.42}$$

as an alternative to Equation 3.40. Note that

$$\sum_{N=0}^{\infty} \int d\mathbf{r}^N \mathcal{P}^{(N)}(\mathbf{r}^N) = 1 \tag{3.43}$$

as required for a probability density.

Systems that are infinitely large at least in some direction are special cases, for instance a bulk fluid that fills the whole space, a fluid in contact with an infinite wall where the fluid fills the whole space on one side of the wall or a fluid between two walls with infinite lateral extent. Such cases can be obtained as the limit where the volume $V \to \infty$ under the condition of constant μ. The grand canonical ensemble is particularly useful for these limits since it is designed to handle any volume for constant μ and T. Such a limit, in particular for a bulk fluid, is usually called the **thermodynamic limit**.

Example 3.3: Inhomogeneous Ideal Gas in the Grand Canonical Ensemble

In Example 3.1, we found that for an ideal gas exposed to an external potential $v(\mathbf{r})$ we have $Z_N = (Z_1)^N$, where $Z_1 = \int d\mathbf{r}\, e^{-\beta v(\mathbf{r})}$, in the canonical ensemble with N particles. We also found that the average density at position $\mathbf{r}$ is $n_N(\mathbf{r}) = N e^{-\beta v(\mathbf{r})}/Z_1$, where we here have put a subscript N on the density to indicate that the expression is valid for a system with N particles. We will now treat the inhomogeneous ideal gas in the grand canonical ensemble.

From Equation 3.37 follows

$$\Xi = \sum_{N=0}^{\infty} \frac{\zeta^N Z_N}{N!} = \sum_{N=0}^{\infty} \frac{\zeta^N (Z_1)^N}{N!} = e^{\zeta Z_1} \quad \text{(ideal gas).} \tag{3.44}$$

and from Equation 3.38, we see that

$$\mathcal{P}(N) = \frac{\zeta^N Z_N}{N!\, \Xi} = \frac{\zeta^N (Z_1)^N}{N!\, e^{\zeta Z_1}}.$$

To obtain the density distribution $n(\mathbf{r})$ in the grand canonical ensemble, we must take the average of $n_N(\mathbf{r})$ for all possible N in the system weighted with the probability $\mathcal{P}(N)$ that the system contains N particles with $N \geq 1$. We thus have

$$n(\mathbf{r}) = \sum_{N=1}^{\infty} \mathcal{P}(N) n_N(\mathbf{r}) = \sum_{N=1}^{\infty} \frac{\zeta^N (Z_1)^N}{N!\, e^{\zeta Z_1}} \cdot \frac{N}{Z_1} e^{-\beta v(\mathbf{r})}$$

$$= \frac{\zeta e^{-\beta v(\mathbf{r})}}{e^{\zeta Z_1}} \sum_{N=1}^{\infty} \frac{\zeta^{N-1}(Z_1)^{N-1}}{(N-1)!} = \zeta e^{-\beta v(\mathbf{r})},$$

where the sum in the last row is equal to $e^{\zeta Z_1}$ (this can be realized by replacing $N-1$ with N' where $N' \geq 0$). Thus, we have

$$n(\mathbf{r}) = \zeta e^{-\beta v(\mathbf{r})} \quad \text{(ideal gas),} \tag{3.45}$$

which can be written as Equation 3.13, if desired. Since $\eta = 1/\Lambda^3$ and hence $\zeta = e^{\beta\mu}/\Lambda^3$, the logarithm of Equation 3.45 can be written as

$$\mu = k_B T \ln(\Lambda^3 n(\mathbf{r})) + v(\mathbf{r}) \quad \text{(ideal gas)}, \tag{3.46}$$

which is the same as Equation 3.34 for the canonical ensemble.

The grand potential can be written as $\Theta = \bar{U} - TS - \mu\bar{N}$ (Equation 2.118), which implies that

$$\Theta = A - \mu\bar{N} = A - \mu \int d\mathbf{r}\, n(\mathbf{r}),$$

where $A = \bar{U} - TS$ is the Helmholtz free energy. From Equation 3.44, it follows that $\ln \Xi = \zeta Z_1 = \zeta \int d\mathbf{r}\, e^{-\beta v(\mathbf{r})} = \int d\mathbf{r}\, n(\mathbf{r})$, where we have used Equation 3.45. This yields

$$\Theta = -k_B T \ln \Xi = -k_B T \int d\mathbf{r}\, n(\mathbf{r}).$$

By eliminating Θ from these two expressions and then inserting μ from Equation 3.46, we obtain

$$A = \int d\mathbf{r}\, n(\mathbf{r})[\mu - k_B T] = \int d\mathbf{r}\, n(\mathbf{r})[k_B T \ln(\Lambda^3 n(\mathbf{r})) + v(\mathbf{r}) - k_B T].$$

This can be written as

$$A = k_B T \int d\mathbf{r}\, n(\mathbf{r}) \left[\ln(\Lambda^3 n(\mathbf{r})) - 1 \right] + \int d\mathbf{r}\, n(\mathbf{r}) v(\mathbf{r}) \quad \text{(ideal gas)}, \tag{3.47}$$

which is the same expression for Helmholtz free energy as Equation 3.21 in the canonical ensemble. The first term is the ideal part of the free energy and the last term is the interaction energy between the gas and the external potential.

3.4 REAL GASES

For **real gases** (also called **imperfect gases**) in the bulk phase, deviations from the ideal gas law can be expressed in terms of a so-called **virial expansion**, that is, a power series expansion of $P/k_B T$ in terms of the density n,

$$\frac{P}{k_B T} = n + B_2(T)n^2 + B_3(T)n^3 + \cdots, \tag{3.48}$$

where B_j are called **virial coefficients** that depend on temperature only. The first term on the rhs yields the ideal gas value and the remaining terms give the deviations from it. This expansion is often used in various practical applications. Then the B_j values have usually been determined from experimental measurements of, for instance, $P/k_B T$ as functions of n. In statistical mechanics, the virial coefficients can be determined from a microscopic descrip-tion of the molecules and their interactions. We will express B_j in terms of Z_N for $N \geq 2$, which can be calculated directly via Equation 3.10 from the interactional energy $\breve{U}_N^{\text{pot}}$ of the molecules in the gas.

The principle that we will use to determine B_j in terms of Z_N is the following. For a homo-geneous bulk fluid, we have $PV = -\Theta = k_B T \ln \Xi$ and since Equation 3.37 gives Ξ as a power series expansion in ζ where the coefficients are $Z_N/N!$, we can quite easily obtain $P/k_B T$ as a power series in ζ. In order to obtain B_j, the latter series then has to be trans-formed into a power series in n instead. The details of the derivation are given in the shaded text box below, which can be skipped in the first reading. The end results for the second and third virial coefficients are

$$B_2(T) = -\frac{Z_2 - V^2}{2V} \tag{3.49}$$

$$B_3(T) = -\frac{VZ_3 - 3Z_2^2 + 3V^2 Z_2 - V^4}{3V^2}. \tag{3.50}$$

The virial coefficients $B_j(T)$ for $j \geq 4$ can likewise be expressed in terms of Z_N with $N \leq j$.

★[5] To derive Equations 3.49 and 3.50, let us start by looking at the general power series expansion of $P/k_B T$ in ζ

$$\frac{P}{k_B T} = b_1 \zeta + b_2 \zeta^2 + b_3 \zeta^3 + \cdots \tag{3.51}$$

The ideal gas law implies that the first coefficient b_1 is equal to one since, as we have seen, $\zeta/n \to 1$ in the limit $n \to 0$ and hence $P/(k_B T n) \to b_1$. All coefficients b_l for $l \geq 2$ depend on T only, $b_l = b_l(T)$.

We will now express b_l in terms of Z_N that occurs in the coefficients for the expansion

$$\Xi = \sum_{N=0}^{\infty} \frac{\zeta^N Z_N}{N!} \tag{3.52}$$

(see Equation 3.37). Note that $Z_1 = V$, as obtained from Equation 3.10 since $\breve{U}_1^{\text{pot}} = 0$ for a bulk phase.

Using Equation 3.51 with $b_1 = 1$, we can express Ξ as

$$\Xi = e^{\beta PV} = e^{V[\zeta + b_2 \zeta^2 + b_3 \zeta^3 + \cdots]} = e^{V\zeta} e^{Vb_2 \zeta^2} e^{Vb_3 \zeta^3} \cdots.$$

---

[5] Text marked with ★ is more advanced than the rest and can be skipped in the first reading.

By expanding the exponential functions $\exp(Vb_l\zeta^l) = 1 + Vb_l\zeta^l + (Vb_l\zeta^l)^2/2! + (Vb_l\zeta^l)^3/3! + \ldots$ and inserting the sum into this expression, we obtain after simplification

$$\Xi = 1 + V\zeta + [V^2/2 + Vb_2]\zeta^2 + [V^3/6 + V^2b_2 + Vb_3]\zeta^3 + \ldots$$

If we identify the coefficients with those in Equation 3.52, we see that $V = Z_1$ (which we already know), $V^2/2 + Vb_2 = Z_2/2$ and $V^3/6 + V^2b_2 + Vb_3 = Z_3/6$. This gives b_2 and b_3 and we have

$$b_1 = 1$$
$$b_2 = [Z_2 - V^2]/2V \qquad (3.53)$$
$$b_3 = [Z_3 - 3VZ_2 + 2V^3]/6V.$$

It remains to convert the expansion (3.51) in ζ into an expansion in n.

From the Gibbs-Duhem Equation 2.124, we have $VdP - \bar{N}d\mu = 0$ when T is constant and hence $n = \bar{N}/V = (\partial P/\partial \mu)_T$. Using the chain rule, we obtain $n = (\partial P/\partial \zeta)_T(\partial \zeta/\partial \mu)_T = (\partial P/\partial \zeta)_T\beta\zeta$, where the latter equality follows from $(\partial \zeta/\partial \mu)_T = (\partial[\eta e^{\beta\mu}]/\partial \mu)_T = \eta\beta e^{\beta\mu} = \beta\zeta$ and we have used the definition (3.27) of ζ. Thus, from the expansion (3.51) with $b_1 = 1$, we obtain

$$n = \zeta\left(\frac{\partial(\beta P)}{\partial \zeta}\right)_T = \zeta + 2b_2\zeta^2 + 3b_3\zeta^3 + \cdots.$$

Next, we will insert this expression for n into Equation 3.48, so we need the corresponding expressions for n^2, n^3 etcetera. If we restrict ourselves to terms in ζ^l with $l \leq 3$, we have $n^2 = [\zeta + 2b_2\zeta^2 + 3b_3\zeta^3 + \cdots]^2 = \zeta^2 + 4b_2\zeta^3 + \cdots$ and, in the same way, $n^3 = \zeta^3 + \cdots$, so we obtain from Equation 3.48

$$\frac{P}{k_BT} = [\zeta + 2b_2\zeta^2 + 3b_3\zeta^3] + B_2[\zeta^2 + 4b_2\zeta^3] + B_3\zeta^3 + \cdots$$
$$= \zeta + [2b_2 + B_2]\zeta^2 + [3b_3 + B_24b_2 + B_3]\zeta^3 + \cdots.$$

By comparing with expansion (3.51) and identifying coefficients in front of ζ^l, we see that $b_2 = 2b_2 + B_2$ and $b_3 = 3b_3 + B_24b_2 + B_3$. This means that

$$B_2 = -b_2$$
$$B_3 = 4b_2^2 - 2b_3.$$

Finally, by inserting Equation 3.53 into these results, we obtain Equations 3.49 and 3.50. In the same manner, one can express all $B_j(T)$ in terms of Z_N with $N \leq j$, but this will not be done here.

We see that $B_2(T)$ only contains V and $Z_2 = \int d\mathbf{r}_1 d\mathbf{r}_2 \exp(-\beta \breve{U}_2^{\text{pot}}(\mathbf{r}_1, \mathbf{r}_2))$, so only interactions between pairs of particles are involved. $B_3(T)$ involves interactions between triplets of particles (in Z_3) and between pairs. Likewise, in $B_j(T)$, interactions between j particles and between fewer ones are involved.

For spherical particles that interact with a pair potential $u(r)$, we have $\breve{U}_2^{\text{pot}}(\mathbf{r}_1, \mathbf{r}_2) = u(r_{12})$, where $r_{12} = |\mathbf{r}_{12}|$ is the distance between the sphere centers. Then

$$B_2(T) = -\frac{\int d\mathbf{r}_1 d\mathbf{r}_2 e^{-\beta u(r_{12})} - \int d\mathbf{r}_1 \int d\mathbf{r}_2}{2V} = -\frac{\int d\mathbf{r}_1 d\mathbf{r}_2 \left[e^{-\beta u(r_{12})} - 1 \right]}{2V}.$$

In the $\mathbf{r}_2$ integration (for constant $\mathbf{r}_1$), we can make the variable substitution $\mathbf{r}_2 \to \mathbf{r}_{12}$, whereby the resulting $\mathbf{r}_{12}$ integral is independent of $\mathbf{r}_1$ and we obtain

$$B_2(T) = -\frac{1}{2} \int d\mathbf{r}_{12} \left[e^{-\beta u(r_{12})} - 1 \right] = -2\pi \int_0^\infty dr_{12} r_{12}^2 \left[e^{-\beta u(r_{12})} - 1 \right] \quad (3.54)$$

since $\int d\mathbf{r}_1 = V$. We have assumed that the volume is so large that we can extend the last integral to infinity and ignore effects of the boundaries of the system.

Let us consider the integrand in Equation 3.54 for large r_{12}, where $u(r_{12})$ is small so $e^{-\beta u(r_{12})} - 1 \approx -\beta u(r_{12})$ and the integrand decays like $-r_{12}^2 \beta u(r_{12})$. For monotonically decaying pair potentials, this means that the integral converges provided that $u(r_{12})$ decays faster than r_{12}^{-3} when $r_{12} \to \infty$. For more slowly decaying potentials, a second virial coefficient does not exist and neither does a virial expansion.

In $B_3(T)$, the interaction potential $\breve{U}_3^{\text{pot}}(\mathbf{r}_1, \mathbf{r}_2, \mathbf{r}_3)$ enters via Z_3. The interaction energy between three particles can, in general, be a complicated function of the coordinates, but in many cases $\breve{U}_3^{\text{pot}}$ can be written as a sum of pair interactions between the three particles

$$\breve{U}_3^{\text{pot}}(\mathbf{r}_1, \mathbf{r}_2, \mathbf{r}_3) = u(r_{12}) + u(r_{13}) + u(r_{23}), \quad (3.55)$$

at least as a reasonable approximation. We then have pairwise additivity of the intermolecular interactions. In such cases, the expression for $B_3(T)$ in Equation 3.50 can be simplified and written as

$$B_3(T) = -\frac{1}{3V} \int d\mathbf{r}_1 d\mathbf{r}_2 d\mathbf{r}_3 \left[e^{-\beta u(r_{12})} - 1 \right] \left[e^{-\beta u(r_{13})} - 1 \right] \left[e^{-\beta u(r_{23})} - 1 \right]. \quad (3.56)$$

The simplest way to verify this expression is to perform the multiplications of the three square brackets in Equation 3.56 and obtain a sum of exponential functions. Their integrals are of the following types: there is one term containing $\int d\mathbf{r}_1 d\mathbf{r}_2 d\mathbf{r}_3 \exp[\beta u(r_{12})] \exp[\beta u(r_{13})] \exp[\beta u(r_{23})]$, three terms of the type $-\int d\mathbf{r}_1 d\mathbf{r}_2 d\mathbf{r}_3 \exp[\beta u(r_{ij})] \exp[\beta u(r_{il})]$ (where i, j and l assume different combinations of values 1, 2, 3), three terms of the type $\int d\mathbf{r}_1 d\mathbf{r}_2 d\mathbf{r}_3 \exp[\beta u(r_{ij})]$ and one term

$- \int d\mathbf{r}_1 d\mathbf{r}_2 d\mathbf{r}_3$. The first term is equal to Z_3, the second group yields $-3Z_2Z_2/V$, the third group $3Z_2V$ and the final term equals $-V^3$. Inserted into Equation 3.56, they will give the rhs of Equation 3.50.

The function $f_M(r_{ij}) \equiv \exp[\beta u(r_{ij})] - 1$ that occurs in Equations 3.54 and 3.56 is called the **Mayer f-function**. These equations can be written as

$$B_2(T) = -\frac{1}{2V} \int d\mathbf{r}_1 d\mathbf{r}_2 f_M(r_{12})$$

$$B_3(T) = -\frac{1}{3V} \int d\mathbf{r}_1 d\mathbf{r}_2 d\mathbf{r}_2 f_M(r_{12}) f_M(r_{13}) f_M(r_{23}).$$

Provided that the interactions between the particles are pairwise additive, $B_j(T)$ for $j \geq 4$ can also be written in terms of integrals of products of $f_M(r_{ij})$ (called *cluster integrals*). The expressions for $B_j(T)$ contain sums of such integrals that become more and more complicated when the value of j increases.

Note that the virial coefficients, which are given in terms of interactions between finite groups (clusters) of molecules, are properties of the gas in the limit of zero density. This is also a mathematical consequence of the fact that the virial expansion is a Taylor series of βPV in n, so the coefficients are given by derivatives of βPV with respect to n evaluated at $n = 0$. The virial expansion (3.48) is useful for gases, at least up to some finite density, but not for liquids.

Example 3.4: Density Expansion for the Chemical Potential

Other thermodynamical quantities can also be expanded in a power series in density. As an example, we will determine the expansion for the excess chemical potential.

The chemical potential of a fluid is (Equation 3.32)

$$\mu = \mu^{\text{ideal}} + \mu^{\text{ex}} = k_B T \ln \left(\frac{n}{\eta} \right) + k_B T \ln \gamma \qquad (3.57)$$

and here we will derive power series expansion in n for the excess chemical potential $\mu^{\text{ex}} = k_B T \ln \gamma$ that is analogous to the virial expansion for the pressure in Equation 3.48.

We will use the Gibbs-Duhem Equation 2.124 that becomes $V dP - \bar{N} d\mu = 0$ when T is constant. This can be written $d\mu = dP/n$, which implies that $(\partial \mu / \partial n)_T = (\partial P / \partial n)_T / n$. From the virial expansion (3.48), we obtain

$$\frac{1}{k_B T} \left(\frac{\partial P}{\partial n} \right)_T = 1 + 2B_2(T)n + 3B_3(T)n^2 + \cdots,$$

which hence implies that

$$\frac{1}{k_B T}\left(\frac{\partial \mu}{\partial n}\right)_T = \frac{1}{n} + 2B_2(T) + 3B_3(T)n + \cdots.$$

Now, Equation 3.57 gives

$$\frac{1}{k_B T}\left(\frac{\partial \mu}{\partial n}\right)_T = \frac{1}{n} + \frac{1}{k_B T}\left(\frac{\partial \mu^{\text{ex}}}{\partial n}\right)_T,$$

so we can conclude that

$$\frac{1}{k_B T}\left(\frac{\partial \mu^{\text{ex}}}{\partial n}\right)_T = 2B_2(T) + 3B_3(T)n + \cdots.$$

By integrating this expression, we obtain the final result

$$\frac{\mu^{\text{ex}}}{k_B T} = \ln \gamma = 2B_2(T)n + \frac{3}{2}B_3(T)n^2 + \cdots, \tag{3.58}$$

where we have used the fact that $\mu^{\text{ex}} = 0$ when $n = 0$.

CHAPTER 4

Illustrative Examples from Some Classical Theories of Fluids

A MAJOR PURPOSE OF this chapter is to give illustrations of some important concepts for liquids and solutions taken from some approximate lattice theories of gases and liquid mixtures. Since these theories have a quite simple mathematical structure, they are pedagogically useful in order to help in the understanding of the concepts via explicit formulas that can quite easily be explained. They also provide very good opportunities to illustrate the statistical mechanical machinery. These theories, however, give quite poor representations of liquids and solutions in general and are included mostly for pedagogical and historical reasons.

4.1 THE ISING MODEL

The Ising model is a model for interacting particles located on lattice sites. Originally, it was used as a model for ferromagnetism in the solid state, but it has been used for many other kinds of problems. The interaction between the particles can be arbitrarily strong but is restricted to neighboring ones, that is, only nearest neighbors interact with each other. The lattice may have one, two or three dimensions and it may have any structure (for instance, square, cubic, hexagonal, orthorhombic). Some examples are shown in Figure 4.1.

The filled circles in the figure may denote sites that are occupied by particles, one particle per site, and open circles denote empty sites. The same model can be used for a mixture of two species of particles on the lattice. All sites are then occupied; the filled circles denote particles of one kind and the open circles denote the other kind. Another use of the model is for particles of different spins on the lattice. All sites are then occupied by the same species of particles but the spin states of the particles are different. In the examples shown, the filled circles may denote particles with spin up and the open circles denote particles with spin

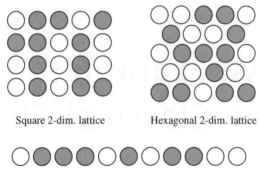

FIGURE 4.1   Some examples of simple lattices.

down. The latter was the original use of the Ising model, which will not be of particular concern here. The case with filled and vacant sites is a model of a one-component fluid, a so-called lattice fluid, and the two-component case is a model of a mixture. For a lattice fluid, the particles move around between all sites whereby the particle configuration changes. In the two-component case, the particle configuration changes when the particles swap places with each other; there are no vacant sites.

In all cases, one may formally regard each lattice site as being in one of two possible states: occupied/empty, species 1/species 2 or spin up/spin down. The interaction between lattice sites which are nearest neighbors depends on their states. There are four possibilities shown in Figure 4.2. The interaction energies for the middle two alternatives are equal and we do not distinguish between them.

We denote the number of lattice sites in one state N_1 (filled circles) and the number of sites in the other state N_2 (open circles). Totally, there are $N_S = N_1 + N_2$ sites, where subscript S stands for "sites." Assume that each site has c nearest neighbors. For each site, one only includes the interactions with these c nearest neighbors ("bonds") as illustrated in Figure 4.3. The total interaction energy of each configuration depends on the number of "bonds" of kind 11, 22 and 12 in the entire lattice. If we denote these numbers N_{11}, N_{22} and N_{12}, respectively, we obviously have

$$N_{11} + N_{22} + N_{12} = cN_S/2$$

since the total number of bonds is $cN_S/2$ and since each bond must be of kind 11, 22 or 12. We have here neglected all boundary effects at the lattice edges. They are negligible when N_S is large.

If we denote the interactional energy for the three kinds of bonds u_{11}, u_{22} and u_{12}, respectively, the total interaction energy equals

$$U^{\text{pot}} = N_{11}u_{11} + N_{22}u_{22} + N_{12}u_{12}. \tag{4.1}$$

FIGURE 4.2   Illustration of neighbors in a lattice.

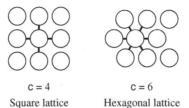

c = 4

Square lattice

c = 6

Hexagonal lattice

FIGURE 4.3 Illustration of lattice "bonds." The lines show which neighbors interact with the middle site. It is customary (and quite illustrative) to use the term bonds for these interactions. The number c is called the coordination number of the lattice.

This energy is, in general, different for different configurations (i.e., distributions of N_1 "filled circles" and N_2 "open circles" on the lattice sites) because N_{11}, N_{22} and N_{12} vary.

Let us consider the N_1 filled circles. They will be engaged in bonds of kinds 11 and 12. Let us calculate how many bonds the filled circles in the lattice participate in, irrespective of what kind of bonds they are. Denote this number N_1^{bonds}. Each lattice site has c nearest neighbors, but N_1^{bonds} is less than cN_1 since each 11 bond is counted twice in cN_1 (each 11 bond involves two filled circles). Therefore, we have to subtract N_{11} in order to correct for the double counting, so we have $N_1^{bonds} = cN_1 - N_{11}$. Alternatively, we can calculate N_1^{bonds} from the observation that each bond to a filled circle is either a 11 or 12 bond. Since each 11 and 12 bond involves at least one filled circle, we realize that $N_1^{bonds} = N_{11} + N_{12}$. By setting the two expression for N_1^{bonds} equal to each other, we obtain $cN_1 - N_{11} = N_{11} + N_{12}$, that is,

$$cN_1 = 2N_{11} + N_{12}.$$

In the same manner, we obtain for open circles

$$cN_2 = 2N_{22} + N_{12}.$$

We therefore have $N_{11} = (cN_1 - N_{12})/2$ and $N_{22} = (cN_2 - N_{12})/2$, so the total interaction energy in Equation 4.1 can be written as

$$
\begin{aligned}
U^{pot} &= u_{11}\frac{cN_1}{2} + u_{22}\frac{cN_2}{2} - [u_{11} + u_{22} - 2u_{12}]\frac{N_{12}}{2} \\
&= u_{11}\frac{cN_1}{2} + u_{22}\frac{cN_2}{2} - w\frac{N_{12}}{2},
\end{aligned}
\tag{4.2}
$$

where[1]

$$w = u_{11} + u_{22} - 2u_{12} \tag{4.3}$$

(distinguish this w from other meanings of w in this treatise). For given values of N_1 and N_2, the two first terms in U^{pot} are constant, so we have $U^{pot} = \text{constant} - wN_{12}/2$. This means

---

[1] Note that w is equal to the change in energy when two bonds are changed according to 2(12) → (11) + (22), see Equation 4.24.

that the energy for each configuration depends only on how many 12-bonds there are for that particular configuration.

The problem is accordingly reduced to finding the number of 12-bonds for each possible configuration. Alternatively, we may determine the number of configurations with a certain N_{12} value and do that for all possible values of N_{12}. Let us select the latter alternative and define $G(N_1, N_2, N_{12})$ as the number of configurations with precisely N_{12} bonds (i.e., nearest neighbor pairs) of kind 12 in a lattice with $N_S = N_1 + N_2$ sites when N_1 sites are in state 1 and N_2 sites in state 2. We will now find an expression for the canonical partition function Q.

Consider a particle that occupies a site, where it can move locally without leaving the site. It can also switch sites, but two particles cannot occupy the same site. Furthermore, the particle can assume various internal states. We assume that the motional and internal states of the particle on a site are independent of the neighboring particles, so these states are independent of the u_{ij} interactions with the neighbors. The states therefore give rise to single-particle partition functions $q_i = q_i(T)$ for each kind of particle. The canonical partition function can hence be written

$$Q(T, N_1, N_2) = q_1^{N_1} q_2^{N_2} \sum_{N_{12}} G(N_1, N_2, N_{12}) e^{-\beta U^{\text{pot}}}$$

$$= \left(q_1 e^{-\beta c u_{11}/2} \right)^{N_1} \left(q_2 e^{-\beta c u_{22}/2} \right)^{N_2}$$

$$\times \sum_{N_{12}} G(N_1, N_2, N_{12}) e^{\beta w N_{12}/2}, \tag{4.4}$$

where we have inserted Equation 4.2. This is a general result for the Ising model. It remains to determine the function $G(N_1, N_2, N_{12})$, which is a very difficult problem in general.

The Ising model for a one-dimensional lattice can be solved exactly without any real difficulty. This was done by Ernst Ising. However, in higher dimensions, the problem is much more difficult. For a two-dimensional square lattice with half the sites in one state and half in the other, an exact, analytic solution was found by Lars Onsager. This was (and still is) regarded as a great breakthrough in the study of the Ising model. No other exact solutions are known when $w \neq 0$. In general, one therefore has to rely on approximate methods.

4.2 THE ISING MODEL APPLIED TO LATTICE GASES AND BINARY LIQUID MIXTURES

The Ising model for a lattice fluid, where the lattice has filled and vacant sites, can be used as a model of a gas when the number of particles is low, a so-called **lattice gas**. The two-component lattice system without vacancies can be used as a model for binary liquid mixtures in bulk – albeit a rather crude model for such a system. The particles in the liquid are then placed on the lattice sites. The lattice does, of course, not exist in reality in the liquid; its role is to approximately take care of the fact that the particles have a finite size, that is, that they strongly repel each other when the distance between them becomes too short. We will consider mixtures of two species filling all lattice sites. This kind of lattice model is reasonable if the particles have about the same size. Usually, the energy parameters u_{11}, u_{12}

and u_{22} mainly describe the attractive contributions to the intermolecular interactions. We will start with the lattice gas and then proceed to liquid mixtures.

4.2.1 Ideal Lattice Gas

Consider a lattice gas where sites of kind 1 are filled and those of kind 2 vacant, so there are N_1 gas particles and N_2 vacant sites, in total $N_1 + N_2 = N_S$ sites. We have $u_{12} = u_{22} = 0$, $w = u_{11}$ and hence

$$Q(T, N_1, N_S - N_1) = \left(q_1 e^{-\beta c u_{11}/2}\right)^{N_1}$$
$$\times \sum_{N_{12}} G(N_1, N_S - N_1, N_{12}) e^{\beta u_{11} N_{12}/2}. \tag{4.5}$$

We will start by treating the special case of a non-interacting lattice gas, where $u_{11} = 0$ and the sum reduces to $\sum_{N_{12}} G(N_1, N_S - N_1, N_{12})$ = number of configurations of N_1 particles on N_S sites, irrespective of the value of N_{12}. If the particles were distinguishable, the number of different ways to distribute the particles on the sites equals $N_S \times (N_S - 1) \times (N_S - 2) \times \cdots \times (N_S - [N_1 - 1]) = N_S!/(N_S - N_1)!$. This follows for the fact the first particle we put in the lattice has N_S places to be located at, the second has $N_S - 1$ places, the third $N_S - 2$, etc. Since we deal with indistinguishable particles we must divide by $N_1!$ and then we obtain the number of configurations

$$\Omega^{\text{conf}} = \sum_{N_{12}} G(N_1, N_S - N_1, N_{12}) = \frac{N_S!}{N_1!(N_S - N_1)!} = \frac{(N_1 + N_2)!}{N_1! N_2!}. \tag{4.6}$$

For the non-interacting lattice gas, Equation 4.5 hence becomes $Q = q_1^{N_1} \Omega^{\text{conf}}$.

The configurational entropy is $S^{\text{conf}} = k_B \ln \Omega^{\text{conf}}$. By using Stirling's formula $\ln N! = N \ln N - N$, we obtain

$$\ln \Omega^{\text{conf}} = \ln \frac{(N_1 + N_2)!}{N_1! N_2!} =$$
$$= (N_1 + N_2) \ln(N_1 + N_2) - N_1 \ln N_1 - N_2 \ln N_2$$
$$= N_1 \ln \frac{N_1 + N_2}{N_1} + N_2 \ln \frac{N_1 + N_2}{N_2}. \tag{4.7}$$

This implies that

$$S^{\text{conf}}(N_1, N_2) = -k_B [N_1 \ln x_1 + N_2 \ln x_2] \tag{4.8}$$

where $x_i = N_i/(N_1 + N_2)$ is the mole fraction of species i. For the current case, x_2 is the fraction of sites in the lattice that are unoccupied.

For a thin ideal gas, $N_1 \ll N_2$ and the last term in the rhs of Equation 4.7 can be written

$$N_2 \ln \frac{N_1 + N_2}{N_2} = N_2 \ln \left[1 + \frac{N_1}{N_2}\right] \approx N_2 \frac{N_1}{N_2} = N_1,$$

where we have used the fact that $\ln(1 + x) \approx x$ for small x. Thus, we have as a very good approximation

$$\ln \Omega^{\text{conf}} = N_1 \ln \frac{N_1 + N_2}{N_1} + N_1 = N_1 \ln \frac{eN_S}{N_1} \quad \text{(ideal lattice gas)}.$$

The free energy is $A = -k_B T \ln Q = -k_B T \ln \left[q_1^{N_1} \Omega^{\text{conf}} \right] = -k_B T N_1 \ln [q_1 eN_S/N_1]$, and by introducing the volume $V = N_S v_S$ where v_S is the volume occupied by a lattice site (the "cell volume"), we obtain

$$A = N_1 k_B T \ln \left[\frac{v_S}{e q_1} \cdot \frac{N_1}{V} \right] \quad \text{(ideal lattice gas)}.$$

This expression is similar to the free energy of an ideal gas in Equation 3.20; the difference is that η is replaced by q_1/v_S, whereby v_S takes the same role as Λ^3.

4.2.2 Ideal Liquid Mixture

Let us consider a binary liquid mixture where the particles are interacting, $u_{ij} \neq 0$. We will start by treating the special case where the interactions happen to be such that $w = u_{11} + u_{22} - 2u_{12} = 0$, that is, we have $u_{12} = [u_{11} + u_{22}]/2$. This means that species 1 and 2 "like" to be surrounded by its own species equally much as the other species. The energy of all Ω^{conf} configurations are the same, with $\Omega^{\text{conf}} = \sum_{N_{12}} G(N_1, N_2, N_{12})$. Since $w = 0$ Equation 4.4 becomes[2]

$$Q(T, N_1, N_2) = \left(q_1 e^{-\beta c u_{11}/2} \right)^{N_1} \left(q_2 e^{-\beta c u_{22}/2} \right)^{N_2} \Omega^{\text{conf}}$$

$$= q_1^{N_1} q_2^{N_2} e^{-\beta U^{\text{pot}}} \frac{(N_1 + N_2)!}{N_1! N_2!}, \tag{4.9}$$

where in this case $U^{\text{pot}} = [cN_1 u_{11} + cN_2 u_{22}]/2$, as follows from Equation 4.2. The free energy is $A = -k_B T \ln Q$ and we have

$$A(T, N_1, N_2) = U^{\text{pot}} - k_B T(N_1 \ln q_1 + N_2 \ln q_2) - TS^{\text{conf}}(N_1, N_2). \tag{4.10}$$

Explicitly, this is

$$A(T, N_1, N_2) = c[N_1 u_{11} + N_2 u_{22}]/2$$

$$+ N_1 k_B T [\ln x_1 - \ln q_1] + N_2 k_B T [\ln x_2 - \ln q_2], \tag{4.11}$$

where we have used Equation 4.8.

---

[2] Equation 4.6 for Ω^{conf} of a one-component gas applies also in the present case since the vacant sites (here corresponding to species 2) were treated as indistinguishable during the derivation of the equation.

The **free energy of mixing** ΔA_{mix} is defined as the change in A when N_1 particles of pure substance 1 are mixed with N_2 particles of pure substance 2. The free energy of pure substance i with N_i particles located at N_i sites is

$$A_i(T, N_i) = -k_B T \ln Q_i(T, N_i) = -k_B T \ln \left[q_i^{N_i} e^{-\beta N_i c u_{ii}/2} \right]$$

$$= N_i \left[\frac{c u_{ii}}{2} - k_B T \ln q_i \right]$$

since $U^{\mathrm{pot}} = N_i c u_{ii}/2$ (there are $N_i c/2$ bonds) and there is just one configuration (the particles are indistinguishable). We have

$$\Delta_{\mathrm{mix}} A = A(T, N_1, N_2) - A_1(T, N_1) - A_2(T, N_2)$$

$$= k_B T \left[N_1 \ln x_1 + N_2 \ln x_2 \right] = -T S^{\mathrm{conf}}(N_1, N_2). \tag{4.12}$$

The entropy is given by $S = -\partial A / \partial T$, which gives the entropy of mixing

$$\Delta_{\mathrm{mix}} S = S^{\mathrm{conf}}(N_1, N_2).$$

We see that there is only the configurational entropy in $\Delta_{\mathrm{mix}} A$ and in $\Delta_{\mathrm{mix}} S$. This originates from the fact that all configurations have the same energy. This kind of mixture is called an **ideal mixture**. Counted per particle in the mixture, we have

$$\Delta_{\mathrm{mix}} \widehat{S} \equiv \frac{\Delta_{\mathrm{mix}} S}{N_1 + N_2} = -k_B \left[x_1 \ln x_1 + x_2 \ln x_2 \right] \quad \text{(ideal mixture)}.$$

This expression is valid for ideal liquid mixtures in general, that is, not only for liquids described by a lattice model. It is also the same as the entropy change when two ideal gases of species 1 and 2 with equal initial number densities are mixed, Equation 2.152.

The chemical potential of species 1 is obtained from the derivative of Equation 4.11 with respect to N_1

$$\mu_1 = \left(\frac{\partial A}{\partial N_1} \right)_{T, N_2} = \frac{c u_{11}}{2} + k_B T \left[\ln x_1 - \ln q_1 \right], \tag{4.13}$$

since

$$\left(\frac{\partial U^{\mathrm{pot}}}{\partial N_1} \right)_{T, N_2} = \frac{c u_{11}}{2} \quad \text{and} \quad \left(\frac{\partial S^{\mathrm{conf}}}{\partial N_1} \right)_{T, N_2} = -k_B \ln x_1,$$

where the last equation can easily be shown from Equations 4.7 and 4.8.

{**EXERCISE 4.1**
Show this result for the S^{conf} derivative.}

An expression analogous to Equation 4.13 applies to μ_2 and we can write these results as

$$\mu_i = \mu_i^0 + k_B T \ln x_i \quad \text{(ideal mixture)}, \tag{4.14}$$

where μ_i^0 is the chemical potential for pure i; in the present case $\mu_i^0 = c u_{ii}/2 - k_B T \ln q_i$, as follows by inserting $x_i = 1$ in the expression for the chemical potential above. This expression for μ_i is valid for ideal liquid mixtures in general.

We see that we have $A = N_1 \mu_1 + N_2 \mu_2$, which means that the free energy A in the present case satisfies a relationship that normally is valid only for the Gibbs free energy, Equation 2.98. In Equation 4.13, we see that we have kept T and N_2 constant during the differentiation, but we have not specified whether volume or (possibly) pressure is constant. If the lattice is incompressible, which we have tacitly assumed, its volume is independent of pressure and we therefore ignore any pressure-volume effect. Since we initially have a lattice filled with particles, we cannot add a new particle without extending the lattice and thereby increasing the volume. Thus, V is not kept constant in Equation 4.13, but this has no real consequences since we ignore pressure-volume effects, which distinguish the Gibbs free energy from Helmholtz free energy. Liquids are, in general, nearly incompressible provided the pressure is not very large, which is one reason why a lattice model for a liquid makes some sense at all.

In the general case of a mixture of species 1 and 2 with $w \neq 0$, we obtain the free energy $A = -k_B T \ln Q$ by using Q from Equation 4.4. Then there are entropy contributions from both particle configurations and distributions of energy; the interaction energy is different depending on the configuration. As mentioned above, the function $G(N_1, N_2, N_{12})$, which gives the number of configurations with the same energy, is very difficult to calculate. To proceed, we will use a simple approximation, as explained in the following section.

4.2.3 The Bragg-William Approximation

So far, nobody has been able to solve three-dimensional lattice statistics problems both completely and exactly. On the other hand, there exist many different approximate methods. One of the simplest methods is called the **Bragg-Williams approximation**, which for the case of mixtures (solutions) is also called the **regular solution theory**. However, it is good enough to give insights into several of the qualitative behaviors of the system. It is a so-called **mean field theory** where one does not consider details regarding, for example, various configurations, but rather considers averages over many configurations as an approximation.

A one-component *non-ideal lattice fluid* in the Bragg-William approximation is the special case where u_{11} gives the interaction between the gas particles and $u_{12} = u_{22} = 0$. As mentioned in Section 4.2.1, in such a system the lattice sites of kind 1 are filled and those of kind 2 vacant. We will not treat such systems explicitly here; the free energy for such a fluid can be deduced from the expression obtained in the regular solution theory in a manner that will be described below in footnote 3.

4.2.3.1 Regular Solution Theory

We will start with the general and exact expression for Q in Equation 4.4 for a lattice fluid mixture with N_1 and N_2 particles of either species and $N_1 + N_2 = N_S$ sites. As mentioned earlier, it is very difficult to apply this expression. The difficulty originates from the fact that the probability for each configuration depends on the total interaction energy between the particles. If we disregard this fact and instead assume that the particles are randomly distributed on the lattice sites irrespective of the intermolecular interactions, the distributions will be the same as for the case $w = 0$. This means that the probability for each site to be occupied by a particle of species i is N_i/N_S. Consider a site occupied by a particle of substance 1. The probability that a neighboring site is occupied by a particle of substance 2 is N_2/N_S. Since there are c such sites, the average number of neighbors of species 2 must be cN_2/N_S in this approximation. Since there are N_1 particles of species 1, there must be on average $N_1 \times cN_2/N_S$ neighbors of species 2 for all of them. The average number of 12-bonds in the lattice is therefore

$$\bar{N}_{12} = N_1 \frac{cN_2}{N_S} = \frac{cN_1N_2}{N_1 + N_2}.$$

Except when $\beta w = w/k_BT \approx 0$, this is a rather rough approximation, but it still gives interesting results. In the expression for Q in Equation 4.4, we accordingly insert $\bar{N}_{12}$ instead of N_{12} in the exponent and obtain

$$Q(T, N_1, N_2) = \left(q_1 e^{-\beta cu_{11}/2}\right)^{N_1} \left(q_2 e^{-\beta cu_{22}/2}\right)^{N_2}$$
$$\times \sum_{N_{12}} G(N_1, N_2, N_{12}) e^{\beta w \bar{N}_{12}/2}.$$

Since $\bar{N}_{12}$ is independent of N_{12}, the exponential function can be taken outside the sum. Using $\sum_{N_{12}} G(N_1, N_2, N_{12}) = \Omega^{\text{conf}}$ and Equation 4.6, we obtain

$$Q(T, N_1, N_2) = \left(q_1 e^{-\beta cu_{11}/2}\right)^{N_1} \left(q_2 e^{-\beta cu_{22}/2}\right)^{N_2} \frac{(N_1 + N_2)!}{N_1! N_2!} e^{\beta w \bar{N}_{12}/2}.$$

It follows that (cf. Equation 4.10)

$$A(T, N_1, N_2) = \bar{U}^{\text{pot}} - k_BT(N_1 \ln q_1 + N_2 \ln q_2) - TS^{\text{conf}}(N_1, N_2)$$

where

$$\bar{U}^{\text{pot}} = u_{11} \frac{cN_1}{2} + u_{22} \frac{cN_2}{2} - w \frac{\bar{N}_{12}}{2}$$

is the average of U^{pot} in Equation 4.2 for a random distribution of the particles on the lattice sites. Explicitly, we have[3]

$$A(T, N_1, N_2) = \frac{c}{2} \left[N_1 u_{11} + N_2 u_{22} - w \frac{N_1 N_2}{N_1 + N_2} \right] \tag{4.15}$$
$$+ N_1 k_BT \left[\ln x_1 - \ln q_1 \right] + N_2 k_BT \left[\ln x_2 - \ln q_2 \right].$$

---

[3] The free energy for a one-component nonideal lattice fluid can be obtained from Equation 4.15 by setting $u_{12} = u_{22} = 0$ and $q_2 = 1$. In this case, N_2 is the number of vacant sites.

It follows that the free energy of mixing is (cf. Equation 4.12)

$$\Delta_{\text{mix}} A = -\frac{cw}{2}(N_1 + N_2)x_1x_2 + k_BT\left[N_1 \ln x_1 + N_2 \ln x_2\right]. \tag{4.16}$$

Per particle in the mixture, we thus have

$$\Delta_{\text{mix}}\widehat{A} \equiv \frac{\Delta_{\text{mix}} A}{N_1 + N_2} = -\frac{cw}{2}x_1x_2 + k_BT\left[x_1 \ln x_1 + x_2 \ln x_2\right] \tag{4.17}$$

and the corresponding entropy and energy of mixing are

$$\Delta_{\text{mix}}\widehat{S} = -k_B\left[x_1 \ln x_1 + x_2 \ln x_2\right]$$

and

$$\Delta_{\text{mix}}\widehat{U} = -\frac{cw}{2}x_1x_2,$$

respectively. Note that the entropy of mixing is the same as for an ideal mixture. This is a consequence of the assumption that the particles are randomly distributed on lattice positions, irrespective of the intermolecular interactions. $\Delta_{\text{mix}}\widehat{S}$ is always positive since $x_i \leq 1$ and hence $\ln x_i \leq 0$. If $w > 0$, that is, when the 12-contacts are energetically preferred over 11 and 22-contacts, $2u_{12} < u_{11} + u_{22}$, we see that $\Delta_{\text{mix}}\widehat{U} < 0$. This is expected since mixing means that we form 12-contacts and the energy is thereby lowered since these contacts are preferred.

Next, we consider the chemical potential. From Equation 4.15, we obtain

$$\mu_1 = \left(\frac{\partial A}{\partial N_1}\right)_{T,N_2} = \frac{cu_{11}}{2} - \frac{cw(1-x_1)^2}{2} + k_BT\left[\ln x_1 - \ln q_1\right], \tag{4.18}$$

where we have used $x_2 = 1 - x_1$. The analogous expression is valid for μ_2. Hence, we have (cf. Equation 4.14)

$$\mu_i = \mu_i^0 + k_BT \ln x_i - \frac{cw(1-x_i)^2}{2}, \tag{4.19}$$

where μ_i^0 is the chemical potential for pure i. Equation 4.19 can be written

$$\mu_i = \mu_i^{\text{ideal}} + \mu_i^{\text{ex}},$$

where μ_i^{ideal} is the chemical potential for an ideal mixture given by Equation 4.14 and $\mu_i^{\text{ex}} = -cw(1-x_i)^2/2$ is the excess chemical potential. Here the term "excess" means "in excess to an *ideal mixture*." Note that this is is different from the excess chemical potential in Equation 3.29, which is in excess to the *ideal gas* chemical potential. One always has to keep in mind what kind of ideal state one talks about. In the rest of the book, we normally deal with excess quantities relative to the ideal gas, but in the present chapter we deal with other ones.

We can introduce an activity coefficient γ_i from $\mu_i^{ex} = k_B T \ln \gamma_i$ (cf. Equation 3.32) and write

$$\mu_i = \mu_i^0 + k_B T \ln x_i + k_B T \ln \gamma_i = \mu_i^0 + k_B T \ln(\gamma_i x_i), \tag{4.20}$$

where we have explicitly

$$\gamma_i = \gamma_i(T, x_i) = e^{-\beta c w (1 - x_i)^2 / 2}.$$

For pure substance i, we have $x_i = 1$, $\gamma_i = 1$ and $\mu_i = \mu_i^0$.

In Equation 4.20, the entity μ_i^0 plays the role of a **standard state chemical potential**, that is, the chemical potential in the **standard state**; here this state is the pure substance i. Earlier, when we discussed ideal gases in connection to Example 2.1 in Section 2.2, the standard state was selected as the pure gas at pressure $P^\ominus = 1$ bar. However, we also saw that any other value of $P^\ominus$ could equally well have been selected as the standard pressure. The common terminology with "standard states," "ideal" and "excess" can be quite confusing. A standard state is, in fact, something one can choose quite freely. To illustrate this, we shall consider the expression for μ_1 in Equation 4.18.

When the pure substance 1 is the standard state, we accordingly set the standard state chemical potential $\mu_i^\ominus$ equal to μ_i^0, so we have

$$\mu_1^\ominus = \mu_1^0 = c u_{11}/2 - k_B T \ln q_1$$
$$\mu_i^{ex} = -c w (1 - x_i)^2 / 2.$$

The term $c u_{11}/2$ in $\mu_1^\ominus$ is the interaction energy per particle in a lattice fully occupied by substance 1 (the factor 1/2 is present in order to avoid double counting each bond) and the term $-k_B T \ln q_1$ is the free energy per particle due to its local motion at the site and its inner degrees of freedom.

The splitting of μ_1 into ideal and excess terms is not unique. We could equally well have taken some other standard and excess chemical potentials

$$\mu_1^{\ominus'} = \mu_1^0 - f(T)$$
$$\mu_1^{ex'} = \mu_1^{ex} + f(T),$$

where $f(T)$ is an arbitrary function of temperature (f is not allowed to depend on the composition since the standard chemical potential must be independent of x_1). The selection of different $f(T)$ implies that one is selecting different standard states with different $\mu_1^{\ominus'}$. In all cases considered here, we have (cf. Equation 4.20)

$$\mu_1 = \mu_1^{\ominus'} + k_B T \ln x_i + \mu_1^{ex'}$$
$$= \mu_1^{\ominus'} + k_B T \ln(\gamma_1' x_1)$$

with different $\mu_1^{\ominus'}$, $\mu_1^{ex'}$ and γ_1'. The value of μ_1 does not change; it is only the splitting of μ_1 into the different terms that changes. Our initial choice with the pure substance as the standard state corresponds to $f(T) = 0$.

Another choice of standard state is to select $f(T) = cu_{11}/2$, that is,

$$\mu_1^{\ominus'} = -k_B T \ln \mathfrak{q}_1$$

$$\mu_1^{ex'} = \frac{c[u_{11} - w(1 - x_1)^2]}{2} = \frac{c[u_{11} - wx_2^2]}{2}.$$

Our standard state then equals an *imagined* state in which the fluid of pure substance 1 behaves as if the particles do not interact with each other, so $\mu_1^{\ominus'}$ does not contain any inter-particle interaction terms (between the particles on the lattice sites). In this case, $\mu_1^{ex'}$ is the contribution to μ_1 in Equation 4.18 that originates from such interactions.

A third choice of standard state is to take $f(T) = cw/2$, that is,

$$\mu_1^{\ominus''} = \frac{c[u_{11} - w]}{2} - k_B T \ln \mathfrak{q}_1 = cu_{12} - \frac{cu_{22}}{2} - k_B T \ln \mathfrak{q}_1$$

$$\mu_1^{ex''} = \frac{cw}{2}[1 - x_2^2] = \frac{cw}{2}x_1[1 + x_2],$$

where we have inserted $w = u_{11} + u_{22} - 2u_{12}$ in the first line. This standard state is an infinitely dilute solution of molecular species 1 in a fluid of molecular species 2 (the solvent). This can be seen from the fact that $\mu_1^{\ominus''}$ equals the sum of $-k_B T \ln \mathfrak{q}_1$ (the interpretation of which we have already given) and $cu_{12} - cu_{22}/2$, that is, the energy required to insert a single particle of species 1 in a lattice filled with molecular species 2.[4] We see that $\mu_1^{ex''} \to 0$ when $x_1 \to 0$, which implies that the activity coefficient $\gamma_1'' = \exp[\beta \mu_1^{ex''}] \to 1$. Note that only for this choice of standard state, the activity coefficient $\to 1$ at infinite dilution.

The latter choice is preferred in practical applications when we are dealing with dilute solutions of substance 1 with nonzero x_1 since we can set $\gamma_1'' = 1$ as a good approximation.[5] For substance 2 (the solvent), on the other hand, the preferred standard state is the pure fluid: our original choice. When is x_1 is small we have $x_2 \approx 1$ and $\gamma_2 \approx 1$, so we can set $\gamma_2 = 1$ as a good approximation. Thus, for both substances, we have the activity coefficient approximately equal to one since we have selected suitable (different) standard states for them.

4.2.3.2 Some Applications of Regular Solution Theory

Let us first treat a system composed of a mixture of two liquids 1 and 2 in equilibrium with a gas phase of these substances. We will treat the liquid mixture according to the regular solution theory and the gas phase as a mixture of ideal gases. The chemical potential for substance i in the liquid mixture is given by Equation 4.19 and in the gas phase it is

---

[4] This follows since cu_{12} is the energy for a particle 1 entirely surrounded by species 2, while $-cu_{22}/2$ is the energy required to make place for the new particle by moving away the particle of species 2 that initially was present on the site where we insert particle 1. The factor of $1/2$ in $-cu_{22}/2$ can be explained as follows. If we remove the particle of species 2 entirely from the system to make an empty site for the new particle, the energy change would be $-cu_{22}$, but since the former particle is kept inside the system in contact with the other particles, the energy change is $-cu_{22} + cu_{22}/2$, where we have added the interaction energy per particle in pure substance 2.

[5] In practice, the value of $\mu_1^{\ominus''}$ has been determined experimentally, often by extrapolating a series of measured μ_1 values to $x_1 = 0$.

$\mu_i = \mu_i^{\ominus} + k_B T \ln(P_i/P^{\ominus})$ (Equation 2.150), where $\mu_i^{\ominus} = k_B T \ln(\beta P^{\ominus}/\eta)$ and $P^{\ominus}$ is the standard pressure, which can be anything we like. The value of μ_i in the gas phase is not affected by the choice of $P^{\ominus}$. The standard state for the liquid is the pure liquid of substance i with chemical potential μ_i^0. For the gas phase, it is suitable here to take the vapor in equilibrium with the pure liquid as the standard state, that is, we select $P^{\ominus} = P_i^0 =$ the vapor pressure of the pure liquid of the same substance. This means that the standard chemical potential for substance i in the gas phase is the same as in the liquid phase: μ_i^0 in both phases. Note that the value of P_i^0 is different for substance 1 and 2.

At equilibrium, the chemical potentials in the liquid and in the gas phases are equal, which means that

$$\mu_i^0 + k_B T \ln x_i - \frac{cw(1-x_i)^2}{2} = \mu_i^0 + k_B T \ln \frac{P_i}{P_i^0}.$$

This implies

$$\frac{P_i}{P_i^0} = x_i^{\text{liq}} e^{-\beta cw(1-x_i^{\text{liq}})^2/2}, \tag{4.21}$$

where we for clarity have set superscript "liq" on x_i in order to emphasize that it is the mole fraction in the liquid phase. For a system where $w = 0$, that is, $u_{12} = [u_{11} + u_{22}]/2$, we see that

$$P_i = x_i^{\text{liq}} P_i^0. \tag{4.22}$$

This is **Raoult's law**. By definition, a liquid mixture that obeys Raoult's law is ideal, which hence is in accordance with our concept of ideal mixture in Section 4.2.2. As mentioned earlier, $u_{12} = [u_{11} + u_{22}]/2$ means that each particle is equally "happy" to be surrounded by particles of the same kind as of the other kind. Therefore, the tendency to escape from the liquid phase for each particle is equal for both species and the amount in the gas phase is simply proportional to the mole fraction in the liquid mixture.

Raoult's law was originally an empirical observation by the French chemist François-Marie Raoult for certain types of liquid mixtures. In some cases, the law holds as a good approximation for all compositions, and in other cases it holds approximately within a limited composition interval. Note that by combining the ideal gas relationship $\mu_i = \mu_i^0 + k_B T \ln(P_i/P_i^0)$ with this law, one obtains the chemical potential μ_i given by Equation 4.14. This is a common route used in classical thermodynamics to derive the formula for the chemical potential in an ideal liquid mixture.

If the partial pressure of a substance in the vapor phase is larger than the value predicted by Equation 4.22, one says that there is a *positive deviation* from Raoult's law. An example is a mixture of ethanol and chloroform as illustrated in Figure 4.4. Instead, when the partial pressure is lower than the value from Equation 4.22, there is a *negative deviation* from Raoult's law. Negative deviation is, for example, observed for mixtures of acetone and chloroform (not shown).

The regular solution theory can predict such behaviors qualitatively. Let us regard the case $w < 0$, that is, $2u_{12} > u_{11} + u_{22}$. This implies that the energy is increased in the process $(11) + (22) \to 2(12)$, that is, 11-contacts and 22-contacts are energetically favored

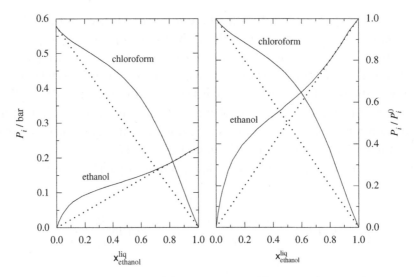

FIGURE 4.4 (a) The partial pressures of the vapor in equilibrium with a liquid mixture of ethanol and chloroform at 45°C for varying compositions.[6] $P_{\text{chloroform}}$ and P_{ethanol} are plotted as functions of the molar fraction of ethanol, $x^{\text{liq}}_{\text{ethanol}}$, of the liquid (we have $x^{\text{liq}}_{\text{chloroform}} = x^{\text{liq}}_{\text{ethanol}} - 1$). The dotted lines show the predictions from Raoult's law. (b) The same data, but plotted as $P_{\text{chloroform}}/P^0_{\text{chloroform}}$ and $P_{\text{chloroform}}/P^0_{\text{chloroform}}$ as functions of $x^{\text{liq}}_{\text{ethanol}}$, where P^0_i is the vapor pressure of pure substance i.

compared to 12-contacts in the liquid, implying that the particles are less "happy" to be mixed with each other in the liquid phase compared to the situation in an ideal mixture. Since 12-contacts cannot be avoided in the mixture, the particles have a larger tendency to leave the liquid phase compared to the ideal case. Thereby, P_1 and P_2 become larger than the prediction from Raoult's law, $P_i > x^{\text{liq}}_i P^0_i$, so there is a positive deviation from Raoult's law. When $w > 0$, there is instead a negative deviation, $P_i < x^{\text{liq}}_i P^0_i$. The partial pressure for each of the two components in a mixture as given by the formula (4.21) is shown in Figure 4.5 for various values of $cw/k_B T$.

Next, we consider the free energy of mixing for a liquid mixture with composition x_1 and x_2 in Equation 4.17

$$\Delta_{\text{mix}}\widehat{A} = \Delta_{\text{mix}}\widehat{U} - T\Delta_{\text{mix}}\widehat{S}$$
$$= -\frac{cw}{2}x_1 x_2 + k_B T \left[x_1 \ln x_1 + x_2 \ln x_2 \right], \qquad (4.23)$$

where $x_2 = 1 - x_1$. The entropy term, which is always negative, favors mixing (negative free energy contribution), while the energy term favors mixing when w is positive and disfavors it when w is negative. When the value of w turns increasingly more negative, one reaches a situation where liquid mixtures with certain compositions are not formed; there is **phase separation**. Such a behavior is predicted by the regular solution theory. To see how, let us investigate $\Delta_{\text{mix}}\widehat{A}/k_B T$ as a function of x_1, which is shown in Figure 4.6a for various negative

---

[6] Figure 4.4 is based on experimental data taken from G. Scatchard and C. L. Raymond, *J. Am. Chem. Soc.*, **60** (1938) 1278.

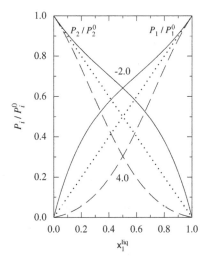

FIGURE 4.5 P_i/P_i^0 for the vapor phase in equilibrium with a binary mixture of two liquids (species 1 and 2) with composition specified by x_1^{liq} as predicted by the regular solution theory for various values of cw/k_BT. The full lines show results for $cw/k_BT = -2.0$ (positive deviation from Raoult's law) and the dashed lines are for $cw/k_BT = 4.0$ (negative deviation). The dotted lines constitute the predictions of Raoult's law ($cw/k_BT = 0$).

values of cw/k_BT. For $cw/k_BT \geq -4$, the curve simply shows a minimum at $x_1 = 0.5$. The second derivative of $\Delta_{mix}\widehat{A}$ with respect to x_1 equals

$$\frac{\partial^2 \Delta_{mix}\widehat{A}}{\partial x_1^2} = cw + \frac{k_BT}{x_1(1-x_1)}.$$

It is positive for all x_1 when $cw/k_BT > -4$, which implies that the curve is concave upwards. For more negative values of cw/k_BT, something else happens. Consider the value of the second derivative at $x_1 = 0.5$, which equals $cw + 4k_BT$. When $cw/k_BT = -4$, the second derivative is zero at $x_1 = 0.5$, and when $cw/k_BT < -4$ it is negative there. In each curve for $cw/k_BT < -4.0$ in Figure 4.6a, there is a "bump" in the middle (the part of the curve between the filled circles). The dotted straight line is a tangent to the curve at both points indicated by the filled circles (i.e., the line that is the common tangent at these points). The system can lower its free energy by splitting into two phases and $\Delta_{mix}\widehat{A}$ of the coexisting phases is, in fact, given by the dotted curve between the filled circles. These circles give the compositions of the two phases. These facts can be understood from the following reasoning in the shaded text box.

Consider two phases with compositions x_1' and x_1'' given by the filled circles and with free energy values $\Delta_{mix}\widehat{A}'$ and $\Delta_{mix}\widehat{A}''$, respectively. Let us have N_1' and N_2' particles of either species in the first phase and N_1'' and N_2'' in the second. The free energy of mixing $\Delta_{mix}A$ (without hat) is $(N_1' + N_2')\Delta_{mix}\widehat{A}'$ for the first phase and $(N_1'' + N_2'')\Delta_{mix}\widehat{A}''$

for the second. The sum of these two numbers is the total free energy of mixing when the two phases are formed side by side from the pure components. Counted per total number of particles in the two phases, we have

$$\Delta_{\mathrm{mix}}\widehat{A} = \frac{(N_1' + N_2')\Delta_{\mathrm{mix}}\widehat{A}' + (N_1'' + N_2'')\Delta_{\mathrm{mix}}\widehat{A}''}{N_1' + N_2' + N_1'' + N_2''}$$

$$= X'\Delta_{\mathrm{mix}}\widehat{A}' + X''\Delta_{\mathrm{mix}}\widehat{A}'',$$

where $X' = (N_1' + N_2')/(N_1' + N_2' + N_1'' + N_2'')$ is the fraction of all particles that belongs to the first phase and $X'' = 1 - X'$ is the fraction that belongs to the second. The average composition of the two phases together is

$$x_1 = \frac{N_1' + N_1''}{N_1' + N_2' + N_1'' + N_2''}$$

$$= \frac{x_1'[N_1' + N_2'] + x_1''[N_1'' + N_2'']}{N_1' + N_2' + N_1'' + N_2''} = x_1'X' + x_1''X''$$

and likewise for $x_2 = 1 - x_1$. When $X' = 1$ there is only the first phase present and when $X' = 0$ there is only the second phase. For intermediate X' values, the mean composition x_1 varies linearly with X' and so does $\Delta_{\mathrm{mix}}\widehat{A}$. This means that the straight line between the filled circles in the figure gives $\Delta_{\mathrm{mix}}\widehat{A}$ for the two phases existing side by side. Thus, $\Delta_{\mathrm{mix}}\widehat{A}$ for the phase-separated system (dotted line) is lower than for the mixture (the part of the curve between the dots), so the mixture will phase separate into two phases with compositions x_1' and x_1''.[7]

Note that the arguments would be true even if the line were not horizontal. In a plot of $\Delta_{\mathrm{mix}}\widehat{A}$ as a function of x_1 (or x_2), a straight line that is the common tangent to the curve at two points hence represents $\Delta_{\mathrm{mix}}\widehat{A}$ for the phase-separated system (provided that the tangent lies below the $\Delta_{\mathrm{mix}}\widehat{A}$ curve). Later, in Figure 4.8, we will see an example of this.

If we have a liquid mixture with a negative w and decrease the temperature, the mixture will phase separate for some compositions since cw/k_BT becomes more negative with decreasing temperature (provided that the system remains liquid at sufficiently low temperatures). In Figure 4.6b, the compositions of the two phase-separated phases at each value of the parameter $cw/k_BT \geq -4$ are plotted (that is, the curve which corresponds to the filled circles in Figure 4.6a). The ordinate in the figure is $[c|w|/k_BT]^{-1}$, which is proportional to temperature. Thus, the figure shows the composition/temperature **phase diagram** of the

---

[7] The free energies of the pure substances vanish in the difference of $\Delta_{\mathrm{mix}}A$ between the mixed and the phase-separated systems and do not contribute.

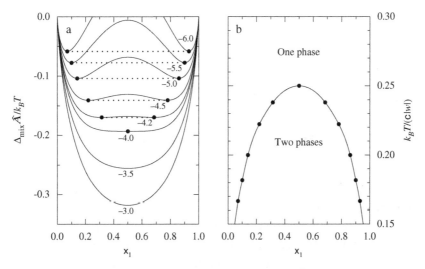

FIGURE 4.6   (a) The regular solution theory predictions of $\Delta_{\mathrm{mix}}\widehat{A}/k_B T$ as a function of the mole fraction x_1 for a two-component mixture at various negative values w. The values of $cw/k_B T$ are shown next to each curve ($cw/k_B T = -3.0, -3.5, -4.0, -4.2, -4.5, -5.0, -5.5,$ and -6.0 counted from bottom to top). For $cw/k_B T \geq -4.0$, there is a minimum in the curve at $x_1 = 0.5$ while for $cw/k_B T < -4.0$ there is a maximum. The dotted straight lines are tangents of each curve at the two points shown by filled circles. (b) The phase diagram for the mixture when w is negative. The ordinate is proportional to temperature. The filled circles correspond to the filled circles in (a). The topmost circle shows the **critical point**, which gives the composition and temperature where phase separation starts to occur when T is decreased.

mixture and the curve shows the phase boundary of the two-phase region for various temperatures. For (x_1, T) values above the curve, the mixture exists as a single phase, while it is split into two phases for (x_1, T) values below the curve.

The highest temperature where phase separation occurs is the **critical temperature**, T^{crit}, of the mixture; in the current theory, this occurs when $cw/k_B T = -4$ (in the figure at ordinate $k_B T/(|c|w) = 1/4$) so we have $T^{\mathrm{crit}} = -cw/(4k_B)$. The composition where phase separation starts to occur is $x_1 = 0.5$, that is, the **critical composition** x_1^{crit}. The value 0.5 for x_1^{crit} is a peculiarity of the current theory and originates from the simplifying assumption that both substances occupy lattice sites of equal cell volumes. In reality, the critical composition can vary quite a lot between different mixtures and is usually not equal to 0.5. In Section 4.2.3.3, we will see examples where the molecules differ in size and where the critical composition varies depending on the sizes.

When T is decreased below the critical temperature, the compositions of the two phases vary (in the current case, as shown by the curve in Figure 4.6b). Immediately below the critical point both phases have x_1 values close to the critical composition, whereafter one phase increases its value of x_1 and the other phase decreases its value.

In regular solution theory and other mean field theories, the phase boundary around the critical point is shaped like a parabola. In reality the boundary is much flatter there and to obtain the correct shape, the phase separation must be treated by more sophisticated theories, which we will not pursue here.

Regular solution theory can be applied to more cases by generalizing it in the following manner. Let us consider the parameter w. Since it is defined as $w = u_{11} + u_{22} - 2u_{12}$, it gives the change in energy when two 12 bonds are replaced by one 11 and one 22 bond

$$
\begin{array}{ccc}
② & ② & ② \quad ② \\
② \; ❶ \; ② \;+\; ② \; ❶ \; ② \;\rightarrow\; ② \; ❶ \; ❶ \; ② \;+\; ② \;\; ② \,. \\
② & ② & ② \quad ②
\end{array}
\tag{4.24}
$$

This illustration can also be regarded as what happens when two molecules of species 1 approach each other in the mixture and thereby lose one contact each with species 2.[8] In our treatment of mixtures, we have so far assumed that w is constant. Considering that w is a combination of the interaction energies between the molecules on the lattice sites, this is quite reasonable. However, the regular solution theory for mixtures can be applied to an even larger number of interesting problems if w is allowed to depend, for instance, on the temperature $w = w(T)$. Thereby, we have generalized w to be an "effective" interaction that includes other effects than "pure" interactional energy when two molecules approach each other. It may, for example, include contributions due to changes in freedom of orientation when molecules interact with other molecules. Therefore, w may contain entropic as well as energetic effects, that is, w is a free energy quantity. The interpretation of w in Equation 4.24 then incorporates these other changes. We have thereby assumed that the possible states of the interacting nearest-neighbor pairs are independent of each other, which is a crude approximation.[9] This kind of approach to the behavior of mixtures is in many respects phenomenological, but as we will see it gives some interesting insights. (There exist much more elaborate manners to treat orientational degrees of freedom in lattice models, but we will not include such theories here.)

We can write

$$
w(T) = u_w - Ts_w,
$$

where

$$
s_w = -\frac{dw}{dT}
$$

and

$$
u_w = \frac{d(w/T)}{d(1/T)} = -T^2 \frac{d(w/T)}{dT}.
$$

The entities u_w and s_w can can also be temperature dependent, $u_w = u_w(T)$ and $s_w = s_w(T)$, so we have a kind of "heat capacity"

$$
c_w = \frac{du_w}{dT} = T\frac{ds_w}{dT}
$$

---

[8] The corresponding argument applies when two molecules of species 2 approach each other and lose one contact each with species 1.

[9] This approximation is in some respects similar to the so-called superposition approximation described in Section 5.12.

that can also depend on temperature. With a temperature dependent w, the expressions for the entropy and energy of mixing, as obtained from differentiation of $\Delta_{mix}\widehat{A}$ in Equation 4.23, equal

$$\Delta_{mix}\widehat{S} = -k_B\left[x_1 \ln x_1 + x_2 \ln x_2\right] - \frac{cs_w}{2}x_1x_2$$

and

$$\Delta_{mix}\widehat{U} = -\frac{cu_w}{2}x_1x_2.$$

Apparently, the entropy no longer is equal to that of an ideal mixture.

We saw above that phase separations can only occur at *decreasing* temperatures when w is temperature independent. This is a consequence of the fact that $w/k_BT < 0$ has to be sufficiently negative (i.e., $-w/k_BT$ sufficiently large) for the phase separation to take place. However, if w is temperature dependent, there are more possibilities. If the T-dependence of w is such that $-w(T)/k_BT > 0$ becomes larger despite that T in the denominator increases, phase separation can occur when the temperature *increases*. In such a case, the system is miscible at fairly low temperatures but phase separates at higher T. One condition for this to happen apparently is that $-w(T)/k_BT$ increases. Thus,

$$\frac{d(-w/T)}{dT} > 0$$

which means that $u_w > 0$. However, for phase separation to be possible at all, it is required that $w = u_w - Ts_w < 0$. Since $u_w > 0$, it follows that

$$Ts_w > u_w > 0,$$

which implies that w *is dominated by the entropy contribution*. We conclude that a phase separation at increasing temperature is only possible when w contains a positive (repulsive) energy contribution u_w which is dominated by an entropy contribution $-Ts_w$ that is attractive. This condition is necessary but not sufficient. One also has to require that the condition $Ts_w > u_w > 0$ is maintained at the increasing T until w/k_BT has reached its critical value, that is, that c_w is not too negative.

Phase separations at increasing T are fairly uncommon, which is not too surprising considering the strict requirements that have to be fulfilled for this to happen. Some cases are known in systems where one of the components is strongly hydrogen bonding, for example, water, glycol and glycerol, and in such systems, large entropy effects can occur when bonds are formed and broken. The presence of hydrogen bonds affects, for example, possible orientations of the molecules and gives entropic contributions to the interaction free energy. Some examples of phase diagrams of this kind are presented in Figure 4.7. Other examples are nicotine-water and polyethylene glycol-water mixtures. Phase separation occurs in these systems when the temperature lies between two values, which are at the top and the bottom of the two-phase region, the "loop" shown in the figure. For compositions within the loop, the system forms two phases. There are two critical points, one at the top point and one at the bottom point of the loop. The upper and lower critical temperatures are often called the **upper critical solution temperature** (UCST) and the **lower critical solution temperature**

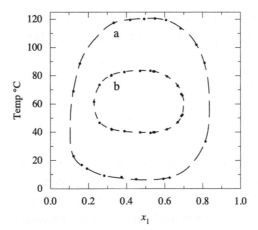

FIGURE 4.7 Examples of phase diagrams of the so-called "closed loop" type: (a) mixtures of glycerol + m-toluidine (i.e., 1-methyl-3-amino-benzene) and (b) mixtures of glycerol + guaiacol (i.e., 1-hydroxy-2-methoxy-benzene). Inside the loop, there are two phases with different compositions in equilibrium with each other.[10]

(LCST), respectively. Note that the experimental phase boundary is very flat close to each critical point. This is in contrast to the regular solution theory prediction as mentioned earlier.

These systems behave more "normally" at high temperatures, that is, the phase separation ceases when the temperature is increased (when the UCST is passed). This is most likely a consequence of the fact that the hydrogen bonds are not important at high temperatures – most of them are broken due to thermal motion. Therefore, s_w decreases and becomes small at increasing T and because of that u_w can eventually dominate. In order to reproduce these phase diagrams *quantitatively*, it is necessary to use much more refined theories than the regular solution theory. However, the *qualitative* behavior can be obtained in this simple theory and, as we have seen, we can thereby gain insight into some aspects of the mechanisms.

In this connection, it is of interest to make some further comments about the illustration of the meaning of $w = u_{11} + u_{22} - 2u_{12}$ given in Equation 4.24. Let subscript 1 denote the solute and 2 the solvent. When two molecules of the solute are brought next to each other, they will interact with energy u_{11} (which can be a free energy, w_{11}, when w is T dependent). However, for this to happen, two solvent molecules (one next to each solute molecule) have to be removed and, as a result, a 22 contact arises in addition to the 11 contact. If $w < 0$, the (free) energy of interaction will decrease when the two solute molecules are brought together, that is, there is an attraction between them. The existence of this attraction does not necessarily imply that u_{11} is attractive; the negative value of w may as well be a consequence of a large negative value of $u_{22} - 2u_{12}$. The negative value of the latter can be of energetic or entropic origin (the latter when w is T dependent). This constitutes a simple example of a

---

[10] Figure 4.7 is based on experimental data taken from R. R. Parvatiker and B. C. McEwen, *J. Chem. Soc., Trans.* **125** (1924) 1484 and B. C. McEwen, *ibid* **123** (1923) 2284.

solvent effect where the process is not driven by the direct 11 interaction but by the change in solvent coordination around the solutes (solvent molecules leaving the neighborhood of the two solute molecules and moving into the surrounding solution).

There is a similarity between this kind of effect and hydrophobic interactions between small nonpolar molecules in aqueous solution. The direct interaction between such molecules is relatively weak, but they are still strongly attracted to each other in aqueous systems. This can be explained essentially as in the previous case. When water molecules are located next to a nonpolar molecule, the free energy is larger than when the water molecules are only surrounded by other water molecules. This increase in free energy is of entropic origin; the orientational freedom of the water molecules near the nonpolar molecule is decreased since they cannot be hydrogen bonded to it but only to other water molecules in the neighborhood – the water molecules thus become constrained. The detailed causes are quite complex, but the final result is that the free energy is increased for the water around the solute due to the entropy decrease.[11] When two nonpolar molecules are brought together, the constrained water is released, the entropy increases and the free energy decreases, which leads to the attraction.

4.2.3.3 Flory-Huggins Theory for Polymer Solutions

An important kind of system that can be approximately treated by lattice theories are polymer solutions and polymer mixtures. Here we will present the basics of the **Flory-Huggins theory** for such systems,[12] which is an approximation on the same level as the regular solution theory for small molecules. Each polymer molecule consists of several polymer units, which we will call "monomers," where each unit is placed on a lattice site. Such a unit is not necessarily equal to the repeat unit (the usual monomer) of the polymer molecule. Each polymer consisting of ℓ monomers (i.e., our units) will occupy a sequence of ℓ sites on the lattice, where consecutive monomers occupy neighboring sites. For a polymer solution, the sites that are not occupied by the polymer molecules are filled with solvent molecules (one molecule per site).

We have seen in Section 4.2.1 that the number of different ways to distribute N_1' distinguishable particles on N_S sites equals $N_S \times (N_S - 1) \times (N_S - 2) \times \cdots \times (N_S - [N_1' - 1]) = N_S!/(N_S - N_1')!$. This result was obtained by placing the particles one by one on the lattice on any sites that are vacant at that instance. Then we corrected for the fact that the particles are indistinguishable by dividing by $N_1'!$ and obtained the number of configurations

$$\Omega^{\text{conf}} = \frac{N_S!}{N_1'! \, (N_S - N_1')!}. \tag{4.25}$$

---

[11] This is valid mainly for small nonpolar molecules like noble gas atoms and short hydrocarbons. The energy for the water around the solute is also changed but this is less important than the change in entropy. However, for long hydrocarbons, the energy contributes to a quite large extent to the increase in free energy.

[12] The Flory-Huggins theory can also be formulated without reference to any lattice model, but here we will use such a model.

We will now investigate how many configurations N_1 polymers with ℓ monomers each can have on N_S sites. Thereby, we place one monomer at a time on the lattice. The first monomer of each polymer can be placed on any vacant site, but each of the following monomers must be placed as a nearest neighbor to the previously placed monomer until all ℓ monomers have been entered. Thereby, we must avoid placing a monomer on a site that is already occupied. A certain fraction of the nearest neighbor sites is already occupied and an attempt to place a monomer at such a site should not be counted. At each instance when we place a monomer as a nearest neighbor, we estimate the fraction of occupied sites by assuming as an approximation that the monomers of the polymers already present on the lattice are randomly distributed. This is in the spirit of regular solution theory and is a fair approximation unless the polymer concentration is very low.[13]

The details of the derivation of an expression for Ω^{conf} for the polymer solution can be found in the shaded text box below. The end result is

$$\Omega^{conf} = \frac{N_S! \, \tau_{conn}^{N_1}}{N_1! \, (N_S - \ell N_1)!}, \tag{4.26}$$

where

$$\tau_{conn} = \frac{c(c-1)^{\ell-2}}{N_S^{\ell-1}}.$$

Apart from the factor $\tau_{conn}^{N_1}$, this expression for Ω^{conf} differs very little from Equation 4.25. The latter equation contains $(N_S - N_1')!$, where $N_S - N_1'$ is the number of vacant sites when N_1' unconnected monomers occupy N_1' sites. When the monomers are connected into N_1 polymer chains of length ℓ, there are still $N_1' = \ell N_1$ occupied sites and therefore $N_S - \ell N_1$ vacant sites. This is the reason why $(N_S - \ell N_1)!$ is present in Equation 4.26. The factor $N_1'!$ in the denominator of Equation 4.25 is present because the N_1' unconnected monomers are indistinguishable, while for N_1 indistinguishable polymers the corresponding factor is $N_1!$ in Equation 4.26. The factor $\tau_{conn}^{N_1}$ contains the main effect of connecting the monomers into polymers; each polymer molecule thereby contributes with a factor τ_{conn} as we will see in the text box.

To find the number of configurations Ω^{conf}, consider the situation somewhat along the way in the process of entering molecules and let us assume that M sites are already occupied when the first monomer of a polymer is entered. This monomer can be placed on any of the $N_S - M$ empty sites. The second monomer in the polymer can be placed on any of the c nearest neighbor sites, provided the site is unoccupied. The M sites that are occupied by the previous polymers are distributed over $N_S - 1$ sites (the site occupied by the first monomer of the present polymer is unavailable), so the fraction of

---

[13] At very low concentrations, the individual polymers are far apart on average. Monomers belonging to different polymers are therefore very distant from each other, while all monomers belonging to the same polymer are located not very far from each other. Then a random distribution of *all* monomers over the whole system gives a very poor estimate of the density of monomers near each monomer and the Flory-Huggins theory is a poor approximation.

occupied sites is $M/(N_S - 1)$ and the fraction $1 - M/(N_S - 1)$ of the sites are empty. Assuming random distribution, this fraction also applies to the c nearest neighbor sites, that is, there are $[1 - M/(N_S - 1)]c = (N_S - 1 - M)c/(N_S - 1)$ empty sites where the second monomer can be placed. This is displayed in the row marked 2 in Table 4.1.

TABLE 4.1 Relevant entities when monomers 1 to ℓ of a polymer are placed on a lattice.

	Number of Sites to Try	Fraction Empty	Number of Empty Sites Where the Monomer Can Be Placed
1	$N_S - M$	1	$N_S - M$
2	c	$1 - \frac{M}{N_S - 1}$	$\left(1 - \frac{M}{N_S - 1}\right)c = (N_S - 1 - M)\frac{c}{N_S - 1}$
3	$c - 1$	$1 - \frac{M}{N_S - 2}$	$\left(1 - \frac{M}{N_S - 2}\right)(c - 1) = (N_S - 2 - M)\frac{c - 1}{N_S - 2}$
4	$c - 1$	$1 - \frac{M}{N_S - 3}$	$\left(1 - \frac{M}{N_S - 3}\right)(c - 1) = (N_S - 3 - M)\frac{c - 1}{N_S - 3}$
$\vdots$	$\vdots$	$\vdots$	$\vdots$
ℓ	$c - 1$	$1 - \frac{M}{N_S - \ell + 1}$	$\left(1 - \frac{M}{N_S - \ell + 1}\right)(c - 1) = (N_S - \ell + 1 - M)\frac{c - 1}{N_S - \ell + 1}$

The third monomer in the polymer can be placed on any of the $c - 1$ nearest neighbor sites (one neighboring site is occupied by the first monomer), provided the site is unoccupied. We know that the M sites occupied by the previous polymers are located on $N_S - 2$ sites (two sites are occupied by the two first monomers of the current polymer), so the fraction of occupied sites among those is $M/(N_S - 2)$. Again, assuming random distribution, the fraction $1 - M/(N_S - 2)$ of the $c - 1$ nearest neighbor sites are empty. We can therefore place the third monomer on any of these $[1 - M/(N_S - 2)](c - 1) = (N_S - 2 - M)(c - 1)/(N_S - 2)$ empty sites, as shown in the row marked 3 in Table 4.1. The same applies for the fourth monomer, except that there are $N_S - 3$ sites that are not occupied by the previous monomers of the current polymer. The fourth monomer can thus be placed on any of $(N_S - 3 - M)(c - 1)/(N_S - 3)$ sites, as shown in the row marked 4 in the table. We have thereby neglected the possibility that one of the $c - 1$ sites happens to be the site where we placed the first monomer, which would occur if the new polymer folds back on itself. Considering that our treatment of the polymers is a rough approximation where overlaps with the other polymers are estimated by assuming random distribution, it would be futile to improve the approximation at this point only. For concentrated solutions of long polymers, the majority of possible overlaps occurs with other polymer chains. We therefore continue in the same fashion and place all ℓ monomers of the polymer in the same manner, neglecting possible overlaps of the polymer with itself; the details for the ℓth monomer are given by the corresponding row in the table.

The total number of possibilities for the whole polymer is given by the product of the entries in the right-hand column in the table. In line l of this column for $2 \leq l \leq \ell$, the denominator $N_S - l + 1$ can be replaced by N_S since N_S is much larger than ℓ. We thereby obtain the number of different ways to distribute the polymer on the lattice as $(N_S - M) \times (N_S - M - 1) \times \cdots \times (N_S - M - \ell + 1)c(c - 1)^{\ell - 2}/N_S^{\ell - 1}$.

Let us compare this with the situation where ℓ monomers are individually placed without being connected to each other. The first monomer can be placed on any of $N_S - M$ sites, the second on any of the remaining $N_S - M - 1$ empty sites etcetera until monomer ℓ that can be placed on any of $(N_S - M - \ell + 1)$ sites. Thus, it is the factor $c(c-1)^{\ell-2}/N_S^{\ell-1} = \tau_{\text{conn}}$ that accounts for the difference in number of possible configurations for the monomers in a polymer compared to ℓ unconnected monomers. This factor arises because the ℓ monomers are connected to each other in a chain. Note that this factor is the same for all polymers of length ℓ.

For the case of $N_1' = \ell N_1$ unconnected monomers, Equation 4.25 is valid. To obtain the total number of configurations for N_1 polymers, we first replace $N_1'!$ in the denominator with $N_1!$ (to account for the fact that the N_1 polymer molecules are indistinguishable). Furthermore, in the last factor in the denominator, we insert $N_1' = \ell N_1$ and multiply by the factor $(\tau_{\text{conn}})^{N_1}$ to account for the fact that each of the N_1 sets of ℓ monomers forms a polymer. As a result, we get

$$\Omega^{\text{conf}} = \frac{N_S!}{N_1! \, (N_S - \ell N_1)!} \left[\frac{c(c-1)^{\ell-2}}{N_S^{\ell-1}} \right]^{N_1}, \tag{4.27}$$

which is the same as Equation 4.26. The solvent molecules (species 2) occupy the remaining $N_2 = N_S - \ell N_1$ sites in the lattice and they contribute with a factor of 1 to Ω^{conf} since there is only one possible way to place them on the lattice sites when the polymers are already present.

By entering $N_S = \ell N_1 + N_2$ in Equation 4.27, we have

$$\Omega^{\text{conf}} = \frac{(\ell N_1 + N_2)!}{N_1! \, N_2!} \left[\frac{c(c-1)^{\ell-2}}{(\ell N_1 + N_2)^{\ell-1}} \right]^{N_1}$$

for a polymer solution. For a pure polymer phase, where $N_S = \ell N_1$, we have

$$\Omega^{\text{conf}} = \frac{(\ell N_1)!}{N_1!} \left[\frac{c(c-1)^{\ell-2}}{(\ell N_1)^{\ell-1}} \right]^{N_1} \quad \text{(pure polymer)}$$

and for a pure solvent phase $\Omega^{\text{conf}} = 1$. The entropy of mixing is

$$\Delta_{\text{mix}} S = S(\text{mixture}) - S(\text{pure polymer}) - S(\text{pure solvent})$$

$$= k_B \ln \left[\frac{\Omega^{\text{conf}}(\text{mixture})}{\Omega^{\text{conf}}(\text{pure polymer}) \Omega^{\text{conf}}(\text{pure solvent})} \right]$$

$$= k_B \ln \left(\frac{(\ell N_1 + N_2)!}{(\ell N_1)! N_2!} \left[\frac{(\ell N_1)^{\ell-1}}{(\ell N_1 + N_2)^{\ell-1}} \right]^{N_1} \right),$$

as obtained by inserting the expressions for Ω^{conf} above and simplifying the result. Compared to a system with $N_1' = \ell N_1$ unconnected monomers and N_2 solvent molecules (where Ω^{conf} for both pure phases equals 1), the expression for $\Delta_{mix}S$ differs by the factor with square bracket in the logarithm (cf. Equation 4.6). The contribution to $\Delta_{mix}S$ due to the polymerization of the monomers into polymers of length ℓ is thus equal to

$$\text{polymerization contribution} = k_B N_1 \ln \left[\frac{(\ell N_1)^{\ell-1}}{(\ell N_1 + N_2)^{\ell-1}} \right]$$

$$= k_B(\ell N_1 - N_1) \ln \frac{\ell N_1}{\ell N_1 + N_2}.$$

For the unconnected case, Equation 4.8 yields

$$\Delta_{mix}S = -k_B \left[\ell N_1 \ln \frac{\ell N_1}{\ell N_1 + N_2} + N_2 \ln \frac{N_2}{\ell N_1 + N_2} \right] \quad \text{(unconnected)}$$

and by adding the polymerization contribution, we finally obtain

$$\Delta_{mix}S = -k_B \left[N_1 \ln \frac{\ell N_1}{\ell N_1 + N_2} + N_2 \ln \frac{N_2}{\ell N_1 + N_2} \right]$$

for a polymer solution. Note that in contrast to Equation 4.8, mole fractions do not occur in this expression when $\ell > 1$. The meaning of the arguments of the logarithms can be realized by noting that if v_S is the volume of a lattice site, the volume occupied by the polymers is $\ell N_1 v_S$, the volume of the solvent is $N_2 v_S$ and the total volume is $(\ell N_1 + N_2)v_S$. We therefore have the *volume fraction* φ_i of species i

$$\varphi_i = \frac{\ell_i N_i v_S}{(\ell_1 N_1 + \ell_2 N_2)v_S} = \frac{\ell_i N_i}{\ell_1 N_1 + \ell_2 N_2},$$

where $\ell_1 = \ell$ and $\ell_2 = 1$ in the present case. Thus, we can write

$$\Delta_{mix}S = -k_B \left[N_1 \ln \varphi_1 + N_2 \ln \varphi_2 \right]. \tag{4.28}$$

Note that this entropy originates only from the configurations (including conformations) of the molecules; it is a *configurational entropy*. Since $\varphi_i \leq 1$, the square bracket is always negative, so the configurational entropy of mixing is always positive.

Equation 4.28 holds also for a blend of two polymers (without solvent), where species 1 has ℓ_1 monomers, species 2 has ℓ_2 monomers and $N_S = \ell_1 N_1 + \ell_2 N_2$. This fact can be realized by entering polymer species 2 monomer by monomer in the same fashion as before when all polymers of species 1 have been entered in the lattice. Thereby, one obtains the same kind of "polymerization factor" for species 2 and Equation 4.27 is replaced by

$$\Omega^{conf} = \frac{N_S!}{N_1! N_2!} \left[\frac{c(c-1)^{\ell_1-2}}{N_S^{\ell_1-1}} \right]^{N_1} \left[\frac{c(c-1)^{\ell_2-2}}{N_S^{\ell_2-1}} \right]^{N_2},$$

where $N_2!$ in the denominator accounts for the indistinguishability of the N_2 molecules of species 2. Note that $(N_S - \ell N_1)!\,)$ in Equation 4.27 is replaced by $(N_S - \ell_1 N_1 - \ell_2 N_2)!$, which is equal to 1.

{**EXERCISE 4.2**
Fill in the details in the proof of Equation 4.28 for a blend of two polymers.}

Let us now turn to $\Delta_{\text{mix}} U$. We use the same kind of random mixing as done previously in regular solution theory and in the calculation of $\Delta_{\text{mix}} S$. For the case of a polymer solution, the interaction energy between two monomers that are nearest neighbors (but not covalently bonded to each other) is denoted by u_{11}, the interaction energy between two solvent molecules is u_{22} and between a monomer and a solvent molecule is u_{12} – in all cases provided they are nearest neighbors. For a blend of two polymers, the interaction energies u_{22} and u_{12} are instead between two monomers (nearest neighbors; not covalently bonded to each other). For simplicity, we will use the notation "unit" for both solvent molecules and monomers in what follows – a unit is the entity that can occupy a site in the lattice. Assuming random mixing, the probability that any site is occupied by a unit of species i is equal to φ_i.

Consider a site occupied by a monomer in a polymer of species j. Neglecting end-of-chain effects, there are $c - 2$ nearest neighboring sites where other units can be present (not covalently bonded). On average, we have $(c - 2)\varphi_i$ units of species i being nearest neighbors to the monomer. Thus, the average number of ij bonds is

$$\bar{N}_{ij} = \begin{cases} c'\varphi_i \ell_j N_j, & i \neq j \\ c'\varphi_i \ell_j N_j/2, & i = j, \end{cases} \tag{4.29}$$

where $c' = c - 2$ and we have multiplied by $1/2$ in the last line to avoid double counting. For a solvent molecule (species 2), we have instead on average $c\varphi_2$ other solvent molecules as nearest neighbors and $\bar{N}_{22} = c\varphi_2 \ell_2 N_2/2$, where $\ell_2 = 1$. For simplicity, we make the further approximation and replace c with $c' = c - 2$ in this case, so Equation 4.29 is valid for this case too (alternatively, we may as an approximation replace c' with c in the latter equation; this is frequently done). The coordination number is not so well determined in practice since the actual liquid system is just modeled using lattice as a rough approximation.

In the approximation we use, we have $U = u_{11}\bar{N}_{11} + u_{22}\bar{N}_{22} + u_{12}\bar{N}_{12}$. Using Equation 4.29 and the fact that $\varphi_i = 1$ for pure substances, we obtain

$$\Delta_{\text{mix}} U = u_{11}\frac{c'\varphi_1 \ell_1 N_1}{2} + u_{22}\frac{c'\varphi_2 \ell_2 N_2}{2} + u_{12}c'\varphi_1 \ell_2 N_2$$

$$- u_{11}\frac{c'\ell_1 N_1}{2} - u_{22}\frac{c'\ell_2 N_2}{2}$$

$$= \frac{c'}{2}(-u_{11} - u_{22} + 2u_{12})N_S\varphi_1\varphi_2$$

since $(1 - \varphi_1)\ell_1 N_1$, $(1 - \varphi_2)\ell_2 N_2$ and $\varphi_1 \ell_2 N_2$ are all equal to $N_S \varphi_1 \varphi_2$. Thus, we have

$$\frac{\Delta_{\mathrm{mix}} U}{k_B T} = -\frac{c'w}{2k_B T} N_S \varphi_1 \varphi_2 = \chi_{\mathrm{FH}} N_S \varphi_1 \varphi_2, \tag{4.30}$$

where $w = u_{11} + u_{22} - 2u_{12}$ as usual and where we have introduced the parameter

$$\chi_{\mathrm{FH}} = -\frac{c'w}{2k_B T}$$

called the **Flory-Huggins interaction parameter**. It depends on temperature even in cases where w is temperature independent, which we assume is the case here.

We can now obtain the free energy of mixing

$$\begin{aligned}
\frac{\Delta_{\mathrm{mix}} A}{k_B T} &= \frac{\Delta_{\mathrm{mix}} U}{k_B T} - \frac{\Delta_{\mathrm{mix}} S}{k_B} \\
&= \chi_{\mathrm{FH}} N_S \varphi_1 \varphi_2 + N_1 \ln \varphi_1 + N_2 \ln \varphi_2 \\
&= N_S \left[\chi_{\mathrm{FH}} \varphi_1 \varphi_2 + \frac{\varphi_1}{\ell_1} \ln \varphi_1 + \frac{\varphi_2}{\ell_2} \ln \varphi_2 \right]
\end{aligned} \tag{4.31}$$

(compare the second line with Equation 4.16). Remember that $N_S = \ell_1 N_1 + \ell_2 N_2$ and that $\ell_2 = 1$ when species 2 is a simple solvent. From Equation 4.31, we obtain the chemical potentials

$$\begin{aligned}
\mu_1 - \mu_1^0 &= \left(\frac{\partial \Delta_{\mathrm{mix}} A}{\partial N_1} \right)_{N_2, T} = k_B T \left[\chi_{\mathrm{FH}} \ell_1 \varphi_2^2 + \ln \varphi_1 + \left(1 - \frac{\ell_1}{\ell_2} \right) \varphi_2 \right] \\
\mu_2 - \mu_2^0 &= \left(\frac{\partial \Delta_{\mathrm{mix}} A}{\partial N_2} \right)_{N_1, T} = k_B T \left[\chi_{\mathrm{FH}} \ell_2 \varphi_1^2 + \ln \varphi_2 + \left(1 - \frac{\ell_2}{\ell_1} \right) \varphi_1 \right].
\end{aligned} \tag{4.32}$$

We can write $\Delta_{\mathrm{mix}} A(\varphi_1)$, $\mu_1(\varphi_1)$ and $\mu_2(\varphi_1)$ explicitly as functions of φ_1 by inserting $\varphi_2 = 1 - \varphi_1$ in these expressions and they can be written as functions of mole fraction by inserting $\varphi_j = x_j \ell_j / (x_1 \ell_1 + x_2 \ell_2)$.

{EXERCISE 4.3
Show Equation 4.32 starting from Equation 4.31.}

The contribution to $\Delta_{\mathrm{mix}} A$ from the configurational entropy (given by Equation 4.28) is negative, so it always favors mixing. The contribution from $\Delta_{\mathrm{mix}} U$ (Equation 4.30) is negative when $\chi_{\mathrm{FH}} < 0$, that is, when $w > 0$ and hence $2u_{12} < u_{11} + u_{22}$. Then 12-contacts are energetically favored compared to 11-contacts and and 22-contacts in the mixture, so the molecules "like" to be mixed, whereas for $\chi_{\mathrm{FH}} > 0$, on the other hand, the 11-contacts and 22-contacts are favored compared to 12-contacts and the molecules "dislike" to be mixed. For polymer solutions, the first situation ($\chi_{\mathrm{FH}} < 0$) is often described as the solvent being a "good" solvent for the polymer, while it is a "poor" solvent in the second case ($\chi_{\mathrm{FH}} > 0$). When the solvent is sufficiently poor (when χ_{FH} is sufficiently large), the solution phase

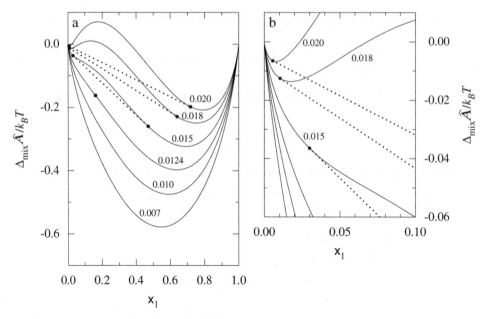

FIGURE 4.8 (a) The free energy of mixing $\Delta_{mix}\widehat{A} = \Delta_{mix}A/(N_1 + N_2)$ as a function of the mole fraction x_1 for a polymer mixture with $\ell_1 = 300$ and $\ell_2 = 100$ at various positive values of χ_{FH} (negative w). The χ_{FH} values are shown next to each curve ($\chi_{FH} = 0.007, 0.010, 0.0124, 0.015, 0.018$ and 0.020 counted from bottom to top). For $\chi_{FH} \leq 0.0124$, the curves are concave upwards, while for $\chi_{FH} > 0.0124$ there is a portion of each curve that is concave downwards. In the latter cases, the dotted straight line is the common tangent of the curve at the two points shown by filled circles. These circles give the composition of the two phases in the two-phase region of the phase diagram. The single circle at the curve for $\chi_{FH} = 0.0124 = (1/\sqrt{\ell_1} + 1/\sqrt{\ell_2})^2/2$ corresponds to the critical point. (b) A magnified view of the same plot for small x_1 values. The left-hand point of tangency for each dotted line is clearly visible in this view.

separates, just like for ordinary solutions when w is sufficiently negative as explored in the previous section on the regular solution theory (remember that χ_{FH} is positive when w is negative).

An example of a system with phase separation is presented in Figure 4.8, which shows $\Delta_{mix}\widehat{A}$ for a polymer mixture as a function of x_1 for some positive values of χ_{FH}. One of the polymers has 300 monomers and the other 100 monomers. For small χ_{FH} values, the $\Delta_{mix}\widehat{A}$ curves are concave upwards and the polymers are completely miscible. When χ_{FH} is sufficiently large, there is a portion of the curve that is concave downwards and, as we did in Figure 4.6, we can draw a straight line that is a common tangent to the curve at two points (dotted line in the figure). The left-hand point of tangency for each dotted line can be seen clearly in Figure 4.8b. The straight line between the two points of tangency lies below the curve and its existence shows that the system can lower its free energy by separating into two phases, as explained in the discussion of Figure 4.6 and in the arguments presented in the shaded text box on page 145f.

The critical point occurs for the χ_{FH} value where the downward concavity starts to appear, that is, there is a value χ_{FH}^{crit} such that downward concavity exists for curves with

$\chi_{FH} > \chi_{FH}^{crit}$, while lower values of χ_{FH} give curves that are solely concave upwards. When $\chi_{FH} > \chi_{FH}^{crit}$, there exist two points where the second derivative of $\Delta_{mix}\widehat{A}(x_1)$ is zero, while for $\chi_{FH} = \chi_{FH}^{crit}$ there is only a single such point. In the figure, this single point with a zero second derivative is shown as the filled circle on the curve for $\chi_{FH} = 0.0124$, which hence corresponds to the critical point. As we will see, the condition for criticality in the Flory-Huggins theory is

$$
\begin{aligned}
\chi_{FH}^{crit} &= \frac{1}{2}\left(\frac{1}{\sqrt{\ell_1}} + \frac{1}{\sqrt{\ell_2}}\right)^2 \\
x_1^{crit} &= \frac{1}{1 + [\ell_1/\ell_2]^{3/2}} \\
\varphi_1^{crit} &= \frac{1}{1 + [\ell_1/\ell_2]^{1/2}},
\end{aligned}
\tag{4.33}
$$

where x_1^{crit} and φ_1^{crit} give the critical composition in mole and volume fractions, respectively. Equation 4.33 is, in fact, the condition for a single point with a zero second derivative of $\Delta_{mix}\widehat{A}(x_1)$, but this condition for criticality can alternatively be found in the following manner:

When there is equilibrium between two phases A and B with different compositions φ_1^A and φ_1^B, the chemical potentials are equal in the two phases

$$
\begin{aligned}
\mu_1(\varphi_1^A) &= \mu_1(\varphi_1^B) \\
\mu_2(\varphi_1^A) &= \mu_2(\varphi_1^B).
\end{aligned}
\tag{4.34}
$$

Let us consider the expression $\mu_j(\varphi_1) = \mu_j^{common}$ for $j = 1, 2$, where μ_j^{common} is the common value of the chemical potential for substance j in the two phases. Equation 4.34 implies that we regard $\mu_j(\varphi_1) = \mu_j^{common}$ as an equation for φ_1; this equation has two solutions $\varphi_1 = \varphi_1^A$ and $\varphi_1 = \varphi_1^B$.[14] At the critical point, the two solutions have merged into a "double solution" $\varphi_1 = \varphi_1^{crit}$ where φ_1^A and φ_1^B are equal (such a solution would correspond to a double root of a polynomial equation). This means that the solution is characterized by

$$
\frac{\partial\mu_j(\varphi_1)}{\partial\varphi_1}\bigg|_{\varphi_1=\varphi_1^{crit}} = \frac{\partial^2\mu_j(\varphi_1)}{\partial\varphi_1^2}\bigg|_{\varphi_1=\varphi_1^{crit}} = 0 \quad \text{for } j = 1, 2
$$

---

[14] When $\chi_{FH} > \chi_{FH}^{crit}$, the equation $\mu_j(\varphi_1) = \mathcal{K}_j$, where $\mathcal{K}_j$ is s constant, has two solutions for various values of $\mathcal{K}_j$, but there is only one pair of solutions φ_1^A and φ_1^B that simultaneously satisfies Equation 4.34 for both μ_1 and μ_2. Our μ_j^{common} is equal to $\mathcal{K}_j$ for this pair of solutions.

for exactly the same reasons as for a double root. Selecting $j = 2$, we find by taking the first and second derivatives of μ_2 in Equation 4.32 that

$$2\chi_{\text{FH}}\,\ell_2\varphi_1 - \frac{1}{1 - \varphi_1} + 1 - \frac{\ell_2}{\ell_1} = 0$$

$$2\chi_{\text{FH}}\,\ell_2 - \frac{1}{(1 - \varphi_1)^2} = 0.$$

By using the last line to eliminate $2\chi_{\text{FH}}\,\ell_2$ in the first line, we obtain a quadratic equation with a solution[15] equal to φ_1^{crit} in Equation 4.33 and by insertion of φ_1^{crit} into the last line, we obtain $\chi_{\text{FH}}^{\text{crit}}$.

{**EXERCISE 4.4**
Show that Equation 4.33 is also obtained from the corresponding derivatives of μ_1 in Equation 4.32 as function of φ_1.}

The composition φ_1^{crit} and x_1^{crit} at the critical point depends, as seen in Equation 4.33, solely on the ratio ℓ_1/ℓ_2 and when $\ell_1 \gg \ell_2$ the critical point occurs for a mixture that is very dilute in species 1. In the two-phase region (for $\chi_{\text{FH}} > \chi_{\text{FH}}^{\text{crit}}$), the phase with a low content of species 1 is even more dilute, $x_1 < x_1^{\text{crit}}$. In a plot like Figure 4.8, the left point of tangency, which gives the composition of this phase, will lie very close to the left ordinate axis in such cases. As we can see in pane b of this figure, $\Delta_{\text{mix}}\widehat{A}(x_1)$ has a minimum close to the point of tangency when χ_{FH} is somewhat larger than $\chi_{\text{FH}}^{\text{crit}}$, see the top two curves. Such a minimum must exist for all larger χ_{FH} because $\Delta_{\text{mix}}\widehat{A}(x_1)$ has an infinite negative derivative at $x_1 = 0$ that originates from the logarithmic term in the entropy contribution. This makes it possible for the equilibrium condition to be satisfied for a nonzero, albeit very small, concentration of species 1.

Let us finally focus on polymer solutions where species 2 is the solvent and we have $\ell_1 = \ell$ and $\ell_2 = 1$. In this case,

$$\varphi_1^{\text{crit}} = 1/(1 + \sqrt{\ell})$$

$$\chi_{\text{FH}}^{\text{crit}} = \tfrac{1}{2}(1 + 1/\sqrt{\ell})^2,$$

so when ℓ is large $\varphi_1^{\text{crit}} \approx 1/\sqrt{\ell}$ and $\chi_{\text{FH}}^{\text{crit}} \approx \tfrac{1}{2}$. For instance, when $\ell = 300$ we have $\varphi_1^{\text{crit}} = 0.0546$ and $\chi_{\text{FH}}^{\text{crit}} = 0.559$ and for $\ell = 30000$ we have $\varphi_1^{\text{crit}} = 0.00574$ and $\chi_{\text{FH}}^{\text{crit}} = 0.506$. The polymer content in the dilute phase is hence very low in the solvent-rich phase for the phase-separated system when ℓ is large.

In the regular solution theory for small molecules of about equal size, we saw that there is a deviation from Raoult's law when $w \neq 0$; the molecules then have a different tendency to escape from the mixture for energetic reasons compared to an ideal mixture where $w = 0$. For polymer solutions, there is also an entropic reason for a deviation from Raoult's law. In

---

[15] The other solution to the quadratic equation gives negative φ_1 or φ_2 and can be ignored.

the vapor phase, we have $\mu_2 - \mu_2^0 = k_B T \ln(P_2/P_2^0)$. At equilibrium, μ_2 is the same in both phases and therefore we obtain using Equation 4.32

$$\frac{P_2}{P_2^0} = \varphi_2^{\text{liq}} \exp\left[\chi_{\text{FH}}(\varphi_1^{\text{liq}})^2 + \left(1 - \frac{1}{\ell}\right) \varphi_1^{\text{liq}} \right], \tag{4.35}$$

where we for clarity have set superscript "liq" on φ_j. Since $\varphi_2^{\text{liq}} = x_2^{\text{liq}}/(x_1^{\text{liq}}\ell + x_2^{\text{liq}})$, we see that $P_i = x_i^{\text{liq}} P_i^0$ is not fulfilled even if $\chi_{\text{FH}} = 0$. In that case, Equation 4.35 can be written

$$\frac{P_2}{P_2^0} = x_2^{\text{liq}} \frac{\exp\left[\frac{\tau_1}{1+\tau_1}\right]}{1 + \tau_1} \quad \text{when } \chi_{\text{FH}} = 0,$$

where $\tau_1 = x_1^{\text{liq}}(\ell - 1)$ and we have used $x_1^{\text{liq}}\ell + x_2^{\text{liq}} = 1 + x_1^{\text{liq}}(\ell - 1)$. It follows that $P_2/P_2^0 < x_2^{\text{liq}}$ when $x_1^{\text{liq}} > 0$ because the function $e^{\tau_1/(1+\tau_1)}/(1 + \tau_1) < 1$ when $\tau_1 > 0$ and decays like $e/(1 + \tau_1)$ for large τ_1. Thus, for polymer solutions, there is a considerable negative deviation from Raoult's law for the solvent when w and hence χ_{FH} are zero.

II

Fluid Structure and Interparticle Interactions

CHAPTER 5

Interaction Potentials and Distribution Functions

To start with, this chapter presents examples of common pair interaction potentials for spherical particles and gives an intuitive introduction to distribution functions in homogeneous bulk fluids like the radial distribution function (pair distribution function), its higher order analogues (triplet and quartet distribution functions) and the relationships between them. For inhomogeneous fluids, where the density varies in space, the density distribution is determined by the forces that act on the fluid particles and the central concept of *potential of mean force* is introduced via a straightforward force balance argument. The potential of mean force describes the density distribution at equilibrium via a simple Boltzmann distribution law and it has a direct relationship to the excess chemical potential, as will be shown. At the next (second) level of the introduction of distribution functions, the relationships between density distributions and pair distributions are described in more detail, and at the third level, the distribution functions of all orders are formally defined from the probability density for particle configurations (the configurational probability density given by Equation 3.9). The concept of structure factor, which is important in scattering experiments, is introduced and its relationship to distribution functions is explained. Such experiments can be used to investigate the structure of liquids. Basic thermodynamic quantities are expressed in terms of distribution functions, whereby important relationships are established that will be central for later chapters. Connections between distribution functions of various orders are explained in more detail and the Born-Green-Yvon equations, which give formal but physically transparent relationships between these functions, are obtained. They will be very useful in later chapters. Some mean field approximations are introduced, which form the basis of commonly used simple approximations. Finally, the chapter contains a section on computer simulations where the basics of Molecular Dynamics and Monte Carlo simulations are explained. Furthermore, various techniques for the determination of thermodynamical quantities and distribution functions (density and pair distributions) by simulation are given for both homogeneous and inhomogeneous fluids.

5.1 BULK FLUIDS OF SPHERICAL PARTICLES. THE RADIAL DISTRIBUTION FUNCTION

Consider a homogeneous bulk fluid composed of particles that interact with a pairwise additive potential, that is, the interaction energy between each pair of particles is independent of the positions of the other particles. The number of particles per unit volume, the **number density**,[1] is denoted by n^b, where superscript b stands for "bulk." We initially restrict ourselves to spherically symmetric particles. The center of particle ν has coordinates $\mathbf{r}_\nu = (x_\nu, y_\nu, z_\nu)$. The interaction potential between each pair of particles, the **pair potential**, is given by $u(\mathbf{r}_1, \mathbf{r}_2) = u(r_{12})$, where $r_{12} = |\mathbf{r}_2 - \mathbf{r}_1|$ is the distance between the particle centers, and it is zero at infinite separation between the particles. When the particles are in a configuration specified by the positions $(\mathbf{r}_1, \mathbf{r}_2, \mathbf{r}_3, \mathbf{r}_4, \ldots)$, the instantaneous total potential energy is given by

$$\check{U}^{\text{pot}} = \sum_\nu \sum_{\substack{\nu' \\ (\nu' > \nu)}} u(\mathbf{r}_\nu, \mathbf{r}_{\nu'}) = \sum_{\nu, \nu' > \nu} u(\mathbf{r}_\nu, \mathbf{r}_{\nu'}),$$

where the sum is over all pairs of particles in the fluid and where we for simplicity have written the two summation symbols as a single one, $\sum_{\nu, \nu' > \nu}$, in the rhs (this will be done from now on). This energy is a function of the particle coordinates, $\check{U}^{\text{pot}} = \check{U}^{\text{pot}}(\mathbf{r}_1, \mathbf{r}_2, \mathbf{r}_3, \mathbf{r}_4, \ldots)$.

Common interaction potentials used for simple modeling of real particles are, for example, the **hard sphere potential**

$$u(r) = \begin{cases} \infty, & r < d^h \\ 0, & r \geq d^h \end{cases} \tag{5.1}$$

where d^h is the diameter of the spheres, the **square well potential**

$$u(r) = \begin{cases} \infty, & r < d^h \\ -\varepsilon^{\text{sw}}, & d^h \leq r < d^{\text{sw}} \\ 0, & r \geq d^{\text{sw}} \end{cases} \tag{5.2}$$

where d^h is the diameter of the hard core of the particles, ε^{sw} the depth of the attractive potential well and $d^{\text{sw}} - d^h$ its width, and the **hard core Yukawa potential**[2]

$$u(r) = \begin{cases} \infty, & r < d^h \\ A^Y \dfrac{e^{-(r-d^h)/l^Y}}{r}, & r \geq d^h \end{cases} \tag{5.3}$$

---

[1] For the general (microscopic) case, we need the following more precise definition: The number density is the number of *particle centers* per unit volume.

[2] The original Yukawa potential $-a \exp(-mr)/r$ used in elementary particle physics was attractive, $a > 0$, but in applications for other systems the potential may be attractive or repulsive.

where l^Y is a decay length and A^Y gives the magnitude of the potential (A^Y may be positive or negative).

For charged hard spheres of diameter d^h and charge q in vacuum, the pair interaction is a **Coulomb potential** (or, more correctly, a Coulomb interaction energy) outside the hard core

$$u(r) = \begin{cases} \infty, & r < d^h \\ \dfrac{q^2}{4\pi\varepsilon_0 r}, & r \geq d^h, \end{cases} \tag{5.4}$$

where ε_0 is the permittivity of vacuum. Sometimes it is useful to write this potential as a sum $u(r) = u^{\text{core}}(r) + u^{\text{el}}(r)$, where

$$u^{\text{core}}(r) = \begin{cases} \infty, & r < d^h \\ 0, & r \geq d^h \end{cases} \tag{5.5}$$

is the same as the hard sphere potential and $u^{\text{el}}(r) = q^2/(4\pi\varepsilon_0 r)$ is the electrostatic part of the pair interaction. If the spheres are placed in a dielectric continuum with a dielectric constant (relative permittivity) ε_r, which is a simple manner to model a solvent, the factor ε_0 in the denominator is replaced by $\varepsilon_0\varepsilon_r$.

So-called "soft spheres" interact with a repulsive potential that becomes progressively larger when r decreases and eventually becomes infinitely large. An example is the simple **soft sphere potential**

$$u(r) = \varepsilon^{\text{ss}}\left(\frac{d^{\text{ss}}}{r}\right)^{\nu},$$

where ν is a positive number (normally an integer). The repulsion becomes sharper and the spheres therefore "harder" when ν is increased. A common model potential which includes dispersion r^{-6} interactions is the **Lennard-Jones potential**

$$u(r) = 4\varepsilon^{\text{LJ}}\left[\left(\frac{d^{\text{LJ}}}{r}\right)^{12} - \left(\frac{d^{\text{LJ}}}{r}\right)^{6}\right], \tag{5.6}$$

where d^{LJ} is the separation where the potential is zero (a measure of the particle diameter) and ε^{LJ} is the depth of the attractive potential well, which occurs at $r = 2^{1/6}d^{\text{LJ}} \approx 1.122d^{\text{LJ}}$. The potential is positive (repulsive) for small r and when r is increased $u(r)$ changes sign at $r = d^{\text{LJ}}$, goes through its minimum value $-\varepsilon^{\text{LJ}}$ and finally decays to zero for large r. A fluid consisting of particles interacting with this pair potential is called a **Lennard-Jones fluid**.

There exist several other simple model interaction potentials for spherical particles, but the ones listed here suffice for now. In general, the particles repel each other at short separations while the interaction for larger separations may be repulsive or attractive or, for some model systems, equal to zero.

Let us select any one of the particles in the fluid and regard the distribution of particles around it; that is, as if we were sitting on the particle and watching the other particles.

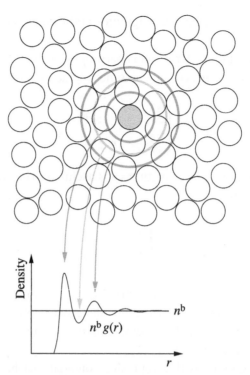

FIGURE 5.1 Sketch showing a snapshot picture of a possible configuration for the particles around a particle (gray) located at the center of the figure. A plot of the average density distribution around this particle is shown in the bottom part of the figure, where n^b is the bulk density and $g(r)$ is the radial distribution function. The dark gray circles in the top part show the approximate locations of the first two "shells" (density peaks in the plot), while the light gray circle indicates the low density region between the shells. (The reason why the light gray circle is a low density region is illustrated in Figure 5.2.)

This particle will be denoted as the "central particle." As time progresses, the positions of the particles in the surroundings change relative to the central one, that is, the configuration of particles changes and may at one instance of time look like in Figure 5.1. The probability to observe a certain configuration depends on the interaction between the particles, including that with the particle at the center. From a long sequence of particle configurations, one can obtain the average density around the central particle by recording how many particles there are at various distances from the central one.

In many cases, the particles in a liquid tend to form a "shell structure" around the central particle, that is, in the average density there are a series of maxima and minima like in Figure 5.1. The density is the number density *of particle centers*, so a maximum corresponds to a region in space where it is very likely to have particle centers, like near the two dark circles in the figure. Likewise, a minimum is a region with few particle centers on average. The alternating maxima and minima have decreasing amplitudes with increasing distance from the central particle, so the "shells" are less and less developed as we proceed outwards. The density approaches the bulk density as the distance increases. This kind of shell structure becomes enhanced when the bulk density of the fluid is increased. For a

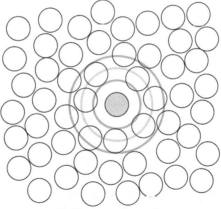

Unlikely configuration

FIGURE 5.2 A configuration with several particle centers near the light gray circle. Such configurations are possible but are very improbable since they require an empty region around the central particle. The particles closest to this region will be forced into it by collisions from the surrounding particles. Therefore, the average density will be low near the light gray circle, as apparent in the diagram in the bottom part of Figure 5.1.

Lennard-Jones fluid, for example, the first maximum near the central particle is associated with the attractive minimum in the interaction with this particle, but no attraction is needed between the particles to form a shell structure. It also appears, for example, in a hard sphere fluid.

The appearance of the oscillatory structure can be understood by examination of why there are few particles with centers near a minimum in the density plot. Let us therefore consider particle configurations where some particles actually have their centers in a low density region, like the light gray circle in Figure 5.1. Such configurations are possible, but they are less likely than other ones. A simple explanation for this is provided in Figure 5.2, which shows an example of a snapshot picture of an unlikely configuration (see figure caption). The empty region (void) around the central particle will very quickly be filled since the particles surrounding the void will experience collisions from the outside but none from the inside – the empty side. The collisions push the particles towards the central one. On the other hand, particle configurations with *no* large voids like that in Figure 5.1 tend to be followed by other similar configurations. There are simply many more configurations of the latter kind in general, which makes them more likely to appear.

Let $\mathbf{r}_2$ denote the midpoint of the central particle and consider the density at distance r_{12} from it, that is, the density at position $\mathbf{r}_1$ when the particle midpoint is located at $\mathbf{r}_2$. We will denote this density by $n(\mathbf{r}_1|\mathbf{r}_2)$, where "$|\mathbf{r}_2$" indicates that the central particle is located at $\mathbf{r}_2$. It is this density that is plotted in the example shown at the bottom of Figure 5.1. The deviation of $n(\mathbf{r}_1|\mathbf{r}_2)$ from the bulk value is described by the so-called **radial distribution function** $g(r_{12})$, which is defined from

$$n(\mathbf{r}_1|\mathbf{r}_2) = n^{\mathrm{b}}g(r_{12}). \tag{5.7}$$

Since $n(\mathbf{r}_1|\mathbf{r}_2) \to n^b$ when r_{12} increases, it follows that $g(r_{12}) \to 1$ when $r_{12} \to \infty$. In this treatise, $g(r_{12})$ is usually denoted by $g^{(2)}(r_{12})$, where superscript (2) indicates that it is a **pair distribution function**, which gives the distribution of pairs of particles in the fluid (we will encounter $g^{(\nu)}$ with $\nu \neq 2$ later). From the point of view of a particle, $g^{(2)}(r_{12})$ gives the distribution of other particles around it.

As we have seen from the discussion of particle configurations above, the probability to find a particle at a certain distance r_{12} from another particle in a fluid does not depend only on the interactions between these two particles. It also depends on the interactions with and between all other particles in their surroundings. The average density is high where the probability to find a particle is high and low where the probability is low. We shall now investigate the probability to find a particle at various places when there is a particle at $\mathbf{r}_2$. As before, we look at the fluid from the point of view of the latter particle and consider what happens with another particle 1 at position $\mathbf{r}_1$. Let us first consider the situation when the two particles are located *very* far away from each other in the fluid. The pair interaction $u(r_{12})$ between the two particles is then negligibly weak, but the interactions between each one of these particles and the particles in its immediate surroundings are strong. The mean density around particle 1 is described by the radial distribution function and is spherically symmetric, see Figure 5.3a, and likewise the mean density around the particle at $\mathbf{r}_2$. Let us investigate the situation for particle 1 a bit closer. At each instance of time, it experiences a net force from the interactions with all particles in its immediate neighborhood. This force fluctuates strongly both in magnitude and direction. On average the force is, however, zero since the average particle density around it is spherically symmetric.[3] Since the mean force

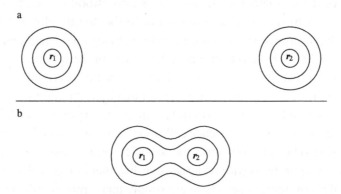

FIGURE 5.3 A simplified sketch of a contour plot for the density distribution around two particles located at $\mathbf{r}_1$ and $\mathbf{r}_2$ in a bulk fluid. The distribution is the average density for all particles whenever one particle is located at $\mathbf{r}_1$ and another at $\mathbf{r}_2$. (a) When the two particles are *very* far away from each other, the average density distribution around each of them is spherically symmetric. (b) When they are close together, the density distribution around them is affected by the interactions with both.

---

[3] This can be understood as follows: Assume that the mean force is nonzero. Then it must have a direction. Assume therefore that it points in direction "A." There must be a reason why it points in this direction and not in another direction, say "B." Due to the spherical symmetry there is, however, no preferred direction in space so it is equally likely that the force points in direction "A" as in direction "B," or for that matter any other direction. The only possibility to avoid a contradiction is that the force is zero.

on the particle is zero, there is no "preferred" direction of the particle's motion, so it is in the long run equally likely to be anywhere (as long as it is far from the particle at $\mathbf{r}_2$; remember that we regard the scene from the point of view of the latter). The same applies to any other particle.

Let us now consider what happens when particle 1 is rather close to the particle at $\mathbf{r}_2$, Figure 5.3b. Then the pair interaction $u(r_{12})$ is not negligible in general (an exception is hard spheres for $r > d^h$), so there is a force $-\nabla_1 u(r_{12})$ acting on particle 1 from this interaction, where $\nabla_1 = (\partial/\partial x_1, \partial/\partial y_1, \partial/\partial z_1)$.[4] If the force is attractive, particle 1 will spend quite some time close to the particle at $\mathbf{r}_2$, which implies that the density around the latter would be higher than the bulk value. The situation is, however, more complex than this since the interactions with the other particles in the neighborhood also contribute to the force on particle 1. Now, this latter force contribution is no longer zero on average because the mean density distribution around particle 1 is not spherically symmetric, as illustrated in Figure 5.3b. For example, there are on average more particles on one side of the particle at $\mathbf{r}_1$ than on the other side, so the force from one side must be different than from the other, that is, the net force is different from zero on average. Thus, a net mean force arises in addition to the direct interaction force $-\nabla_1 u(r_{12})$. This constitutes an *indirect contribution* to the mean force between the two particles. The total mean force is the sum of the direct force and this indirect contribution. For symmetry reasons, the force will be directed along the connection line between the particle centers. It may be attractive or repulsive depending on the details of the average density distribution around the two particles for each value of the separation r_{12} between the two particles. The mean force on each particle (for instance, particle 1) that surrounds the particle at $\mathbf{r}_2$ as a function of distance from the latter determines the density distribution $n(\mathbf{r}_1|\mathbf{r}_2)$ and thereby the radial distribution function $g(r_{12})$. This will be elaborated and more fully explained in the following sections. Here it suffices to say that it is the complex behavior of the indirect mean force that gives rise to structures like the oscillating density illustrated in Figure 5.1.

For a hard sphere fluid there exists only an indirect mean force between each pair of particles when their separation r is larger than d_h since the direct force is zero there. If two

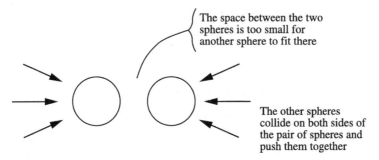

The space between the two spheres is too small for another sphere to fit there

The other spheres collide on both sides of the pair of spheres and push them together

FIGURE 5.4 Two hard spheres at short separation from each other in a hard sphere fluid are pushed together by collisions from the surrounding spheres.

---

[4] An equal but opposite force will of course act on particle 2, but since we are "sitting on" this particle we only consider the force on particle 1 here.

spheres are at a short separation from each other like in Figure 5.4, they are pushed together by collisions from the surrounding hard spheres. Much fewer collisions act in the opposite direction ("outwards") since there is not enough space between the two spheres to fit another sphere, so there is a void there. No collisions happen in the void. Thus, there is an indirect force[5] that gives an attraction between the two spheres at short separations. The indirect force is actually alternatingly attractive and repulsive when the separation between the two spheres is increased. When spheres can efficiently fill the space between the two spheres, the force from collisions in the outward direction (from the "inside") can be larger than in the inward direction (from the "outside"), which leads to net repulsion between the two spheres. In contrast, when spheres cannot efficiently fill the space and voids must be present there, the force can be attractive. This leads to an oscillatory radial distribution function similar to that in Figure 5.1.

Before we close this section, we may add that Figure 5.3 and the discussion above illustrate that it is necessary, at least in principle, to know the distribution of particles around two particles in order to calculate the radial distribution function $g^{(2)}(r)$, which gives the distribution around one particle. The latter is a *pair* distribution function and the distribution around two particles is called a **triplet distribution function**. Thus, one needs to know the triplet distribution function to calculate the pair distribution, that is, a more complex entity is needed to calculate a simpler one! Likewise, in order to calculate the density distribution around two particles, one needs in principle to know the mean force acting on a third particle. The latter force is determined by the density distribution around the three particles, the **quartet distribution function**, in complete analogy to the discussion above. Thus, to calculate the triplet distribution function, one needs to know the quartet distribution function, an even more complicated entity. This continues in the same fashion to higher and higher distribution functions, which may seem quite depressing. This kind of chain of interdependencies between the distribution functions of higher and higher orders is called a **hierarchy**. We will give examples later how this can be formulated mathematically. A major task in the theory of liquids is to find ways to handle this situation without explicit consideration of all these complicated functions.

To introduce the necessary tools and concepts needed to develop the theory, we will now proceed to consider density distributions in inhomogeneous fluids. This is actually more efficient and conceptually more transparent than to proceed with a treatment of homogeneous bulk fluids only.

5.2 NUMBER DENSITY DISTRIBUTIONS: DENSITY PROFILES

Let us now treat the general case when the particles in a fluid are exposed to an **external potential** $v(\mathbf{r})$, that is, an interaction potential that originates from a source that is not part of the fluid like the walls of the vessel that contain the fluid or some body immersed in the fluid. Initially, we restrict ourselves to the case when the fluid consists of spherically symmetric particles. A particle located with its center at position $\mathbf{r}$ is acted on by the force $-\nabla v(\mathbf{r})$ from

---

[5] Note that the momentum transfer between particles during a collision is equivalent to the action of a force according to Newton's equation $\mathbf{F} = m\mathbf{a} = d\mathbf{p}/dt$, where $\mathbf{F}$ is a force, m the mass, $\mathbf{a}$ the acceleration, $\mathbf{p}$ the momentum and t the time.

the external potential. The presence of such a potential makes the fluid inhomogeneous, that is, the density of the fluid varies in space. The interaction energy of the particles with each other is given by a potential $\breve{U}^{\text{intr}}(\mathbf{r}_1, \mathbf{r}_2, \mathbf{r}_3, \mathbf{r}_4, \ldots)$, where superscript "intr" stands for intrinsic (intrinsic for the fluid itself). The total interaction energy equals

$$\breve{U}^{\text{pot}} = \breve{U}^{\text{intr}}(\mathbf{r}_1, \mathbf{r}_2, \mathbf{r}_3, \mathbf{r}_4, \ldots) + \sum_{\nu} v(\mathbf{r}_\nu), \tag{5.8}$$

where the sum is over all particles in the system. A special case is when the particles interact with a pair potential[6] $u(\mathbf{r}_\nu, \mathbf{r}_{\nu'})$ that presently is assumed to be spherically symmetric, $u(\mathbf{r}_\nu, \mathbf{r}_{\nu'}) = u(r_{\nu\nu'})$.[7] The total interaction energy then equals

$$\breve{U}^{\text{pot}} = \sum_{\nu, \nu' > \nu} u(\mathbf{r}_\nu, \mathbf{r}_{\nu'}) + \sum_{\nu} v(\mathbf{r}_\nu), \tag{5.9}$$

where the sum in the first term is over all pairs of particles in the system.

The number density, $n(\mathbf{r})$, of the fluid at point $\mathbf{r}$ is defined as the average number of *particle centers* per volume there, or more precisely: if we consider a small volume element $d\mathbf{r} = dx\, dy\, dz$ located at the point $\mathbf{r}$ (see Figure 3.1), then $n(\mathbf{r})d\mathbf{r}$ is equal to the average number of particle centers inside $d\mathbf{r}$.[8] For a macroscopic system far from the source of the external interaction potential (for instance, the container walls), the fluid is homogeneous and the density there is n^{b}. On the other hand, in the inhomogeneous region close to this source, the number density depends on $\mathbf{r}$. When the distance from the source is increased, $n(\mathbf{r})$ approaches the bulk density n^{b}. Far from the source, the force $-\nabla v(\mathbf{r})$ is very small or equal to zero;[9] it generally goes to zero with increasing distance from the source and it is zero in the bulk. It is common to select the value of the potential $v(\mathbf{r})$ in bulk to be equal to zero, but any constant value would do since the force is not changed by the addition of a constant to the potential. We will select the value in bulk equal to zero in what follows unless otherwise stated.

---

[6] The assumption that the interaction potential is pairwise additive is not needed for the arguments in, for example, Sections 5.2 through 5.4, so the results there are valid for a general interaction potential between the particles. For the numerical examples in Figures 5.5 and 5.8, a pair potential is, however, assumed.

[7] An exception to spherical symmetry of $u(\mathbf{r}_\nu, \mathbf{r}_{\nu'})$ for spherical particles is, for example, ions in the neighborhood of a wall that is polarized by them. Then the absolute positions $\mathbf{r}_\nu$ and $\mathbf{r}_{\nu'}$ need to be specified since they contain information about the distances from the wall.

[8] When calculating the average density $n(\mathbf{r})$, we may make a long series of discrete observations of the particle positions (at instances evenly spaced in time) and count the number of particle centers in $d\mathbf{r}$. We may then use a counter originally set to zero. Each time we see a particle center located in the volume element $d\mathbf{r}$, we increase the counter by one. The counter value divided by the total number of observations is equal to the average density n times the volume $d\mathbf{r}$. Note that if the volume element is smaller than the particle size, only one center at a time can be inside $d\mathbf{r}$ and then it is very common to observe that *no* center is located there. To make $n(\mathbf{r})$ precisely defined, we must make the time interval between the observations to go to zero and also make the total time for the observations to be very long (in principle infinitely long). Furthermore, to make $n(\mathbf{r})$ specified point-wise microscopically, we must take the limit where dx, dy and dz all go to zero (an "infinitesimal" volume element $d\mathbf{r}$). All these limits are assumed to be taken for the number density we will use in what follows.

[9] A planar sheet of charge as an external source is a special case since the force $-\nabla v(\mathbf{r})$ then is not small even far away (for an infinitely large sheet, the electrostatic force on a charged particle is constant irrespective of the distance from the sheet). This situation will be dealt with separately later. For a charged planar surface in contact with a bulk electrolyte, the presence of counterions near the surface make the total interaction force (from the surface charge and the ions) decay to zero with distance from the surface.

A special case is when the external potential depends only on z but is independent of x and y, that is, $v(\mathbf{r}) = v(z)$, for instance in the neighborhood of a perfectly planar wall[10] where z is the coordinate perpendicular to the wall surface. Then the density also varies in the z direction only, that is, $n(\mathbf{r}) = n(z)$. In such a case, the density distribution is often denoted by "**density profile**," $n(z)$. This term is also used for other simple geometries like the density distribution around a spherically symmetric object when $n = n(r)$ and r is the distance from the sphere center.

An example of a density profile for a hard sphere fluid in a slit between two planar hard walls is shown in Figure 5.5a. The wall surfaces are located at coordinates $z = \pm 2.5 d^h$ (i.e., at the left and right edges of the plot). The fluid in the slit is in equilibrium with at bulk fluid with density n^b. The oscillations seen in the figure reflect the packing of the particles in layers near the walls; the layers become less defined at increasing distance from the surface, so the amplitude decreases. When the distance between the walls is much larger than in the figure, the region in the middle harbors a bulk fluid with a constant density n^b. Figure 5.5b

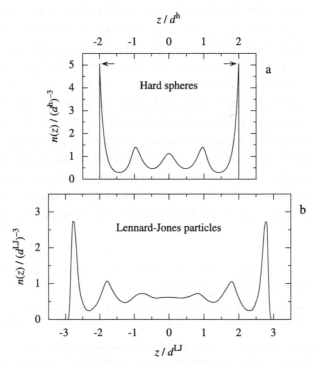

FIGURE 5.5 (a) The density profile $n(z)$ for an inhomogeneous fluid of hard spheres in a slit of width $5.0\ d^h$ between two hard walls. The fluid in the slit is in equilibrium with a hard sphere bulk fluid of density $n^b = 0.75\ (d^h)^{-3}$. The arrows show the value of the density for spheres in contact with the wall surfaces, the so-called contact density. (b) The density profile of an inhomogeneous Lennard-Jones fluid between two soft walls and in equilibrium with a bulk fluid of density $n^b = 0.59\ (d^{LJ})^{-3}$ calculated by Monte Carlo simulations. The fluid film thickness is $6.8\ d^{LJ}$ and $k_B T / \varepsilon^{LJ} = 1.2$.[11]

---

[10] A perfectly planar wall, that is, a wall with a smooth surface without atomic structure, is a commonly used model in the study of liquids near a wall.

[11] The calculations of the data in Figure 5.5a are described in more detail in Section 7.2.1.2. Figure 5.5b is based on data from I. K. Snook and V. van Megen, *J. Chem. Phys.* **72** (1980) 2907.

shows an example of a Lennard-Jones fluid in the slit between two planar soft walls. The fluid film thickness is 6.8 d^{LJ}.[12] In pane (a) of the figure, the hard walls cause the density to have a discontinuous drop to zero at the distance $d^h/2$ from each wall surface, that is, at the point of closest approach of the sphere centers to the wall. The value of the density at this point (just before the drop) is called the **contact density**, which is marked with arrows in the figure. For the Lennard-Jones fluid in pane (b), the soft walls give a more gradual drop to zero near the walls.

5.3 FORCE BALANCE AND THE BOLTZMANN DISTRIBUTION FOR DENSITY: POTENTIAL OF MEAN FORCE

Let us now consider what determines the density distribution of an inhomogeneous fluid at equilibrium. In Figure 5.6, there is depicted a case with a density profile, $n(z)$. The reason why there is a nonzero density gradient is that the net interactional force $\mathbf{F}(z)$ that acts on each particle at coordinate z is nonzero on average (the interactions from both the external source and the particles in the fluid contribute to $\mathbf{F}$). The nonzero average force from interactions counteracts the flow that otherwise would occur due to the density gradient and is directed to the right in the figure. In part b of the figure, we have focused on a slice of the fluid of thickness dz lying between the coordinates z and $z + dz$. Particles are all the time entering and leaving the slice through the planes at z and at $z + dz$. For the case depicted in the figure, there are on average more particles at $z + dz$ than at z since the density is higher to the right. In the *absence* of a net interactional force $\mathbf{F}$, more particles would enter the slice

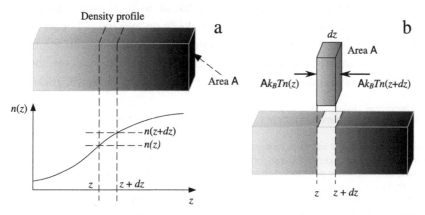

FIGURE 5.6 (a) A sketch of the density profile $n(z)$ for an inhomogeneous fluid in an external potential $v(z)$. A portion of the fluid with cross-section area A is shown in the top part, where the shade of gray indicates the density. The curve at the bottom part shows the corresponding plot of the density $n(z)$ as a function of z. (b) A slice of the system between coordinates z and $z + dz$ is considered separately. This slice of thickness dz experiences a mean force $Ak_BTn(z)$ from momentum transfer on its left-hand side and a corresponding opposing force $Ak_BTn(z + dz)$ on its right-hand side, see text. The net force from momentum transfer is balanced by the mean interactional force on the particles in the slice at equilibrium, so the total net force is zero on average.

---

[12] The measure of film thickness for this system is defined in Section 7.2.1.1.

than leave it (at least initially), but in the *presence* of a nonzero $\mathbf{F}$ the net change in number of particles at equilibrium must on average be equal to zero everywhere despite the presence of a density gradient.

The mean force $\mathbf{F}(z)$ that acts on each particle at z can be calculated by following the time evolution of the particle locations in space and by considering the instantaneous force[13] $\check{\mathbf{F}}$ that acts on each particle whenever its center is located at z. One then obtains $\mathbf{F}(z)$ as the average of $\check{\mathbf{F}}$ for all particles that are located at z at any time, that is, among all particle configurations one only considers the configurations that have a particle at z and only these contribute to the average. As will be discussed in more detail later, this implies that when one calculates $\mathbf{F}(z)$, one includes all possible configurations of particles around a particle located at z, that is, as if there is a particle fixed at z and $\check{\mathbf{F}}$ is the instantaneous force that acts on it. Due to the planar symmetry, the x and y components of the *mean* force must be zero and hence only the z component contributes, $\mathbf{F}(z) = \langle \check{\mathbf{F}} \rangle = (0, 0, F_z(z))$, while the *instantaneous* force $\check{\mathbf{F}}$ can point in any direction.

How large must the mean interactional force $\mathbf{F}$ be to maintain equilibrium? The *total* net force on each volume element of the fluid must at equilibrium be equal to zero on average, otherwise that part of the fluid will move which cannot happen at equilibrium. There are two contributions to the total force: the interactional force and the force from momentum transfer due to the motion of the particles. When calculating the latter, we can use the fact that in classical statistical mechanics the momentum distribution for the particles in a fluid at equilibrium is independent of the interactions.[14] As we have seen, Equation 3.7, the momenta of the particles are Maxwell-Boltzmann distributed for both a strongly coupled liquid and an ideal gas. Thus, the force from momentum transfer is the same for a liquid as for an ideal gas at the same density, that is, the force per unit area is $k_B T n$ on average.[15] For an inhomogeneous fluid, the number density to be used here is the local one at the position in question.

Accordingly, the momentum transfer force acting at coordinate z on the slice in Figure 5.6b with area A is on average equal to $A k_B T\, n(z)$ and it is directed inwards (to the

---

[13] Like before, the symbol "$\check{}$" over a quantity, like in $\check{\mathbf{F}}$, indicates that the value is the instantaneous one, cf. footnote 1 in Chapter 3.

[14] The slice of fluid depicted in Figure 5.6b gains momentum when a particle passes its boundary. If a particle with mass m has a positive momentum p_z and hence velocity p_z/m in the z direction when it passes the boundary at coordinate z, the increase in momentum of the slice is p_z. A particle that passes *out* from the slice through this boundary (in the negative z direction) makes the slice to lose a negative momentum $p_z = -|p_z|$. Hence, this kind of event implies a change $-(-|p_z|)$ in momentum, that is, it contributes to an *increase* in momentum of the slice that is equal to the contribution from a particle that passes inwards with the same absolute velocity. Let us consider the change during an infinitesimally small time interval dt for an ideal gas. Any particle with absolute velocity $|p_z|/m$ that is located within a distance $[|p_z|/m]dt$ from the boundary (inside or outside, respectively) will pass through it during this time interval provided it moves in the appropriate direction. Halfway through the time interval (after $dt/2$), all such particles will be located within a slice of width $[|p_z|/m]dt$ centered at the boundary and no other particles with the same absolute velocity will be located there. If the number density of these particles is n_{p_z} there are $n_{p_z}[|p_z|/m]dt$ particles per unit area that pass during the interval dt. Since each particle contributes with momentum change $|p_z|$, the total change is $n_{p_z}[(p_z)^2/m]dt$. Hence, the rate of momentum change is $n_{p_z}(p_z)^2/m$ per unit area from these particles, which according to Newton's second law $F = dp/dt$ corresponds to a force acting on the slice. We now sum the contributions from all particles irrespective of their velocities. The total number density at z is $n(z)$ and we have a fraction $n_{p_z}/n(z)$ of particles that has absolute velocity $|p_z|/m$ in the z direction. Hence, we obtain the total force per unit area $n(z) \sum_{p_z} [n_{p_z}/n(z)](p_z)^2/m = n(z)\langle(p_z)^2\rangle/m$ from momentum transfer.

[15] From our result in footnote 14, we see that the force per unit area is equal to $n(z)\langle(p_z)^2\rangle/m = n(z)\langle 2\epsilon_z\rangle$ where $\epsilon_z = (p_z)^2/2m$ is the contribution to the kinetic energy from translational motion in the z direction. Since according to the equipartition theorem $\langle\epsilon_z\rangle = k_B T/2$, we obtain the mean force $k_B T n(z)$ per unit area.

right). Likewise, the corresponding force at coordinate $z + dz$ is equal to $Ak_BT\, n(z + dz)$ and it is also directed inwards (here to the left). The net force from momentum transfer on the slice is accordingly $Ak_BT[n(z) - n(z + dz)] = -Ak_BT(dn(z)/dz)dz$ in the limit where dz is infinitesimally small. This net force is directed to the left (negative). It must be balanced by the interactional force on the slice, which is equal to $F_z(z)n(z)Adz$, since there are on average $n(z)Adz$ molecules in the slice. Thus, we have the equilibrium condition

$$\left[F_z(z)n(z) - k_BT\frac{dn(z)}{dz}\right]Adz = 0,$$

which can be written

$$F_z(z) = k_BT\frac{d\ln n(z)}{dz}. \tag{5.10}$$

This gives the interactional force F_z required to counteract the net flow of particles due to the density gradient at equilibrium.

The mean force can be written in terms of a potential function $w(z)$, called the **potential of mean force**, as $F_z(z) = -dw(z)/dz$ (alternative names are the **average force** and the **potential of average force**, respectively). The equilibrium condition (5.10) then becomes

$$-\frac{dw(z)}{dz} = k_BT\frac{d\ln n(z)}{dz},$$

which can be integrated to give $-w(z) = k_BT\ln n(z) + C_1$ where C_1 is an integration constant. Thus, we can write the equilibrium condition as

$$k_BT\ln n(z) + w(z) = \text{constant} \tag{5.11}$$

where the constant $(-C_1)$ is independent of z. As we shall see later, this is equivalent to the condition that the chemical potential μ is equal everywhere at equilibrium; the first term in Equation 5.11 is the local value of the ideal part of μ (apart from a constant $k_BT\ln\Lambda^3$, see Equation 3.34) and $w(z)$ is the nonideal part (possibly also apart from a constant). Alternatively, we can write the condition as $n(z) = C_2\exp(-w(z)/k_BT)$ where $C_2 = \exp(-C_1/k_BT)$ is another constant. This yields

$$n(z) = n^0 e^{-\beta w(z)},$$

where n^0 is the number density at a place where we have selected $w = 0$, for instance in a bulk phase far from the external source (the place where w is zero can be selected freely since we can add an arbitrary constant to the potential without changing the force). We recognize this as a form of Boltzmann's distribution law.

These results can easily be generalized to an arbitrary density distribution $n(\mathbf{r})$ that depends also on x and y. Here one considers the force balance on a volume element bounded by $[x, x + dx]$, $[y, y + dy]$ and $[z, z + dz]$, Figure 3.1. Let us regard the configurations of all particles as a function of time and consider the instantaneous interactional force $\check{\mathbf{F}}$ on each particle whenever it happens to have its center inside this volume element. To obtain $\mathbf{F}$, we take the average of $\check{\mathbf{F}}$ for all particles that are located in this volume element at any time. In the limit when dx, dy and dz all go to zero, we obtain the mean force $\mathbf{F}(\mathbf{r}) = \langle \check{\mathbf{F}} \rangle = (F_x(\mathbf{r}), F_y(\mathbf{r}), F_z(\mathbf{r}))$ acting on a particle when its center is located at position $\mathbf{r}$.

The argument for the force balance above is now done in the three separate directions x, y and z.

{EXERCISE 5.1

Fill in the details of the following arguments leading to Equations 5.16 and 5.17.}

When, for example, one considers the force balance in the z direction, one uses $A = dx\, dy$ in the derivation above and obtains the equilibrium condition

$$F_z(\mathbf{r}) = k_B T \frac{\partial \ln n(\mathbf{r})}{\partial z}$$

and analogously for the x and y direction. The mean force therefore fulfills

$$\mathbf{F}(\mathbf{r}) = k_B T \, \nabla \ln n(\mathbf{r}) \tag{5.12}$$

at equilibrium. The potential of mean force w is related to the mean force by

$$\mathbf{F}(\mathbf{r}) = -\nabla w(\mathbf{r}) \tag{5.13}$$

and one performs a line integral of $\nabla[kT \ln n(\mathbf{r}) + w(\mathbf{r})] = 0$ between two arbitrary points $\mathbf{r}$ and $\mathbf{r}_0$ to conclude that

$$k_B T \ln n(\mathbf{r}) + w(\mathbf{r}) = k_B T \ln n(\mathbf{r}_0) + w(\mathbf{r}_0). \tag{5.14}$$

This implies that

$$k_B T \ln n(\mathbf{r}) + w(\mathbf{r}) = \text{constant}, \tag{5.15}$$

which is equivalent to the condition of equal chemical potential everywhere. From Equation 5.14, one obtains the Boltzmann distribution

$$n(\mathbf{r}) = n(\mathbf{r}_0) e^{-\beta[w(\mathbf{r}) - w(\mathbf{r}_0)]} \tag{5.16}$$

or, when $w(\mathbf{r}_0) = 0$,

$$n(\mathbf{r}) = n^0 e^{-\beta w(\mathbf{r})}. \tag{5.17}$$

Note that the force $\mathbf{F}(\mathbf{r})$ is the *actual average force* that on acts on a particle located at $\mathbf{r}$, a force that, at least in principle, can be measured in experiments.[16]

The force $\mathbf{F}(\mathbf{r})$ on a particle at $\mathbf{r}$ originates from both the external potential $v(\mathbf{r})$ and the interactions with the other particles in the fluid,

$$\mathbf{F}(\mathbf{r}) = -\nabla v(\mathbf{r}) + \mathbf{F}^{\text{intr}}(\mathbf{r}), \tag{5.18}$$

where the last term is the mean force from the other particles, the "intrinsic" part of the force. Consequently, the potential of mean force can likewise be divided into two parts

$$w(\mathbf{r}) = v(\mathbf{r}) + w^{\text{intr}}(\mathbf{r}) \tag{5.19}$$

and we have $\mathbf{F}^{\text{intr}}(\mathbf{r}) = -\nabla w^{\text{intr}}(\mathbf{r})$. To calculate the intrinsic part of the potential of mean force, $w^{\text{intr}}(\mathbf{r})$, for a given system, that is, for given interaction potentials $u(r)$ and $v(\mathbf{r})$, is one of the major tasks in the theory of inhomogeneous fluids. Once it is known, the density distribution is given by Equations 5.17 and 5.19, which yield

$$n(\mathbf{r}) = n^0 e^{-\beta[v(\mathbf{r}) + w^{\text{intr}}(\mathbf{r})]}. \tag{5.20}$$

In general, we have

$$n(\mathbf{r}) = n(\mathbf{r}_0) e^{-\beta[v(\mathbf{r}) - v(\mathbf{r}_0) + w^{\text{intr}}(\mathbf{r}) - w^{\text{intr}}(\mathbf{r}_0)]}, \tag{5.21}$$

which is obtained from Equation 5.16.

Let us now consider the physical meaning of $w^{\text{intr}}(\mathbf{r})$ and the associated force $\mathbf{F}^{\text{intr}}(\mathbf{r})$. It is illustrative to look at this force from a slightly different perspective. Previously, when we calculated $\mathbf{F}$, we regarded the configurations of all particles as a function of time and considered the instantaneous force $\check{\mathbf{F}}$ on each particle whenever it happens to be located with its center at $\mathbf{r}$ (or rather in an infinitesimal volume element at $\mathbf{r}$). Then we took the average for all particles that are located there at any time to obtain $\mathbf{F}(\mathbf{r})$. From the sequence of particle configurations, we thereby picked out those that actually have a particle at $\mathbf{r}$ and it is only this subset that contributes in the calculation of the mean $\mathbf{F}$. The intrinsic part of the force, $\mathbf{F}^{\text{intr}}(\mathbf{r})$, is the force from interactions with the other particles of the fluid averaged over this subset of configurations. It does not matter how this subset is generated: from a time sequence or from a set of configurations in the appropriate ensemble (we assume that the system is ergodic). The main thing is that there is a particle with its center at $\mathbf{r}$ for every

---

[16] One can in several cases measure the mean force acting on a particle in an inhomogeneous liquid near a surface by using an *Atomic Force Microscope* (AFM), where the particle is attached to a fine tip on a cantilever spring and placed at various distances from the surface. A technique to measure the mean force acting on freely moving particles (like colloidal particles) near a planar surface is *Total Internal Reflection Microscopy*, where the density profile $n(z)$ is measured and then converted to $w(z)$ via the Boltzmann relationship (5.17) and to $\mathbf{F}(z)$ via Equation 5.13. These techniques are, for example, described in J. N. Israelachvili, *Intermolecular and surface forces*, 3rd edition (Academic Press, 2011). One can also manipulate colloidal particles in liquids by optical tweezers and by first positioning particles and then releasing them, one can obtain the potential of mean force via statistical analysis of their subsequent motion (Crocker J. C. and Grier D. G., *Phys. Rev. Lett.* **73** (1994) 352 and *Phys. Rev. Lett.* **77** (1996) 1897).

configuration in the subset and that each of these configurations occurs with the correct probability, considering the interactions between all particles and with the external potential. This subset of configurations also represents the situation when we have a particle located at $\mathbf{r}$ and is fixed there,[17] while all other particles are free to move. It does not make any difference whether a moving particle is recorded at the instances when it is at $\mathbf{r}$ or if the particle is stationary there; the particle configurations we use for the average are the same in principle. Remember that the distribution of particle velocities is not dependent on the interactions and is independent of the spatial configurations.

Thus, we can identify $\mathbf{F}^{\mathrm{intr}}(\mathbf{r})$ with the mean force that acts on a particle held fixed at $\mathbf{r}$ due to the interactions with the moving particles in its environment. The distribution of these particles in the neighborhood of $\mathbf{r}$ is, of course, affected by the interactions with the fixed particle and with the external potential. We will denote the number density of particles at position $\mathbf{r}'$ when a particle is held fixed at $\mathbf{r}$ as $n(\mathbf{r}'|\mathbf{r})$. When we shift the fixed particle to a different position $\mathbf{r}$, the density distribution around it will change, which gives rise to the $\mathbf{r}$ dependence of $\mathbf{F}^{\mathrm{intr}}$. This $\mathbf{r}$ dependent distribution is nontrivial to obtain since it depends on the interactions between all mobile particles. This is the key difficulty in the determination of $\mathbf{F}^{\mathrm{intr}}(\mathbf{r})$ and hence of $w^{\mathrm{intr}}(\mathbf{r})$ and $n(\mathbf{r})$.

As a concrete example of how the density distribution $n(\mathbf{r}'|\mathbf{r})$ around a particle located at $\mathbf{r}$ can look like, let us consider an inhomogeneous fluid between two planar walls as sketched in Figure 5.7a. We take the example from calculations[18] for the same Lennard-Jones fluid as in Figure 5.5b but for a shorter distance between the two walls, where there are only four maxima of the density profile in the slit. Plots of $n(\mathbf{r}'|\mathbf{r})$ around a fixed Lennard-Jones particle in the fluid between the two walls are shown in Figure 5.8, where the density plot is made for the cross-section plane shown in Figure 5.7b. The point $\mathbf{r}$ where the particle is located is marked by a cross in the figures and $n(\mathbf{r}'|\mathbf{r})$ for two different positions $\mathbf{r}$ is shown in Figure 5.8. For each position, $n(\mathbf{r}'|\mathbf{r})$ is plotted as a function of $\mathbf{r}'$ in two different manners shown in the top and bottom frames, respectively, as explained in the figure caption.

The fluid is affected by both the planar walls and the fixed spherical particle and the resulting density distribution is a "compromise" between planar and spherical symmetry, as can be seen most clearly in the contour plots in frames b and d of Figure 5.8. $\mathbf{F}^{\mathrm{intr}}(\mathbf{r})$ is the force on the particle at $\mathbf{r}$ due to interactions with this rather complicated distribution of particles and by comparing the plots for the two different positions of $\mathbf{r}$, we can see that the variation of the distribution with $\mathbf{r}$ is quite intricate. These issues will be treated in detail in Section 7.2. Inhomogeneous Lennard-Jones fluids are treated in Section 7.2.1.1, where Figure 5.8 will be discussed further.

---

[17] We may hold the particle fixed at $\mathbf{r}$ by some suitable means, for instance by invoking a counter-force that *exactly* cancels $\check{\mathbf{F}}$ at each instance of time so the particle does not move. This is, of course, an artificial way of doing it, like a "Maxwell's demon" device, and the presence of a fixed particle can be regarded as a "gedanken experiment," which helps to bring out important principles of the theory. In practice one may, however, be able to keep one particle fixed without affecting the surrounding particles in the liquid too much by using some variant of the optical tweezers technique, for which Arthur Ashkin was awarded the 2018 Nobel Prize in Physics. When a particle is trapped in a deep and narrow potential well, the counter-force $\check{\mathbf{F}}$ mentioned above will (approximately) arise and keep the particle near the potential minimum.

[18] In these calculations, an accurate integral equation theory for inhomogeneous fluids has been used, whereby the pair distributions and the density profiles are calculated in a self-consistent manner. Such theories are introduced in Chapter 9 (in the 2nd volume).

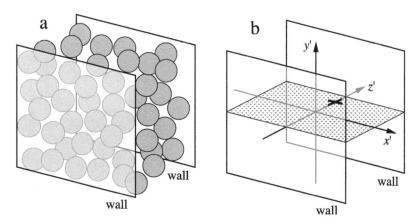

FIGURE 5.7 (a) A sketch of a fluid consisting of spherical particles between two planar walls. An instantaneous configuration of the particles is shown. (b) A cross-section plane of the system made perpendicularly to the surfaces; here it is the (x', z') plane at $y' = 0$. In illustrations of the density $n(\mathbf{r}'|\mathbf{r})$ at position $\mathbf{r}'$, like in Figure 5.8, the density values in such a plane are plotted as a function of (x', z') when a particle is located with its center at $\mathbf{r}$. One possible position of this particle is marked by a cross in the figure.

5.4 THE RELATIONSHIP TO FREE ENERGY AND CHEMICAL POTENTIAL

Let us shift a particle very slowly from one position, $\mathbf{r}_1$, to another, $\mathbf{r}_2$, so slowly that the particle distribution around it is fully relaxed all the time (a quasistatic change in position). To do this, we have to counteract the force $\mathbf{F}(\mathbf{r})$, so we must act with the force $-\mathbf{F}(\mathbf{r})$ on the particle (or rather a force infinitesimally different from $-\mathbf{F}(\mathbf{r})$). The work performed on the system during this process is given by the line integral

$$-\int_{\mathbf{r}_1}^{\mathbf{r}_2} \mathbf{F}(\mathbf{r}) \cdot d\mathbf{r} = \int_{\mathbf{r}_1}^{\mathbf{r}_2} \nabla w(\mathbf{r}) \cdot d\mathbf{r} = w(\mathbf{r}_2) - w(\mathbf{r}_1). \qquad (5.22)$$

Note that the reversible work performed on a system is equal to the change in free energy.[19] Accordingly, $w(\mathbf{r})$ is a free energy function: its variation gives the change in free energy due to interactions when shifting the position of a particle. The change does not depend on the path chosen between $\mathbf{r}_1$ and $\mathbf{r}_2$ since the free energy is a state function. Note that there is a change in entropy in this process; it is associated with the change in distribution of the mobile particles during the shift. The free energy change has two parts: $v(\mathbf{r}_2) - v(\mathbf{r}_1)$ from the work against the external potential (this is usually a pure energy change) and $w^{\text{intr}}(\mathbf{r}_2) - w^{\text{intr}}(\mathbf{r}_1)$ from the work against the interparticle interactions (see Equation 5.19).

---

[19] In the canonical ensemble, the Helmholtz free energy $A = \bar{U} - TS$ is the relevant entity, where $\bar{U}$ is the internal energy and S the entropy. For a reversible process with work w^{rev} and heat $q^{\text{rev}} = T\Delta S$ at constant temperature, we have $\Delta A = \Delta \bar{U} - T\Delta S = q^{\text{rev}} + w^{\text{rev}} - T\Delta S = w^{\text{rev}}$, a standard thermodynamic result.

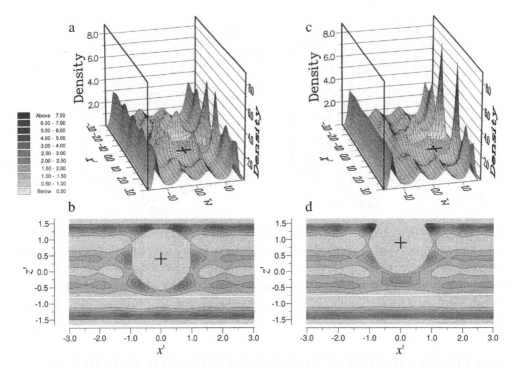

FIGURE 5.8 A plot of the average number density distribution around a Lennard-Jones particle in an inhomogeneous Lennard-Jones fluid between two walls[20] (see Figure 5.7a for the geometry of the system). The system is the same as in Figure 5.5b, except that the slit is more narrow; the fluid film thickness here is 4.0 d^{LJ}. The plots at the top and bottom show different representations of the number density $n(\mathbf{r}'|\mathbf{r})$ at position $\mathbf{r}'$ around a particle located with its center at the cross (= the point $\mathbf{r}$). Each contour plot at the bottom shows a cross-section perpendicular to the wall surfaces (the x', z' plane depicted in Figure 5.7b). The wall surfaces are perpendicular to the plane of the paper and located at $z' = \pm 2.35\ d^{LJ}$. In three dimensions, the density distribution has cylindrical symmetry around a vertical axis through the cross. The surface plots at the top show a different representation of the same data as the contour plots. In the figures, the origin of the coordinate system is placed at the midplane between the surfaces. The coordinates are given in units of d^{LJ} and the density in units of $(d^{LJ})^{-3}$. The particle is placed with its center at (a, b) 0.4 d^{LJ} and (c, d) 0.9 d^{LJ} from the midplane.

We can now interpret Equation 5.16 in a new manner. If we take a particle from a point $\mathbf{r}_0$ and put it at $\mathbf{r}$, the reversible work done is $w(\mathbf{r}) - w(\mathbf{r}_0)$. When this work is positive, the density at $\mathbf{r}$ is lower than at $\mathbf{r}_0$ and when the work is negative, the density is higher. Thus, the density is low at places where it requires positive work to place a particle and high where the work is negative. Furthermore, the free energy of the system is decreased if we take a particle from a place with high $w(\mathbf{r})$ to one with low, so if we release the particle it would have a higher probability to be at the latter place than at the former. This is expressed in terms of densities by Equation 5.17. Note that if we add an arbitrary constant to w, this does not

---

[20] The results in Figure 5.8 are taken from integral equation calculations presented in "A study of anisotropic pair distribution theories for Lennard-Jones fluids in narrow slits. Part II. Pair correlations and solvation forces" by R. Kjellander and S. Sarman, *Mol. Phys.* **74** (1991) 665. The figure is reprinted from this work by permission of the publisher (Taylor & Francis Ltd., http://www.tandfonline.com).

affect the difference $w(\mathbf{r}) - w(\mathbf{r}_0)$ in Equation 5.16, so $w(\mathbf{r})$ is so far defined up to an additive constant that does not affect the physics. In Equation 5.17, we have selected a certain value of the constant since we have set the value of w to zero at $\mathbf{r}_0$.

As mentioned in connection to Equations 5.11 and 5.15 above, $w(\mathbf{r})$ has a close relationship to the chemical potential. By definition, the chemical potential is the change in free energy when one increases the number of particles in the system by one. One can do this by taking a particle from a place (fixed in space) where it does not interact with anything ("infinitely far away") and moving it into the system. Alternatively, one can do this (at least in theory) by "creating" a new particle in the system. Initially, the system has, say, N particles and finally $N + 1$ particles in both cases. One can imagine doing this addition in two steps: one first places the new particle (or creates it) at some point $\mathbf{r}$ where it is held fixed and then one releases it so it becomes freely mobile like the other particles (so it cannot be distinguished from them). The chemical potential μ is the *total* change in free energy for these two steps. From the discussion above, it follows that the free energy change during the first step is equal to $w(\mathbf{r})$ (apart possibly from a constant). This is the contribution to μ from interactions, the **excess chemical potential**, $\mu^{\text{ex}}(\mathbf{r})$, at location $\mathbf{r}$ in an inhomogeneous system.[21] The remainder is the change in free energy associated with the release of the particle from the point $\mathbf{r}$ (due to the ideal entropy of mixing) and is equal to the ideal part of μ. It is the sum of these two contributions that is constant everywhere at equilibrium, as expressed in Equation 5.15, while each of these contributions is $\mathbf{r}$ dependent. We conclude that *the potential of mean force and the excess chemical potential are essentially the same thing*; they differ at most by a constant: $\mu^{\text{ex}}(\mathbf{r}) = w(\mathbf{r}) + \text{const}$. If we by convention set $w(\mathbf{r})$ to zero in bulk, we have

$$\mu^{\text{ex}}(\mathbf{r}) = w(\mathbf{r}) + \mu^{\text{ex,b}} \quad (\text{when } w^{\text{b}} = 0), \tag{5.23}$$

where $\mu^{\text{ex,b}}$ is the reversible work against interactions to insert (or create) a particle in the bulk phase. The total chemical potential is given by

$$\mu = k_B T \ln(\Lambda^3 n(\mathbf{r})) + \mu^{\text{ex}}(\mathbf{r}) = k_B T \ln n(\mathbf{r}) + w(\mathbf{r}) + \text{const} \tag{5.24}$$

where $k_B T \ln(\Lambda^3 n)$ is the ideal (non-interactional) part of the chemical potential.

For an ideal gas in an external field, $\mu^{\text{ex}}(\mathbf{r}) = v(\mathbf{r})$ since the gas particles do not interact with each other ($w^{\text{intr}} = 0$) and we have (cf. Equations 3.34 and 3.46)

$$\mu = k_B T \ln(\Lambda^3 n(\mathbf{r})) + v(\mathbf{r}) \quad (\text{ideal gas}).$$

We can write this as (cf. Equation 3.45)

$$n(\mathbf{r}) = \zeta e^{-\beta v(\mathbf{r})} \quad (\text{ideal gas})$$

---

[21] In an alternative definition of the excess (nonideal) chemical potential, which is used in some treatises, the external potential is not included. The excess part then only contains the intrinsic contributions. In the present treatise, μ^{ex} includes contributions from *all* kinds of interactions, both external and interparticle interactions.

where the activity $\zeta = e^{\beta\mu}/\Lambda^3$. Alternatively, this can be written (cf. Equation 3.13)

$$n(\mathbf{r}) = n(\mathbf{r}_0)e^{-\beta[v(\mathbf{r})-v(\mathbf{r}_0)]} \quad \text{(ideal gas)},$$

which also can be obtained from Equation 5.16 with $w(\mathbf{r}) = v(\mathbf{r})$.

5.5 DISTRIBUTION FUNCTIONS OF VARIOUS ORDERS FOR SPHERICAL PARTICLES

5.5.1 Singlet Distribution Function

Let us consider an inhomogeneous fluid in equilibrium with a bulk fluid. The interaction energy is pairwise additive, Equation 5.9. The density distribution is given by Equation 5.20, which can be written

$$n(\mathbf{r}_1) = n^b e^{-\beta[v(\mathbf{r}_1)+w^{\text{intr}}(\mathbf{r}_1)]}, \tag{5.25}$$

when we select $w = 0$ in the bulk. For a given n^b and for any external potential $v(\mathbf{r}_1)$, the density $n(\mathbf{r}_1)$ is expressed by this equation.

When v is varied, the intrinsic potential of mean force w^{intr} and the density distribution n will change. Different $v(\mathbf{r}_1)$ occur, for example, when we study a fluid near various surfaces in contact with a bulk fluid; different surfaces give rise to different v. Another example is a fluid between two walls for various surface separations (as in Figure 5.5 for diverse slit widths). In the latter case, $v(\mathbf{r}_1)$ depends on the surface separation since the external potential originates from both walls and the interaction depends on the distances between $\mathbf{r}_1$ and both surfaces. For large separations, the density in the middle approaches n^b when the fluid in the slit is in equilibrium with a bulk fluid.

We can write Equation 5.25 as

$$n(\mathbf{r}_1) = n^b g^{(1)}(\mathbf{r}_1),$$

where $g^{(1)}(\mathbf{r}_1) = \exp(-\beta[v(\mathbf{r}_1) + w^{\text{intr}}(\mathbf{r}_1)])$ is called the **singlet distribution function** (a singlet function depends on one coordinate point, also called a one-point function). In the bulk phase, $g^{(1)}(\mathbf{r}_1) = 1$. In order to emphasize that $n(\mathbf{r}_1)$ is a singlet density distribution, we can alternatively write it as $n^{(1)}(\mathbf{r}_1)$, but in most cases we will not show the superscript (1) for the density.

Let us apply Equation 5.25 for a particular kind of choices of $v(\mathbf{r}_1)$, namely the density distribution in the absence or presence of an additional body immersed in the fluid at some fixed location. This body contributes to the external potential that is changed from $v(\mathbf{r}_1)$ to $v(\mathbf{r}_1) + v^{\text{body}}(\mathbf{r}_1)$ when the additional body is added, where $v^{\text{body}}(\mathbf{r}_1)$ is due to the presence of the latter. The particle distributions are affected by the change in external potential, which leads to a change in the intrinsic part of the potential of mean force from $w^{\text{intr}}(\mathbf{r}_1)$ to, say, $w^{\text{intr}}(\mathbf{r}_1) + \Delta w^{\text{intr}}(\mathbf{r}_1)$, where $\Delta w^{\text{intr}}(\mathbf{r}_1)$ is as yet unknown. If we insert the new values of these functions in Equation 5.25, it follows that the particle density becomes

$$n(\mathbf{r}_1)\Big|_{\text{with body present}} = n^b e^{-\beta[v(\mathbf{r}_1)+v^{\text{body}}(\mathbf{r}_1)+w^{\text{intr}}(\mathbf{r}_1)+\Delta w^{\text{intr}}(\mathbf{r}_1)]} \tag{5.26}$$

$$= n(\mathbf{r}_1)e^{-\beta[v^{\text{body}}(\mathbf{r}_1)+\Delta w^{\text{intr}}(\mathbf{r}_1)]}$$

where $n(\mathbf{r}_1)$ is the density in the absence of the body and Equation 5.25 has been used to obtain the last equality.

5.5.2 Pair Distribution Function

A particularly important case is when the additional body in Equation 5.26 is a particle of the same species as those of the fluid. Let us place this particle with its center at $\mathbf{r}_2$. Then the density $n(\mathbf{r}_1)\big|_{\text{with body present}}$ is equal to $n(\mathbf{r}_1|\mathbf{r}_2)$ like the case depicted in Figure 5.8 (with $\mathbf{r}' = \mathbf{r}_1$ and $\mathbf{r} = \mathbf{r}_2$). The mobile particles feel the interaction with the fixed particle, that is, they feel the pair potential $u(\mathbf{r}_1, \mathbf{r}_2)$ in addition to $v(\mathbf{r}_1)$ so $v^{\text{body}}(\mathbf{r}_1)$ is equal to $u(\mathbf{r}_1, \mathbf{r}_2)$ in this case. Furthermore, $\Delta w^{\text{intr}}(\mathbf{r}_1)$ is given by

$$\Delta w^{\text{intr}}(\mathbf{r}_1|\mathbf{r}_2) = w^{\text{intr}}(\mathbf{r}_1|\mathbf{r}_2) - w^{\text{intr}}(\mathbf{r}_1),$$

where we use the notation $w^{\text{intr}}(\mathbf{r}_1|\mathbf{r}_2)$ to mark that the additional particle is placed at $\mathbf{r}_2$. Equation 5.26 for this case becomes

$$n(\mathbf{r}_1|\mathbf{r}_2) = n(\mathbf{r}_1)e^{-\beta[u(\mathbf{r}_1,\mathbf{r}_2)+\Delta w^{\text{intr}}(\mathbf{r}_1|\mathbf{r}_2)]}. \tag{5.27}$$

This is the same kind of conditional number density as shown in Figure 5.8 (with $\mathbf{r}_1 = \mathbf{r}'$ and $\mathbf{r}_2 = \mathbf{r}$) and discussed for a bulk fluid in connection with Figure 5.1. The **pair distribution function**[22] $g^{(2)}(\mathbf{r}_1, \mathbf{r}_2)$ is defined from

$$n(\mathbf{r}_1|\mathbf{r}_2) = n(\mathbf{r}_1)g^{(2)}(\mathbf{r}_1, \mathbf{r}_2). \tag{5.28}$$

It gives the fraction $n(\mathbf{r}_1|\mathbf{r}_2)/n(\mathbf{r}_1)$ that tells how much a fixed particle at $\mathbf{r}_2$ influences the density distribution on a relative scale (relative to the density before we added the particle). Here and in what follows, we will always use the superscript "(2)" to distinguish the pair distribution function $g^{(2)}$ from the distribution functions $g^{(1)}$, $g^{(3)}$, etc. of other orders. By comparing with Equation 5.27, we see that

$$g^{(2)}(\mathbf{r}_1, \mathbf{r}_2) = e^{-\beta[u(\mathbf{r}_1,\mathbf{r}_2)+w^{\text{intr}}(\mathbf{r}_1|\mathbf{r}_2)-w^{\text{intr}}(\mathbf{r}_1)]}. \tag{5.29}$$

Note that $g^{(2)}(\mathbf{r}_1, \mathbf{r}_2)$ constitutes the pair distribution function for a system *without* any fixed particles. That the expression includes $w^{\text{intr}}(\mathbf{r}_1|\mathbf{r}_2)$ for a fixed particle at $\mathbf{r}_2$ is just a theoretical device to calculate $g^{(2)}$. As we shall see later, the pair distribution function is symmetric $g^{(2)}(\mathbf{r}_1, \mathbf{r}_2) = g^{(2)}(\mathbf{r}_2, \mathbf{r}_1)$, although this is not immediately apparent from Equation 5.29.

---

[22] The function $g^{(2)}(\mathbf{r}_1, \mathbf{r}_2)$ is sometimes called the "pair correlation function," but in this treatise this name is reserved for $h^{(2)} = g^{(2)} - 1$, see Equation 5.32.

It is illuminating to obtain Equation 5.27 in a slightly different manner, whereby we will also see the link to the excess chemical potential. In the absence of the fixed particle at $\mathbf{r}_2$ the density of the fluid is according to Equation 5.17 with $n^0 = n^b$ given by

$$n(\mathbf{r}_1) = n^b e^{-\beta w(\mathbf{r}_1)} = n^b e^{-\beta[\mu^{\mathrm{ex}}(\mathbf{r}_1) - \mu^{\mathrm{ex,b}}]}, \tag{5.30}$$

where we have inserted Equation 5.23 to obtain the last equality. When we place a fixed particle at $\mathbf{r}_2$, the density becomes instead

$$n(\mathbf{r}_1|\mathbf{r}_2) = n^b e^{-\beta w(\mathbf{r}_1|\mathbf{r}_2)} = n^b e^{-\beta[\mu^{\mathrm{ex}}(\mathbf{r}_1|\mathbf{r}_2) - \mu^{\mathrm{ex,b}}]}. \tag{5.31}$$

In both cases the fluid is, as before, in equilibrium with a bulk phase with density n^b. By taking the ratio between these equations, we have

$$\frac{n(\mathbf{r}_1|\mathbf{r}_2)}{n(\mathbf{r}_1)} = e^{-\beta[w(\mathbf{r}_1|\mathbf{r}_2) - w(\mathbf{r}_1)]} = e^{-\beta[\mu^{\mathrm{ex}}(\mathbf{r}_1|\mathbf{r}_2) - \mu^{\mathrm{ex}}(\mathbf{r}_1)]},$$

where the first equality is the same as Equation 5.27 since $w(\mathbf{r}_1) = v(\mathbf{r}_1) + w^{\mathrm{intr}}(\mathbf{r}_1)$ and $w(\mathbf{r}_1|\mathbf{r}_2) = v(\mathbf{r}_1) + u(\mathbf{r}_1, \mathbf{r}_2) + w^{\mathrm{intr}}(\mathbf{r}_1|\mathbf{r}_2)$.

A fixed particle at $\mathbf{r}_2$ has a vanishing influence on the density distribution at positions $\mathbf{r}_1$ far from it since $u(\mathbf{r}_1, \mathbf{r}_2) \to 0$ when $r_{12} \to \infty$ and hence $g^{(2)}(\mathbf{r}_1, \mathbf{r}_2) = n(\mathbf{r}_1|\mathbf{r}_2)/n(\mathbf{r}_1) \to 1$ far from the particle. The **pair correlation function**, also called the **total correlation function**, $h^{(2)}(\mathbf{r}_1, \mathbf{r}_2)$ is defined as

$$h^{(2)}(\mathbf{r}_1, \mathbf{r}_2) = g^{(2)}(\mathbf{r}_1, \mathbf{r}_2) - 1. \tag{5.32}$$

It gives the relative deviation of the density $n(\mathbf{r}_1|\mathbf{r}_2)$ from $n(\mathbf{r}_1)$

$$h^{(2)}(\mathbf{r}_1, \mathbf{r}_2) = h^{(2)}(\mathbf{r}_2, \mathbf{r}_1) = \frac{n(\mathbf{r}_1|\mathbf{r}_2) - n(\mathbf{r}_1)}{n(\mathbf{r}_1)} \tag{5.33}$$

and $h^{(2)}(\mathbf{r}_1, \mathbf{r}_2) \to 0$ when r_{12} becomes very large. Particles at $\mathbf{r}_1$ and $\mathbf{r}_2$ are then uncorrelated to each other.

The pair distribution function for a bulk fluid can be obtained as a special case of the results obtained above. In this case, $n(\mathbf{r}_1) = n^b$ is independent of $\mathbf{r}_1$ and $g^{(2)}(\mathbf{r}_1, \mathbf{r}_2) = g^{(2)}(|\mathbf{r}_2 - \mathbf{r}_1|) = g^{(2)}(r_{12})$ due to spherical symmetry. Equation 5.28 becomes

$$n(\mathbf{r}_1|\mathbf{r}_2) = n^b g^{(2)}(r_{12}) \quad \text{(bulk phase)}$$

in accordance with Equation 5.7, where $g^{(2)}(r)$ is denoted by $g(r)$. From Equation 5.27, it follows that

$$g^{(2)}(r_{12}) = e^{-\beta[u(r_{12}) + w^{(2)\mathrm{intr}}(r_{12}) - w^{\mathrm{intr,b}}]} \quad \text{(bulk phase)}, \tag{5.34}$$

where $w^{(2)\text{intr}}(r_{12}) = w^{\text{intr}}(\mathbf{r}_1|\mathbf{r}_2)$ has superscript "(2)" to indicate that it is a two-particle function (to distinguish it from $w^{\text{intr}}(\mathbf{r})$ in Equation 5.19) and where $w^{\text{intr,b}}$ is the value of w^{intr} in bulk. The difference $w^{(2)\text{intr}}(r_{12}) - w^{\text{intr,b}}$ in the exponent is the *indirect contribution to the potential of mean force* between a pair of particles at $\mathbf{r}_1$ and $\mathbf{r}_2$, that is, the reversible work against the interactions with the surrounding particles when one of the particles is brought from far away in the bulk to the distance r_{12} from the other one; compare with Figure 5.3 and the discussion in Section 5.1 (the indirect force on particle 1 considered there is equal to $-\nabla_1 w^{(2)\text{intr}}(r_{12})$ in the present notation). The direct contribution from the interaction between the two particles is expressed by the term $u(r_{21})$ in the exponent. If we by convention set w equal to zero in the bulk phase, we have $w^{\text{intr,b}} = 0$, which simplifies Equation 5.34. The total pair potential of mean force is $w^{(2)}(r_{12}) = u(r_{12}) + w^{(2)\text{intr}}(r_{12})$ and we have

$$g^{(2)}(r_{12}) = e^{-\beta w^{(2)}(r_{12})} \quad \text{(bulk phase).} \tag{5.35}$$

The mean force on a particle at $\mathbf{r}_1$ when a particle is located at $\mathbf{r}_2$ is $\mathbf{F}_1(r_{12}) = -\nabla_1 w^{(2)}(r_{12})$ and the force on the other particle is $\mathbf{F}_2(r_{12}) = -\nabla_2 w^{(2)}(r_{12})$.

The experimental pair distribution function $g^{(2)}(r)$ for liquid argon at 85 K (close to the triple point) is shown in Figure 5.9a. It has been obtained by neutron scattering experiments. The pair potential of mean force given by $\beta w^{(2)}(r) = -\ln g^{(2)}(r)$ is also shown in the figure. Note that $w^{(2)}$ has minima where $g^{(2)}$ has maxima and vice versa.

Figure 5.9b shows the structure factor $S(k)$, which is an entity obtained from the scattering intensity in the experiments as will be explained later in Section 5.6. The structure factor is directly related to the pair correlation function $h^{(2)}(r) = g^{(2)}(r) - 1$ via the relationship $S(k) = 1 + n^b \tilde{h}^{(2)}(k)$, where $\tilde{h}^{(2)}(k)$ is the Fourier transform of $h^{(2)}(r)$ defined as $\tilde{h}^{(2)}(k) = \int d\mathbf{r}\, h^{(2)}(r) \sin(kr)/(kr)$; for details, see Section 5.6.

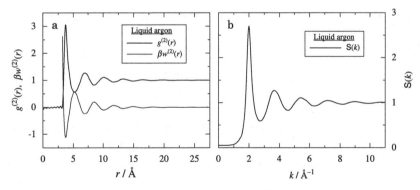

FIGURE 5.9 (a) Experimental pair distribution function $g^{(2)}(r)$ for liquid argon close to the triple point and the corresponding pair potential of mean force $w^{(2)}(r)$. The rippled feature in $g^{(2)}(r)$ for $r \lesssim 3$ Å is an artifact caused by the processing of the experimental neutron scattering data. This part of the data is left out from $w^{(2)}(r)$. (b) The structure factor $S(k)$ for the same system. The curve shows smoothed and somewhat extended experimental data.[23]

[23] Figure 5.9 is based on data from J. L. Yarnell, M. J. Katz, R. G. Wenzel, and S. H. Koening, *Phys. Rev. A* **7** (1973) 2130.

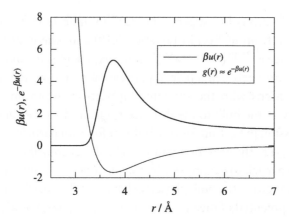

FIGURE 5.10 The pair distribution function $g(r) \approx e^{-\beta u(r)}$ for argon gas at low densities calculated with an accurate pair potential $u(r)$ [see text], which is also plotted. The temperature is 85 K; same as in Figure 5.9.

For a thin gas phase $w^{(2)\text{intr}}(r_{12})$ in Equation 5.34 is small since the particle at $\mathbf{r}_2$ interacts weakly with the other particle for most of the time since they are usually far away. When $n^b \to 0$, both $w^{(2)\text{intr}}$ and $w^{\text{intr,b}}$ go to zero. Therefore, for sufficiently small n^b, we have

$$g^{(2)}(r_{12}) \approx e^{-\beta u(r_{12})} \quad (\text{small } n^b) \tag{5.36}$$

as a good approximation. An example is shown in Figure 5.10, where the distribution function function for a low density argon gas is plotted. The pair potential $u(r)$ used has been obtained[24] from an adaption of computer simulation results to experimental data.

5.5.3 Distribution Functions in the Canonical Ensemble

Let us consider a system with temperature T, volume V and N particles in the canonical ensemble. The total interaction energy is given by

$$\check{U}_N^{\text{pot}} = \check{U}_N^{\text{intr}}(\mathbf{r}_1, \ldots, \mathbf{r}_N) + \sum_{\nu=1}^{N} v(\mathbf{r}_\nu), \tag{5.37}$$

(cf. Equation 5.8). We do not assume pairwise additivity of the interaction between the fluid particles, so $\check{U}_N^{\text{intr}}$ can have any form. As we have seen in Equation 3.9, the probability density to observe a certain particle configuration is equal to

$$\mathcal{P}_N^{(N)}(\mathbf{r}^N) = \frac{1}{Z_N} e^{-\beta \check{U}_N^{\text{pot}}(\mathbf{r}^N)}, \tag{5.38}$$

where $\mathbf{r}^N \equiv \mathbf{r}_1, \mathbf{r}_2, \ldots, \mathbf{r}_N$ and Z_N is the configurational partition function defined in Equation 3.10. $\mathcal{P}_N^{(N)}(\mathbf{r}^N) d\mathbf{r}^N$ gives the probability that the center of particle 1 is in the volume

[24] J. A. Barker, R. A. Fisher, and R. O. Watts, *Mol. Phys.* **21** (1971) 657.

element $d\mathbf{r}_1$ located at the point $\mathbf{r}_1$ at the same time as particle 2 is in $d\mathbf{r}_2$ at $\mathbf{r}_2$ ... and particle N is in $d\mathbf{r}_N$ at $\mathbf{r}_N$. Since we do not care about the numbering of the molecules (they are indistinguishable), we have to consider that any of the N molecules can be located at $\mathbf{r}_1$, any of the remaining $N - 1$ molecules at $\mathbf{r}_2, \ldots$ and analogously any of $N - (l - 1)$ particles at $\mathbf{r}_l, \ldots$, any of two particles at $\mathbf{r}_{N-1}$ and the only remaining one at $\mathbf{r}_N$. To obtain a physically relevant function for indistinguishable particles, we therefore multiply $\mathcal{P}_N^{(N)}$ by $N(N - 1) \times \cdots \times 2 \times 1 = N!$ and define the **N-particle density distribution function**

$$n_N^{(N)}(\mathbf{r}^N) = N! \mathcal{P}_N^{(N)}(\mathbf{r}^N) = \frac{N!}{Z_N} e^{-\beta \check{U}_N^{\mathrm{pot}}(\mathbf{r}^N)}. \tag{5.39}$$

It has the dimension (number density)^N, that is, (length)$^{-3N}$.

The functions $\mathcal{P}_N^{(N)}$ and $n_N^{(N)}$ give full information about the distribution of all N particles. However, we are seldom – if ever – interested in all of this detail. We are mostly interested in distributions of a few of them. Say that we are interested in the probability to find a particle in a certain volume element irrespective of where the other particles are. The probability $\mathcal{P}_N^{(1)}(\mathbf{r}_1) d\mathbf{r}_1$ that particle 1 is in $d\mathbf{r}_1$ at the point $\mathbf{r}_1$, irrespective of the locations of the other particles, is given by

$$\mathcal{P}_N^{(1)}(\mathbf{r}_1) d\mathbf{r}_1 = \left[\int d\mathbf{r}_2 d\mathbf{r}_3 \ldots d\mathbf{r}_N \mathcal{P}_N^{(N)}(\mathbf{r}^N) \right] d\mathbf{r}_1. \tag{5.40}$$

In the square brackets, we have summed (integrated) the probability for all possible positions for particles 2, 3, ...,N. Since the particles are indistinguishable, the relevant entity is not the probability to find particle 1 in $d\mathbf{r}_1$, but rather the probability to find *any* particle there. All N particles have equal probability to be in the volume element $d\mathbf{r}_1$; each of them with probability $\mathcal{P}_N^{(1)}(\mathbf{r}_1) d\mathbf{r}_1$. Therefore, $N\mathcal{P}_N^{(1)}(\mathbf{r}_1) d\mathbf{r}_1$ gives the mean number of particles (i.e., particle centers) in the volume element. By dividing by $d\mathbf{r}_1$ (the volume of the volume element), we obtain the number density of particles at $\mathbf{r}_1$, that is,

$$n_N^{(1)}(\mathbf{r}_1) = N\mathcal{P}_N^{(1)}(\mathbf{r}_1) = \frac{N}{Z_N} \int d\mathbf{r}_2 d\mathbf{r}_3 \ldots d\mathbf{r}_N e^{-\beta \check{U}_N^{\mathrm{pot}}(\mathbf{r}^N)}. \tag{5.41}$$

This is the usual **number density distribution** (also called the **singlet density distribution**) and we can alternatively write $n_N^{(1)}(\mathbf{r}_1)$ as $n_N(\mathbf{r}_1)$, where we have dropped superscript (1) and where subscript N indicates, as before, that there are N particles in the system. Equation 5.41 gives the density distribution explicitly in terms of the interaction energy $\check{U}_N^{\mathrm{pot}}$ for the particles in the system.

By inserting $\check{U}_N^{\mathrm{pot}}$ from Equation 5.37, we can write Equation 5.41 as

$$n_N(\mathbf{r}_1) = e^{-\beta v(\mathbf{r}_1)} \frac{N}{Z_N} \int d\mathbf{r}_2 d\mathbf{r}_3 \ldots d\mathbf{r}_N e^{-\beta \left[\check{U}_N^{\mathrm{intr}}(\mathbf{r}^N) + \sum_{\nu > 1} v(\mathbf{r}_\nu) \right]}$$

$$= n^0 e^{-\beta [v(\mathbf{r}_1) + w_N^{\mathrm{intr}}(\mathbf{r}_1)]} = n^0 e^{-\beta w_N(\mathbf{r}_1)} \tag{5.42}$$

(cf. Equation 5.20), where w_N^{intr} is defined from

$$e^{-\beta w_N^{intr}(\mathbf{r}_1)} = \frac{N}{n^0 Z_N} \int d\mathbf{r}_2 d\mathbf{r}_3 \ldots d\mathbf{r}_N e^{-\beta \left[\breve{U}_N^{intr}(\mathbf{r}^N) + \sum_{\nu > 1} v(\mathbf{r}_\nu)\right]} \tag{5.43}$$

and $w_N = v + w_N^{intr}$. Here n^0 is arbitrary, but as before we can take $n^0 = n_N(\mathbf{r}_0)$ as the density at a point $\mathbf{r}_0$ where we have selected $w_N(\mathbf{r}_0) = 0$.

If we have a part of the system far from the source of the external potential, $v(\mathbf{r})$ is negligible there and $n(\mathbf{r})$ has a constant value; the fluid is in its bulk state in this part. Let us select $\mathbf{r}_0$ to be located somewhere in this part of the system, where the constant density is, say, n^b. Then we have $n_N(\mathbf{r}_0) = n^b$ and

$$n_N(\mathbf{r}_1) = n^b e^{-\beta[v(\mathbf{r}_1) + w_N^{intr}(\mathbf{r}_1)]}$$

(cf. Equation 5.25).

Let us now consider two of the particles in the system. The probability that particle 1 is in $d\mathbf{r}_1$ at the point $\mathbf{r}_1$ simultaneously as particle 2 is in $d\mathbf{r}_2$ at $\mathbf{r}_2$, irrespective of the locations of the other particles, is given by

$$\mathcal{P}_N^{(2)}(\mathbf{r}_1, \mathbf{r}_2) d\mathbf{r}_1 d\mathbf{r}_2 = \left[\int d\mathbf{r}_3 \ldots d\mathbf{r}_N \mathcal{P}_N^{(N)}(\mathbf{r}^N)\right] d\mathbf{r}_1 d\mathbf{r}_2. \tag{5.44}$$

Again, since the particles are indistinguishable, we have to consider that any of the N particles can be in $d\mathbf{r}_1$ and any of the $N-1$ remaining ones in $d\mathbf{r}_2$, so the physically relevant entity is

$$n_N^{(2)}(\mathbf{r}_1, \mathbf{r}_2) = N(N-1) \mathcal{P}_N^{(2)}(\mathbf{r}_1, \mathbf{r}_2)$$

$$= \frac{N(N-1)}{Z_N} \int d\mathbf{r}_3 \ldots d\mathbf{r}_N e^{-\beta \breve{U}_N^{pot}(\mathbf{r}^N)}. \tag{5.45}$$

This is called the **pair density distribution** function. It has the dimension (number density)^2, that is, (length)$^{-6}$.

We now introduce the conditional probability $\mathcal{P}_N^{(1)}(\mathbf{r}_1|\mathbf{r}_2)d\mathbf{r}_1$ to find the center of particle 1 in $d\mathbf{r}_1$ at the point $\mathbf{r}_1$ provided that particle 2 is located at $\mathbf{r}_2$. When particle 2 is at $\mathbf{r}_2$, there is an equal probability for any of the other $N-1$ particles to be in the volume element $d\mathbf{r}_1$; each of them has a probability $\mathcal{P}_N^{(1)}(\mathbf{r}_1|\mathbf{r}_2)d\mathbf{r}_1$ to be there. Thus, $(N-1)\mathcal{P}_N^{(1)}(\mathbf{r}_1|\mathbf{r}_2)d\mathbf{r}_1$ gives the mean number of particles (particle centers) in the volume element. Dividing by the volume $d\mathbf{r}_1$, we obtain the number density of particles at $\mathbf{r}_1$

$$n_N^{(1)}(\mathbf{r}_1|\mathbf{r}_2) = (N-1)\mathcal{P}_N^{(1)}(\mathbf{r}_1|\mathbf{r}_2) \tag{5.46}$$

when particle 2 is located at $\mathbf{r}_2$. We have

$$\mathcal{P}_N^{(2)}(\mathbf{r}_1, \mathbf{r}_2) = \mathcal{P}_N^{(1)}(\mathbf{r}_1|\mathbf{r}_2)\mathcal{P}_N^{(1)}(\mathbf{r}_2),$$

where $\mathcal{P}_N^{(1)}(\mathbf{r}_2)$ is, as before, the probability density that particle 2 is at $\mathbf{r}_2$, irrespective of the locations of the other particles. From this relationship together with Equations 5.45 and 5.46, we obtain

$$n_N^{(2)}(\mathbf{r}_1, \mathbf{r}_2) = n_N^{(1)}(\mathbf{r}_1|\mathbf{r}_2)n_N^{(1)}(\mathbf{r}_2), \tag{5.47}$$

where we have used Equation 5.41 to obtain the last factor.

Analogously to Equation 5.28, the pair distribution function $g_N^{(2)}(\mathbf{r}_1, \mathbf{r}_2)$ is defined from

$$n_N^{(1)}(\mathbf{r}_1|\mathbf{r}_2) = n_N^{(1)}(\mathbf{r}_1)g_N^{(2)}(\mathbf{r}_1, \mathbf{r}_2) \tag{5.48}$$

and we have from Equation 5.47

$$n_N^{(2)}(\mathbf{r}_1, \mathbf{r}_2) = n_N^{(1)}(\mathbf{r}_1)n_N^{(1)}(\mathbf{r}_2)g_N^{(2)}(\mathbf{r}_1, \mathbf{r}_2). \tag{5.49}$$

From this relationship, it is obvious that $g_N^{(2)}$ is symmetric, $g_N^{(2)}(\mathbf{r}_1, \mathbf{r}_2) = g_N^{(2)}(\mathbf{r}_2, \mathbf{r}_1)$, since $n_N^{(2)}$ is symmetric as follows directly from its definition, Equation 5.45. For a bulk fluid, we have

$$n_N^{(1)}(\mathbf{r}_1|\mathbf{r}_2) = n^{\mathrm{b}}g_N^{(2)}(r_{12}) \quad \text{(bulk phase)}$$

and

$$n_N^{(2)}(\mathbf{r}_1, \mathbf{r}_2) = n_N^{(2)}(r_{12}) = (n^{\mathrm{b}})^2 g_N^{(2)}(r_{12}) \quad \text{(bulk phase)}.$$

We can express the pair distribution function explicitly in terms of the interaction energy $\check{U}_N^{\mathrm{pot}}$ for the particles in the system by using Equations 5.41, 5.45 and 5.49.

In an analogous manner, we can introduce the triplet, quartet and higher order distribution functions, which were mentioned at the end of Section 5.1. For the general case of l particles, the probability that particle 1 is located in $d\mathbf{r}_1$ at the point $\mathbf{r}_1$, particle 2 in $d\mathbf{r}_2$ at $\mathbf{r}_2, \ldots$ and particle l in $d\mathbf{r}_l$ at $\mathbf{r}_l$, irrespective of the locations of the remaining molecules $l+1, \ldots, N$ equals

$$\mathcal{P}_N^{(l)}(\mathbf{r}_1, \ldots, \mathbf{r}_l)d\mathbf{r}_1 \ldots d\mathbf{r}_l = \left[\int d\mathbf{r}_{l+1} \ldots d\mathbf{r}_N \mathcal{P}_N^{(N)}(\mathbf{r}^N)\right] d\mathbf{r}_1 \ldots d\mathbf{r}_l.$$

This means that

$$\mathcal{P}_N^{(l)}(\mathbf{r}_1, \ldots, \mathbf{r}_l) = \int d\mathbf{r}_{l+1} \ldots d\mathbf{r}_N \mathcal{P}_N^{(N)}(\mathbf{r}^N)$$

$$= \frac{1}{Z_N}\int d\mathbf{r}_{l+1} \ldots d\mathbf{r}_N e^{-\beta \check{U}_N^{\mathrm{pot}}(\mathbf{r}^N)}. \tag{5.50}$$

Again, we have to consider that any of the N molecules can be at $\mathbf{r}_1$, any of the remaining $N-1$ molecules at $\mathbf{r}_2 \ldots$ and any of the other $N-(l-1)$ particles at $\mathbf{r}_l$. Thus, we have to multiply $\mathcal{P}_N^{(l)}$ by

$$N \cdot (N-1) \cdot \ldots \cdot [N-(l-1)] = \frac{N!}{(N-l)!}$$

and define the *l*-point density distribution function (also called the *l*-particle density distribution function) as

$$n_N^{(l)}(\mathbf{r}_1,\ldots,\mathbf{r}_l) = \frac{N!}{(N-l)!} \mathcal{P}_N^{(l)}(\mathbf{r}_1,\ldots,\mathbf{r}_l)$$

$$= \frac{N!}{(N-l)!Z_N} \int d\mathbf{r}_{l+1}\ldots d\mathbf{r}_N e^{-\beta \check{U}_N^{\text{pot}}(\mathbf{r}^N)}. \qquad (5.51)$$

It has the dimension (number density)^l, that is, (length)$^{-3l}$. The triplet function is the case $l = 3$ and quartet function is $l = 4$.

5.6 THE STRUCTURE FACTOR FOR HOMOGENEOUS AND INHOMOGENEOUS FLUIDS ★

The pair distribution function of a bulk fluid can be experimentally determined by X-ray or neutron scattering experiments and for inhomogeneous fluids one can determine the structure on both singlet and pair distribution levels by such techniques. In these experiments, a sample of the fluid is exposed to an incident beam of X-rays or neutrons and the intensity of the scattered wave is measured in various directions. For bulk fluids, the intensity is usually measured as a function of the scattering angle ($2\theta_k$) as counted from the direction of an incident beam with wavelength λ_b, where b stands for beam (see Figure 5.11). From these measurements, one can determine an entity called the *structure factor* $S(k)$, where k, the length of the so-called *scattering vector* $\mathbf{k}$, is related to the scattering angle via $k = (4\pi/\lambda_b)\sin(\theta_k)$. $S(k)$ describes, as we will see, the influence of the fluid structure on the intensity of the scattered wave at the scattering angle; $S(k)$ is closely related to the pair

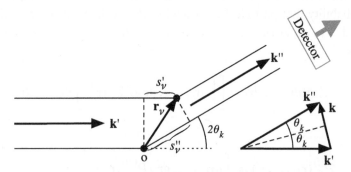

FIGURE 5.11 An illustration of some important entities in scattering measurements. An incident beam (wave) with wavelength λ_b and wave vector $\mathbf{k}'$ hits two point scatterers (particles), one at the origin (o) and one at location $\mathbf{r}_\nu$. The intensity of the scattered wave at scattering angle $2\theta_k$ is measured by a detector that lies far away and $\mathbf{k}''$ is the wave vector of the scattered wave in that direction. For elastic scattering, which is assumed, we have $|\mathbf{k}''| = |\mathbf{k}'| = 2\pi/\lambda_b$. The scattering vector $\mathbf{k}$ is defined as $\mathbf{k} = \mathbf{k}'' - \mathbf{k}'$ and from the Pythagoras theorem follows $k/2 = (2\pi/\lambda_b)\sin(\theta_k)$ as illustrated in the inset to the right. The two paths (full straight lines) travelled by the waves to the detector via the position $\mathbf{r}_\nu$ and via the origin, respectively, differ in lengths by $s'_\nu - s''_\nu$, where $s'_\nu = \hat{\mathbf{k}}' \cdot \mathbf{r}_\nu$ and $s''_\nu = \hat{\mathbf{k}}'' \cdot \mathbf{r}_\nu$ are the projections of the vector $\mathbf{r}_\nu$ in the directions $\hat{\mathbf{k}}'$ and $\hat{\mathbf{k}}''$.

correlation function of the fluid. While the structure factor depends only on k for a bulk fluid, it also depends on the direction of $\mathbf{k}$ for an inhomogeneous fluid, so we have $S = S(\mathbf{k})$ in the latter case. In this section, we will give a simplified presentation of the theory of scattering for simple fluids. An example of the experimental structure factor for a liquid argon in the bulk phase is given in Figure 5.9b and in Section 7.2.1.2 we will present some results for an inhomogeneous fluid.

We start by considering the scattering from a distribution of point scatterers, that is, particles that scatter from a single point only. Say that we have a set of N equal point scatterers located at $\mathbf{r}_\nu$, $1 \leq \nu \leq N$. The incident beam is a planar wave[25] proportional to $\exp(i\,\mathbf{k}' \cdot \mathbf{r}) = \cos(\mathbf{k}' \cdot \mathbf{r}) + i\sin(\mathbf{k}' \cdot \mathbf{r})$, where i is the imaginary unit and $\mathbf{k}'$ is the wave vector with length $k' = |\mathbf{k}'| = 2\pi/\lambda_b$, that is, k' is the wave number of the beam. The beam is thereby represented by its wave function, which is complex-valued.

The scattered wave from each particle is spherical, but at the detector which lies far away, the scattered wave can be approximated by a plane wave proportional to $\exp(i\,\mathbf{k}'' \cdot \mathbf{r})$ (see Figure 5.11). We therefore use a simple argument that gives the correct result using the planar wave approximation near the detector. The scattering is assumed to be elastic, so we have $k'' = k' = 2\pi/\lambda_b$, and we neglect multiple scattering (i.e., scattering of a wave that has been scattered by some other particle).

The scattered waves from the particles give rise to an interference pattern at the detector. The path lengths travelled by the waves are different for the various particles, so there is a difference in phase of the wave due to each of them. Say that the particle located at the origin gives a scattered wave $B\exp(i\,\mathbf{k}'' \cdot \mathbf{r})$ near the detector. Then the particle located at $\mathbf{r}_\nu$ gives rise to a wave with a phase shift since, as can be seen in Figure 5.11, the path travelled differs by $s'_\nu - s''_\nu$ compared to the path through the origin, where $s'_\nu = \hat{\mathbf{k}}' \cdot \mathbf{r}_\nu$ and $s''_\nu = \hat{\mathbf{k}}'' \cdot \mathbf{r}_\nu$. The phase shift is therefore $(s'_\nu - s''_\nu)2\pi/\lambda_b = (\mathbf{k}' - \mathbf{k}'') \cdot \mathbf{r}_\nu = -\mathbf{k} \cdot \mathbf{r}_\nu$, where $\mathbf{k} = \mathbf{k}'' - \mathbf{k}'$ is the **scattering vector**. The scattered wave due to the particle located at $\mathbf{r}_\nu$ therefore gives a scattered wave $B\exp(i\,[\mathbf{k}'' \cdot \mathbf{r} - \mathbf{k} \cdot \mathbf{r}_\nu])$ near the detector, where we have added the phase shift to the exponent.

We now sum over all N particles and obtain the total scattered wave near the detector

$$\sum_{\nu=1}^{N} Be^{i\,[\mathbf{k}'' \cdot \mathbf{r} - \mathbf{k} \cdot \mathbf{r}_\nu]} = Be^{i\,\mathbf{k}'' \cdot \mathbf{r}} \sum_{\nu=1}^{N} e^{-i\,\mathbf{k} \cdot \mathbf{r}_\nu}. \tag{5.52}$$

The intensity of the wave is given by

$$\text{Intensity} = \left| Be^{i\,\mathbf{k}'' \cdot \mathbf{r}} \sum_{\nu=1}^{N} e^{-i\,\mathbf{k} \cdot \mathbf{r}_\nu} \right|^2 = |B|^2 \sum_{\nu=1}^{N} \sum_{\nu'=1}^{N} e^{-i\,\mathbf{k} \cdot [\mathbf{r}_\nu - \mathbf{r}_{\nu'}]} \tag{5.53}$$

---

[25] The incident wave is proportional to $\exp(i\,[\mathbf{k}' \cdot \mathbf{r} - \omega'_b t])$, where $\mathbf{k}'$ is the wave vector of length $|\mathbf{k}'| = 2\pi/\lambda_b$ and ω'_b is the angular frequency. We will only consider the spatial coordinate dependence of the wave. The time dependence is given by the factor $\exp(-i\,\omega'_b t)$.

since for a complex number a we have $|a|^2 = a\underline{a}$, where $\underline{a}$ is the complex conjugate of a. Each of the terms with $\nu = \nu'$ in the sum is equal to 1, so we have

$$\text{Intensity} = |B|^2 \left[N + \sum_{\nu=1}^{N} \sum_{\substack{\nu'=1 \\ (\nu' \neq \nu)}}^{N} e^{-i\mathbf{k}\cdot[\mathbf{r}_\nu - \mathbf{r}_{\nu'}]} \right].$$

Note that the intensity depends on the separation vectors $\mathbf{r}_\nu - \mathbf{r}_{\nu'}$ between all pairs of particles in the system.

In a fluid, the particles are mobile so we need to take the ensemble average over all particle configurations, which yields

$$\text{Intensity} = |B|^2 \left[N + \left\langle \sum_{\nu=1}^{N} \sum_{\substack{\nu'=1 \\ (\nu' \neq \nu)}}^{N} e^{-i\mathbf{k}\cdot[\mathbf{r}_\nu - \mathbf{r}_{\nu'}]} \right\rangle \right].$$

The total **structure factor** $S_{\text{tot}}(\mathbf{k})$ is defined as[26]

$$S_{\text{tot}}(\mathbf{k}) = 1 + \frac{1}{N} \left\langle \sum_{\nu=1}^{N} \sum_{\substack{\nu'=1 \\ (\nu' \neq \nu)}}^{N} e^{-i\mathbf{k}\cdot[\mathbf{r}_\nu - \mathbf{r}_{\nu'}]} \right\rangle. \tag{5.54}$$

For an isotropic and homogeneous fluid, $S_{\text{tot}}(\mathbf{k})$ does not depend on the direction of $\mathbf{k}$, so we have the intensity $I(2\theta_k)$ at scattering angle $2\theta_k$ equal to

$$I(2\theta_k) = N |B|^2 S_{\text{tot}}(k). \tag{5.55}$$

Thus, the intensity is proportional to $S_{\text{tot}}(k)$, in agreement with the statements earlier in this section. For an inhomogeneous fluid, on the other hand, the intensity of the scattered wave depends on the direction of $\mathbf{k}$, so one has to consider not only the angle $2\theta_k$, but also the azimuthal angle ϕ_k of $\mathbf{k}''$ around the direction of the beam and we have

$$I(2\theta_k, \phi_k) = N |B|^2 S_{\text{tot}}(\mathbf{k}). \tag{5.56}$$

This result will be of use in Section 7.2.1.2 that deals with a narrowly confined and therefore inhomogeneous fluid.

---

[26] This entity is often denoted by $S(\mathbf{k})$, but we will, in congruence with the notation often used in liquid state theory, reserve $S(\mathbf{k})$ to the part that originates from the pair correlations in the fluid.

To find the relationship between $S_{tot}(\mathbf{k})$ and the pair distribution function, we write Equation 5.54 as

$$S_{tot}(\mathbf{k}) = 1 + \frac{1}{N} \int d\mathbf{r}^N \mathcal{P}_N^{(N)}(\mathbf{r}^N) \sum_{\nu=1}^{N} \sum_{\substack{\nu'=1 \\ (\nu' \neq \nu)}}^{N} e^{-i\mathbf{k}\cdot[\mathbf{r}_\nu - \mathbf{r}_{\nu'}]}, \qquad (5.57)$$

where the probability density $\mathcal{P}_N^{(N)}(\mathbf{r}^N)$ is defined in Equation 5.38. Let us investigate one of the terms in the sum in Equation 5.57, say, the one with subscripts $\nu = 1$ and $\nu' = 2$. It gives the contribution to the integral in $S_{tot}(\mathbf{k})$

$$\int d\mathbf{r}^N \mathcal{P}_N^{(N)}(\mathbf{r}^N) e^{-i\mathbf{k}\cdot[\mathbf{r}_1 - \mathbf{r}_2]} = \int d\mathbf{r}_1 d\mathbf{r}_2 \left[\int d\mathbf{r}_3 \dots d\mathbf{r}_N \mathcal{P}_N^{(N)}(\mathbf{r}^N) \right] e^{-i\mathbf{k}\cdot[\mathbf{r}_1 - \mathbf{r}_2]}$$

$$= \int d\mathbf{r}_1 d\mathbf{r}_2 \mathcal{P}_N^{(2)}(\mathbf{r}_1, \mathbf{r}_2) e^{-i\mathbf{k}\cdot[\mathbf{r}_1 - \mathbf{r}_2]}$$

$$= \frac{1}{N(N-1)} \int d\mathbf{r}_1 d\mathbf{r}_2 n_N^{(2)}(\mathbf{r}_1, \mathbf{r}_2) e^{-i\mathbf{k}\cdot[\mathbf{r}_1 - \mathbf{r}_2]},$$

where we have used Equations 5.44 and 5.45 to obtain the last two lines. Since the result is independent of the subscripts 1 and 2, we would obtain the same result for any of the $N(N-1)$ terms in the sum of Equation 5.57. Therefore,

$$S_{tot}(\mathbf{k}) = 1 + \frac{1}{N} \int d\mathbf{r}_1 d\mathbf{r}_2 n_N^{(2)}(\mathbf{r}_1, \mathbf{r}_2) e^{-i\mathbf{k}\cdot[\mathbf{r}_1 - \mathbf{r}_2]}$$

$$= 1 + \frac{1}{N} \int d\mathbf{r}_1 d\mathbf{r}_2 n_N^{(1)}(\mathbf{r}_1) n_N^{(1)}(\mathbf{r}_2) \left[1 + h_N^{(2)}(\mathbf{r}_1, \mathbf{r}_2) \right] e^{-i\mathbf{k}\cdot[\mathbf{r}_1 - \mathbf{r}_2]},$$

where we have inserted Equation 5.49 and used $g_N^{(2)}(\mathbf{r}_1, \mathbf{r}_2) = 1 + h_N^{(2)}(\mathbf{r}_1, \mathbf{r}_2)$. The contribution from the first term in the square brackets is

$$\frac{1}{N} \int d\mathbf{r}_1 d\mathbf{r}_2 n_N^{(1)}(\mathbf{r}_1) n_N^{(1)}(\mathbf{r}_2) e^{-i\mathbf{k}\cdot[\mathbf{r}_1 - \mathbf{r}_2]}$$

$$= \frac{1}{N} \left| \int d\mathbf{r}_1 n_N^{(1)}(\mathbf{r}_1) e^{-i\mathbf{k}\cdot\mathbf{r}_1} \right|^2 \equiv S_{singlet}(\mathbf{k}),$$

which defines $S_{singlet}(\mathbf{k})$ that depends on the singlet density distribution only. The pair correlations contribute to the rest

$$S_{pair}(\mathbf{k}) = S(\mathbf{k}) \equiv 1 + \frac{1}{N} \int d\mathbf{r}_1 d\mathbf{r}_2 n_N^{(1)}(\mathbf{r}_1) n_N^{(1)}(\mathbf{r}_2) h_N^{(2)}(\mathbf{r}_1, \mathbf{r}_2) e^{-i\mathbf{k}\cdot[\mathbf{r}_1 - \mathbf{r}_2]} \qquad (5.58)$$

and we have $S_{tot}(\mathbf{k}) = S_{singlet}(\mathbf{k}) + S_{pair}(\mathbf{k})$. We will use the notation $S(\mathbf{k})$ instead of $S_{pair}(\mathbf{k})$ since the former is common in liquid state theory. For an inhomogeneous fluid, $S_{singlet}(\mathbf{k})$

gives information about the density distribution; a concrete example will be given later in Section 7.2.1.2. $S(\mathbf{k})$ gives information about the pair correlation function for both homogeneous and inhomogeneous fluids.

At $\mathbf{k} = \mathbf{0}$, we have $S_{\text{singlet}}(\mathbf{0}) = N$ since $\int d\mathbf{r}_1 n_N^{(1)}(\mathbf{r}_1) = N$, so this contribution is huge for a macroscopic system; it corresponds experimentally to a significant part of the unscattered radiation that passes on a straight line along $\mathbf{k}'$ through the system. For bulk systems, $S_{\text{singlet}}(\mathbf{k})$ is negligible for $\mathbf{k} \neq \mathbf{0}$. To see this, let us consider a macroscopic bulk fluid with density n^b enclosed in a container of volume $V = L_x L_y L_z$ with sides oriented along the coordinate axes. We have

$$S_{\text{singlet}}(\mathbf{k}) = \frac{1}{N} \left| \int_{-L_x/2}^{L_x/2} dx_1 \int_{-L_y/2}^{L_y/2} dy_1 \int_{-L_z/2}^{L_z/2} dz_1 n^b e^{-i\mathbf{k}\cdot\mathbf{r}_1} \right|^2$$

$$= N \left[\frac{\sin(k_x L_x/2)}{k_x L_x/2} \frac{\sin(k_y L_y/2)}{k_y L_y/2} \frac{\sin(k_z L_z/2)}{k_z L_z/2} \right]^2$$

where we have used $\int_{-a}^{a} dt \, e^{-ibt} = 2\sin(ba)/b$ and $n^b = N/V$. Since L_x, L_y, and L_z have macroscopic dimensions and $\sin(t)/t \approx 0$ when t is large, $S_{\text{singlet}}(\mathbf{k})$ is very small except in a very small region of $\mathbf{k}$ values near $\mathbf{k} = \mathbf{0}$.

This can be substantiated further by introducing

$$\mathfrak{D}_{L_\alpha}(k_\alpha) = \frac{L_\alpha}{2\pi} \left[\frac{\sin(k_\alpha L_\alpha/2)}{k_\alpha L_\alpha/2} \right]^2, \tag{5.59}$$

and writing

$$S_{\text{singlet}}(\mathbf{k}) = \frac{N(2\pi)^3}{L_x L_y L_z} \mathfrak{D}_L^{(3)}(\mathbf{k}) = n^b (2\pi)^3 \mathfrak{D}_L^{(3)}(\mathbf{k}),$$

where $\mathfrak{D}_L^{(3)}(\mathbf{k}) = \mathfrak{D}_{L_x}(k_x) \mathfrak{D}_{L_y}(k_y) \mathfrak{D}_{L_z}(k_z)$. The function $\mathfrak{D}_{L_\alpha}(k_\alpha)$ has the properties

$$\lim_{L_\alpha \to \infty} \mathfrak{D}_{L_\alpha}(k_\alpha) \to 0, \quad k_\alpha \neq 0$$

$$\int dk_\alpha \, \mathfrak{D}_{L_\alpha}(k_\alpha) = 1,$$

as follows from the fact that $\int_{-\infty}^{\infty} dt \, [\sin(bt/2)/(bt/2)]^2 = 2\pi/b$. Since L_α is very large, this means that $\mathfrak{D}_{L_\alpha}(k_\alpha)$ is virtually equal to the Dirac delta function $\delta(k_\alpha)$, which is defined in Appendix 5A. In practice, we therefore have for a macroscopic system

$$S_{\text{singlet}}(\mathbf{k}) = n^b (2\pi)^3 \delta^{(3)}(\mathbf{k}) \quad \text{(bulk fluid)}, \tag{5.60}$$

where $\delta^{(3)}(\mathbf{k}) = \delta(k_x)\delta(k_y)\delta(k_z)$ is the three-dimensional Dirac delta function (cf. Appendix 5A). The function $\mathfrak{D}_L^{(3)}(\mathbf{k})$ goes to $\delta^{(3)}(\mathbf{k})$ in the limit where L_x, L_y, and L_z all go to infinity.

For a bulk phase, Equation 5.58 can be written

$$S(\mathbf{k}) = 1 + \frac{1}{N} \int d\mathbf{r}_1 d\mathbf{r}_2 (n^b)^2 h_N^{(2)}(r_{12}) e^{-i\mathbf{k}\cdot[\mathbf{r}_1 - \mathbf{r}_2]}$$

$$= 1 + n^b \int d\mathbf{r} h_N^{(2)}(r) e^{-i\mathbf{k}\cdot\mathbf{r}} \quad \text{(bulk fluid)}, \tag{5.61}$$

where the last equality follows since for a fixed $\mathbf{r}_2$ one can make the variable substitution $\mathbf{r} = \mathbf{r}_1 - \mathbf{r}_2$ whereby $d\mathbf{r} = d\mathbf{r}_1$. The final integration over $\mathbf{r}_2$ then simply gives a factor $\int d\mathbf{r}_2 = V$. The result in Equation 5.61 depends only on k and we can write

$$S(k) = 1 + n^b \tilde{h}_N^{(2)}(k) \quad \text{(bulk fluid)}, \tag{5.62}$$

where $\tilde{h}_N^{(2)}(k)$ is the Fourier transform of $h_N^{(2)}(r)$. The **Fourier transform** $\tilde{f}(\mathbf{k})$ of a function $f(\mathbf{r})$ is defined as

$$\tilde{f}(\mathbf{k}) = \int d\mathbf{r} f(\mathbf{r}) e^{-i\mathbf{k}\cdot\mathbf{r}} \tag{5.63}$$

and for a function $f(r)$ that only depends on r, we can write this as

$$\tilde{f}(k) = \frac{4\pi}{k} \int_0^\infty dr f(r) r \sin(kr) = \int d\mathbf{r} f(r) \frac{\sin(kr)}{kr},$$

as follows from an integration over the spherical polar coordinate angles of $\mathbf{r}$. Conversely, if $\tilde{f}(\mathbf{k})$ is known, $f(\mathbf{r})$ can be obtained from an **inverse Fourier transformation** given by

$$f(\mathbf{r}) = \frac{1}{(2\pi)^3} \int d\mathbf{k} \tilde{f}(\mathbf{k}) e^{i\mathbf{k}\cdot\mathbf{r}}$$

Thus, $h_N^{(2)}(r)$ is given by the inverse Fourier transform of $\tilde{h}_N^{(2)}(k)$.

The theory presented so far is adequate for neutron scattering[27] of fluids composed of single atoms like liquid argon, since the neutrons are scattered by the atoms due to very short-range interactions with the nuclei. The particles of the fluid are point scatterers to a good approximation. In *X-ray scattering*, on the other hand, the scattering occurs due to interactions with the electrons and since they are distributed in the whole atom, the particles are not point scatterers. Therefore, the theory must be somewhat modified to account for the electron distribution inside the particles.

To start with, let us treat scattering from an electron distribution with number density $n_e(\mathbf{r}')$ in some part of space. Consider an infinitesimal volume element $d\mathbf{r}'$ at the point $\mathbf{r}'$ (cf. Figure 3.1). It contains $n_e(\mathbf{r}')d\mathbf{r}'$ electrons that scatter the incident X-rays, just like the situation with point scatterers described earlier. If an electron at the origin gives a scattered

---

[27] This applies to the coherent part of the neutron scattering.

wave $B\exp(i\,\mathbf{k}'' \cdot \mathbf{r})$ near the detector (with different value of B than earlier), each of the electrons in the volume element at $\mathbf{r}'$ gives a scattered wave $B\exp(i\,[\mathbf{k}'' \cdot \mathbf{r} - \mathbf{k} \cdot \mathbf{r}'])$ near the detector, so all of these electrons give $B\exp(i\,[\mathbf{k}'' \cdot \mathbf{r} - \mathbf{k} \cdot \mathbf{r}'])n_e(\mathbf{r}')d\mathbf{r}'$.

Let us now sum all contributions from the electrons in the entire distribution $n_e(\mathbf{r}')$. In analogy to Equation 5.52, we obtain the scattered wave near the detector as

$$\int Be^{i\,[\mathbf{k}''\cdot\mathbf{r}-\mathbf{k}\cdot\mathbf{r}']}n_e(\mathbf{r}')d\mathbf{r}' = Be^{i\,\mathbf{k}''\cdot\mathbf{r}}\int d\mathbf{r}'\,n_e(\mathbf{r}')e^{-i\,\mathbf{k}\cdot\mathbf{r}'} = Be^{i\,\mathbf{k}''\cdot\mathbf{r}}\tilde{n}_e(\mathbf{k}),$$

where $\tilde{n}_e(\mathbf{k})$ is the Fourier transform of $n_e(\mathbf{r}')$. The intensity of the wave is given by

$$\text{Intensity} = \left|Be^{i\,\mathbf{k}''\cdot\mathbf{r}}\tilde{n}_e(\mathbf{k})\right|^2 = |B|^2\left|\tilde{n}_e(\mathbf{k})\right|^2 = |B|^2\tilde{n}_e(\mathbf{k})\tilde{n}_e(-\mathbf{k}), \qquad (5.64)$$

where the last equality follows from the fact that the complex conjugate of, say, $\tilde{f}(\mathbf{k})$ in Equation 5.63 is equal to $\int d\mathbf{r}f(\mathbf{r})e^{i\mathbf{k}\cdot\mathbf{r}} = \tilde{f}(-\mathbf{k})$ when $f(\mathbf{r})$ is real.

Let us now assume that the electron density $n_e(\mathbf{r}')$ originates from N spherically symmetric atoms located at $\mathbf{r}_\nu$, $1 \le \nu \le N$. Each of these atoms has an internal electron density distribution given by a number density that we will denote as $f_e(s)$, where s is the distance from the center of the atom. This means that

$$n_e(\mathbf{r}') = \sum_{\nu=1}^{N} f_e(|\mathbf{r}' - \mathbf{r}_\nu|)$$

and

$$\tilde{n}_e(\mathbf{k}) = \int d\mathbf{r}' \sum_{\nu=1}^{N} f_e(|\mathbf{r}' - \mathbf{r}_\nu|)e^{-i\,\mathbf{k}\cdot\mathbf{r}'}$$

$$= \int d\mathbf{r}' \sum_{\nu=1}^{N} f_e(|\mathbf{r}' - \mathbf{r}_\nu|)e^{-i\,\mathbf{k}\cdot[\mathbf{r}'-\mathbf{r}_\nu]}e^{-i\,\mathbf{k}\cdot\mathbf{r}_\nu}$$

$$= \tilde{f}_e(k) \sum_{\nu=1}^{N} e^{-i\,\mathbf{k}\cdot\mathbf{r}_\nu}, \qquad (5.65)$$

where $\tilde{f}_e(k)$ is the Fourier transform of $f_e(s)$ and the last equality follows from the fact that for each ν we have

$$\int d\mathbf{r}'f_e(|\mathbf{r}' - \mathbf{r}_\nu|)e^{-i\,\mathbf{k}\cdot[\mathbf{r}'-\mathbf{r}_\nu]} = \int d\mathbf{r}''f_e(|\mathbf{r}''|)e^{-i\,\mathbf{k}\cdot\mathbf{r}''} = \tilde{f}_e(k)$$

with $\mathbf{r}'' = \mathbf{r}' - \mathbf{r}_\nu$ and $d\mathbf{r}'' = d\mathbf{r}'$ since $\mathbf{r}_\nu$ is constant during the integration. By inserting Equation 5.65 into Equation 5.64, we obtain[28]

$$\text{Intensity} = |B|^2 [\tilde{\mathfrak{f}}_e(k)]^2 \sum_{\nu=1}^{N} \sum_{\nu'=1}^{N} e^{-i\,\mathbf{k}\cdot[\mathbf{r}_\nu - \mathbf{r}_{\nu'}]},$$

which differs from Equation 5.53 solely by the factor $[\tilde{\mathfrak{f}}_e(k)]^2$. All results obtained for point scatterers are therefore valid in the present case provided one multiplies by $[\tilde{\mathfrak{f}}_e(k)]^2$, for example Equation 5.56 is replaced by

$$I(2\theta_k, \phi_k) = N\,|B|^2\,[\tilde{\mathfrak{f}}_e(k)]^2 S_{\text{tot}}(\mathbf{k}). \tag{5.66}$$

The factor $\tilde{\mathfrak{f}}_e(k)$ is called the **form factor** since it describes the individual atomic form.

As another example, let us consider a spherical macroparticle of radius r_p with a uniform internal electron density n_e^{int} inside, $\mathfrak{f}_e(s) = n_e^{\text{int}}$ for $s \leq r_p$. In this case, we have

$$\tilde{\mathfrak{f}}_e(k) = \frac{4\pi}{k} \int_0^{r_p} dr\, n_e^{\text{int}} r \sin(kr) = 3 n_e^{\text{int}} V_p \frac{\sin(kr_p) - kr_p \cos(kr_p)}{(kr_p)^3}, \tag{5.67}$$

where $V_p = 4\pi r_p^3/3$ is the particle volume and $n_e^{\text{int}} V_p$ is the number of electrons per particle. This result, together with Equation 5.66, will be of interest in Section 7.2.1.2, where we will discuss X-ray scattering by inhomogeneous colloidal dispersions. For neutron scattering of such systems, a similar result applies. For instance, for spherical macroparticles with interior uniform density n_a^{int} of atoms, there is a form factor $\tilde{\mathfrak{f}}_a(k)$ given by an expression analogous to Equation 5.67.

5.7 THERMODYNAMICAL QUANTITIES FROM DISTRIBUTION FUNCTIONS

The Helmholtz free energy A_N for a system with given T, V and N is given by Equation 3.19. We will not consider internal degrees of freedom of the particles, so we have $\eta = 1/\Lambda^3$ and

$$A_N = k_B T \ln \frac{N! \Lambda^{3N}}{Z_N} = -k_B T \ln \left[\frac{1}{N! \Lambda^{3N}} \int d\mathbf{r}^N e^{-\beta \check{U}_N^{\text{pot}}(\mathbf{r}^N)} \right]. \tag{5.68}$$

The interactions between the particles are assumed to be are pairwise additive, so we have

$$\check{U}_N^{\text{pot}} = \sum_{\nu,\nu'>\nu} u(r_{\nu\nu'}) + \sum_{\nu=1}^{N} v(\mathbf{r}_\nu). \tag{5.69}$$

As we will see, various thermodynamical quantities for a homogeneous bulk fluid with density n^b can be calculated when the pair distribution function $g_N^{(2)}(r_{12})$ is known. When the

---

[28] We neglect quantum-mechanical corrections, which are beyond the scope of the present discussion. As before, multiple scattering is ignored.

fluid is inhomogeneous, such quantities can be calculated when the density distribution $n_N(\mathbf{r}_1)$ and the pair distribution function $g_N^{(2)}(\mathbf{r}_1, \mathbf{r}_2)$ are known.

As a first example, we shall express the internal energy $\bar{U}$ for a bulk fluid in terms of the pair distribution. Let us first make an intuitive derivation. The total energy of the molecules consists of the sum of the kinetic and potential energies, $\bar{U}_N \equiv \langle U_N \rangle = \langle U_N^{\text{kin}} \rangle + \langle U_N^{\text{pot}} \rangle$ (Equation 3.15). Since the velocities of the molecules are Maxwell-Boltzmann distributed irrespective of the interactions, we easily realize that the average kinetic energy is independent of the interactions. This means that $\langle U_N^{\text{kin}} \rangle$ must have the same value as if the gas were ideal. Thus, we have $\langle U_N^{\text{kin}} \rangle = 3Nk_BT/2$ in all cases. Alternatively, this also follows from the equipartition theorem.

It remains to determine $\langle U_N^{\text{pot}} \rangle$. Let us regard a particular but arbitrary particle as a "central" particle (we will position the origin of the coordinate system at its midpoint). The average density of other molecules around this molecule equals $n^b g_N^{(2)}(r)$. The total intermolecular interaction energy between the central molecule and other molecules located at distances between r and $r + dr$ (i.e., inside a spherical shell of thickness dr) equals $u(r) \cdot n^b g_N^{(2)}(r) \cdot 4\pi r^2 dr$, where $4\pi r^2 dr$ is the volume of the spherical shell and $n^b g_N^{(2)}(r) \cdot 4\pi r^2 dr$ is the average number of particles there. $\langle U_N^{\text{pot}} \rangle$ can now be obtained by integrating over all r values and by multiplying by $N/2$ (since any molecule can be the central one). The factor $1/2$ arises because we must avoid counting the pair interaction twice for each molecule. In total, we have

$$\bar{U}_N = \frac{3}{2}Nk_BT + \frac{Nn^b}{2} \int_0^\infty u(r) g_N^{(2)}(r) 4\pi r^2 dr \quad \text{(bulk phase)}, \tag{5.70}$$

where we take the integral to infinity despite that our system has a finite size (this makes no difference since $u(r)$ becomes negligible for large r; here we are only considering cases when $u(r)$ decays faster to zero than r^{-3}). The fluid is enclosed in a container of volume V with macroscopic dimensions in all directions and we assume that both N and V are very large. Only a vanishingly small part of the fluid is near the walls of the container and the overwhelming part of the fluid is homogeneous. We can therefore neglect the influence of the wall-particle interactions on A_N and $\bar{U}_N$.

We shall now prove this expression for $\bar{U}_N$ and simultaneously generalize it to inhomogeneous fluids, so the effects of any external potential will be included, for instance from the interactions with the container walls. The treatment includes cases where a substantial part of the fluid is influenced by interactions with the walls, like when the fluid is enclosed in a narrow slit. At least one of the dimensions of the container is, however, assumed to be macroscopic. To obtain the expression for the internal energy, we use Equation 2.47, which can be written $\bar{U}_N = (\partial[\beta A_N]/\partial\beta)_{V,N}$. By inserting Equation 5.68, we obtain

$$\bar{U}_N = \left(\frac{\partial \ln(N!\Lambda^{3N})}{\partial\beta} \right)_{V,N} - \frac{1}{Z_N} \left(\frac{\partial Z_N}{\partial\beta} \right)_{V,N}.$$

The first term gives the kinetic energy $3Nk_BT/2$ and the second term is

$$\left\langle U_N^{\text{pot}} \right\rangle = \frac{1}{Z_N} \int d\mathbf{r}^N \, \breve{U}_N^{\text{pot}}(\mathbf{r}^N) e^{-\beta \breve{U}_N^{\text{pot}}(\mathbf{r}^N)}$$

$$= \sum_{\nu,\nu'>\nu} \int d\mathbf{r}^N u(r_{\nu\nu'}) \mathcal{P}_N^{(N)}(\mathbf{r}^N) + \sum_{\nu=1}^{N} \int d\mathbf{r}^N v(\mathbf{r}_\nu) \mathcal{P}_N^{(N)}(\mathbf{r}^N),$$

where we have used Equations 5.38 and 5.69 to obtain the last equality. The first sum can be written

$$\int d\mathbf{r}_1 d\mathbf{r}_2 u(r_{12}) \int d\mathbf{r}_3 \dots d\mathbf{r}_N \mathcal{P}_N^{(N)}(\mathbf{r}^N)$$

$$+ \int d\mathbf{r}_1 d\mathbf{r}_3 u(r_{13}) \int d\mathbf{r}_2 d\mathbf{r}_4 \dots d\mathbf{r}_N \mathcal{P}_N^{(N)}(\mathbf{r}^N)$$

$$+ \dots + \int d\mathbf{r}_{N-1} d\mathbf{r}_N u(r_{N-1,N}) \int d\mathbf{r}_1 \dots d\mathbf{r}_{N-2} \mathcal{P}_N^{(N)}(\mathbf{r}^N)$$

and the second sum is

$$\int d\mathbf{r}_1 v(\mathbf{r}_1) \int d\mathbf{r}_2 \dots d\mathbf{r}_N \mathcal{P}_N^{(N)}(\mathbf{r}^N)$$

$$+ \int d\mathbf{r}_2 v(\mathbf{r}_2) \int d\mathbf{r}_1 d\mathbf{r}_3 \dots d\mathbf{r}_N \mathcal{P}_N^{(N)}(\mathbf{r}^N)$$

$$+ \dots + \int d\mathbf{r}_N v(\mathbf{r}_N) \int d\mathbf{r}_1 \dots d\mathbf{r}_{N-1} \mathcal{P}_N^{(N)}(\mathbf{r}^N).$$

We see that all terms in the first sum are numerically equal (they differ only by the *name* of the integration variables) and the same applies to the second sum. Thus, all terms are equal to the first one in each of the sums. It is easy to see that there are $N(N-1)/2$ terms in the first sum and N terms in the second sum, so therefore we have

$$\left\langle U_N^{\text{pot}} \right\rangle = \frac{N(N-1)}{2} \int d\mathbf{r}_1 d\mathbf{r}_2 u(r_{12}) \mathcal{P}_N^{(2)}(\mathbf{r}_1, \mathbf{r}_2) + N \int d\mathbf{r}_1 v(\mathbf{r}_1) \mathcal{P}_N^{(1)}(\mathbf{r}_1),$$

where we have used Equations 5.40 and 5.44. Finally, by utilizing the definitions of $n_N^{(1)}(\mathbf{r}_1) \equiv n_N(\mathbf{r}_1)$ from Equation 5.41 and $n_N^{(2)}(\mathbf{r}_1, \mathbf{r}_2)$ from Equation 5.45, it follows that the total energy is

$$\bar{U}_N = \frac{3}{2} N k_B T + \frac{1}{2} \int d\mathbf{r}_1 d\mathbf{r}_2 u(r_{12}) n_N^{(2)}(\mathbf{r}_1, \mathbf{r}_2) + \int d\mathbf{r}_1 v(\mathbf{r}_1) n_N(\mathbf{r}_1). \qquad (5.71)$$

Here we can insert $n_N^{(2)}(\mathbf{r}_1, \mathbf{r}_2) = n_N(\mathbf{r}_1) n_N(\mathbf{r}_2) g_N^{(2)}(\mathbf{r}_1, \mathbf{r}_2)$ if required. For a bulk liquid where $n_N(\mathbf{r}) = n^b$, the result (5.71) reduces to Equation 5.70 since $N = n^b V = \int d\mathbf{r} n_N(\mathbf{r})$. Note that $d\mathbf{r} = 4\pi r^2 dr$ due to spherical symmetry in the latter case.

Let us now determine the pressure P for a bulk fluid (inhomogeneous fluids will be treated later in Chapter 8). From Equations 2.51 and 5.68, we have

$$P = -\left(\frac{\partial A_N}{\partial V}\right)_{T,N} = \frac{k_B T}{Z_N}\left(\frac{\partial Z_N}{\partial V}\right)_{T,N}.$$

Let us for simplicity assume that the fluid is contained in a cubic container with sides of length L which are aligned with the coordinate axes. The volume of the fluid is hence $V = L^3$. Since the macroscopic properties of a homogeneous fluid do not depend on the shape of the box that contains it, the final result is valid in general for the fluid.

We have

$$Z_N = \int_0^L dx_1 \int_0^L dy_1 \int_0^L dz_1 \ldots \int_0^L dx_N \int_0^L dy_N \int_0^L dz_N \, e^{-\beta \breve{U}_N^{\text{pot}}(\mathbf{r}^N)}$$

and we see that the L dependence and therefore the V dependence of Z_N arise from the integration limits. In order to handle this dependence in a convenient manner, let us do a variable substitution $x_\nu = Lx_\nu^*$, $y_\nu = Ly_\nu^*$ and $z_\nu = Lz_\nu^*$ for all ν, whereby we obtain

$$Z_N = L^{3N} \int_0^1 dx_1^* \int_0^1 dy_1^* \int_0^1 dz_1^* \ldots \int_0^1 dx_N^* \int_0^1 dy_N^* \int_0^1 dz_N^* \, e^{-\beta \breve{U}_N^{\text{pot}}(L\mathbf{r}^{*N})}$$

$$= V^N \int d\mathbf{r}^{*N} e^{-\beta \breve{U}_N^{\text{pot}}(L\mathbf{r}^{*N})},$$

where the integration limits now are constants independent of V and where

$$\breve{U}_N^{\text{pot}}(L\mathbf{r}^{*N}) = \sum_{\nu,\nu'>\nu} u(Lr_{\nu\nu'}^*)$$

since $L\mathbf{r}_\nu^* = (Lx_\nu^*, Ly_\nu^*, Lz_\nu^*) = \mathbf{r}_\nu$ and $Lr_{\nu\nu'}^* = |L\mathbf{r}_{\nu'}^* - L\mathbf{r}_\nu^*| = |\mathbf{r}_{\nu'} - \mathbf{r}_\nu| = r_{\nu\nu'}$.

By taking the derivative with respect to V, we obtain

$$\frac{\partial Z_N}{\partial V} = NV^{N-1} \int d\mathbf{r}^{*N} e^{-\beta \breve{U}_N^{\text{pot}}(L\mathbf{r}^{*N})} + V^N \frac{\partial}{\partial L}\left[\int d\mathbf{r}^{*N} e^{-\beta \breve{U}_N^{\text{pot}}(L\mathbf{r}^{*N})}\right]\frac{dL}{dV}$$

$$= \frac{N}{V}Z_N - \frac{V^N}{3L^2} \int d\mathbf{r}^{*N} \beta \frac{\partial \breve{U}_N^{\text{pot}}(L\mathbf{r}^{*N})}{\partial L} e^{-\beta \breve{U}_N^{\text{pot}}(L\mathbf{r}^{*N})}, \tag{5.72}$$

since $dL/dV = 1/[dV/dL] = 1/[dL^3/dL] = 1/(3L^2)$. Next, we insert $\breve{U}_N^{\text{pot}}$ in terms of the pair interaction potential of the particles and use the fact that

$$\frac{\partial \breve{U}_N^{\text{pot}}(L\mathbf{r}^{*N})}{\partial L} = \sum_{\nu,\nu'>\nu} \frac{\partial u(Lr_{\nu\nu'}^*)}{\partial L} = \sum_{\nu,\nu'>\nu} r_{\nu\nu'}^* u'(Lr_{\nu\nu'}^*) = \sum_{\nu,\nu'>\nu} \frac{r_{\nu\nu'}}{L} u'(r_{\nu\nu'}),$$

where $u'(r) = du(r)/dr$. Thus, Equation 5.72 yields

$$\frac{1}{Z_N}\frac{\partial Z_N}{\partial V} = \frac{N}{V} - \frac{\beta}{3V}\sum_{v,v'>v}\int d\mathbf{r}^N r_{vv'}u'(r_{vv'})\mathcal{P}_N^{(N)}(\mathbf{r}^N)$$

where we have have gone back to the original integration variables and used Equation 5.38. Exactly like in the calculation of $\bar{U}_N$, we note that the sum consists of $N(N-1)/2$ numerically identical terms, all equal to the first one where $v = 1$ and $v' = 2$. Finally, we therefore obtain

$$P = \frac{Nk_BT}{V} - \frac{1}{6V}\int d\mathbf{r}_1 d\mathbf{r}_2 r_{12}u'(r_{12})n_N^{(2)}(\mathbf{r}_1,\mathbf{r}_2)$$

$$= n^b k_B T - \frac{(n^b)^2}{6}\int d\mathbf{r}\, r\frac{du(r)}{dr}g_N^{(2)}(r) \quad \text{(bulk phase)}, \tag{5.73}$$

where we have used Equation 5.45. This result is known as the **virial equation** for the pressure.

The first term in this expression for P constitutes the kinetic contribution to the pressure, which equals the pressure P^{ideal} of an ideal gas of the same density. It is the momentum transfer per unit area across an arbitrary (mathematical) surface in the fluid. Since the molecular velocities are Maxwell-Boltzmann distributed irrespective of the interactions, this term is equal for all fluids (and is therefore identical to that for an ideal gas). The second term gives the pressure contribution due to interactions between the particles of the fluid (the force per unit area), the excess pressure P^{ex}. In a liquid, these two contributions counteract each other, so the net pressure is normally far smaller than each of the terms alone.

We can get a feeling for the magnitudes of these contributions to P by noting that, for liquid argon at its boiling point at 1 atm ($T = 87.3$ K), the density in molar units is 34.9 mol dm^{-3} (1.394 kg dm^{-3}), which implies that $n^b k_B T = 250$ atm. Since the net pressure is 1 atm, the interactional contribution therefore must equal –249 atm. It is this very strongly attractive interaction that makes the fluid to be in its liquid state. The pressure of the vapor phase at this temperature is, of course, also 1 atm and it originates mostly from the kinetic term. Argon is nearly an ideal gas under these conditions (see Example 5.1 below). For liquid water the density is 55.4 mol dm^{-3} (at 20°C), which implies that $n^b k_B T = 1334$ atm. At normal atmospheric net pressure, the interactional contribution is therefore –1333 atm (for water the expression for the pressure is more complicated than in Equation 5.73, but this does not make any difference for the argument).

Example 5.1: Pressure for a Gas at Low Densities

Our expression for P is valid for both simple liquids and gases. For gases at very low densities, we have $g_N^{(2)}(r) \approx e^{-\beta u(r)}$, that is, a simple Boltzmann distribution in the pair potential $u(r)$ (see Equation 5.36 and Figure 5.10). If we insert this and $d\mathbf{r} = 4\pi r^2 dr$

in our expression (5.73) for P, we obtain

$$P \approx n^b k_B T - \frac{2\pi (n^b)^2}{3} \int_0^\infty dr \, r^3 \frac{du(r)}{dr} e^{-\beta u(r)}. \tag{5.74}$$

Utilizing $d \left(\exp[-\beta u(r)] - 1 \right) / dr = -\beta \exp[-\beta u(r)] \, du(r)/dr$, we can write the integral as

$$\int_0^\infty dr \, r^3 \frac{du(r)}{dr} e^{-\beta u(r)} = -\beta^{-1} \int_0^\infty dr \, r^3 \frac{d}{dr} \left(e^{-\beta u(r)} - 1 \right)$$

$$= -\beta^{-1} \left[r^3 \left(e^{-\beta u(r)} - 1 \right) \right]_0^\infty + \beta^{-1} \int_0^\infty dr \, 3r^2 \left(e^{-\beta u(r)} - 1 \right),$$

where we have done a partial integration to obtain the last equality. The first term on the rhs is equal to zero (as before, we limit ourselves to cases where $u(r)$ decays faster to zero than r^{-3}). Therefore, we obtain from Equation 5.74

$$\frac{P}{k_B T} \approx n^b - 2\pi (n^b)^2 \int_0^\infty dr \, r^2 \left[e^{-\beta u(r)} - 1 \right] = n^b + B_2(T) \, (n^b)^2,$$

where we have used the expression (3.54) for the second virial coefficient $B_2(T)$. This result consists of the two first terms in the virial expansion (3.48) for the pressure of a real gas. In molar units, we have $P/RT \approx n_m^b + B_2(T) \, (n_m^b)^2$, where $n_m^b = n^b/N_{Av}$ is the molar density and B_2 is expressed in such units.

For argon gas at $T = 87.3$ K, we have $B_2 = -0.24$ dm^3 mol^{-1}.[29] When $P = 1$ atm, we have $P/RT = 0.140$ mol dm^{-3}. The density of the gas at this pressure can be determined from the virial expansion, which implies that $0.140 = n_m^b - 0.24(n_m^b)^2$ (the third order virial term is negligible here). This equation has the solution $n_m^b = 0.145$ mol dm^{-3}, which implies that the first term in the virial expansion is 0.145 and the second term is -0.005 mol dm^{-3}. The pressure of an ideal gas at this density is 1.04 atm at $T = 87.3$ K. Thus, the attractive interactions between the argon atoms reduce the pressure by 0.04 atm compared to that of an ideal gas under these conditions, so the gas is nearly ideal. This should be compared to the huge attraction in the liquid phase that is in equilibrium with the gas at this pressure and temperature, as discussed earlier.

For a fluid of particles with a hard core interactions, we have $u(r) = \infty$ when $r < d^h$ and $g_N^{(2)}(r) = 0$ there. In the shaded text box below, which can be skipped in the first reading,

---

[29] B_2 is obtained from accurate fits to experimental data, see R. B. Stewart and R. T. Jacobsen, *J. Phys. Chem. Ref. Data* **18** (1989) 639.

we will see how one can handle the product $[du(r)/dr]g_N^{(2)}(r)$ that appears in Equation 5.73 for the bulk pressure in this case. The end result is

$$\frac{P}{k_B T} = n^b + \frac{2\pi (d^h)^3}{3}(n^b)^2 g_N^{(2)}(d^{h+}) - \frac{\beta (n^b)^2}{6} \int\limits_{d^h}^{\infty} dr\, 4\pi r^3 \frac{du(r)}{dr} g_N^{(2)}(r), \qquad (5.75)$$

where $g_N^{(2)}(d^{h+})$ is the so-called **contact value** of the pair distribution function ($r = d^h$ corresponds to two particles in contact with each other). It is defined as

$$g_N^{(2)}(d^{h+}) = \lim_{r \to d^{h+}} g_N^{(2)}(r), \qquad (5.76)$$

where $r \to d^{h+}$ means that we take the limit by approaching d^h from r values that are larger than d^h. This kind of limit is taken because $g_N^{(2)}(r)$ has a discontinuity at $r = d^h$ and drops to the value zero there. The product $n^b g_N^{(2)}(d^{h+})$ constitutes the **contact density**, which is the average density of particles in contact with the central particle. This density is analogous to the contact density at a hard wall surface indicated by arrows in Figure 5.5a.

For a hard sphere fluid, $u(r) = 0$ when $r \geq d^h$, so the integral is zero and we have

$$\frac{P}{k_B T} = n^b + \frac{2\pi (d^h)^3}{3}(n^b)^2 g_N^{(2)}(d^{h+})$$

$$= n^b + B_2(n^b)^2 g_N^{(2)}(d^{h+}) \quad \text{(hard sphere fluid)}, \qquad (5.77)$$

where $B_2 = 2\pi (d^h)^3/3$ is the value of the second virial coefficient for a hard sphere fluid. Thus, the pressure of the fluid in excess to the ideal part is proportional to the contact density $n^b g_N^{(2)}(d^{h+})$ at the hard sphere surface.

★ Let us derive Equation 5.75 from Equation 5.73 for a hard core fluid. From Equation 5.34, we see that we can write

$$g_N^{(2)}(r) = e^{-\beta u(r)} y_N^{(2)}(r), \qquad (5.78)$$

where $y_N^{(2)}(r) = \exp[-\beta\{w_N^{(2)\text{intr}}(r) - w_N^{\text{intr,b}}\}]$. It is the factor $\exp[-\beta u(r)]$ that makes $g_N^{(2)}(r)$ to be zero when $r < d^h$. The function $y_N^{(2)}(r)$ is continuous at $r = d^h$,[30] while $g_N^{(2)}(r)$ makes a step from 0 to $g_N^{(2)}(d^{h+})$ there.

---

[30] This is a consequence of the fact that $w_N^{(2)\text{intr}}(r)$ is continuous there since it originates from the average interactions due to all particles that surround the central particle. The continuity will be demonstrated in more detail in Chapter 9 (in the 2nd volume).

Let us initially treat a hard sphere fluid with pair interaction potential given by Equation 5.1. Using Equation 5.78, we can write the integrand in Equation 5.73 as

$$\mathcal{I}(r) \equiv r \frac{du(r)}{dr} g_N^{(2)}(r) = r \frac{du(r)}{dr} e^{-\beta u(r)} y_N^{(2)}(r) = r \frac{de^{-\beta u(r)}}{dr} (-\beta)^{-1} y_N^{(2)}(r).$$

For $r < d^h$, where $u(r) = \infty$, the function $\exp[-\beta u(r)]$ is zero and for $r > d^h$, where $u(r) = 0$, it is equal to one. Let us introduce the so-called **Heaviside step function** $\mathcal{H}(x)$, which is equal to 0 for $x < 0$ and 1 for $x > 0$. We see that $\exp[-\beta u(r)] = \mathcal{H}(r - d^h)$ and the integrand can therefore be written as

$$\mathcal{I}(r) = -\beta^{-1} r \frac{d\mathcal{H}(r - d^h)}{dr} y_N^{(2)}(r),$$

which is zero everywhere except at $r = d^h$ where $\mathcal{H}(r - d^h)$ makes a step from 0 to 1. The derivative of a function with such a step does not exist in usual differential calculus, but one can generalize the concept of the derivative so one can handle this situation.

The derivative $d\mathcal{H}(x)/dx$ is zero everywhere except at $x = 0$, where it is "infinitely large." By introducing the so-called **Dirac delta function** $\delta(x)$ as this derivative, $\delta(x) = d\mathcal{H}(x)/dx$, we have a "function" that is zero everywhere except at $x = 0$ where it has an infinite peak. The width of this peak is zero but the "area" of the peak is set equal to one, $\int dx\, \delta(x) = 1$. With this "definition" of $\delta(x)$, we can see that

$$\int_{-\infty}^{x} dx'\, \delta(x') = \int_{-\infty}^{x} dx' \frac{d\mathcal{H}(x')}{dx'} = \mathcal{H}(x)$$

since the value of the integral is zero for $x < 0$ and then, at increasing x, it obtains a contribution equal to 1 at $x = 0$ and stays at 1 thereafter since $\delta(x') = 0$ for $x' > 0$. Thus, it is appropriate to say that $\delta(x)$ is the derivative of $\mathcal{H}(x)$. In general, the delta function has the property $\int_{-\infty}^{\infty} dx\, \delta(x) f(x) = f(0)$ for any (well-behaved) function $f(x)$. This follows since the integrand is zero everywhere except at $x = 0$ where the integral obtains the contribution $1 \cdot f(x)|_{x=0} = f(0)$. Likewise, the delta function $\delta(x - a)$ has an "infinitely sharp peak" at $r = a$ and we have $\int_{-\infty}^{\infty} dx\, \delta(x - a) f(x) = f(a)$. The Dirac delta function is explained in more detail in Appendix 5A, where it is treated in a more formal manner.

Let us return to the hard sphere fluid. We see that $\mathcal{I}(r) = -\beta^{-1} r \delta(r - d^h) y_N^{(2)}(r)$, so Equation 5.73 can be written

$$\frac{P}{k_B T} = n^b + \frac{(n^b)^2}{6} \int_0^{\infty} dr\, 4\pi r^3 \delta(r - d^h) y_N^{(2)}(r)$$

$$= n^b + \frac{2\pi (d^h)^3}{3} (n^b)^2 g_N^{(2)}(d^{h+}) \quad \text{(hard sphere fluid)}$$

since $y_N^{(2)}(d^{h+}) = g_N^{(2)}(d^{h+})$. This is Equation 5.77.

The generalization of this result to a fluid with hard core particles is straightforward. In such cases, we write $u(r) = u^{\text{core}}(r) + u^{\text{rest}}(r)$, where $u^{\text{core}}(r)$ is the hard core potential (that is, the hard sphere potential) and $u^{\text{rest}}(r)$ is the same as $u(r)$ for $r \geq d^h$ (for $r < d^h$, the values of $u^{\text{rest}}(r)$ do not matter). We have $\exp[-\beta u(r)] = \mathcal{H}(r - d^h)\exp[-\beta u^{\text{rest}}(r)]$ and we obtain

$$
\begin{aligned}
\mathcal{I}(r) &= -\beta^{-1} r \frac{d[\mathcal{H}(r - d^h)e^{-\beta u^{\text{rest}}(r)}]}{dr} y_N^{(2)}(r) \\
&= -\beta^{-1} r \left[\delta(r - d^h)e^{-\beta u^{\text{rest}}(r)} + \mathcal{H}(r - d^h) \frac{de^{-\beta u^{\text{rest}}(r)}}{dr} \right] y_N^{(2)}(r) \\
&= -\beta^{-1} r \left[\delta(r - d^h)g_N^{(2)}(r) - \beta \mathcal{H}(r - d^h) \frac{du(r)}{dr} g_N^{(2)}(r) \right],
\end{aligned}
$$

where we have used $u^{\text{rest}}(r) = u(r)$ for $r \geq d^h$ and reintroduced $g_N^{(2)}$ from Equation 5.78. Equation 5.75 is obtained from Equation 5.73 after integration of $\mathcal{I}(r)$.

Compared to the examples of thermodynamic quantities we have already seen, it is more complicated to express the free energy A_N or the chemical potential μ in terms of the distribution functions. Let us consider a bulk fluid (like before we neglect boundary effects). As we have seen in Equation 3.22, the free energy can be written (with $\eta = 1/\Lambda^3$)

$$
A_N = A_N^{\text{ideal}} + A_N^{\text{ex}} = k_B T \ln \frac{N! \Lambda^{3N}}{V^N} + k_B T \ln \frac{V^N}{Z_N}. \tag{5.79}
$$

The ideal free energy A_N^{ideal} is the non-interactional contribution to A_N and the excess free energy A_N^{ex} arises from the interparticle interactions as expressed in Equation 3.23. If $g_N(r)$ is known or can be calculated for all temperatures, we can obtain A_N^{ex} by utilizing the relationship (compare with Equation 2.49)

$$
\left(\frac{\partial (A_N^{\text{ex}}(T)/T)}{\partial (1/T)} \right)_{V,N} = \bar{U}_N^{\text{ex}}(T) = \frac{Nn^b}{2} \int_0^{\infty} u(r)g_N^{(2)}(r; T) 4\pi r^2 dr, \tag{5.80}
$$

where $\bar{U}_N^{\text{ex}}$ is the interactional part of $\bar{U}_N$ (the last term in Equation 5.70) and we have shown explicitly that $g_N^{(2)}$ depends on temperature. By integrating this relationship between the temperatures T to ∞ (that is, the inverse temperature varies between $1/T$ and 0), we obtain

$$
\frac{A_N^{\text{ex}}(T)}{T} = \int_0^{1/T} \frac{\partial (A_N^{\text{ex}}/T')}{\partial (1/T')} d\left(\frac{1}{T'}\right) = \int_T^{\infty} \bar{U}_N^{\text{ex}}(T') \frac{dT'}{(T')^2}. \tag{5.81}
$$

Then one utilizes the fact that g_N approaches one when T increases to infinity (random distribution of the particles).[31] The total free energy A_N is obtained by adding A_N^{ideal} to A_N^{ex} thus obtained. We can see that it is not sufficient to know g_N for only one temperature in order to calculate the free energy; here we need to know it for a range of temperatures. This manner of calculating the free energy is called a **thermodynamic integration** since one integrates one thermodynamic entity as a function of another thermodynamic entity – here the average energy as function temperature T.

Alternatively, by integrating $P^{\text{ex}} = -\left(\partial A_N^{\text{ex}}/\partial V\right)_{T,N}$ over V and using $P^{\text{ex}} = P^{\text{ex}}(V)$ as determined from the second term in Equation 5.73 for a bulk fluid, we can determine free energy differences

$$\Delta A_N^{\text{ex}} = A_N^{\text{ex}}(V_2) - A_N^{\text{ex}}(V_1) = -\int_{V_1}^{V_2} dV \, P^{\text{ex}}(V). \tag{5.82}$$

In this case, $g_N^{(2)}$ must be known as a function of the density N/V. For example, for a real gas we can determine $A_N^{\text{ex}}(V)$ by performing an integration of $P^{\text{ex}}(V')$ between $V' = V$ and $V' = \infty$ and using the fact that A_N^{ex} is zero at $V' = \infty$ (that is, at zero pressure). This is also a kind of thermodynamic integration.

To obtain the chemical potential of a bulk fluid, we should calculate $\mu = A_N - A_{N-1}$, which can be written $\mu = \mu^{\text{ideal}} + \mu^{\text{ex}}$, where $\mu^{\text{ideal}} = A_N^{\text{ideal}} - A_{N-1}^{\text{ideal}} = k_B T \ln \left(\Lambda^3 n^{\text{b}}\right)$ is the ideal contribution and $\mu^{\text{ex}} = k_B T \ln \gamma = A_N^{\text{ex}} - A_{N-1}^{\text{ex}}$ is the excess part, compare with Equation 3.29 (with $\eta = 1/\Lambda^3$). Since the change in free energy is equal to reversible work done at constant temperature,[32] μ^{ex} is equal to the isothermal reversible work we have to do on the system *against the intermolecular interactions* when we add a particle to the system.

Since A_N is a state function, it does not matter in what manner we add the new particle. The only essential conditions are that this particle initially does not interact with the other particles and that it finally interacts fully with them (i.e., it then interacts like any other particle in the system). One way to perform this is to let the particle initially be located infinitely far away from the other particles (at $\mathbf{r}_\infty$) and then to move it into their midst (to $\mathbf{r}$). The work done is equal to the integral of *(the force that acts on the particle in the direction of motion)* × *(the distance moved)*, that is,

$$w^{\text{rev}} = \int_{\mathbf{r}_\infty}^{\mathbf{r}} \mathbf{F}(\mathbf{r}') \cdot d\mathbf{r}',$$

where $\mathbf{r}'$ is the position of the particle. The process is reversible provided the particle is moved so slowly (quasistatically) that the system at all times is relaxed to its equilibrium state. This is assumed here. The force originates from the interactions with the other particles and from any external potential. Since the density distribution of these particles depends on

---

[31] For particles with a hard core, this is valid outside the core.
[32] See footnote 19.

the position of the particle (because of the interactions with it), $\mathbf{F}$ depends in general on $\mathbf{r}'$ and must be determined at each point $\mathbf{r}'$ on the way from $\mathbf{r}_\infty$ to $\mathbf{r}$. This is a fact that has to be considered when calculating $\mathbf{F}$.

The procedure is rather tricky to perform, and therefore we are going to use a smarter way to calculate μ. We will let the new particle remain all the time at one point in the midst of all other particles, and instead we will vary the strength of the interaction between it and the surrounding particles. This can be done by introducing a coupling parameter ξ that varies between $\xi = 0$ and $\xi = 1$. We let the pair potential that involves the new particle depend on ξ and we can, for example, take $u(r; \xi) = \xi u(r)$ as the interaction between this particle and all other particles.[33] The latter interact among themselves with the full potential $u(r)$ irrespective of the value of ξ. When $\xi = 0$, the other particles do not perceive the new particle (and vice versa), so from their point of view it does not exist. We then increase ξ gradually and when $\xi = 1$, the new particle interacts equally strongly as all other particles, $u(r; 1) = u(r)$.

Let us now calculate the reversible work done against the intermolecular forces during this process. Let us place the origin of the coordinate system at the new particle. For each value of ξ, $0 \le \xi \le 1$, we introduce a radial distribution function $g_N^{(2)}(r; \xi)$ such that the density of the other particles around the central, partially coupled particle is $n^b g_N^{(2)}(r; \xi)$ when $u(r; \xi) = \xi u(r)$. When $\xi = 0$, we have, of course, $g_N^{(2)}(r; 0) = 1$ and when $\xi = 1$, we have $g_N^{(2)}(r; 1) = g_N^{(2)}(r)$ (since the central particle then interacts like all other particles). Figure 5.12 illustrates how the density around the central particle (placed at the cross) may change with ξ.

The interaction energy between the central molecule and a molecule at distance r is equal to $\xi u(r)$. When ξ is increased from ξ to $\xi + d\xi$, we obtain the following contribution from this particular interaction to the reversible work done on the system

$$\frac{\partial u(r; \xi)}{\partial \xi} d\xi = \frac{\partial [\xi u(r)]}{\partial \xi} d\xi = u(r) d\xi.$$

The contribution from the interactions between the central particle and all particles in a spherical shell between r and $r + dr$ therefore equals

$$u(r) d\xi \times n^b g_N^{(2)}(r; \xi) \times 4\pi r^2 dr$$

when ξ is increased by $d\xi$. The total work when ξ is increased from 0 to 1 simply is equal to the integral of this expression from $\xi = 0$ to $\xi = 1$ and we have

$$\mu = k_B T \ln \left(\Lambda^3 n^b \right) + \int_0^1 d\xi \int_0^\infty dr \, 4\pi r^2 u(r) n^b g_N^{(2)}(r; \xi).$$

---

[33] When the particle has a hard core, one must also increase the repulsion in the core region from zero to the final value during the coupling process.

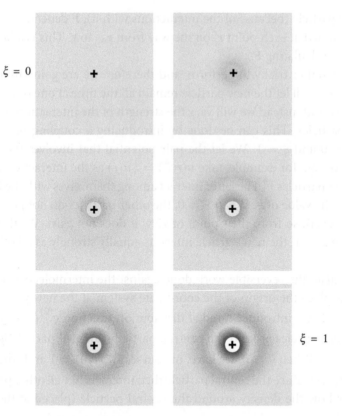

$\xi = 0$

$\xi = 1$

FIGURE 5.12 A sketch of how the density around a particle (placed with its center at the cross) may change when the coupling parameter ξ varies between 0 and 1. The density distribution is the equilibrium one at every stage of the process. The shade of gray indicates the value of the density.

We here see that it is necessary to determine $g_N^{(2)}$ for a series of states with different ξ values in order to calculate μ. The last term is $\mu^{\text{ex}} = k_B T \ln \gamma$ and, as we have seen, it equals the work that must be done on the system against the interactions when adding a new particle, so the activity factor $\gamma = \exp(\beta \mu^{\text{ex}})$ depends on this work.

Since free energy is a state function, the coupling of the new particle can be done in any manner provided $u(r; \xi)$ satisfies

$$u(r; \xi) = \begin{cases} 0 & \text{when } \xi = 0 \\ u(r) & \text{when } \xi = 1. \end{cases} \tag{5.83}$$

Any dependence of $u(r; \xi)$ on the parameter ξ can be used provided that this condition is fulfilled (i.e., there is no need to have precisely $u(r; \xi) = \xi u(r)$). The general formula for μ is

$$\mu = k_B T \ln \left(\Lambda^3 n^b \right) + \int_0^1 d\xi \int d\mathbf{r} \frac{\partial u(r; \xi)}{\partial \xi} n^b g_N^{(2)}(r; \xi), \tag{5.84}$$

where $n^b g_N^{(2)}(r; \xi)$ is the density distribution around the partially coupled particle when the interaction potential with it is $u(r; \xi)$. This manner of calculating the chemical potential is called a **coupling parameter integration**. Sometimes it is inappropriately also called a thermodynamic integration.

{**EXERCISE 5.2**

Consider two simple fluids: one (fluid I) consisting of particles that interact with a pair potential $u_I(r)$ and another (fluid II) in which the particles interact with the pair potential $u_{II}(r) = \alpha u_I(r)$, where α is a positive number. Each fluid is enclosed in a container of volume V and consists of N particles.

a. Show that A^{ex}/T, $g_N^{(2)}(r)$ and U^{ex}/T of fluid I at temperature T are identical to the corresponding quantities of fluid II at another temperature T', where $T' = \alpha T$.

b. By letting α go to zero at constant T', you may relate the properties of one system at $T \to \infty$ with the properties of the other at constant temperature when the pair potential of its particles goes to zero for all r (it is assumed that the pair potential is finite for all r values). Use this observation to conclude what are the values of A^{ex}/T, $g_N^{(2)}(r)$ and U^{ex}/T at infinite temperature.}

{**EXERCISE 5.3**

The pressure of a simple fluid, where the particles interact with the pair potential $u(r)$, is given by (Equation 5.73)

$$P = nk_B T - \frac{n^2}{6} \int d\mathbf{r}\, r \frac{du(r)}{dr} g^{(2)}(r; n), \tag{5.85}$$

where we have written n instead of n^b and have dropped subscript N on $g_N^{(2)}$. We have shown explicitly that $g^{(2)}$ depends on the density $n = N/V$. The pair distribution function can be written as in Equation 5.78

$$g^{(2)}(r; n) = e^{-\beta u(r)} y^{(2)}(r; n). \tag{5.86}$$

a. For each value of r, we can expand $y^{(2)}$ in a Taylor series in n

$$y^{(2)}(r; n) = 1 + n f_1(r) + n^2 f_2(r) + n^3 f_3(r) + \cdots,$$

where

$$f_\nu(r) = \frac{1}{\nu!} \left. \frac{\partial^\nu y^{(2)}(r; n)}{\partial n^\nu} \right|_{n=0}$$

is independent of n. Show that one obtains a virial expansion for P if this expansion of $y^{(2)}$ is inserted in Equations 5.85 and 5.86.

b. Identify the second virial coefficient B_2 expressed in $u(r)$ from the result in (a) and simplify the expression for B_2 so it does not contain any derivatives.

c. Use the Gibbs-Duhem equation $-SdT + VdP - Nd\mu = 0$ (Equation 2.124) to show that

$$\left(\frac{\partial \mu}{\partial n}\right)_T = \left(\frac{\partial (P/n)}{\partial n}\right)_T + \frac{P}{n^2}.$$

By inserting the excess chemical potential μ^{ex} and the excess pressure P^{ex} defined from $\mu = k_B T \ln(\Lambda^3 n) + \mu^{\text{ex}}$ and $P = nk_B T + P^{\text{ex}}$, respectively, show that

$$\left(\frac{\partial \mu^{\text{ex}}}{\partial n}\right)_T = \left(\frac{\partial (P^{\text{ex}}/n)}{\partial n}\right)_T + \frac{P^{\text{ex}}}{n^2}. \tag{5.87}$$

Show that

$$\mu^{\text{ex}} = \frac{P^{\text{ex}}}{n} + \int_0^n dn' \frac{P^{\text{ex}}}{[n']^2} \tag{5.88}$$

by integrating Equation 5.87 with respect to density between the values 0 and n.

d. Determine P^{ex} from Equation 5.85 and use Equation 5.88 to derive an expression for μ^{ex} that contains $g^{(2)}(r; n)$ and $\int_0^n dn' \, g^{(2)}(r; n')$ in the integrands. The result is an alternative to Equation 5.84 for μ. Note that we still must calculate $g^{(2)}$ for many different states to determine μ.}

5.8 MICROSCOPIC DENSITY DISTRIBUTIONS AND DENSITY-DENSITY CORRELATIONS ★

First, we need to generalize the concept of Dirac's delta function to three-dimensional space. The **three-dimensional Dirac delta function** is defined as $\delta^{(3)}(\mathbf{r}) = \delta(x)\delta(y)\delta(z)$, where each of the factors in the rhs is a one-dimensional delta function that we have encountered earlier. The function $\delta^{(3)}(\mathbf{r})$ has an "infinitely sharp peak" at $\mathbf{r} = \mathbf{0}$, is zero everywhere else and has unit integral $\int d\mathbf{r}\delta^{(3)}(\mathbf{r}) = 1$. It is spherically symmetric so we can write $\delta^{(3)}(\mathbf{r}) = \delta^{(3)}(r)$. Moreover, $\delta^{(3)}(\mathbf{r} - \mathbf{a})$ has a peak at $\mathbf{r} = \mathbf{a}$ and we have $\int d\mathbf{r}\delta^{(3)}(\mathbf{r} - \mathbf{a})f(\mathbf{r}) = f(\mathbf{a})$. Dirac's delta function is treated in more detail in Appendix 5A.

Let us consider N particles in a configuration with spatial coordinates $\mathbf{r}^N \equiv \mathbf{r}_1, \mathbf{r}_2, \ldots, \mathbf{r}_N$ and define the function

$$\breve{n}^{(1)}_{\{\mathbf{r}^N\}}(\mathbf{r}) \equiv \sum_{\nu=1}^N \delta^{(3)}(\mathbf{r} - \mathbf{r}_\nu), \tag{5.89}$$

which has infinitely sharp peaks at the locations of the N particles and is zero elsewhere (the subscript $\{\mathbf{r}^N\}$ indicates that the function depends on the particle coordinates). Since the dimension of the delta function $\delta^{(3)}(\mathbf{r} - \mathbf{r}_\nu)$ is (volume)$^{-1}$, which follows from the fact that $\int d\mathbf{r}\delta^{(3)}(\mathbf{r} - \mathbf{r}_\nu) = 1$, the function $\breve{n}^{(1)}_{\{\mathbf{r}^N\}}(\mathbf{r})$ also has the dimension (volume)$^{-1}$. Its integral is

$$\int d\mathbf{r}\, \breve{n}^{(1)}_{\{\mathbf{r}^N\}}(\mathbf{r}) = \int d\mathbf{r} \sum_{\nu=1}^{N} \delta^{(3)}(\mathbf{r} - \mathbf{r}_\nu) = \sum_{\nu=1}^{N} 1 = N,$$

so it is a number density. The function $\breve{n}^{(1)}_{\{\mathbf{r}^N\}}(\mathbf{r})$ is called the **microscopic number density distribution** of the system and it contains information about the instantaneous locations of all particles. The ensemble average of this microscopic density is the usual average number density, $\left\langle \breve{n}^{(1)}_{\{\mathbf{r}^N\}}(\mathbf{r}) \right\rangle = n_N^{(1)}(\mathbf{r})$, which can be realized from (cf. Equations 5.40 and 5.41)

$$\left\langle \breve{n}^{(1)}_{\{\mathbf{r}^N\}}(\mathbf{r}) \right\rangle = \int d\mathbf{r}^N \breve{n}^{(1)}_{\{\mathbf{r}^N\}}(\mathbf{r}) \mathcal{P}_N^{(N)}(\mathbf{r}^N)$$

$$= \sum_{\nu=1}^{N} \int d\mathbf{r}^N \delta^{(3)}(\mathbf{r} - \mathbf{r}_\nu) \mathcal{P}_N^{(N)}(\mathbf{r}^N) \tag{5.90}$$

$$= \sum_{\nu=1}^{N} \int d\mathbf{r}^N \Big|_{\text{not } d\mathbf{r}_\nu} \mathcal{P}_N^{(N)}(\mathbf{r}^N)\Big|_{\mathbf{r}_\nu=\mathbf{r}} = N\mathcal{P}_N^{(1)}(\mathbf{r}) \equiv n_N^{(1)}(\mathbf{r}),$$

where, for example, $d\mathbf{r}^N\big|_{\text{not } d\mathbf{r}_1} = d\mathbf{r}_2 \ldots d\mathbf{r}_N$ (i.e., all differentials except $d\mathbf{r}_1$) and where we have used the fact that all N terms in the sum are identical (only the names of the integration variables differ).

Next, we define the function

$$\breve{n}^{(2)}_{\{\mathbf{r}^N\}}(\mathbf{r}, \mathbf{r}') \equiv \sum_{\nu=1}^{N} \sum_{\substack{\nu'=1 \\ (\nu' \neq \nu)}}^{N} \delta^{(3)}(\mathbf{r} - \mathbf{r}_\nu)\delta^{(3)}(\mathbf{r}' - \mathbf{r}_{\nu'}), \tag{5.91}$$

which contains information about the instantaneous locations of all pairs of particles, where the particles in each pair are labeled ν and ν', respectively. Its ensemble average is the usual pair density distribution function $\left\langle \breve{n}^{(2)}_{\{\mathbf{r}^N\}}(\mathbf{r}, \mathbf{r}') \right\rangle = n_N^{(2)}(\mathbf{r}, \mathbf{r}')$, so $\breve{n}^{(2)}_{\{\mathbf{r}^N\}}(\mathbf{r}, \mathbf{r}')$ is the **microscopic pair density distribution** function. This can be shown as follows

$$\left\langle \breve{n}^{(2)}_{\{\mathbf{r}^N\}}(\mathbf{r}, \mathbf{r}') \right\rangle = \sum_{\nu=1}^{N} \sum_{\substack{\nu'=1 \\ (\nu' \neq \nu)}}^{N} \int d\mathbf{r}^N \delta^{(3)}(\mathbf{r} - \mathbf{r}_\nu)\delta^{(3)}(\mathbf{r}' - \mathbf{r}_{\nu'})\mathcal{P}_N^{(N)}(\mathbf{r}^N)$$

$$= N(N-1)\mathcal{P}_N^{(2)}(\mathbf{r}, \mathbf{r}') \equiv n_N^{(2)}(\mathbf{r}, \mathbf{r}'), \tag{5.92}$$

where we have used the fact the $N(N-1)$ terms in the sum are equal numerically.

{**EXERCISE 5.4**
Fill in the details in the proof of Equation 5.92.}

The difference $\check{n}^{(1)}_{\{r^N\}}(\mathbf{r}) - n^{(1)}_N(\mathbf{r})$ is the instantaneous deviation of the microscopic number density at $\mathbf{r}$ from its average value. It is of considerable interest to investigate how this deviation at one point $\mathbf{r}$ correlates with the deviation at another point $\mathbf{r}'$, that is, how fluctuations in density at the two points correlate with each other. Therefore, we define the **density-density correlation function** $H^{(2)}_N(\mathbf{r}, \mathbf{r}')$ as the ensemble average

$$H^{(2)}_N(\mathbf{r}, \mathbf{r}') = \left\langle \left[\check{n}^{(1)}_{\{r^N\}}(\mathbf{r}) - n^{(1)}_N(\mathbf{r}) \right] \left[\check{n}^{(1)}_{\{r^N\}}(\mathbf{r}') - n^{(1)}_N(\mathbf{r}') \right] \right\rangle$$

$$= \left\langle \check{n}^{(1)}_{\{r^N\}}(\mathbf{r}) \check{n}^{(1)}_{\{r^N\}}(\mathbf{r}') \right\rangle - n^{(1)}_N(\mathbf{r}) n^{(1)}_N(\mathbf{r}'). \tag{5.93}$$

We have

$$\check{n}^{(1)}_{\{r^N\}}(\mathbf{r}) \check{n}^{(1)}_{\{r^N\}}(\mathbf{r}') = \sum_{\nu=1}^{N} \delta^{(3)}(\mathbf{r} - \mathbf{r}_\nu) \sum_{\nu'=1}^{N} \delta^{(3)}(\mathbf{r}' - \mathbf{r}_{\nu'})$$

$$= \check{n}^{(2)}_{\{r^N\}}(\mathbf{r}, \mathbf{r}') + \sum_{\nu=1}^{N} \delta^{(3)}(\mathbf{r} - \mathbf{r}_\nu) \delta^{(3)}(\mathbf{r}' - \mathbf{r}_\nu) \tag{5.94}$$

and since we already know the ensemble average of the first term on the rhs, we only need to calculate the average of the last term, which is

$$\int d\mathbf{r}^N \sum_{\nu=1}^{N} \delta^{(3)}(\mathbf{r} - \mathbf{r}_\nu) \delta^{(3)}(\mathbf{r}' - \mathbf{r}_\nu) \mathcal{P}^{(N)}_N(\mathbf{r}^N) = \left\langle \check{n}^{(1)}_{\{r^N\}}(\mathbf{r}) \right\rangle \delta^{(3)}(\mathbf{r}' - \mathbf{r}), \tag{5.95}$$

where we have used the fact that[34]

$$\delta^{(3)}(\mathbf{r} - \mathbf{r}_\nu) \delta^{(3)}(\mathbf{r}' - \mathbf{r}_\nu) = \delta^{(3)}(\mathbf{r} - \mathbf{r}_\nu) \delta^{(3)}(\mathbf{r}' - \mathbf{r}).$$

Hence, we can conclude that

$$\left\langle \check{n}^{(1)}_{\{r^N\}}(\mathbf{r}) \check{n}^{(1)}_{\{r^N\}}(\mathbf{r}') \right\rangle = \left\langle \check{n}^{(2)}_{\{r^N\}}(\mathbf{r}, \mathbf{r}') \right\rangle + \left\langle \check{n}^{(2)}_{\{r^N\}}(\mathbf{r}) \right\rangle \delta^{(3)}(\mathbf{r}' - \mathbf{r}) \tag{5.96}$$

$$= n^{(2)}_N(\mathbf{r}, \mathbf{r}') + n^{(1)}_N(\mathbf{r}) \delta^{(3)}(\mathbf{r}' - \mathbf{r}), \tag{5.97}$$

so we have

$$H^{(2)}_N(\mathbf{r}, \mathbf{r}') = n^{(2)}_N(\mathbf{r}, \mathbf{r}') + n^{(1)}_N(\mathbf{r}) \delta^{(3)}(\mathbf{r}' - \mathbf{r}) - n^{(1)}_N(\mathbf{r}) n^{(1)}_N(\mathbf{r}')$$

$$= n^{(1)}_N(\mathbf{r}) \left[n^{(1)}_N(\mathbf{r}') h^{(2)}_N(\mathbf{r}, \mathbf{r}') + \delta^{(3)}(\mathbf{r}' - \mathbf{r}) \right], \tag{5.98}$$

---

[34] This fact follows from $\int d\mathbf{r}_\nu f(\mathbf{r}_\nu) \delta^{(3)}(\mathbf{r} - \mathbf{r}_\nu) \delta^{(3)}(\mathbf{r}' - \mathbf{r}_\nu) = [f(\mathbf{r}_\nu) \delta^{(3)}(\mathbf{r}' - \mathbf{r}_\nu)]_{\mathbf{r}_\nu = \mathbf{r}} = f(\mathbf{r}) \delta^{(3)}(\mathbf{r}' - \mathbf{r}) = \int d\mathbf{r}_\nu f(\mathbf{r}_\nu) \delta^{(3)}(\mathbf{r} - \mathbf{r}_\nu) \delta^{(3)}(\mathbf{r}' - \mathbf{r}).$

where we have inserted $n_N^{(2)}(\mathbf{r}, \mathbf{r}') = n_N^{(1)}(\mathbf{r})n_N^{(1)}(\mathbf{r}')[h_N^{(2)}(\mathbf{r}, \mathbf{r}') + 1]$. The delta function term describes the perfect correlation of a particle at $\mathbf{r}$ with itself and the pair correlation function $h_N^{(2)}$ carries the information about the correlations between different particles. For a bulk fluid, we obtain

$$H_N^{(2)}(\mathbf{r}, \mathbf{r}') = n^b \left[n^b h_N^{(2)}(|\mathbf{r} - \mathbf{r}'|) + \delta^{(3)}(\mathbf{r}' - \mathbf{r}) \right],$$

which we can write as

$$H_N^{(2)}(r) = n^b \left[n^b h_N^{(2)}(r) + \delta^{(3)}(r) \right]. \tag{5.99}$$

5.9 DISTRIBUTION FUNCTION HIERARCHIES AND CLOSURES, PRELIMINARIES

As mentioned at the end of Section 5.1, in order to calculate the pair distribution function one needs in principle to know the triplet distribution function, that is, density distribution around two particles. According to Equation 5.34 with $w^{\text{intr,b}} = 0$, we have

$$g^{(2)}(r_{12}) = e^{-\beta w^{(2)}(r_{12})} = e^{-\beta[u(r_{12})+w^{(2)\text{intr}}(r_{12})]} \quad \text{(bulk phase)} \tag{5.100}$$

and one needs to know the triplet distribution function to calculate $w^{(2)\text{intr}}$, that is, a more complex entity is needed to calculate a simpler one. Likewise, in order to calculate the triplet distribution function, one needs in principle to know the quartet distribution function etcetera.

For a bulk system with N particles, this means that in order to know $n_N^{(2)}(\mathbf{r}_1, \mathbf{r}_2)$, we require $n_N^{(3)}(\mathbf{r}_1, \mathbf{r}_2, \mathbf{r}_3)$, which in turn requires $n_N^{(4)}(\mathbf{r}_1, \mathbf{r}_2, \mathbf{r}_3, \mathbf{r}_4)$ etcetera all the way to $n_N^{(N)}(\mathbf{r}_1, \mathbf{r}_2, \ldots, \mathbf{r}_N)$, which we know explicitly in terms of the interaction potentials, Equation 5.39. This chain of dependencies is called a **distribution function hierarchy**. The N-particle function $n_N^{(N)}$ and its associated probability density $\mathcal{P}_N^{(N)}(\mathbf{r}^N)$ are extremely complex entities with $3N$ variables and they require that we know Z_N, which means that we need to know the free energy of the system. To obtain any l-point distribution function $n_N^{(l)}(\mathbf{r}_1, \ldots, \mathbf{r}_l)$ and the associated $\mathcal{P}_N^{(l)}(\mathbf{r}_1, \ldots, \mathbf{r}_l)$ (Equation 5.51) by evaluating the integral of $\mathcal{P}_N^{(N)}(\mathbf{r}^N)$ in Equation 5.50 is very complex and we still need to know Z_N. Thus, the problem to calculate the low order distribution functions like $n_N^{(2)}$ and hence $g_N^{(2)}$ accordingly seem quite unsolvable.

In practice it is, however, possible to use this kind of hierarchy to obtain low-order distribution functions. In order to determine $n_N^{(2)}$ to a good approximation, it is not necessary to have complete information about all higher order distribution functions. It is sufficient to have an approximative solution for them. An error in a high order distribution function normally gives rise to a much smaller error for a lower order function.

In one particularly important technique, one makes an approximation that implies that the distribution functions of higher order than, say, $n_N^{(l)}$ are not treated explicitly. The entire problem is then expressed solely in distribution functions of order $\leq l$. This kind

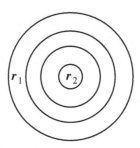

FIGURE 5.13 A sketch of the mean field approximation where the density around a particle at $\mathbf{r}_2$ is assumed not to be affected by the presence of another particle at $\mathbf{r}_1$.

of approximation is called a **closure approximation**, implying that one cuts the chain of dependencies between the distribution functions of different orders in such a manner that one obtains a closed, solvable problem for the lower order functions (of order $\leq l$). To obtain a very good approximation it is, however, necessary to reformulate the problem quite extensively before the chain of dependencies is cut. This will be done later on, but we will illustrate the principle with a simple example.

In Figure 5.3b, we have seen a sketch of the fact that the density around one particle at $\mathbf{r}_1$ affects the density around another particle at $\mathbf{r}_2$, provided the two particles are quite close to each other (this is described by the triplet distribution). The simplest closure approximation for the calculation of $g^{(2)}(r_{12})$ is to entirely neglect that the presence of one particle affects the density around the other (i.e., we neglect the triplet distributions). This is illustrated in Figure 5.13, where the density around the particle at $\mathbf{r}_2$ is unaffected by the particle at $\mathbf{r}_1$ and, likewise, by any other particle in the surroundings. We thus assume that all surrounding particles correlate with the particle at $\mathbf{r}_2$ but not with each other when we calculate $g^{(2)}(r_{12})$. This means that we make an artificial distinction between the particles. (We will, however, use the $g^{(2)}(r_{12})$ function thus obtained for all particles.)

Let us consider the potential that is felt by the particle at $\mathbf{r}_1$ when we neglect the influence from this particle on the density. The potential is equal to the sum of $u(r_{12})$ from the particle at $\mathbf{r}_2$ and the potential from all other particles when their density distribution is given by $n^b g^{(2)}(r_{23})$ as in Figure 5.13, where $\mathbf{r}_3$ is any point in the neighborhood of $\mathbf{r}_2$. The sum of $u(r_{12})$ and latter contribution is called the mean potential, $u^{\text{mean}}(r_{12})$, and is given by

$$u^{\text{mean}}(r_{12}) = u(r_{12}) + \int d\mathbf{r}_3\, n^b g^{(2)}(r_{23}) u(r_{31}).$$

It is so called because it is calculated from the mean density $n^b g^{(2)}(r_{23})$ around the particle at $\mathbf{r}_2$ in the absence of other fixed particles (i.e., when *all* other particles are mobile). We can use the mean potential to obtain an approximation for $w^{(2)}(r_{12})$ in Equation 5.100 by taking

$$w^{(2)}(r_{12}) \approx u^{\text{mean}}(r_{12}) + \text{constant},$$

where the constant is selected such that $w^{(2)}(r_{12}) \rightarrow 0$ when r_{12} becomes large, as required. (Note that one can always add a constant to a potential without changing the force since the latter is obtained from the derivative of the potential.)

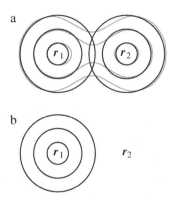

FIGURE 5.14 (a) The density distribution around two particles, as illustrated in Figure 5.3b (here shown as gray contours), is approximated by two superimposed spherical distributions around $\mathbf{r}_1$ and $\mathbf{r}_2$, respectively (shown as black contours). (b) The spherical density distribution around $\mathbf{r}_1$ is shown separately. The potential at $\mathbf{r}_1$ due to this distribution alone is independent of the distance r_{12} between $\mathbf{r}_1$ and $\mathbf{r}_2$.

This approximation for $w^{(2)}(r_{12})$ is quite rough, but it is not as bad as it may seem to disregard the influence that the particle at $\mathbf{r}_1$ has on the density in its neighborhood. If the particles at $\mathbf{r}_1$ and $\mathbf{r}_2$ are not too close to each other, the correct density distribution around them is similar to two superimposed spherical distributions as illustrated in Figure 5.14a. Let us consider the potential felt by the particle at $\mathbf{r}_1$ and the force that acts on this particle. The contribution to the potential at $\mathbf{r}_1$ due to a spherical distribution around this point, shown in Figure 5.14b, is simply a constant that does not depend on r_{12} and vanishes in the force on the particle at $\mathbf{r}_1$. On the other hand, the contribution to the potential at $\mathbf{r}_1$ from the spherical distribution around the particle at $\mathbf{r}_2$ *does* depend on r_{12} and *is* included in the approximation.

Let us now determine the constant we added to $u^{\text{mean}}(r_{12})$ to obtain our approximate $w^{(2)}(r_{12})$. It can be determined by noting that

$$u^{\text{mean}}(r_{12}) = u(r_{12}) + \int d\mathbf{r}_3\, n^{\text{b}} g^{(2)}(r_{23}) u(r_{31}) \rightarrow \int d\mathbf{r}_3\, n^{\text{b}} u(r_{31})$$

for large r_{12}. This follows from the facts that $g^{(2)}(r_{23}) \rightarrow 1$ when r_{23} becomes large, while $u(r_{31}) \rightarrow 0$ when r_{31} becomes large. Thus, we take

$$w^{(2)}(r_{12}) \approx u(r_{12}) + \int d\mathbf{r}_3\, n^{\text{b}} g^{(2)}(r_{23}) u(r_{31}) - \int d\mathbf{r}_3\, n^{\text{b}} u(r_{31}).$$

The sum of the integrals is

$$\int d\mathbf{r}_3\, n^{\text{b}} [g^{(2)}(r_{23}) - 1] u(r_{31}) = \int d\mathbf{r}_3\, n^{\text{b}} h^{(2)}(r_{23}) u(r_{31}),$$

and from Equation 5.100 we thus obtain

$$g^{(2)}(r_{12}) \approx e^{-\beta\left[u(r_{12})+n^b \int d\mathbf{r}_3 \, h^{(2)}(r_{23})u(r_{31})\right]}, \tag{5.101}$$

where the last term in the square bracket of the exponent constitutes our approximation for $w^{(2)\text{intr}}(r_{12})$. This equation only involves the pair distribution as unknown since $g^{(2)}(r_{12}) = h^{(2)}(r_{12}) + 1$. The equation can be solved, at least numerically, and gives $g^{(2)}(r_{12})$. This kind of approximation is called a **mean field approximation** (MFA). The "mean field" originates from the potential u^{mean}. The equation is a so-called *integral equation*, since the unknown, the pair distribution function, occurs both inside and outside an integral.

The approximation is useful as it stands for some kinds of point-like particles. For particles with highly singular interaction potentials, for example those with a hard core inter-action, the integral in Equation 5.101 does not exist. As you can see, for particles with hard cores, the integral $\int d\mathbf{r}_3 \, h^{(2)}(r_{23})u(r_{31}))$ is infinite for nearly all r_{12} since $u(r_{31})$ is infinite whenever $r_{31} < d^h$, that is, inside the hard core diameter d^h. The equation therefore has to be modified somewhat. To avoid the problem, one splits the pair potential into two parts, $u(r) = u^{\text{core}}(r) + u^{\text{rest}}(r)$, where u^{core} is the hard core potential defined in Equation 5.5 and u^{rest} is the remainder (that is well-behaved), and one takes

$$w^{(2)\text{intr}}(r_{12}) \approx \int d\mathbf{r}_3 \, n^b h^{(2)}(r_{23})u^{\text{rest}}(r_{31}) + w^{(2)\text{core}}(r_{12}),$$

where $w^{(2)\text{core}}$ describes the effects of the hard cores. One then obtains

$$g^{(2)}(r_{12}) \approx e^{-\beta\left[u(r_{12})+n^b \int d\mathbf{r}_3 \, h^{(2)}(r_{23})u^{\text{rest}}(r_{31})+w^{(2)\text{core}}(r_{12})\right]}. \tag{5.102}$$

Note that the first term in the exponent is the full pair interaction potential. Since $u(r_{12}) = \infty$ when $r_{12} < d^h$, we obtain $g^{(2)}(r_{12}) = 0$ there as required. A reasonable approximation for $w^{(2)\text{core}}(r_{12})$ is, for instance, to set it equal to $w^{(2)}(r_{12})$ obtained from some approximate treatment of a hard sphere fluid. The simplest possible approximation is, however, to set $w^{(2)\text{core}}(r_{12}) \equiv 0$. One then assumes that the particles around the central particle at $\mathbf{r}_2$ are treated like point particles in their interactions with each other. They are, however, not treated like point particles in their interactions with the central particle. This latter approximation is common in mean field approximations for electrolytes (see Section 5.12).

5.10 DISTRIBUTION FUNCTIONS IN THE GRAND CANONICAL ENSEMBLE

For a system with N particles, the probability density to observe l particles in the configuration $\mathbf{r}^l$ is given by $\mathcal{P}_N^{(l)}(\mathbf{r}_1, \ldots, \mathbf{r}_l)$ defined in Equation 5.50. To obtain the probability for an open system to have N particles *and* be in a certain configuration, we multiply by $\mathscr{P}(N)$ given by Equation 3.38 and obtain (cf. Equation 3.41) $\mathscr{P}(N)\mathcal{P}_N^{(l)}(\mathbf{r}_1, \ldots, \mathbf{r}_l)$. To account for the fact that the particles are indistinguishable, we must multiply with $N!/(N-l)!$, as we did in Section 5.5.3, and consider

$$\mathscr{P}(N)\mathcal{P}_N^{(l)}(\mathbf{r}_1,\ldots,\mathbf{r}_l)\frac{N!}{(N-l)!} = \mathscr{P}(N)n_N^{(l)}(\mathbf{r}_1,\ldots,\mathbf{r}_l),$$

where we have used the definition of $n_N^{(l)}$ (Equation 5.51) To obtain the **l-point density distribution function** $n^{(l)}$ for an open system, we must include all occurrences of at least l particles in the system, so we define

$$n^{(l)}(\mathbf{r}_1,\ldots,\mathbf{r}_l) = \sum_{N=l}^{\infty} \mathscr{P}(N)n_N^{(l)}(\mathbf{r}_1,\ldots,\mathbf{r}_l)$$

$$= \sum_{N=l}^{\infty} \mathscr{P}(N)\frac{N!}{(N-l)!Z_N} \int d\mathbf{r}_{l+1}\ldots d\mathbf{r}_N e^{-\beta \breve{U}_N^{\text{pot}}(\mathbf{r}^N)}$$

$$= \frac{1}{\Xi} \sum_{N=l}^{\infty} \frac{\zeta^N}{(N-l)!} \int d\mathbf{r}_{l+1}\ldots d\mathbf{r}_N e^{-\beta \breve{U}_N^{\text{pot}}(\mathbf{r}^N)}, \qquad (5.103)$$

where we have inserted $\mathscr{P}(N)$ from Equation 3.38 to obtain the last equality. The first term in the sum (for $N=l$) is equal to

$$\mathscr{P}(l)n_l^{(l)}(\mathbf{r}_1,\ldots,\mathbf{r}_l) = \mathscr{P}(l)\frac{l!}{Z_l}e^{-\beta \breve{U}_l^{\text{pot}}(\mathbf{r}^l)} = \frac{\zeta^l}{\Xi}e^{-\beta \breve{U}_l^{\text{pot}}(\mathbf{r}^l)},$$

which means that the rhs of Equation 5.103 should be interpreted such that the integral is absent from the first term.

The case $l=1$ gives the number density distribution $n^{(1)}(\mathbf{r}_1) \equiv n(\mathbf{r}_1)$, so we have

$$n^{(1)}(\mathbf{r}_1) = \sum_{N=1}^{\infty} \mathscr{P}(N)n_N^{(1)}(\mathbf{r}_1)$$

$$= \frac{1}{\Xi} \sum_{N=1}^{\infty} \frac{\zeta^N}{(N-1)!} \int d\mathbf{r}_2\ldots d\mathbf{r}_N e^{-\beta \breve{U}_N^{\text{pot}}(\mathbf{r}^N)}, \qquad (5.104)$$

while the pair density distribution is obtained for $l=2$

$$n^{(2)}(\mathbf{r}_1,\mathbf{r}_2) = \frac{1}{\Xi} \sum_{N=2}^{\infty} \frac{\zeta^N}{(N-2)!} \int d\mathbf{r}_3\ldots d\mathbf{r}_N e^{-\beta \breve{U}_N^{\text{pot}}(\mathbf{r}^N)}. \qquad (5.105)$$

The pair distribution function $g^{(2)}$ is, like before, obtained from

$$n^{(2)}(\mathbf{r}_1,\mathbf{r}_2) = n^{(1)}(\mathbf{r}_1)\, n^{(1)}(\mathbf{r}_2)\, g^{(2)}(\mathbf{r}_1,\mathbf{r}_2) \qquad (5.106)$$

and the density distribution around a particle located at $\mathbf{r}_2$ is given by

$$n^{(1)}(\mathbf{r}_1|\mathbf{r}_2) = n^{(1)}(\mathbf{r}_1)\,g^{(2)}(\mathbf{r}_1,\mathbf{r}_2), \tag{5.107}$$

which implies

$$n^{(2)}(\mathbf{r}_1,\mathbf{r}_2) = n^{(1)}(\mathbf{r}_1|\mathbf{r}_2)n^{(1)}(\mathbf{r}_2) \tag{5.108}$$

(cf. Equations 5.47 and 5.48 for the canonical ensemble).

The triplet density distribution function $n^{(3)}(\mathbf{r}_1,\mathbf{r}_2,\mathbf{r}_3)$ is likewise obtained from Equation 5.103 with $l = 3$ and we can define the **triplet distribution function** $g^{(3)}$ from

$$n^{(3)}(\mathbf{r}_1,\mathbf{r}_2,\mathbf{r}_3) = n^{(1)}(\mathbf{r}_1)\,n^{(1)}(\mathbf{r}_2)\,n^{(1)}(\mathbf{r}_3)\,g^{(3)}(\mathbf{r}_1,\mathbf{r}_2,\mathbf{r}_3). \tag{5.109}$$

The function $g^{(3)}$ is an analogue of the pair distribution function $g^{(2)}$, but it expresses the correlations between three particles instead of two. It is symmetric with respect to permutations of the three positions $\mathbf{r}_1$, $\mathbf{r}_2$ and $\mathbf{r}_3$, as can be seen from the definition of $n^{(3)}$.

The pair density distribution in the system when a particle is placed fixed at $\mathbf{r}_2$ is given by

$$n^{(2)}(\mathbf{r}_1,\mathbf{r}_2|\mathbf{r}_3) = n^{(1)}(\mathbf{r}_1)\,n^{(1)}(\mathbf{r}_2)\,g^{(3)}(\mathbf{r}_1,\mathbf{r}_2,\mathbf{r}_3) \tag{5.110}$$

as shown in the shaded text box on page 222, which can be skipped in the first reading. The function $n^{(2)}(\mathbf{r}_1,\mathbf{r}_2|\mathbf{r}_3)$ is the same as the pair density obtained when the interaction with a fixed particle at $\mathbf{r}_3$ is included in the external potential for the fluid. Therefore, this function can also be obtained from Equation 5.108 by inserting a particle of the same species at $\mathbf{r}_3$, whereby the latter equation becomes

$$n^{(2)}(\mathbf{r}_1,\mathbf{r}_2|\mathbf{r}_3) = n^{(1)}(\mathbf{r}_1|\mathbf{r}_2,\mathbf{r}_3)n^{(1)}(\mathbf{r}_2|\mathbf{r}_3). \tag{5.111}$$

By multiplying this expression with $n^{(1)}(\mathbf{r}_3)$ and using Equations 5.109 and 5.110, we obtain

$$n^{(3)}(\mathbf{r}_1,\mathbf{r}_2,\mathbf{r}_3) = n^{(1)}(\mathbf{r}_1|\mathbf{r}_2,\mathbf{r}_3)n^{(1)}(\mathbf{r}_2|\mathbf{r}_3)n^{(1)}(\mathbf{r}_3),$$

which expresses $n^{(3)}$, which has dimension (number density)^3, as a product of three densities. The first factor, $n^{(1)}(\mathbf{r}_1|\mathbf{r}_2,\mathbf{r}_3)$, is a density like that sketched in Figure 5.3, which depicts $n^{(1)}(\mathbf{r}|\mathbf{r}_1,\mathbf{r}_2)$.

The average potential energy for the system can be obtained by inserting $\breve{U}_N^{\mathrm{pot}} = \breve{U}_N^{\mathrm{intr}}(\mathbf{r}^N) + \sum_{\nu=1}^{N} v(\mathbf{r}_\nu)$ into Equation 3.42, which yields

$$\langle U^{\mathrm{pot}}\rangle = \sum_{N=0}^{\infty} \int d\mathbf{r}^N \breve{U}_N^{\mathrm{pot}}(\mathbf{r}^N)\mathcal{P}^{(N)}(\mathbf{r}^N)$$

$$= \langle U^{\mathrm{intr}}\rangle + \sum_{N=1}^{\infty}\sum_{\nu=1}^{N} \int d\mathbf{r}^N v(\mathbf{r}_\nu)\mathcal{P}^{(N)}(\mathbf{r}^N)$$

(the $N = 0$ term is zero). For $\nu = 1$, we have in the last sum

$$\int d\mathbf{r}^N v(\mathbf{r}_1) \mathcal{P}^{(N)}(\mathbf{r}^N) = \int d\mathbf{r}^N v(\mathbf{r}_1) \frac{\zeta^N}{N!\,\Xi} e^{-\beta \check{U}_N^{\text{pot}}(\mathbf{r}^N)}$$

$$= \frac{\zeta^N}{N!\,\Xi} \int d\mathbf{r}_1 v(\mathbf{r}_1) \int d\mathbf{r}_2 \ldots d\mathbf{r}_N e^{-\beta \check{U}_N^{\text{pot}}(\mathbf{r}^N)}$$

where we have inserted $\mathcal{P}^{(N)}(\mathbf{r}^N)$ from Equation 3.41. For all other ν values, we obtain the same numerical result, but with a permutation of the integration variables (cf. the corresponding argumentation for $\langle U_N^{\text{pot}} \rangle$ in Section 5.7). Since the sum over ν therefore contains N numerically equal terms, the entire last term in $\langle U^{\text{pot}} \rangle$ equals

$$\sum_{N=1}^{\infty} N \left[\frac{\zeta^N}{N!\,\Xi} \int d\mathbf{r}_1 v(\mathbf{r}_1) \int d\mathbf{r}_2 \ldots d\mathbf{r}_N e^{-\beta \check{U}_N^{\text{pot}}(\mathbf{r}^N)} \right] = \int d\mathbf{r}_1 v(\mathbf{r}_1) n^{(1)}(\mathbf{r}_1)$$

and we obtain

$$\langle U^{\text{pot}} \rangle = \langle U^{\text{intr}} \rangle + \int d\mathbf{r}_1 v(\mathbf{r}_1) n^{(1)}(\mathbf{r}_1). \tag{5.112}$$

Incidentally, we note that the grand canonical average of any quantity that depends solely on one-particle coordinates, say, $f(\mathbf{r}_\nu)$ for any particle ν, is given by

$$\langle f \rangle = \sum_{N=1}^{\infty} \int d\mathbf{r}^N \mathcal{P}^{(N)}(\mathbf{r}^N) \sum_{\nu=1}^{N} f(\mathbf{r}_\nu) = \int d\mathbf{r}_1 f(\mathbf{r}_1) n^{(1)}(\mathbf{r}_1), \tag{5.113}$$

as can be shown in the same manner.

If $\check{U}_N^{\text{intr}}(\mathbf{r}^N)$ is pairwise additive so Equation 5.69 applies, one can show in a similar manner that

$$\langle U^{\text{intr}} \rangle = \frac{1}{2} \int d\mathbf{r}_1 d\mathbf{r}_2 u(r_{12}) n^{(2)}(\mathbf{r}_1, \mathbf{r}_2).$$

{EXERCISE 5.5
Show this result for $\langle U^{\text{intr}} \rangle$ in the grand canonical ensemble.}

The total average energy equals

$$\bar{U} = \frac{3}{2} \bar{N} k_B T + \frac{1}{2} \int d\mathbf{r}_1 d\mathbf{r}_2 u(r_{12}) n^{(2)}(\mathbf{r}_1, \mathbf{r}_2) + \int d\mathbf{r}_1 v(\mathbf{r}_1) n^{(1)}(\mathbf{r}_1) \tag{5.114}$$

in this case (cf. Equation 5.71 for the canonical ensemble). If the particle interacts with a potential that contains contributions that depend on three particle coordinates, $u^{(3)}(\mathbf{r}_1, \mathbf{r}_2, \mathbf{r}_3)$, and U^{intr} therefore cannot be written solely in terms of a pair potential $u^{(2)}(\mathbf{r}_1, \mathbf{r}_2)$, the average energy contains also a similar term with the triplet density distribution function $n^{(3)}$.

★ Let us now show Equation 5.110. We take an open system with density distribution functions $n^{(l)}$ for any l described by Equation 5.103 and add a particle of the same species placed fixed at position $\mathbf{r}_0$. The potential energy for $N + 1$ particles in the new system (N mobile particles and one fixed) is denoted by $\check{U}_{N+1}^{\text{pot}}(\mathbf{r}_0, \mathbf{r}^N)$ for any N. Since we have a particle fixed at $\mathbf{r}_0$, the pair density distribution in this system is denoted by $n^{(2)}(\mathbf{r}_1, \mathbf{r}_2 | \mathbf{r}_0)$ and Equation 5.105 becomes

$$n^{(2)}(\mathbf{r}_1, \mathbf{r}_2 | \mathbf{r}_0) = \frac{1}{\Xi^{(1)}(\mathbf{r}_0)} \sum_{N=2}^{\infty} \frac{\zeta^N}{(N-2)!} \int d\mathbf{r}_3 \dots d\mathbf{r}_N e^{-\beta \check{U}_{N+1}^{\text{pot}}(\mathbf{r}_0, \mathbf{r}^N)},$$

where

$$\Xi^{(1)}(\mathbf{r}_0) = \sum_{N=0}^{\infty} \frac{\zeta^N}{N!} \int d\mathbf{r}_1 \dots d\mathbf{r}_N e^{-\beta \check{U}_{N+1}^{\text{pot}}(\mathbf{r}_0, \mathbf{r}^N)}$$

is the grand canonical partition function for the system with a particle kept fixed at $\mathbf{r}_0$. By replacing N by $N' - 1$ in the last equation, we can write it as

$$\Xi^{(1)}(\mathbf{r}_0) = \sum_{N'=1}^{\infty} \frac{\zeta^{N'-1}}{(N'-1)!} \int d\mathbf{r}_1 \dots d\mathbf{r}_{N'-1} e^{-\beta \check{U}_{N'}^{\text{pot}}(\mathbf{r}_0, \mathbf{r}^{N'-1})} = \frac{\Xi\, n^{(1)}(\mathbf{r}_0)}{\zeta},$$

where the last equality is obtained from Equation 5.104 after renaming the integration variables (by setting new indices). If we do the same thing in the equation for $n^{(2)}$ above, we obtain

$$n^{(2)}(\mathbf{r}_1, \mathbf{r}_2 | \mathbf{r}_0) = \frac{1}{\Xi^{(1)}(\mathbf{r}_0)} \sum_{N'=3}^{\infty} \frac{\zeta^{N'-1}}{(N'-3)!} \int d\mathbf{r}_3 \dots d\mathbf{r}_{N'-1} e^{-\beta \check{U}_{N'}^{\text{pot}}(\mathbf{r}_0, \mathbf{r}^{N'-1})}$$

$$= \frac{1}{\Xi^{(1)}(\mathbf{r}_0)} \frac{\Xi\, n^{(3)}(\mathbf{r}_0, \mathbf{r}_1, \mathbf{r}_2)}{\zeta}$$

where Equation 5.103 with $l = 3$ is used to obtain the last equality. By combining these results, we obtain $n^{(2)}(\mathbf{r}_1, \mathbf{r}_2 | \mathbf{r}_0) = n^{(3)}(\mathbf{r}_1, \mathbf{r}_2, \mathbf{r}_0)/n^{(1)}(\mathbf{r}_0)$, which is equal to $n^{(1)}(\mathbf{r}_1)\, n^{(1)}(\mathbf{r}_2)\, g^{(3)}(\mathbf{r}_1, \mathbf{r}_2, \mathbf{r}_0)$ and hence yields Equation 5.110 with $\mathbf{r}_3 = \mathbf{r}_0$.

5.11 THE BORN-GREEN-YVON EQUATIONS

Let us consider an inhomogeneous fluid in equilibrium with a bulk fluid of density n^{b} (the latter constitutes an infinitely large reservoir of particles), so we treat the system in the grand canonical ensemble. In most cases, we will assume that the interactions between the particles are pairwise additive, Equation 5.9. We found earlier, Equation 5.25, that the density distribution is given by

$$n(\mathbf{r}_1) = n^b e^{-\beta w(\mathbf{r}_1)} = n^b e^{-\beta[v(\mathbf{r}_1)+w^{intr}(\mathbf{r}_1)]},$$

where we have selected $w = 0$ in the bulk. The potential of mean force $w(\mathbf{r}_1)$ has been divided into two parts, the external potential $v(\mathbf{r}_1)$ and the contribution from the interactions between the particles of the system $w^{intr}(\mathbf{r}_1)$, the intrinsic part of w, as explained in connection to Equation 5.19.

While it is general in difficult to obtain w^{intr}, it is easier to obtain the derivative ∇w^{intr} or small variations δw^{intr}. This is analogous to a fact that we touched upon in Section 5.7 for the canonical ensemble, namely that it is easier to determine a free energy derivative like $(\partial A/\partial V)_{T,N} = -P$ or the variation $dA = -P\,dV$ than to determine a finite variation ΔA or A itself. As we have seen, to obtain ΔA from the pressure, the latter must be known as a function of volume $P = P(V)$ in a finite volume interval. On the other hand, P can be determined from the variation of A in an infinitesimal interval of width dV, a much easier task. One may say that A contains more information about the system than P. Here we shall investigate ∇w^{intr}, while we shall deal with δw^{intr} and other variations in Chapter 9 (in the 2nd volume).

The average force $\mathbf{F}$ on a particle at position $\mathbf{r}_1$ is given by (Equation 5.18)

$$\mathbf{F}(\mathbf{r}_1) = -\nabla_1 v(\mathbf{r}_1) + \mathbf{F}^{intr}(\mathbf{r}_1) = -\nabla_1 v(\mathbf{r}_1) - \nabla_1 w^{intr}(\mathbf{r}_1). \tag{5.115}$$

The force $\mathbf{F}^{intr}$ is the sum of the forces from all particles in the surroundings of the particle at $\mathbf{r}_1$. These particles have an average density distribution at position $\mathbf{r}_3$ equal to $n(\mathbf{r}_3|\mathbf{r}_1)$ and each one of them gives the force $-\nabla_1 u(\mathbf{r}_3,\mathbf{r}_1)$ on the particle. The number of particles in the volume element $d\mathbf{r}_3$ at the point $\mathbf{r}_3$ is $n(\mathbf{r}_3|\mathbf{r}_1)d\mathbf{r}_3$, so the total average force $\mathbf{F}^{intr}$ from interparticle interactions, the sum from all volume elements, equals

$$\mathbf{F}^{intr}(\mathbf{r}_1) = \int d\mathbf{r}_3 \, n(\mathbf{r}_3|\mathbf{r}_1)\left(-\nabla_1 u(\mathbf{r}_3,\mathbf{r}_1)\right). \tag{5.116}$$

(Incidentally, this is the force that was discussed at the end of Section 5.3 in connection to Figure 5.8 and $n(\mathbf{r}_3|\mathbf{r}_1)$ is the density distribution of the kind that is depicted in the figure.) It follows that

$$\nabla_1 w^{intr}(\mathbf{r}_1) = -\mathbf{F}^{intr}(\mathbf{r}_1) = \int d\mathbf{r}_3 \, n(\mathbf{r}_3)g^{(2)}(\mathbf{r}_1,\mathbf{r}_3)\nabla_1 u(\mathbf{r}_3,\mathbf{r}_1). \tag{5.117}$$

From the fact that $\mathbf{F}(\mathbf{r}_1) = k_B T \nabla_1 \ln n(\mathbf{r}_1)$ (Equation 5.12) and the relationships above, we see that

$$\nabla_1 \ln n(\mathbf{r}_1) = \beta \mathbf{F}(\mathbf{r}_1)$$
$$= -\beta\left[\nabla_1 v(\mathbf{r}_1) + \int d\mathbf{r}_3 \, n(\mathbf{r}_3)g^{(2)}(\mathbf{r}_1,\mathbf{r}_3)\nabla_1 u(\mathbf{r}_3,\mathbf{r}_1)\right]. \tag{5.118}$$

This is known as the **first Born-Green-Yvon (BGY) equation** and is often written as

$$\nabla_1 n(\mathbf{r}_1) = -\beta n(\mathbf{r}_1) \left[\nabla_1 v(\mathbf{r}_1) + \int d\mathbf{r}_3 \, n(\mathbf{r}_3) g^{(2)}(\mathbf{r}_1, \mathbf{r}_3) \nabla_1 u(\mathbf{r}_3, \mathbf{r}_1) \right]. \tag{5.119}$$

It is an integro-differential equation for the density distribution $n(\mathbf{r})$, but it requires that $g^{(2)}$, which is a more complicated function, is known. One possible strategy to determine $n(\mathbf{r})$ is to obtain an approximate $g^{(2)}$ and then solve the equation.

To determine $g^{(2)}$, one can follow analogous arguments to obtain an equation for w^{intr} when a particle is placed at $\mathbf{r}_2$, that is, $w^{\text{intr}}(\mathbf{r}_1|\mathbf{r}_2)$, and to thereby obtain an equation for the pair distribution function. The force $\mathbf{F}^{\text{intr}}$ on a particle at $\mathbf{r}_1$ when there is a particle at $\mathbf{r}_2$ will be denoted as $\mathbf{F}^{\text{intr}}(\mathbf{r}_1|\mathbf{r}_2)$. It can be obtained by applying Equation 5.116 to this situation. We obtain

$$\mathbf{F}^{\text{intr}}(\mathbf{r}_1|\mathbf{r}_2) = \int d\mathbf{r}_3 \, n(\mathbf{r}_3|\mathbf{r}_1, \mathbf{r}_2) \left(-\nabla_1 u(\mathbf{r}_3, \mathbf{r}_1) \right),$$

where $n(\mathbf{r}_3|\mathbf{r}_1, \mathbf{r}_2)$ is the density at $\mathbf{r}_3$ when there is one particle at $\mathbf{r}_1$ and one at $\mathbf{r}_2$, that is, the kind of density distribution sketched in Figure 5.3 for a bulk fluid. We can take over the results that we derived earlier in the present section by including the interaction from the particle at $\mathbf{r}_2$ in the external potential so it becomes $v(\mathbf{r}_1|\mathbf{r}_2) = v(\mathbf{r}_1) + u(\mathbf{r}_1, \mathbf{r}_2)$ instead of $v(\mathbf{r}_1)$. By going through the same steps as above (that lead to Equation 5.118), we obtain

$$\nabla_1 \ln n(\mathbf{r}_1|\mathbf{r}_2) = -\beta \left[\nabla_1 v(\mathbf{r}_1|\mathbf{r}_2) + \int d\mathbf{r}_3 \, n(\mathbf{r}_3|\mathbf{r}_1, \mathbf{r}_2) \nabla_1 u(\mathbf{r}_3, \mathbf{r}_1) \right]. \tag{5.120}$$

This is the **second Born-Green-Yvon equation**. It is usually written in a different form, which can be obtained as follows.

Equation 5.111 can be written as $n^{(2)}(\mathbf{r}_3, \mathbf{r}_1|\mathbf{r}_2) = n(\mathbf{r}_3|\mathbf{r}_1, \mathbf{r}_2) n(\mathbf{r}_1|\mathbf{r}_2)$ From this relationship and Equation 5.110, we obtain

$$n(\mathbf{r}_3|\mathbf{r}_1, \mathbf{r}_2) = \frac{n^{(2)}(\mathbf{r}_3, \mathbf{r}_1|\mathbf{r}_2)}{n(\mathbf{r}_1|\mathbf{r}_2)} = \frac{n(\mathbf{r}_3) n(\mathbf{r}_1) g^{(3)}(\mathbf{r}_1, \mathbf{r}_2, \mathbf{r}_3)}{n(\mathbf{r}_1|\mathbf{r}_2)}. \tag{5.121}$$

By inserting this into Equation 5.120 and subtracting Equation 5.118 from the result, we obtain

$$\nabla_1 \ln g^{(2)}(\mathbf{r}_1, \mathbf{r}_2) = -\beta \nabla_1 u(\mathbf{r}_1, \mathbf{r}_2) \tag{5.122}$$

$$-\beta \int d\mathbf{r}_3 \, n(\mathbf{r}_3) \left[\frac{g^{(3)}(\mathbf{r}_1, \mathbf{r}_2, \mathbf{r}_3)}{g^{(2)}(\mathbf{r}_1, \mathbf{r}_2)} - g^{(2)}(\mathbf{r}_1, \mathbf{r}_3) \right] \nabla_1 u(\mathbf{r}_3, \mathbf{r}_1),$$

where we have used Equation 5.107. This is the second BGY equation written in a more common form than in Equation 5.120.

Equation 5.120 is an equation for $n(\mathbf{r}_1|\mathbf{r}_2)$ (from which $g^{(2)}$ can be determined), but it requires that $n(\mathbf{r}_3|\mathbf{r}_1, \mathbf{r}_2)$, which is a more complicated function, is known. The increasing complexity when we go from the first to the second BGY equation is an explicit example of the situation we discussed in Section 5.9, and these two equations are just the first two members of a hierarchy of equations for higher and higher order distribution functions. The next member, which is an equation for the quartet distribution function (the third BGY equation), can be derived in a straightforward manner by including yet another fixed particle and going through the same kind of procedure. It requires the knowledge of the next higher distribution function and so it goes on for higher members.

5.12 MEAN FIELD APPROXIMATIONS FOR BULK SYSTEMS

As we saw in the previous section, one can determine the pair correlation function $g^{(2)}$ via Equation 5.122 provided one knows the triplet distribution function $g^{(3)}$, which is a more complicated function than $g^{(2)}$. To obtain an equation that one can solve, it is necessary to do some kind of approximation, and we shall, as an example, discuss a bulk fluid with density n^b. For a bulk fluid, only the second and higher BGY equations are relevant; here we will only consider the second. We will approximate $g^{(3)}$ somehow to obtain a solvable equation and a possible strategy is to express $g^{(3)}$ in terms of the pair distribution function. Thereby, we obtain an equation that only involves the pair function as unknown, so it is solvable in principle. This strategy is an example of the concept of *closure approximation* that we introduced in Section 5.9. It implies that one makes an approximation that transforms the set of equations (or, in the present case, one single equation) into a closed set that has an equal number of equations as unknown entities. The set of equations is then solvable.

When one wants to determine the distribution around a particle, the simplest approximation is to entirely neglect the correlations between particles in its surroundings. Let us see what this means for the second BGY equation and the triplet distribution function. The latter can according to Equation 5.121 be written as

$$g^{(3)}(\mathbf{r}_1, \mathbf{r}_2, \mathbf{r}_3) = \frac{n(\mathbf{r}_3|\mathbf{r}_1, \mathbf{r}_2)n(\mathbf{r}_1|\mathbf{r}_2)}{n(\mathbf{r}_1)n(\mathbf{r}_3)}.$$

We consider a particle placed at $\mathbf{r}_2$. To neglect correlations between particles in its surroundings means that correlations between particles at positions $\mathbf{r}_1$ and $\mathbf{r}_3$ are ignored. This implies that a particle at $\mathbf{r}_3$ does not "feel" a particle at $\mathbf{r}_1$, so $g^{(2)}(\mathbf{r}_1, \mathbf{r}_3) \approx 1$ and likewise $n(\mathbf{r}_3|\mathbf{r}_1, \mathbf{r}_2) \approx n(\mathbf{r}_3|\mathbf{r}_2) = n(\mathbf{r}_3)g^{(2)}(\mathbf{r}_3, \mathbf{r}_2)$. Using the latter equation, we see that

$$g^{(3)}(\mathbf{r}_1, \mathbf{r}_2, \mathbf{r}_3) \approx g^{(2)}(\mathbf{r}_3, \mathbf{r}_2)g^{(2)}(\mathbf{r}_1, \mathbf{r}_2),$$

where we have used $n(\mathbf{r}_1|\mathbf{r}_2) = n(\mathbf{r}_1)g^{(2)}(\mathbf{r}_1, \mathbf{r}_2)$. By inserting these approximations into Equation 5.122, we obtain

$$\nabla_1 \ln g^{(2)}(r_{12}) \approx -\beta \left[\nabla_1 u(r_{12}) + n^b \int d\mathbf{r}_3 \left(g^{(2)}(r_{23}) - 1 \right) \nabla_1 u(r_{31}) \right], \tag{5.123}$$

which can be integrated, yielding

$$g^{(2)}(r_{12}) = e^{-\beta \left[u(r_{12}) + n^b \int d\mathbf{r}_3 \, h^{(2)}(r_{23})u(r_{31}) \right]} \quad \text{(MFA).} \tag{5.124}$$

This is the same as Equation 5.101 and it is appropriate provided the pair interaction potential is sufficiently well-behaved.[35] The approximation we have made to obtain Equation 5.124 is exactly the same as we did when we obtained the Equation 5.101, so this is the *mean field approximation* (MFA) for pair correlations that we introduced in a simpler fashion in Section 5.9. For highly singular interaction potentials, like those with a hard core interaction, we replace this equation by

$$g^{(2)}(r_{12}) = e^{-\beta\left[u(r_{12}) + n^b \int d\mathbf{r}_3\, h^{(2)}(r_{23})u^{\text{rest}}(r_{31}) + w^{(2)\text{core}}(r_{12})\right]} \quad \text{(MFA)}, \tag{5.125}$$

that is, Equation 5.102 as discussed in Section 5.9, where the definitions of u^{rest} and $w^{(2)\text{core}}$ can be found.

For systems with soft particles, like a LJ fluid, one can make a similar approximation by splitting the pair potential into a short-range repulsive potential $u^{\text{rep}}(r)$ and the rest, $u(r) = u^{\text{rep}}(r) + u^{\text{rest}}(r)$, whereby Equation 5.125 still applies as an approximation, but with $w^{(2)\text{core}}$ replaced by some other function. It is, however, common to use $w^{(2)\text{core}}$ from a hard sphere fluid also in this case.

An approximation that is somewhat more refined than the mean field approximation is obtained if we set $g^{(3)}(\mathbf{r}_1, \mathbf{r}_2, \mathbf{r}_3) \approx g^{(2)}(\mathbf{r}_3, \mathbf{r}_2)g^{(2)}(\mathbf{r}_1, \mathbf{r}_2)g^{(2)}(\mathbf{r}_1, \mathbf{r}_3)$, which is called **Kirkwood's superposition approximation**. By inserting this into Equation 5.122, we obtain

$$\nabla_1 \ln g^{(2)}(r_{12}) \approx -\beta\left[\nabla_1 u(r_{12}) + n^b \int d\mathbf{r}_3\, \left(g^{(2)}(r_{23}) - 1\right)g^{(2)}(r_{31})\nabla_1 u(r_{31})\right].$$

Here the factor $g^{(2)}(r_{31})$ in the integrand, which is not present in the MFA Equation 5.123, assures that a hard-core interaction does not cause any problems. The integrand does not have any contribution from the hard-core region in $u(r_{31})$ since $g^{(2)}(r_{31})$ is zero there. For a bulk fluid, it is appropriate to define the potential of mean force for pair and triplet interactions from $g^{(3)} = \exp(-\beta w^{(3)})$ and $g^{(2)} = \exp(-\beta w^{(2)})$, so the superposition approximation means that $w^{(3)}(\mathbf{r}_1, \mathbf{r}_2, \mathbf{r}_3) \approx w^{(2)}(\mathbf{r}_3, \mathbf{r}_2) + w^{(2)}(\mathbf{r}_1, \mathbf{r}_2) + w^{(2)}(\mathbf{r}_1, \mathbf{r}_3)$. This approximation works well when the density of the fluid is quite low (it becomes exact in the limit $n^b \to 0$), but fails at higher densities.

To find more accurate approximations for $g^{(2)}$ and other quantities, one needs to make a deeper analysis of the statistical mechanical theory. It is thereby necessary to introduce some more fundamental concepts in liquid state theory. This will be done done in Chapter 9 (in the 2nd volume), whereafter such approximations are introduced like more advanced integral equation theories and density functional theory.[36] We have already encountered results from such theories and more will appear in the following chapters. A key element in most integral equation theories is to make a closure approximation for some higher order distribution function in order to obtain approximate lower order functions, as discussed

---

[35] If the potential is infinite at the origin, it must have an integrable singularity there. This is the case for the Coulomb potential.
[36] In density functional theory, one starts from an approximate expression for the free energy of a system and then obtains distribution functions and thermodynamic properties by taking various derivatives.

in Section 5.9. To be able to make very clever closure approximation, one needs the more fundamental concepts mentioned earlier.

The most accurate manner to calculate statistical mechanical quantities for a given Hamiltonian is – at least in principle but usually also in practice – via computer simulations and we will give a brief introduction to this topic in the following section.

5.13 COMPUTER SIMULATIONS AND DISTRIBUTION FUNCTIONS

5.13.1 General Background

A **computer simulation** of a fluid is a numerical method to calculate various statistical mechanical properties of the system. In this treatise, we will only treat a few aspects of this subject[37] and, furthermore, we will limit ourselves to spherical particles. We will only be concerned with equilibrium properties.

One can distinguish two principally different manners to do such calculations, namely Molecular Dynamics and Monte Carlo simulations. In both methods, one generates a very large number of configurations for the particles in the system, like those sketched in Figures 5.1, 5.2 and 5.7a. More precisely, for a system with N particles, one generates sets of particle positions $\mathbf{r}_\nu$ with $1 \leq \nu \leq N$, where each set constitutes a configuration $\mathbf{r}^N \equiv \mathbf{r}_1, \mathbf{r}_2, \ldots, \mathbf{r}_N$.

In a **Molecular Dynamics (MD) simulation**, the particle coordinates are generated by numerical solution of Newton's equations of motion for the particles and, as a result, one obtains $\mathbf{r}_\nu = \mathbf{r}_\nu(t)$ as functions of time and, at the same time, the momenta $\mathbf{p}_\nu = \mathbf{p}_\nu(t)$. Thus, one obtains a part of a trajectory for the phase space point $\mathbf{\Gamma} = \mathbf{\Gamma}(t)$ of the system (cf. Section 3.1). As will be described later, the values of $\mathbf{r}_\nu$ and $\mathbf{p}_\nu$ are obtained for a discrete set of M instances of time, $t = t_\alpha$, for $1 \leq \alpha \leq M$, so we have the set of configurations $\mathbf{r}^N_\alpha \equiv \mathbf{r}_{1,\alpha}, \mathbf{r}_{2,\alpha}, \ldots, \mathbf{r}_{N,\alpha}$, where $\mathbf{r}_{\nu,\alpha} = \mathbf{r}_\nu(t_\alpha)$, and the corresponding set of momentum values. The time step $\Delta t = t_{\alpha+1} - t_\alpha$ is assumed in this treatise to be the same for all α, but this is not necessary in general. The various equilibrium properties of the system are calculated as time averages, whereby the momentum values are not needed (except in the calculation of the average kinetic energy, see below). The average of a quantity X, which depends on the coordinates $\mathbf{r}_1, \mathbf{r}_2, \ldots, \mathbf{r}_N$ and thereby on time, $X = X(\mathbf{r}^N_\alpha(t)) \equiv X(t)$, is given by

$$\langle X \rangle = \frac{1}{t_f - t_i} \int_{t_i}^{t_f} dt\, X(t), \tag{5.126}$$

where t_i is the initial and t_f the final time.

When we have calculated the trajectory for a very long time interval (in principle infinitely long, $t_f \to \infty$), the time average thus obtained gives the equilibrium value of $\langle X \rangle$ provided that the trajectory is representative of the system. In practice, one can obtain the average to

---

[37] For a comprehensive treatise, see, for example, D. Frenkel and B. Smit, *Understanding Molecular Simulation - From Algorithms to Applications*, 2nd Edition, Academic Press, San Diego, 2002 and M. P. Allen and D. J. Tildesley, *Computer Simulations of Liquids*, 2nd Edition, Oxford Science Publications, Oxford, 2017.

a high degree of accuracy by calculating the particle configurations for a sufficiently long time interval. For a system with given N and V, the total energy U is constant, which is a consequence of Newton's equations of motion (provided they are solved exactly, which is not the case for numerical solutions). In addition, the total momentum $\mathbf{p}_{\text{tot}} = \sum_{\nu=1}^{N} \mathbf{p}_\nu$ is constant, which is also a consequence of these equations. Both U and $\mathbf{p}_{\text{tot}}$ are so-called constants of motion.

As discussed in Chapter 3, we may alternatively calculate properties of an equilibrium system by taking an ensemble average rather than a time average. The two averages are the same provided the system is ergodic, that is, it fulfills the ergodic hypothesis (cf. Section 2.1). By generating a long sequence of particle configurations $\mathbf{r}_\alpha^N$, where $1 \le \alpha \le M$, the ensemble average of the quantity $X = X(\mathbf{r}_\alpha^N)$ is in the limit $M \to \infty$ given by

$$\langle X \rangle = \frac{1}{M} \sum_{\alpha=1}^{M} X(\mathbf{r}_\alpha^N) \tag{5.127}$$

provided the different configurations occur with probabilities given by the ensemble and are representative of the system. In the ensemble average, each configuration thereby represents an ensemble member and corresponds to a microstate of the system. In practice, one has a finite M and one can obtain the average to a high degree of accuracy by calculating a sufficiently large number of particle configurations. This is the principle behind a **Monte Carlo (MC) simulation**, where one generates particle configurations by moving the particles around in a more or less random manner and singling out configurations in such a manner that they occur with the correct probabilities. The sequence of configurations does not follow a temporal order in this case and one does not generate any momenta. The latter are not needed since we know their distribution analytically; for example, in the case of the canonical ensemble, each momentum vector $\mathbf{p}_\nu$ is Maxwell-Boltzmann distributed as shown in Equation 3.7. Once the average over configuration space is calculated according to Equation 5.127, one can add the contributions due to the momenta. The latter are the same as for an ideal gas and are known explicitly. One can do MC simulations for various ensembles, for example the NVT, NPT and μVT ensembles.

Initially, we will treat simulations of a confined fluid of N particle in a container of volume V, and later we will see how one can treat bulk fluids and fluids between two infinitely large, flat surfaces.

5.13.1.1 Basics of Molecular Dynamics Simulations

Let us start by describing MD simulations in more detail. The average of a quantity $X(\mathbf{r}_\alpha^N)$ can be calculated from Equation 5.127 with $M = (t_f - t_i)/\Delta t$, so the same expression can be used for both kinds of simulations. The difference from MC simulations is the manner used to generate the sequence of configurations.

A simple, but very efficient, manner to numerically solve Newton's equations of motion is the **Verlet algorithm**,[38] which can be derived as follows. Let us consider the particle coordinates $\mathbf{r}_\nu(t)$ before and after a time step, that is, at t and $t + \Delta t$. We will also be concerned

---

[38] L. Verlet, *Phys. Rev.* **159** (1967) 98.

with its previous value $\mathbf{r}_\nu(t - \Delta t)$. The Taylor expansion of $\mathbf{r}_\nu(t \pm \Delta t)$ is given by

$$\mathbf{r}_\nu(t + \Delta t) = \mathbf{r}_\nu(t) + \dot{\mathbf{r}}_\nu(t)\Delta t + \frac{\ddot{\mathbf{r}}_\nu(t)}{2!}\Delta t^2 + \frac{\dddot{\mathbf{r}}_\nu(t)}{3!}\Delta t^3 + \mathcal{O}(\Delta t^4)$$

$$\mathbf{r}_\nu(t - \Delta t) = \mathbf{r}_\nu(t) - \dot{\mathbf{r}}_\nu(t)\Delta t + \frac{\ddot{\mathbf{r}}_\nu(t)}{2!}\Delta t^2 - \frac{\dddot{\mathbf{r}}_\nu(t)}{3!}\Delta t^3 + \mathcal{O}(\Delta t^4),$$

$$(5.128)$$

where $\dot{\mathbf{r}}_\nu = d\mathbf{r}_\nu/dt = \mathbf{v}_\nu$, $\ddot{\mathbf{r}}_\nu = d^2\mathbf{r}_\nu/dt^2 = d\mathbf{v}_\nu/dt$, $\dddot{\mathbf{r}}_\nu = d^3\mathbf{r}_\nu/dt^3$ and the symbol $\mathcal{O}(\Delta t^4)$ stands for some function that decays to zero like Δt^4 (or faster) when $\Delta t \to 0$. From Newton's second law, we have

$$\ddot{\mathbf{r}}_\nu(t) = \frac{d\mathbf{v}_\nu(t)}{dt} = \frac{1}{m}\frac{d\mathbf{p}_\nu(t)}{dt} = \frac{\mathbf{F}_\nu(t)}{m},$$

$$(5.129)$$

where $\mathbf{F}_\nu(t)$ is the force that acts on particle ν at time t when it is located at position $\mathbf{r}_\nu(t)$. This force originates from all other particles located at $\mathbf{r}_{\nu'}(t)$ for $\nu' \neq \nu$ and from the walls of the container that contains the N particle system. It is obtained from the instantaneous potential energy $\check{U}_N^{\text{pot}}$ as

$$\mathbf{F}_\nu(t) = -\nabla_\nu \check{U}_N^{\text{pot}}(\mathbf{r}^N(t)) = -\nabla_\nu \check{U}_N^{\text{pot}}(\mathbf{r}_1(t), \mathbf{r}_2(t), \mathbf{r}_3(t), \dots, \mathbf{r}_N(t)),$$

where $\nabla_\nu = (\partial/\partial x_\nu, \partial/\partial y_\nu, \partial/\partial z_\nu)$. For example, for a fluid with pair potential $u(r)$ between the particles and external potential $v(\mathbf{r})$ from the container walls, we have

$$\mathbf{F}_\nu(t) = -\sum_{\nu' \neq \nu} \nabla_\nu u(|\mathbf{r}_\nu(t) - \mathbf{r}_{\nu'}(t)|) - \nabla_\nu v(\mathbf{r}_\nu(t)).$$

Usually, $u(r)$ and $v(\mathbf{r})$ are model potentials like the Lennard-Jones potential or similar and then the force is given by an explicit expression that can be evaluated analytically.

By adding the two equations in Equation 5.128 and inserting Equation 5.129, we obtain

$$\mathbf{r}_\nu(t + \Delta t) + \mathbf{r}_\nu(t - \Delta t) = 2\mathbf{r}_\nu(t) + \frac{\mathbf{F}_\nu(t)}{m}\Delta t^2 + O(\Delta t^4).$$

If we know the particle positions at both $t - \Delta t$ and t and the force at t, we can accordingly calculate the position at time $t + \Delta t$ from

$$\mathbf{r}_\nu(t + \Delta t) \approx 2\mathbf{r}_\nu(t) - \mathbf{r}_\nu(t - \Delta t) + \frac{\mathbf{F}_\nu(t)}{m}\Delta t^2,$$

$$(5.130)$$

where we have made an error of the order Δt^4. This is the Verlet algorithm, which can be used as follows during an MD simulation: One starts from some initial values of $\mathbf{r}_\nu(t_0) \equiv \mathbf{r}_\nu(t_1 - \Delta t)$ and $\mathbf{r}_\nu(t_1)$ for all ν, which we denote as the "old" and "current" positions, respectively, of the particles. New positions $\mathbf{r}_\nu(t_2) = \mathbf{r}_\nu(t_1 + \Delta t)$ for all ν are then calculated from Equation 5.130. In the next step, we first make the current positions become "old" and the new ones become "current." The previous "old" positions are not needed anymore and can be discarded (unless one wants to keep a record of them). The force $\mathbf{F}_\nu(t_2)$ can now be calculated for each particle and Equation 5.130 can be used to calculate new coordinates $\mathbf{r}_\nu(t_3) = \mathbf{r}_\nu(t_2 + \Delta t)$ for all ν. This is repeated and $\mathbf{r}_\nu(t_{\alpha+1}) = \mathbf{r}_\nu(t_\alpha + \Delta t)$ for $\alpha > 3$ is calculated in the same manner until α have become sufficiently large and the trajectory between t_0 and $t_M = t_0 + M\Delta t$ has been determined approximately.

By decreasing the time step Δt, the accuracy of the approximation increases quite rapidly since the errors are of the order Δt^4. There are, however, practical limits for the minimal value of Δt that one can use, since overly small values make the time evolution very inefficient – it takes many steps to get anywhere. Furthermore, the evaluation of the forces constitutes in practice a much more time-consuming part of the simulations than the solution of the equations of motion, so in order to minimize the number of force calculations one should not use a step size that is too small. Note that Equation 5.130 is invariant when Δt is changed to $-\Delta t$, so the trajectory is time-reversible, which is one of the criteria for accurate trajectories. One can show that the Verlet algorithm also fulfills important criteria for accurate long-time evolution of the trajectories. These features make it very successful, which may appear surprising considering its simplicity.

We need to set the initial values for the coordinates of the particles to start the simulation. The values of $\mathbf{r}_\nu(t_1)$ can, for example, be set up by placing the N particles at random positions in the system or by placing them on lattice sites of a suitable lattice. For the first option, random placements of particles too close together should be avoided since this gives very high energies and large forces. If the particles have hard cores, one must avoid overlaps of any core regions, which give infinite energy. The option with lattice sites is usually more convenient for dense systems since it may be difficult to avoid configurations with very high energies/overlaps when particles are placed randomly. The initial velocities of the particles can, for example, be selected randomly and then shifted so that the total momentum $\mathbf{p}_{\text{tot}}$ is zero and scaled so that the equipartition of translational energy $\langle U_N^{\text{tr}} \rangle = 3Nk_BT/2$ is fulfilled. The coordinates at $t = t_0$ for each particle can then be set equal to $\mathbf{r}_\nu(t_0) = \mathbf{r}_\nu(t_1) - \mathbf{v}_\nu \Delta t$.

The system is initially far from equilibrium since the starting configuration is unrealistic and the total energy is usually far from the correct average value. The details of the initialization are not so important because the forces that act on the particles at $t = t_1$ bring the kinetic energy away from the initial value anyway. One needs to run the simulation for a substantial period of time in order to reach more realistic configurations. Thereby, one has to make adjustments to the kinetic energies of the particles at several instances, which is most easily done by multiplying the speed of all particles by a suitable factor so that one gets $\langle U_N^{\text{tr}} \rangle = 3Nk_BT/2$. This *equilibration phase* of the simulation is continued at least until the total average energy of the system is virtually constant in time and the average kinetic energy has reached the correct value, $3Nk_BT/2$.

The total energy U should be constant as a consequence of Newton's equations of motion, but it is not exactly constant in the simulation because the trajectory is approximate. Even if the time step Δt is very small, there are rounding errors in the numerical calculations, so it is unavoidable that U eventually drifts off. Therefore, there is a change in temperature as obtained from the average kinetic energy, which may need further corrections. There exist various methods to keep the temperature constant, like various "thermostats," but it would lead too far here to describe these details.[39] By using thermostats one can, in fact, run MD simulations in other ensembles than the microcanonical one.

The total energy is obtained from

$$\langle U \rangle = \left\langle \breve{U}_N^{\text{pot}} \right\rangle + \left\langle \breve{U}_N^{\text{kin}} \right\rangle = \frac{1}{M} \sum_{\alpha=1}^{M} \breve{U}_N^{\text{pot}}(\mathbf{r}_\alpha^N) + \left\langle \sum_\nu \frac{p_\nu^2}{2m_\nu} \right\rangle, \tag{5.131}$$

where the kinetic term is (approximately) equal to $3Nk_BT/2$ at equilibrium. These averages are obtained during the simulation by collecting the value of the instantaneous energy $\breve{U}_N^{\text{pot}}$ and $\breve{U}_N^{\text{kin}}$ at each time step and, finally, calculating the averages at the end of the simulation run. The same is done for averages of other entities of the type in Equation 5.127. When the initialization phase is completed, one discards the averages obtained so far and restarts the collection of data for the averages in order to obtain the final average value of each quantity that one wants. The simulation is run at least until the calculation of these quantities has converged to constant values.

5.13.1.2 Basics of Monte Carlo Simulations

We will primarily describe MC simulations in the *canonical ensemble* where T, V and N are constant. One can start the simulation from an initial configuration of particles like the one described earlier for MD simulations (the velocities are not needed here). New configurations are generated by consecutive random movements of the particles in the system. One thereby makes random trial moves and accepts a move if it satisfies certain conditions to be described. If the move is not accepted, one stays with the old configuration, which is counted as a new one. In each trial move, one can move several particles or one of the particles at a time; the particle may thereby be selected randomly or in numerical sequence.

The probability for the occurrence of an accepted configuration must at equilibrium be proportional to the canonical probability density (Equation 3.9)

$$\mathcal{P}_N^{(N)}(\mathbf{r}^N) = \frac{1}{Z_N} e^{-\beta \breve{U}_N^{\text{pot}}(\mathbf{r}^N)}. \tag{5.132}$$

To illustrate what this means, let us represent each accepted configuration as a black point at $\mathbf{r}_\alpha^N$ in the $3N$-dimensional configuration space and we imagine that we have made an *extremely* long simulation. These black points then form a swarm of points distributed in this space (a "cloud" of points). The swarm is denser where the probability is high than where the probability is low. By calculating averages using Equation 5.127, one thereby samples the different parts of the configuration space in proportion to the density of black points. The important thing here is that the sampling thereby is done more frequently in the more probable regions than the less probable regions; if a region is, for example, three times more probable than another, the former should be sampled three times as much as the latter in

---

[39] For a description of such methods, see, for example, Chapter 6 in Frenkel and Smit, *loc. cit.* in footnote 37.

order to obtain the correct average in Equation 5.127. Thus, what matters is the *ratio* of probability density between different configurations $\mathbf{r}_A^N$ and $\mathbf{r}_B^N$

$$\frac{\mathcal{P}_N^{(N)}(\mathbf{r}_B^N)}{\mathcal{P}_N^{(N)}(\mathbf{r}_A^N)} = \frac{e^{-\beta \breve{U}_N^{\text{pot}}(\mathbf{r}_B^N)}}{e^{-\beta \breve{U}_N^{\text{pot}}(\mathbf{r}_A^N)}} = e^{-\beta[\breve{U}_N^{\text{pot}}(\mathbf{r}_B^N) - \breve{U}_N^{\text{pot}}(\mathbf{r}_A^N)]}. \tag{5.133}$$

(it is this ratio that is three in our example). Thus, we do not need to know the value of Z_N in order to calculate the average. It is sufficient to know the difference in energy for the configurations.

Let us consider a given but arbitrary configuration $\mathbf{r}_A^N$. Say that the probability for an accepted move from $\mathbf{r}_A^N$ to, say, configuration $\mathbf{r}_B^N$ is equal to $\mathfrak{p}^{\text{move}}(\mathbf{r}_A^N \rightarrow \mathbf{r}_B^N)$, which we for simplicity write as $\mathfrak{p}^{\text{move}}(A \rightarrow B)$. Likewise, we will, for example, write $\mathcal{P}_N^{(N)}(\mathbf{r}_A^N)$ as $\mathcal{P}_N^{(N)}(A)$. How can we make sure that $\mathfrak{p}^{\text{move}}(A \rightarrow B)$ for all configurations A and B is in agreement with the canonical distribution $\mathcal{P}_N^{(N)}$? A very important requirement is that if we sum over all accepted moves *to* configuration B coming from all possible configurations A, the former should have a probability $\mathcal{P}_N^{(N)}(B)$. It is necessary that this holds at equilibrium, that is, when each of the configurations A occurs with a probability $\mathcal{P}_N^{(N)}(A)$. This means that we require that

$$\sum_A \mathcal{P}_N^{(N)}(A)\mathfrak{p}^{\text{move}}(A \rightarrow B) = \mathcal{P}_N^{(N)}(B), \tag{5.134}$$

which implies that once the equilibrium distribution $\mathcal{P}_N^{(N)}$ is attained, it is maintained by the accepted moves. One simple manner to make sure that the moves satisfy this condition is to require that an equal number of moves occur on average from A to B as reverse moves from B to A, that is,

$$\mathcal{P}_N^{(N)}(A)\mathfrak{p}^{\text{move}}(A \rightarrow B) = \mathcal{P}_N^{(N)}(B)\mathfrak{p}^{\text{move}}(B \rightarrow A) \tag{5.135}$$

at equilibrium. This condition, which is called the **detailed balance condition**, is stronger than Equation 5.134 – in fact unnecessarily strong – but it is easier to use. Equation 5.135 satisfies Equation 5.134, which one can show by summing the former over all configurations A and obtain

$$\sum_A \mathcal{P}_N^{(N)}(A)\mathfrak{p}^{\text{move}}(A \rightarrow B) = \sum_A \mathcal{P}_N^{(N)}(B)\mathfrak{p}^{\text{move}}(B \rightarrow A) = \mathcal{P}_N^{(N)}(B).$$

The last equality follows from $\sum_A \mathfrak{p}^{\text{move}}(B \rightarrow A) = 1$, which expresses the fact that moves that start from B have some destinations; they either give some other configuration A (for $A \neq B$) or the same configuration (for $A = B$). We can write Equation 5.135 as

$$\mathfrak{p}^{\text{move}}(A \rightarrow B) = \mathfrak{p}^{\text{move}}(B \rightarrow A)\frac{\mathcal{P}_N^{(N)}(B)}{\mathcal{P}_N^{(N)}(A)}$$

$$= \mathfrak{p}^{\text{move}}(B \rightarrow A)e^{-\beta[\breve{U}_N^{\text{pot}}(B) - \breve{U}_N^{\text{pot}}(A)]}, \tag{5.136}$$

where we have inserted Equation 5.133. If $\mathfrak{p}^{move}$ satisfies this condition, the resulting set of configurations is in agreement with the canonical distribution.

As mentioned earlier, one usually first makes a random trial move and then accepts the move if it satisfies the appropriate condition. This means that we have

$$\mathfrak{p}^{move}(A \to B) = \mathfrak{p}^{trial}(A \to B)\mathfrak{p}^{acc}(A \to B), \tag{5.137}$$

where $\mathfrak{p}^{trial}(A \to B)$ is the probability to make the random move from A to B and $\mathfrak{p}^{acc}(A \to B)$ is the probability that this move is accepted, that is, that the move is such that $\mathfrak{p}^{move}(A \to B)$ satisfies Equation 5.136.

Let us first consider the trial moves. In each such move, one can, for instance, select a random particle, say particle ν, and change $\mathbf{r}_\nu$ to $\mathbf{r}_\nu + \Delta\mathbf{r}_\nu$, where $\Delta\mathbf{r}_\nu$ is a random displacement. One possibility is to select $\Delta\mathbf{r}_\nu = (\alpha_x, \alpha_y, \alpha_z)\delta_{max}$, where α_x, α_y and α_z are three different random numbers with uniform distribution between -1 and 1, and where δ_{max} is the maximal absolute change of each coordinate. A trial move is thereby made from, say, configuration A to configuration B with a probability $\mathfrak{p}^{trial}(A \to B)$ that depends on δ_{max}. A move from, say, $\mathbf{r}'_\nu$ to $\mathbf{r}''_\nu$ has equal probability as one from $\mathbf{r}''_\nu$ to $\mathbf{r}'_\nu$ since $\Delta\mathbf{r}_\nu$ is equally likely as $-\Delta\mathbf{r}_\nu$. It therefore follows that if we instead had started at configuration B, the probability to move to A would have been the same, that is, $\mathfrak{p}^{trial}(B \to A) = \mathfrak{p}^{trial}(A \to B)$. Of course, one can design trial moves that are not symmetrical in A and B, but here we will only consider symmetrical ones.

Let us now find an acceptance rule that is so designed that $\mathfrak{p}^{move} = \mathfrak{p}^{trial}\mathfrak{p}^{acc}$ satisfies Equation 5.136. By inserting Equation 5.137 into Equation 5.136, we can write the latter as

$$\mathfrak{p}^{acc}(A \to B) = \mathfrak{p}^{acc}(B \to A)e^{-\beta[\breve{U}_N^{pot}(B) - \breve{U}_N^{pot}(A)]} \tag{5.138}$$

since $\mathfrak{p}^{trial}(B \to A) = \mathfrak{p}^{trial}(A \to B)$. There exist many different manners to design acceptance rules that satisfy this condition. Here we will limit ourselves to the **Metropolis scheme**, which is very common in applications and gives an efficient sampling of configuration space.[40] In this scheme, we make the following choice: If the energy decreases or stays constant in the trial move from A to B, we *always accept the move*, that is,

$$\mathfrak{p}^{acc}(A \to B) = 1 \quad \text{when } \breve{U}_N^{pot}(A) \geq \breve{U}_N^{pot}(B).$$

What should we do if the energy instead goes up in the trial move from A to B? This is, in fact, dictated by the condition (5.138) and the choice we just have made. To find out what to do if the energy increases, we first make the observation that for the *reverse* move from B to A the energy must decrease or stay constant and such a move should therefore be accepted, $\mathfrak{p}^{acc}(B \to A) = 1$. Equation 5.138 therefore implies that

$$\mathfrak{p}^{acc}(A \to B) = 1 \cdot e^{-\beta[\breve{U}_N^{pot}(B) - \breve{U}_N^{pot}(A)]} \quad \text{when } \breve{U}_N^{pot}(A) < \breve{U}_N^{pot}(B),$$

---

[40] Other schemes can be found in the books cited in footnote 37.

Hence, the trial move from A to B must in this case be *accepted with a probability equal to* $\exp(-\beta[\check{U}_N^{\text{pot}}(\text{B}) - \check{U}_N^{\text{pot}}(\text{A})])$, which is a number between 0 and 1 since $\check{U}_N^{\text{pot}}(\text{B}) - \check{U}_N^{\text{pot}}(\text{A}) > 0$. For any trial move that thereby is not accepted, the system stays in the same configuration A, which then is counted as the new configuration.

An easy manner to decide whether a move should be accepted or not is to generate a random number α with uniform distribution between 0 and 1 and accept the move if α is smaller than $\exp(-\beta[\check{U}_N^{\text{pot}}(\text{B}) - \check{U}_N^{\text{pot}}(\text{A})])$. This gives the correct probability since if the exponential function is, say, 0.75, the probability that $\alpha < 0.75$ is precisely 0.75. We can write the acceptance rule as a single expression valid for both decreasing and increasing energies, namely

$$\mathfrak{p}^{\text{acc}}(\text{A} \to \text{B}) = \min\left[1,\, e^{-\beta[\check{U}_N^{\text{pot}}(\mathbf{r}_{\text{B}}^N) - \check{U}_N^{\text{pot}}(\mathbf{r}_{\text{A}}^N)]}\right],$$

where $\min[x, y]$ is the smallest of x and y.

To summarize, the Metropolis scheme can be expressed as

$$\mathfrak{p}^{\text{move}}(\mathbf{r}_{\text{A}}^N \to \mathbf{r}_{\text{B}}^N) = \mathfrak{p}^{\text{trial}}(\mathbf{r}_{\text{A}}^N \to \mathbf{r}_{\text{B}}^N) \min\left[1,\, e^{-\beta[\check{U}_N^{\text{pot}}(\mathbf{r}_{\text{B}}^N) - \check{U}_N^{\text{pot}}(\mathbf{r}_{\text{A}}^N)]}\right]$$

$$\mathfrak{p}^{\text{move}}(\mathbf{r}_{\text{A}}^N \to \mathbf{r}_{\text{A}}^N) = 1 - \sum_{\text{B} \neq \text{A}} \mathfrak{p}^{\text{move}}(\mathbf{r}_{\text{A}}^N \to \mathbf{r}_{\text{B}}^N), \tag{5.139}$$

where the last line deals with the cases where the configuration is kept. Note that in the energy difference $\check{U}_N^{\text{pot}}(\mathbf{r}_{\text{B}}^N) - \check{U}_N^{\text{pot}}(\mathbf{r}_{\text{A}}^N)$, the only interactions that contribute are those that involve the moved particle(s) for the trial move from A to B. All the other interactions are unchanged and cancel in this difference.

Like in MD simulations, one needs to carry through an *equilibration phase* for the simulation. The system is initially far from equilibrium since the starting configuration is unrealistic and one therefore runs the simulation for a large number of moves before starting to collect the data one wants to calculate. Once equilibrium is reached, averages calculated from Equation 5.127 give the appropriate values for the conditions of constant T, V and N since the configurations are distributed according to the canonical ensemble. The simulation is run at least until the averages have converged for the quantities one is interested in. It is thereby important that the sampling of configuration space is representative of the system. It is, for example, crucial that the sampling is not stuck in a part of the configuration space due to, for example, high energetic barriers that prevent parts of the space to be sampled during a reasonable number of moves. There exist methods to overcome such problems, but it will lead too far here to describe this.[41]

The rate of convergence and the efficiency of the simulation depend on the parameter δ_{max}. If δ_{max} is too large, the number of accepted moves can be very few in a dense fluid and one needs many trial moves to get anywhere, and if δ_{max} is very small, many moves are accepted but one needs a huge number of moves in order to sample the configuration space to a sufficient degree. It is sometimes claimed that a reasonable value of δ_{max} yields

---

[41] Some methods can be found in the books cited in footnote 37.

an acceptance rate of about 50% of the trial moves, but there are, in fact, no simple, general relationships between the acceptance rate and the simulation efficiency.

Simulations can also be done, for example, in the *isobaric-isothermal* and the *grand canonical* ensembles. The particle moves that sample the configuration space are thereby complemented by volume changes and variations in the number of particles in the system, respectively. We will not go into any details here[42] and limit ourselves to the following brief remarks. During MC simulations in the isobaric-isothermal ensemble, one makes random trial changes in volume at more or less regular intervals among the particle moves. The volume changes are accepted or rejected in such a manner that one obtains the appropriate distribution $\mathscr{P}(V)$ of volume values for the system at constant T, P and N. Likewise, for MC simulations in the grand canonical ensemble, one now and then makes trial insertions/removals of particles in the system. They are accepted or rejected according to rules that make the distribution $\mathscr{P}(N)$ of the number of particles appropriate for the system at constant T, V and μ. Between the volume changes or particle insertions/removals, respectively, the particles are usually moved around like in the canonical ensemble.

5.13.2 Bulk Fluids

5.13.2.1 Boundary Conditions
In the description of MD and MC simulations in the previous sections, we assumed that the system was enclosed in a container. The presence of container walls makes the fluid inhomogeneous at least in a region close to the wall surfaces. This is the case even if the walls contain the fluid by simply interacting like a hard wall, that is, the external potential is $v(\mathbf{r}) = \infty$ inside the walls and $v(\mathbf{r}) = 0$ in the interior of the container. A particle does not interact with a hard wall unless it is touching its surface, but it *does* experience a force which originates from the interactions with the other particles in the fluid. When a particle is near a wall surface, it has many particles on the side away from the surface and few, if any, particles on the other side that faces the surface because no particles can enter into the wall. The sum of the interparticle forces from the former particles is not the same as the sum of the forces from the latter ones, which makes the net average force on the particle nonzero. As we have seen, this makes the density distribution vary with the distance from the surface, at least near the wall.

If the system is sufficiently large, the fluid is homogeneous away from the walls,[43] so in the interior of the system we have bulk conditions to a very good approximation. If we are interested in the bulk properties of the fluid, we can in many cases calculate averages of various quantities for the particles in this interior region and thereby obtain the bulk values. However, unless we have a very large number of particles in the system, a substantial fraction of the particles are located in the region near the walls where the density varies. This means that we are spending a large part of the computer power to simulate both the homogeneous

---

[42] For details, see the books cited in footnote 37.
[43] Provided, of course, that the bulk fluid is in a single-phase state in the phase diagram.

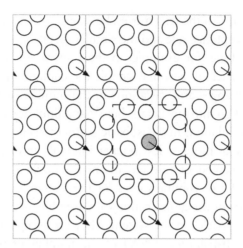

FIGURE 5.15 Illustration of periodic boundary conditions in two dimensions, see text. One of the particles is shown in gray solely for the purpose of illustrating some aspects of the figure.

and inhomogeneous parts of the fluid. This is a waste of effort unless we are also interested in the inhomogeneous region of the fluid.

In many cases, an efficient manner to calculate bulk properties is to use so-called **periodic boundary conditions,** which is a simple trick to mimic bulk conditions in simulations with a limited number of particles. One thereby replaces the walls by replicas of the same fluid system in the manner illustrated in Figure 5.15, which shows a two-dimensional example. The part shown in the central square constitutes the system (the simulation box) and the six surrounding squares in the figure are replicas of the central square. All particles (circles) in each replica have exactly the same positions as in the central part. This pattern is repeated throughout space so each square is surrounded by eight squares with equal content. When a particle leaves the simulation box, for instance the gray one, a particle from a surrounding replica enters from the opposite side as illustrated by the arrows in the figure. The same occurs in all other boxes. By surrounding the simulation box by replicas in this manner, one avoids the boundary walls and each particle is surrounded by other particles irrespective of where it is located in the box. In three dimensions, a cubic simulation box is repeated periodically in the corresponding manner in all directions throughout the entire space.

When the interaction energy is calculated, one can include the interactions of each particle with *all* other particles – both in the central box and in all replicas in the whole space. If the interactions between the particles decay sufficiently fast with distance, the contributions from distant particles can be neglected so one does not need to include all of them (we here assume that the central simulation box contains a large number of particles – much larger than in the figure). A possibility when one calculates pair interactions for a certain particle, for instance the gray one in the figure, is that one only includes contributions from particles with centers inside the dashed box. It is centered at the particle and has equal dimensions as the central box. Thereby, it contains one "representative" of each particle in the simulation box (the representative is either a particle in the central simulation box or in one of the replicas). This is called the *minimal image convention.* Provided that the interactions have a sufficiently short range, one can further restrict the calculation to only include particles with centers inside a cutoff distance $\mathcal{R}_{cut}$, which is less than half the length of a side of the simulation box. This limit of $\mathcal{R}_{cut}$ is used in order to avoid including more than one representative of a particle.

When the particles have long-range interactions, like Coulomb interactions, a simple cutoff is not a viable option. For an ionic system, the total charge of the ions inside a cutoff distance $\mathcal{R}_{cut}$ is in general not zero, which causes a large error. Sometimes, it is sufficient to include interactions for an ion with the other anions and cations according to the minimal image convention to achieve sufficient accuracy. The ions inside a minimal image form an electroneutral entity – exactly like the central box. However, since the $1/r$ decay of the potential is very slow, the sum of the contributions from distant anions and cations is often substantial and cannot be neglected. One therefore commonly includes the interactions with all particles in all replicas of the simulation box. It is thereby not necessary to do sums of pair interactions between each particle and the infinite number of particles in all replicas; there exist mathematical techniques to calculate the entire interaction energy to a sufficient accuracy by using sums with a finite number of terms. In the *Ewald summation technique*, one uses the periodicity of the simulated system (the central box and the replicas) to write the interaction energy into two parts: one from a periodic charge distribution in the entire space and one from a charge distribution in the central simulation box. The periodic charge distribution can be written as a Fourier series. We will not go into any details here;[44] suffices it to say that by a suitable choice of these charge distributions, one can calculate the energy to a very good approximation in terms of finite sums, one sum over the Fourier components of the periodic distribution and the rest as sums over the particle positions in the central box.

In the general case, a drawback of periodic boundary conditions is that they impose an unnatural periodicity in the system. To decrease the effect of this, one can, as a reasonable approximation in the calculation of the interaction energy between a particle and the surrounding particles, replace the particles outside a cutoff radius $\mathcal{R}_{cut}$ by a continuous medium. One thereby uses a suitable model of this medium such that it responds to the interactions with the particles and gives a feedback to them, that is, the medium mimics the response of a fluid outside the cutoff. This is the basis of the so-called *reaction field methods* to calculate the interaction energies. Again, we will not consider any details.

5.13.2.2 Distribution Functions

Having described the most common manners to do simulations for bulk systems, let us now turn to the relationship between simulations and various distribution functions. As we have seen, a simulation for N particles in the canonical ensemble gives a distribution of particle configurations according to the canonical N-particle probability density $\mathcal{P}_N^{(N)}(\mathbf{r}^N)$ in Equation 5.132.[45] Therefore, it gives configurations described by the N-particle density distribution function $n_N^{(N)}(\mathbf{r}^N) = N!\mathcal{P}_N^{(N)}(\mathbf{r}^N)$ of Equation 5.39.

---

[44] For details, see, for example, the literature cited in footnote 37.

[45] The simulation does not determine $\mathcal{P}_N^{(N)}(\mathbf{r}^N)$ itself because the factor $1/Z_N$ is not calculated, but the probabilities of the configurations are in accordance with $\mathcal{P}_N^{(N)}$. One can say that they describe $\mathcal{P}_N^{(N)}$ up to a multiplicative factor.

In Section 5.5.3, we have seen how one can extract the lower order distribution functions like $n_N^{(2)}(\mathbf{r}_1, \mathbf{r}_2)$ and $n_N^{(3)}(\mathbf{r}_1, \mathbf{r}_2, \mathbf{r}_3)$ once the function $\mathcal{P}_N^{(N)}(\mathbf{r}^N)$ is given. For example, from Equations 5.44 and 5.45, we have

$$n_N^{(2)}(r_{12}) = n_N^{(2)}(\mathbf{r}_1, \mathbf{r}_2) = N(N-1) \int d\mathbf{r}_3 \ldots d\mathbf{r}_N \mathcal{P}_N^{(N)}(\mathbf{r}^N).$$

In $\mathcal{P}_N^{(N)}$, it is particle 1 that is located at $\mathbf{r}_1$ and particle 2 at $\mathbf{r}_2$, while in $n_N^{(2)}$ it can be any two particles that are located at these coordinates; the particles are indistinguishable. The prefactor $N(N-1)$ makes the necessary "correction" to insure the latter. The integration means that we are summing over all possibilities, that is, irrespective of the positions of all particles, apart from the ones at $\mathbf{r}_1$ and $\mathbf{r}_2$.

In a simulation, one should consider particle configurations without caring about which particle is located at what position in order to treat the particles as being indistinguishable. Thus, when one has a large number of configurations that are distributed according to the canonical ensemble, one can extract $n_N^{(2)}(\mathbf{r}_1, \mathbf{r}_2)$ by recording the locations of any pairs of particles, irrespective of where all other particles are (we will describe later in what manner this can be done). Likewise, we can extract $n_N^{(3)}(\mathbf{r}_1, \mathbf{r}_2, \mathbf{r}_3)$ by recording the locations of any three particles irrespective of the positions of the others. The N-particle density distribution function $n_N^{(N)}(\mathbf{r}^N)$ is obtained by recording the positions of all particles.

In a simulation of N particles, the set of configurations obtained gives a very lousy description of $n^{(N)}(\mathbf{r}^N)$ for a bulk fluid since the latter should describe the probability of distributions of N particles in the bulk phase, that is, in the presence of *many* more particles than N. In this situation, is there any hope that, for example, the extracted $n_N^{(2)}(r_{12})$ from a simulation of N particles is a fair representation of the pair function for a bulk fluid? An answer can be found in the discussion about distribution function hierarchies in Section 5.9, where we discussed how $n_N^{(2)}$ depends on $n_N^{(3)}$, which in turn depends on $n_N^{(4)}$ etc. all the way up to $n_N^{(N)}$. By making a closure approximation for a higher order distribution function, one can in most cases obtain the lower order functions to a fair or good approximation (as mentioned earlier, this is the technique that is used to set up most approximate integral equation theories). The quality depends on how clever the approximation is that one uses.

When one extracts the pair distribution function or any other *low-order* distribution function from a simulation with N particles, one may say that one has made a closure approximation: one entirely neglects all higher order distribution functions $n^{(N')}$ with $N' > N$ for the bulk fluid and uses the result to calculate approximate lower order functions. Thus, *a simulation for a bulk fluid can be regarded as a kind of closure approximation*. The approximation one thereby performs is very accurate for the low-order distributions in most cases, but not always, and the same is true for any entity that depends only on distributions of low orders. Exceptions occur, for instance, when there are long-range correlations in the system like near critical points or other collective phenomena that involve coupled correlations between a huge number of particles. One can usually check whether or not the number of particles N is sufficient by making simulations for larger systems with the same n^b, that is, for larger N and $V = N/n^b$. If each quantity one is interested in does not change significantly, one can usually trust the result. Otherwise, one should increase N further. Sometimes it may, however, be difficult to judge whether one would obtain the same result if N is made even larger. A quantity may have a slow N dependence that persists till large system sizes so its value does not change significantly until N is *very* large.

The most straightforward manner to determine $n_N^{(2)}(r) = (n^b)^2 g_N^{(2)}(r)$ is the following, which focuses on $n^b g_N^{(2)}(r)$, that is, the number density at distance r from the center of a particle as depicted in Figure 5.1. For each particle configuration, one counts how many other particles there are at various distances from the particle center. Having divided the distance r into a series of small intervals $r^{(\alpha)} \leq r < r^{(\alpha+1)}$ for $\alpha = 1, 2, 3, \ldots$, one thereby counts how many particle centers are located in each spherical shell between $r^{(\alpha)}$ and $r^{(\alpha+1)}$ for each α (see Figure 5.16a). One has a counter for each shell that increases by one whenever a particle is located there. This is done for all N particles of the same species taking the role of the central particle, whereby each particle acts as a representative for all of them (the same set of counters is used throughout). One repeats this for all configurations. The numbers stored in different counters can be represented by a histogram that shows the occurrence of particle centers in the different shells. In the end, one takes the resulting histogram and divides each entry by N to get the average per particle. Then one divides by the number of configurations to get the average over the simulation. The resulting numbers are the average number of particles inside each spherical shell between $r^{(\alpha)}$ and $r^{(\alpha+1)}$ around a particle. By dividing these numbers by the volume of the respective shell, one obtains the average density in each shell, that is, the average of $n^b g_N^{(2)}(r)$. An example of such a histogram is shown in Figure 5.16b. The density values of this histogram are used as an approximation for $n^b g_N^{(2)}(r)$ evaluated at the midpoint of each shell, that is, at $r = [r^{(\alpha)} + r^{(\alpha+1)}]/2$ for $\alpha = 1, 2, 3, \ldots$. Thereby, we have determined $n^b g_N^{(2)}(r)$ and hence $g_N^{(2)}(r)$ for this grid of points on the r axis.

For particles with a hard core, for instance a hard sphere fluid, it is important to determine the contact value $g_N^{(2)}(d^h)$, which appears in the expression (5.75) for the pressure. In the histogram method just outlined to determine $g_N^{(2)}(r)$, the first shell outside the core has a

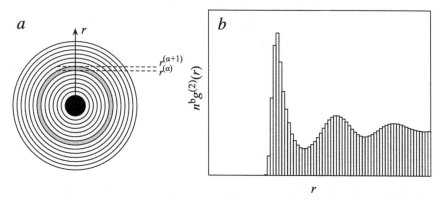

FIGURE 5.16 (a) The space around a particle (black sphere) is divided into spherical shells, each of them being located between radii $r^{(\alpha)}$ and $r^{(\alpha+1)}$ for some value of the integer α. One of the shells is shown in gray in the figure. (b) Example of a histogram for the pair distribution density $n^b g^{(2)}(r)$. The histogram values equal the average number of particles in each shell divided by the shell volume.

finite width (like all other shells). Therefore, one does not obtain the value at $r = d^h$, but instead at the midpoint of the shell which lies bit further out. Since the density often varies very rapidly with r near the core, the value obtained at this midpoint can differ a lot from $n^b g_N^{(2)}(d^h)$. If one uses a thinner first shell outside the core, the midpoint lies closer to $r = d^h$, but then the statistics deteriorate since there are on average fewer particle centers in a thinner shell.

One way to remedy this problem is to use two counters per shell: one that records any particle center that lies inside the shell in each configuration (like before) and one that *only* records the center for the particle that lies *closest* to the central particle.[46] At the end of the simulation, the latter counter will store a number that is smaller than the other counter for the same shell. Since a particle in contact with the central particle must be the closest one, the two types of counters approach the same value when r decreases. In this manner, one obtains two functions of r that tend to the same value when $r \to d^h$. By extrapolating the curves slightly from the midpoint of the first shell to $r = d^h$, we can obtain a good estimate for $n^b g_N^{(2)}(d^h)$ from the intersection of the two curves.

Another way to determine the contact value is to use the relationship (Equation 5.31)

$$n(\mathbf{r}_1|\mathbf{r}_2) = n^b e^{-\beta[\mu^{ex}(\mathbf{r}_1|\mathbf{r}_2) - \mu^{ex,b}]}, \tag{5.140}$$

which says that the density at $\mathbf{r}_1$, that is $n^b g_N^{(2)}(r_{12})$, can be calculated if we know the excess chemical potential at the same place and in the bulk. We therefore need to find out how to calculate these chemical potentials, in particular at $r_{12} = d^h$, to find the contact density $n^b g_N^{(2)}(d^h)$. The evaluation of chemical potentials from simulations is an important topic in itself, so we will investigate this in some detail.

First, we will consider $\mu^{ex,b}$ for the bulk fluid modeled as an N particle system. As we have seen earlier, the excess chemical potential is the change in excess free energy when we add a particle to the system at some point, say at $\mathbf{r}_0$, and keep it fixed there. For a bulk fluid, it does not matter where $\mathbf{r}_0$ is located since the system is homogeneous. We start with an N particle system and end up with an $N + 1$ particle system, where the extra particle is placed at $\mathbf{r}_0$. The excess free energy of the former system is equal to $A_N^{ex} = k_B T \ln(V^N/Z_N)$ (Equation 5.79) and the excess free energy of the latter system is

$$A_{N+1}^{ex}(\mathbf{r}_0) = k_B T \ln \frac{V^N}{Z_{N+1}(\mathbf{r}_0)},$$

where

$$Z_{N+1}(\mathbf{r}_0) = \int d\mathbf{r}_1 \dots d\mathbf{r}_N e^{-\beta \check{U}_{N+1}^{pot}(\mathbf{r}_0, \mathbf{r}^N)} \tag{5.141}$$

is the configurational partition function for the $N + 1$ particle system with one particle kept fixed at $\mathbf{r}_0$ and the other ones being mobile. $\check{U}_{N+1}^{pot}(\mathbf{r}_0, \mathbf{r}^N)$ is the potential energy for

---

[46] If two particles qualify as the closest one since they have exactly the same distance to the central particle, both can be recorded as being closest.

$N + 1$ particles located at $(\mathbf{r}_0, \mathbf{r}_1, \ldots \mathbf{r}_N)$. Note that we have V^N instead of V^{N+1} since only N particles are mobile. We assume that the particles are interacting with a pairwise additive potential, so we have

$$\check{U}_{N+1}^{\text{pot}}(\mathbf{r}_0, \mathbf{r}^N) = \check{U}_N^{\text{pot}}(\mathbf{r}^N) + \sum_{\nu=1}^{N} u(r_{0\nu}), \tag{5.142}$$

where the sum is the total interaction energy that involves the added particle located at $\mathbf{r}_0$.

The excess chemical potential at $\mathbf{r}_0$ is given by[47]

$$\mu^{\text{ex}}(\mathbf{r}_0) = A_{N+1}^{\text{ex}}(\mathbf{r}_0) - A_N^{\text{ex}} = -k_B T \ln \frac{Z_{N+1}(\mathbf{r}_0)}{Z_N}. \tag{5.143}$$

For a bulk fluid, $\mu^{\text{ex}}(\mathbf{r}_0)$ has the same value everywhere and equals $\mu^{\text{ex,b}}$. Using the definition (5.141) of $Z_{N+1}(\mathbf{r}_0)$ with Equation 5.142 inserted, we obtain

$$\mu^{\text{ex}}(\mathbf{r}_0) = -k_B T \ln \frac{\int d\mathbf{r}_1 \ldots d\mathbf{r}_N e^{-\beta \check{U}_N^{\text{pot}}(\mathbf{r}^N)} e^{-\beta \sum_{\nu=1}^{N} u(r_{0\nu})}}{Z_N}$$

$$= -k_B T \ln \int d\mathbf{r}_1 \ldots d\mathbf{r}_N \mathcal{P}_N^{(N)}(\mathbf{r}^N) e^{-\beta \sum_{\nu=1}^{N} u(r_{0\nu})},$$

where $\mathcal{P}_N^{(N)}(\mathbf{r}^N)$ is the probability density for the N mobile particles *in the absence* of the particle at $\mathbf{r}_0$. The integral in the rhs is equal to the canonical average of the exponential function, so we can write

$$\mu^{\text{ex}}(\mathbf{r}_0) = -k_B T \ln \left\langle e^{-\beta \sum_{\nu=1}^{N} u(r_{0\nu})} \right\rangle^{\circ} \bigg|_{\mathbf{r}_0 \text{ fix}} \tag{5.144}$$

where superscript "∘" on the average indicates that the probability density for the N mobile particles is not affected by the particle at $\mathbf{r}_0$. This means that the average is taken for a situation where the fixed particle is treated as a "ghost" (often called a "test particle") that the mobile particles do not interact with. This result, which we can write as

$$e^{-\beta \mu^{\text{ex}}(\mathbf{r}_0)} = \left\langle e^{-\beta \sum_{\nu=1}^{N} u(r_{0\nu})} \right\rangle^{\circ} \bigg|_{\mathbf{r}_0 \text{ fix}},$$

is known as **Widom's formula** for the excess chemical potential.[48]

The average in Widom's formula can be evaluated from a simulation in the manner shown in Equation 5.127 with $X = e^{-\beta \sum_{\nu=1}^{N} u(r_{0\nu})}$ and to obtain $\mu^{\text{ex,b}}$ it does not matter where $\mathbf{r}_0$ is located. In practice, it is not optimal to have a fixed $\mathbf{r}_0$ and a much better option is to take

---

[47] In the notation of Chapter 10 (in the 2nd volume), this quantity is written as μ_{N+1}^{ex}.

[48] B. Widom, *J. Chem. Phys.*, **39** (1963) 2808.

an average over many different random positions $\mathbf{r}_0$ that are uniformly distributed over the simulation cell. This is done at regular intervals during the simulation and we have

$$e^{-\beta\mu^{\mathrm{ex},b}} = \left\langle \left\langle e^{-\beta\sum_{\nu=1}^{N} u(r_{0\nu})} \right\rangle_{\{\mathbf{r}_0\}} \right\rangle^{\circ}, \tag{5.145}$$

where $\langle\cdot\rangle_{\{\mathbf{r}_0\}}$ denotes the average over a set of $\mathbf{r}_0$ values for each configuration involved.

Widom's formula works well in practice provided the density is not too high. To determine the excess chemical potential in this manner is called the **Widom insertion method**. Note, however, that the test particle is never inserted into the system on an equal footing to the particles already present. It is only used as a means to measure μ_N^{ex} and is never a part of the system. The derivation of Widom's formula was done in the canonical ensemble, but an analogous formula is valid in other ensembles as will be shown in Chapter 10 (in the 2nd volume).

In order to obtain $\mu^{\mathrm{ex}}(\mathbf{r}_1|\mathbf{r}_2)$ one can use a similar procedure, but instead of letting $\mathbf{r}_0$ to be randomly distributed over the simulation cell, one places it at positions located at distance r_{12} from centers of randomly selected mobile particles. Thereby, the sampled values of $\exp\left(-\beta\sum_{\nu=1}^{N} u(r_{0\nu})\right)$ are appropriate for the conditions near a particle. By selecting $r_{12} = d^{\mathrm{h}}$, one thereby finds the excess chemical potential at contact, which gives the contact density via Equation 5.140.

Note that one can also use this method to determine $n^{\mathrm{b}}g^{(2)}(r_{12})$ for all r_{12} as an alternative to the histogram method based on spherical shells depicted in Figure 5.16. Thereby, one can determine the density at each precise distance r_{12} rather than as an average over the width of a shell. This can be useful when the density varies rapidly with r. For dense fluids, where Widom's formula is inaccurate in practice, the chemical potential can be determined with other methods presented in the next section.

5.13.2.3 Thermodynamical Quantities

To calculate thermodynamical quantities like the energy and pressure of the fluid, one can in principle use equations that express these quantities in terms of the pair distribution function, Equations 5.70 and 5.73. In practice it is, however, preferable in simulations to calculate the energy directly from the particle configurations $\{\mathbf{r}_\alpha^N\}_{\alpha=1}^{M}$ via

$$\langle U \rangle = \frac{3}{2}Nk_BT + \frac{1}{M}\sum_{\alpha=1}^{M}\breve{U}_N^{\mathrm{pot}}(\mathbf{r}_\alpha^N) = \frac{3}{2}Nk_BT + \left\langle \sum_{\nu,\nu'>\nu} u(r_{\nu\nu'}) \right\rangle \tag{5.146}$$

rather than to first calculate $g_N^{(2)}(r)$ and then use this function to obtain $\langle U \rangle$ via Equation 5.70. Likewise, one can calculate the pressure from

$$P = n^{\mathrm{b}}k_BT - \frac{1}{3V}\left\langle \sum_{\nu,\nu'>\nu} r_{\nu\nu'}u'(r_{\nu\nu'}) \right\rangle, \tag{5.147}$$

where, as before, $u'(r) = du(r)/dr$ and where the last term gives the excess pressure. Equation 5.147 gives the same result as Equation 5.73 since the average of any function $f(r)$ summed over the distances $r_{\nu\nu'}$ between all pairs of particles, $\sum_{\nu,\nu'>\nu} f(r_{\nu\nu'})$, can be written as

$$\left\langle \sum_{\nu,\nu'>\nu} f(r_{\nu\nu'}) \right\rangle = \frac{Nn^b}{2} \int_0^\infty f(r)g^{(2)}(r)4\pi r^2 dr. \tag{5.148}$$

This can be proven in a similar manner as the derivation of the second term in Equation 5.71, which can be written for a bulk phase as in Equation 5.70. Both Equations 5.73 and 5.147 are often called the **virial equation**.

{**EXERCISE 5.6**

Prove Equation 5.148 in the canonical ensemble.}

We have already seen from Equation 5.145 how the excess chemical potential for a bulk fluid can be calculated with the Widom insertion method. The total chemical potential is

$$\mu = \mu^{\text{ideal,b}} + \mu^{\text{ex,b}}$$
$$= k_B T \ln\left(\Lambda^3 n^b\right) - k_B T \ln\left\langle\left\langle e^{-\beta \sum_{\nu=1}^N u(r_{0\nu})}\right\rangle_{\{\mathbf{r}_0\}}\right\rangle^\circ, \tag{5.149}$$

which can be used in practice provided the density is not too high.

The average in Equation 5.149 has a particularly interesting interpretation for a hard sphere fluid. In this case, $u(r_{0\nu}) = \infty$ when particle ν (hard sphere ν) overlaps with the test particle at $\mathbf{r}_0$ (a hard sphere), and $u(r_{0\nu}) = 0$ otherwise. For a given configuration, this means that $\sum_{\nu=1}^N u(r_{0\nu}) = \infty$ if *any* of the N mobile spheres overlaps with the test particle at $\mathbf{r}_0$ and zero if *none* of the spheres do this, so $\exp[-\beta \sum_{\nu=1}^N u(r_{0\nu})]$ is zero when any overlap occurs and one otherwise. Notice that when no overlap occurs for a certain configuration of the N spheres, the point $\mathbf{r}_0$ must be surrounded by a vacant space that is large enough to fit an additional sphere at $\mathbf{r}_0$ – otherwise, some overlap(s) would occur with the latter. For each configuration, we now let $\mathbf{r}_0$ be anywhere in space and take the average over all positions $\mathbf{r}_0$

$$\left\langle e^{-\beta \sum_{\nu=1}^N u(r_{0\nu})}\right\rangle_{\{\mathbf{r}_0\}} = \frac{\int d\mathbf{r}_0\, e^{-\beta \sum_{\nu=1}^N u(r_{0\nu})}}{\int d\mathbf{r}_0} = \frac{V_{\text{free}}}{V} \quad \text{(hard sphere fluid),}$$

where V_{free} is the volume of the part of space where the exponential function is equal to one.[49] Anywhere in this part of space one can insert a sphere without overlaps – or, more precisely, one can have the center of an additional sphere anywhere in this part of space. V_{free} is called the **free volume** for the configuration and it is accordingly the total volume available for the center of an additional sphere that does not overlap with any of the N spheres. The

---

[49] We here disregard the boundary effects due to the container that holds the fluid. This can be done when the volume V is sufficiently large.

average of V_{free} over all configurations is the ensemble average $\langle V_{\text{free}}\rangle^\circ = \bar{V}_{\text{free}}$, and the last term in Equation 5.149 is therefore equal to

$$\mu^{\text{ex,b}} = -k_B T \ln \frac{\bar{V}_{\text{free}}}{V} = -k_B T \ln \mathfrak{p}^{\text{hs}} \quad \text{(hard sphere fluid)}, \qquad (5.150)$$

where $\mathfrak{p}^{\text{hs}}$ is the average probability to succeed with an insertion of an additional hard sphere into the bulk fluid. The last equality follows from the fact that the probability is one inside the free volume and zero otherwise. Since the activity coefficient γ defined in Equation 3.30 is given by $\mu^{\text{ex,b}} = k_B T \ln \gamma^{\text{b}}$ for a bulk fluid, we see that for a hard sphere fluid $\gamma^{\text{b}} = 1/\mathfrak{p}^{\text{hs}}$. Since $\mathfrak{p}^{\text{hs}} \leq 1$, we see that $\gamma^{\text{b}} \geq 1$ for hard spheres. When the density $n^{\text{b}} \to 0$, we have $\mathfrak{p}^{\text{hs}} \to 1$ and hence $\gamma^{\text{b}} \to 1$ as expected.

From Equation 5.150, it is easy to understand why the Widom formula does not provide a suitable way to calculate the chemical potential of dense fluids. The probability is then low to find some vacant space in the fluid where a particle can fit, so it requires a large number of random trials at various places to find such a space. Hence, Widom's formula provides an inefficient manner to obtain the chemical potential for a dense fluid. This is true in general, not only for hard spheres.

For dense fluids, one therefore needs to calculate $\mu^{\text{ex,b}}$ in some other manner. One possibility is to perform a *coupling parameter integration* as in Equation 5.84.[50] Let us recall the use of the coupling parameter. One starts with a particle located inside the fluid at some fixed position, say $\mathbf{r}_0$. It initially does not interact with the mobile particles of the fluid and then one gradually increases its interaction from zero to the full interaction potential $u(r)$. The pair potential between this particle and the other ones is thereby $u(r;\xi)$, where ξ is the coupling parameter that varies between 0 and 1. This potential must satisfy Equation 5.83. The other particles are fully interacting with each other at all times and the average distribution of particles varies as depicted in Figure 5.12.

When the coupling parameter method is used in simulations, one needs to do a full simulation for each value of ξ in order to obtain the appropriate particle distribution around the partially coupled particle. This is in contrast to the Widom insertion method, where one only does a single simulation and does not actually insert the test particle. In the present case, one really adds a new particle to the system and it thereby interacts with the other particles when $\xi > 0$.

To calculate the chemical potential from the series of simulations, one does not have to explicitly calculate the pair distribution function $g(r;\xi)$ in Equation 5.84. Instead, one calculates

$$\mu^{\text{ex}}(\mathbf{r}_0) = \int_0^1 d\xi \left\langle \sum_{\nu=1}^N \frac{\partial u(r_{0\nu};\xi)}{\partial \xi} \right\rangle_{(\xi)}\Bigg|_{\mathbf{r}_0 \text{ fix}}, \qquad (5.151)$$

---

[50] This is often called *thermodynamic integration*, which is inappropriate. As mentioned in Section 5.7, in a thermodynamic integration, one integrates one thermodynamical entity as a function of another thermodynamical entity. A coupling parameter is not such an entity.

where subscript (ξ) on the average means that the particle at $\mathbf{r}_0$ is coupled to the extent ξ. This gives the same result as the last term in Equation 5.84 since, similar to Equation 5.148, we have

$$\left\langle \sum_{\nu=1}^{N} f(r_{0\nu}) \right\rangle_{(\xi)}\Bigg|_{\mathbf{r}_0\text{ fix}} = n^{\mathrm{b}} \int_0^{\infty} f(r) g^{(2)}(r;\xi) 4\pi r^2 dr, \tag{5.152}$$

which can be used for each ξ value by setting $f(r_{0\nu}) = \partial u(r_{0\nu};\xi)/\partial\xi$.

{EXERCISE 5.7

Prove Equation 5.152 in the canonical ensemble with a particle coupled to the extent ξ placed at $\mathbf{r}_0$. This can be done by setting the external potential for a system with N mobile particles equal to $v(\mathbf{r}_\nu) = u(|\mathbf{r}_\nu - \mathbf{r}_0|;\xi) = u(r_{0\nu};\xi)$ when evaluating the average in the lhs of Equation 5.152. Thereby the average density distribution of the system is $n(\mathbf{r}) = n^{\mathrm{b}} g^{(2)}(|\mathbf{r} - \mathbf{r}_0|;\xi)$ for each ξ value.}

In practice, one does simulations for a finite number of ξ values. Thereby, one can use the final configuration for one ξ value as the starting configuration for the next ξ value, and after a relatively short equilibration (if needed) one can start collecting data for the average in Equation 5.151 for the latter ξ.

The rhs of Equation 5.151 can, in fact, be used for an inhomogeneous fluid to calculate the contribution to $\mu^{\mathrm{ex}}(\mathbf{r}_0)$ from pair interactions at any point $\mathbf{r}_0$. This will be utilized later. For a bulk fluid, one does not have to keep $\mathbf{r}_0$ fixed; the partially coupled particle can move around like the other particles and $\mathbf{r}_0$ then denotes its position in each separate configuration. Since $\mu^{\mathrm{ex}}(\mathbf{r}_0)$ in a bulk phase does not depend on position, one obtains $\mu^{\mathrm{ex,b}}$ irrespective of whether the particle is fixed or not.

Another method to obtain $\mu^{\mathrm{ex}}(\mathbf{r}_0)$ is to use the **free energy perturbation** technique, which is based on the general equality

$$\Delta A_{\mathrm{A}\to\mathrm{B}}^{\mathrm{ex}} = A_{\mathrm{B}}^{\mathrm{ex}} - A_{\mathrm{A}}^{\mathrm{ex}} = -k_B T \ln \left\langle e^{-\beta[\breve{U}_{\mathrm{B}}^{\mathrm{pot}} - \breve{U}_{\mathrm{A}}^{\mathrm{pot}}]} \right\rangle_{(\mathrm{A})}, \tag{5.153}$$

where A and B denote two different equilibrium states where the interaction potentials differ. Note that the ensemble average is taken for state A, which is indicated by subscript (A) on the average. This equality can be derived as follows. For a system with N mobile particles, we have from Equation 3.23 (for simplicity in notation, we here suppress subscript N for all entities)

$$\begin{aligned} A_{\mathrm{B}}^{\mathrm{ex}} &= -k_B T \ln \left[\frac{1}{V^N} \int d\mathbf{r}^N e^{-\beta \breve{U}_{\mathrm{B}}^{\mathrm{pot}}(\mathbf{r}^N)} \right] \\ &= -k_B T \ln \left[\frac{1}{V^N} \int d\mathbf{r}^N e^{-\beta \breve{U}_{\mathrm{A}}^{\mathrm{pot}}(\mathbf{r}^N)} e^{-\beta[\breve{U}_{\mathrm{B}}^{\mathrm{pot}}(\mathbf{r}^N) - \breve{U}_{\mathrm{A}}^{\mathrm{pot}}(\mathbf{r}^N)]} \right] \\ &= -k_B T \ln \left[\frac{Z_{\mathrm{A}}}{V^N} \int d\mathbf{r}^N \mathcal{P}_{\mathrm{A}}^{(N)}(\mathbf{r}^N) e^{-\beta[\breve{U}_{\mathrm{B}}^{\mathrm{pot}}(\mathbf{r}^N) - \breve{U}_{\mathrm{A}}^{\mathrm{pot}}(\mathbf{r}^N)]} \right] \\ &= A_{\mathrm{A}}^{\mathrm{ex}} - k_B T \ln \left\langle e^{-\beta[\breve{U}_{\mathrm{B}}^{\mathrm{pot}} - \breve{U}_{\mathrm{A}}^{\mathrm{pot}}]} \right\rangle_{(\mathrm{A})}. \end{aligned}$$

We now let states A and B differ by the extent of the coupling of an additional fixed particle at $\mathbf{r}_0$, so we have $\breve{U}_B^{pot}(\mathbf{r}^N) - \breve{U}_A^{pot}(\mathbf{r}^N) = \sum_{\nu=1}^{N}[u(r_{0\nu};\xi_B) - u(r_{0\nu};\xi_A)]$, where ξ_A and ξ_B are the values of the coupling parameter for the additional particle in A and B, respectively. Then we obtain

$$\Delta A_{A \to B}^{ex} = -k_B T \ln \left\langle e^{-\beta \sum_{\nu=1}^{N}[u(r_{0\nu};\xi_B) - u(r_{0\nu};\xi_A)]} \right\rangle_{(A)} \Bigg|_{\mathbf{r}_0 \text{ fix}}.$$

Note that the Widom insertion method is the special case where $\xi_A = 0$, $\xi_B = 1$ and the average is taken for the fluid without any fixed particle ($\xi_A = 0$). We then have $\Delta A_{A \to B}^{ex} = \mu^{ex}(\mathbf{r}_0)$. In the present case, we will do a real insertion of the fixed particle in several steps, whereby the other particles interact with this particle when it is partially coupled. This makes it possible to calculate $\mu^{ex}(\mathbf{r}_0)$ also for dense systems. If we divide the interval $0 \leq \xi \leq 1$ into, $\mathcal{L}$ parts $\xi_\ell \leq \xi \leq \xi_{\ell+1}$ where $\xi_1 = 0$ and $\xi_{\mathcal{L}+1} = 1$, we have the change in free energy for interval ℓ

$$\Delta A_\ell^{ex} = -k_B T \ln \left\langle e^{-\beta \sum_{\nu=1}^{N}[u(r_{0\nu};\xi_{\ell+1}) - u(r_{0\nu};\xi_\ell)]} \right\rangle_{(\xi_\ell)} \Bigg|_{\mathbf{r}_0 \text{ fix}},$$

where $\langle \cdot \rangle_{(\xi_\ell)}$ means that the average is to be taken for the system with the coupling parameter $\xi = \xi_\ell$. The total excess chemical potential equals

$$\mu^{ex}(\mathbf{r}_0) = \sum_{\ell=1}^{\mathcal{L}} \Delta A_\ell^{ex}. \tag{5.154}$$

Thereby, one needs to do a simulation for each ξ_ℓ value, $1 \leq \ell \leq \mathcal{L}$, that is, for all $\xi_\ell < 1$.

Simulation results for $\mu^{ex}(\mathbf{r}_0)$ calculated with this method can be improved as follows without much extra effort. By setting $\xi_A = \xi_\ell$ and $\xi_B = \xi_{\ell-1}$, we have

$$-k_B T \ln \left\langle e^{-\beta \sum_{\nu=1}^{N}[u(r_{0\nu};\xi_{\ell-1}) - u(r_{0\nu};\xi_\ell)]} \right\rangle_{(\xi_\ell)} \Bigg|_{\mathbf{r}_0 \text{ fix}} = -\Delta A_{\ell-1}^{ex},$$

so a simulation for $\xi = \xi_\ell$ can be used to obtain an independent estimate of $\Delta A_{\ell-1}^{ex}$, which complements the estimate of $\Delta A_{\ell-1}^{ex}$ from the simulation with $\xi = \xi_{\ell-1}$. No extra simulations have to be done except for one with $\xi = 1$. The latter must be done in order to obtain an additional estimate of $\Delta A_{\mathcal{L}}^{ex}$ as obtained from a simulation with $\xi = \xi_{\mathcal{L}+1} = 1$. This manner of obtaining a pair of estimates of ΔA_ℓ^{ex} for each ℓ is called "double wide sampling" and the final value of ΔA_ℓ^{ex} is the average of the pair. The procedure allows control of the consistency of the results. If the two estimates are too different from each other, one can achieve a better result by, for example, performing a longer simulation and/or decreasing the size of the intervals between consecutive ξ values. In the present brief introduction to free energy methods, we will, however, not go into any details about how to obtain high accuracy in the simulation results.

Instead of doing full simulations for each of the separate values of the coupling parameter, one can do a so-called **expanded ensemble simulation**.[51] In such a simulation, in addition

---

[51] A. P. Lyubartsev, A. A. Martsinovski, S. V. Shevkunov, P. N. Vorontsov-Velyaminov, *J. Chem. Phys.* **96** (1992) 1776.

to particle moves, one also does random changes in some other variable like the temperature or a coupling parameter by treating this variable as an additional degree of freedom. Like the free energy perturbation technique, the expanded ensemble simulation technique is a general method to calculate free energy differences between various states of a system. Here we will only consider the same case as before, namely the insertion of an additional particle coupled to the extent ξ.

In the expanded ensemble method, one handles the system for the various values of ξ for $0 \leq \xi \leq 1$ in a single simulation, where apart from particle moves one lets ξ spontaneously vary and thereby pass between the end point values 0 and 1 many times. Like before, one divides the interval $0 \leq \xi \leq 1$ into $\mathcal{L}$ parts $\xi_\ell \leq \xi \leq \xi_{\ell+1}$ where $\xi_1 = 0$ and $\xi_{\mathcal{L}+1} = 1$. The total potential energy of the system depends on the coordinates $\mathbf{r}^N$ of the N particles, the position $\mathbf{r}_0$ of the inserted particle and the value of the coupling parameter for the latter, so we have $\breve{U}^{\text{pot}}(\mathbf{r}^N, \mathbf{r}_0, \xi_\ell)$.

The principle of the simulation is to do several particle moves while keeping ξ_ℓ constant, then change the value of ξ_ℓ and continue with more particle moves at constant ξ_ℓ, change ξ_ℓ yet again followed by more particle moves etc. Each of the changes in ξ_ℓ is made to a neighboring value, in general from ξ_ℓ to $\xi_{\ell-1}$ or $\xi_{\ell+1}$ (the selection of direction to higher or lower ξ is random).[52] Thus, among the particle moves, one now and then attempts to change the value of ξ_ℓ. The acceptance or rejection of both particle moves and changes in ξ is thereby performed according to the standard Metropolis algorithm based on the variations in $\breve{U}^{\text{pot}}(\mathbf{r}^N, \mathbf{r}_0, \xi_\ell)$.[53]

During an expanded ensemble simulation, the average probability $\mathcal{P}(\xi_\ell)$ to observe the system at a certain ξ_ℓ value is given by

$$\mathcal{P}(\xi_\ell) = \frac{Z(\xi_\ell)}{\sum_{\ell'=1}^{\mathcal{L}+1} Z(\xi_{\ell'})}, \tag{5.155}$$

where $Z(\xi_\ell)$ is the configurational partition function for the system when $\xi = \xi_\ell$. This can be realized from the fact that the probability for a certain configuration of particles when $\xi = \xi_\ell$ is proportional to the Boltzmann factor $\exp[-\beta \breve{U}^{\text{pot}}(\mathbf{r}^N, \mathbf{r}_0; \xi_\ell)]$. The probability that $\xi = \xi_\ell$ irrespective of configuration $\mathbf{r}^N$ is therefore

$$\mathcal{P}(\xi_\ell) = \mathcal{K} \int d\mathbf{r}^N e^{-\beta \breve{U}^{\text{pot}}(\mathbf{r}^N, \mathbf{r}_0; \xi_\ell)} = \mathcal{K} Z(\xi_\ell),$$

where $\mathcal{K}$ is the proportionality factor, which can be determined from the fact that $\sum_{\ell'=1}^{\mathcal{L}+1} \mathcal{P}(\xi_\ell) = 1$ and Equation 5.155 follows. We have here assumed that the inserted particle is kept fixed at $\mathbf{r}_0$, so the functions $Z(\xi_\ell)$ and $\mathcal{P}(\xi_\ell)$ depend also on $\mathbf{r}_0$ in the general case. For a homogeneous fluid, we can let the inserted particle be mobile and then one also includes an integration over $\mathbf{r}_0$ in $Z(\xi_\ell)$, which becomes independent of $\mathbf{r}_0$.

---

[52] At the end points of ξ, changes in both directions are generated at random, but all changes to the outside of $0 \leq \xi \leq 1$ are rejected.

[53] Alternatively, the particles can be moved by using the MD simulation technique, but changes in ξ are performed as a MC step according to the Metropolis algorithm.

The excess free energy of the system for any ξ_ℓ is given by

$$A^{\mathrm{ex}}(\xi_\ell) = A^{\mathrm{ex}}(0) - k_B T \ln \frac{Z(\xi_\ell)}{Z(0)} = A^{\mathrm{ex}}(0) - k_B T \ln \frac{\mathcal{P}(\xi_\ell)}{\mathcal{P}(0)}.$$

In particular, μ^{ex} is equal to $A^{\mathrm{ex}}(1) - A^{\mathrm{ex}}(0) = -k_B T \ln[\mathcal{P}(1)/\mathcal{P}(0)]$. Thus, by recording the probability for the system to be at the end points $\xi = 0$ and $\xi = 0$ during the simulation, we can determine the difference in free energy and thereby the excess chemical potential. These probabilities are proportional to the length of time that the system spends at the respective end point; in an MC simulation, the "time" is measured as the number of particle moves (both accepted and rejected).

In general, $\breve{U}^{\mathrm{pot}}(\mathbf{r}^N, \mathbf{r}_0, \xi_\ell)$ can vary a lot as a function of ξ_ℓ, in particular when the surroundings of the particle at $\mathbf{r}_0$ are crowded. Changes in ξ_ℓ in the direction of increasing potential energy then have low probability to be accepted, while all changes in the opposite direction are accepted. The simulation may therefore in practice spend a very long time at ξ_ℓ values where the system has a low energy. For the method to be efficient and the value of $A^{\mathrm{ex}}(1) - A^{\mathrm{ex}}(0)$ reliable, the time that the system spends at the different ξ_ℓ values should not be too different during the simulation. One can accomplish this by introducing a bias that makes transitions between consecutive ξ_ℓ values likely in both directions. A suitable manner to do this is to add a "bias potential" $\varphi_{\mathrm{bias}}(\xi_\ell)$ to the potential energy.[54] This potential is independent of the particle coordinates but dependent of ξ_ℓ. Thus, φ_{bias} does not affect the forces between the molecules and hence not the transitions between various particle configurations for fixed ξ_ℓ, but it does affect the transitions between the ξ_ℓ values. In this approach, one makes a simulation with potential energy $\breve{U}^{\mathrm{pot}}(\mathbf{r}^N, \mathbf{r}_0, \xi_\ell) + \varphi_{\mathrm{bias}}(\xi_\ell)$. Thereby, one obtains a new value of the probability $\mathcal{P}^{[\varphi]}(\xi_\ell) = \mathcal{P}(\xi_\ell) e^{-\beta \varphi_{\mathrm{bias}}(\xi_\ell)}$, where superscript $[\varphi]$ means that the simulation is performed with the bias potential. We now have

$$A^{\mathrm{ex}}(\xi_\ell) = A^{\mathrm{ex}}(0) - k_B T \ln \frac{\mathcal{P}^{[\varphi]}(\xi_\ell)}{\mathcal{P}^{[\varphi]}(0)} + \varphi_{\mathrm{bias}}(0) - \varphi_{\mathrm{bias}}(\xi_\ell) \tag{5.156}$$

and hence

$$\mu^{\mathrm{ex}}(\mathbf{r}_0) = A^{\mathrm{ex}}(1) - A^{\mathrm{ex}}(0) = -k_B T \ln \frac{\mathcal{P}^{[\varphi]}(1)}{\mathcal{P}^{[\varphi]}(0)} + \varphi_{\mathrm{bias}}(0) - \varphi_{\mathrm{bias}}(1). \tag{5.157}$$

The optimal choice of $\varphi_{\mathrm{bias}}(\xi_\ell)$ makes $\mathcal{P}^{[\varphi]}(\xi_\ell)$ equal for all ℓ. From Equation 5.156 follows that in this case $\varphi_{\mathrm{bias}}(\xi_\ell) = A^{\mathrm{ex}}(\xi_\ell) + \text{constant}$. This means that in order to achieve an optimal procedure one needs to know the free energy, that is, the quantity that we wanted to determine in the first place. However, it is unnecessary to have an optimal choice of bias potential; it is sufficient to make the transition probability between the different ξ_ℓ values reasonable. There exist techniques to construct a suitable bias potential on the run during

---

[54] The quantity corresponding to $-\beta\varphi_{\mathrm{bias}}$ is denoted η in the paper cited in footnote 51. It is sometimes called a "balancing factor."

the expanded ensemble simulation from a study of the acceptance probabilities[55] and these techniques can be automatized. Here we will limit ourselves to give an example of a direct manner to obtain a suitable $\varphi_{\text{bias}}(\xi_\ell)$.

Since the optimal bias potential is known once we have $A^{\text{ex}}(\xi_\ell)$, we can obtain a good $\varphi_{\text{bias}}(\xi_\ell)$ from an approximate $A^{\text{ex}}(\xi_\ell)$. This can be obtained without undue effort from, for example, a series of *short* simulations for various ξ_ℓ values where one uses, for example, the free energy perturbation technique or coupling parameter integration to determine reasonable estimates of $A^{\text{ex}}(\xi_\ell)$ as a function of ξ_ℓ. One does not need to do this approximate calculation for all ξ_ℓ values if $A^{\text{ex}}(\xi)$ varies quite smoothly for $0 < \xi < 1$.[56] One may use some interpolation method to obtain $\varphi_{\text{bias}}(\xi_\ell)$ for all ξ_ℓ. Once this is done, one can spend the large majority of computing time to obtain accurate values of $A^{\text{ex}}(\xi_\ell)$ and hence $\mu^{\text{ex}}(\mathbf{r}_0)$ in an expanded ensemble simulation.

Finally, we turn to determinations of the total free energy of the fluid by simulation. One can do this by, for example, calculating the average excess energy $\langle U^{\text{ex}}\rangle$ given by the last term in Equation 5.146 for various temperatures and then performing the thermodynamic integration in Equation 5.81. An alternative is to integrate $P^{\text{ex}} = -(\partial A^{\text{ex}}/\partial V)_{T,N}$ as a function of V, as discussed in connection to Equation 5.82.

To determine A_N^{ex}, one can also introduce a coupling parameter α for *all* particles, such that

$$\check{U}_N^{\text{pot}}(\mathbf{r}^N; \alpha) = \begin{cases} 0, & \alpha = 0 \\ \check{U}_N^{\text{pot}}(\mathbf{r}^N), & \alpha = 1 \end{cases} \tag{5.158}$$

and utilize the relationship

$$\left(\frac{\partial A_N^{\text{ex}}(\alpha)}{\partial \alpha}\right)_{T,V,N} = -k_B T \left(\frac{\partial \ln Z(T,V,N;\alpha)}{\partial \alpha}\right) = \left\langle \frac{\partial \check{U}_N^{\text{pot}}(\mathbf{r}^N; \alpha)}{\partial \alpha}\right\rangle. \tag{5.159}$$

By integrating the rhs with respect to α between 0 and 1, one obtains A_N^{ex}.

{**EXERCISE 5.8**

Fill in the details of the proof of this method to calculate the excess free energy. Do this by proving Equation 5.159 starting from Equation 3.23 and then using this equation to obtain A_N^{ex}.}

One can also use the free energy perturbation technique based on Equation 5.153 or the expanded ensemble simulation technique[57] to calculate the free energy. Other methods to obtain free energy by simulations can be found in, for example, the literature cited in footnote 37.

---

[55] Examples are given in the paper cited in footnote 51, where the expanded ensemble has $1/k_B T$ as the extra degree of freedom rather than ξ.

[56] Here the choice of ξ parameterization of $u(r; \xi)$ is crucial, so one avoids rapid variations of $A^{\text{ex}}(\xi)$ as a function of ξ.

[57] For details, see the paper cited in footnote 51.

{**EXERCISE 5.9**

Sketch a procedure where one uses any coupling parametrization that fulfills Equation 5.158 and the free energy perturbation technique to obtain A_N^{ex}.}

5.13.3 Inhomogeneous Fluids

5.13.3.1 Density Profiles Outside Macroparticles or Near Planar Surfaces

We first consider an inhomogeneous fluid in the surroundings of a spherical macroparticle that interacts with the fluid via a potential $v(r)$. The macroparticle is placed at the center of a spherical simulation box that contains N fluid particles. The outer boundary of the simulation box may be a hard surface (this is assumed in the following unless otherwise specified). This is often called the **spherical cell model** of a macroparticle/fluid system. If one wants to study a macroparticle immersed in a bulk fluid, N must be so large that the density profile $n(r)$ outside the macroparticle approximately reaches the bulk value n^b when r increases. At even larger r values – in the outer region of the simulation box – the density deviates from n^b due to the influence of the outer boundary of the cell, but this region is normally ignored in the sampling for calculations of the properties of the system.[58] One can alternatively use cubic simulation cells and periodic boundary conditions provided the replicas of the macroparticle influence the density in the simulation box to a sufficiently small extent.

To determine the density profile $n(r)$ outside the macroparticle, one can use a method with a division of space into spherical shells around the macroparticle, in analogy to Figure 5.16. One then makes histograms by counting the number of particle centers that fall within each shell and finally one takes the average over all particle configurations during the simulations. This is similar to the procedure used for $n^b g^{(2)}(r)$ earlier; the difference is that we do this only for one particle – in the current case, the macroparticle surrounded by the fluid particles. Therefore, one usually needs longer simulations in order to obtain sufficient accuracy for the density distribution in the present case. One can also use the method where the excess chemical potential at various r values is determined and used to calculate the density, as described earlier.

In the special case where the central particle is not a macroparticle, but instead a particle of the same species as the constituent particles of the fluid, we have $v(r) = u(r)$. Then the density profile $n(r)$ is the same as $n^b g^{(2)}(r)$ (except close to the edges of the simulation box). If the central particle belongs to a different species j than the constituent particles of the fluid (species i), one obtains the pair distribution function $n_i^b g_{ij}^{(2)}(r)$ for a solution of species j in a solvent consisting of i evaluated in the limit of high (infinite) dilution of j.

Let us now turn to inhomogeneous fluids in planar geometry like a fluid outside a planar wall in contact with a bulk phase or a fluid in the slit between two such parallel walls like in Figure 5.5. A smooth planar wall interacts with the fluid particles via a potential $v(z)$, where

---

[58] There exist techniques to minimize the extent of this outer region, for example by imposing an external potential from the outer boundary that makes the density profile to deviate as little as possible from n^b there. Furthermore, one can surround the simulation cell with a continuum phase that mimics a bulk fluid and use a reaction field method to account for the response of it.

z is the coordinate in the normal direction. One can use a simulation box with rectangular sides, of which four are perpendicular and two parallel to the wall(s). One uses periodic boundary conditions in the lateral directions only (the x and y directions).

For the case of one wall in contact with a bulk fluid, the far side of the simulation box parallel to the wall (the "outer side") plays the same role as the outer boundary of the spherical cell discussed previously. The density profile $n(z)$ then should reach the bulk density when z is increased (before the outer region is reached). To evaluate the density profile, one divides the space into thin planar layers parallel to the wall(s). One then makes a histogram by counting the number of particle centers that fall within each layer and at the end one takes the average over all particle configurations, analogously to the previous cases.

For the system with two parallel walls, wall I and wall II, the interaction potential between the fluid particles and each wall is in general different, $v_{I}(\ell)$ and $v_{II}(\ell)$, respectively, where ℓ is the distance from the respective wall. The fluid is exposed to the external potential $v(z) = v_{I}(z) + v_{II}(z - D^S)$, where D^S is the separation between the walls and z is measured from wall I. In cases where the fluid in the slit is in equilibrium with a bulk phase and D^S is very large, the density in the middle of the slit is virtually equal to the bulk density. Then each half on either side of the middle is like a one-wall system in contact with the bulk. For smaller D^S, one must maintain the equilibrium with the bulk phase by requiring that the chemical potential μ for the fluid in the slit is the same as μ in the bulk. For a canonical ensemble simulation, this can be done by adjusting the number of particles per unit area in the slit until μ has the appropriate value. One can also do a Grand Canonical Ensemble simulation, where μ is prescribed. This is usually preferred. In cases where the fluid in the slit is *not* in equilibrium with a bulk phase and the system is closed, the number of particles per unit area is the same for all D^S. Then one simply does simulations with a given number of particles for all separations.

One can also treat cases where there is an internal structure of the wall(s) and the external potential $v(\mathbf{r})$ from the wall(s) depends on the x and y coordinates in addition to z. Then the wall structure should preferably have a lateral periodicity in agreement with the periodic boundary conditions used. In this case, the density profile $n(\mathbf{r})$ depends on the x, y and z coordinates and one must subdivide each layer into small cells with sides of lengths, say, Δx and Δy, in the lateral directions. One then counts the particle centers that fall within each little cell in order to obtain $n(\mathbf{r})$. Long simulations are usually needed in order to achieve sufficient accuracy, in particular if $v(\mathbf{r})$ depends on x and y in a complicated manner. When $v(\mathbf{r})$ has some symmetry, $n(\mathbf{r})$ must have the same kind of symmetry after an infinitely long simulation, so in the evaluation of $n(\mathbf{r})$ in a finite simulation one can in practice improve the accuracy by making $n(\mathbf{r})$ satisfy the same symmetry as $v(\mathbf{r})$ by averaging.

The density distribution $n(\mathbf{r})$ can alternatively be determined from the expression (Equation 5.30)

$$n(\mathbf{r}) = n^b e^{-\beta[\mu^{ex}(\mathbf{r}) - \mu^{ex,b}]}, \tag{5.160}$$

so by determining the excess chemical potential at $\mathbf{r}$, one can obtain the density there. This can in particular be of use at points where $n(\mathbf{r})$ is difficult to determine accurately by the histogram method, like places where the density varies rapidly in space. In the presence of an external potential $v(\mathbf{r})$, Equation 5.144 is replaced by the Widom expression

$$\mu^{\text{ex}}(\mathbf{r}_0) = v(\mathbf{r}_0) - k_B T \ln \left\langle e^{-\beta \sum_{\nu=1}^{N} u(r_{0\nu})} \right\rangle^{\circ} \Bigg|_{\mathbf{r}_0 \text{ fix}} \qquad (5.161)$$

and Equation 5.151 is replaced by

$$\mu^{\text{ex}}(\mathbf{r}_0) = v(\mathbf{r}_0) + \int_0^1 d\xi \left\langle \sum_{\nu=1}^{N} \frac{\partial u(r_{0\nu}; \xi)}{\partial \xi} \right\rangle_{(\xi)} \Bigg|_{\mathbf{r}_0 \text{ fix}} . \qquad (5.162)$$

The same applies to the other ways of calculating $\mu^{\text{ex}}(\mathbf{r}_0)$; there is always a contribution $v(\mathbf{r}_0)$ in addition to the part that originates from interactions between the fluid particles (the intrinsic part).

For systems where the Widom insertion technique works in practice, the determination of the density at $\mathbf{r}$ via Equations 5.160 and 5.161 can be advantageous compared to the histogram method. While the latter method relies on that a sufficient number of particles are found very near $\mathbf{r}$ during the simulation, the Widom method determines $n(\mathbf{r})$ for any $\mathbf{r}$ by measuring the influence of the correlations in the entire surroundings of $\mathbf{r}$ via pair interactions.

5.13.3.2 Pair Distribution Functions

The pair distribution function $g^{(2)}(\mathbf{r}_1, \mathbf{r}_2)$ for an inhomogeneous fluid can *in principle* be obtained in a simulation as follows. We recall that $n(\mathbf{r}_1|\mathbf{r}_2) = n(\mathbf{r}_1)g^{(2)}(\mathbf{r}_1, \mathbf{r}_2)$ is the density at $\mathbf{r}_1$ when there is a particle fixed at $\mathbf{r}_2$ as in the example shown in Figure 5.8 with $\mathbf{r}' = \mathbf{r}_1$ and $\mathbf{r} = \mathbf{r}_2$. Thus, if one performs a simulation of a system where a particle is fixed at $\mathbf{r}_2$, the density distribution obtained is $n(\mathbf{r}_1|\mathbf{r}_2)$ and can in principle be determined by the histogram method. This is, in fact, a special case of a general external potential $v(\mathbf{r})$ discussed previously; the external potential used here is $v(\mathbf{r}|\mathbf{r}_2) = v(\mathbf{r}) + u(\mathbf{r}, \mathbf{r}_2)$. The basis is the same as before, only the geometry is more complicated.[59] Note that one must perform one simulation for each particle position $\mathbf{r}_2$ where one wants to determine $n(\mathbf{r}_1|\mathbf{r}_2)$. If one in addition performs a simulation of the same system in the absence of a fixed particle (i.e., with external potential $v(\mathbf{r})$ only) and determine $n(\mathbf{r}_1)$, one can obtain the pair distribution from $g^{(2)}(\mathbf{r}_1, \mathbf{r}_2) = n(\mathbf{r}_1|\mathbf{r}_2)/n(\mathbf{r}_1)$. This kind of simulation to determine the entire $g^{(2)}(\mathbf{r}_1, \mathbf{r}_2)$ is (currently) prohibitively expensive in computation time because the accuracy is too low unless one performs *very* long simulations.

As an alternative, one can for any given particle position $\mathbf{r}_2$ evaluate $n(\mathbf{r}_1|\mathbf{r}_2)$ point-wise at various $\mathbf{r}_1$ by calculating $\mu^{\text{ex}}(\mathbf{r}_1|\mathbf{r}_2)$ with the same techniques as in the evaluation of $\mu^{\text{ex}}(\mathbf{r}_1)$ for the density distribution in the previous section. When the density is not too high, one can thereby utilize the Widom insertion method in Equation 5.161 with $v(\mathbf{r}_0)$ replaced by $v(\mathbf{r}_0) + u(\mathbf{r}_0, \mathbf{r}_2)$. This gives a practical means to obtain $g^{(2)}(\mathbf{r}_1, \mathbf{r}_2)$ for various $\mathbf{r}_1$ from a

---

[59] The simulation box must have sufficiently large lateral dimensions so the replicas of the particle at $\mathbf{r}_2$ influence the density in the simulation box to a sufficiently small extent.

simulation with a fixed particle at $\mathbf{r}_2$. The other methods for the calculation of μ^{ex} can also be used, but they are much more expensive in computer time.

For conceptual reasons, it is of interest to note that one in principle can obtain $n(\mathbf{r}_1|\mathbf{r}_2)$ from *one single* simulation in the absence of a fixed particle in the following manner. One generates a huge number of particle configurations for the fluid in the presence of the external potential $v(\mathbf{r})$ only. From all these configurations, one selects those where a particle happens to have its center located at $\mathbf{r}_2$ (or, in practice, located within a small volume element that surrounds the point $\mathbf{r}_2$). In this subset of configurations, there is accordingly a particle at $\mathbf{r}_2$ and all other particles are therefore exposed to the potential $v(\mathbf{r}) + u(\mathbf{r}, \mathbf{r}_2)$. This means that this subset of configurations is statistically distributed in the same manner as if there were a particle fixed at $\mathbf{r}_2$,[60] which implies that the average density at $\mathbf{r}_1$ of these particles is equal to $n(\mathbf{r}_1|\mathbf{r}_2)$ and can be calculated from the configurations in the subset in the same manner as before.[61] Next, from the original set of configurations, one can select another subset where a particle happens to have its center located at another point $\mathbf{r}_2$. This gives in the same manner $n(\mathbf{r}_1|\mathbf{r}_2)$ for this other value of $\mathbf{r}_2$. By proceeding in the same manner with different points $\mathbf{r}_2$, one can map the entire $n(\mathbf{r}_1|\mathbf{r}_2)$ and hence $g^{(2)}(\mathbf{r}_1, \mathbf{r}_2)$ from the configurations generated in one single simulation. To achieve any reasonable accuracy, there must exist a sufficient number of configurations with a particle at (or near) each point $\mathbf{r}_2$ where $n(\mathbf{r}_1|\mathbf{r}_2)$ is to be obtained. Therefore, the number of generated configurations ought to be enormous, so this option has (currently) little practical value in general.

There exists however, a simple manner to determine $n^{(2)}(\mathbf{r}_1, \mathbf{r}_2)$ and

$$g^{(2)}(\mathbf{r}_1, \mathbf{r}_2) = n^{(2)}(\mathbf{r}_1, \mathbf{r}_2)/[n(\mathbf{r}_1)n(\mathbf{r}_2)]$$

at any coordinates $\mathbf{r}_1$ and $\mathbf{r}_2$ from a single simulation in the presence of the external potential $v(\mathbf{r})$ only. One thereby applies a variant of Widom's formula that can be used in practice, provided the density is not too high. This is outlined in the following text box.

The expression for $\mu^{\text{ex}}(\mathbf{r}_0)$ in Equation 5.161 was derived by inserting a test particle at $\mathbf{r}_0$ in a fluid of N mobile particles. We can derive a corresponding expression for the insertion of two test particles, one at $\mathbf{r}_0$ and one at $\mathbf{r}_{0'}$, by defining a **two-particle excess chemical potential** $\mu^{(2)\text{ex}}$ from[62] (cf. Equation 5.143)

$$\mu^{(2)\text{ex}}(\mathbf{r}_0, \mathbf{r}_{0'}) = A_{N+2}^{\text{ex}}(\mathbf{r}_0, \mathbf{r}_{0'}) - A_N^{\text{ex}} = -k_B T \ln \frac{Z_{N+2}^{(2)}(\mathbf{r}_0, \mathbf{r}_{0'})}{Z_N}, \tag{5.163}$$

---

[60] Since the distributions of positions and velocities are independent of each other in classical statistical mechanics, it makes no difference whether a particle is fixed or mobile when being located at $\mathbf{r}_2$.

[61] By including configurations where there is a particle located at any other point that is equivalent to $\mathbf{r}_2$ by symmetry, one can increase the number of configurations in the subset provided all particle coordinates are measured relative to the position of this particle.

[62] In the notation of Chapter 10 (in the 2nd volume), this quantity is written as $\mu_{N+2}^{(2)\text{ex}}$.

where

$$Z_{N+2}^{(2)}(\mathbf{r}_0, \mathbf{r}_{0'}) = \int d\mathbf{r}_1 \dots d\mathbf{r}_N e^{-\beta \breve{U}_{N+2}^{\text{pot}}(\mathbf{r}_0, \mathbf{r}_{0'}, \mathbf{r}^N)}$$

is the configurational partition function and A_{N+2}^{ex} is the free energy for the $N+2$ particle system with N mobile particles and two particles kept fixed at $\mathbf{r}_0$ and $\mathbf{r}_{0'}$, respectively. The potential energy for the $N+2$ particles located at $(\mathbf{r}_0, \mathbf{r}_{0'}, \mathbf{r}_1, \dots \mathbf{r}_N)$ is

$$\breve{U}_{N+2}^{\text{pot}}(\mathbf{r}_0, \mathbf{r}_{0'}, \mathbf{r}^N) = \breve{U}_N^{\text{pot}}(\mathbf{r}^N) + \sum_{\nu=1}^{N} [u(r_{0\nu}) + u(r_{0'\nu})] + u(r_{00'}) + v(\mathbf{r}_0) + v(\mathbf{r}_{0'})$$

and we readily obtain (cf. Equation 5.161)

$$\mu^{(2)\text{ex}}(\mathbf{r}_0, \mathbf{r}_{0'}) = u(r_{00'}) + v(\mathbf{r}_0) + v(\mathbf{r}_{0'})$$

$$- k_B T \ln \left\langle e^{-\beta \sum_{\nu=1}^{N} [u(r_{0\nu}) + u(r_{0'\nu})]} \right\rangle^{\circ} \Bigg|_{\mathbf{r}_0, \mathbf{r}_{0'} \text{ fix}}. \tag{5.164}$$

Like before, superscript "o" on the average indicates that the probability density for the N mobile particles is not affected by the fixed particles, in the present case located at $\mathbf{r}_0$ and $\mathbf{r}_{0'}$, respectively. This average is evaluated in the same fashion as the corresponding average in Equations 5.144 and 5.161.

{**EXERCISE 5.10**
Show the result in Equation 5.164 by starting from the definition (5.163) of $\mu^{(2)\text{ex}}(\mathbf{r}_0, \mathbf{r}_{0'})$.}

The link to the pair distribution function can be obtained from Equation 5.45, which can be written for a system with $N+2$ particles as

$$n^{(2)}(\mathbf{r}_0, \mathbf{r}_{0'}) = \mathcal{C} \int d\mathbf{r}_1 \dots d\mathbf{r}_N e^{-\beta \breve{U}_{N+2}^{\text{pot}}(\mathbf{r}_0, \mathbf{r}_{0'}, \mathbf{r}^N)} \equiv \mathcal{C} Z_{N+2}^{(2)}(\mathbf{r}_0, \mathbf{r}_{0'}), \tag{5.165}$$

where $n^{(2)} = n_{N+2}^{(2)}$ and $\mathcal{C} = (N+2)(N+1)/Z_{N+2}$. When N is a large number, this $n^{(2)}(\mathbf{r}_0, \mathbf{r}_{0'})$ is a very good approximation for the distribution function in our system with N particles. By using Equation 5.163, we find from Equation 5.165 that

$$n^{(2)}(\mathbf{r}_0, \mathbf{r}_{0'}) = \mathcal{C}' e^{-\beta \mu^{(2)\text{ex}}(\mathbf{r}_0, \mathbf{r}_{0'})},$$

where $\mathcal{C}' = \mathcal{C} Z_N = (N+2)(N+1)Z_N/Z_{N+2}$. Next, we express the constant $\mathcal{C}'$ in terms of activities by using the relationship $\zeta_N = N Z_{N-1}/Z_N$ (Equation 3.28). By applying this relationship for two systems with $N+1$ and $N+2$ particles, respectively, we see that $\zeta_{N+2} \zeta_{N+1} = \mathcal{C}'$. When N is large, the quantities ζ_N, ζ_{N+1} and ζ_{N+2} are virtually the same and we can set $C' \approx \zeta_N^2 = \zeta^2$ as a very good approximation, where ζ is the activity of a bulk fluid with the same chemical potential as the fluid in the slit. Thus, we have

$$n^{(2)}(\mathbf{r}_0, \mathbf{r}_{0'}) = \zeta^2 e^{-\beta \mu^{(2)\text{ex}}(\mathbf{r}_0, \mathbf{r}_{0'})} = (n^{\text{b}})^2 e^{-\beta[\mu^{(2)\text{ex}}(\mathbf{r}_0, \mathbf{r}_{0'}) - 2\mu^{\text{ex,b}}]}, \qquad (5.166)$$

where we have used the relationship $\zeta = n^{\text{b}} \gamma^{\text{b}} = n^{\text{b}} e^{\beta \mu^{\text{ex,b}}}$ for the bulk fluid. This result can be compared to the expression for $n(\mathbf{r}_1 | \mathbf{r}_2)$ in Equation 5.140.

One utilizes Equation 5.166 in the evaluation of $n^{(2)}(\mathbf{r}_1, \mathbf{r}_2)$ in an analogous manner as Equation 5.140 is used in the evaluation of $n(\mathbf{r}_1 | \mathbf{r}_2)$. The only difference is that two fixed particles are involved in the evaluation of the average in Equation 5.164 instead of only one such particle as in Equation 5.161. By using these formulas, one can obtain $n^{(2)}(\mathbf{r}_1, \mathbf{r}_2)$ and hence $g^{(2)}(\mathbf{r}_1, \mathbf{r}_2)$ for any pair of points $\mathbf{r}_1$ and $\mathbf{r}_2$ from a single simulation of the inhomogeneous fluid. Thereby, one can map out these functions point-wise in any finite region of space. The other methods for calculation of excess chemical potential can be generalized in an analogous fashion, but they require that many simulations are performed.

APPENDIX 5A: THE DIRAC DELTA FUNCTION

The **Dirac delta function** $\delta(x)$ is an important and very useful concept in physics.[63] It has the property $\int dx\, \delta(x) f(x) = f(0)$ for any (well-behaved) function $f(x)$. One can think of $\delta(x)$ to be zero everywhere except at the origin, where it has an infinitely sharp peak with a unit integral, that is, $\int dx\, \delta(x) = 1$. Dirac's delta function is an example of what is called a "generalized function," which can be defined as certain limits of ordinary functions. One can, for example, define $\delta(x)$ as the limit when $X \to 0$ of a function $D_X(x) = 1/(2X)$ for $|x| \leq X$ and $= 0$ elsewhere. This function has a unit integral, $\int dx\, D_X(x) = 1$, and we have $\lim_{X \to 0} \int dx\, D_X(x) f(x) = f(0)$ if $f(x)$ is sufficiently well-behaved. Note that the delta function is an even function, $\delta(-x) = \delta(x)$, and that $\int dx\, \delta(x) f(x' - x) = f(x - 0) = f(x')$ and $\int dx\, \delta(x - x') f(x) = f(x')$.

There are, in fact, a multitude of functions that go to the delta function in some limit. Each of them has a unit integral and goes to zero in this limit for all nonzero values of x. One example that is more complicated than $D_X(x)$ is the function $\mathfrak{D}_{L_x}(x)$ defined in Equation 5.59. It satisfies $\int dx\, \mathfrak{D}_{L_x}(x) = 1$ and $\lim_{L_x \to \infty} \mathfrak{D}_{L_x}(x) \to 0$ for $x \neq 0$. We have

$$\lim_{L_x \to \infty} \int dx\, \mathfrak{D}_{L_x}(x) f(x) = f(0),$$

so $\mathfrak{D}_{L_x}(x)$ goes to the delta function in the limit $L_x \to \infty$.

An advantage with the introduction of $\delta(x)$, which we will make use of, is that it allows us to take the derivative of a function with a step discontinuity, like the **Heaviside step function**

---

[63] In mathematics, it is called the *Dirac distribution*, which is the linear mapping of a so-called test function $f(x)$ to its value $f(0)$ at the origin.

$\mathcal{H}(x)$, which is equal to 0 for $x < 0$ and 1 for $x > 0$. In fact, we have $d\mathcal{H}(x)/dx = \delta(x)$, which can be realized by considering the function

$$H_X(x) = \begin{cases} 0 & x < -X \\ \frac{1+x/X}{2} & |x| \leq X \\ 1 & x > X, \end{cases}$$

which is continuous and has a derivative $dH_X(x)/dx = D_X(x)$. In the limit $X \to 0$, the latter expression goes to $d\mathcal{H}(x)/dx = \delta(x)$ since $H_X(x) \to \mathcal{H}(x)$ for $x \neq 0$. In general, the derivative of a function has the contribution $a\,\delta(x - x_a)$ from a discontinuity with step height a at $x = x_a$.

In three dimensions, we have the **three-dimensional delta function** $\delta^{(3)}(\mathbf{r}) = \delta(x)\delta(y)\delta(z)$, which is zero everywhere except at the origin and has the property $\int d\mathbf{r}\,\delta^{(3)}(\mathbf{r}) = 1$. It is spherically symmetric, so we can write $\delta^{(3)}(\mathbf{r}) = \delta^{(3)}(r)$. For any well-behaved function $f(\mathbf{r})$, we have $\int d\mathbf{r}\,\delta^{(3)}(r)f(\mathbf{r}) = f(\mathbf{0})$. Note that $\int d\mathbf{r}\,\delta^{(3)}(\mathbf{r})f(\mathbf{r}' - \mathbf{r}) = f(\mathbf{r}' - \mathbf{0}) = f(\mathbf{r}')$ and that $\int d\mathbf{r}\,\delta^{(3)}(\mathbf{r} - \mathbf{r}')f(\mathbf{r}) = f(\mathbf{r}')$.

One can also define $\delta^{(3)}(r)$ as the limit when $R \to 0$ of, for example, the function $D_R^{(3)}(\mathbf{r}) = 3/[4\pi R^3]$ when $|\mathbf{r}| \leq R$ and $= 0$ elsewhere. This function has a unit integral, $\int d\mathbf{r}\,D_R^{(3)}(\mathbf{r}) = 1$, and we have $\lim_{R \to 0} \int d\mathbf{r}\,D_R^{(3)}(\mathbf{r})f(\mathbf{r}) = f(\mathbf{0})$. The dimension of both $D_R^{(3)}(\mathbf{r})$ and $\delta^{(3)}(r)$ is (volume)$^{-1}$, same as for a number density. Another, more complicated example of a function that goes to $\delta^{(3)}(r)$ is $\mathfrak{D}_L^{(3)}(\mathbf{r}) = \mathfrak{D}_{L_x}(x)\mathfrak{D}_{L_y}(y)\mathfrak{D}_{L_z}(z)$ in the limit where L_x, L_y, and L_z all go to infinity (cf. Section 5.6).

CHAPTER 6

Interactions and Correlations in Simple Bulk Electrolytes

W HEN WE APPLY THE mean field approximation introduced in Section 5.12 to ionic fluids, we obtain a theory called the Poisson-Boltzmann (PB) approximation. It is the traditional theory for electrolytes, which still is quite extensively used in many applications today and constitutes an approximate basis for an understanding of the special properties of such systems, in particular regarding the electrostatic interactions in ionic fluids. A large part of the current chapter deals with this theory and includes some aspects of it for nonspherical particles that have been realized not so long ago[1] (Sections 6.1.2 and 6.1.3). An advantage with the Poisson-Boltzmann approximation is that it can be used to illustrate many important aspects of statistical mechanics for fluids in a quite concrete manner, which motivates a quite extensive treatment in the present book. For this and other reasons, the chapter contains a presentation of the Debye-Hückel theory, which is the classical approximate theory for bulk electrolytes.

The concept of effective charge, as defined in a strict manner, is introduced and we investigate the decay behavior of the electrostatic potential from spherical and nonspherical particles immersed in simple electrolytes in the PB approximation. The direction dependence of the electrostatic potential from a nonspherical particle differs in a fundamental manner from that in a polar medium and in vacuum. This leads to the introduction of a "multipolar effective charge" (a direction dependent entity) of the particle. The treatment includes the electrostatic interactions between nonspherical particles in simple electrolytes.

We will also investigate what happens when we take a little step beyond the PB approximation by demanding that all ions of the same kind should be treated in the same manner,

---

[1] The screened electrostatic interactions between anisotropic colloidal particles like finite rods or disks with arbitrary orientations in the PB approximation were, for example, treated in D. Chapot, L. Bocquet, and E. Trizac, *J. Chem. Phys.* **120** (2004) 3969 and R. Agra, E. Trizac, and L. Bocquet, *Eur. Phys. J. E* **15** (2004) 345.

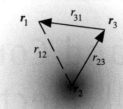

FIGURE 6.1 Illustration of the application of Coulomb's law for the calculation of the electrostatic potential at $\mathbf{r}_1$ from the charge density distribution $\rho(r_{23})$ (shown with a gray shading) around an ion located at $\mathbf{r}_2$. The potential at $\mathbf{r}_1$ from a charge at $\mathbf{r}_3$ is $1/(4\pi\varepsilon_0 r_{31})$ times the value of the charge, in this case given by the charge $\rho(r_{23})d\mathbf{r}_3$ in a volume element $d\mathbf{r}_3$. By integrating with respect to $\mathbf{r}_3$ over the whole space, one obtains the potential from the entire charge distribution.

which is not done in this approximation. The consequences are, as we will see, rather dramatic since there arise multiple decay lengths and oscillatory electrostatic interactions, while the PB approximation only gives monotonic exponential decay with a single decay length (called the Debye length).

Thereafter, in Section 6.2, we will use formally exact theory in a reasonably elementary manner to analyze the nature of the interactions and correlations in simple bulk electrolytes. We will thereby find that the structure of the PB theory in several respects "survives" and can be extended to the general, exact case. This gives tools to understand several aspects of electrostatic interactions and correlations in simple electrolytes ranging from dilute electrolyte solutions/thin ionic gases to dense liquids like molten NaCl. Several concrete explicit examples are given. The concepts of density-density, charge-density and charge-charge correlations are introduced and used in the analysis.

A key concept both in the PB approximation and the formally exact theory is the (unit) *screened Coulomb potential*, which takes a similar role as the unscreened Coulomb potential $1/(4\pi\varepsilon_0 r)$ and is given a strict, general definition. We will see that the electrostatic potential due to a charged particle in an electrolyte can be expressed in terms of a simple generalization of Coulomb's law with the screened instead of the unscreened Coulomb potential. This gives a basis for the understanding of screened electrostatic interactions in the general case between particles in electrolytes.

6.1 THE POISSON-BOLTZMANN (PB) APPROXIMATION

6.1.1 Bulk Electrolytes, Basic Treatment

We will start with homogeneous bulk phases consisting of **classical plasmas**, that is, gaseous electrolytes treated in classical statistical mechanics.

Example 6.1: One-component Plasma
As an application of the mean field approximation, we will consider the so-called **one-component plasma** (OCP), which is the simplest model of an ionic fluid. It consists of charged point particles (point ions) of charge q in vacuum. They interact with each other via the pair interaction in Equation 5.4 with $d^h = 0$

$$u(r) = u^{\text{el}}(r) \equiv \frac{q^2}{4\pi \varepsilon_0 r}, \tag{6.1}$$

where superscript el stands for electrostatic. The fluid has a uniform density n^b. To maintain electroneutrality, the system also contains a uniform background charge density $\rho^B = -qn^b$ that permeates the whole space. This charge density acts as the countercharges to the ions, that is, the charges of the opposite sign that make the system electroneutral.

The ions that surround an ion at $\mathbf{r}_2$ (the central ion) are distributed with an average density at point $\mathbf{r}_3$ equal to $n^b g^{(2)}(r_{23})$. The charge density from these ions is $qn^b g(r_{23})$ and the total charge density (including the background) is

$$\rho(r_{23}) = qn^b g^{(2)}(r_{23}) + \rho^B = qn^b h^{(2)}(r_{23}). \tag{6.2}$$

The ion at $\mathbf{r}_2$ and the surrounding charge distribution $\rho(r_{23})$ give rise to an electrostatic potential ψ. The potential at position $\mathbf{r}_1$ is according to Coulomb's law given by (see illustration in Figure 6.1)

$$\psi(r_{12}) = \frac{q}{4\pi \varepsilon_0 r_{12}} + \int d\mathbf{r}_3 \frac{\rho(r_{23})}{4\pi \varepsilon_0 r_{31}}$$

$$= \frac{q}{4\pi \varepsilon_0 r_{12}} + \int d\mathbf{r}_3 \frac{qn^b h^{(2)}(r_{23})}{4\pi \varepsilon_0 r_{31}}. \tag{6.3}$$

The potential ψ is called the **mean electrostatic potential** since the charge distribution $\rho(r_{23})$ is mean charge density around the central ion at $\mathbf{r}_2$. Comparison with the expression (6.1) for $u(r)$ shows that we can write

$$q\psi(r_{12}) = u(r_{12}) + \int d\mathbf{r}_3 \, n^b h^{(2)}(r_{23}) u(r_{31}), \tag{6.4}$$

which is exactly the expression in the square brackets of the exponent in Equation 5.124, so in the mean field approximation we have

$$g^{(2)}(r_{12}) = e^{-\beta q\psi(r_{12})}. \tag{6.5}$$

By comparison with Equation 5.35, we see that the pair potential of mean force is $w^{(2)}(r_{12}) = q\psi(r_{12})$ in this approximation (this is the *Poisson-Boltzmann approximation* which will be described in more detail later). The mean force on an ion at $\mathbf{r}_1$ when an ion is located at $\mathbf{r}_2$ is accordingly

$$\mathbf{F}_1(r_{12}) = -\nabla_1 w^{(2)}(r_{12}) = -q\nabla_1\psi(r_{12}) = q\mathbf{E}_1(r_{12}), \tag{6.6}$$

where $\mathbf{E}_1 = -\nabla_1\psi$ is the electrostatic field from the charge distribution that gives rise to ψ. The integral in Equation 6.4 constitutes the indirect contribution to the potential of mean force, that is, from the interactions with all ions surrounding the two ions.

The first equality in Equation 6.3 is a relationship between the charge density $\rho(r)$ and the potential $\psi(r)$ that it gives rise to. The same potential also satisfies the Poisson's equation of electrostatics

$$-\varepsilon_0 \nabla^2 \psi(r) = \rho(r), \tag{6.7}$$

which is also a relationship between these entities. We will apply this equation for $r > 0$ with the boundary conditions that there is a charge q at the origin and that $\psi(r) \to 0$ when $r \to \infty$. Indeed, $\psi(r)$ from Coulomb's law in Equation 6.3 is the solution to this equation with the given boundary conditions.

It is often more convenient to solve Poisson's equation in order to obtain $\psi(r)$ than to use Coulomb's law directly. By multiplying Equation 6.5 with qn^b, inserting this in Equation 6.2 and combining the result with Poisson's equation, we obtain

$$-\varepsilon_0 \nabla^2 \psi(r) = qn^b e^{-\beta q\psi(r)} + \rho^B \tag{6.8}$$

for $r > 0$. This the called the *Poisson-Boltzmann equation* (the last term, ρ^B, is is present only in cases where there is a background charge density).

When the electrostatic potential is small, so $|\beta q\psi| \ll 1$, we can linearize the exponential function,

$$e^{-\beta q\psi(r)} \approx 1 - \beta q\psi(r) \tag{6.9}$$

and obtain for $r > 0$

$$\nabla^2 \psi(r) = \kappa_D^2 \psi(r), \tag{6.10}$$

where we have introduced the parameter

$$\kappa_D^2 = \frac{\beta q^2 n^b}{\varepsilon_0}. \tag{6.11}$$

This differential equation is called the *linearized Poisson-Boltzmann equation*. It can be solved analytically as follows.

Since $\psi(r)$ depends only on r, we can use the radial part of ∇^2, which can be written $r^{-1}(\partial^2/\partial r^2)r$. Thus, we have $(\partial^2/\partial r^2)[r\psi(r)] = \kappa_D^2[r\psi(r)]$, which has the general solution $r\psi(r) = C_1 \exp(-\kappa_D r) + C_2 \exp(\kappa_D r)$, where C_1 and C_2 are constants.

From the boundary condition at $r \to \infty$, it follows that $C_2 = 0$ and from the boundary condition at the origin it follows that $C_1 = q/(4\pi \varepsilon_0)$. Therefore, we have the solution

$$\psi(r) = \frac{q}{4\pi \varepsilon_0} \cdot \frac{e^{-\kappa_D r}}{r}. \tag{6.12}$$

This potential is a *screened Coulomb potential* and it has very much shorter range than the usual Coulomb potential due to the appearance of the exponential function. The charge q at the center is screened by the surrounding charge of the opposite sign, which in this case originates from the repulsion of ions of the same sign, so the background charge dominates near the ion at the center. From the definition of κ_D, it follows that when $n^b \to 0$ or $T \to \infty$ we have $\kappa_D \to 0$, so the screened Coulomb potential goes over to the usual one, $q/(4\pi \varepsilon_0 r)$ in these limits.

A function with the functional form $\exp(-ar)/r$, where a is a constant, will be denoted as a **Yukawa function**. Thus, the electrostatic potential in the plasma decays as a function of distance like a Yukawa function (here with $a = \kappa_D$). As we will see, this is a common behavior for electrostatic interactions in electrolytes.

The pair distribution function is given by Equation 6.5 and by inserting the same linearization (6.9) as before, we obtain

$$g^{(2)}(r) = 1 - \frac{\beta q^2}{4\pi \varepsilon_0} \cdot \frac{e^{-\kappa_D r}}{r}, \tag{6.13}$$

which is valid as an approximation at low ionic density since $|\beta q\psi| \ll 1$ where most of the surrounding ions are located. The charge density around an ion as given by Equation 6.2 then is

$$\rho(r) = -\frac{q\kappa_D^2}{4\pi} \cdot \frac{e^{-\kappa_D r}}{r}, \tag{6.14}$$

where we have used Equation 6.11 to simplify the coefficient. The total charge of this surrounding charge density is $\int d\mathbf{r}\, \rho(r) = -q$, so the sum of the charge q of the ion and the charge of its surrounding charge density is equal to zero. This is called the **"local electroneutrality condition,"** which actually holds in the general case.

{**EXERCISE 6.1**
Show that $\rho(r)$ given by Equation 6.14 satisfies $\int d\mathbf{r}\, \rho(r) = \int_0^\infty dr\, 4\pi r^2 \rho(r) = -q.$}

In this example, we found that the force $\mathbf{F}_1$ on an ion at $\mathbf{r}_1$ is given by Equation 6.6 when an ion is located at $\mathbf{r}_2$ (the central ion). This may seem all right at first glance, but $\mathbf{F}_1$ is actually not the force on an ion when it is located at $\mathbf{r}_1$ because $\mathbf{E}_1$ is not the electrostatic field that this ion feels. $\mathbf{E}_1$ is the electrostatic field from the ion at $\mathbf{r}_2$ and the charge distribution $\rho(r_{32})$, but the latter *is not the average charge distribution around two ions* located at $\mathbf{r}_1$ and $\mathbf{r}_2$, respectively. We have seen that the indirect contribution to the potential of mean force originates from a density distribution like that depicted in Figure 5.3b, but $\rho(r_{32})$ is the charge distribution around a single particle at $\mathbf{r}_2$ when no other particle is fixed, like that illustrated in Figure 5.13. This feature is a consequence of the fact that we in the mean field approximation have *neglected the correlations between the ions* in the surroundings of the ion at $\mathbf{r}_2$.

Despite that the approximation has this fault, it constitutes a classic cornerstone in the theory of electrolyte systems under the names of *Poisson-Boltzmann theory* for diffuse electric double-layers,[2] colloidal dispersions and other electrolyte systems and the *Debye-Hückel theory* of electrolyte solutions. We shall therefore continue to investigate this approach here. The key ingredient in these theories is the **Poisson-Boltzmann approximation** for the potential of mean force

$$w^{(2)}(r_{12}) = q\psi(r_{12}) \quad \text{(PB)} \tag{6.15}$$

as used in Equation 6.5 (the notation (PB) next to an equation means that it is valid in the PB approximation). For ions with hard cores, this approximation is used outside the core region, as we are going to see soon.

{EXERCISE 6.2

a. For the classical one-component plasma of Example 6.1, the internal energy for a system of N particles is $U_N = 3Nk_BT/2 + U_N^{\text{ex}}$, where

$$U_N^{\text{ex}} = \frac{Nn^b}{2} \int_0^\infty u(r)[g_N^{(2)}(r) - 1]4\pi r^2 dr. \tag{6.16}$$

The reason why there is $g_N^{(2)}(r) - 1$ in the formula instead of $g_N^{(2)}(r)$, as in Equation 5.70, is that the interaction with the uniform background $\rho^B = -qn^b$ also contributes to the energy.

Use the pair distribution function in Equation 6.13, which is useful for low ionic densities, to show that

$$U_N^{\text{ex}} = -\frac{Nq^2\kappa_D}{8\pi\varepsilon_0}. \tag{6.17}$$

b. The Helmholtz free energy is $A_N = A_N^{\text{ideal}} + A_N^{\text{ex}}$, where A_N^{ideal} is given in Equation 3.20 with $\eta = 1/\Lambda^3$. The excess free energy $A_N^{\text{ex}}(T)$ at temperature T can be calculated by thermodynamic integration (Equation 5.81)

$$\frac{A_N^{\text{ex}}(T)}{T} = \int_T^\infty dT' \frac{\bar{U}_N^{\text{ex}}(T')}{(T')^2}$$

---

[2] A diffuse electric double-layer is, for example, a charged surface and the distribution of countercharges outside it. The charge distribution of the countercharges (normally from the surrounding ions) is "diffuse," which explains the name.

provided we know how the excess internal energy varies with temperature. Use this relationship and Equation 6.17, where κ_D depends on T, to show that

$$A_N = Nk_BT\ln\left(\frac{\Lambda^3 n^b}{e}\right) - \frac{Nq^2\kappa_D}{12\pi\varepsilon_0}. \tag{6.18}$$

c. By using A_N in Equation 6.18 and taking the V and N derivatives to obtain P and μ (Equations 2.51 and 2.53), show that

$$P = n^b k_B T - \frac{n^b q^2 \kappa_D}{24\pi\varepsilon_0} \tag{6.19}$$

$$\mu = k_B T\ln(\Lambda^3 n^b) - \frac{q^2 \kappa_D}{8\pi\varepsilon_0}. \tag{6.20}$$

d. Verify the result (6.19) for P by using the pair distribution function from Equation 6.13 and the relationship (cf. Equation 5.73)

$$P = n^b k_B T - \frac{(n^b)^2}{6}\int d\mathbf{r}\, r\frac{du(r)}{dr}[g_N^{(2)}(r) - 1],$$

where we have $g_N^{(2)}(r) - 1$ in the formula for the same reason as in Equation 6.16.

e. The Gibbs-Duhem Equation 2.124 can be written $dP = n^b d\mu$ when the temperature is constant. It implies that

$$\left(\frac{\partial P}{\partial n^b}\right)_T = n^b\left(\frac{\partial \mu}{\partial n^b}\right)_T.$$

Verify that P and μ from Equations 6.19 and 6.20 satisfy this relationship.}

The results in Example 6.1 can readily be generalized to ions with size, say, ions with a hard core diameter d^h. Then the pair potential $u(r)$ is given by Equation 5.4. Since we have only one charged species, we must have the uniform background charge density ρ^B present. For ions with hard cores, the Poisson-Boltzmann theory is obtained by applying Equation 5.125 instead of Equation 5.124. We thereby set $u^{rest}(r) = u^{el}(r) \equiv q^2/(4\pi\varepsilon_0 r)$ and take $w^{(2)core}(r) = 0$. In this approximation, the ions that surround the central ion are treated as uncorrelating point-like particles (point ions) in their interactions with each other, despite that they have a finite size and are charged. They only correlate with the central ion. Due to the hard core interactions with the latter, $g^{(2)}$ for the ion distribution that surrounds the central ion is zero inside the core region, $g^{(2)}(r_{12}) = 0$ for $r_{12} < d^h$, since other ions cannot enter there. This is ensured in Equation 5.125 since $u(r_{12}) = \infty$ for $r_{12} < d^h$ in the exponent. The charge density ρ around the central ion is still given by Equation 6.2 and the mean electrostatic potential is given by (compare with Equation 6.4 where $u(r) = u^{el}(r)$)

$$q\psi(r_{12}) = u^{el}(r_{12}) + \int d\mathbf{r}_3\, n^b h^{(2)}(r_{23})u^{el}(r_{31}).$$

According to Equation 5.125 with $w^{(2)\text{core}} = 0$, we thus have

$$g^{(2)}(r_{12}) = \begin{cases} 0, & r_{12} < d^{\text{h}} \\ e^{-\beta q \psi(r_{12})}, & r_{12} \geq d^{\text{h}} \end{cases} \quad \text{(PB)} \quad (6.21)$$

so the PB approximation (6.15) is used outside the core region as we set out to show. In this case, the Poisson-Boltzmann equation takes the form Equation 6.8 for $r \geq d^{\text{h}}$ and $-\epsilon_0 \nabla^2 \psi(r) = \rho^{\text{B}}$ for $0 < r < d^{\text{h}}$. The boundary conditions at $r = 0$ and for $r \to \infty$ are the same as before.

The one-component plasma in the bulk phase is rather artificial. More realistic bulk systems consist of at least two species of ions, anions and cations, like in bulk electrolyte solutions of simple salts. Let us for the time being restrict ourselves to binary electrolytes, which consist of two species of ions only. We then have cations with density n_+^{b} and positive charge q_+ and anions with density n_-^{b} and negative charge q_-. Consider a solution with concentration c_{salt} (in molar units, mol m^{-3}) of a binary salt $C_{l_+} A_{l_-}$, each unit consisting of l_+ cations C^{z+} and l_- anions A^{z-} (z_i is the valency of ions of species i), for example $CaCl_2$ with one ion Ca^{2+} and two ions Cl^- where $l_+ = 1$, $z_+ = +2$, $l_- = 2$, and $z_- = -1$. We then have $n_+^{\text{b}} = l_+ N_{Av} c_{\text{salt}}$ and $n_-^{\text{b}} = l_- N_{Av} c_{\text{salt}}$ and the ionic charges are $q_+ = z_+ q_{\text{e}}$ and $q_- = z_- q_{\text{e}}$, where q_{e} is the elementary charge (the protonic charge).

The average charge density from ions of species i in the bulk solution is equal to $n_i^{\text{b}} q_i$. Since $\sum_i n_i^{\text{b}} q_i = n_+^{\text{b}} q_+ + n_-^{\text{b}} q_- = 0$, the net charge density of the solution is zero (no background charge density is present) and the system is electroneutral. We will specify the valencies of the ions by giving them in the form $z_+ : z_-$; for example, NaCl is denoted as a 1:-1 electrolyte and $CaCl_2$ a 2:-1 electrolyte. The cases of $z : -z$ electrolytes, for instance 1:-1 and 2:-2, are called **symmetric electrolytes**. The anions and cations then have the same absolute valency $z = z_+ = |z_-|$. They may, however, have different sizes, so "symmetric" only refers to the absolute valency.

In a common simple model of salt solutions, the ions are modeled as charged hard spheres (with a point charge at the sphere center) and the solvent is modeled as a dielectric continuum with the dielectric constant ε_r. This is called the **primitive model** of electrolyte solutions. The pair interaction potential is

$$u_{ij}(r_{12}) = u_{ij}^{\text{el}}(r_{12}) + u_{ij}^{\text{core}}(r_{12}), \quad (6.22)$$

where u_{ij}^{el} is the electrostatic and u_{ij}^{core} is the hard core potential, which is infinitely large when two ions overlap and zero otherwise. We shall for the time being assume that all ions have the same hard core diameter d^{h}. The interaction potential $u_{ij}(r)$ between two ions of species i and j is then given by

$$u_{ij}(r) = \begin{cases} \infty, & r < d^{\text{h}} \\ \frac{q_i q_j}{4\pi \varepsilon_r \varepsilon_0 r}, & r \geq d^{\text{h}}. \end{cases} \quad (6.23)$$

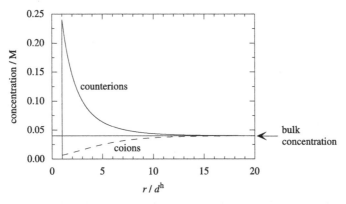

FIGURE 6.2 A sketch of the counterion and coion distributions around an ion in a bulk electrolyte as a function of distance r from the ion center. The figure shows a quite typical example of the distributions for a restricted primitive model electrolyte at low bulk concentration.

The case of *symmetric* electrolytes in the primitive model with the *same size* of anions and cations is called the **restricted primitive model** (RPM). The anions and cations then differ only by the signs of their charges.

The **coions** to a charged particle have charges of the same sign as the particle and the **counterions** have the opposite sign. An example of quite typical counterion and coion distributions around an ion in a bulk electrolyte solution at low concentration is given in Figure 6.2. The counterions are attracted to the central ion and their density close to the ion is therefore higher than in the bulk, while the coions are repelled from the ion and have lower density there than in bulk.

The number density of ions of species j around an ion of species i (the central ion placed at $\mathbf{r}_2$) is $n_j^b g_{ij}^{(2)}(r_{23})$, where $g_{ij}^{(2)}$ is the pair distribution function for these two species. The charge density $\rho_i(r_{32})$ around this ion is (compare with Equation 6.2)

$$\rho_i(r_{23}) = \sum_j q_j n_j^b g_{ij}^{(2)}(r_{23}) = \sum_j q_j n_j^b \left[1 + h_{ij}^{(2)}(r_{23})\right] \tag{6.24}$$

$$= \sum_j q_j n_j^b h_{ij}^{(2)}(r_{23}), \tag{6.25}$$

where subscript i on ρ_i indicates that this is the charge density around an ion of species i and where the sum is over all species in the solution. The last equality follows from the fact that $\sum_i n_i^b q_i = 0$ due to electroneutrality of the bulk phase. For example, the charge density ρ_- around an anion is $\rho_-(r_{23}) = q_+ n_+^b g_{-+}^{(2)}(r_{23}) + q_- n_-^b g_{--}^{(2)}(r_{23})$, where the first term on the rhs is the change density from cations and the second term is that from anions. A positive charge density around the anion is usually due to an attraction of cations (the counterions to the central anion) as well as a repulsion of anions (the coions to the central anion). As illustrated in Figure 6.2, the repulsion of the coions makes their density lower than the bulk density, so in the present case this contributes with a positive charge (a depletion of negative charge gives a positive charge).

The mean electrostatic potential ψ_i at distance r_{12} from the central ion of species i is

$$\psi_i(r_{12}) = \frac{q_i}{4\pi\varepsilon_r\varepsilon_0 r_{12}} + \int d\mathbf{r}_3 \frac{\rho_i(r_{23})}{4\pi\varepsilon_r\varepsilon_0 r_{31}}. \tag{6.26}$$

In the Poisson-Boltzmann approximation, an ion of species j at distance r_{12} from the central ion feels the potential of mean force

$$w_{ij}^{(2)}(r_{12}) = q_j\psi_i(r_{12}) \quad \text{(PB)} \tag{6.27}$$

(outside the core region) and we obtain (compare with Equation 6.21)

$$g_{ij}^{(2)}(r_{12}) = \begin{cases} 0, & r_{12} < d^h \\ e^{-\beta q_j\psi_i(r_{12})}, & r_{12} \geq d^h \end{cases} \quad \text{(PB)}. \tag{6.28}$$

The potential ψ_i in Equation 6.26 and the charge density ρ_i satisfy Poisson's equation $-\varepsilon_r\varepsilon_0\nabla^2\psi_i(r) = \rho_i(r)$ for $r > 0$, and by inserting Equations 6.24 and 6.28 we obtain the (nonlinear) **Poisson-Boltzmann equation** for this case:

$$-\varepsilon_r\varepsilon_0\nabla^2\psi_i(r) = \begin{cases} 0, & 0 < r < d^h \\ \sum_j q_j n_j^b e^{-\beta q_j\psi_i(r_{12})}, & r \geq d^h \end{cases} \tag{6.29}$$

with the boundary conditions that there is a point charge q_i at $r = 0$ and that $\psi_i(r) \to 0$ when $r \to \infty$.

Before continuing, let us point out a deficiency of the nonlinear PB approximation. Since the central ion is treated differently than the other ions, which are treated like point ions, the resulting $g_{ij}^{(2)}(r)$ does not have the correct symmetry in this approximation. In a correct theory, one must have $g_{ij}^{(2)}(r) = g_{ji}^{(2)}(r)$ for all i and j, but this is not the case in general for this approximation. Consider a case with $q_+ \neq |q_-|$. Then $\psi_+(r)/q_+ \neq \psi_-(r)/q_-$ since $\rho_+(r)/q_+ \neq \rho_-(r)/q_-$ due to the nonlinearity in Equations 6.28 and 6.29.[3] Thus, $\psi_+(r)q_- \neq \psi_-(r)q_+$, which implies that $g_{+-}^{(2)}(r) \neq g_{-+}^{(2)}(r)$. This is also the case for any q_+ and q_- (including $q_+ = |q_-|$) when the ions have different sizes.

To determine $\psi_i(r)$, the nonlinear PB equation has to be solved numerically (it does not have any solutions that can be expressed analytically in terms of elementary functions). If the exponent $\beta q_j\psi_i(r_{12})$ is small for the most important values of r_{12}, we can, however, linearize the exponential function as we did in Example 6.1 (we shall see later what "the most important values of r_{12}" mean). We thereby obtain the **linearized PB equation**, also called

---

[3] This can easily be seen by expanding the exponential function in Equation 6.29 in a power series. For $r \geq d^h$, we have $-\varepsilon_r\varepsilon_0\nabla^2\psi_i(r) = -\beta\sum_j q_j^2 n_j^b\psi_i(r) + \beta^2\sum_j q_j^3 n_j^b\psi_i^2(r)/2 + \cdots$, which can be written $-\varepsilon_r\varepsilon_0\nabla^2 f_i(r) = -\beta\sum_j q_j^2 n_j^b f_i(r) + q_i\beta^2\sum_j q_j^3 n_j^b[f_i(r)]^2/2 + \cdots$, where $f_i(r) = \psi_i(r)/q_i$. The presence of the last term makes it impossible for $f_+(r)$ to be equal to $f_-(r)$ when $q_+ \neq |q_-|$. (This term vanishes when $q_+ = |q_-|$.)

the **Debye-Hückel (DH) equation,**

$$\nabla^2 \psi_i(r) = \begin{cases} 0, & 0 < r < d^h \\ \kappa_D^2 \psi_i(r), & r \geq d^h \end{cases}, \tag{6.30}$$

where

$$\kappa_D^2 = \frac{\beta}{\varepsilon_r \varepsilon_0} \sum_j q_j^2 n_j^b. \tag{6.31}$$

As we will see, the linearized approximation gives pair distributions with the correct symmetry $g_{ij} = g_{ji}$.

The entity κ_D will be denoted as the **Debye parameter** and has the dimension (length)$^{-1}$. Its inverse $1/\kappa_D$ is called the **Debye length**. By inserting $q_i = q_e z_i$, we can alternatively write

$$\kappa_D^2 = \frac{\beta q_e^2}{\varepsilon_r \varepsilon_0} \sum_j z_j^2 n_j^b = \frac{2\beta q_e^2 N_{Av}}{\varepsilon_r \varepsilon_0} I, \tag{6.32}$$

where quantity $I = \sum_j z_j^2 c_j^b / 2$ is called the **ionic strength** of the electrolyte solution and $c_j^b = n_j^b / N_{Av}$ is the concentration of species j in molar units (mol m^{-3}). For a 1:-1 electrolyte, we have $I = c_{salt}$, for a 2:-1 electrolyte $I = 3c_{salt}$ and for a 2:-2 electrolyte, for instance MgSO$_4$, $I = 4c_{salt}$. We can see that κ_D goes to zero when the concentration goes to zero proportionally to $\sqrt{I}$ and hence also proportionally to $\sqrt{c_{salt}}$. The linearized PB equation is the cornerstone in the **Debye-Hückel theory** of electrolyte solutions, which we will treat in some detail.

Equation 6.30 has analytic solutions and in the same fashion as in Example 6.1 we find that for $r \geq d^h$ we have $r\psi_i(r) = C_1 \exp(-\kappa_D r) + C_2 \exp(\kappa_D r)$, where C_2 must be zero due to the boundary condition at $r \to \infty$. For $0 < r < d^h$, we have the solution $\psi_i(r) = q_i/(4\pi\varepsilon_r\varepsilon_0 r) + C_3$, where the first term is the potential from q_i at the center and C_3 is the constant potential from the spherically symmetric charge distribution outside the core. From continuity of $\psi_i(r)$ at $r = d^h$, it follows that $C_3 = \psi_i(d^h) - q_i/(4\pi\varepsilon_r\varepsilon_0 d^h)$. The constant C_1 can be determined from the boundary conditions too, but instead we are going to determine it in a different manner.

We first note that Poisson's equation $\rho_i(r) = -\varepsilon_r\varepsilon_0 r^{-1}(\partial^2/\partial r^2)[r\psi_i(r)]$ with $r\psi_i(r) = C_1 \exp(-\kappa_D r)$ inserted implies that $\rho_i(r) = -C_1\varepsilon_r\varepsilon_0\kappa_D^2 \exp(-\kappa_D r)/r$ for $r \geq d^h$. The system must be electroneutral. If we have an ion of charge q_i at the origin, the total charge in the distribution $\rho_i(r)$ must, according to the local electroneutrality condition, therefore be equal $-q_i$, that is,

$$-q_i = \int_{d^h}^{\infty} dr\, 4\pi r^2 \rho_i(r) = -C_1\varepsilon_r\varepsilon_0\kappa_D^2 \int_{d^h}^{\infty} dr\, 4\pi r e^{-\kappa_D r}$$

$$= -C_1\varepsilon_r\varepsilon_0 4\pi (1 + \kappa_D d^h) e^{-\kappa_D d^h}$$

and we obtain $C_1 = q_i \exp(\kappa_D d^h)/[4\pi \varepsilon_r \varepsilon_0 (1 + \kappa_D d^h)]$. Accordingly, we have

$$\psi_i(r) = \frac{q_i}{4\pi \varepsilon_r \varepsilon_0 (1 + \kappa_D d^h)} \cdot \frac{e^{-\kappa_D(r-d^h)}}{r} \quad \text{for } r \geq d^h \quad \text{(DH theory)} \tag{6.33}$$

and

$$\rho_i(r) = -\frac{q_i \kappa_D^2}{4\pi (1 + \kappa_D d^h)} \cdot \frac{e^{-\kappa_D(r-d^h)}}{r} \quad \text{for } r \geq d^h \quad \text{(DH theory).} \tag{6.34}$$

Note that in the linearized theory we have $\psi_+(r)q_- = \psi_-(r)q_+$, so $w_{ij}^{(2)}(r) = w_{ji}^{(2)}(r)$ and hence $g_{ij}^{(2)}(r) = g_{ji}^{(2)}(r)$ for all i and j. This symmetry is, however, fortuitous, as will be explained in Section 6.1.2. It exists *despite* that we have treated the central ion differently from the other ions.

The charge density $\rho_i(r)$ is often called the charge density of the "ion atmosphere" or "ion cloud" around an ion in the electrolyte solution. This function is proportional to a Yukawa function $f_1(r) = e^{-\kappa r}/r$ with $\kappa = \kappa_D$ and in Figure 6.3 we have plotted $f_1(r)/\kappa$ as a function of κr (the dashed curve). The charge density decays quickly to zero with increasing r. The potential $\psi_i(r)$ for $r \geq d^h$ decays in the same manner with r.

To see how the charge in the ion atmosphere is distributed around an ion, let us investigate the amount of charge in a spherical shell of radius r, that is, $4\pi r^2 \rho_i(r)dr$, where dr is the shell thickness. This quantity is proportional to $f_2(r) = re^{-\kappa r}$ with $\kappa = \kappa_D$. In Figure 6.3, we have plotted $\kappa f_2(r)$ as a function of κr (the full curve). The maximum is located at

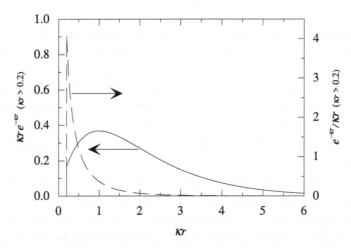

FIGURE 6.3 The Debye-Hückel prediction of the charge density $\rho_i(r)$ surrounding an ion of species i is proportional to the function shown by the dashed curve (when $\kappa = \kappa_D$). The figure shows the plot for a case with $d^h = 0.2/\kappa_D$, so the density is zero for $\kappa r < 0.2$ in the figure, which corresponds to $r < d^h$. The mean electrostatic potential $\psi_i(r)$ for $r \geq d^h$ is also proportional to this function. The full curve shows a function that is proportional to the relative net amount of charge between r and $r + dr$ in the ion atmosphere (when $\kappa = \kappa_D$). The ordinate scale for the dashed curve is to the right in the plot and that for the full curve is to the left.

$\kappa r = 1$, that is, at $r = 1/\kappa$. The "most important distances," where most charges of $\rho_i(r)$ are found, are obviously located around $1/\kappa_D$, that is, the Debye length. When the electrolyte solution is diluted, $1/\kappa_D$ becomes larger, that is, the most important region moves to larger r.

To give a feeling for the order of magnitude of the Debye length, we can insert numbers in Equation 6.32, whereby we obtain for an aqueous electrolyte solution at $25°C$

$$\frac{1}{\kappa_D} = \frac{3.04}{\sqrt{I}} \quad \text{(in Å)}, \tag{6.35}$$

where the ionic strength I is measured in M (mol dm^{-3}). This means, for example, that the Debye length for a 0.01 M NaCl solution is 30.4 Å.

Let us now determine the chemical potential for a bulk electrolyte solution as predicted by the Debye-Hückel theory. In the canonical ensemble, we derived Equation 5.84 for μ in a single-component system. We obtained this equation from a calculation of the reversible work done to gradually insert a new particle in the system. The gradual insertion was described by a coupling parameter ξ that varied between 0 and 1. When we have a two-component electrolyte, the work to insert a particle of species i against interactions with the other particles will involve both $g_{i+}(r; \xi)$ and $g_{i-}(r; \xi)$; otherwise, the derivation is identical. This also applies when we are not using the canonical ensemble. Thus, we can readily generalize Equation 5.84 to the present case and obtain

$$\mu_i = k_B T \ln\left(\Lambda_i^3 n_i^b\right) + \sum_j \int_0^1 d\xi \int d\mathbf{r} \frac{\partial u_{ij}(r; \xi)}{\partial \xi} n_j^b g_{ij}(r; \xi), \tag{6.36}$$

where Λ_i is the thermal de Broglie wavelength for species i and

$$u_{ij}(r; \xi) = \begin{cases} 0 & \text{when } \xi = 0 \\ u_{ij}(r) & \text{when } \xi = 1. \end{cases}$$

The last term on the rhs in Equation 6.36 is the excess chemical potential for species i, $\mu_i^{ex} = k_B T \ln \gamma_i$, where γ_i is the activity coefficient for this species.

When adding an ion of diameter d^h, one first has to make a "hole" in the solution where one can place the ion. This hole must have a radius equal to d^h, since this is the distance of closest approach of two particles of diameter d^h. The work done when pushing away the ions in the solution when making the hole is easy to calculate in the current mean field approximation. Since we neglect the three-particle correlations in this approximation (that is, the correlations between the ions around the hole are neglected), these ions behave like an ideal gas when we create the hole. Therefore, the work done = (the volume of the hole) × (the ideal pressure of the solution), which means that this contribution to the excess chemical potential is equal to $[4\pi (d^h)^3/3] \times k_B T \sum_j n_j^b$. We denote this contribution as the hard core part of μ_i^{ex} and write

$$\mu_i^{\text{ex,core}} = \frac{4\pi (d^{\text{h}})^3 k_B T n_{\text{tot}}^{\text{b}}}{3}, \tag{6.37}$$

where $n_{\text{tot}}^{\text{b}} = \sum_j n_j^{\text{b}}$ is the total number density of ions.

The remaining part of μ_i^{ex} is the work done when charging up the added ion from zero charge to q_i. We do this by setting the charge of this ion equal to $\xi' q_i$ (where $0 \le \xi' \le 1$), that is, the interaction between this ions and the surrounding ions outside the hole equals $\xi' u_{ij}^{\text{el}}(r)$, where

$$u_{ij}^{\text{el}}(r) = \frac{q_i q_j}{4\pi \varepsilon_r \varepsilon_0 r}.$$

Then we calculate the work done by performing an integral like in Equation 6.36. Since $\partial[\xi' u_{ij}^{\text{el}}(r)]/\partial \xi' = u_{ij}^{\text{el}}(r)$, we obtain the electrostatic contribution to μ_i^{ex}

$$\mu_i^{\text{ex,el}} = \sum_j \int_0^1 d\xi' \int_{d^{\text{h}}}^\infty dr \, 4\pi r^2 u_{ij}^{\text{el}}(r) n_j^{\text{b}} g_{ij}(r; \xi')$$

$$= \int_0^1 d\xi' \int_{d^{\text{h}}}^\infty dr \, 4\pi r^2 \frac{q_i \sum_j q_j n_j^{\text{b}} g_{ij}(r; \xi')}{4\pi \varepsilon_r \varepsilon_0 r}.$$

We have $\sum_j q_j n_j^{\text{b}} g_{ij}(r; \xi') = \rho_i(r; \xi') =$ the charge density around the ion with charge $\xi' q_i$ (compare with Equation 6.24) and we can obtain $\rho_i(r; \xi')$ by simply inserting $\xi' q_i$ instead of q_i in Equation 6.34. Hence, the expression for $\mu_i^{\text{ex,el}}$ can be written

$$\mu_i^{\text{ex,el}} = -\frac{q_i^2 \kappa_D^2}{4\pi \varepsilon_r \varepsilon_0 (1 + \kappa_D d^{\text{h}})} \int_0^1 \xi' d\xi' \int_{d^{\text{h}}}^\infty dr \, e^{-\kappa_D(r - d^{\text{h}})}$$

$$= -\frac{q_i^2 \kappa_D}{8\pi \varepsilon_r \varepsilon_0 (1 + \kappa_D d^{\text{h}})}.$$

By adding $\mu_i^{\text{ex,core}}$, we finally obtain

$$\mu_i^{\text{ex}} = k_B T \ln \gamma_i$$

$$= \frac{4\pi (d^{\text{h}})^3 k_B T n_{\text{tot}}^{\text{b}}}{3} - \frac{q_i^2 \kappa_D}{8\pi \varepsilon_r \varepsilon_0 (1 + \kappa_D d^{\text{h}})} \quad \text{(DH theory)}. \tag{6.38}$$

At high dilution, when $n_{\text{tot}}^{\text{b}} \to 0$ and hence $\kappa_D \to 0$ proportionally to $\sqrt{n_{\text{tot}}^{\text{b}}}$, we have $\kappa_D d^{\text{h}} \ll 1$, $n_{\text{tot}}^{\text{b}} \ll \sqrt{n_{\text{tot}}^{\text{b}}}$ and hence

$$\mu_i^{\text{ex}} = k_B T \ln \gamma_i \sim -\frac{q_i^2 \kappa_D}{8\pi \varepsilon_r \varepsilon_0} \quad \text{when } n_{\text{tot}}^{\text{b}} \to 0. \tag{6.39}$$

The latter relationship is called the **Debye-Hückel limiting law for the activity coefficient** and we will see later that it is an *exact result* in the sense that μ_i^{ex} for electrolyte solutions always approaches this expression at very high dilution. It follows that μ_i^{ex} goes to zero like $\sqrt{n_{tot}^b}$ when the concentration goes to zero.

Instead of dealing with individual ion activity factors, it is customary to introduce an average chemical potential, which is the change in free energy counted per added ion when one adds an electroneutral salt. An important reason for this is the huge difficulties to determine individual ion activity factors experimentally. For example, if one adds only one kind of ion at a time to a solution, a very large voltage difference is built up quickly between the solution and the surroundings.

If we add one unit of salt $C_{l_+}A_{l_-}$ (l_+ cations and l_- anions) into the electrolyte solution, the free energy changes by the amount $l_+\mu_+ + l_-\mu_-$. We define the average chemical potential $\mu_\pm$ such that the change in free energy when adding the salt equals $[l_+ + l_-]\mu_\pm$, that is,

$$[l_+ + l_-]\mu_\pm = l_+\mu_+ + l_-\mu_-.$$

The excess free energy changes by $l_+\mu_+^{ex} + l_-\mu_-^{ex}$ and we likewise define $\mu_\pm^{ex}$ from

$$[l_+ + l_-]\mu_\pm^{ex} = l_+\mu_+^{ex} + l_-\mu_-^{ex}. \tag{6.40}$$

It is relatively straightforward to experimentally determine $\mu_\pm^{ex}$, but this does not apply to the individual μ_+^{ex} and μ_-^{ex}. The former is obtained in experiments where one adds salt, which is electroneutral.

For a solution containing N units of salt, we have $N_i = l_i N$ and hence

$$\mu_\pm = \frac{N[l_+\mu_+ + l_-\mu_-]}{N[l_+ + l_-]} = x_+\mu_+ + x_-\mu_-,$$

where $x_i = N_i/(N_+ + N_-) = n_i/(n_+ + n_-)$. An analogous expression applies to $\mu_\pm^{ex}$.

In analogy to $\mu_i^{ex} = k_B T \ln \gamma_i$, we define the average activity factor $\gamma_\pm$ from $\mu_\pm^{ex} = k_B T \ln \gamma_\pm$ and it follows from Equation 6.40 that

$$k_B T[l_+ + l_-] \ln \gamma_\pm = k_B T[l_+ \ln \gamma_+ + l_- \ln \gamma_-]$$

which implies that

$$\ln \gamma_\pm = \frac{l_+ \ln \gamma_+ + l_- \ln \gamma_-}{l_+ + l_-}$$

(a weighted arithmetic average) and hence

$$\gamma_\pm^{l_+ + l_-} = \gamma_+^{l_+} \gamma_-^{l_-}$$

(a "geometric average" for $\gamma_\pm$).

By inserting the Debye-Hückel result (6.38) for $\ln \gamma_i$ in the expression for $\ln \gamma_\pm$, we obtain

$$k_B T \ln \gamma_\pm = \frac{4\pi (d^h)^3 k_B T n_{tot}^b}{3} - \frac{\kappa_D}{8\pi \varepsilon_r \varepsilon_0 (1 + \kappa_D d^h)} \cdot \frac{l_+ q_+^2 + l_- q_-^2}{l_+ + l_-}.$$

The last factor can be simplified. Electroneutrality implies $l_+ q_+ + l_- q_- = 0$, and hence we have $l_+ q_+ = -l_- q_-$. This means that $l_+ q_+ q_+ + l_- q_- q_- = -(l_- q_-)q_+ - (l_+ q_+)q_- = -(l_- + l_+)q_+ q_-$. Since q_+ and q_- have opposite signs, we have $-q_+ q_- = |q_+ q_-|$. Thus, we can write our result for $\ln \gamma_\pm$ as

$$\ln \gamma_\pm = \frac{4\pi (d^h)^3 n_{tot}^b}{3} - \frac{\kappa_D |q_+ q_-|}{8\pi \varepsilon_r \varepsilon_0 k_B T (1 + \kappa_D d^h)} \quad \text{(DH theory).} \quad (6.41)$$

In the literature, it is common to forget the core contribution (the first term) in the Debye-Hückel chemical potential, but, as we shall see, this term is quite important. The limiting law takes the form

$$\ln \gamma_\pm \sim -\frac{\kappa_D |q_+ q_-|}{8\pi \varepsilon_r \varepsilon_0 k_B T} \quad \text{when } n_{tot}^b \to 0.$$

In Figure 6.4, $\ln \gamma_\pm$ for a 1:-1 electrolyte in aqueous solution calculated in the Debye-Hückel approximation is compared with the corresponding primitive model results from Monte Carlo simulations. The latter are very accurate results for an electrolyte solution where the ions are charged hard spheres of diameter 4.25 Å and the solvent is a continuum with dielectric constant $\varepsilon_r = 78.7$. We see that the Debye-Hückel approximation works quite

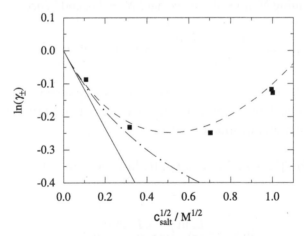

FIGURE 6.4 The logarithm of the activity coefficient as a function of $\sqrt{c_{salt}}$, where c_{salt} is the concentration in M. The Debye-Hückel (DH) prediction (curves) is compared to the result from Monte Carlo simulations[4] (symbols) for a (1:-1) salt in aqueous solution. The full curve is the DH limiting law, the dashed-dotted curve is the DH electrostatic contribution (i.e., without the core contribution) and the dashed curve is the DH prediction of the total $\ln \gamma_\pm$. The ion diameter is $d^h = 4.25$ Å.

---

[4] The Monte Carlo data in Figure 6.4 are taken from J. P. Valleau and L. K. Cohen, *J. Chem. Phys.* **72** (1980) 5935.

well for the activity coefficient for this model 1:-1 electrolyte. Obviously, it is important to include the core contribution to achieve good results for higher concentrations. For divalent electrolytes, however, the Debye-Hückel approximation is insufficient and more refined theories must be used.

From the limiting law, it follows that both μ_i^{ex} and $\mu_{\pm}^{\text{ex}}$ go to zero like $\sqrt{n_{\text{tot}}^{\text{b}}}$ when the concentration goes to zero. From Equation 6.41, we can obtain an expansion of $\mu_{\pm}^{\text{ex}}$ in powers of $n_{\text{tot}}^{\text{b}}$. The first few terms can be written as

$$\frac{\mu_{\pm}^{\text{ex}}}{k_B T} = \ln \gamma_{\pm} = b_{\frac{1}{2}} \sqrt{n_{\text{tot}}^{\text{b}}} + b_1 n_{\text{tot}}^{\text{b}} + b_{\frac{3}{2}} \left(\sqrt{n_{\text{tot}}^{\text{b}}}\right)^3 + \cdots,$$

where the coefficients b_{α} can readily be obtained from the results earlier in this section. The appearance of the leading square root term implies that the nonideal contributions remain important even in very dilute solutions; $\sqrt{n_{\text{tot}}^{\text{b}}}$ decays much slower to zero than $n_{\text{tot}}^{\text{b}}$. This is in contrast to solutions of electroneutral solutes, where the nonideal contributions go to zero like the density, that is, in the same manner as for real gases as we have seen in Equation 3.58. A virial expansion like Equation 3.58 does not exist for ionic fluids. As we saw in Section 3.4, the virial coefficient expressions converge for monotonically decaying pair potentials provided that $u(r)$ decays faster than r^{-3} when $r \to \infty$. This is not the case for ionic fluids where the pair potential decays like $1/r$. It is the very long range of the Coulomb interaction that sets ionic fluids apart from other fluids.

Historically, the Debye-Hückel theory for electrolyte solutions, which was published in 1923 by Peter Debye and Erich Hückel, had a huge impact, partly because it explained the experimental observation that the nonideal contributions remain important at high dilution. This theory has had a major role in the physics and chemistry of electrolytes for a very long time. Debye and Hückel were, however, not the first scientists to use the Poisson-Boltzmann equation for electrolytes. It was used earlier by Louis George Gouy in 1910 and David L. Chapman in 1913 to theoretically describe the behavior of electrolyte solutions near charged surfaces, so-called diffuse electric double-layer systems, in planar geometry. In Section 7.1, we will turn to the PB treatment of such systems, often called the *Gouy-Chapman theory*, which long has been a cornerstone in Surface and Colloid Science and other related fields of science. This theory is closely connected to the PB treatment of interactions between macroions, that is, large charged particles of various sizes and shapes. Such systems are treated in Sections 6.1.3 and 6.1.4, where we investigate interactions between particles immersed in electrolytes.

The **osmotic pressure**, that is, the pressure of a solution in excess of the pressure of the pure solvent (here a dielectric continuum), is in the primitive model of electrolytes given by

$$P^{\text{osm}} = k_B T \sum_i n_i^{\text{b}} - \sum_{i,j} \frac{n_i^{\text{b}} n_j^{\text{b}}}{6} \int d\mathbf{r}\, r \frac{du_{ij}(r)}{dr} g_{ij}^{(2)}(r), \tag{6.42}$$

which is the generalization of Equation 5.73 to many-component systems. (A common alternative notation for P^{osm} is Π.) In the limit of zero concentration, where the second term in the rhs is negligible, we have $P^{osm} = k_B T n^b_{tot}$, which is called **van't Hoff's law** and corresponds to the ideal gas law. At finite concentrations, there are deviations from van's Hoff's law and we have instead

$$P^{osm} = \phi^{osm} k_B T n^b_{tot}, \tag{6.43}$$

where the entity ϕ^{osm} is called the **osmotic coefficient**. The deviation of this coefficient from one describes the nonideality, that is, the influence of the last term in Equation 6.42. The following exercise deals with the Debye-Hückel limiting law for the osmotic coefficient and that for the average activity coefficient in the general case (i.e., not only in binary electrolytes).

{**EXERCISE 6.3**

a. The average activity coefficient can in the general case be defined from

$$\left(\sum_i l_i\right) \ln \gamma_\pm = \sum_i l_i \ln \gamma_i \tag{6.44}$$

or, equivalently, $\left(\sum_i n^b_i\right) \ln \gamma_\pm = \sum_i n^b_i \ln \gamma_i$. Use Equation 6.39 to show that the Debye-Hückel limiting law for the average activity coefficient can be written

$$\ln \gamma_\pm = -\frac{\kappa^3_D}{8\pi n^b_{tot}}, \tag{6.45}$$

which is valid at very low concentrations.

b. For an electrolyte solution with very low concentration (so $\kappa_D d^h \ll 1$), we have from Equation 6.34

$$\rho_i(r) = \sum_j q_j n^b_j g^{(2)}_{ij}(r) = -\frac{q_i \kappa^2_D}{4\pi} \cdot \frac{e^{-\kappa_D r}}{r}, \tag{6.46}$$

where we can neglect the size of the ions since the concentration is very low and therefore can apply the equation for all r. In Equation 6.42, we can for the same reason set $u_{ij}(r)$ equal to the electrostatic pair potential $u^{el}_{ij}(r)$. From Equations 6.42 and 6.46, derive

$$\phi^{osm} = 1 - \frac{\kappa^3_D}{24\pi n^b_{tot}}, \tag{6.47}$$

which is valid at very low concentrations and is called the **Debye-Hückel limiting law for the osmotic coefficient**.

c. When T is constant, the Gibbs-Duhem Equation 2.127 can be written $dP = \sum_i n^b_i d\mu_i$, where we have $P = P^{osm}$ in the current case. As before, the chemical potential can be written as $\mu_i = k_B T \ln(\Lambda^3_i n^b_i) + k_B T \ln \gamma_i$. Use these expressions and Equations 6.43

and 6.44 to show the following general relationship between the average activity coefficient and the osmotic coefficient for an electrolyte solution

$$\left(\frac{\partial[n_{\text{tot}}^{\text{b}}(\phi^{\text{osm}} - 1)]}{\partial n_{\text{tot}}^{\text{b}}}\right)_T = n_{\text{tot}}^{\text{b}}\left(\frac{\partial \ln \gamma_\pm}{\partial n_{\text{tot}}^{\text{b}}}\right)_T. \tag{6.48}$$

Note that you shall show Equation 6.48 *for the general case*, so you may not use the Debye-Hückel results (6.45) and (6.47) in the derivation.

d. Equation 6.48 implies that the average activity coefficient at some concentration given by, say, $n_{\text{tot}}^{\text{b*}}$, can be determined from the osmotic coefficient by integration with respect to $n_{\text{tot}}^{\text{b}}$ from zero to $n_{\text{tot}}^{\text{b*}}$. Discuss how this could be done.

e. Verify that $\ln \gamma_\pm$ and ϕ^{osm} from the Debye-Hückel limiting laws (6.45) and (6.47), respectively, satisfy relationship (6.48).}

6.1.2 Decay of Electrostatic Potential and Effective Charges of Particles

6.1.2.1 The Concept of Effective Charge

A charged particle immersed in a bulk electrolyte is, as we have seen, surrounded by an ion cloud with a net charge opposite to that of the particle. The charge density of this cloud originates from deviations in the ion densities around the particle from their bulk values. When analyzing the decay of the electrostatic potential from a particle and its cloud, like $\psi_i(r)$ in Equation 6.33, it is often conceptually useful to utilize the notion of an **effective charge** of the particle.

Let us consider the magnitude of the potential in the tail region far away from a particle placed at the origin (in general for large distances r). A small value of the potential at a certain r can arise for different reasons, for example that the particle has a small charge (i.e., its bare charge is small) or that it has attracted a large amount of countercharge near its surface which nearly neutralizes the particle (the net charge of the particle, including the charge in its immediate neighborhood, can thereby be small even if the bare charge is large). In both cases, the potential far from the particle is small, and in the latter case one can say that "the effective charge" is small.

Likewise, a particle can have a large effective charge because it attracts coions and/or repels counterions. This increases the net charge of the same sign as the particle in the immediate neighborhood and makes the potential large. The potential can be large also for other reasons. Loosely speaking, the effective charge is the charge that the particle "appears to have" when seen from a distance (that is, when one does not know any details about the particle and the ion cloud distribution in its immediate neighborhood). Before we make a general treatment of the issue of effective charges in the PB approximation, we will give an example of how one can relate an effective charge to the magnitude of the potential.

The electrostatic potential $\psi_i(r)$ from an ion of diameter d^{h} is in the DH theory given by (Equation 6.33)

$$\psi_i(r) = \frac{q_i}{4\pi \varepsilon_r \varepsilon_0 (1 + \kappa_{\text{D}} d^{\text{h}})} \cdot \frac{e^{-\kappa_{\text{D}}(r - d^{\text{h}})}}{r}$$

outside the core region. If we let the diameter go to zero, we obtain

$$\psi_i(r) = \frac{q_i}{4\pi \varepsilon_r \varepsilon_0} \cdot \frac{e^{-\kappa_D r}}{r},$$

which differs from the previous $\psi_i(r)$ by a factor $\exp[\kappa_D d^{h}]/(1 + \kappa_D d^{h})$. The potential from an ion with diameter d^{h} is therefore the same as the potential from a point ion (diameter zero) with a bare charge that is larger by this factor. The ion with $d^{h} \neq 0$ can therefore be assigned an effective charge q_i^{eff} equal to

$$q_i^{\text{eff}} = q_i \cdot \frac{e^{\kappa_D d^{h}}}{1 + \kappa_D d^{h}} \quad \text{(DH theory)} \tag{6.49}$$

and we have

$$\psi_i(r) = \frac{q_i^{\text{eff}}}{4\pi \varepsilon_r \varepsilon_0} \cdot \frac{e^{-\kappa_D r}}{r} \quad \text{(DH theory)} \tag{6.50}$$

for large r (outside the core region). An ion with effective charge q_i^{eff} accordingly has the same potential for large r as a point ion with a bare charge that is numerically equal to q_i^{eff}, so they cannot be distinguished solely by the magnitude of the potential there.

Note that there is an amount of freedom in the definition of the value of an effective charge: this charge is defined relative to a reference case where the charge is taken equal to the bare charge: here the point ion case. A different choice of reference gives a different value of the effective charge. With the current choice, q_i^{eff} for an ion in the DH theory goes to the bare charge q_i, the limit of zero ionic diameter.

As shown in the following shaded text box, these matters illustrate the fortuitous symmetry $w_{ij}^{(2)}(r) = w_{ji}^{(2)}(r)$ in the DH theory (linearized PB approximation) for ions of equal sizes, which we started to discuss earlier in connection to Equation 6.33. As we will see later in Section 6.1.5, there are important lessons to be learned from this.

The condition $w_{ij}^{(2)} = w_{ji}^{(2)}$, which is required by an exact theory, is, as we have seen, not fulfilled in general in the PB approximation, but it happens to be fulfilled in the DH theory. Let us see how and why. In the PB relationship, $w_{ij}^{(2)}(r) = q_j \psi_i(r)$ in Equation 6.27, the central ion of species i has a diameter d^{h}, while we have treated the ion of species j as a point ion. By inserting Equation 6.50, we see that

$$w_{ij}^{(2)}(r) = \frac{q_j q_i^{\text{eff}}}{4\pi \varepsilon_r \varepsilon_0} \cdot \frac{e^{-\kappa_D r}}{r} \quad \text{(DH theory)} \tag{6.51}$$

for large r (outside the core region in this case). Here it is clear that the j ion and the central i ion are treated in an unsymmetrical manner. The apparent symmetry for

$w_{ij}^{(2)}$ is due to the fact that q_i^{eff} is proportional to q_i in the linearized PB approximation (see Equation 6.49). In a truly symmetrical $w_{ij}^{(2)}$, we should have $q_j^{\text{eff}} q_i^{\text{eff}}$ instead
of $q_j q_i^{\text{eff}}$ in the numerator, that is, the i ion and the j ion should appear in the same
manner.

In fact, if one solves the DH equation $\nabla^2 \psi(r) = \kappa_D^2 \psi(r)$ for two ions with nonzero
diameters separated by distance r, one ion of species i and one of species j, the potential
of mean force between these two ions satisfies

$$w_{ij}^{(2)}(r) \sim \frac{q_j^{\text{eff}} q_i^{\text{eff}}}{4\pi \varepsilon_r \varepsilon_0} \cdot \frac{e^{-\kappa_D r}}{r} \quad \text{when } r \to \infty \quad \text{(DH, two ions)}, \qquad (6.52)$$

where both q_i^{eff} and q_j^{eff} are given by the expression (6.49).[5] The two ions have thereby
been treated in an equal manner, which leads to a truly symmetric $w_{ij}^{(2)}$. However, all
other ions in the electrolyte solutions (those that surround these two ions) are still
treated like point ions that do not correlate with each other. Only the selected two ions
are treated in the same manner; all other ions (of the same and opposite species as these
two) are treated in a different manner. Thus, the approximation used is still quite rough.

A relationship like Equation 6.52, which says that the function $w_{ij}^{(2)}(r)$ decays like the
rhs for large r, is called an **asymptotic expression**. The symbol $\sim$ means "approaches
asymptotically." The statement $f_1(r) \sim f_2(r)$ when $r \to \infty$ means that $f_1(r)/f_2(r) \to 1$ in
this limit.

Let us consider the issue of effective charges treated in the (nonlinear) PB approximation.
We will see that the result in Equation 6.50 holds for large r,

$$\psi_i(r) \sim \frac{q_i^{\text{eff}}}{4\pi \varepsilon_r \varepsilon_0} \cdot \frac{e^{-\kappa_D r}}{r} \quad \text{when } r \to \infty \quad \text{(PB)}, \qquad (6.53)$$

in the nonlinear case, but with a different value of the effective charge q_i^{eff}. Thus, the potential still decays like a Yukawa function, but with a different coefficient than in the DH theory.
Note that we are only considering the behavior for large r here; the functional dependence
of r is different for small r values in the nonlinear case. As regards the value of q_i^{eff}, we
will find that it depends on the charge distribution close to the ion as anticipated in the
introduction of the current section. We will also find that it is not the full charge distribution $\rho_i(r)$ of the entire ion cloud that determines q_i^{eff}, but rather *a part of* $\rho_i(r)$ close to
the ion. These results are shown in the following section, which treats the potential from
nonspherical particles. Thereby the spherical case referred to here constitutes a special case
(Example 6.3 below).

---

[5] This fact has been shown by H. Ohshima (*J. Colloid Interface Sci.*, **170** (1995) 432). It can also be shown by superposition of
the electrostatic potentials from two separate ions at large separations as given by Equation 6.33, see G. M. Bell, S. Levine,
and L. N. McCartney, *J. Colloid Interface Sci.*, **33** (1970) 335.

6.1.2.2 Electrostatic Potential from Nonspherical Particles

We now turn to the case with a nonspherical particle P immersed in a bulk electrolyte solution. The particle can be of *any shape and have an arbitrary internal charge distribution* $\sigma_P(\mathbf{s})$ *in its interior*, where the coordinate vector $\mathbf{s}$ is counted from the center of mass of P. The ions in the surrounding electrolyte are assumed to be charged hard spheres. The particle P takes the role of the "central ion" discussed earlier and it has a fixed orientation relative to the frame of reference.[6] We place the origin at the center of the particle. Of course, P can be an ion of species i as in the previous cases and then the internal charge is a point charge at the center of mass. Thus, our treatment includes the cases described earlier. The techniques we are going to use here *can quite easily be generalized to more refined theories than the PB approximation*, so the treatment here has general interest (the generalization will be done in Section 6.2).

The particle P interacts via a pair potential $u_{Pj}(\mathbf{r})$ with the ions of species j in the surrounding solution. This potential consists of an electrostatic interaction $u_{Pj}^{el}(\mathbf{r})$ and a nonspherical hard core potential $u_{Pj}^{core}(\mathbf{r})$, which is infinitely large when the j-ion at $\mathbf{r}$ overlaps with the core region of P and zero otherwise (compare with Equation 6.22). The interactions make ion densities of the various species density around the particle to deviate from the bulk, giving rise to an ion cloud of density $\rho_P(\mathbf{r})$ in the neighborhood. This charge density can be described as the **polarization response** of the electrolyte due to the interactions with P. The total charge density associated with the particle – from the internal charges and the ion cloud – is

$$\rho_P^{tot}(\mathbf{r}) = \sigma_P(\mathbf{r}) + \rho_P(\mathbf{r}).$$

Note that the density $\rho_P(\mathbf{r})$ is zero inside the particle, while $\sigma_P(\mathbf{r})$ is zero outside it. The mean electrostatic potential ψ_P due to the particle is obtained from Coulomb's law as

$$\psi_P(\mathbf{r}_1) = \int d\mathbf{r}_2 \frac{\rho_P^{tot}(\mathbf{r}_2)}{4\pi \varepsilon_r \varepsilon_0 r_{21}} \tag{6.54}$$

and Poisson's equation is

$$-\varepsilon_r \varepsilon_0 \nabla^2 \psi_P(\mathbf{r}) = \rho_P^{tot}(\mathbf{r}) \tag{6.55}$$

for all $\mathbf{r}$. The boundary condition is $\psi_P(\mathbf{r}) \to 0$ infinitely far away from the particle. Equation 6.54 constitutes the solution to Poisson's equation with this boundary condition.

To start with, we assume, as before, that the ions in the surrounding electrolyte have the same diameter d^h. In the PB approximation, the potential of mean force between P and an ion of species j is equal to $q_j \psi_P(\mathbf{r})$ outside the particle, so we have

$$\rho_P(\mathbf{r}) = \begin{cases} 0, & \text{inside P} \\ \sum_j q_j n_j^b e^{-\beta q_j \psi_P(\mathbf{r})}, & \text{outside P} \end{cases} \quad \text{(PB)} \tag{6.56}$$

---

[6] The orientation of a particle can be specified by the orientational variable ω introduced in Appendix 6A. For simplicity in notation, we will not write this variable explicitly here. This will, however, be done later in Section 6.1.3 where interactions between two nonspherical particle are treated.

and the PB equation is

$$-\varepsilon_r \varepsilon_0 \nabla^2 \psi_P(\mathbf{r}) = \rho_P^{tot}(\mathbf{r}) = \begin{cases} \sigma_P(\mathbf{r}), & \text{inside P} \\ \sum_j q_j n_j^b e^{-\beta q_j \psi_P(\mathbf{r})}, & \text{outside P} \end{cases} \quad \text{(PB)}. \qquad (6.57)$$

Let us expand the exponential function in a Taylor series. For positions $\mathbf{r}$ outside particle P, we have

$$\rho_P(\mathbf{r}) = \sum_j q_j n_j^b [1 - \beta q_j \psi_P(\mathbf{r}) + (\beta q_j \psi_P(\mathbf{r}))^2/2 - \cdots]$$

$$= -\beta \sum_j q_j^2 n_j^b \psi_P(\mathbf{r}) + \text{nonlinear terms} \quad \text{(PB)}. \qquad (6.58)$$

The term that is linear in ψ_P will be denoted as $\rho_P^{lin}(\mathbf{r})$

$$\rho_P^{lin}(\mathbf{r}) = -\beta \sum_j q_j^2 n_j^b \psi_P(\mathbf{r}) = -\varepsilon_r \varepsilon_0 \kappa_D^2 \psi_P(\mathbf{r}) \quad \text{(PB)}, \qquad (6.59)$$

where the last equality is obtained from the definition (6.31) of κ_D. The entity $\rho_P(\mathbf{r})$ is the polarization response of the bulk electrolyte to the electrostatic field due to particle P. This response is nonlinear and $\rho_P^{lin}(\mathbf{r})$ is the linear part of it. We may denote $\rho_P^{lin}(\mathbf{r})$ as the *linear polarization response* of the bulk electrolyte to this field.

By subtracting $\rho_P^{lin}(\mathbf{r})$ from both sides of Poisson's Equation 6.55 and introducing

$$\rho_P^\star(\mathbf{r}) = \rho_P^{tot}(\mathbf{r}) - \rho_P^{lin}(\mathbf{r}), \qquad (6.60)$$

we can write Poisson's equation for all $\mathbf{r}$ values as

$$-\varepsilon_r \varepsilon_0 [\nabla^2 \psi_P(\mathbf{r}) - \kappa_D^2 \psi_P(\mathbf{r})] = \rho_P^\star(\mathbf{r}) \quad \text{(PB)}. \qquad (6.61)$$

This is, of course, equivalent to Equation 6.55.

We can write Equation 6.60 as

$$\rho_P^\star(\mathbf{r}) = \rho_P^{tot}(\mathbf{r}) + \varepsilon_r \varepsilon_0 \kappa_D^2 \psi_P(\mathbf{r})$$

$$= \sigma_P(\mathbf{r}) + \rho_P(\mathbf{r}) + \varepsilon_r \varepsilon_0 \kappa_D^2 \psi_P(\mathbf{r}) \quad \text{(PB)} \qquad (6.62)$$

and from Equation 6.58 follows

$$\rho_P^\star(\mathbf{r}) = \begin{cases} \sigma_P(\mathbf{r}) + \varepsilon_r \varepsilon_0 \kappa_D^2 \psi_P(\mathbf{r}), & \text{inside P} \\ \frac{1}{2} \sum_j q_j^3 n_j^b \beta^2 \psi_P^2(\mathbf{r}) - \frac{1}{3!} \sum_j q_j^4 n_j^b \beta^3 \psi_P^3(\mathbf{r}) + \cdots, & \text{outside P}. \end{cases} \quad \text{(PB)} \qquad (6.63)$$

Thus, outside particle P $\rho_P^\star(\mathbf{r})$ is equal to the *nonlinear* terms of $\rho_P(\mathbf{r})$. The function $\rho_P^\star(\mathbf{r})$ defined in Equation 6.60 will be denoted as the charge density of a "**dressed particle**," in this case particle P.[7] Equations 6.62 and 6.63 show how this density is expressed in the PB approximation.

To proceed, we need some additional tools and concepts, *Dirac's delta function* and *Green's function*. The latter is described in Example 6.2 below, while we have already briefly encountered Dirac's delta function in the shaded text box on page 205ff; it is treated in more detail in Appendix 5A. As explained in the appendix, one can say that the three-dimensional delta function $\delta^{(3)}(\mathbf{r})$ is zero everywhere except at the origin, $\mathbf{r} = \mathbf{0}$, where it has an infinitely sharp peak with integral equal to one. A key property of Dirac's delta function is that $\int d\mathbf{r}\, f(\mathbf{r})\delta^{(3)}(\mathbf{r}) = f(\mathbf{0})$. Since the delta function is spherically symmetric, we can write $\delta^{(3)}(\mathbf{r}) = \delta^{(3)}(r)$. The function $\delta^{(3)}(\mathbf{r} - \mathbf{r}_\nu)$ is zero everywhere except at the point $\mathbf{r} = \mathbf{r}_\nu$ where it has a peak with integral 1, that is, for any finite volume V, we have $\int_V d\mathbf{r}\, \delta^{(3)}(\mathbf{r} - \mathbf{r}_\nu) = 1$ whenever $\mathbf{r}_\nu$ is inside V and zero otherwise. It follows that $\int d\mathbf{r}\, f(\mathbf{r})\delta^{(3)}(\mathbf{r} - \mathbf{r}_\nu) = f(\mathbf{r}_\nu)$.

Example 6.2: Coulomb's Law and the Concept of Green's Function

Let us consider a point charge q placed at the origin in a dielectric continuum with dielectric constant ε_r (with no electrolyte present). The Coulomb potential at distance r from this charge is $q\phi_{\mathrm{Coul}}(r)$, where

$$\phi_{\mathrm{Coul}}(r) = \frac{1}{4\pi\varepsilon_r\varepsilon_0 r} \quad \text{(dielectric continuum)} \tag{6.64}$$

is the "unit" Coulomb potential in the dielectric continuum. The point charge can be represented by a charge density $q\delta^{(3)}(r)$, which has the dimension charge (volume)$^{-1}$. The total charge of this density is $\int d\mathbf{r}\, q\delta^{(3)}(r) = q$. Furthermore, Coulomb's law gives the electrostatic potential from this charge density as

$$\int d\mathbf{r}' \frac{q\delta^{(3)}(r')}{4\pi\varepsilon_r\varepsilon_0|\mathbf{r} - \mathbf{r}'|} = \frac{q}{4\pi\varepsilon_r\varepsilon_0|\mathbf{r} - \mathbf{0}|} = \frac{q}{4\pi\varepsilon_r\varepsilon_0 r},$$

that is, the same as the potential from the point charge q. Poisson's equation for this case is $-\varepsilon_r\varepsilon_0\nabla^2[q\phi_{\mathrm{Coul}}(r)] = q\delta^{(3)}(r)$, or, in other words, the unit Coulomb potential satisfies

$$-\varepsilon_r\varepsilon_0\nabla^2\phi_{\mathrm{Coul}}(r) = \delta^{(3)}(r) \tag{6.65}$$

with the boundary condition $\phi_{\mathrm{Coul}}(r) \to 0$ when $r \to \infty$.

We will now look at Equation 6.65 from a slightly different perspective. We know that the potential $\psi(\mathbf{r})$ from a charge density $\rho(\mathbf{r})$ satisfies Poisson's equation

$$-\varepsilon_r\varepsilon_0\nabla^2\psi(\mathbf{r}) = \rho(\mathbf{r}) \tag{6.66}$$

---

[7] By defining the charge density $\rho_{\mathrm{P}}^{\mathrm{dress}}(\mathbf{r}) \equiv \rho_{\mathrm{P}}(\mathbf{r}) - \rho_{\mathrm{P}}^{\mathrm{lin}}(\mathbf{r})$, we can write $\rho_{\mathrm{P}}^{\star}(\mathbf{r}) = \sigma_{\mathrm{P}}(\mathbf{r}) + \rho_{\mathrm{P}}^{\mathrm{dress}}(\mathbf{r})$, where the latter term contains the nonlinear contributions to $\rho_{\mathrm{P}}(\mathbf{r})$. The latter are shown explicitly in Equation 6.63. The function $\rho_{\mathrm{P}}^{\mathrm{dress}}$ gives the charge density of the "dress" of particle P.

and that Coulomb's law can be written

$$\psi(\mathbf{r}) = \int d\mathbf{r}' \rho(\mathbf{r}') \phi_{\text{Coul}}(|\mathbf{r} - \mathbf{r}'|). \tag{6.67}$$

Equation 6.67, which is a solution to Equation 6.66, contains the same function $\phi_{\text{Coul}}(r)$ as Equation 6.65. As we will see, this is no coincidence. Note that the left-hand sides of Equations 6.65 and 6.66 have the same form, but the equations have different right-hand sides; the former has a delta function as rhs. The solution $\phi_{\text{Coul}}(r)$ to Equation 6.65 is the so-called **Green's function** associated with Equation 6.66. This concept can be explained as follows.

Consider an inhomogeneous differential equation $\mathscr{L}F(\mathbf{r}) = f(\mathbf{r})$, where $\mathscr{L}$ is a linear differential operator and $f(\mathbf{r})$ is a given function (the Laplace operator ∇^2 is a simple example of such an operator). A Green's function $G(\mathbf{r} - \mathbf{r}')$ for this equation is, by definition, a solution of $\mathscr{L}G(\mathbf{r}) = \delta^{(3)}(\mathbf{r})$.[8] The function $F(\mathbf{r}) = \int d\mathbf{r}' f(\mathbf{r}') G(\mathbf{r} - \mathbf{r}')$ is then a solution of the original differential equation. Let us investigate why, whereby we start with our case where $\mathscr{L} = \nabla^2$.

We must show that the solution $\psi(\mathbf{r})$ to Equation 6.66 is given by Equation 6.67 with the Green's function $\phi_{\text{Coul}}(r)$ inserted. Let us hence verify that $\psi(\mathbf{r})$ from Equation 6.67 satisfies Equation 6.66. We insert $\psi(\mathbf{r})$ from the former equation into the latter whereby we can write the lhs as

$$\begin{aligned}
\text{lhs} &= -\varepsilon_r \varepsilon_0 \nabla^2 \int d\mathbf{r}' \rho(\mathbf{r}') \phi_{\text{Coul}}(|\mathbf{r} - \mathbf{r}'|) \\
&= \int d\mathbf{r}' \rho(\mathbf{r}') [-\varepsilon_r \varepsilon_0 \nabla^2 \phi_{\text{Coul}}(|\mathbf{r} - \mathbf{r}'|)] \\
&= \int d\mathbf{r}' \rho(\mathbf{r}') \delta^{(3)}(|\mathbf{r} - \mathbf{r}'|) = \rho(\mathbf{r}),
\end{aligned}$$

where we have used Equation 6.65 to obtain the 3rd line from the 2nd. This shows the assertion. The proof for the general case with operator $\mathscr{L}$ can be done in the same manner.

Let us specialize to a spherical ion of species i with a point charge at its center. The internal charge density $\sigma_i(r)$ is equal to $q_i \delta^{(3)}(r)$, where r is counted from the center. In an electrolyte solution, the Poisson equation is $-\varepsilon_r \varepsilon_0 \nabla^2 \psi_i(r) = \rho_i^{\text{tot}}(r)$ for all r, where the total charge density for the ion and its surrounding ion cloud is $\rho_i^{\text{tot}}(r) = q_i \delta^{(3)}(r) + \rho_i(r)$. By inserting ρ_i^{tot} as the charge density in Equation 6.67, we obtain

$$\begin{aligned}
\psi_i(r) &= \int d\mathbf{r}' [q_i \delta^{(3)}(r') + \rho_i(r')] \phi_{\text{Coul}}(|\mathbf{r} - \mathbf{r}'|) \\
&= q_i \phi_{\text{Coul}}(r) + \int d\mathbf{r}' \rho_i(r') \phi_{\text{Coul}}(|\mathbf{r} - \mathbf{r}'|),
\end{aligned}$$

which is the same as Equation 6.26.

---

[8] In mathematics, a solution of $\mathscr{L}G(\mathbf{r}) = \delta^{(3)}(\mathbf{r})$ is called a **fundamental solution** of $\mathscr{L}F(\mathbf{r}) = f(\mathbf{r})$.

We are now prepared to continue with our investigation of the potential from the particle P in an electrolyte solution. Let us consider the function $\phi^{\star}_{\text{Coul}}(r)$ which we define as the solution of

$$-\varepsilon_r\varepsilon_0[\nabla^2\phi^{\star}_{\text{Coul}}(r) - \kappa_D^2\phi^{\star}_{\text{Coul}}(r)] = \delta^{(3)}(r) \quad \text{(PB)} \tag{6.68}$$

with the boundary condition $\phi^{\star}_{\text{Coul}}(r) \to 0$ when $r \to \infty$. This function is the appropriate Green's function for Equation 6.61. Note that Equation 6.68 for $r > 0$, where $\delta^{(3)}(r) = 0$, is equivalent to the linearized PB equation [the DH equation] in Equation 6.10. The solution to Equation 6.68 is

$$\phi^{\star}_{\text{Coul}}(r) = \frac{e^{-\kappa_D r}}{4\pi\varepsilon_r\varepsilon_0 r}, \quad \text{(PB)} \tag{6.69}$$

which we will denote as the **unit screened Coulomb potential** of the PB approximation. It gives the spatial propagation of electrostatic interactions in the electrolyte as given by this approximation. Since $\phi^{\star}_{\text{Coul}}(r)$ is Green's function for Equation 6.61, it follows that the solution ψ_P of Equation 6.61 for a given $\rho_P^{\star}$ is given by

$$\psi_P(\mathbf{r}_1) = \int d\mathbf{r}_2 \rho_P^{\star}(\mathbf{r}_2)\phi^{\star}_{\text{Coul}}(r_{21}), \tag{6.70}$$

$$\text{i.e., } \psi_P(\mathbf{r}_1) = \int d\mathbf{r}_2 \rho_P^{\star}(\mathbf{r}_2)\frac{e^{-\kappa_D r_{21}}}{4\pi\varepsilon_r\varepsilon_0 r_{21}} \quad \text{(PB)} \tag{6.71}$$

for all $\mathbf{r}_1$. One can show this by inserting Equation 6.70 in Equation 6.61 and using Equation 6.68 in an analogous manner as in Example 6.2.

{**EXERCISE 6.4**

Show that $\psi_P(\mathbf{r})$ given by Equation 6.70 satisfies Equation 6.61.}

Equation 6.70 can be interpreted physically as follows. The potential ψ_P originates from the internal charges of particle P and the polarization it induces in the surroundings. From the identification of the linear response term in Equation 6.58 and the consequent construction of $\rho_P^{\star}$ in Equations 6.60 and 6.62, we see that $\rho_P^{\star}$ contains the nonlinear part of the polarization response of the electrolyte. The effects of the linear part of the response, which is not included in $\rho_P^{\star}$, is provided in Equation 6.70 by $\phi^{\star}_{\text{Coul}}$. This can most clearly be seen in Equation 6.61, where the linear response in contained in the lhs, which has the same form as the lhs of Equation 6.68 that defines $\phi^{\star}_{\text{Coul}}$.

Note that Equation 6.70 has the same mathematical form as Coulomb's law in Equation 6.54, which we can write as

$$\psi_P(\mathbf{r}_1) = \int d\mathbf{r}_2 \rho_P^{\text{tot}}(\mathbf{r}_2)\phi_{\text{Coul}}(r_{21}).$$

The potential $\psi_P(\mathbf{r}_1)$ is exactly the same for all $\mathbf{r}_1$ in both cases. In Equation 6.70, we have written $\psi_P(\mathbf{r}_1)$ in terms of the screened Coulomb potential $\phi^{\star}_{\text{Coul}}(r)$ instead of the ordinary (unscreened) Coulomb potential $\phi_{\text{Coul}}(r)$. The source charge distribution ρ_P^{tot} in Coulomb's law has thereby been replaced by another source charge distribution $\rho_P^{\star}$ in Equation 6.70, that is, the *charge density of the dressed particle*. The dressed particle charge density $\rho_P^\star(\mathbf{r}_2)$, which is used in conjunction with the screened Coulomb potential, has a shorter range than $\rho_P(\mathbf{r}_2)$ of the ion cloud; the latter decays for large r_2 like $\psi_P(\mathbf{r}_2)$ while $\rho_P^\star(\mathbf{r}_2)$ decays like $\psi_P^2(\mathbf{r}_2)$ or faster, as can be seen in Equation 6.63.

6.1.2.3 The Decay of Electrostatic Potential from Spherical and Nonspherical Particles

In this section, we will use the result in Equation 6.71 to investigate the decay of the electrostatic potential due to a particle immersed in an electrolyte solution in the PB approximation. We will first consider the potential from a spherically symmetric particle and then we will generalize the treatment to nonspherical particles.

Example 6.3: Effective Charge for a Spherical Particle in an Electrolyte

Let us consider a spherically symmetric particle P with internal charge density $\sigma_P(r')$ in an electrolyte solution. It can be a point charge at the center, a uniform surface charge density at the periphery of P or any other spherically symmetric charge distribution. The charge density $\rho_P^\star$ associated with P also has this symmetry, $\rho_P^\star = \rho_P^\star(r')$.

We will first consider a part of the screened electrostatic potential $\psi_P(r)$ due to the particle, namely that from $\sigma_P(r')$ in the interior of P. It is given by

$$\psi_P^\sigma(r) = \int d\mathbf{r}'\sigma_P(r')\phi_{Coul}^\star(|\mathbf{r}-\mathbf{r}'|) = \frac{1}{4\pi\varepsilon_r\varepsilon_0}\int d\mathbf{r}'\sigma_P(r')\frac{e^{-\kappa_D|\mathbf{r}-\mathbf{r}'|}}{|\mathbf{r}-\mathbf{r}'|},$$

where superscript σ indicates that ψ_P^σ is associated with σ_P according to this formula. By introducing spherical polar coordinates $(r',\varphi_r',\theta_r')$ for $\mathbf{r}'$, we can write this as

$$\psi_P^\sigma(r) = \frac{1}{4\pi\varepsilon_r\varepsilon_0}\int dr'(r')^2\sigma_P(r')\left[\int d\hat{\mathbf{r}}'\frac{e^{-\kappa_D|\mathbf{r}-\mathbf{r}'|}}{|\mathbf{r}-\mathbf{r}'|}\right], \tag{6.72}$$

where $\hat{\mathbf{r}}' = \mathbf{r}'/r'$ is the unit vector with polar angles (φ_r',θ_r') and $d\hat{\mathbf{r}}' = \sin(\theta_r')d\theta_r'd\varphi_r'$ stands for the angle differential. The angular integration in the square brackets can be performed analytically and we have[9]

$$\int d\hat{\mathbf{r}}'\frac{e^{-\kappa_D|\mathbf{r}-\mathbf{r}'|}}{|\mathbf{r}-\mathbf{r}'|} = 2\pi\frac{e^{-\kappa_D|r-r'|} - e^{-\kappa_D|r+r'|}}{\kappa_D r r'}.$$

If $r > r'$, we have $|r \pm r'| = r \pm r'$, whereby this becomes

$$\int d\hat{\mathbf{r}}'\frac{e^{-\kappa_D|\mathbf{r}-\mathbf{r}'|}}{|\mathbf{r}-\mathbf{r}'|} = 4\pi\frac{e^{-\kappa_D r}}{r}\cdot\frac{\sinh(\kappa_D r')}{\kappa_D r'} \quad \text{when } r > r'. \tag{6.73}$$

---

[9] By selecting a z axis that points in the direction of $\mathbf{r}$ and using the law of cosines in trigonometry $|\mathbf{r}-\mathbf{r}'|^2 = r^2 + (r')^2 - 2rr'\cos\theta_r'$, where θ_r' is the angle between the vectors $\mathbf{r}$ and $\mathbf{r}'$, one can can easily perform the integration after the change of variable $t^2 = r^2 + (r')^2 - 2rr'\cos\theta_r'$, whereby $t\,dt = rr'\sin(\theta_r')d\theta_r'$ and the limits $|r-r'| \leq t \leq |r+r'|$ correspond to $0 \leq \theta_r' \leq \pi$.

The condition $r > r'$ is fulfilled in Equation 6.72 provided that the point $\mathbf{r}$ is outside the particle (the integration over r' can be limited to inside P since $\sigma_P(r') = 0$ outside). We then obtain from these equations

$$\psi_P^\sigma(r) = \frac{e^{-\kappa_D r}}{4\pi \varepsilon_r \varepsilon_0 r} \int d\mathbf{r}' \sigma_P(r') \frac{\sinh(\kappa_D r')}{\kappa_D r'} \quad \text{(outside P)},$$

where we have used $d\mathbf{r}' = dr'(r')^2 4\pi$ (due to spherical symmetry). We can write this as $\psi_P^\sigma(r) = q_P^{\sigma\,\text{eff}} \phi_{\text{Coul}}^\star(r)$, where

$$q_P^{\sigma\,\text{eff}} = \int d\mathbf{r}' \sigma_P(r') \frac{\sinh(\kappa_D r')}{\kappa_D r'} \tag{6.74}$$

defines a kind of effective charge for σ_P.

Next, we investigate the full electrostatic potential due to particle P given by

$$\psi_P(r) = \int d\mathbf{r}' \rho_P^\star(r') \phi_{\text{Coul}}^\star(|\mathbf{r} - \mathbf{r}'|).$$

Since $\rho_P^\star(r')$ is not zero outside P (cf. Equation 6.63), the results for $\psi_P^\sigma(r)$ above cannot be taken over exactly as they are, but we can obtain a corresponding result for large r. The function $\rho_P^\star(r')$ has a shorter range than $\phi_{\text{Coul}}^\star(r)$, so the contributions to the integral from points with $r' > r$ for large r are negligible since $\rho_P^\star(r')$ is very small there. Thus, the condition $r > r'$ is fulfilled in practice, and by following the same steps as for ψ_P^σ above, we obtain

$$\psi_P(r) \sim \frac{q_P^{\text{eff}}}{4\pi \varepsilon_r \varepsilon_0} \cdot \frac{e^{-\kappa_D r}}{r} = q_P^{\text{eff}} \phi_{\text{Coul}}^\star(r) \quad \text{when } r \to \infty \quad \text{(PB)}, \tag{6.75}$$

where

$$q_P^{\text{eff}} = \int d\mathbf{r}' \rho_P^\star(r') \frac{\sinh(\kappa_D r')}{\kappa_D r'} \tag{6.76}$$

is the *effective charge* of P. Equation 6.75 can be compared with Equation 6.50, which can be written (with a different value of the effective charge)

$$\psi_i(r) = \frac{q_i^{\text{eff}}}{4\pi \varepsilon_r \varepsilon_0} \cdot \frac{e^{-\kappa_D r}}{r} = q_i^{\text{eff}} \phi_{\text{Coul}}^\star(r) \quad \text{(DH theory)}.$$

Note that the effective charge q_P^{eff} in an electrolyte is in general different from the particle's actual (bare) charge q_P, which is given by $q_P = \int d\mathbf{r}' \sigma_P(\mathbf{r}')$. In the limit of infinite dilution of the electrolyte, where $\kappa_D \to 0$, we have $q_P^{\text{eff}} \to q_P$ since $\rho_P^\star(r') \to \sigma_P(r')$ and $\sinh(\kappa_D r')/(\kappa_D r') \to 1$ in that limit.

Next, we turn to a simple example of a nonspherical particle, and to start with, we investigate the screened electrostatic potential from the particle itself in Example 6.4. We will thereby find an important difference between electrostatics in electrolytes compared to the "usual" electrostatics in a polar medium or in vacuum. After that, we will include the potential from the surrounding ion cloud and we will see that this difference remains.

Example 6.4: Screened Potential from a Dipole in an Electrolyte

There is a "peculiarity" of the screened Coulomb potential that originates from its r dependence, $\exp(-\kappa_D r)/r$. Let us consider a dipole consisting of two charges δq and $-\delta q$ separated by a distance a, see Figure 6.5. The dipole has the dipole moment vector $\mathbf{m}_\delta^d = \delta q\, \mathbf{a}$ (in this treatise we use the notation $\mathbf{m}^d$ for dipole moment, where d stands for "dipole").

Consider first the potential from the dipole in pure solvent ($\kappa_D = 0$). The Coulomb potential at a point located at the distance r' from the positive charge ($+\delta q$) and distance r from the negative ($-\delta q$) is proportional to

$$\frac{\delta q}{r'} - \frac{\delta q}{r} \sim \frac{\delta q\, \mathbf{a} \cdot \hat{\mathbf{r}}}{r^2} = \frac{\mathbf{m}_\delta^d \cdot \hat{\mathbf{r}}}{r^2} \quad \text{when } r \to \infty, \tag{6.77}$$

where $\hat{\mathbf{r}} = \mathbf{r}/r$ is the unit vector in the direction of $\mathbf{r}$. The rhs is the normal expression for the potential from a dipole. In electrolytes ($\kappa_D \neq 0$), where the potential from a point charge decays like $\exp(-\kappa_D r)/r$, the potential from the dipole is instead proportional to

$$\frac{\delta q e^{-\kappa_D r'}}{r'} - \frac{\delta q e^{-\kappa_D r}}{r} \sim \frac{\delta q e^{-\kappa_D r}}{r}\left[e^{\kappa_D \mathbf{a}\cdot\hat{\mathbf{r}}} - 1 \right] \quad \text{when } r \to \infty. \tag{6.78}$$

To derive this result, we need to express r' in terms of r for large distances from the dipole. The law of cosines in trigonometry says that

$$|\mathbf{r}'|^2 = |\mathbf{r} - \mathbf{a}|^2 = r^2 + a^2 - 2ra\cos\theta,$$

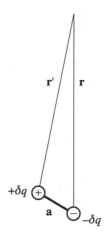

FIGURE 6.5 A dipole consisting of two charges $+\delta q$ and $-\delta q$ separated by the vector $\mathbf{a}$ that points towards the positive charge. Its dipole moment is $\mathbf{m}_\delta^d = \delta q\, \mathbf{a}$. The electrostatic potential is calculated at a point that lies at distances r' and r, respectively, from the charges (see text). The vectors $\mathbf{r}$ and $\mathbf{r}'$ are directed towards this point.

where θ is the angle between the vectors $\mathbf{r}$ and $\mathbf{a}$. We can write the last term as the scalar product $-2\mathbf{a} \cdot \mathbf{r}$ and we have

$$r' = [r^2 + a^2 - 2\mathbf{a} \cdot \mathbf{r}]^{1/2} = r[1 + (a/r)^2 - 2\mathbf{a} \cdot \hat{\mathbf{r}}/r]^{1/2} = r - \mathbf{a} \cdot \hat{\mathbf{r}} + \mathcal{O}(1/r), \quad (6.79)$$

where we have expanded the square root in a Taylor series to obtain the last equality; the sum of the higher order terms in the series is written as $\mathcal{O}(1/r)$, which means $1/r$ times a function that stays finite when $r \to \infty$, implying that this sum goes to zero in this limit. To the leading order, we therefore have

$$\frac{e^{-\kappa_D r'}}{r'} \sim \frac{e^{-\kappa_D(r - \mathbf{a} \cdot \hat{\mathbf{r}})}}{r - \mathbf{a} \cdot \hat{\mathbf{r}}} \sim \frac{e^{-\kappa_D r}}{r} e^{\kappa_D \mathbf{a} \cdot \hat{\mathbf{r}}} \quad \text{when } r \to \infty$$

and the rhs of Equation 6.78 is readily obtained. Incidentally, we note that when $\kappa_D = 0$, we have instead

$$\frac{1}{r'} \sim \frac{1}{r - \mathbf{a} \cdot \hat{\mathbf{r}}} = \frac{1}{r(1 - \mathbf{a} \cdot \hat{\mathbf{r}}/r)} \sim \frac{1}{r}\left(1 + \frac{\mathbf{a} \cdot \hat{\mathbf{r}}}{r}\right) \quad \text{when } r \to \infty$$

and we obtain the result in Equation 6.77.

The essential difference between the results in Equations 6.77 and 6.78 is the following. In the former equation, the inverse distance contributions from the two charges cancel exactly for large r, which implies that the potential decays faster than that from a single charge. In Equation 6.78, the two contributions do *not* cancel and they give a contribution that decays *equally fast* as the potential from a charge, that is, proportionally to $\exp(-\kappa_D r)/r$. The potential from a simple charge and that from a dipole thus decays like a Yukawa function with the same decay length. This highlights an important difference between electrostatics in electrolytes and "usual" electrostatics.

Let us investigate the result in Equation 6.78 further. The rhs of the equation depends on the direction of $\hat{\mathbf{r}}$, but there is a direction independent part which can be found by taking the average $\langle \cdot \rangle_{\hat{\mathbf{r}}}$ over all directions of $\hat{\mathbf{r}}$. The average $\langle \exp(\kappa_D \mathbf{a} \cdot \hat{\mathbf{r}}) \rangle_{\hat{\mathbf{r}}}$ is equal to $\sinh(\kappa_D a)/(\kappa_D a)$,[10] so we have

$$\left\langle \frac{\delta q e^{-\kappa_D r'}}{r'} - \frac{\delta q e^{-\kappa_D r}}{r} \right\rangle_{\hat{\mathbf{r}}} = \delta q \left[\frac{\sinh(\kappa_D a)}{\kappa_D a} - 1\right] \frac{e^{-\kappa_D r}}{r} \equiv q_\delta^{\sigma \text{ eff}} \frac{e^{-\kappa_D r}}{r}, \quad (6.80)$$

---

[10] The average of $\exp(\kappa_D \mathbf{a} \cdot \hat{\mathbf{r}})$ over all directions $\hat{\mathbf{r}}$ can be obtained by integrating over the direction angles of $\hat{\mathbf{r}}$ in spherical polar coordinates. This spherical average is independent of coordinate system, so we can select the direction of $\mathbf{a}$ as the z axis, whereby the angle θ between the vectors $\mathbf{r}$ and $\mathbf{a}$ is the polar angle. The average of $\exp(\kappa_D \mathbf{a} \cdot \hat{\mathbf{r}}) = \exp(\kappa_D a \cos\theta)$ is

$$\int_0^\pi d\theta \sin\theta e^{\kappa_D a \cos\theta} \Big/ \int_0^\pi d\theta \sin\theta = \sinh(\kappa_D a)/(\kappa_D a),$$

since the integration over the azimuthal angle contributes by 2π and cancels in the average. The integration in the numerator is readily performed after the variable substitution $t = \cos\theta$ and the rhs follows.

which defines $q_\delta^{\sigma\,\text{eff}}$. This means that the dipole gives the same direction independent contribution to the potential as a point particle with charge $q_\delta^{\sigma\,\text{eff}}$, so the dipole in general has an effective charge(!) in an electrolyte. Since the "internal" charge distribution of the dipole is $\sigma_\delta(\mathbf{r}) = \delta q\left[\delta^{(3)}(\mathbf{r} - \mathbf{a}) - \delta^{(3)}(\mathbf{r})\right]$, we can write (cf. Equation 6.74)

$$q_\delta^{\sigma\,\text{eff}} = \int d\mathbf{r}'\sigma_\delta(\mathbf{r}')\frac{\sinh(\kappa_D r')}{\kappa_D r'} \tag{6.81}$$

(superscript σ on $q_\delta^{\sigma\,\text{eff}}$ means that the latter originates from the internal charge density $\sigma_\delta(\mathbf{r})$).

The direction *dependent* part of the potential decays also like $\exp(-\kappa_D r)/r$, but it has a different direction dependence compared to the ordinary dipole potential. From Equation 6.78, it follows that the total potential from the dipole decays like

$$Q_\delta^\sigma(\hat{\mathbf{r}})\frac{e^{-\kappa_D r}}{r} \quad \text{when } r \to \infty$$

where we have introduced

$$Q_\delta^\sigma(\hat{\mathbf{r}}) = \delta q\left[e^{\kappa_D \mathbf{a}\cdot\hat{\mathbf{r}}} - 1\right].$$

Thus, $Q_\delta^\sigma(\hat{\mathbf{r}})$, which occurs in the prefactor of the potential for large r, describes the direction dependence. This prefactor can be written as

$$Q_\delta^\sigma(\hat{\mathbf{r}}) = \int d\mathbf{r}'\,\sigma_\delta(\mathbf{r}')e^{\kappa_D \mathbf{r}'\cdot\hat{\mathbf{r}}} \tag{6.82}$$

and we have

$$\left\langle Q_\delta^\sigma(\hat{\mathbf{r}})\right\rangle_{\hat{\mathbf{r}}} = \int d\mathbf{r}'\,\sigma_\delta(\mathbf{r}')\left\langle e^{\kappa_D \mathbf{r}'\cdot\hat{\mathbf{r}}}\right\rangle_{\hat{\mathbf{r}}} = q_\delta^{\sigma\,\text{eff}}.$$

Note that $Q_\delta^\sigma(\hat{\mathbf{r}})$ and $q_\delta^{\sigma\,\text{eff}}$ depend on where the center of the molecule is located relative to the charges (in the present example, the center is located at the negative charge).

After these preparatory examples, let us now investigate the full mean electrostatic potential ψ_P due to a particle P with an arbitrary internal charge distribution $\sigma_P(\mathbf{s})$ in its interior, that is, the potential from the particle itself and its surrounding ion cloud. A general formula for the decay of $\psi_P(\mathbf{r}_1)$ for large r_1 can easily be obtained from Equation 6.71. By using Equation 6.79 with $\mathbf{r} = \mathbf{r}_1$, $\mathbf{a} = \mathbf{r}_2$ and $\mathbf{r}' = \mathbf{r}_{21}$, we have

$$\frac{e^{-\kappa_D r_{21}}}{r_{21}} \sim \frac{e^{-\kappa_D(r_1 - \mathbf{r}_2\cdot\hat{\mathbf{r}}_1)}}{r_1 - \mathbf{r}_2\cdot\hat{\mathbf{r}}_1} \sim \frac{e^{-\kappa_D r_1}}{r_1}e^{\kappa_D \mathbf{r}_2\cdot\hat{\mathbf{r}}_1} \quad \text{when } r_2 \ll r_1 \to \infty.$$

We insert this result in Equation 6.71 and obtain

$$\psi_P(\mathbf{r}_1) \sim \frac{e^{-\kappa_D r_1}}{4\pi\varepsilon_r\varepsilon_0 r_1}\int d\mathbf{r}_2\rho_P^\star(\mathbf{r}_2)e^{\kappa_D \mathbf{r}_2\cdot\hat{\mathbf{r}}_1} \quad \text{when } r_1 \to \infty,$$

where the integral converges since $\rho_P^\star(\mathbf{r}_2)$ decays fast to zero. We can write this as

$$\psi_P(\mathbf{r}_1) \sim \phi_{\text{Coul}}^\star(r_1) \int d\mathbf{r}_2 \rho_P^\star(\mathbf{r}_2) e^{\kappa_D \mathbf{r}_2 \cdot \hat{\mathbf{r}}_1} \quad \text{when } r_1 \to \infty \tag{6.83}$$

$$= \phi_{\text{Coul}}^\star(r_1) Q_P^{\text{eff}}(\hat{\mathbf{r}}_1), \tag{6.84}$$

where (cf. Equation 6.82)

$$Q_P^{\text{eff}}(\hat{\mathbf{r}}_1) = \int d\mathbf{r}_2 \rho_P^\star(\mathbf{r}_2) e^{\kappa_D \mathbf{r}_2 \cdot \hat{\mathbf{r}}_1} \quad \text{(PB)}, \tag{6.85}$$

and we see that the distance dependence of the potential is accordingly the same as for the screened Coulomb potential $\phi_{\text{Coul}}^\star(r_1)$. Note, however, that the value of the coefficient $Q_P^{\text{eff}}(\hat{\mathbf{r}}_1)$ is in general dependent on the direction of the vector $\mathbf{r}_1$. Particle P has an **effective charge** equal to (cf. Equation 6.81)

$$q_P^{\text{eff}} = \left\langle Q_P^{\text{eff}}(\hat{\mathbf{r}}_1) \right\rangle_{\hat{\mathbf{r}}_1} = \int d\mathbf{r}_2 \rho_P^\star(\mathbf{r}_2) \frac{\sinh(\kappa_D r_2)}{\kappa_D r_2} \quad \text{(PB)}, \tag{6.86}$$

which gives the *magnitude of the direction independent part* of $\psi_P(\mathbf{r}_1)$ far from the particle

$$\langle \psi_P(\mathbf{r}_1) \rangle_{\hat{\mathbf{r}}_1} \sim q_P^{\text{eff}} \phi_{\text{Coul}}^\star(r_1) \quad \text{when } r_1 \to \infty.$$

We see that the principal features of the potential from a dipole in Example 6.4 apply in general for a particle in an electrolyte.

For the special case when particle P is spherically symmetric, for example when it is a simple ion, $\rho_P^\star$ and ψ_P are spherically symmetric. Then we have

$$\psi_P(r_1) \sim q_P^{\text{eff}} \phi_{\text{Coul}}^\star(r_1) \quad \text{when } r_1 \to \infty, \tag{6.87}$$

which we derived earlier in a different manner, Equation 6.75. This result for the potential is valid at some distance from the particle – in the tail region for large r_1. Closer in to the particle, the distance dependence of $\psi_P(r_1)$ is not given by $\phi_{\text{Coul}}^\star(r_1)$ in the nonlinear PB case.

For a particle that is not spherical, Equation 6.84 implies, as we have seen, that the direction dependent part of $\psi_P(\mathbf{r}_1)$ decays with distance in exactly the same manner for large r_1 as the direction independent part. This shows that there is a profound difference between the direction dependence of the electrostatic potential due to a charge distribution $\sigma_P(\mathbf{r}_2)$ in a particle immersed in an electrolyte compared to the same particle in vacuum or in a pure dielectric medium. In the latter cases, the potential far away from the particle does not depend on the direction and is given by $q_P/(4\pi\varepsilon_r\varepsilon_0 r_1)$, that is, proportional to the charge q_P of the particle. When the distance r_1 is decreased, the potential eventually becomes dependent on the direction of $\hat{\mathbf{r}}_1$; the potential due to the particle's dipole moment $\mathbf{m}_P^d = \int d\mathbf{r}_2 \mathbf{r}_2 \sigma_P(\mathbf{r}_2)$ then starts to give substantial contributions (it decays like $1/r_1^2$ with distance). Further in, the potential due to the particle's quadrupole moment starts to set in and for even shorter distances the higher multipole moments set in one after the other (their contributions are negligible for large r_1). In contrast to this, *in an electrolyte the direction dependence of the electrostatic potential due to a charge distribution $\sigma_P(\mathbf{r}_2)$ extends to infinity* when σ_P is not spherically symmetric. As we will see later in Chapter 13 (in the 2nd volume), one can define effective dipole, quadrupole and higher multipole moment contributions to $\psi_P(\mathbf{r}_1)$ that all decay with distance as $\exp(-\kappa_D r_1)/r_1$. The sum of all these contributions, together with the direction independent part, constitutes $\phi_{\text{Coul}}^{\star}(r_1)Q_P^{\text{eff}}(\hat{\mathbf{r}}_1)$ in Equation 6.84.

The entity $Q_P^{\text{eff}}(\hat{\mathbf{r}}_1)$ is accordingly the prefactor that gives both the magnitude and the direction dependence of the decay of the electrostatic potential from a nonspherical particle P for large r_1. It has the unit of charge and can be described as a "**multipolar effective charge**" (i.e., direction dependent) for electrostatic interactions involving particle P.

We have assumed that all ions in the electrolyte have the same diameter, but we can easily generalize the results we have obtained so far to ions of different diameters. In the PB approximation, the potential of mean force between P and an ion of species j is equal to $u_{Pj}^{\text{core}}(\mathbf{r}) + q_j\psi_P(\mathbf{r})$ for all $\mathbf{r}$, so we have

$$\rho_P(\mathbf{r}) = \sum_j q_j n_j^{\text{b}} e^{-\beta[u_{Pj}^{\text{core}}(\mathbf{r})+q_j\psi_P(\mathbf{r})]} = \sum_j q_j n_j^{\text{b}} O_{Pj}(\mathbf{r}) e^{-\beta q_j\psi_P(\mathbf{r})} \quad \text{(PB)}, \qquad (6.88)$$

where $O_{Pj}(\mathbf{r}) = \exp(-\beta u_{Pj}^{\text{core}}(\mathbf{r}))$ is zero when the j ion overlaps with P and one otherwise. Since small ions can approach P closer than large ions, $O_{Pj}(\mathbf{r})$ depends on j for positions $\mathbf{r}$ near the surface. The PB equation is (compare with Equation 6.57)

$$-\varepsilon_r\varepsilon_0\nabla^2\psi_P(\mathbf{r}) = \begin{cases} \sigma_P(\mathbf{r}), & \text{inside P} \\ \sum_j q_j n_j^{\text{b}} O_{Pj}(\mathbf{r}) e^{-\beta q_j\psi_P(\mathbf{r})}, & \text{outside P} \end{cases} \quad \text{(PB)}. \qquad (6.89)$$

For positions $\mathbf{r}$ away from P, where $O_{Pj} = 1$ for all j, we have

$$\rho_P(\mathbf{r}) = \sum_j q_j n_j^{\text{b}} e^{-\beta q_j\psi_P(\mathbf{r})},$$

which can be expanded as in Equation 6.58. Thereafter, the arguments follow in the same manner leading to the results in Equations 6.83 through 6.87.

{**EXERCISE 6.5**

For a spherical particle of species i with a charge q_i at its center, the internal charge density is $\sigma_i(r) = q_i\delta^{(3)}(r)$. The particle has diameter d^{h}. In the Debye-Hückel theory (linearized PB approximation), it follows from Equation 6.63 that the charge density of the dressed particle is equal to

$$\rho_i^{\star}(r) = \begin{cases} q_i\delta^{(3)}(r) + \varepsilon_r\varepsilon_0\kappa_D^2\psi(r), & r \leq d^{\text{h}} \\ 0, & r > d^{\text{h}} \end{cases} \quad \text{(DH)} \qquad (6.90)$$

since the nonlinear terms are absent in the linearized case. The electrostatic potential $\psi(r)$ is equal to the potential from q and that from the surrounding ion cloud. Inside the particle, the potential from the latter is constant due to spherical symmetry, so we have

$$\psi(r) = \frac{q_i}{4\pi\varepsilon_r\varepsilon_0}\left[\frac{1}{r} - \frac{1}{d^h}\right] + \psi(d^h) \quad \text{for } r \le d^h,$$

where $\psi(d^h)$ can be obtained from Equation 6.33.

Show that the effective charge of the particle, as obtained from

$$q_i^{\text{eff}} = \int d\mathbf{r}'\rho_i^\star(r')\frac{\sinh(\kappa_D r')}{\kappa_D r'}$$

(Equation 6.76), is equal to

$$q_i^{\text{eff}} = q_i \cdot \frac{e^{\kappa_D d^h}}{1 + \kappa_D d^h},$$

that is, the same as in Equation 6.49.}

As seen in Equation 6.90 for the linearized PB (DH) approximation, the only contribution to $\rho_i^\star$, apart from the charge q at the center, is the term $\varepsilon_r\varepsilon_0\kappa_D^2\psi(r)$ for $r \le d^h$. This charge density for $r \le d^h$ is nominal (there are no actual charges inside the particle apart from q) and it arises in Equation 6.60 from the contribution $-\rho_P^{\text{lin}}$, where ρ_P^{lin} is given by Equation 6.59. The subtraction of ρ_P^{lin} represents the removal of the linear part of the polarization response of the surrounding electrolyte, but since there is no such polarization response inside the particle, there is nothing to remove there. One can say that the term $\varepsilon_r\varepsilon_0\kappa_D^2\psi(r)$ for $r \le d^h$ is equal to minus the charge density that linear polarization response would have placed inside the particle if it had been able to. This term also appears for the nonlinear PB case, as seen in Equation 6.63, but then there are contributions for $r > d^h$ as well (the nonlinear part of the response). A nominal charge density of this kind inside the particles always exists in the dressed particle charge density $\rho_i^\star$ and is necessary in order to give the correct potential from Equation 6.70.

6.1.3 Interaction between two Particles Treated on an Equal Basis

6.1.3.1 Background

In the PB approximation for electrolyte solutions, the central ion is, as we have seen, treated in a different manner than the surrounding ions (in its ion cloud), including those of the same species as the central one. While the correlations between the central ion and the surrounding ions are taken into account, the correlations between the ions in the cloud are neglected; they are thereby treated like uncorrelating point ions that do not have any ion clouds of their own. The ion cloud around a central ion of species j gives rise to the screening of the potential from it, which leads to an exponential decay of $\psi_j(r)$ that is given by $\psi_j(r) \sim q_j^{\text{eff}}\phi_{\text{Coul}}^\star(r)$ when $r \to \infty$ (Equation 6.87), where the magnitude (and sign) of the potential is expressed in terms of the effective charge q_j^{eff}. An ion of species i in the cloud interacts with this potential via its bare charge q_i in the PB approximation, leading to $w_{ij}^{(2)}(r) = q_i \psi_j(r) \sim q_i q_j^{\text{eff}} \phi_{\text{Coul}}^\star(r)$ for large r. As noted earlier, we have $w_{ij}^{(2)} \neq w_{ji}^{(2)}$ in general in this approximation.

In the linearized PB approximation (the DH theory), we have seen that if we consider two ions treated on an equal basis where both have ion clouds, their interaction decays according to Equation 6.52, that is, as $w_{ij}^{(2)}(r) \sim q_i^{\text{eff}} q_j^{\text{eff}} \phi_{\text{Coul}}^\star(r)$ for large r, which has the correct symmetry. This result can also be applied to any spherically symmetric particles immersed in an electrolyte treated in the linearized DH theory, for example two spherical colloid particles.

Large charged particles, macroions, in electrolyte solutions are of interest in, for example, colloid and interface science. When we have two macroions, P_I and P_{II}, immersed in an electrolyte and treat them on an equal basis, the potential of mean force between them fulfills the requirement $w_{I,II}^{(2)} = w_{II,I}^{(2)}$. When we use the PB approximation for the electrolyte solution to calculate the interactions between the macroions, the small ions in the surrounding electrolyte solution are, however, still treated like uncorrelating point ions. The potential of mean force between a small ion of species i and, say, macroion I is approximated as $w_{i,I}^{(2)} = q_i \psi_I$, which decays like $q_i q_I^{\text{eff}} \phi_{\text{Coul}}^\star(r)$ when the macroion is spherical. An asymmetry in the treatment of the different kinds of charged species thus remains since the two macroions have ion clouds while the small ions do not. In many cases, the results of this approximation are, however, quite reasonable, in particular for colloid particles in 1:-1 aqueous electrolyte solutions. We will therefore continue with the PB approximation for the interaction between macroions. In Section 6.1.5, we will, however, see in a very simple approximation what happens when all small ions in a bulk electrolyte are treated on the same basis and all have ion clouds of their own.

6.1.3.2 The Decay of Interaction between Two Nonspherical Macroions

Let us investigate a system with two particles P_I and P_{II} immersed in an electrolyte solution. Thereby, we will treat these two particles on an equal basis, just like we did for two ions in Equation 6.52, but now in the *nonlinear* PB approximation. Initially, we will assume that the particles are spherical and have internal charge distributions $\sigma_I(s)$ and $\sigma_{II}(s)$, where s is counted from the particle center. The particles are placed with their centers at $\mathbf{r}_I$ and $\mathbf{r}_{II}$, respectively. The main application is for two macroions P_I and P_{II}, but the treatment includes, of course, also the case when P_I and P_{II} are small ions.

It is quite simple to determine the decay of the potential of mean force $w_{I,II}(r_{I,II})$ between P_I and P_{II} for large separations $r_{I,II}$, where $r_{I,II} = |\mathbf{r}_{I,II}|$ and $\mathbf{r}_{I,II} = \mathbf{r}_{II} - \mathbf{r}_I$. The mean electrostatic potential ψ_J from a particle J is, according to Equation 6.70, given by

$$\psi_J(\mathbf{r}_1) = \int d\mathbf{r}_2 \rho_J^\star(\mathbf{r}_2 - \mathbf{r}_J) \phi_{\text{Coul}}^\star(r_{21})$$

(we have $\mathbf{r}_2 - \mathbf{r}_J$ rather than $\mathbf{r}_2$ as the argument since the origin of the coordinate system in the current case is not placed at the center of particle J). In connection to Equation 6.70, we noted that $\phi^\star_{\text{Coul}}$ describes polarization response effects of the electrolyte to linear order while the nonlinear parts are contained in $\rho^\star_J$. If we place P_I and P_{II} very far from each other, the ion clouds of each particle are not affected by the other particle. When we let the two particles approach each other, the electrostatic field from one particle will start to polarize the surroundings of the other, but for large separations the influence from one to the other is weak. Therefore, the polarization between them can be treated to linear order. The nonlinear part of the response of the electrolyte due to one of the particles can then be neglected near the other ion; the nonlinear polarization due to the former particle is important only in its own immediate neighborhood. In other words, $\rho^\star_J(\mathbf{r}_2)$ for $J = I$ and II are virtually unchanged if the separation between the particles is sufficiently large. Since $\phi^\star_{\text{Coul}}$ describes the polarization response to linear order, the interaction free energy for large separations will go like[11]

$$w^{(2)}_{I,II}(\mathbf{r}_{I,II}) \sim \int d\mathbf{r}_1 d\mathbf{r}_2 \rho^\star_I(\mathbf{r}_2 - \mathbf{r}_I)\phi^\star_{\text{Coul}}(r_{21})\rho^\star_{II}(\mathbf{r}_1 - \mathbf{r}_{II})$$

$$= \int d\mathbf{r}_1 \psi_I(\mathbf{r}_1 - \mathbf{r}_I)\rho^\star_{II}(\mathbf{r}_1 - \mathbf{r}_{II}) = \int d\mathbf{r}_2 \rho^\star_I(\mathbf{r}_2 - \mathbf{r}_I)\psi_{II}(\mathbf{r}_2 - \mathbf{r}_{II}) \quad (6.91)$$

when $r_{I,II} \to \infty$. The potential of mean force accordingly decays like the screened electrostatic interaction between the dressed particle charge densities $\rho^\star_I$ and $\rho^\star_{II}$ of particles I and II.

Since $\rho^\star_I$, $\rho^\star_{II}$, ψ_I and ψ_{II} are spherically symmetric, it follows that

$$w^{(2)}_{I,II}(r_{I,II}) \sim \int d\mathbf{r}_1 \psi_I(r_{1,1})\rho^\star_{II}(r_{II,1}) \sim \int d\mathbf{r}_1 q^{\text{eff}}_I \phi^\star_{\text{Coul}}(r_{1,I})\rho^\star_{II}(r_{II,1})$$

$$= q^{\text{eff}}_I \psi_{II}(r_{I,II}) \sim q^{\text{eff}}_I q^{\text{eff}}_{II} \phi^\star_{\text{Coul}}(r_{I,II}),$$

where q^{eff}_J is defined in Equation 6.86 and where we have used Equation 6.87 and the fact that $\rho^\star_J$ decays quickly. Thus, we have

$$w^{(2)}_{I,II}(r) \sim q^{\text{eff}}_I q^{\text{eff}}_{II} \phi^\star_{\text{Coul}}(r) \quad \text{when } r \to \infty, \tag{6.92}$$

which means that Equation 6.52 holds also in the nonlinear PB approximation, but with q^{eff}_J given by Equation 6.86 instead of Equation 6.49. For spherical colloid particles with uniform surface charge density, the effective charge can be expressed in terms of an effective surface charge density by dividing q^{eff}_J by the surface area of the particle.

We can generalize these results to *nonspherical particles* P_I and P_{II} with internal charge distributions $\sigma_I(\mathbf{s})$ and $\sigma_{II}(\mathbf{s})$ placed with their centers of mass at $\mathbf{r}_I$ and $\mathbf{r}_{II}$ and separated by $\mathbf{r}_{I,II} = \mathbf{r}_{II} - \mathbf{r}_I$ (Figure 6.6a). The vector $\mathbf{s}$ is counted from the center of mass of each particle. So far, we have implicitly assumed that particles P have fixed orientation in space.

---

[11] This asymptotic relationship is also valid for the screened electrostatic part of the potential of mean force in other cases – not only in the PB approximation – as shown in Section 6.2.

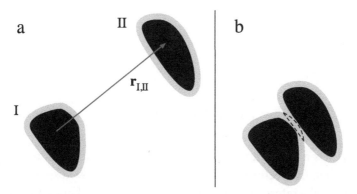

FIGURE 6.6 (a) A sketch of two macroions I and II with orientations ω_I and ω_{II}, respectively, separated by the vector $\mathbf{r}_{I,II}$ between their centers of mass which are located at $\mathbf{r}_I$ and $\mathbf{r}_{II}$. The gray region around each macroion indicates the exclusion zone where the centers of the ions in the electrolyte cannot enter because of core-core interactions between ions and the macroion. (b) At short separations, the exclusion zones of the macroions overlap. The overlap region is indicated by the dashed lines. The excluded volume around the two macroions is less than that in (a) by the volume of the overlap.

Since the interaction between P_I and P_{II} in general depend on both the distance and the particle orientations, it is, however, important to show the latter. In order to explicitly do this, we will use orientational variable ω introduced in Appendix 6A. This variable gives the orientation of a particle relative to the laboratory frame of reference. The internal charge distribution of particle P_J with orientation ω_J is then $\sigma_J(\mathbf{s}, \omega_J)$. We also have, for example, $\rho_J^\star = \rho_J^\star(\mathbf{s}, \omega_J)$ and $\psi_J = \psi_J(\mathbf{s}, \omega_J)$. For the pair potential of mean force, which depends on the orientations of the two particles, we have $w_{I,II} = w_{I,II}^{(2)}(\mathbf{r}_{I,II}, \omega_I, \omega_{II})$ and the same applies to other pair entities. Equation 6.91 should thus be written

$$w_{I,II}^{(2)}(\mathbf{r}_{I,II}, \omega_I, \omega_{II}) \sim \int d\mathbf{r}_1 d\mathbf{r}_2 \rho_I^\star(\mathbf{r}_{I,2}, \omega_I)\phi_{Coul}^\star(r_{21})\rho_{II}^\star(\mathbf{r}_{II,1}, \omega_{II})$$

$$= \int d\mathbf{r}_1 \psi_I(\mathbf{r}_{I,1}, \omega_I)\rho_{II}^\star(\mathbf{r}_{II,1}, \omega_{II}) = \int d\mathbf{r}_2 \rho_I^\star(\mathbf{r}_{I,2}, \omega_I)\psi_{II}(\mathbf{r}_{II,2}, \omega_{II}), \quad (6.93)$$

which applies when $r_{I,II} \to \infty$.

The asymptotic result that corresponds to Equation 6.92 is

$$w_{I,II}^{(2)}(\mathbf{r}_{I,II}, \omega_I, \omega_{II}) \sim Q_I^{eff}(\hat{\mathbf{r}}_{I,II}, \omega_I)Q_{II}^{eff}(-\hat{\mathbf{r}}_{I,II}, \omega_{II})\phi_{Coul}^\star(r_{I,II}) \quad \text{when } r_{I,II} \to \infty, \quad (6.94)$$

where Q_J^{eff} for $J = I$ and II are defined in Equation 6.85. This is derived in the following text box, which can be skipped in the first reading. Note that the unit vector in the argument of Q_I^{eff} and Q_{II}^{eff}, respectively, in Equation 6.94 points to the other particle along the connection line between the centers of mass of the particles. The asymptotic result in Equation 6.94 shows that the strength of the interaction between the two particles depends on their relative orientations also for very large separations.

★ From Equations 6.93 and 6.84, we obtain when $r_{I,II} \to \infty$

$$w_{I,II}^{(2)}(\mathbf{r}_{I,II}, \boldsymbol{\omega}_I, \boldsymbol{\omega}_{II}) \sim \int d\mathbf{r}_1 \psi_I(\mathbf{r}_{I,1}, \boldsymbol{\omega}_I) \rho_{II}^\star(\mathbf{r}_{II,1}, \boldsymbol{\omega}_{II})$$

$$\sim \int d\mathbf{r}_1 Q_I^{\text{eff}}(\hat{\mathbf{r}}_{I,1}, \boldsymbol{\omega}_I) \phi_{\text{Coul}}^\star(r_{I,1}) \rho_{II}^\star(\mathbf{r}_{II,1}, \boldsymbol{\omega}_{II}).$$

Since $\rho_{II}^\star(\mathbf{r}_{II,1}, \boldsymbol{\omega}_{II})$ has a short range, the integrand in the rhs has its main contribution for large $r_{I,II}$ when the vector $\mathbf{r}_{I,1}$ from particle I points towards particle II, that is, towards the point $\mathbf{r}_{II}$. Therefore, in the limit $r_{I,II} \to \infty$, we can replace $Q_I^{\text{eff}}(\hat{\mathbf{r}}_{I,1}, \boldsymbol{\omega}_I)$ by $Q_I^{\text{eff}}(\hat{\mathbf{r}}_{I,II}, \boldsymbol{\omega}_I)$, which means that the rhs goes like

$$Q_I^{\text{eff}}(\hat{\mathbf{r}}_{I,II}, \boldsymbol{\omega}_I) \int d\mathbf{r}_1 \phi_{\text{Coul}}^\star(r_{I,1}) \rho_{II}^\star(\mathbf{r}_{II,1}, \boldsymbol{\omega}_{II}) = Q_I^{\text{eff}}(\hat{\mathbf{r}}_{I,II}, \boldsymbol{\omega}_I) \psi_{II}(\mathbf{r}_{II,I}, \boldsymbol{\omega}_{II})$$

when $r_{I,II} \to \infty$. Now, since $\psi_{II}(\mathbf{r}_{II,I}, \boldsymbol{\omega}_{II}) \sim Q_{II}^{\text{eff}}(\hat{\mathbf{r}}_{II,I}, \boldsymbol{\omega}_{II}) \phi_{\text{Coul}}^\star(r_{II,I}) = Q_{II}^{\text{eff}}(-\hat{\mathbf{r}}_{I,II}, \boldsymbol{\omega}_{II}) \phi_{\text{Coul}}^\star(r_{I,II})$, we finally obtain the result in Equation 6.94.

6.1.4 The Interaction between Two Macroions for all Separations

6.1.4.1 Poisson-Boltzmann Treatment

The asymptotic formulas that we have obtained in the previous section (Section 6.1.3.2) for the decay of $w_{I,II}^{(2)}$ are conceptually important since they give a qualitative picture of how the interaction between two particles behaves in an electrolyte. To be useful in a quantitative manner, one needs to be able to calculate the values of the effective charges q_J^{eff} for $J = I$ and II or, for particles that are not spherically symmetric, the coefficients Q_J^{eff}. To do this calculation, it is sufficient to solve the PB Equation 6.89 for a system with *only one* particle J present in the electrolyte solution. From the charge density distribution $\rho_J(\mathbf{r})$ given by Equation 6.88, which gives $\rho_J^\star(\mathbf{r})$ via Equation 6.62, one can then calculate q_J^{eff} or Q_J^{eff} from the latter via Equations 6.76 or 6.85, respectively.

To obtain the potential of mean force $w_{I,II}^{(2)}$ for all separations between the particles P_I and P_{II} in the PB approximation, one must solve the PB equation for the system with both particles present in the electrolyte solution[12]

$$-\varepsilon_r \varepsilon_0 \nabla^2 \psi(\mathbf{r}) = \begin{cases} \sigma_I(\mathbf{r} - \mathbf{r}_I), & \text{inside } P_I \\ \sigma_{II}(\mathbf{r} - \mathbf{r}_{II}), & \text{inside } P_{II} \\ \sum_i q_j n_j^b e^{-\beta q_j \psi(\mathbf{r})}, & \text{otherwise} \end{cases} \quad \text{(PB)},$$

where "inside" P_I and P_{II} includes the exclusion zone of ions around the particles where the charge density is zero (the gray regions in Figure 6.6). The equation will in general have to

---

[12] For simplicity, we have assumed that all ions in the electrolyte solution have the same size.

be solved numerically and in the present section we describe how one can obtain $w_{\mathrm{I,II}}^{(2)}$ from the solution. This is done for the general case of nonspherical particles P_I and P_II, but for simplicity in notation we do not explicitly show the orientation variables. Therefore, we write $\sigma_J(\mathbf{s})$ instead of $\sigma_J(\mathbf{s}, \boldsymbol{\omega}_J)$ for $J = \mathrm{I}$ and II and, likewise, we leave out the orientation variable for entities like $w_{\mathrm{I,II}}$ and ψ that depend on the orientations of both P_I and P_II. Consequently, we assume that P_I and P_II have fixed orientations in space all the time.

To evaluate $w_{\mathrm{I,II}}^{(2)}$, one can calculate the change in free energy when particle II is inserted at various positions $\mathbf{r}_\mathrm{II}$ in a system with particle I present at $\mathbf{r}_\mathrm{I}$, that is, one inserts particle II in the electrolyte that is inhomogeneous due to the presence of particle I. Thereby, one calculates the excess chemical potential $\mu_\mathrm{II}^{\mathrm{ex}}$ for various positions of II relative to I and obtains $\mu_\mathrm{II}^{\mathrm{ex}}(\mathbf{r}_\mathrm{I,II})$. The potential of mean force $w_{\mathrm{I,II}}^{(2)}(\mathbf{r}_\mathrm{I,II})$ is equal to the difference between $\mu_\mathrm{II}^{\mathrm{ex}}(\mathbf{r}_\mathrm{I,II})$ and the bulk value $\mu_\mathrm{II}^{\mathrm{ex,b}}$, as follows from Equation 5.23 applied to the current situation (the bulk value is independent of orientation due to the homogeneity and isotropy of the bulk phase).

Alternatively, one can evaluate the change in free energy of the system when both particles are simultaneously inserted at various positions $\mathbf{r}_\mathrm{I}$ and $\mathbf{r}_\mathrm{II}$ in the bulk electrolyte solution. The two particles are initially, before the insertion, neither interacting with the system nor with each other. The free energy change when two particles are inserted is, by definition, a **two-particle excess chemical potential**[13] $\mu_{\mathrm{I,II}}^{(2)\mathrm{ex}}(\mathbf{r}_\mathrm{I,II})$ and we have

$$w_{\mathrm{I,II}}^{(2)}(\mathbf{r}_\mathrm{I,II}) = \mu_{\mathrm{I,II}}^{(2)\mathrm{ex}}(\mathbf{r}_\mathrm{I,II}) - \mu_{\mathrm{I,II}}^{(2)\mathrm{ex}}(\infty) = \mu_{\mathrm{I,II}}^{(2)\mathrm{ex}}(\mathbf{r}_\mathrm{I,II}) - \mu_\mathrm{I}^{\mathrm{ex,b}} - \mu_\mathrm{II}^{\mathrm{ex,b}}, \qquad (6.95)$$

where we have used the fact that the change in free energy is $\mu_\mathrm{I}^{\mathrm{ex,b}} + \mu_\mathrm{II}^{\mathrm{ex,b}}$ when the particles are inserted infinitely far from each other. We will proceed with this alternative and obtain $w_{\mathrm{I,II}}^{(2)}$ via Equation 6.95.

The evaluation of $\mu_{\mathrm{I,II}}^{(2)\mathrm{ex}}(\mathbf{r}_\mathrm{I,II})$ can be done, as will be described in detail later (Equation 6.100), by using a generalization of Equation 5.84, where the last term is μ^{ex}. One gradually changes the interactions between the two particles and the surrounding electrolyte from zero to the full value by varying a coupling parameter ξ from 0 to 1 (analogously to Equation 5.83). One does this by altering the internal charge distributions of the particles from zero to the final values, whereby the electrostatic interactions with the surrounding ions vary in the same manner. As we will see, one thereby obtains the electrostatic contribution $\mu_{\mathrm{I,II}}^{(2)\mathrm{ex,el}}$ to the two-particle excess chemical potential $\mu_{\mathrm{I,II}}^{(2)\mathrm{ex}}$.

Like in our treatment of μ^{ex} in the Debye-Hückel theory in Section 6.1.1, there is also a contribution from making "holes" for the particles in the electrolyte before they are charged; the holes are the hard core exclusion zones where the ions of the electrolyte cannot enter. In the PB approximation, this contribution, $\mu_{\mathrm{I,II}}^{(2)\mathrm{ex,core}}$, is simply equal to the product of the volume of the holes (the excluded volume) times the ideal pressure of the bulk phase (cf. Equation 6.37 for the creation of a spherical hole). The excluded volume does not depend on the distance between the two particles except when they are very close to each other and

---

[13] In Section 5.13.3.2, the same kind of excess chemical potential was used for a similar purpose in a different context.

the exclusion zones overlap as illustrated in Figure 6.6b. Therefore, $\mu_{I,II}^{(2)ex,core}$ is independent of $\mathbf{r}_{I,II}$ for most separations. Since we take the difference between $\mu_{I,II}^{(2)ex}$ and its value at infinite separation, the core contribution vanishes for $w_{I,II}^{(2)}$ except when overlap occurs, so in the PB approximation we only need to include the electrostatic part, $\mu_{I,II}^{(2)ex,el}$, for most separations $\mathbf{r}_{I,II}$.

We accordingly have

$$w_{I,II}^{(2)}(\mathbf{r}_{I,II}) = w_{I,II}^{(2)el}(\mathbf{r}_{I,II}) + w_{I,II}^{(2)core}(\mathbf{r}_{I,II}) \tag{6.96}$$

with

$$w_{I,II}^{(2)el}(\mathbf{r}_{I,II}) = \mu_{I,II}^{(2)ex,el}(\mathbf{r}_{I,II}) - \mu_{I,II}^{(2)ex,el}(\infty), \tag{6.97}$$

where $\mu_{I,II}^{(2)ex,el}(\infty)$ is the electrostatic part of $\mu_I^{ex,b} + \mu_{II}^{ex,b}$, and

$$w_{I,II}^{(2)core}(\mathbf{r}_{I,II}) = \mu_{I,II}^{(2)ex,core}(\mathbf{r}_{I,II}) - \mu_{I,II}^{(2)ex,core}(\infty), \tag{6.98}$$

which is zero in the PB approximation except for very small $r_{I,II}$ where overlaps of the exclusion zones occur.

In the next section (Section 6.1.4.2), which can be skipped in the first reading, we derive the following result, which is valid in general, that is, not only in the PB approximation

$$\mu_{I,II}^{(2)ex,el}(\mathbf{r}_{I,II}) = \int_0^1 d\xi \int d\mathbf{r}[\sigma_I(\mathbf{r} - \mathbf{r}_I)\psi(\mathbf{r};\xi)$$
$$+ \sigma_{II}(\mathbf{r} - \mathbf{r}_{II})\psi(\mathbf{r};\xi)] + \text{constant}, \tag{6.99}$$

where $\psi(\mathbf{r};\xi)$ is the mean electrostatic potential in the system when the particles are charged to the extent ξ, that is, when their internal charge distribution is $\sigma_J(\mathbf{s};\xi) = \xi\sigma_J(\mathbf{s})$ for $J = I$ and II. The constant[14] does not depend on the separation vector $\mathbf{r}_{I,II}$ between the particles, so it cancels $w_{I,II}^{(2)el}(\mathbf{r}_{I,II})$. Equation 6.99 says that solely the interaction between the mean electrostatic potential and the internal charge distributions σ_I and σ_{II} of the particles contributes to $\mu_{I,II}^{(2)ex,el}$ for all values of the coupling parameter ξ.

For simplicity in notation in Equation 6.99, we have suppressed that $\psi(\mathbf{r};\xi)$ depends on $\mathbf{r}_I$ and $\mathbf{r}_{II}$, but in reality we have $\psi = \psi(\mathbf{r}|\mathbf{r}_I, \mathbf{r}_{II};\xi)$. This potential originates from the two particles I and II and the surrounding charge density of the ion cloud, where the density of ion species j is equal to $n_j(\mathbf{r}|\mathbf{r}_I, \mathbf{r}_{II};\xi)$. In the PB approximation, we have $n_j(\mathbf{r}|\mathbf{r}_I, \mathbf{r}_{II};\xi) = n_j^b \exp[-\beta q_j\psi(\mathbf{r};\xi)]$ and $\psi(\mathbf{r};\xi)$ is the solution to the PB equation when the particles are charged to the extent ξ.

---

[14] As we will see in Section 6.1.4.2, this constant contains the self-energies of the two particles I and II, that is, the interaction of their respective internal charge distribution σ_J with itself.

6.1.4.2 Electrostatic Part of Pair Potential of Mean Force, General Treatment ★

In this section, we will derive Equation 6.99 for $\mu_{I,II}^{(2)ex,el}$, which *is valid in general*. By inserting this result in Equation 6.97, we obtain a general expression for $w_{I,II}^{(2)el}$ in terms of the mean electrostatic potential $\psi(\mathbf{r}''; \xi)$. The latter is calculated from the charge distribution around the partially charged particles (the macroions) placed with their centers at $\mathbf{r}_I$ and $\mathbf{r}_{II}$, that is, when the density of ion species j at $\mathbf{r}'$ is equal to $n_j(\mathbf{r}'|\mathbf{r}_I, \mathbf{r}_{II}; \xi)$. In the general case, the densities must be obtained either in some approximate theory that yields the density distributions or by computer simulation.

To obtain the full potential of mean force, $w_{I,II}^{(2)}(\mathbf{r}_{I,II})$ in Equation 6.96, one must also include the contribution $w_{I,II}^{(2)el,core}$ from hard-core exclusions, which can be obtained by calculating $\mu_{I,II}^{(2)ex,core}$ for uncharged particles when they are separated by $\mathbf{r}_{I,II}$ and when they are infinitely far apart, Equation 6.98. This excess chemical potential is equal to the reversible work done during a gradual creation of the holes for the two particles from zero to the final volume in the bulk electrolyte. It is only in the PB approximation that the hard core contribution is trivial and cancels in $w_{I,II}^{(2)}$ for most separations $r_{I,II}$.

By generalizing Equation 5.84 to the insertion of two particles I and II, we can write the electrostatic part of the excess chemical potential as

$$
\mu_{I,II}^{(2)ex,el}(\mathbf{r}_{I,II}) = u_{I,II}^{el}(\mathbf{r}_{I,II})
$$
$$
+ \int_0^1 d\xi \sum_j \int d\mathbf{r}' \left[\frac{\partial u_{j,I}^{el}(\mathbf{r}', \mathbf{r}_I; \xi)}{\partial \xi} + \frac{\partial u_{j,II}^{el}(\mathbf{r}', \mathbf{r}_{II}; \xi)}{\partial \xi} \right] n_j(\mathbf{r}'|\mathbf{r}_I, \mathbf{r}_{II}; \xi), \quad (6.100)
$$

where $u_{I,II}^{el}$ is electrostatic pair potential between the two particles and $u_{j,I}^{el}$ and $u_{j,II}^{el}$ are the corresponding potentials between the respective particle and an ion of species j. Coulomb's law yields

$$
u_{J,J'}^{el}(\mathbf{r}_{J,J'}) = \int d\mathbf{r}' d\mathbf{r}'' \frac{\sigma_J(\mathbf{r}' - \mathbf{r}_J)\sigma_{J'}(\mathbf{r}'' - \mathbf{r}_{J'})}{4\pi \varepsilon_r \varepsilon_0 |\mathbf{r}' - \mathbf{r}''|} \quad (6.101)
$$

for $J = I$ and $J' = II$. Likewise, we have

$$
u_{j,J}^{el}(\mathbf{r}', \mathbf{r}_J) = q_j \int d\mathbf{r}'' \frac{\sigma_J(\mathbf{r}'' - \mathbf{r}_J)}{4\pi \varepsilon_r \varepsilon_0 |\mathbf{r}' - \mathbf{r}''|}
$$

for $J = I$ and II and we have $u_{j,J}^{el}(\mathbf{r}', \mathbf{r}_J; \xi) = \xi u_{j,J}^{el}(\mathbf{r}', \mathbf{r}_J)$.

The last term in Equation 6.100 can be written as

$$
\int_0^1 d\xi \int d\mathbf{r}' \int d\mathbf{r}'' \frac{\sigma_I(\mathbf{r}'' - \mathbf{r}_I) + \sigma_{II}(\mathbf{r}'' - \mathbf{r}_{II})}{4\pi \varepsilon_r \varepsilon_0 |\mathbf{r}' - \mathbf{r}''|} \rho(\mathbf{r}'|\mathbf{r}_I, \mathbf{r}_{II}; \xi),
$$

where $\rho(\mathbf{r}'|\mathbf{r}_I, \mathbf{r}_{II}; \xi) = \sum_j q_j n_j(\mathbf{r}'|\mathbf{r}_I, \mathbf{r}_{II}; \xi)$ is the charge density around the two partially charged particles. The electrostatic potential ψ_ρ from this charge distribution is

$$\psi_\rho(\mathbf{r}''; \xi) \equiv \psi_\rho(\mathbf{r}''|\mathbf{r}_I, \mathbf{r}_{II}; \xi) = \int d\mathbf{r}' \frac{\rho(\mathbf{r}'|\mathbf{r}_I, \mathbf{r}_{II}; \xi)}{4\pi \varepsilon_r \varepsilon_0 |\mathbf{r}' - \mathbf{r}''|},$$

where we for simplicity in notation write $\psi_\rho(\mathbf{r}''; \xi)$ whereby the dependence on $\mathbf{r}_I$ and $\mathbf{r}_{II}$ is not explicitly shown. From these results, we see that Equation 6.100 can be written

$$\mu_{I,II}^{(2)ex,el}(\mathbf{r}_{I,II}) = u_{I,II}^{el}(\mathbf{r}_{I,II})$$

$$+ \int_0^1 d\xi \int d\mathbf{r}'' \left[\sigma_I(\mathbf{r}'' - \mathbf{r}_I) + \sigma_{II}(\mathbf{r}'' - \mathbf{r}_{II}) \right] \psi_\rho(\mathbf{r}''; \xi). \quad (6.102)$$

Note that the potential $\psi_\rho(\mathbf{r}''; \xi)$ is not equal to the total mean potential $\psi(\mathbf{r}''; \xi)$, which in addition includes the potentials from particles I and II.

Given Equation 6.102, it is, as we will see shortly, straightforward to show the final result Equation 6.99, which contains the interaction of particles I and II with ψ instead of ψ_ρ. To show Equation 6.99, let us consider its integrand which we can write as $[\sigma_I + \sigma_{II}]\psi(\xi)$, where we for simplicity in notation have skipped the coordinates. If ψ_I denotes the potential from particle I and ψ_{II} that from particle II when they are fully charged, we have $\psi(\xi) = \psi_\rho(\xi) + \xi\psi_I + \xi\psi_{II}$ and we obtain $[\sigma_I + \sigma_{II}]\psi(\xi) = [\sigma_I + \sigma_{II}]\psi_\rho(\xi) + \xi[\sigma_I\psi_I + \sigma_I\psi_{II} + \sigma_{II}\psi_I + \sigma_{II}\psi_{II}]$. The first term on the rhs is the integrand of Equation 6.102, so it only remains to investigate the rest, which equals ξ times the four terms in the square brackets. These four terms are independent of ξ, so when we do the ξ integration in Equation 6.99 for this latter part, we obtain the factor $\int_0^1 \xi \, d\xi = 1/2$. Let us now perform the integration over $\mathbf{r}$ of these four terms. We have $\int d\mathbf{r}\sigma_I\psi_{II} = \int d\mathbf{r}\sigma_{II}\psi_I = u_{I,II}^{el}$, so the sum of these two terms multiplied by 1/2 gives the first term of the rhs in Equation 6.102. There remain $\int d\mathbf{r}\sigma_I\psi_I/2$ and $\int d\mathbf{r}\sigma_{II}\psi_{II}/2$ which equal $u_{I,I}^{el}(0)/2$ and $u_{II,II}^{el}(0)/2$, respectively, with $u_{J,J'}^{el}$ defined in Equation 6.101. Each of these is a "self-energy" of the internal charge distributions of the particle and is independent of the separation vector $\mathbf{r}_{I,II}$. The constant in Equation 6.99 is minus the sum of these terms for the two particles. If one wishes, one can include the self-energies in the chemical potentials[15] for both $\mu_{I,II}^{(2)ex}$ and $\mu_J^{ex,b}$, and then Equation 6.99 is valid without the "const" term. This would, however, not work for ions with point charges,[16] wherefore we have selected to define chemical potentials without such self-energies.

---

[15] The excess chemical potential would then be equal to the reversible work to *assemble* the particle by gradually bringing infinitesimal charges from infinity and then insert the particle into the system or, alternatively, to likewise assemble it in place in the system. We have instead selected to start with a ready-made particle that does not interact with anything else and we define the excess chemical potential as the reversible work to insert it into the system.

[16] A point charge has an infinite self-energy. In the difference $w_{I,II}^{(2)}(\mathbf{r}_{I,II}) = \mu_{I,II}^{(2)ex}(\mathbf{r}_{I,II}) - \mu_I^{ex,b} - \mu_{II}^{ex,b}$, this infinity would be cancelled, but we want to deal with each of these terms separately.

6.1.5 One Step beyond PB: What Happens When *all* Ions are Treated on an Equal Basis?

In Section 6.1.3, we used the PB approximation to investigate the decay of the screened electrostatic interaction between two particles, I and II, in an electrolyte solution. Both particles were treated on an equal basis and we considered how the surrounding ion clouds change when the particles approach each other. When they are very far from each other, they have unperturbed ion clouds of their own, but at shorter separations the interaction from one particle to the other distorts the clouds. For spherical particles, the potential of mean force decays like (Equation 6.92)

$$w_{I,II}^{(2)}(r) \sim q_I^{\text{eff}} q_{II}^{\text{eff}} \phi_{\text{Coul}}^{\star}(r) = \frac{q_I^{\text{eff}} q_{II}^{\text{eff}}}{4\pi \varepsilon_r \varepsilon_0} \cdot \frac{e^{-\kappa_D r}}{r}$$

when $r \to \infty$. Furthermore, we have $w_{I,II}^{(2)}(r) \sim q_I^{\text{eff}} \psi_{II}(r)$, $w_{I,II}^{(2)}(r) \sim q_{II}^{\text{eff}} \psi_I(r)$ and $\psi_J(r) \sim q_J^{\text{eff}} \phi_{\text{Coul}}^{\star}(r)$ for $J = $ I and II.

An inherent feature of the PB approximation is, as we have seen, that the ions in the ion cloud are treated in a different manner – like point ions without ion clouds of their own. The potential of mean force between an ion of species i in the ion cloud and particle J, for $J = $ I or II, is assumed to be $w_{iJ}^{(2)}(r) = q_i \psi_J(r)$, which decays like

$$w_{iJ}^{(2)}(r) \sim \frac{q_i q_J^{\text{eff}}}{4\pi \varepsilon_r \varepsilon_0} \cdot \frac{e^{-\kappa_D r}}{r}.$$

These relationships hold, of course, also when particles I and II are of the same species as ions in the surrounding electrolyte solution, that is, those in the ion cloud. As we have discussed earlier, the unequal treatment of the ions in the PB approximation leads to the asymmetric feature $w_{ij}^{(2)}(r) \neq w_{ji}^{(2)}(r)$, which is incorrect. In the linearized PB approximation, the DH theory, we discussed these aspects in some detail in connection to Equations 6.51 and 6.52, which correspond to the two equations above. Then the value of the effective charge of an ion is given by (Equation 6.49)

$$q_i^{\text{eff}} = q_i \cdot \frac{e^{\kappa_D d^h}}{1 + \kappa_D d^h} \quad \text{(DH theory)}.$$

As we have seen, the DH theory is a reasonable approximation when $\beta q \psi$ is weak, like for monovalent ions in sufficiently dilute aqueous solutions or at sufficiently high temperatures.

We will now investigate what happens when *all ions* are treated on an equal basis, not only the two ions we considered earlier but also the ions in the ion cloud. We will do this by introducing a very simple but illustrative approximation that avoids the inherent asymmetry in the PB approximation, so we will have $w_{ij}^{(2)}(r) = w_{ji}^{(2)}(r)$ for the interaction between all ions. Before doing this, we note that the previous analysis shows two key features of effective charges of ions:

i. The relationship $\psi_j(r) \sim q_j^{\text{eff}} \phi_{\text{Coul}}^{\star}(r)$ when $r \to \infty$ means that the electrostatic potential from a j-ion has a magnitude and sign in the tail region (large r) that is described by the value of the effective charge q_j^{eff}. In this region, the potential $\psi_j(r)$ is weak. If $\beta q_i \psi_j$ is weak everywhere outside the ion, the potential is to a good approximation equal to $q_j^{\text{eff}} \phi_{\text{Coul}}^{\star}(r)$ for small r as well, Equation 6.50.

ii. The expression $w_{ij}^{(2)}(r) \sim q_i^{\text{eff}} \psi_j(r)$ shows that q_i^{eff} gives the strength of the interaction of the i-ion with the potential $\psi_j(r)$ from the j-ion in the tail region where ψ_j is small. This holds provided the i-ion is treated on an equal basis as the j-ion.

These points show that the effective charge has a dual role: it gives the strength of the electrostatic potential where the latter is weak as well as the strength of the interaction of an ion with this potential.

Let us limit ourselves to cases where $\beta q_i \psi_j$ and $\beta q_i^{\text{eff}} \psi_j$ are small. The product $q_i^{\text{eff}} \psi_j(r)$ can then be taken as an approximation for $w_{ij}^{(2)}(r)$ not only in the tail region, which means we have $w_{ij}^{(2)}(r) = q_i^{\text{eff}} \psi_j(r)$ everywhere outside the ions. We assume all ions have the same diameter d^{h} so we take

$$g_{ij}^{(2)}(r) = \begin{cases} 0, & r < d^{\text{h}} \\ e^{-\beta q_i^{\text{eff}} \psi_j(r)}, & r \geq d^{\text{h}} \end{cases}$$

for all ions in the bulk electrolyte. This is done instead of the PB approximation (6.28), so in this improved approximation we have q_i^{eff} instead of q_i in the exponent. Since $\beta q_i^{\text{eff}} \psi_j$ is small, we can expand the exponential and obtain for $r \geq d^{\text{h}}$

$$\rho_j(r) = \sum_i q_i n_i^{\text{b}} e^{-\beta q_i^{\text{eff}} \psi_j(r)} \approx \sum_j q_i n_i^{\text{b}} - \beta \sum_i q_i q_i^{\text{eff}} n_i^{\text{b}} \psi_i(r),$$

where the first term on the rhs is zero, as always. By inserting this in the Poisson equation, we obtain the approximation $-\varepsilon_r \varepsilon_0 \nabla^2 \psi_j(r) = -\beta \sum_i q_i q_i^{\text{eff}} n_i^{\text{b}} \psi_j(r)$ and we have

$$\nabla^2 \psi_j(r) = \begin{cases} 0, & 0 < r < d^{\text{h}} \\ \kappa^2 \psi_j(r), & r \geq d^{\text{h}} \end{cases}, \tag{6.103}$$

where

$$\kappa^2 = \frac{\beta}{\varepsilon_r \varepsilon_0} \sum_i q_i q_i^{\text{eff}} n_i^{\text{b}}. \tag{6.104}$$

The parameter κ differs from the Debye parameter κ_{D} solely by the occurrence of $q_i q_i^{\text{eff}}$ instead of q_i^2 (compare with Equation 6.31). Thus, the bare charge q_i is replaced by the effective charge in one of the factors of q_i and all other factors in the expression remain the same. The solution of Equation 6.103 is the same as that of the DH Equation 6.30 except that we have κ instead of κ_D, that is, the Yukawa function $\psi_j(r) = C_1 \exp(-\kappa r)/r$. We can determine C_1 in the same manner as when we obtained Equation 6.33, which results in $C_1 = q_j \exp(\kappa d^h)/[4\pi \varepsilon_r \varepsilon_0 (1 + \kappa d^h)]$. Thus, we have

$$\psi_j(r) = \frac{q_j e^{\kappa d^h}}{1 + \kappa d^h} \cdot \frac{e^{-\kappa r}}{4\pi \varepsilon_r \varepsilon_0 r} = q_j^{\text{eff}} \phi_{\text{Coul}}^{\star}(r) \quad \text{for } r \geq d^h,$$

where we have identified

$$q_j^{\text{eff}} = \frac{q_j e^{\kappa d^h}}{1 + \kappa d^h} \tag{6.105}$$

and defined

$$\phi_{\text{Coul}}^{\star}(r) = \frac{e^{-\kappa r}}{4\pi \varepsilon_r \varepsilon_0 r} \quad \text{(improved DH approx.)} \tag{6.106}$$

as we did when we obtained Equation 6.69, but with κ_D replaced by κ. Since these results are valid for all ions in the electrolyte, we have $w_{ij}^{(2)}(r) = q_i^{\text{eff}} \psi_j(r) = q_i^{\text{eff}} q_j^{\text{eff}} \phi_{\text{Coul}}^{\star}(r)$, which has the correct symmetry. In this manner, all ions in the electrolyte have been treated on the same basis and all ions of the same species have the same effective charge, including the ions in the ion cloud of each ion, and we have $q_j^{\text{eff}} \neq q_j$. This is in contrast to the PB and DH approximations for bulk electrolytes, where only the central ion has $q_j^{\text{eff}} \neq q_j$.

An immediate consequence of this simple improvement of the DH theory is that the decay parameter κ in Equation 6.104 is different from κ_D. Thus, *when all ions are treated on an equal basis, the screening decay length $1/\kappa$ is different from the Debye length $1/\kappa_D$.* We can easily determine this new decay length by combining Equation s 6.104 and 6.105, which yields

$$\kappa^2 = \frac{\beta}{\varepsilon_0} \sum_i n_i^b q_i \left[q_i \frac{e^{\kappa d^h}}{1 + \kappa d^h} \right].$$

This is an equation for κ since the rhs depends on κ and it can be written as

$$\left[\frac{\kappa}{\kappa_D} \right]^2 = \frac{e^{\kappa d^h}}{1 + \kappa d^h} \tag{6.107}$$

or as

$$\alpha^2 = \frac{e^{\alpha \tau}}{1 + \alpha \tau}, \tag{6.108}$$

where we have introduced $\alpha = \kappa/\kappa_D$ and $\tau = \kappa_D d^h$. By solving this equation, which has to be done numerically, we can obtain α as a function of τ and thereby determine the decay length $1/\kappa$ as a function of the salt concentration (τ is proportional to $\sqrt{c_{\text{salt}}}$ via the factor κ_D). When the density goes to zero, we have $\tau \to 0$ and from Equation 6.108 it follows that $\alpha \to 1$, that is, $\kappa \sim \kappa_D$ at low densities. Thus, the DH theory is recovered in this limit. Likewise, $q_j^{\text{eff}} \to q_j$ in the same limit.

Let us compare the results from this simple approximation with results from Monte Carlo (MC) simulations for symmetric electrolytes with spherical hard ions of diameter d^h. If distances are measured in units of d^h, such a system can be fully characterized in terms of

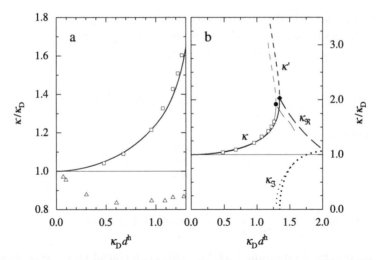

FIGURE 6.7 (a) The decay parameter κ divided by the Debye parameter κ_D obtained from the simple approximation in Equation 6.107 for a binary symmetric electrolyte with ions of diameter d^h plotted as functions of $\tau = \kappa_D d^h$ (which is proportional to the square root of the total ion density). The symbols show the corresponding results from Monte Carlo (MC) simulations for $\beta_R = 1.55$ (open squares) and 6.22 (open triangles). (b) Comparison between results from Equation 6.107 and accurate calculations for $\beta_R = 1.55$ showing the crossover between monotonic and oscillatory decay of the electrostatic potential. Thick curves show results from Equation 6.107, thin curves show results from the Hypernetted Chain (HNC)[17] integral equation theory and open squares show MC results. Various decay parameters (see text) divided by the Debye parameter κ_D are displayed: full curves and symbols, κ; short dashed curves, κ' (both κ and κ' are real solutions of the equation for κ); long dashed curves, $\kappa_{\Re}$; dotted curves, $\kappa_{\Im}$ (for complex solutions of the κ equation, we have $\kappa = \kappa_{\Re} \pm i\kappa_{\Im}$). The filled symbols show the crossover point between monotonic exponential (Yukawa function) and exponentially damped oscillatory decays, that is, real and complex κ, respectively, as calculated in the HNC approximation and from Equation 6.107.[18]

two reduced (dimensionless) parameters, for example $\tau = \kappa_D d^h$ and $\beta_R = \beta q^2/(4\pi\varepsilon_r\varepsilon_0 d^h)$, where $q = q_+ = -q_-$ (the physical meaning of β_R is the magnitude of the interaction energy of two ions in contact measured in $k_B T$ units). The choice of dimensionless parameters is not unique; alternative parameters can be selected that are combinations of these two.[19] The MC calculations that we will compare with were performed at $\beta_R = 1.55$ and 6.22, which corresponds, for example, to 1:-1 and 2:-2 aqueous electrolyte solutions, respectively, at temperature $T = 298$ K and with $d^h = 4.6$ Å. These values of β_R also correspond to a 1:-1 classical plasma in vacuum ($\varepsilon_r = 1$) at temperatures 23 400 and 5840 K, respectively.

The ratio κ/κ_D ($= \alpha$) obtained from Equation 6.107 and from MC simulations is shown in Figure 6.7a as a function of $\kappa_D d^h$ ($= \tau$). For $\beta_R = 1.55$, which corresponds both to the 1:-1 aqueous electrolyte solution and the high temperature plasma, the values of κ/κ_D obtained

---

[17] The HNC approximation is introduced in Chapter 9 (in the 2nd volume).

[18] The MC and HNC data in Figure 6.7 are taken from J. Ulander, and R. Kjellander, *J. Chem. Phys.* **114** (2001) 4893.

[19] Alternative parameters that can be used to characterize the system can be written as combinations of τ and β_R, for example, the reduced density $n_R^b = n^b(d^h)^3 = \tau^2/(8\pi\beta_R)$, where $n^b = n_+^b = n_-^b$, and the so-called plasma parameter $(\beta_R\tau)^{-1}$, which is often denoted by Λ (which is different from the thermal de Broglie wavelength Λ in this treatise).

from Equation 6.107 are surprisingly accurate considering the simplicity of the approximation. On the other hand, for $\beta_R = 6.22$, corresponding to the 2:-2 electrolyte solution and the plasma at the lower temperature, the approximation does not work, as apparent from the figure. This is not surprising since it was derived under the condition of small values of $\beta q_i \psi_j$ (small β_R), for example a low valency electrolyte in a solvent with high ε_r (for water at room temperature $\varepsilon_r = 78.4$) or a plasma at high temperatures. Since Equation 6.108 gives a κ/κ_D that only depends on τ and is independent of β_R, it cannot give correct results unless β_R is small (of the order of 1 or less).

Let us take a closer look at how much the screening parameter κ deviates from the Debye parameter κ_D. For the case $\beta_R = 1.55$, the deviation of κ/κ_D from 1 is appreciable for large densities (large $\kappa_D d^h$), but for $\kappa_D d^h \lesssim 0.5$, which corresponds to a 1:-1 electrolyte solution with concentration less than about 0.1 M, the deviation of κ/κ_D from 1 is small – less than 5%. For electrolyte solutions in the primitive model, the results for higher concentrations are quite irrelevant since the realism of this model is low when the decay length is comparable to the molecular size. The solvent is represented by a continuum without any structure and is not modeled as discrete solvent molecules. The results for all $\kappa_D d^h$ are, however, relevant for classical plasmas at high temperatures, where we have ions in vacuum.

For $\beta_R = 6.22$, the deviation of κ/κ_D from 1 obtained in the MC simulations is up to around 17 % in the range shown. In this case, $\kappa < \kappa_D$ in the displayed range, while for $\beta_R = 1.55$ we have $\kappa > \kappa_D$.[20] Equation 6.107 always gives $\kappa > \kappa_D$.

Equation 6.107 has, in fact, some features that reflect important properties of electrolytes that lie beyond the simple assumptions made in its derivation, most notably a crossover from monotonic to oscillatory decay of the electrostatic potential at an increased density. This is described in the following shaded text box. For small β_R, we will find that the results of Equation 6.107 are remarkably accurate considering its humble origin. The subjects discussed in the box constitute an introduction to the behavior of simple electrolytes in the general case, which is treated in detail in the next section (Section 6.2).

Equation 6.108 has two solutions for α, one of which goes to 1 when $\tau \to 0$, for instance when the electrolyte density goes to zero. This solution, $\alpha = \kappa/\kappa_D$, thus approaches the ordinary DH case, $\kappa \sim \kappa_D$, for low densities. The other solution, $\alpha' = \kappa'/\kappa_D$, corresponds to an additional term in the potential that is a Yukawa function with a shorter decay length, so we have

$$\psi_i(r) \sim C_i \frac{e^{-\kappa r}}{r} + C_i' \frac{e^{-\kappa' r}}{r}, \tag{6.109}$$

where C_i and C_i' are constants. Thus, there are two decay lengths $1/\kappa$ and $1/\kappa'$.

---

[20] In accurate calculations, there is, in fact, a slight dip of κ/κ_D below 1 for very low densities, in the case $\beta_R = 1.55$. For low densities, κ/κ_D is always below 1 in accurate theories for symmetric electrolytes, but the interval with values below 1 is very small in this case. For the $\beta_R = 6.22$ case, the curve lies below 1 for the whole range shown in the figure.

When τ is increased, the two solutions to Equation 6.108 approach each other and at $\tau \approx 1.35$ they merge. For $\tau > 1.35$, the solutions are complex-valued; the two solutions are then complex conjugates to each other: $\kappa = \kappa_\Re + i\kappa_\Im$ and $\kappa' = \kappa_\Re - i\kappa_\Im$, where i is the imaginary unit and $\kappa_\Re$ and $\kappa_\Im$ are the real numbers. The prefactors in Equation 6.109 become complex-valued too (complex conjugates) $C_i = |C_i| \exp(-i\vartheta_i^c)$ and $C_i' = |C_i'| \exp(i\vartheta_i^c)$, where ϑ_i^c is a real number and $|C_i| = |C_i'| = B_i/2$, which defines B_i. The sum of the two terms in Equation 6.109 is therefore a real valued function

$$\psi_i(r) \sim |C_i| e^{-i\vartheta_i^c} \frac{e^{-(\kappa_\Re + i\kappa_\Im)r}}{r} + |C_i'| e^{i\vartheta_i^c} \frac{e^{-(\kappa_\Re - i\kappa_\Im)r}}{r}$$

$$= \frac{B_i}{2} \left[e^{-i(\kappa_\Im r + \vartheta_i^c)} + e^{i(\kappa_\Im r + \vartheta_i^c)} \right] \frac{e^{-\kappa_\Re r}}{r}$$

$$= B_i \cos(\kappa_\Im r + \vartheta_i^c) \frac{e^{-\kappa_\Re r}}{r}, \tag{6.110}$$

where $1/\kappa_\Re$ is the decay length, $2\pi/\kappa_\Im$ is the wavelength and ϑ_i^c is a phase shift. This kind of behavior is what is found in more refined treatments.

The results from Equation 6.108 are compared in Figure 6.7b with those from Monte Carlo (MC) simulations and Hypernetted Chain (HNC) calculations[21] for symmetric electrolytes at $\beta_R = 1.55$. The figure shows the decay parameters κ and κ' when Equation 6.108 has real solutions and $\kappa_\Re$ and $\kappa_\Im$ when the solutions are complex. From the comparison with the MC and HNC results, we see that this simple approximation is remarkably accurate. This shows that the characteristic behavior of electrolytes with several decay lengths and a transition to oscillatory behavior can arise from the interplay between electrostatics and excluded volume *provided all ions are treated on the same basis.*

Very similar results are obtained in various other linear approaches which also treat all ions on an equal basis like the Mean Spherical Approximation (MSA) introduced in Chapter 9 (in the 2nd volume). The crossover to oscillatory decay is often called the **Kirkwood crossover point**, named after John G. Kirkwood who already in the 1930s showed the existence of a transition to oscillatory behavior for electrolytes. This type of crossover is defined as a change where two real-valued decay parameters κ and κ' merge and become complex-valued at a certain density.

6.2 ELECTROSTATIC SCREENING IN SIMPLE BULK ELECTROLYTES, GENERAL CASE

In the PB treatment of electrostatic interactions in electrolytes, one neglects the correlation between the ions in the ion cloud around the central ion or around any other kind of particle

---

[21] The HNC approximation, which is introduced in Chapter 9, is accurate for primitive model electrolytes under a reasonably wide range of conditions.

one considers, for instance macroions. As we have seen repeatedly, this unequal treatment of the particles in the system leads in general to inconsistencies like $g_{ij} \neq g_{ji}$. In Section 6.1.5, we saw that a treatment of all ions on an equal basis has quite dramatic consequences – much larger consequences than a simple rectification of the symmetry $g_{ij} = g_{ji}$. To build a correct (exact) theory for electrolytes may appear to be a formidable task, which it indeed is in many respects, but there are some general, exact features of electrostatic screening in electrolytes that can be brought forward without undue effort. In fact, there are elements in the PB theory, as presented earlier in this treatise, that can be generalized in a straightforward manner and thereby can give important aspects of the correct theory.

Screening of electrostatic interactions is a central feature of electrolytes and gives rise to the exponential decay of the electrostatic potential as expressed in the unit screened Coulomb potential

$$\phi^\star_{\text{Coul}}(r) = \frac{e^{-\kappa_D r}}{4\pi \varepsilon_r \varepsilon_0 r} \quad \text{(PB)}. \tag{6.111}$$

This leads to the decay law $w^{(2)}_{\text{I,II}}(r) \sim q^{\text{eff}}_{\text{I}} q^{\text{eff}}_{\text{II}} \phi^\star_{\text{Coul}}(r)$ when $r \to \infty$ for the interaction between two spherical macroions I and II and the corresponding orientation dependent law for nonspherical macroions in Equation 6.94. As we have seen in Section 6.1.5, screening is a more complex phenomenon than what the PB theory encompasses since there is in general more than one decay length and the decay can also be oscillatory. In order to include these additional aspects of screening, it is, for example, vital to generalize $\phi^\star_{\text{Coul}}(r)$, which describes the spatial propagation of electrostatic interactions in the electrolyte and was defined as a Green's function in Equation 6.68.

In the following sections, we will also investigate the decay of the electrostatic interactions for large separations and we will see that the obtained decay laws can be valid down to quite short separations, sometimes surprisingly short. We will see that these decay laws are very useful as tools in the analysis of the properties of the electrolyte systems. To obtain the potential of mean force w_{ij} accurately for *all* separations and thereby the pair distribution g_{ij}, one has to resort to simulations, Section 5.13, or use theoretical methods that will be treated in Parts III and IV in the 2nd volume of this treatise. Incidentally, we may remind ourselves that the method in Section 6.1.4.2 for calculation of the electrostatic part of the potential of mean force from the electrostatic mean potential can also be used in the general case.

In the present treatment, we restrict ourselves to bulk electrolytes that consist solely of spherical ions in vacuum (no solvent) or spherical ions with a dielectric continuum solvent. We will also, like before, consider electrolytes where we have inserted one or a few fixed particles with arbitrary size, shape and any internal charge distribution. The simple spherical ions interact between themselves with electrostatic u^{el}_{ij} and short-range u^{sh}_{ij} interactions, where the latter do not need to be hard-core potentials, so we have

$$u_{ij}(r) = u^{\text{el}}_{ij}(r) + u^{\text{sh}}_{ij}(r).$$

Likewise, the immersed particles interact with the small ions and with each other via electrostatic and arbitrary short-range potentials. The latter and u_{ij}^{sh} are here assumed to decay to zero with distance faster than any power law.[22]

Before proceeding, let us make the following remark. In various formulas that we will consider in the following sections, the dielectric constant ε_r is included. This is correct for systems where a solvent is modeled as a dielectric continuum, for instance the primitive model of electrolyte solutions. When dealing with systems without a solvent, we must set $\varepsilon_r = 1$ in the formulas (i.e., the dielectric constant of vacuum). The theory for electrolytes with a molecular solvent is postponed until Chapter 13 (in the 2nd volume), where dielectric effects of both the solvent and ions are treated properly.

6.2.1 Electrostatic Interaction Potentials

The mean electrostatic potential from a particle P immersed in a bulk electrolyte and held fixed at some orientation is given in terms of Coulomb's law as

$$\psi_P(\mathbf{r}_1) = \int d\mathbf{r}_2 \rho_P^{tot}(\mathbf{r}_2)\phi_{Coul}(r_{21}), \tag{6.112}$$

where $\rho_P^{tot}(\mathbf{r}) = \sigma_P(\mathbf{r}) + \rho_P(\mathbf{r})$ is the total charge density associated with the particle; $\sigma_P(\mathbf{r})$ is the internal charge density of P and $\rho_P(\mathbf{r})$ is the charge density of the surrounding ion cloud. The origin is here placed at the center of mass of P. In the PB approximation, we found in Section 6.1.2 that ψ_P can be expressed in terms of the unit screened Coulomb potential $\phi_{Coul}^\star(r)$ and the charge density $\rho_P^\star(\mathbf{r})$ of the dressed particle P as (Equation 6.70)

$$\psi_P(\mathbf{r}_1) = \int d\mathbf{r}_2 \rho_P^\star(\mathbf{r}_2)\phi_{Coul}^\star(r_{21}), \tag{6.113}$$

which has the same form as Coulomb's law. The dressed particle charge density is defined as $\rho_P^\star(\mathbf{r}) = \rho_P^{tot}(\mathbf{r}) - \rho_P^{lin}(\mathbf{r})$ (Equation 6.60), where ρ_P^{lin} is the *linear part of the polarization response* of the bulk electrolyte from the electrostatic field due to particle P. This part is given by (Equation 6.59)

$$\rho_P^{lin}(\mathbf{r}) = -\beta \sum_j q_j^2 n_j^b \psi_P(\mathbf{r}) = -\varepsilon_r \varepsilon_0 \kappa_D^2 \psi_P(\mathbf{r}) \quad \text{(PB)}$$

in the PB approximation.

Furthermore, we found in the PB approximation that the potential of mean force between two particles I and II immersed in a bulk electrolyte decays like in Equation 6.93, which can be written as

$$w_{I,II}^{(2)}(r_{I,II}) \sim \int d\mathbf{r}_1 d\mathbf{r}_2 \rho_I^\star(r_{1,2})\phi_{Coul}^\star(r_{21})\rho_{II}^\star(r_{II,1}) \quad \text{when } r_{I,II} \to \infty, \tag{6.114}$$

---

[22] The important case where u_{ij}^{sh} decays like $1/r^6$ (dispersion interactions) is discussed in Section 7.1.5 (see the shaded text box on page 390f at the end of Section 7.1.5); a more complete treatment of this case will be found in Part IV (in the 2nd volume).

for spherical particles I and II. These two particles are here treated in exactly the same manner and both correlate with the surrounding ions, while correlations between the latter ions are ignored in the PB approximation. Our task is to generalize these results so *all* correlations are considered and *all* particles are treated on the same basis. We will thereby find out how to generalize $\phi^\star_{\text{Coul}}(r)$ and how $\rho_P^{\text{lin}}(\mathbf{r})$ should be defined so that Equation 6.113 remains valid. Finally, we will generalize Equations 6.93 and 6.114.

6.2.1.1 Polarization Response and Nonlocal Electrostatics

We can regard particle P as an external body that we immerse into the bulk electrolyte, whereby the ion density distribution of the latter becomes changed and, as a result, the total electrostatic potential around P becomes equal to ψ_P. This density change is the response of the electrolyte to the interaction of the ions with P – an interaction that from the point of view of the fluid constitutes an external potential. The electrostatic interaction of the ions with $\sigma_P(\mathbf{r})$ is a part of this external potential; the rest is equal to the short-range interactions with P. We will start by considering the response of a bulk electrolyte to a small external electrostatic potential, whereby the response is linear. We will later identify the *linear part* of the response to the total electrostatic potential $\psi_P(\mathbf{r})$ due to P. This potential is not small and the full response is therefore nonlinear. Remember that $\psi_P(\mathbf{r})$ is the sum of the electrostatic potential from $\sigma_P(\mathbf{r})$, here considered as an external potential, and the potential from the charge density $\rho_P(\mathbf{r})$ of the surrounding ion cloud that arises as a response of the interactions of the ions with P.

In the general case of an inhomogeneous electrolyte that is exposed to an external electrostatic potential $\Psi^{\text{ext}}(\mathbf{r}')$ from a source outside the system, one may ask what happens when the latter is varied by $\delta\Psi^{\text{ext}}(\mathbf{r}')$, that is, when the fluid instead is exposed to the external potential $\Psi^{\text{ext}}(\mathbf{r}') + \delta\Psi^{\text{ext}}(\mathbf{r}')$ where $\delta\Psi^{\text{ext}}(\mathbf{r}')$ is assumed to be small everywhere in the electrolyte. In particular, one may ask how much the density distribution is changed: for species i, it varies from $n_i(\mathbf{r})$ to, say, $n_i(\mathbf{r}) + \delta n_i(\mathbf{r})$ and one would like to know the variation $\delta n_i(\mathbf{r})$ for a given variation $\delta\Psi^{\text{ext}}(\mathbf{r}')$.[23]

In the present case, we are dealing with a bulk electrolyte where, initially, the external potential $\Psi^{\text{ext}}(\mathbf{r}') = 0$, so a small variation $\delta\Psi^{\text{ext}}(\mathbf{r}')$ in this potential means that the density for any species i goes from the bulk value n_i^{b} to $n_i^{\text{b}} + \delta n_i(\mathbf{r})$. In what follows, the symbol δ in front of an entity therefore simply means that the latter is small.

The density distribution is in general given by

$$n_i(\mathbf{r}) = n_i^{\text{b}} e^{-\beta w_i(\mathbf{r})}$$

and the potential of mean force w_i for the bulk electrolyte is set equal to zero. When a weak external electrostatic potential $\delta\Psi^{\text{ext}}$ is turned on, the potential of mean force deviates from zero and becomes, say, $\delta w_i(\mathbf{r})$, which is small too. If we know $\delta w_i(\mathbf{r})$, we can obtain the density $n_i(\mathbf{r})$ from

$$n_i(\mathbf{r}) = n_i^{\text{b}} e^{-\beta \delta w_i(\mathbf{r})} \approx n_i^{\text{b}}[1 - \beta \delta w_i(\mathbf{r})] = n_i^{\text{b}} - \beta n_i^{\text{b}} \delta w_i(\mathbf{r}),$$

---

[23] The answers to these questions can be found in Appendix 6B.

so we see that $\delta n_i(\mathbf{r}) = -\beta n_i^b \delta w_i(\mathbf{r})$ in the limit when $\delta \Psi^{ext}$ and δw_i go to zero. In Appendix 6B, it is shown that $\delta w_i(\mathbf{r}_1)$ for an i-ion at $\mathbf{r}_1$ is (Equation 6.210)

$$\delta w_i(\mathbf{r}_1) = \int d\mathbf{r}_2 \, \rho_i^{tot}(r_{12}) \delta \Psi^{ext}(\mathbf{r}_2), \tag{6.115}$$

which means that the potential of mean force is given by the electrostatic interaction energy between the weak $\delta \Psi^{ext}$ and the total charge distribution ρ_i^{tot} associated with the ion. This result is based on the so-called *first Yvon equation*, which is an exact relationship between the density distribution and variations in external potential derived in Appendix 6B. The Yvon equation itself and the details of its derivation are not needed for the applications here and can be skipped in the first reading.

Since $\rho_i^{tot}(r_{12}) = q_i \delta^{(3)}(r_{12}) + \rho_i(r_{12})$, we see from Equation 6.115 that

$$\delta w_i(\mathbf{r}_1) = q_i \delta \Psi^{ext}(\mathbf{r}_1) + \int d\mathbf{r}_2 \, \rho_i(r_{12}) \delta \Psi^{ext}(\mathbf{r}_2),$$

where the first term is the interaction energy with the ion itself and the second term originates from the ion-ion correlations, that is, the distribution of ions in the ion cloud around the i-ion. The external field interacts with the ions in the ion cloud and they influence the i-ion at $\mathbf{r}_1$ via the correlations. From $\delta n_i(\mathbf{r}) = -\beta n_i^b \delta w_i(\mathbf{r})$, it follows that the density distribution caused by $\delta \Psi^{ext}$ is

$$\delta n_i(\mathbf{r}_1) = -\beta n_i^b \int d\mathbf{r}_2 \, \rho_i^{tot}(r_{12}) \delta \Psi^{ext}(\mathbf{r}_2). \tag{6.116}$$

Note that these exact results are valid in the limit where $\delta \Psi^{ext}(\mathbf{r}_2) \to 0$ for all $\mathbf{r}_2$ inside the electrolyte.

The charge density caused by the influence of $\delta \Psi^{ext}$ on the bulk electrolyte, that is, the **polarization charge density**, is hence given by

$$\delta \rho^{pol}(\mathbf{r}_1) = \sum_i q_i \delta n_i(\mathbf{r}_1) = -\beta \sum_i q_i n_i^b \int d\mathbf{r}_2 \, \rho_i^{tot}(r_{12}) \delta \Psi^{ext}(\mathbf{r}_2).$$

We can write this as

$$\delta \rho^{pol}(\mathbf{r}_1) = \int d\mathbf{r}_2 \, \chi^\rho(r_{12}) \delta \Psi^{ext}(\mathbf{r}_2), \tag{6.117}$$

where

$$\chi^\rho(r_{12}) \equiv -\beta \sum_i q_i n_i^b \rho_i^{tot}(r_{12}) \tag{6.118}$$

is a function that links the polarization charge density at one point, $\mathbf{r}_1$, to the external electrostatic potential at another point, $\mathbf{r}_2$. More precisely, $\chi^\rho(r_{12})$ gives the contribution to $\delta \rho^{pol}$

at $\mathbf{r}_1$ from the influence of the external electrostatic potential at $\mathbf{r}_2$. This kind of function is called a **polarization response function**. It is a *linear response function*, which means that it gives the polarization caused by the potential when the latter is weak and the response therefore is linear. Furthermore, since we are dealing with equilibrium systems, $\chi^\rho(r)$ is a *static response function* which does not involve any time dependence.[24] The external potential is independent of time and no currents flow in the system.

The polarization charge density gives rise to an electrostatic potential $\delta\Psi^{pol}$ given by

$$\delta\Psi^{pol}(\mathbf{r}_2) = \int d\mathbf{r}_3 \delta\rho^{pol}(\mathbf{r}_3)\phi_{Coul}(r_{32}), \qquad (6.119)$$

so the total electrostatic potential $\delta\Psi$ caused by $\delta\Psi^{ext}$ is

$$\delta\Psi(\mathbf{r}) = \delta\Psi^{ext}(\mathbf{r}) + \delta\Psi^{pol}(\mathbf{r}). \qquad (6.120)$$

Since δn_i and $\delta\rho^{pol}$ are averages, $\delta\Psi(\mathbf{r})$ is the *electrostatic mean potential* at $\mathbf{r}$. Remember that it is the presence of an external source (some stationary charge density) that gives rise to $\delta\Psi^{ext}$ and hence $\delta\Psi^{pol}$ and $\delta\Psi$ in a bulk system.

Note that $\delta\Psi^{ext}$ is the potential due to the source in the *absence of the electrolyte*, while $\delta\Psi$ is the total potential in the *presence of the electrolyte*. Since $\delta\Psi$ is the mean potential that actually exists in the system we are interested in, we have good reason to obtain an expression of $\delta\rho^{pol}$ in terms of $\delta\Psi$, that is, an expression that we can use as an alternative to Equation 6.117. The functions $\delta\Psi^{ext}$, $\delta\rho^{pol}$, and $\delta\Psi$ are, as we have seen, linearly related to each other, so analogously to Equation 6.117 there exists a function $\chi^\star(r)$ that expresses the linear relationship between $\delta\rho^{pol}$ and $\delta\Psi$ as

$$\delta\rho^{pol}(\mathbf{r}_1) = \int d\mathbf{r}_2 \, \chi^\star(r_{12})\delta\Psi(\mathbf{r}_2). \qquad (6.121)$$

The function $\chi^\star(r)$ will also be denoted as a *polarization response function*.[25] We know $\chi^\rho(r)$ from Equation 6.118 and in the shaded text box below, which can be skipped in the first reading, it is shown how $\chi^\star(r)$ can be determined from $\chi^\rho(r)$.

---

[24] Frequency dependent response functions that are common in the study of polarization of fluids are therefore not included in the treatment.

[25] Response functions will be treated in great detail in Chapters 12 and 13 in the second volume of this treatise. A response function normally gives the response of a system due to an external potential, that is, the potential from the source itself (here $\delta\Psi^{ext}$), which is a quantity that we can control directly by external means. Thus, $\chi^\rho(r)$ is a response function in this sense. The function $\chi^\star(r)$ is a slightly different kind of function since it describes a response in terms of the total potential $\delta\Psi$, which contains the potential from the polarization charge density that we cannot control directly. We will still call it a response function. In Chapters 12 and 13, we will see that $\chi^\star(r)$ is closely related to the dielectric susceptibility of the bulk fluid.

★ For a given $\chi^P(r)$, the function $\chi^\star(r)$ is the solution to the equation[26]

$$\chi^P(r_{14}) = \chi^\star(r_{14}) + \int d\mathbf{r}_2 d\mathbf{r}_3 \, \chi^\star(r_{12}) \phi_{Coul}(r_{23}) \chi^P(r_{34}), \qquad (6.122)$$

which can be derived as follows. Using Equations 6.119 and 6.120, we can write Equation 6.121 as

$$\delta\rho^{pol}(\mathbf{r}_1) = \int d\mathbf{r}_2 \, \chi^\star(r_{12})[\delta\Psi^{ext}(\mathbf{r}_2) + \delta\Psi^{pol}(\mathbf{r}_2)]$$

$$= \int d\mathbf{r}_4 \, \chi^\star(r_{14}) \delta\Psi^{ext}(\mathbf{r}_4) + \int d\mathbf{r}_2 \, \chi^\star(r_{12}) \int d\mathbf{r}_3 \, \delta\rho^{pol}(\mathbf{r}_3) \phi_{Coul}(r_{32}),$$

where we have changed the integration variable in the first integral on the rhs from $\mathbf{r}_2$ to $\mathbf{r}_4$. By inserting Equation 6.117, we obtain

$$\delta\rho^{pol}(\mathbf{r}_1) = \int d\mathbf{r}_4 \left[\chi^\star(r_{14}) + \int d\mathbf{r}_2 d\mathbf{r}_3 \, \chi^\star(r_{12}) \chi^P(r_{34}) \phi_{Coul}(r_{32}) \right] \delta\Psi^{ext}(\mathbf{r}_4).$$

Since the lhs is equal to $\int d\mathbf{r}_4 \, \chi^P(r_{14}) \delta\Psi^{ext}(\mathbf{r}_4)$ and $\delta\Psi^{ext}(\mathbf{r}_4)$ is arbitrary, Equation 6.122 follows.

Let us see what the PB approximation says in this case. The potential of mean force is then given by $\delta w_i(\mathbf{r}_1) = q_i \delta\Psi(\mathbf{r}_1)$ and hence $\delta n_i(\mathbf{r}_1) = -\beta n_i^b q_i \delta\Psi(\mathbf{r}_1)$, so we have

$$\delta\rho^{pol}(\mathbf{r}_1) = \sum_i q_i \delta n_i(\mathbf{r}_1) = -\beta \sum_j q_j^2 n_j^b \delta\Psi(\mathbf{r}_1)$$

$$= -\beta \sum_j q_j^2 n_j^b \int d\mathbf{r}_2 \, \delta^{(3)}(r_{12}) \delta\Psi(\mathbf{r}_2) \quad (PB). \qquad (6.123)$$

By comparison with the expression (6.121) for $\chi^\star$, it follows that we have

$$\chi^\star(r_{12}) = -\beta \sum_j q_j^2 n_j^b \delta^{(3)}(r_{12}) \quad (PB). \qquad (6.124)$$

From the first line in Equation 6.123, we see that $\delta\rho^{pol}(\mathbf{r}_1) = -\beta \sum_j q_j^2 n_j^b \delta\Psi(\mathbf{r}_1)$ and, accordingly, the polarization charge density $\delta\rho^{pol}(\mathbf{r}_1)$ is in this case proportional to the total mean electrostatic potential $\delta\Psi(\mathbf{r}_1)$ *at the same point* $\mathbf{r}_1$. In $\chi^\star(r_{12})$, this fact is expressed by

---

[26] In Fourier space, the solution is $\tilde{\chi}^\star(k) = \tilde{\chi}^P(k)/[1 + \tilde{\chi}^P(k)\tilde{\phi}_{Coul}(k)]$. This relationship can be obtained directly from Equation 6.122 by using the convolution theorem which says that the Fourier transform of a convolution integral, $\int d\mathbf{r}' f_1(\mathbf{r} - \mathbf{r}') f_2(\mathbf{r}')$, is equal to $\tilde{f}_1(\mathbf{k})\tilde{f}_2(\mathbf{k})$. All integrals that occur in this text box are convolution integrals and the derivation in the box can alternatively (and more easily) be done in Fourier space.

the appearance of the Dirac delta function $\delta^{(3)}(r_{12})$. We can therefore say that electrostatics is *local* in the PB approximation: $\delta\rho^{pol}(\mathbf{r}_1)$ is not influenced by the values of $\delta\Psi(\mathbf{r}_2)$ at other points $\mathbf{r}_2 \neq \mathbf{r}_1$.

In reality, we have **nonlocal electrostatics**, meaning that the polarization at one point $\mathbf{r}_1$ is influenced by the electrostatic potential at other points in the surroundings. This is shown in the general case in Equation 6.121 since the polarization response function $\chi^\star(r_{12})$ is nonzero for $r_{12} \neq 0$, a feature that is caused by the correlations. The nonlocal nature of the response can be understood as follows. Any ion located at $\mathbf{r}_2$ interacts with the electrostatic potential and will affect the probability for ions to be at $\mathbf{r}_1$ via the correlations. Therefore, the density of ions at $\mathbf{r}_1$ depends on the potential elsewhere, which means that the charge density $\delta\rho^{pol}(\mathbf{r}_1)$ and the potential of mean force $\delta w_i(\mathbf{r}_1)$ depend on the values of $\delta\Psi(\mathbf{r}_2)$ for all points $\mathbf{r}_2$ in the neighborhood of $\mathbf{r}_1$. *This fact is expressed by the nonlocality in the general exact relationships we have obtained.*

6.2.1.2 The Potential of Mean Force and Dressed Particles

Next, let us express the potential of mean force δw_i in terms of $\delta\Psi$. Since $\delta\Psi^{ext}(\mathbf{r}) = \delta\Psi(\mathbf{r}) - \delta\Psi^{pol}(\mathbf{r})$, Equation 6.115 yields

$$\delta w_i(\mathbf{r}_1) = \int d\mathbf{r}_2\, \rho_i^{tot}(r_{12})\delta\Psi(\mathbf{r}_2) - \int d\mathbf{r}_2\, \rho_i^{tot}(r_{12})\delta\Psi^{pol}(\mathbf{r}_2) \qquad (6.125)$$

and it remains to express the last term in terms of $\delta\Psi$. Using Equation 6.119, we have

$$\int d\mathbf{r}_2\, \rho_i^{tot}(r_{12})\delta\Psi^{pol}(\mathbf{r}_2) = \int d\mathbf{r}_2\, \rho_i^{tot}(r_{12}) \int d\mathbf{r}_3 \delta\rho^{pol}(\mathbf{r}_3)\phi_{Coul}(r_{32})$$

$$= \int d\mathbf{r}_3 \left[\int d\mathbf{r}_2\, \rho_i^{tot}(r_{12})\phi_{Coul}(r_{32}) \right] \delta\rho^{pol}(\mathbf{r}_3)$$

$$= \int d\mathbf{r}_3\, \psi_i(r_{13})\delta\rho^{pol}(\mathbf{r}_3),$$

where $\psi_i(r_{13})$ is the mean electrostatic potential due to the i-ion at $\mathbf{r}_1$ as obtained from Equation 6.112 with P $=i$. By inserting Equation 6.121, we can express the rhs in terms of $\delta\Psi$ and, finally, by inserting the result into Equation 6.125, we obtain

$$\delta w_i(\mathbf{r}_1) = \int d\mathbf{r}_2 \left[\rho_i^{tot}(r_{12}) - \int d\mathbf{r}_3\, \psi_i(r_{13})\chi^\star(r_{32}) \right] \delta\Psi(\mathbf{r}_2). \qquad (6.126)$$

Comparing with Equation 6.115, we see that when δw_i is expressed in terms of $\delta\Psi$ instead of $\delta\Psi^{ext}$, the square bracket takes the role corresponding to that of $\rho_i^{tot}(r_{12})$ in Equation 6.115. The bracket differs from $\rho_i^{tot}(r_{12})$ by the integral, so let us interpret the latter.

To do this, let us first consider a more general case, namely a particle P that we immerse in a bulk electrolyte and keep fixed in space. We thereby regard P as external to the system and its internal charge density $\sigma_P(\mathbf{r})$ constitutes the source of the external electrostatic potential, that is, $\Psi^{\text{ext}}(\mathbf{r}')$ is the potential from $\sigma_P(\mathbf{r})$ in this case. Since $\Psi^{\text{ext}}(\mathbf{r}')$ is not small everywhere, we do not have a δ in front of Ψ^{ext}. The polarization response is nonlinear in regions where $\Psi^{\text{ext}}(\mathbf{r}')$ is large. We have placed the origin at P, so coordinates $\mathbf{r}$ and $\mathbf{r}'$ are counted from this particle.

The charge density $\rho_P(\mathbf{r})$ of the ion cloud around P is induced in the bulk electrolyte by the interactions between P and the electrolyte ions. The polarization charge density induced by the *electrostatic* interactions between $\sigma_P(\mathbf{r})$ and the ions is just a part of $\rho_P(\mathbf{r})$; all kinds of interactions between the ions and P influence the density around P. The mean electrostatic potential $\psi_P(\mathbf{r}')$ from charge density $\rho_P(\mathbf{r})$, as given by Equation 6.112, takes the role of $\Psi(\mathbf{r}')$ in the present case.

When $\psi_P(\mathbf{r}')$ is weak and linear response applies, the electrostatically induced polarization charge density is given by[27]

$$\rho_P^{\text{pol}}(\mathbf{r}) = \int d\mathbf{r}' \, \psi_P(\mathbf{r}') \chi^\star(|\mathbf{r} - \mathbf{r}'|) \quad \text{(weak } \psi_P\text{)} \tag{6.127}$$

as follows from Equation 6.121, which is valid in the general case for a weak mean potential Ψ: in the present case ψ_P. When $\psi_P(\mathbf{r}')$ is not weak, like close to P, the rhs of Equation 6.127 instead gives the *linear part of the electrostatic polarization response* due to ψ_P and we therefore define in the general case

$$\rho_P^{\text{lin}}(\mathbf{r}) = \int d\mathbf{r}' \, \psi_P(\mathbf{r}') \chi^\star(|\mathbf{r} - \mathbf{r}'|). \tag{6.128}$$

The rest of the polarization of the bulk electrolyte caused by the interactions between the ions and P is contained in $\rho_P^{\text{tot}}(\mathbf{r}) - \rho_P^{\text{lin}}(\mathbf{r})$. Like in the PB approximation, we define the **dressed particle charge density** $\rho_P^\star$ for P as (cf. Equation 6.60)

$$\rho_P^\star(\mathbf{r}) = \rho_P^{\text{tot}}(\mathbf{r}) - \rho_P^{\text{lin}}(\mathbf{r}) \tag{6.129}$$

so we have

$$\rho_P^\star(\mathbf{r}) = \rho_P^{\text{tot}}(\mathbf{r}) - \int d\mathbf{r}' \, \psi_P(\mathbf{r}') \chi^\star(|\mathbf{r} - \mathbf{r}'|) \tag{6.130}$$

in the general case. Since $\rho_P^{\text{tot}}(\mathbf{r}) = \sigma_P(\mathbf{r}) + \rho_P(\mathbf{r})$, we can write

$$\rho_P^\star(\mathbf{r}) = \sigma_P(\mathbf{r}) + \rho_P^{\text{dress}}(\mathbf{r}), \tag{6.131}$$

where $\rho_P^{\text{dress}}(\mathbf{r}) \equiv \rho_P(\mathbf{r}) - \rho_P^{\text{lin}}(\mathbf{r})$ is the charge density of the "dress" of particle P.[28]

---

[27] More precisely, Equation 6.127 gives $\rho_P^{\text{pol}}(\mathbf{r})$ provided $\psi_P(\mathbf{r}')$ is small within a region where $\chi^\star(|\mathbf{r} - \mathbf{r}'|)$ is non-negligible. In the cases that we will deal with, $\chi^\star(r)$ decays with distance faster than $\psi_P(\mathbf{r}')$ so Equation 6.127 applies, for example, far from P where $\psi_P(\mathbf{r}')$ is weak.

[28] Compare with footnote 7, where ρ_P^{dress} is defined for the PB approximation.

Let us now go back to Equation 6.126 and the case with a weak external potential $\delta\Psi^{ext}$ that causes a polarization charge density $\delta\rho^{pol}$ and a total potential $\delta\Psi$ in a bulk electrolyte. From Equation 6.130 with $P = i$, it follows that the square bracket in Equation 6.126 constitutes the dressed charge density $\rho_i^{\star}(r)$ of an i-ion, so we have

$$\delta w_i(\mathbf{r_1}) = \int d\mathbf{r_2}\, \rho_i^{\star}(r_{12})\delta\Psi(\mathbf{r_2}). \qquad (6.132)$$

Thereby, we have extended the concept of dressed particles to all kinds of particles present in the system and all particles are treated on the same basis as required. Equation 6.132 means that the potential of mean force δw_i for small $\delta\Psi$ is given by electrostatic interaction energy between $\rho_i^{\star}$ and $\delta\Psi$. By comparing with Equation 6.115, we see that *the dressed charge density $\rho_i^{\star}$ has the same role vis-à-vis the total potential $\delta\Psi$ as the total charge density ρ_i^{tot} has vis-à-vis the external potential $\delta\Psi^{ext}$*.

The induced polarization charge density $\delta\rho^{pol}$ is

$$\delta\rho^{pol}(\mathbf{r_1}) = \sum_i q_i \delta n_i(\mathbf{r_1}) = -\beta \sum_i q_i n_i^b \delta w_i(\mathbf{r_1})$$

$$= -\beta \sum_i q_i n_i^b \int d\mathbf{r_2}\, \rho_i^{\star}(r_{12})\delta\Psi(\mathbf{r_2})$$

and by comparison with Equation 6.121, we see that

$$\chi^{\star}(r_{12}) = -\beta \sum_i q_i n_i^b \rho_i^{\star}(r_{12}), \qquad (6.133)$$

which can be compared with the analogous expression for $\chi^\rho(r_{12})$ in Equation 6.118.

6.2.1.3 Screened Electrostatic Interactions

We have now established the necessary concepts and relationships in the general case in order to proceed with our objective to generalize the expressions for the interaction between particles in electrolytes. Let us first investigate the mean electrostatic potential $\psi_P(\mathbf{r'})$ due to a particle P in the electrolyte. Like before, P has arbitrary size, shape and any internal charge density $\sigma_P(\mathbf{r})$. P can be an ion of the same kind as one of the constituent species in the electrolyte, a macroion or any other kind of rigid particle. It has orientation ω, but we will initially keep the particle in an arbitrary but fixed orientation so we will suppress this variable in the formulas.

The mean electrostatic potential satisfies Poisson's equation

$$-\varepsilon_r\varepsilon_0 \nabla^2 \psi_P(\mathbf{r}) = \rho_P^{tot}(\mathbf{r}). \qquad (6.134)$$

By subtracting $\rho_P^{lin}(\mathbf{r})$ from both sides of Poisson's Equation 6.134, we obtain

$$-\varepsilon_r\varepsilon_0 \nabla^2 \psi_P(\mathbf{r}) - \rho_P^{lin}(\mathbf{r}) = \rho_P^{tot}(\mathbf{r}) - \rho_P^{lin}(\mathbf{r}) \equiv \rho_P^{\star}(\mathbf{r})$$

so we have, using Equation 6.128,

$$-\varepsilon_r\varepsilon_0\nabla^2\psi_P(\mathbf{r}) - \int d\mathbf{r}'\,\psi_P(\mathbf{r}')\chi^\star(|\mathbf{r}-\mathbf{r}'|) = \rho_P^\star(\mathbf{r}). \tag{6.135}$$

Let us consider Green's function $\phi_{Coul}^\star(r)$ for Equation 6.135, that is, the solution to the equation

$$-\varepsilon_r\varepsilon_0\nabla^2\phi_{Coul}^\star(r) - \int d\mathbf{r}'\,\phi_{Coul}^\star(r')\chi^\star(|\mathbf{r}-\mathbf{r}'|) = \delta^{(3)}(r) \tag{6.136}$$

[this equation corresponds to Equation 6.68 in the PB case]. It follows from the properties of Green's functions in general (see Example 6.2) that the solution ψ_P of Equation 6.135 is given by

$$\psi_P(\mathbf{r}) = \int d\mathbf{r}'\,\rho_P^\star(\mathbf{r}')\phi_{Coul}^\star(|\mathbf{r}-\mathbf{r}'|), \tag{6.137}$$

which is the same as Equation 6.113 and has the same form as Coulomb's law. The function $\phi_{Coul}^\star(r)$ defined here is the generalization of the corresponding function in the PB approximation and it is the **unit screened Coulomb potential** for the present general case. It has a key role in the spatial propagation of electrostatic interactions in the electrolyte.

{**EXERCISE 6.6**
Show that $\psi_P(\mathbf{r})$ given by Equation 6.137 satisfies Equation 6.135 considering that $\phi_{Coul}^\star(r)$ satisfies Equation 6.136.}

We have seen in Equation 6.132 that the potential of mean force for an ion can be obtained from the mean electrostatic potential provided the latter is small. Consider a j-ion located at $\mathbf{r}_4$ when particle P is placed at $\mathbf{r}_3$. Equation 6.132 corresponds to

$$w_{Pj}^{(2)}(\mathbf{r}_{34}) = \int d\mathbf{r}_1\,\psi_P(\mathbf{r}_{31})\rho_j^\star(r_{41}) \quad \text{(when } \psi_P \text{ is weak)}, \tag{6.138}$$

where $w_{Pj}^{(2)}$ is the pair potential of mean force between P and the ion. By inserting $\rho_j^\star(r) = q_j\delta^{(3)}(r) + \rho_j^{dress}(r)$ (cf. Equation 6.131), we obtain

$$w_{Pj}^{(2)}(\mathbf{r}_{34}) = \psi_P(\mathbf{r}_{34})q_j + \int d\mathbf{r}_1\,\psi_P(\mathbf{r}_{31})\rho_j^{dress}(r_{41}) \quad \text{(weak } \psi_P\text{)}$$

and we recognize the first term on the rhs as the sole contribution to $w_{Pj}^{(2)}$ in the PB approximation. The last term is the *correlation contribution* to $w_{Pj}^{(2)}$ from the ions in the dress of the j-ion. When the ion density goes to zero, there are on average fewer and fewer ions that influence $w_{Pj}^{(2)}(r_{34})$ via correlations, so the magnitude of this term decreases and eventually it becomes negligible. This means that at sufficiently low ion density, $w_{Pj}^{(2)}(\mathbf{r}) = \psi_P(\mathbf{r})q_j$ is a reasonably good approximation – at least when ψ_P is small.

The polarization response function $\chi^\star(r)$ given by Equation 6.133 can likewise be written as

$$\chi^\star(r) = -\beta \sum_j q_j n_j^{\mathrm{b}} \left[q_j \delta^{(3)}(r) + \rho_j^{\mathrm{dress}}(r) \right]$$

$$= -\varepsilon_r \varepsilon_0 \kappa_D^2 \delta^{(3)}(r) - \beta \sum_j q_j n_j^{\mathrm{b}} \rho_j^{\mathrm{dress}}(r), \qquad (6.139)$$

where we have used the definition of κ_D in Equation 6.31. When the ion density goes to zero, the last term in the rhs become less and less important and one can therefore set $\chi^\star(r) \approx -\varepsilon_r \varepsilon_0 \kappa_D^2 \delta^{(3)}(r)$ at low densities, which should be compared with Equation 6.124. By inserting this approximation in Equation 6.136 for $\phi_{\mathrm{Coul}}^\star(r)$, we obtain the corresponding equation in the PB approximation, Equation 6.68, which has the solution given by Equations 6.69. Having established the connection to the PB approximation, let us now continue with the general case.

We can make use of Equation 6.138 whenever $\psi_{\mathrm{P}}(\mathbf{r}_{31})$ is weak. Since this applies far from P (ψ_{P} goes to zero when $r_{31} \to \infty$), we have

$$w_{\mathrm{P}j}^{(2)}(\mathbf{r}_{34}) \sim \int d\mathbf{r}_1 \, \psi_{\mathrm{P}}(\mathbf{r}_{31}) \rho_j^\star(r_{41}) \quad \text{when } r_{34} \to \infty, \qquad (6.140)$$

at least if the screened electrostatic forces dominate for large distances in the system.[29] Furthermore, the integrand must not have any significant contributions from regions where ψ_{P} is not small. This is fulfilled if $\rho_j^\star$ is a short-range function (compared to ψ_{P}).[30] In fact, $\rho_j^\star$ has a shorter range than ψ_{P} and $\phi_{\mathrm{Coul}}^\star(r)$ at least if the ion density is not too high; in the PB approximation, we have seen that $\rho_{\mathrm{P}}^\star$ decays like the square of the mean potential, so $\rho_{\mathrm{P}}^\star$ then has half the decay length of ψ_{P} and $\phi_{\mathrm{Coul}}^\star(r)$. The same applies to $\rho_j^\star$ and ψ_j, which is the special case where P $= j$. In the general case, one can show that the decay length of $\rho_j^\star$ (and $\rho_{\mathrm{P}}^\star$) approaches half the decay length of $\phi_{\mathrm{Coul}}^\star(r)$ (and of ψ_j and ψ_{P}) in the limit of infinite dilution.[31]

Here and in what follows, it is assumed that the conditions are such that $\rho_j^\star$ has a shorter range than $\phi_{\mathrm{Coul}}^\star(r)$ and ψ_j. The assumption is done here in order to simplify the theoretical treatment; the full statistical mechanical analysis can be found in Part IV in the 2nd volume. In fact, most results hold true even if all of these conditions are not satisfied. For example, the decay laws of the electrostatic potential ψ_i and charge distribution ρ_i outside an ion or other particle that we will find here are valid in practice for a larger range of cases than those considered here.

---

[29] It is therefore required that the short-range non-electrostatic interaction u_{ij}^{sh} decays at least exponentially fast to zero with distance: faster than $\phi_{\mathrm{Coul}}^\star(r)$ (cf. footnote 22). Furthermore, the conditions near critical points require special treatment (cf. Section 6.2.3).

[30] When $\mathbf{r}_4$ lies in a region far from P where ψ_{P} is small, a short-range function $\rho_j^\star(r_{41})$ has significant contributions only at points $\mathbf{r}_1$ in the same region, that is, where $\psi_{\mathrm{P}}(\mathbf{r}_{31})$ is small.

[31] Note that Equation 6.133 implies that $\chi^\star(r)$ decays equally fast as $\rho_j^\star(r)$ for any j (or possibly faster). This is relevant for the comment in footnote 27.

By inserting Equation 6.137, which holds for all $\mathbf{r}$, into Equation 6.140, we obtain

$$w_{Pj}^{(2)}(r_{34}) \sim \int d\mathbf{r}_1 d\mathbf{r}_2 \, \rho_P^\star(\mathbf{r}_{32}) \phi_{Coul}^\star(r_{21}) \rho_j^\star(r_{41}) \quad \text{when } r_{34} \to \infty, \tag{6.141}$$

where the rhs is the screened electrostatic interaction between the dressed particle charge distribution of particle P and j-ions. For the special case when particle P is an i-ion, this equation becomes

$$w_{ij}^{(2)}(r_{34}) \sim \int d\mathbf{r}_1 d\mathbf{r}_2 \rho_i^\star(\mathbf{r}_{32}) \phi_{Coul}^\star(r_{21}) \rho_j^\star(r_{41})$$

$$= \int d\mathbf{r}_1 \psi_i(r_{31}) \rho_j^\star(r_{41}) = \int d\mathbf{r}_2 \rho_i^\star(\mathbf{r}_{32}) \psi_j(r_{42}) \tag{6.142}$$

when $r_{34} \to \infty$, where we have used

$$\psi_i(r_{31}) = \int d\mathbf{r}_2 \rho_i^\star(\mathbf{r}_{32}) \phi_{Coul}^\star(r_{21}) \tag{6.143}$$

and the analogous equation for ψ_j. These equations are valid for all ions in the electrolyte and we see that all particles are treated on the same basis. The dressed particle charge distributions $\rho_i^\star$ and $\rho_j^\star$ are hence entities that have the roles as (i) the charge density $\rho_i^\star$ that via the screened Coulomb potential $\phi_{Coul}^\star$ gives rise to the electrostatic potential $\psi_i(r_{31})$ for all r_{31} (Equation 6.143), and (ii) the charge density $\rho_j^\star$ that interacts with ψ_i giving the potential of mean force w_{ij} in the tail region (for large separations, Equation 6.140 with $P = j$). The corresponding is true when $\rho_i^\star$ and $\rho_j^\star$ take the reverse roles, so they *have the dual roles of* both (i) and (ii).

Equation 6.142 is valid for any spherical particles of species i and j, including all constituent ions in a bulk electrolyte, so thereby we have managed to show that Equation 6.114 has a general validity, as we set out to do. For two immersed nonspherical particles P_I and P_{II} at distance $r_{I,II}$ from each other, we have when $r_{I,II} \to \infty$

$$w_{I,II}^{(2)}(\mathbf{r}_{I,II}, \boldsymbol{\omega}_I, \boldsymbol{\omega}_{II}) \sim \int d\mathbf{r}_1 d\mathbf{r}_2 \rho_I^\star(\mathbf{r}_{I,2}, \boldsymbol{\omega}_I) \phi_{Coul}^\star(r_{21}) \rho_{II}^\star(\mathbf{r}_{II,1}, \boldsymbol{\omega}_{II})$$

$$= \int d\mathbf{r}_1 \psi_I(\mathbf{r}_{I,1}, \boldsymbol{\omega}_I) \rho_{II}^\star(\mathbf{r}_{II,1}, \boldsymbol{\omega}_{II}) = \int d\mathbf{r}_2 \rho_I^\star(\mathbf{r}_{I,2}, \boldsymbol{\omega}_I) \psi_{II}(\mathbf{r}_{II,2}, \boldsymbol{\omega}_{II}),$$

$$\tag{6.144}$$

where we explicitly have shown the orientations $\boldsymbol{\omega}_I$ and $\boldsymbol{\omega}_{II}$ of the particles. This is the same as Equation 6.93 for the PB case, but in the present case $\phi_{Coul}^\star$ is the generalized screened Coulomb potential defined by Equation 6.136. The distance dependence of the decay for $w_{I,II}^{(2)}$ at large $r_{I,II}$ is determined by $\phi_{Coul}^\star$ and is the same as that for ψ_I and ψ_{II} for the vast majority of cases we deal with here as will be discussed in detail later.

One can also argue in the following manner to obtain the result in Equation 6.144 (cf. the corresponding argument for the PB case in Section 6.1.3.2). If we let the two particles approach each other from very far away, the polarization of the surroundings of one particle due to the other is initially weak and can be treated to linear order. Therefore, $\rho_j^\star(\mathbf{r}_2, \boldsymbol{\omega}_J)$

for the two particles, $J = I$ and II, are virtually unchanged if the separation between the particles is sufficiently large. Since $\phi^\star_{Coul}$ describes polarization response to linear order, the interaction free energy for large separations will go like in the first line of Equation 6.144.

For future reference, we write Equations 6.129, 6.131 and 6.137 with the orientation variable explicit, whereby they read

$$\rho^\star_P(\mathbf{r}, \omega) = \rho^{tot}_P(\mathbf{r}, \omega) - \rho^{lin}_P(\mathbf{r}, \omega), \tag{6.145}$$

$$\rho^\star_P(\mathbf{r}, \omega) = \sigma_P(\mathbf{r}, \omega) + \rho^{dress}_P(\mathbf{r}, \omega), \tag{6.146}$$

where $\rho^{dress}_P \equiv \rho_P - \rho^{lin}_P$, and

$$\psi_P(\mathbf{r}, \omega) = \int d\mathbf{r}' \rho^\star_P(\mathbf{r}', \omega) \phi^\star_{Coul}(|\mathbf{r} - \mathbf{r}'|). \tag{6.147}$$

The latter has the same form as Coulomb's law

$$\psi_P(\mathbf{r}, \omega) = \int d\mathbf{r}' \rho^{tot}_P(\mathbf{r}', \omega) \phi_{Coul}(|\mathbf{r} - \mathbf{r}'|)$$

with the ordinary Coulomb potential ϕ_{Coul}. Like in the PB case, $\phi^\star_{Coul}(r)$ determines the distance dependence of the decay of ψ_P for large r.

Equation 6.144 applies in the general case, for example when P_I and P_{II} are two macroions or any other particles, provided that the electrostatic interactions dominate for large separations $r_{I,II}$. The three integral expressions in the rhs, which are equivalent to each other, give the screened electrostatic contribution to $w^{(2)}_{I,II}$, but there can exist other contributions that dominate when $r_{I,II} \to \infty$, for example from van der Waals interactions between the particles (cf. Sections 7.1.5 and 8.6). When such interactions are *not* included, the potential of mean force normally decays for large distances like screened electrostatic interactions, but it may in rare cases happen that it does not. This can, for example, happen under some conditions for model electrolyte systems where the anions and cations differ *only* by the sign of their charge. The most common model of this kind is the *restricted primitive model*, where the ions in a symmetric $z : -z$ electrolyte are charged hard spheres of equal diameters. As we will see later in Section 6.2.2.3, these types of model systems show some exceptional behaviors because of their extreme symmetry which decouples density and charge correlations from each other. Equations 6.142 and 6.144 usually apply also for these systems, but it *can* happen at high ionic densities that the leading term in $w^{(2)}_{ij}$ for large distances is not given by the rhs of these equations. The leading term is then instead due to density-density correlations that often are dominated by ionic core interactions. Such a term can dominate in $w^{(2)}_{ij}$, for example under conditions where the electrostatic interactions are very efficiently screened and therefore have a very short range.[32] These exceptional cases will be treated in more detail in Section 6.2.2.3.

Any asymmetry between the anions and cations, for example different sizes, makes this exceptional behavior to cease. For $z : -z$ electrolytes where the non-electrostatic

---

[32] The dominance of density-density correlations also happens near critical points where long-range density fluctuations appear.

interactions $u^{sh}_{++}(r)$, $u^{sh}_{+-}(r)$ and $u^{sh}_{--}(r)$ are different from each other, the electrostatic contributions and the density correlation contributions to $w^{(2)}_{ij}$ are coupled to each other. They thereby acquire a common decay behavior for large r that is included in the analysis made earlier in this section. This will be explored in some detail in Sections 6.2.2.3 and 6.2.3.

6.2.2 The Decay Behavior and the Screening Decay Length

6.2.2.1 Oscillatory and Monotonic Exponential Decays: Explicit Examples

The unit screened Coulomb potential $\phi^\star_{Coul}(r)$ has a key role for the electrostatic interactions in an electrolyte. It determines the spatial propagation of the electrostatic potential due to an ion or other particle in the electrolyte for all distances, Equations 6.143 and 6.147, and in general it determines the decay of the potential of mean force between two particles for large separations. Thereby, it determines the decay of $g^{(2)}_{ij}(r)$, which goes like $1 - \beta w^{(2)}_{ij}(r)$ for large r. In the DH and PB approximations, we have seen that $\phi^\star_{Coul}(r)$ decays like a Yukawa function with decay length $1/\kappa_D$, Equation 6.69, while in the improved mean field theory of Section 6.1.5 it still decays like a Yukawa function but with a different decay length $1/\kappa$, Equation 6.106.

As illustrated in Figure 6.7b, there can be a crossover between monotonic and oscillatory decays: when the electrostatic coupling is weak, the potential has a Yukawa function decay and when it is strong, the potential decays in an exponentially damped oscillatory manner. Such an oscillatory decay is ubiquitous for molten salts and it appears, for example, also in room temperature ionic liquids. The monotonic decay occurs at low ionic densities like dilute electrolyte solutions and thin plasmas, but it also occurs in ionic liquids and concentrated electrolyte solutions. The situation is accordingly multifaceted. Ionic liquids and electrolyte solutions with molecular solvent will be treated in Chapter 13; here we restrict ourselves to simple electrolytes with spherically symmetric ions.

We start with systems where the electrostatic potential and the pair distribution functions are oscillatory. As an example, the pair distributions $g^{(2)}_{ij}(r)$ for molten NaCl near the freezing point are shown in Figure 6.8. A realistic model potential $u_{ij}(r)$ for the ion-ion pair interactions in molten NaCl was used in the calculations. We see in the figure that the equal-species function $g^{(2)}_{--}(r)$ and the unequal-species function $g^{(2)}_{-+}(r)$ are out of phase with each other; one has minima where the other has maxima and vice versa. The same is true for the pair $g^{(2)}_{++}(r)$ and $g^{(2)}_{+-}(r)$ [the latter is the same as $g^{(2)}_{-+}(r)$]. Around an ion, the ionic density for the opposite species is larger that that of the same species for certain distances and vice versa for other distances. This means that the ions in the surroundings form a "shell structure" around the ion, where consecutive shells are richer in either anions or cations in an alternating manner. The charge density $\rho_i(r) = \sum_j q_j n^b_j g^{(2)}_{ij}(r)$ around an ion of species i hence alternates between positive and negative values with the same wavelength as the oscillations of $g^{(2)}_{ij}(r)$. Since electroneutrality demands that $q_+ n^b_+ + q_- n^b_- = 0$, we have

$$\begin{aligned} \rho_+(r) &= q_+ n^b_+ [g^{(2)}_{++}(r) - g^{(2)}_{+-}(r)] = \frac{q_\pm n^b_{tot}}{2}[g^{(2)}_{++}(r) - g^{(2)}_{+-}(r)] \\ \rho_-(r) &= -q_+ n^b_+ [g^{(2)}_{--}(r) - g^{(2)}_{-+}(r)] = -\frac{q_\pm n^b_{tot}}{2}[g^{(2)}_{--}(r) - g^{(2)}_{-+}(r)] \end{aligned} \tag{6.148}$$

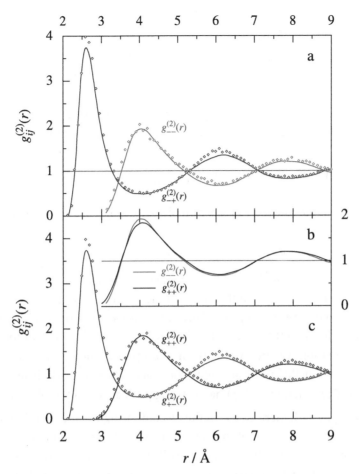

FIGURE 6.8 **(See color insert.)** Pair distribution function $g_{ij}^{(2)}(r)$ for molten NaCl at 1073 K (near the freezing point).[33] (a) Pair distributions around a Cl^- ion. Symbols are results from MD simulations and curves are results from an integral equation theory (MHNC).[34] (b) A comparison of MHNC curves for $g_{++}^{(2)}(r)$ and $g_{--}^{(2)}(r)$. (c) Pair distributions around an Na^+ ion. Symbols show MD and curves show MHNC results.

where $n_{tot}^b = n_+^b + n_-^b$ and $q_{\pm} = q_+ = -q_- = q$ (this definition of $q_{\pm}$ applies only to symmetric electrolytes). Since $g_{++}^{(2)}(r) \approx g_{--}^{(2)}(r)$ for the present system (see pane (b) in Figure 6.8), we have $\rho_+(r) \approx -\rho_-(r)$ in this case.

The oscillatory feature of the functions depends on an optimization of the packing of spheres combined with electrostatic interactions between the different species for high densities of the liquid. The highest peak of $g_{+-}^{(2)}(r)$ corresponds to anions and cations that are brought close together by the strong attraction; in the first shell, the oppositely charged species dominates greatly. The attraction is to a large part of electrostatic origin, but there

[33] Figure 6.8 is based on data from B. Ballone, G. Pastore and M. P. Tosi, *J. Chem. Phys.* **81** (1984) 3174 with MD data taken from J. W. E. Lewis and K. Singer, *J. Chem. Soc., Faraday Trans. 2*, **71** (1975) 41.

[34] The Modified HNC (MHNC) approximation is an integral equation theory for pair distributions introduced in Chapter 9 (in the 2nd volume).

is also a steric component: the ions in the first shell are exposed to collisions by ions from the outside; these collisions push the former ions towards the central ion. Also in the other shells, there are important contributions from ionic collisions and other steric correlations in addition to the electrostatic interactions, but the alternation in sign of ρ_i occurs mainly for electrostatic reasons.

For future reference, we add the following comment. As noted earlier, the definition of $q_\pm$ above is valid only for z : −z electrolytes like NaCl. In the general binary case, we define

$$q_\pm = x^b_+ q_+ + x^b_- |q_-| = \frac{2q_+|q_-|}{q_+ + |q_-|}, \tag{6.149}$$

where $x^b_i = n^b_i/n^b_{tot}$ is a mole fraction that only involves the ionic species. Also in this general case, we have

$$q_+ n^b_+ = -q_- n^b_- = q_\pm n^b_{tot}/2, \tag{6.150}$$

which implies that Equation 6.148 holds for any binary electrolyte.

Next, we turn to an example where the electrostatic potential and the pair distribution function have a Yukawa function decay. In Figure 6.9, we see $g^{(2)}_{ij}(r)$ for an aqueous solution of a 2:-1 electrolyte in the primitive model obtained by MC simulation. The cations are divalent and the anions are monovalent. We see from a comparison of $g^{(2)}_{++}(r)$ and $g^{(2)}_{--}(r)$ that the divalent cations are repelled more strongly from each other at short distances than the anions, as expected. The function $g^{(2)}_{+-}(r)$ has a strong peak at contact, $r/d^h = 1$. All $g^{(2)}_{ij}(r)$ decay monotonically to 1 with increased r. In the inset of the figure, the function $|rh^{(2)}_{ij}(r)|$ is plotted on a log-linear scale. We see that this function is close to a straight line in a quite wide range of distances for the ++ and the −− cases, while the +− curve tends to a straight line much further out where simulation accuracy is low since $h^{(2)}_{ij}$ is very small there. A straight line means that $rh^{(2)}_{ij}(r)$ decays like an exponential function and hence $h^{(2)}_{ij}(r)$ indeed decays like a Yukawa function for large r. The curves that show the asymptotic decay in the figure will be discussed later in Section 6.2.2.3.

6.2.2.2 Roles of Effective Charges, Effective Dielectric Permittivities and the Decay Parameter κ

Let us return to the analysis of the general case where the decay of the screened Coulomb potential $\phi^*_{Coul}(r)$ for large r can be like a Yukawa function or like an exponentially damped oscillatory function. When the electrostatic coupling is sufficiently weak, we have[35]

$$\phi^*_{Coul}(r) \sim A^\star \frac{e^{-\kappa r}}{r} \quad r \to \infty, \tag{6.151}$$

where $A^\star$ is a constant that will be specified later (see Equation 6.166) and κ is the decay parameter which in general is different from κ_D. A difference from the PB approximation

---

[35] The derivation of this decay formula is postponed till Chapter 13 since we currently do not have the theoretical tools needed for this (see also R. Kjellander, Soft Matter, 2019, DOI: 10.1039/C9SM00712A for a derivation based on quite simple statistical mechanics and physical reasoning). We will, however, obtain expressions for κ and $A^\star$ in the current chapter.

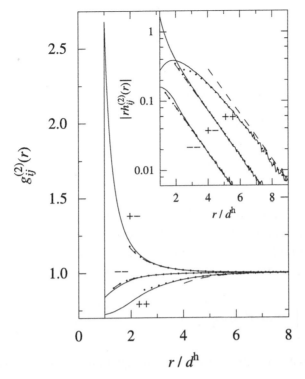

FIGURE 6.9 Pair distribution function $g_{ij}^{(2)}(r)$ for a 60 mM aqueous solution of a 2:-1 electrolyte at room temperature in the primitive model ($\varepsilon_r = 78.4$).[36] Alternatively, the system is a gas of the 2:-1 electrolyte at 23400 K ($\varepsilon_r = 1$). The ionic hard core diameter is $d^h = 4.6$ Å. The insert shows the same data plotted as $|rh_{ij}^{(2)}(r)|$ on a log-linear scale. Full curves show results from MC simulation and the other curves show the asymptotic decay of the functions as obtained with: (dashed curves) one term with an exponential decay length $1/\kappa$ and (dotted curves) two terms with decay lengths $1/\kappa$ and $1/\kappa'$, respectively. These asymptotic curves are *not* fitted; they are calculated from the simulation data as explained in the text.

is that the decay is given by a Yukawa function *only for large r*. The distance dependence of $\phi_{Coul}^*(r)$ for small r is more complicated.[37] At higher electrostatic coupling, $\phi_{Coul}^*(r)$ is often an exponentially damped oscillatory function for large r, but we will start with cases where the decay is given in Equation 6.151.

Many qualitative results that we obtained earlier for the asymptotic decay of the electrostatic potential and the potential of mean force in the PB approximation remain, in fact, valid in the present case. For example, the electrostatic potential $\psi_P(r)$ due to a spherically symmetric particle P satisfies

$$\psi_P(r) = \int d\mathbf{r}' \rho_P^\star(r')\phi_{Coul}^\star(|\mathbf{r} - \mathbf{r}'|) \sim A^\star \int d\mathbf{r}' \rho_P^\star(r')\frac{e^{-\kappa|\mathbf{r}-\mathbf{r}'|}}{|\mathbf{r}-\mathbf{r}'|} \quad \text{when } r \to \infty,$$

---

[36] Figure 6.9 is based on data from J. Ulander and R. Kjellander, *J. Chem. Phys.* **114** (2001) 4893.

[37] As we saw in Equation 6.109 and Figure 6.7b, there are at least two terms with Yukawa function form that have different decay lengths and there are also other contributions with other functional r dependences. In Equation 6.151, we deal with the contribution with the longest decay length (smallest κ) that dominates for large r.

where the rhs follows by insertion of Equation 6.151. By going through the same steps for the rhs of this equation as those performed in Example 6.3 for the PB case, we obtain also in the present case

$$\psi_P(r) \sim q_P^{\text{eff}} \phi_{\text{Coul}}^\star(r) \quad \text{when } r \to \infty \quad \text{(spherical particle)}, \tag{6.152}$$

where the **effective charge** q_P^{eff} of particle P is

$$q_P^{\text{eff}} = \int d\mathbf{r}' \rho_P^\star(r') \frac{\sinh(\kappa r')}{\kappa r'}. \tag{6.153}$$

The decay behavior and definition of an effective charge are accordingly exactly the same as in the PB approximation apart from the facts that the function $\rho_P^\star(r')$ is not the same here and the values of κ and q_P^{eff} are different. Likewise, for a nonspherical particle P with an arbitrary internal charge distribution $\sigma_P(\mathbf{s}, \boldsymbol{\omega})$ in its interior, we have

$$\psi_P(\mathbf{r}, \boldsymbol{\omega}) = \int d\mathbf{r}' \rho_P^\star(\mathbf{r}', \boldsymbol{\omega}) \phi_{\text{Coul}}^\star(|\mathbf{r} - \mathbf{r}'|) \sim A^\star \int d\mathbf{r}' \rho_P^\star(\mathbf{r}', \boldsymbol{\omega}) \frac{e^{-\kappa|\mathbf{r}-\mathbf{r}'|}}{|\mathbf{r} - \mathbf{r}'|}$$

when $r \to \infty$. In the same manner as we obtained Equation 6.84 in the PB case, we obtain also in the present case

$$\psi_P(\mathbf{r}, \boldsymbol{\omega}) \sim Q_P^{\text{eff}}(\hat{\mathbf{r}}, \boldsymbol{\omega}) \phi_{\text{Coul}}^\star(r) \quad \text{when } r \to \infty,$$

where

$$Q_P^{\text{eff}}(\hat{\mathbf{r}}, \boldsymbol{\omega}) = \int d\mathbf{r}' \rho_P^\star(\mathbf{r}', \boldsymbol{\omega}) e^{\kappa \mathbf{r}' \cdot \hat{\mathbf{r}}} \tag{6.154}$$

is the same kind of entity with the same definition as in the PB case (Equation 6.85). Its value and functional dependencies on $\hat{\mathbf{r}}$ and $\boldsymbol{\omega}$ are, however, not the same since $\rho_P^\star$ is a different function and κ has a different value.

Like in the PB case, entity $Q_P^{\text{eff}}(\hat{\mathbf{r}}, \boldsymbol{\omega})$ can be described as the "**multipolar effective charge**" (i.e., direction dependent) of particle P that gives both the magnitude and the direction dependence for electrostatic interactions at large distances. For two particles P_I and P_{II} placed with their centers of mass at $\mathbf{r}_I$ and $\mathbf{r}_{II}$, we have

$$w_{I,II}^{(2)}(\mathbf{r}_{I,II}, \boldsymbol{\omega}_I, \boldsymbol{\omega}_{II}) \sim Q_I^{\text{eff}}(\hat{\mathbf{r}}_{I,II}, \boldsymbol{\omega}_I) Q_{II}^{\text{eff}}(-\hat{\mathbf{r}}_{I,II}, \boldsymbol{\omega}_{II}) \phi_{\text{Coul}}^\star(r_{I,II}) \quad \text{when } r_{I,II} \to \infty, \tag{6.155}$$

which can be derived in the same manner as (the identical) Equation 6.94 for the PB case.

We will now derive a general equation for the decay parameter κ. When $r > 0$, Equation 6.136 is equal to

$$-\varepsilon_r \varepsilon_0 \frac{1}{r} \frac{\partial^2}{\partial r^2} \left[r \phi_{\text{Coul}}^\star(r) \right] = \int d\mathbf{r}' \phi_{\text{Coul}}^\star(r') \chi^\star(|\mathbf{r} - \mathbf{r}'|).$$

Let us investigate the consequences of this equation for large r by inserting the limiting form Equation 6.151 for $\phi_{\text{Coul}}^{\star}(r)$ on both sides (note that $\chi^{\star}$, which is a sum of $\rho_j^{\star}$ functions, has a shorter range than $\phi_{\text{Coul}}^{\star}$). We obtain, after performing the derivatives in the lhs,

$$-\varepsilon_r\varepsilon_0\kappa^2 A^{\star}\frac{e^{-\kappa r}}{r} \sim \int d\mathbf{r}' A^{\star}\frac{e^{-\kappa r'}}{r'}\chi^{\star}(|\mathbf{r}-\mathbf{r}'|)$$

in the limit $r \to \infty$. By making the variable substitution $\mathbf{r}'' = \mathbf{r} - \mathbf{r}'$ in the rhs and going over to spherical polar coordinates $(r'', \varphi_r'', \theta_r'')$ for $\mathbf{r}''$, we obtain

$$-\varepsilon_r\varepsilon_0\kappa^2 A^{\star}\frac{e^{-\kappa r}}{r} \sim \int dr''(r'')^2\left[\int d\hat{\mathbf{r}}'' A^{\star}\frac{e^{-\kappa|\mathbf{r}-\mathbf{r}''|}}{|\mathbf{r}-\mathbf{r}''|}\right]\chi^{\star}(r'') \qquad (6.156)$$

where $d\hat{\mathbf{r}}''$ stands for the angle differential $\sin(\theta_r'')d\theta_r''d\varphi_r''$. The angular integration in the square brackets is the same as in Equation 6.73. The latter equation applies in the present case since we have $r > r''$ (or even $r \gg r''$ since we are interested in the limit $r \to \infty$). By inserting the angular integration result from Equation 6.73 into Equation 6.156, we obtain

$$-\varepsilon_r\varepsilon_0\kappa^2 A^{\star}\frac{e^{-\kappa r}}{r} \sim 4\pi A^{\star}\frac{e^{-\kappa r}}{r}\int dr''(r'')^2\frac{\sinh(\kappa r'')}{\kappa r''}\chi^{\star}(r'')$$

when $r \to \infty$. An identification of the coefficients for $\exp(-\kappa r)/r$ on either side gives after simplification

$$-\varepsilon_r\varepsilon_0\kappa^2 = \int d\mathbf{r}''\frac{\sinh(\kappa r'')}{\kappa r''}\chi^{\star}(r''), \qquad (6.157)$$

where $d\mathbf{r}'' = dr''(r'')^2 4\pi$ (due to spherical symmetry). If we know the function $\chi^{\star}(r)$, this equation can be used to determine κ, which appears on both sides. In general, one has to solve the equation numerically for the unknown variable κ.

Equation 6.157 is a general equation for κ. We can write it in some equivalent manners that are quite illuminating. By inserting Equation 6.139, we obtain after simplification

$$\kappa^2 = \kappa_D^2 + \frac{\beta}{\varepsilon_r\varepsilon_0}\sum_j q_j n_j^b \int d\mathbf{r}''\frac{\sinh(\kappa r'')}{\kappa r''}\rho_j^{\text{dress}}(r),$$

which clearly shows that the reason why κ deviates from κ_D is the presence of ρ_j^{dress}. This entity contains the ion-ion correlation effect discussed earlier, for example in connection to Equation 6.139, where we saw that the influence of ρ_j^{dress} vanishes in the limit of zero density of the ions, so $\kappa \sim \kappa_D$ in this limit.

If we instead insert Equation 6.133 into 6.157, we obtain

$$\kappa^2 = \frac{\beta}{\varepsilon_r\varepsilon_0}\sum_j q_j n_j^b \int d\mathbf{r}''\frac{\sinh(\kappa r'')}{\kappa r''}\rho_j^{\star}(r'')$$

and by using Equation 6.153 for an ion of species j (setting $P = j$), it follows that

$$\kappa^2 = \frac{\beta}{\varepsilon_r \varepsilon_0} \sum_j q_j q_j^{\text{eff}} n_j^{\text{b}}, \quad \text{(spherical ions).} \tag{6.158}$$

This expression differs from the definition of the Debye parameter κ_D in Equation 6.31 solely by the replacement of q_j^2 in the latter equation by $q_j q_j^{\text{eff}}$. Incidentally, we note that this expression for κ^2 is the same as Equation 6.104 obtained in the approximate theory of Section 6.1.5, which thereby is verified in this respect. Since the effective charge q_j^{eff} according to its definition Equation 6.153 depends on κ, we have $q_j^{\text{eff}} = q_j^{\text{eff}}(\kappa)$. This means that Equation 6.158 is actually an equation for κ, which it must since Equations 6.157 and 6.158 are equivalent to each other.

Although Equation 6.158 is both exact and appealing, it is valid only for electrolytes that consist of spherically symmetric ions. As we will find out in Chapter 13, where we will generalize the current analysis to electrolytes consisting of nonspherical ions, the expression for κ that corresponds to Equation 6.158 is more complicated (it contains, for example, Q_j^{eff} instead of q_j^{eff}) and is therefore much less appealing. For this reason, we are going to derive an alternative expression for κ that is valid irrespective of the shape and internal charge distributions of the ions. Again, we start from Equation 6.157, which is, in fact, valid also in the nonspherical case. The definition of $\chi^\star(r)$ in Equation 6.133 is, however, valid only for spherical ions.

Equation 6.157 can be written as

$$-\varepsilon_r \varepsilon_0 \kappa^2 = \int d\mathbf{r} \frac{\sinh(\kappa r)}{\kappa r} \chi^\star(r)$$

$$= \int d\mathbf{r} \chi^\star(r) + \kappa^2 \int d\mathbf{r}\, r^2 \left[\frac{\sinh(\kappa r) - \kappa r}{(\kappa r)^3} \right] \chi^\star(r) \tag{6.159}$$

(note that the square bracket remains finite when $\kappa r \to 0$). By moving the last term in the rhs to the lhs, we obtain

$$-\varepsilon_0 \kappa^2 \left(\varepsilon_r + \frac{1}{\varepsilon_0} \int d\mathbf{r}\, r^2 \left[\frac{\sinh(\kappa r) - \kappa r}{(\kappa r)^3} \right] \chi^\star(r) \right) = \int d\mathbf{r} \chi^\star(r). \tag{6.160}$$

From Equation 6.133, it follows that

$$\int d\mathbf{r} \chi^\star(r) = -\beta \sum_j q_j n_j^{\text{b}} \int d\mathbf{r} \rho_j^\star(r) = -\beta \sum_j q_j n_j^{\text{b}} q_j^\star, \tag{6.161}$$

where

$$q_j^\star = \int d\mathbf{r} \rho_j^\star(r) \tag{6.162}$$

is the *total charge* of the dressed particle charge density $\rho_j^\star$. We will call $q_j^\star$ the **dressed particle charge** of the j-ion. By inserting this into Equation 6.160, we obtain the appealing expression for the decay parameter κ

$$\kappa^2 = \frac{\beta}{\mathcal{E}_r^\star(\kappa)\varepsilon_0} \sum_j q_j q_j^\star n_j^b, \qquad (6.163)$$

where we have defined

$$\mathcal{E}_r^\star(\kappa) = \varepsilon_r + \frac{1}{\varepsilon_0} \int d\mathbf{r}\, r^2 \left[\frac{\sinh(\kappa r) - \kappa r}{(\kappa r)^3} \right] \chi^\star(r), \qquad (6.164)$$

which will be denoted as the **dielectric factor**. Equation 6.163 *is an exact expression for κ that holds in general.*[38] Note that the dressed particle charge $q_j^\star$ is independent of κ. The κ dependence in the rhs of Equation 6.163 arises solely from $\mathcal{E}_r^\star(\kappa)$. This is in contrast to Equation 6.158, where the κ dependence of the rhs arises solely from $q_j^{\text{eff}} = q_j^{\text{eff}}(\kappa)$.[39]

The general and exact expression (6.163) for κ differs from the definition (6.31) of the Debye parameter κ_D solely by the occurrence of $q_j q_j^\star$ instead of q_j^2 and and $\mathcal{E}_r^\star$ instead of ε_r. The dielectric factor $\mathcal{E}_r^\star$ for the electrolyte thus replaces the dielectric constant of the solvent. This difference is conceptually important since $\mathcal{E}_r^\star$ is a property that depends on the presence of ions, which is apparent from Equation 6.164 since $\chi^\star$ contains ionic contributions.

The coefficient $A^\star$ in the decay of $\phi_{\text{Coul}}^\star(r)$ in Equation 6.151 can be determined from the function $\chi^\star(r)$. It will be shown in Chapter 13 (see also *loc. cit.* in footnote 35) that this coefficient is given by $A^\star = 1/(4\pi \mathcal{E}_r^{\text{eff}} \varepsilon_0)$ with

$$\mathcal{E}_r^{\text{eff}} = \mathcal{E}_r^{\text{eff}}(\kappa) = \varepsilon_r + \frac{1}{2\varepsilon_0} \int d\mathbf{r}\, r^2 \left[\frac{\kappa r \cosh(\kappa r) - \sinh(\kappa r)}{(\kappa r)^3} \right] \chi^\star(r), \qquad (6.165)$$

---

[38] Equation 6.163 holds also for systems with nonspherical ions which may contain polar molecules. As we will find in Chapter 13 (see also *loc. cit.* in footnote 35), the general definition of the polarization response function is $\chi^\star(r_{12}) = -\beta \sum_j \int d\mathbf{r}_3 n_j^b \left\langle \rho_j^\star(\mathbf{r}_{31}, \omega_3) \sigma_j(\mathbf{r}_{32}, \omega_3) \right\rangle_{\omega_3}$, where $\langle \cdot \rangle_{\omega_3}$ denotes the average over orientations and the sum is over all species in the system (including the polar molecules, if any). In this case, the result in Equation 6.161 is still valid because

$$\int d\mathbf{r}_2 \chi^\star(r_{12}) = -\beta \sum_j \int d\mathbf{r}_3 n_j^b \left\langle \rho_j^\star(\mathbf{r}_{31}, \omega_3) \int d\mathbf{r}_2 \sigma_j(\mathbf{r}_{32}, \omega_3) \right\rangle_{\omega_3}$$
$$= -\beta \sum_j \int d\mathbf{r}_3 n_j^b \left\langle \rho_j^\star(\mathbf{r}_{31}, \omega_3) q_j \right\rangle_{\omega_3} = -\beta \sum_j n_j^b q_j^\star q_j,$$

where we have used the fact that $q_j = \int d\mathbf{r}\, \sigma_j(\mathbf{r}, \omega)$ and $q_j^\star \equiv \int d\mathbf{r}\, \rho_j^\star(\mathbf{r}, \omega)$ are independent of orientation (note that electroneutral polar molecules do not contribute to the sum since $q_j = 0$ for them). The expression for $\mathcal{E}_r^\star$ in Equation 6.164 is valid for an electrolyte in a dielectric continuum with dielectric constant ε_r. For an electrolyte solution with a molecular solvent or for an ionic liquid, ε_r in the formula should be replaced by 1 (ε_r for vacuum). Then the dielectric properties of the system are included in $\chi^\star(r_{12})$ and enter automatically in $\mathcal{E}_r^\star$. In fact, for an electrolyte solution with a molecular polar solvent in the limit of infinite dilution, $\mathcal{E}_r^\star$ goes to the dielectric constant of the pure polar solvent as shown in Chapter 13. For finite electrolyte concentrations, the contributions from the molecular solvent is, however, different from ε_r for the pure solvent. This effect is, of course, totally missing from a model where a dielectric continuum replaces the molecular solvent.

[39] From a comparison of Equations 6.158 and 6.163, it may appear as if $q_j^{\text{eff}}(\kappa)/\varepsilon_r$ is equal to $q_j^\star/\mathcal{E}_r^\star(\kappa)$, but *this is not the case* in general.

where ε_r is the dielectric constant of the solvent modeled as a dielectric continuum.[40] Thus, we have

$$\phi_{\text{Coul}}^*(r) \sim \frac{1}{4\pi \mathcal{E}_r^{\text{eff}} \varepsilon_0} \cdot \frac{e^{-\kappa r}}{r} \quad \text{when } r \to \infty, \tag{6.166}$$

which should be compared with the corresponding expression in the PB approximation (Equation 6.69)

$$\phi_{\text{Coul}}^\star(r) = \frac{e^{-\kappa_D r}}{4\pi \varepsilon_r \varepsilon_0 r}, \quad \text{(PB)}.$$

As apparent from this comparison, $\mathcal{E}_r^{\text{eff}}$ is an entity that *plays a role similar to a dielectric constant*, so it is a kind of **effective dielectric permittivity** of the electrolyte. Thus, for electrolytes, there are two entities, $\mathcal{E}_r^\star$ and $\mathcal{E}_r^{\text{eff}}$, that replace the dielectric constant of the solvent in the expression for κ and in the magnitude of the electrostatic potential, respectively. We can see from the definitions (6.164) and (6.165) that $\mathcal{E}_r^\star$ and $\mathcal{E}_r^{\text{eff}}$ have different values in general, but both go ε_r in the limit of zero ion density.

For a spherical ion of species i, Equation 6.152 can be written

$$\psi_i(r) \sim \frac{q_i^{\text{eff}}}{4\pi \mathcal{E}_r^{\text{eff}} \varepsilon_0} \cdot \frac{e^{-\kappa r}}{r} \quad \text{when } r \to \infty,$$

which should be compared with the decay formula obtained for the PB case in Equation 6.53. The only differences are the occurrence of $\mathcal{E}_r^{\text{eff}}$ instead of ε_r and the unequal values of the effective charge q_i^{eff} and the decay parameter for the two cases. For the potential of mean force of the ions in the electrolyte, we have

$$w_{ij}^{(2)}(r) \sim \frac{q_i^{\text{eff}} q_j^{\text{eff}}}{4\pi \mathcal{E}_r^{\text{eff}} \varepsilon_0} \cdot \frac{e^{-\kappa r}}{r} \quad \text{when } r \to \infty, \tag{6.167}$$

which is a special case of Equation 6.155.

We thus have the equivalent Equations 6.157 and 6.163 for the decay parameter κ, which are exact and general. Any of them can be used, but let us select the second equation for the time being. So far, we have focused on the solution κ, which, according to Equation 6.151, gives the decay length $1/\kappa$ of the leading term in $\phi_{\text{Coul}}^*(r)$ for large r. In fact, Equation 6.163 has more than one solution; the solution κ discussed up to now is the one that has the smallest value (largest decay length). There is (at least) another solution κ' as well, that is, we have

$$(\kappa')^2 = \frac{\beta}{\mathcal{E}_r^\star(\kappa') \varepsilon_0} \sum_j q_j q_j^\star n_j^{\text{b}}. \tag{6.168}$$

---

[40] As explained in footnote 38, the term ε_r should be replaced by 1 (ε_r for vacuum) when there is a molecular solvent or in cases without a solvent.

The two real-valued solutions κ and κ' to Equation 6.163 with $\kappa < \kappa'$ give rise to two Yukawa function terms in $\phi^*_{\text{Coul}}(r)$, which hence decays like[41]

$$\phi^*_{\text{Coul}}(r) \sim \frac{1}{4\pi\varepsilon_0} \left[\frac{e^{-\kappa r}}{\mathcal{E}^{\text{eff}}_r(\kappa)r} + \frac{e^{-\kappa' r}}{\mathcal{E}^{\text{eff}}_r(\kappa')r} \right] \quad \text{when } r \to \infty, \tag{6.169}$$

where $\mathcal{E}^{\text{eff}}_r$ is evaluated at κ' in the last term. This is very much different from the PB approximation where there is only one decay parameter κ_D and only one term in $\phi^*_{\text{Coul}}(r)$.

The consequences of these two solutions are discussed in the following shaded text box, which can be skipped in the first reading. The main points are: The existence of the two solutions implies that there are to two Yukawa terms in the mean electrostatic potential ψ_i, just like in Equation 6.109 of the approximate theory of Section 6.1.5 (see the discussion of these two terms in the shaded text box on page 303f). Likewise, there are two such terms in various distribution functions, for instance the charge density $\rho_i(r)$ around an ion. Furthermore, the solutions to Equation 6.163 can become complex-valued, $\kappa = \kappa_{\Re} + i\kappa_{\Im}$, where $\kappa_{\Re}$ and $\kappa_{\Re}$ are the real and imaginary parts, respectively. Then the other solution κ' is the complex conjugate $\kappa' = \kappa_{\Re} - i\kappa_{\Im}$. The appearance of such solutions corresponds to the crossover from a monotonic to an oscillatory decay that was illustrated in Figure 6.7b and discussed in the shaded text box on page 303f. The potential in the oscillatory case decays like in Equation 6.110 and the same is true for $\rho_i(r)$ and various other distribution functions.[42]

In the next section, we will discuss these issues further for the systems dealt with in Figures 6.8 and 6.9.

★ The two terms in $\phi^*_{\text{Coul}}(r)$ give rise to two corresponding terms in the decay of the electrostatic potential from a particle in the electrolyte

$$\psi_i(r) \sim \frac{1}{4\pi\varepsilon_0} \left[\frac{q^{\text{eff}}_i(\kappa)}{\mathcal{E}^{\text{eff}}_r(\kappa)} \cdot \frac{e^{-\kappa r}}{r} + \frac{q^{\text{eff}}_i(\kappa')}{\mathcal{E}^{\text{eff}}_r(\kappa')} \cdot \frac{e^{-\kappa' r}}{r} \right] \quad \text{when } r \to \infty. \tag{6.170}$$

(cf. Equation 6.109 in the approximate theory of Section 6.1.5). Note that $\mathcal{E}^{\text{eff}}_r$ and q^{eff}_i in the two terms have different values. This means that there is not one single value of the "effective charge" of an ion – there is one value for each decay term (decay mode) with different decay parameters κ and κ'. The same applies for the "effective permittivity"

---

[41] The contributions to the asymptotic expression for $\phi^*_{\text{Coul}}(r)$ will be discussed in some detail in Chapter 13 in the second volume (see also *loc. cit.* in footnote 35). In addition to the terms shown, there are in general terms with a different r dependence than Yukawa functions, terms like $e^{-2\kappa r}f(r)/r^2$ where $f(r)$ is a slowly varying function. Such terms decay faster than the leading term when $r \to \infty$, but they can make significant contributions for small r. When $\kappa' > 2\kappa$, such a term can, however, be the second leading term in $\phi^*_{\text{Coul}}(r)$ for large r, but we limit the discussion here to cases where Yukawa terms with real or complex decay parameters are the leading ones.

[42] In the general case, there are several terms, each with its own value of the decay parameter that is a solution to Equation 6.163. There are also terms with other functional dependence of r, but in the cases considered here the leading term is a Yukawa function or the corresponding oscillatory function. The second leading term is often such a term, but see footnote 41.

$\mathcal{E}_r^{\text{eff}}$ for the electrolyte. Likewise, the value of dielectric factor $\mathcal{E}_r^{\star}$ in Equation 6.163 for κ is different from the value in Equation 6.168 for κ'. Thus, each decay mode has its own values of these entities.

The charge density $\rho_i(r)$ around an i-ion can be obtained from Poisson's equation, which reads $\rho_i(r) = -\varepsilon_r \varepsilon_0 (\partial^2 \left[r\psi_i(r) \right] / \partial r^2)/r$ outside the ion, and we obtain from Equation 6.170

$$\rho_i(r) \sim -\frac{\varepsilon_r}{4\pi} \left[\frac{\kappa^2 q_i^{\text{eff}}(\kappa)}{\mathcal{E}_r^{\text{eff}}(\kappa)} \cdot \frac{e^{-\kappa r}}{r} + \frac{(\kappa')^2 q_i^{\text{eff}}(\kappa')}{\mathcal{E}_r^{\text{eff}}(\kappa')} \cdot \frac{e^{-\kappa' r}}{r} \right] \quad \text{when } r \to \infty. \quad (6.171)$$

Finally, the potential of mean force decays when $r \to \infty$ as

$$w_{ij}^{(2)}(r) \sim \frac{1}{4\pi \varepsilon_0} \left[\frac{q_i^{\text{eff}}(\kappa) q_j^{\text{eff}}(\kappa)}{\mathcal{E}_r^{\text{eff}}(\kappa)} \cdot \frac{e^{-\kappa r}}{r} + \frac{q_i^{\text{eff}}(\kappa') q_j^{\text{eff}}(\kappa')}{\mathcal{E}_r^{\text{eff}}(\kappa')} \cdot \frac{e^{-\kappa' r}}{r} \right] \quad (6.172)$$

and since $g_{ij}^{(2)}(r) - 1 = h_{ij}^{(2)}(r) \approx -\beta w_{ij}^{(2)}(r)$ when $w_{ij}^{(2)}$ is small, $h_{ij}^{(2)}(r)$ decays like $-\beta$ times the rhs of Equation 6.172.

The corresponding relationships are true for the electrostatic potential, potential of mean force and charge density for nonsymmetrical particles. In such cases, $Q_P^{\text{eff}}(\hat{\mathbf{r}}, \boldsymbol{\omega})$ enters in the relationships instead of q_i^{eff}. Note that Q_P^{eff} is dependent on the value of κ and takes on different values for κ and κ'.

In Figure 6.7b, we saw that when the ionic density is increased, the solutions κ and κ' can merge and become complex-valued; the two solutions are then complex conjugates to each other: $\kappa = \kappa_{\Re} + i\kappa_{\Im}$ and $\kappa' = \kappa_{\Re} - i\kappa_{\Im}$. Obviously, the factors $\mathcal{E}_r^{\text{eff}}(\kappa)$ and $\mathcal{E}_r^{\text{eff}}(\kappa')$ in Equation 6.169 also become complex conjugates to each other, $\mathcal{E}_r^{\text{eff}}(\kappa_{\Re} \pm i\kappa_{\Im})$, and the same is true for $q_i^{\text{eff}}(\kappa)$ and $q_i^{\text{eff}}(\kappa')$. Therefore, in the formulas for the decay of $\phi_{\text{Coul}}^*(r)$, $\psi_i(r)$, $\rho_i(r)$ and $w_{ij}(r)$ above, the two terms become complex conjugates to each other. Since the sum of a complex number A and its complex conjugate $\underline{A}$ is given by $A + \underline{A} = 2\Re(A)$, where $\Re(A)$ stands for the real part of A, we obtain

$$\phi_{\text{Coul}}^*(r) \sim 2\Re \left[\frac{e^{-(\kappa_{\Re} + i\kappa_{\Im})r}}{4\pi |\mathcal{E}_r^{\text{eff}}| e^{-i\vartheta_{\mathcal{E}}} \varepsilon_0 r} \right] \quad \text{when } r \to \infty, \quad (6.173)$$

where we have written $\mathcal{E}_r^{\text{eff}}(\kappa_{\Re} + i\kappa_{\Im}) = |\mathcal{E}_r^{\text{eff}}| \exp(-i\vartheta_{\mathcal{E}})$.[43] This asymptotic relationship equals

$$\phi_{\text{Coul}}^*(r) \sim \frac{1}{2\pi |\mathcal{E}_r^{\text{eff}}| \varepsilon_0} \cdot \frac{e^{-\kappa_{\Re} r}}{r} \cos(\kappa_{\Im} r - \vartheta_{\mathcal{E}}) \quad \text{when } r \to \infty, \quad (6.174)$$

---

[43] We may always select a phase ϑ in the interval $-\pi < \vartheta \leq \pi$ for it to be definite.

where $\vartheta_{\mathcal{E}}$ is a phase shift, $1/\kappa_{\mathfrak{R}}$ is the decay length and $2\pi/\kappa_{\mathfrak{I}}$ is the wavelength. In the same manner, we obtain

$$\psi_i(r) \sim \frac{|q_i^{\text{eff}}|}{2\pi|\mathcal{E}_r^{\text{eff}}|\varepsilon_0} \cdot \frac{e^{-\kappa_{\mathfrak{R}}r}}{r} \cos(\kappa_{\mathfrak{I}}r + \vartheta_i - \vartheta_{\mathcal{E}}) \quad \text{when } r \to \infty,$$

where $q_i^{\text{eff}}(\kappa_{\mathfrak{R}} + i\kappa_{\mathfrak{I}}) = |q_i^{\text{eff}}| \exp(-i\vartheta_i)$. Note that any difference in sign for $\psi_+(r)$ and $\psi_-(r)$ is incorporated in the phase ϑ_i. For instance, if $\psi_+(r) \approx -\psi_+(r)$ for a range of r values, we normally have $\vartheta_+ - \vartheta_- \approx \pi$.

Likewise,

$$w_{ij}^{(2)}(r) \sim \frac{|q_i^{\text{eff}} q_j^{\text{eff}}|}{2\pi|\mathcal{E}_r^{\text{eff}}|\varepsilon_0} \cdot \frac{e^{-\kappa_{\mathfrak{R}}r}}{r} \cos(\kappa_{\mathfrak{I}}r + \vartheta_i + \vartheta_j - \vartheta_{\mathcal{E}}) \quad \text{when } r \to \infty, \quad (6.175)$$

where each ion contributes with a phase shift ϑ_l and a magnitude $|q_l^{\text{eff}}|$ for $l = i, j$. The decay of the charge density around an i-ion can similarly be obtained from Equation 6.171 and we have when $r \to \infty$

$$\rho_i(r) \sim -\frac{\varepsilon_r|\kappa|^2 |q_i^{\text{eff}}|}{2\pi|\mathcal{E}_r^{\text{eff}}|} \cdot \frac{e^{-\kappa_{\mathfrak{R}}r}}{r} \cos(\kappa_{\mathfrak{I}}r + \vartheta_i - \vartheta_{\mathcal{E}} + 2\vartheta_\kappa), \quad (6.176)$$

where we have written $\kappa^2 = (\kappa_{\mathfrak{R}} + i\kappa_{\mathfrak{I}})^2 = |\kappa|^2 \exp(-i2\vartheta_\kappa)$ in the numerator of Equation 6.171. For simplicity, we will write this as

$$\rho_i(r) \sim b_i \frac{e^{-\kappa_{\mathfrak{R}}r}}{r} \cos(\kappa_{\mathfrak{I}}r + \vartheta_{b,i}), \quad (6.177)$$

where the constants b_i and $\vartheta_{b,i}$ are defined by identification with the previous equation. The oscillations in sign of $\rho_i(r)$ and $\psi_i(r)$ as functions of r depend on the fact that the surrounding ions form a shell structure of the kind that exists in molten NaCl as illustrated by $g_{ij}^{(2)}(r)$ in Figure 6.8.

Let us also consider the function $\rho_Q(r) = [\rho_+(r) - \rho_-(r)]/2$, which will be of interest in what follows. For the case of real κ and κ', the decay of $\rho_Q(r)$ when $r \to \infty$ is given by

$$\rho_Q(r) \sim a_Q \frac{e^{-\kappa r}}{r} + a_Q' \frac{e^{-\kappa'r}}{r} \quad (6.178)$$

where $a_Q = -\varepsilon_r\kappa^2[q_+^{\text{eff}}(\kappa) - q_-^{\text{eff}}(\kappa)]/[8\pi\mathcal{E}_r^{\text{eff}}(\kappa)]$, as obtained from the first term in Equation 6.171, and where a_Q' is given by the same expression but with κ' inserted instead of κ. For the oscillatory case, the decay is obtained by inserting Equation 6.177

into the definition of $\rho_Q(r)$ and we have

$$\rho_Q(r) \sim \frac{e^{-\kappa_{\Re} r}}{r} \left[b_+ \cos(\kappa_{\Im} r + \vartheta_{b,+}) - b_- \cos(\kappa_{\Im} r + \vartheta_{b,-}) \right]/2$$

$$= b_Q \frac{e^{-\kappa_{\Re} r}}{r} \cos(\kappa_{\Im} r + \vartheta_Q), \qquad (6.179)$$

where the last equality follows since one can always write the difference of two cosines in this manner with suitable values of b_Q and ϑ_Q.[44]

6.2.2.3 The Significance of the Asymptotic Decays: Concrete Examples

Let us see how results for the asymptotic decay obtained in the previous section work out in practice for the systems presented earlier in Section 6.2.2.1. For molten NaCl, the charge density $\rho_i(r)$ around an ion of species i is oscillatory as discussed in connection to Figure 6.8. As mentioned there, we have $\rho_+(r) \approx -\rho_-(r)$ in this particular case. For large r, the asymptotic decay of $\rho_i(r)$ is, as shown in the shaded text box above (Equations 6.176 and 6.177),

$$\rho_i(r) \sim b_i \frac{e^{-\kappa_{\Re} r}}{r} \cos(\kappa_{\Im} r + \vartheta_{b,i}) \quad \text{when } r \to \infty,$$

where b_i is a constant, $\vartheta_{b,i}$ is a phase shift, $1/\kappa_{\Re}$ is the decay length and $2\pi/\kappa_{\Im}$ is the wavelength. The parameters b_i and $\vartheta_{b,i}$ can be expressed in terms of effective charges and other entities as shown in Equation 6.176.

We will investigate the function $\rho_Q(r) = \left[\rho_+(r) - \rho_-(r)\right]/2$, which gives a kind of "average" charge density around an ion in the electrolyte; the minus sign in front of ρ_- takes into account that the ion cloud has an opposite sign for an anion compared to a cation. It is shown in the shaded text box above that this function decays in the oscillatory case as (Equation 6.179)

$$\rho_Q(r) \sim b_Q \frac{e^{-\kappa_{\Re} r}}{r} \cos(\kappa_{\Im} r + \vartheta_Q) \quad \text{when } r \to \infty, \qquad (6.180)$$

where the constants b_Q and ϑ_Q can be expressed in terms of b_i and $\vartheta_{b,i}$.[45] For molten NaCl where $\rho_+(r) \approx -\rho_-(r)$, we have $\rho_Q(r) \approx \rho_+(r) \approx -\rho_-(r)$.

---

[44] The parameters b_Q and ϑ_Q satisfy $b_Q^2 = b_+^2 + b_-^2 - 2b_+ b_- \cos(\vartheta_{b,+} - \vartheta_{b,-})$ and $\tan \vartheta_Q = [b_+ \sin(\vartheta_{b,+}) - b_- \sin(\vartheta_{b,-})]/[b_+ \cos(\vartheta_{b,+}) - b_- \cos(\vartheta_{b,-})]$.

[45] In the current case, $b_Q \approx b_+ \approx b_-$ and $\vartheta_Q \approx \vartheta_{b,+} \approx \vartheta_{b,-} - \pi$, where the term $-\pi$ takes care of the different signs for the anions.

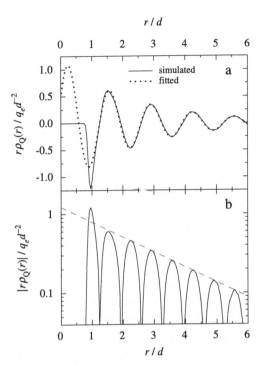

FIGURE 6.10 The charge density $\rho_Q(r)$, obtained from MD simulations[46] for molten NaCl with a bulk density $n_{tot}^b = n_+^b + n_-^b = 0.28\,d^{-3}$, plotted as $r\rho_Q(r)$. The average ion size is $d = (d_{Na^+} + d_{Cl^-})/2 = 2.76$ Å. (a) Results at 1000 K (full curve) compared with a fit to an exponentially decaying oscillatory function (dotted curve), see text. (b) Absolute value of the same function $r\rho_Q(r)$ as in pane a plotted on a log-linear scale. The deep "minima" occur where the oscillatory function passes zero. The dashed line show the tangent at the maximal values for large r and its negative slope gives the decay parameter $\kappa_\Re$.

In Figure 6.10, taken from computer simulations by Keblinski and coworkers,[47] the function $r\rho_Q(r)$ for molten NaCl near the freezing point is shown. A realistic model potential for the ion-ion pair interactions in the molten salt was used in the calculations. Pane a of the figure shows $r\rho_Q(r)$ plotted together with a curve fit of this function to the asymptotic expression in the rhs of Equation 6.180. We see that the fitted curve (dotted) is very close to $r\rho_Q(r)$ for $r/d \gtrsim 1.2$, where $d = (d_{Na^+} + d_{Cl^-})/2$ is the average ionic size. Beyond $r/d = 2$, the fit is essentially indistinguishable from $r\rho_Q(r)$. This means that $\rho_Q(r)$ is accurately represented by its asymptotic expression beyond a couple of ionic diameters. In Figure 6.10b, the absolute value of $r\rho_Q(r)$ is plotted on a log-linear scale. The tangent to the curve for large r at the maximal values is also shown; these maximal values correspond to the maxima and minima of $r\rho_Q(r)$. The slope of the tangent (with inverted sign) gives the decay length $1/\kappa_\Re$, which is about 2.3 ionic diameters in this case.

[46] Figure 6.10 is based on data from P. Keblinski, J. Eggebrecht, D. Wolf, and S. R. Phillpot, *J. Chem. Phys.* **113** (2000) 282. Prof. Keblinski is acknowledged for kindly providing the original data.

[47] Keblinski et al. *loc. cit.* in footnote 46. The function $Q(r)$ in that work is in our notation equal to $\rho_Q(r)4/q_\pm$ with $q_\pm = q_+ = |q_-|$.

Keblinski et al. found similarly good agreement for fits to the asymptotic expression also at other temperatures and densities. It can be concluded that the asymptotic estimate of $\rho_Q(r)$ and hence of $\rho_+(r)$ and $\rho_-(r)$ is very useful for molten NaCl. The asymptotic formulas give an accurate description of these functions for a *quite wide range of distances* and not only for very large r as one may have thought considering the formal derivations that are valid in the limit $r \to \infty$.

Asymptotic expressions are accordingly very useful, but in order to have quantitative results one needs in practice to know the parameters in the asymptotic formulas, like the effective charges and decay parameters. The determination of these parameter values can be an equally demanding task as the determination of the distribution functions themselves. However, a knowledge of the relevant *qualitative decay behavior* of these functions is in many cases essential for the correct interpretation of theoretical and experimental results. Furthermore, the asymptotic formulas can constitute the basis for the construction of efficient approximations that have the correct behavior built in from the start (the approximate theory of Section 6.1.5 is a very simple example of this).

Keblinski et al. also studied the crossover between monotonic and oscillatory decay behavior for the NaCl system at higher temperatures. In Figure 6.11, the absolute values of the function $r\rho_Q(r)$ are plotted on a log-linear scale for NaCl gas at 10000 K for various densities as shown in panes a–c in the figure. At high density (pane a), the function is oscillatory as before, while at low density (pane c) the decay is monotonic. In the latter case, the

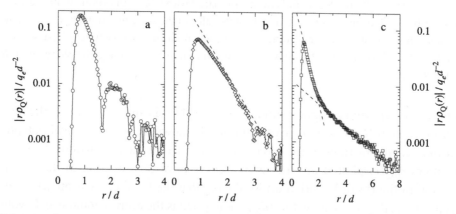

FIGURE 6.11 Same as in Figure 6.10b, but for NaCl gas at 10 000 K for various densities n^b_{tot}: (a) 0.125, (b) 0.028 and (c) 0.0013 d^{-3}.[48] The curve in pane (a) shows an oscillatory decay like in Figure 6.10b, but with a longer wavelength and much faster decay, so the curve is dominated by numerical noise after only two maxima. Pane (b) shows the results for a system near the crossover point between oscillatory and monotonic decay. The system in pane (c) has an exponential decay with two decay lengths $1/\kappa'$ and $1/\kappa$ as indicated by the dashed lines. These two decay lengths have merged into one single decay length at the crossover point as illustrated by the dashed curve in pane (b).

---

[48] Figure 6.11 is based on data from Keblinski et al. *loc. cit.* in footnote 46.

decay for $\rho_Q(r)$ is given by

$$\rho_Q(r) \sim a_Q \frac{e^{-\kappa r}}{r} + a_Q' \frac{e^{-\kappa' r}}{r}$$

(see Equation 6.178, where the coefficients are specified in the text). In pane c, two straight parts of the curve are clearly visible (shown by the dashed lines) with decay lengths $1/\kappa'$ and $1/\kappa$, which correspond to the two terms in the asymptotic decay formula. For large r, the first asymptotic term with the largest decay length $1/\kappa$ dominates since the second term with a smaller decay length $1/\kappa'$ is negligible there. For smaller r, the second term contributes significantly and if the coefficient a_Q' is appreciably larger than a_Q in absolute values, this term can dominate there, as seen in the figure. We see that $\rho_Q(r)$ is accurately represented by its asymptotic limiting form with two terms down to quite small distances.

When the density is increased, the values of κ and κ' approach each other and at around the density $n_{tot}^b = 0.028 \, d^{-3}$ they merge. The function $|r\rho_Q(r)|$ at the point of merger is shown in Figure 6.11b, which is the borderline case between monotonic and oscillatory decay behavior. The change between the monotonic and oscillatory behaviors constitutes a *Kirkwood crossover point*, which was defined at the end of Section 6.1.5. When the density is further increased beyond this point, the wavelength $2\pi/\kappa_\Im$ of the oscillations is initially very large, but it decreases with increasing density (at the crossover point, the wavelength is infinite). This corresponds qualitatively to the behavior of $\kappa_\Im$ as shown in Figure 6.7b.

In general, there can be both monotonic and oscillatory terms in the decay of $g_{ij}^{(2)}(r)$ for large r. As we have seen, the appearance of $\mathcal{E}_r^\star(\kappa)$ in Equation 6.163 for κ leads to the possibility to have decay behaviors with more than one decay length and/or oscillatory decay. It depends on the circumstances which one of these possibilities takes place, and it is, in fact, possible to have both at the same time (i.e., there are simultaneously both real and complex roots to the equation). We may, for example, have

$$g_{ij}^{(2)}(r) - 1 \sim -\beta w_{ij}^{(2)}(r) \sim b_{ij} \frac{e^{-\kappa_\Re r}}{r} \cos(\kappa_\Im r + \vartheta_{b,ij}) + a_{ij}'' \frac{e^{-\kappa'' r}}{r} \qquad (6.181)$$

when $r \to \infty$. Incidentally, we note that by using Equation 6.175 one can express the phase shift $\vartheta_{b,ij}$ and the coefficient b_{ij} in terms of q_i^{eff}, q_j^{eff} and $\mathcal{E}_r^{eff}$. A corresponding expression for the coefficient a_{ij}'' can be obtained from Equation 6.167.

In Equation 6.181, it is the term with the slowest decay (the longest decay length) that eventually takes over and dominates for large r. An example from the MD simulations by Keblinski et al. for NaCl at 3000 K at various densities is given in Figure 6.12, which shows the decay lengths $1/\kappa_\Re$ and $1/\kappa''$ as functions of density that ranges from a relatively low density gas ($n_{tot}^b = 0.0012 \, d^{-3}$) to a dense melt ($n_{tot}^b = 0.28 \, d^{-3}$). At high densities, the oscillatory term has the longest decay length and is the leading term for large r, while for lower densities the monotonic term is leading; the crossover point occurs at around $n_{tot}^b = 0.1 \, d^{-3}$. This is another type of crossover between leading oscillatory or monotonic decays than the Kirkwood crossover point in Figure 6.11b. In the present case, a complex-valued solution,

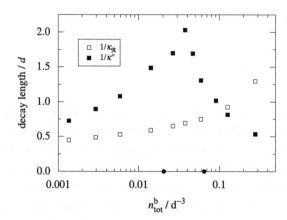

FIGURE 6.12 The decay lengths $1/\kappa_{\mathfrak{R}}$ for the oscillatory term (open squares) and $1/\kappa''$ for the monotonic term (filled squares) of $g_{ij}^{(2)}(r)$ given by Equation 6.181 for NaCl at 3000 K as functions of density obtained from MD simulations.[49] The two small points at the density axis mark the limits of the two-phase region for a large system (cf. footnote 51). The density is measured in units of d^{-3} with $d = 2.76$ Å (cf. Figure 6.10).

κ, and a real-valued one, κ'', exist at the same time and the crossover occurs at the density where $\kappa_{\mathfrak{R}}$ and κ'' become equal. This is called a **Fisher-Widom crossover point**.

At decreasing densities, there ought to be a Kirkwood crossover point for the oscillatory contribution at some low density: a point where two monotonic terms with real-valued decay parameters κ and κ', respectively, will appear (outside the range of Figure 6.12). The monotonic term with the smallest decay parameter (κ) will have a decay length ($1/\kappa$) that must be larger than $1/\kappa''$ at very low densities since $1/\kappa$ goes (as required) to the Debye length $1/\kappa_D$ in the limit of zero ionic density. The Debye-Hückel result for $g_{ij}^{(2)}(r)$ is always the leading term in this limit.

Let us now consider the functions

$$\rho_Q(r) = q_\pm n_{tot}^b \left[g_{++}^{(2)}(r) + g_{--}^{(2)}(r) - 2g_{+-}^{(2)}(r) \right]/4$$

$$n_N(r) = n_{tot}^b \left[g_{++}^{(2)}(r) + g_{--}^{(2)}(r) + 2g_{+-}^{(2)}(r) \right]/4,$$

where $\rho_Q(r)$ is the same function as we have encountered earlier (note that n_N [with N in roman] is different from n_N used earlier [with N in italics]; in the latter N means number of particles). These functions are treated in more detail in the next section (Section 6.2.3), which can be skipped in the first reading (cf. Equations 6.184 and 6.188; the expression for $n_N(r)$ above is valid only for $z : -z$ electrolytes). In that section, it is shown that the function $n_N(r)$ for $r > 0$ is proportional to the density-density correlation function $H_{NN}(r)$, which describes the correlations in density fluctuations between any two points at separation r from each other,[50] while $\rho_Q(r)$ is proportional to the corresponding charge-charge

---

[49] Figure 6.12 is based on data from Keblinski et al. *loc. cit.* in footnote 46.

[50] For a one-component system, $H_{NN}(r)$ is equal to the function $H^{(2)}(r)$ introduced in Section 5.8.

correlation function $H_{QQ}(r)$. Section 6.2.3 deals also with $H_{QN}(r)$, the charge-density correlation function, which describes correlations between fluctuations in charge at one point and fluctuations in density at another point separated by r. All three correlation functions describe the correlations in fluctuations in number density and/or charge density at two points irrespective of the ion species.

As shown in the shaded text box below, for the NaCl systems in Figure 6.12 the first term in the rhs of Equation 6.181 mainly contributes to $\rho_Q(r)$ and $H_{QQ}(r)$, while the second term mainly contributes to $n_N(r)$ and $H_{NN}(r)$. The main term in the decay of the charge-charge correlations therefore has decay length $1/\kappa_\Re$ and the main term in the density-density correlations has decay length $1/\kappa''$. There is, however, also a small contribution to $\rho_Q(r)$ with decay length $1/\kappa''$ (from the second term in Equation 6.181), and likewise a small contribution to $n_N(r)$ with decay length $1/\kappa_\Re$ (from the first term). These contributions have small prefactors and can be ignored except, possibly, for very large r, where the term with the longest decay length – the leading term – ultimately dominates. When the leading term has a small prefactor compared to the other terms, one has to go to very large r for this to happen.

Like in the case of molten NaCl shown in Figure 6.8, $\rho_Q(r)$ is oscillatory for the present NaCl systems – a behavior that originates from the first term in Equation 6.181. Both in Figure 6.8 and in the present case, we have $g_{++}^{(2)}(r) \approx g_{--}^{(2)}(r)$ and the functions $g_{+-}^{(2)}(r)$ and $g_{++}^{(2)}(r)$ have opposite signs but nearly equal magnitudes for large r (the same applies to $g_{+-}^{(2)}(r)$ and $g_{--}^{(2)}(r)$). Therefore, $b_{++} \approx b_{--} \approx b_{+-}$ and $\vartheta_{b,++} \approx \vartheta_{b,--} \approx \vartheta_{b,+-} + \pi$, where the phase contribution π gives the difference in signs of $g_{+-}^{(2)}(r)$ on the one hand and $g_{++}^{(2)}(r)$ and $g_{--}^{(2)}(r)$ on the other. Due to this sign difference and the nearly equal magnitudes of b_{ij} for all i and j, the first term in Equation 6.181 is nearly canceled in the combinations $g_{++}^{(2)}(r) + g_{+-}^{(2)}(r)$ and $g_{--}^{(2)}(r) + g_{+-}^{(2)}(r)$, so this term gives a very small contribution to $n_N(r)$, that is, a contribution with a small prefactor. On the other hand, the first term gives a large contribution to $g_{++}^{(2)}(r) - g_{+-}^{(2)}(r)$ and $g_{--}^{(2)}(r) - g_{+-}^{(2)}(r)$ for the same reasons. Thus, the first term mainly contributes to $\rho_Q(r)$ and hence to $H_{QQ}(r)$.

For the second term in Equation 6.181, we have, in fact, $a'_{++} \approx a'_{--} \approx a'_{+-}$ in the present case, so $g_{++}^{(2)}(r)$, $g_{--}^{(2)}(r)$ and $g_{+-}^{(2)}(r)$ have equal signs and nearly equal magnitudes. This implies that the second term gives only a small contribution to $\rho_Q(r)$. It contributes mainly to $n_N(r)$ and hence to $H_{NN}(r)$. All these matters are explained in more detail in Section 6.2.3.

The maximum of $1/\kappa''$ in Figure 6.12 occurs for conditions not far from the critical point, where the density-density correlations have a long range.[51] The decay length diverges to

---

[51] For densities inside the two-phase region between the points at the density axis of Figure 6.12, some of the finite system simulations by Keblinski et al. can represent metastable gas or liquid. The maximum of $1/\kappa''$ must, however, lie in the unstable region of the phase diagram for a large system.

infinity at the critical point, which lies at a higher temperature. As regards the charge-charge correlations, the decay length $1/\kappa_\Re$ does not diverge and these correlations remain limited when criticality is approached.[52]

Let us turn to the general case. Then the functions $g_{ij}^{(2)}(r)$, $\rho_Q(r)$ and $n_N(r)$ can be dominated either by oscillatory or monotonically decaying contributions depending on the conditions. Normally, all three functions have *the same decay length* and, when appropriate, *the same wavelength* for large r. When $g_{ij}^{(2)}(r)$ has a term with a certain decay length, there normally appears, as we have seen, a term with the same decay length in both $n_N(r)$ and $\rho_Q(r)$.

There exist, however, some exceptional cases where $\rho_Q(r)$ and $n_N(r)$ have different decay lengths and then the same applies to $H_{QQ}(r)$ and $H_{NN}(r)$. As will be explained in Section 6.2.3, this occurs for $z : -z$ bulk electrolytes where the anions and cations differ *only* by the signs of their charges, like the *restricted primitive model* (RPM) where anions and cations have the same diameters. This kind of model has an extreme symmetry because all properties remain the same when the signs of all charges are reversed from $+$ to $-$ and vice versa. Therefore, $g_{++}^{(2)}(r) \equiv g_{--}^{(2)}(r)$ and the charge-density correlation function $H_{QN}(r)$ is identically zero by symmetry. The unequal decay lengths of the charge-charge and density-density correlations are a consequence of the absence of charge-density correlations. In these exceptional cases, we have

$$\rho_Q(r) \sim a_Q \frac{e^{-\kappa r}}{r}$$
$$n_N(r) \sim B_N \frac{e^{-\alpha r}}{r} \qquad \text{(RPM and similar models)}, \qquad (6.182)$$

where a_Q is the same as before, B_N is another constant and where we have $\kappa \neq \alpha$ in general (analogous formulae applies for oscillatory cases). Here the decay behavior of $n_N(r)$ involves ionic core interactions to a large extent, while the decay of $\rho_Q(r)$ involves the screened electrostatic interactions in a similar manner as in the previous systems. The decay length $1/\kappa$ is always larger than $1/\alpha$ when the ionic density is sufficiently small,[53] but it can happen at high densities that $n_N(r)$ has a longer range than $\rho_Q(r)$. This also happens near critical points where fluctuations in density are large and $H_{NN}(r)$ and $n_N(r)$ have a very long range.

The decay parameter κ is a solution to Equation 6.163 as before, but α is not a solution to this equation. The value of α is instead determined by another equation, which we presently do not have the statistical mechanical tools to obtain.[54] These matters are treated in detail in Chapter 13 of the second volume of this treatise, where the decay behaviors of

---

[52] As explained in Section 6.2.3, the small contribution to $\rho_Q(r)$ with decay length $1/\kappa''$ [from the second term in Equation 6.181] has a coefficient that goes to zero when $1/\kappa''$ diverges. Hence, this contribution does not contribute under these conditions.

[53] One can show that $\alpha \sim 2\kappa$ when $n_{tot}^b \to 0$.

[54] The decay parameter α satisfies the equation $n_{tot}^b \int d\mathbf{r}\, c_N^{(2)}(r) \sinh(\alpha r)/(\alpha r) = 1$ where $c_N^{(2)}(r) = [c_{++}^{(2)}(r) + c_{+-}^{(2)}(r)]/2$ and $c_{ij}^{(2)}(r)$ is the so-called *direct correlation function*, which is introduced in the second volume of this treatise. Equation 6.163 for κ can also be expressed solely in terms of the direct correlation functions as will be shown in Chapter 13.

the distribution functions in electrolytes in general are treated more fully than in the present section.

Next, we will investigate the monotonic decay of the pair distribution function $g_{ij}^{(2)}(r)$ for the 1:-2 electrolyte shown in Figure 6.9. From Equation 6.167, it follows that

$$g_{ij}^{(2)}(r) \sim 1 - \beta w_{ij}^{(2)}(r) \sim 1 - \beta \frac{q_i^{\text{eff}} q_j^{\text{eff}}}{4\pi \mathcal{E}_r^{\text{eff}} \varepsilon_0} \cdot \frac{e^{-\kappa r}}{r} \quad \text{when } r \to \infty. \quad (6.183)$$

The values of the effective charges can be calculated from Equation 6.153 applied for $P = i, j$ provided $\rho_i^\star(r)$ and $\rho_j^\star(r)$ are known. $\mathcal{E}_r^{\text{eff}}$ can likewise be calculated from Equation 6.165 (with $\chi^*(r)$ obtained from Equation 6.133). In Chapter 13, it will be shown how $\rho_l^\star(r')$ for $l = i, j$ can be calculated directly from the pair distribution functions (the values of the latter for small r are the most important ones for this task).[55] Since the coefficient in front of $\exp(-\kappa r)/r$ in the decay formula can be calculated from $\rho_l^\star(r')$ and κ can be determined as a solution to Equation 6.157 or 6.163, *one does not need to make a curve fit to* $g_{ij}^{(2)}(r)$ *for large r* in order to obtain the parameters in the decay formula. Such a fit can be very tricky to do since it is difficult to know in the general case whether one has obtained $g_{ij}^{(2)}(r)$ accurately for large r, where this function is very close to 1.

There is also a contribution to the decay from the solution κ' in Equation 6.168, so a more complete formula for the decay is

$$g_{ij}^{(2)}(r) \sim 1 + a_{ij} \frac{e^{-\kappa r}}{r} + a_{ij}' \frac{e^{-\kappa' r}}{r} \quad \text{when } r \to \infty,$$

where a_{ij} is equal to the coefficient in the previous equation and a_{ij}' likewise can be written in terms of (different) effective charges and an effective permittivity that can be determined in the same fashion from $\rho_l^\star(r')$ (cf. Equation 6.172). We have $\kappa < \kappa'$, so like in Figure 6.11c the first term dominates for very large r, while the second contributes significantly for smaller r values. In the present case, of the 60 mM 1:-2 electrolyte, the decay length $1/\kappa'$ is about half of $1/\kappa$.

In Figure 6.9, the asymptotic decay given by the first term alone is shown as dashed curves and that from the full expression with two terms is shown as dotted curves. Starting with the former, we see that $g_{--}^{(2)}(r)$ and $g_{+-}^{(2)}(r)$ are very close to the results from the first term for $r/d^{\text{h}} \gtrsim 2.3$ and $r/d^{\text{h}} \gtrsim 3.0$, respectively. Even for r/d^{h} values that are about one unit smaller, these functions remain well represented by this term. The dotted curve is very close to the dashed curve in these cases, so the addition of the second term does not make any appreciable difference. This is a consequence of the fact that the coefficients a_{--}' and a_{+-}' are relatively small. We can also see these features in the insert, where a dominant first term yields a straight line for $|rh_{ij}^{(2)}(r)|$ for large r in these two cases.

For the divalent-divalent distribution function $g_{++}^{(2)}(r)$, however, the situation is different. The first term is clearly insufficient and from the insert it can be seen that $|rh_{++}^{(2)}(r)|$ does not

---

[55] In principle, it is also possible to calculate $\rho_l^\star(r)$ for $l = i, j$ from the system of equations consisting of Equations 6.133, 6.136 and 6.143 with input of $\psi_l(r)$ determined from the pair distributions via the charge densities $\rho_l(r)$ for $l = i, j$.

agree with the value of this term in the entire range plotted, although this is hard to judge for $r/d^h \gtrsim 8$ due to numerical noise in the simulation data in the region where $g_{++}^{(2)}(r)$ is very small. However, the entire asymptotic formula with two terms agrees well with the distribution function for $r/d^h \gtrsim 3.5$ and it does not deviate much for about one unit smaller values of r/d^h. Thus, the entire asymptotic formula represents the pair distribution function $g_{ij}^{(2)}(r)$ well beyond about 2–3 ionic diameters for all species.

Before we close the current section, we will mention some additional aspects of Equation 6.163 for the decay parameter κ. The appearance of $q_j^\star$ in the expression (6.163) for κ necessitates a different thinking as regards electrostatic screening compared to the PB approximation where the decay is determined by the corresponding expression (6.31) for the Debye parameter κ_D. For a binary electrolyte, we can write the latter expression as

$$\kappa_D^2 = \frac{\beta q_{\pm} n_{tot}^b}{2\varepsilon_0} \cdot \frac{q_+ - q_-}{\varepsilon_r}$$

and the former as

$$\kappa^2 = \frac{\beta q_{\pm} n_{tot}^b}{2\varepsilon_0} \cdot \frac{q_+^\star - q_-^\star}{\mathcal{E}_r^\star(\kappa)},$$

where we have used $q_+ n_+^b = -q_- n_-^b = q_{\pm} n_{tot}^b/2$ ($q_{\pm}$ is defined in Equation 6.149). While the Debye parameter for a given system at a certain temperature depends only on the ion density, the actual decay parameter κ depends in an intricate manner on the state of the system via the last factor since $q_j^\star$ and $\mathcal{E}_r^\star(\kappa)$ are state-dependent entities. As we have seen in Equation 6.164, $\mathcal{E}_r^\star(\kappa)$ depends on the ion-ion correlations via the polarization response function $\chi^\star$ of the electrolyte. The charge $q_j^\star$, which is defined in Equation 6.162, equals

$$q_j^\star = \int d\mathbf{r} \rho_j^\star(r) = q_j + \int d\mathbf{r} \rho_j^{dress}(r)$$

and depends on the ion-ion correlations via the last term.

In order to give a concrete example, let us consider the case with real κ and hence a Yukawa function decay. The magnitude of κ compared to κ_D depends on how much $q_j^\star$ for $j = +$ and $-$ deviates from q_j and how much $\mathcal{E}_r^\star$ deviates from ε_r. As we have seen in Figure 6.7a, κ can be both larger or smaller than κ_D depending on the state of the system. There exist several different reasons for such deviations in either direction – the situation is in general quite involved.

Here we shall solely discuss one interesting case, namely when there is strong ion pairing in a $z : -z$ electrolyte, that is, an appreciable number of anions and cations form an electroneutral pair. Such a pair is a transient phenomena, but a pair can have a considerable lifetime. For an ion of species j in a pair, there is an ion of opposite sign located near it. The contribution of the latter to ρ_j^{dress} and hence to $q_j^\star$ has a sign opposite to that of q_j. This makes the magnitude of $q_j^\star$ smaller than q_j for both anions and cations. The ion pairing also influences $\mathcal{E}_r^\star(\kappa)$. By inserting the expression for $\chi^\star(r)$ given by Equation 6.139 into Equation 6.164, we obtain

$$\mathcal{E}_r^\star(\kappa) = \varepsilon_r - \frac{\beta q_{\pm} n_{\text{tot}}^b}{2\varepsilon_0} \int d\mathbf{r}\, r^2 \left[\frac{\sinh(\kappa r) - \kappa r}{(\kappa r)^3} \right] [\rho_+^{\text{dress}}(r) - \rho_-^{\text{dress}}(r)].$$

As we saw, the ion pairing makes a contribution to ρ_+^{dress} that is negative and a contribution to ρ_-^{dress} that is positive. Since the rest of the integrand is positive, the pairing gives a positive contribution to $\mathcal{E}_r^\star$. Thus, the appearance of ion pairs acts like an increase in the dielectric constant, which is very reasonable since an ion pair is in many respects similar to a polar molecule. The net effect of pairing is like a decrease of the concentration of "free" ions, that is, those that are not in pairs. Both effects – the decrease in magnitude of $q_j^\star$ and the increase in $\mathcal{E}_r^\star$ – contribute to make κ smaller and hence the decay length $1/\kappa$ larger. In fact, it is not necessary to have actual pairing for this to happen. It is sufficient that anions and cations have a strong tendency to be close together without actually forming pairs. The ions of opposite charge near an ion then give contributions like those we have just discussed.

6.2.3 Density-Density, Charge-Density and Charge-Charge Correlations ★

The definition of the function $\rho_Q(r)$ is

$$\rho_Q(r) = \tfrac{1}{2} \left[\rho_+(r) - \rho_-(r) \right] = \frac{q_{\pm} n_{\text{tot}}^b}{4} \left[g_{++}^{(2)}(r) + g_{--}^{(2)}(r) - 2 g_{+-}^{(2)}(r) \right], \tag{6.184}$$

where the last equality is obtained from $\rho_i(r) = \sum_j q_j n_j^b g_{ij}^{(2)}(r)$ (cf. Equation 6.148). The closely related function

$$H_{QQ}(r) \equiv \frac{q_{\pm} n_{\text{tot}}^b}{q_e^2} \left[\rho_Q(r) + q_Q \delta^{(3)}(r) \right], \tag{6.185}$$

where $q_Q = (q_+ - q_-)/2 = (q_+ + |q_-|)/2$, is called the **charge-charge correlation function**. It has the unit (number density)^2 (i.e., m^{-6}). As shown in Appendix 6C, $H_{QQ}(r)$ gives the correlations in charge fluctuations at two points separated by distance r in the electrolyte (irrespective of which species the charges are sitting on). The delta function contribution in H_{QQ} comes from self-correlations, that is, from the fact that when we look at distribution of charges from the point of view of a charge, there is a charge sitting in the same place, namely the charge itself.

For binary electrolytes, the general expression for the function $n_N(r)$ is

$$n_N(r) = x_+^b n_{\text{tot};+}(r) + x_-^b n_{\text{tot};-}(r)$$

$$= n_{\text{tot}}^b \left[(x_+^b)^2 g_{++}^{(2)}(r) + (x_-^b)^2 g_{--}^{(2)}(r) + 2 x_+^b x_-^b g_{+-}^{(2)}(r) \right], \tag{6.186}$$

where $n_{\text{tot};i}(r) = \sum_j n_j^b g_{ij}^{(2)}(r) = n_{\text{tot}}^b \sum_j x_j^b g_{ij}^{(2)}(r)$ is the total density distribution around an ion of species i. This function occurs in the **density-density correlation function**, which is given by

$$H_{NN}(r) \equiv n_{\text{tot}}^b \left[n_N(r) - n_{\text{tot}}^b + \delta^{(3)}(r) \right] \tag{6.187}$$

(see Appendix 6C). It gives the correlations in density fluctuations in the electrolyte. For a $z : -z$ electrolyte, we have

$$n_N(r) = \tfrac{1}{2} \left[n_{\text{tot};+}(r) + n_{\text{tot};-}(r) \right]$$

$$= \frac{n_{\text{tot}}^b}{4} \left[g_{++}^{(2)}(r) + g_{--}^{(2)}(r) + 2g_{+-}^{(2)}(r) \right] \quad \text{(for } z : -z\text{).} \tag{6.188}$$

There is also the **charge-density correlation function** given by

$$H_{QN}(r) \equiv \frac{q_{\pm} n_{\text{tot}}^b}{q_e} n_Q(r), \tag{6.189}$$

where

$$n_Q(r) = \tfrac{1}{2} \left[n_{\text{tot};+}(r) - n_{\text{tot};-}(r) \right]$$

$$= \frac{n_{\text{tot}}^b}{2} \left[x_+^b g_{++}^{(2)}(r) - x_-^b g_{--}^{(2)}(r) + (x_-^b - x_+^b) g_{+-}^{(2)}(r) \right]. \tag{6.190}$$

As shown in Appendix 6C, $H_{QN}(r)$ gives the correlations between charge fluctuations on the one hand and density fluctuations on the other. It can also be written as

$$H_{QN}(r) \equiv \frac{n_{\text{tot}}^b}{q_e} \rho_N(r), \tag{6.191}$$

where

$$\rho_N(r) = x_+^b \rho_+(r) + x_-^b \rho_-(r),$$

which, in fact, is equal to $q_{\pm} n_Q(r)$. For a $z : -z$ electrolyte, we have

$$n_Q(r) = \frac{n_{\text{tot}}^b}{4} \left[g_{++}^{(2)}(r) - g_{--}^{(2)}(r) \right] \quad \text{(for } z : -z\text{).} \tag{6.192}$$

and

$$\rho_N(r) = \tfrac{1}{2} \left[\rho_+(r) + \rho_-(r) \right] = \frac{q_{\pm} n_{\text{tot}}^b}{4} \left[g_{++}^{(2)}(r) - g_{--}^{(2)}(r) \right] \quad \text{(for } z : -z\text{).}$$

Note that we can write $\rho_+(r) = \rho_N(r) + \rho_Q(r)$ and $\rho_-(r) = \rho_N(r) - \rho_Q(r)$ in this case.

In the general case, we have for a binary electrolyte[56]

$$\rho_i(r) = \rho_N(r) + \frac{q_i}{q_Q}\rho_Q(r). \tag{6.193}$$

Thus, the charge density $\rho_i(r)$ around an ion species i as the sum of two contributions: $\rho_N(r)$ that is the same for both anions and cations and $q_i\rho_Q(r)/q_Q$ that depends on the charge of the ion via the factor q_i, but is otherwise equal for both kinds of ions. Likewise, we have

$$n_{\text{tot};i}(r) = n_N(r) + \frac{q_i}{q_Q}n_Q(r), \tag{6.194}$$

so the total ion density around an ion can be divided up in an analogous manner for a binary electrolyte. Normally, the functions $\rho_Q(r)$, $n_N(r)$ and $\rho_N(r) = q_\pm n_Q(r)$ all decay with the same decay length and, when appropriate, the same wavelength. The same therefore applies for $\rho_i(r)$, $n_{\text{tot};i}(r)$, $H_{NN}(r)$, $H_{QN}(r)$ and $H_{QQ}(r)$. The exceptions to this will be treated at the end of the current section.

The dressed particle charge density $\rho_i^\star(r)$ can be written in the same manner

$$\rho_i^\star(r) = \rho_N^\star(r) + \frac{q_i}{q_Q}\rho_Q^\star(r),$$

where $\rho_N^\star(r) = x_+^b\rho_+^\star(r) + x_-^b\rho_-^\star(r)$ and $\rho_Q^\star(r) = [\rho_+^\star(r) - \rho_-^\star(r)]/2$. This implies via the definitions of the dressed particle charge $q_i^\star$ and the effective charge q_i^{eff} of an i-ion (Equations 6.162 and 6.153 with P $= i$) that

$$q_i^\star = q_N^\star + \frac{q_i}{q_Q}q_Q^\star \quad \text{and} \quad q_i^{\text{eff}} = q_N^{\text{eff}} + \frac{q_i}{q_Q}q_Q^{\text{eff}},$$

where $q_N^\star$, $q_Q^\star$, q_N^{eff} and q_Q^{eff} are defined analogously to the other entities, for example $q_Q^\star = [q_+^\star - q_-^\star]/2$ and $q_N^\star(r) = x_+^b q_+^\star + x_-^b q_-^\star$. Recall that $q_i^{\text{eff}} = q_i^{\text{eff}}(\kappa)$ is a function of κ while $q_i^\star$ is independent of κ. Likewise, $q_N^{\text{eff}} = q_N^{\text{eff}}(\kappa)$ and $q_Q^{\text{eff}} = q_Q^{\text{eff}}(\kappa)$, while $q_N^\star$ and $q_Q^\star$ are independent of κ.

Let us consider a case where we have a real-valued decay parameter κ and $g_{ij}^{(2)}$ decays like (cf. Equation 6.167)

$$g_{ij}^{(2)}(r) \sim 1 - \frac{\beta\, q_i^{\text{eff}}(\kappa)\, q_j^{\text{eff}}(\kappa)}{4\pi\, \mathcal{E}_r^{\text{eff}}(\kappa)\varepsilon_0} \cdot \frac{e^{-\kappa r}}{r}$$

---

[56] This can be shown by using the definitions of ρ_N and ρ_Q together with the identities $q_+/q_Q = 2x_-^b$ and $q_-/q_Q = -2x_+^b$, which are consequences of electroneutrality $q_+x_+^b + q_-x_-^b = 0$. For instance, we have $2q_Qx_-^b = q_+x_-^b - q_-x_-^b = q_+x_-^b + q_+x_+^b = q_+$.

when $r \to \infty$ [recall that $g_{ij}^{(2)}(r) \sim 1 - \beta w_{ij}^{(2)}(r)$]. By inserting this into Equation 6.184 and defining the constant $\mathcal{B}_1 = \beta n_{\text{tot}}^{\text{b}}/[4\pi \mathcal{E}_r^{\text{eff}}(\kappa)\varepsilon_0]$, we obtain

$$\frac{\rho_Q(r)}{q_{\pm}} \sim -\frac{\left[(q_+^{\text{eff}})^2 + (q_-^{\text{eff}})^2 - 2q_+^{\text{eff}}q_-^{\text{eff}}\right]\mathcal{B}_1}{4} \cdot \frac{e^{-\kappa r}}{r}$$

$$= -(q_Q^{\text{eff}})^2 \, \mathcal{B}_1 \, \frac{e^{-\kappa r}}{r} \quad \text{when } r \to \infty \tag{6.195}$$

with $q_i^{\text{eff}} = q_i^{\text{eff}}(\kappa)$. Likewise, from Equation 6.186, it follows that

$$n_N(r) \sim n_{\text{tot}}^{\text{b}} - (q_N^{\text{eff}})^2 \, \mathcal{B}_1 \, \frac{e^{-\kappa r}}{r} \quad \text{when } r \to \infty \tag{6.196}$$

and from Equation 6.190 is obtained

$$n_Q(r) = \frac{\rho_N(r)}{q_{\pm}} \sim -q_Q^{\text{eff}} q_N^{\text{eff}} \, \mathcal{B}_1 \, \frac{e^{-\kappa r}}{r} \quad \text{when } r \to \infty. \tag{6.197}$$

From the relationships of these functions to $H_{QQ}(r)$, $H_{NN}(r)$ and $H_{QN}(r)$, it follows that

$$H_{NN}(r) \sim -(q_N^{\text{eff}})^2 \, \mathcal{B}_2 \, \frac{e^{-\kappa r}}{r}$$

$$\frac{H_{QQ}(r)}{(q_{\pm}/q_{\text{e}})^2} \sim -(q_Q^{\text{eff}})^2 \, \mathcal{B}_2 \, \frac{e^{-\kappa r}}{r} \quad \text{when } r \to \infty, \tag{6.198}$$

$$\frac{H_{QN}(r)}{q_{\pm}/q_{\text{e}}} \sim -q_Q^{\text{eff}} q_N^{\text{eff}} \, \mathcal{B}_2 \, \frac{e^{-\kappa r}}{r}$$

where $\mathcal{B}_2 = \beta \, (n_{\text{tot}}^{\text{b}})^2/[4\pi \mathcal{E}_r^{\text{eff}}(\kappa)\varepsilon_0]$. We see that the Q and N components of the effective charges enter in the prefactor in all of these functions. The same is true for any other term in $H_{QQ}(r)$, $H_{NN}(r)$ and $H_{QN}(r)$ with a different decay parameter value, say κ', and in that case $q_N^{\text{eff}}(\kappa')$, $q_Q^{\text{eff}}(\kappa')$ and $\mathcal{E}_r^{\text{eff}}(\kappa')$ appear in the formulae.

Equation 6.163 for κ can be written

$$\kappa^2 = \frac{\beta}{\mathcal{E}_r^{\star}(\kappa)\varepsilon_0} \left[q_+ n_+^{\text{b}} q_+^{\star} + q_- n_-^{\text{b}} q_-^{\star}\right] = \frac{\beta q_{\pm} n_{\text{tot}}^{\text{b}}}{\mathcal{E}_r^{\star}(\kappa)\varepsilon_0} \cdot \frac{q_+^{\star} - q_-^{\star}}{2},$$

where we have used Equation 6.150 (recall that $q_{\pm}$ was defined in Equation 6.149). Thus, we have

$$\kappa^2 = \frac{\beta q_{\pm} q_Q^{\star} n_{\text{tot}}^{\text{b}}}{\mathcal{E}_r^{\star}(\kappa)\varepsilon_0}, \tag{6.199}$$

and we see that *the N-component $q_N^\star$ does not influence κ* since it is canceled in $q_+^\star - q_-^\star$. Likewise, Equation 6.158 yields

$$\kappa^2 = \frac{\beta q_\pm q_Q^{\text{eff}}(\kappa) n_{\text{tot}}^{\text{b}}}{\varepsilon_r \varepsilon_0} \quad \text{(spherical ions)}, \tag{6.200}$$

which is equivalent to the previous equation (when the ions are spherical) and we have $q_Q^{\text{eff}}(\kappa)/\varepsilon_r = q_Q^\star/\mathcal{E}_r^\star(\kappa)$ (cf. footnote 39).

Let us now specialize to $z : -z$ electrolytes, in the case of which we have

$$q_+^{\text{eff}}(\kappa) = q_N^{\text{eff}}(\kappa) + q_Q^{\text{eff}}(\kappa), \quad q_-^{\text{eff}}(\kappa) = q_N^{\text{eff}}(\kappa) - q_Q^{\text{eff}}(\kappa). \tag{6.201}$$

Different solutions κ to Equation 6.200 can give very different values of q_Q^{eff} and q_N^{eff}. The magnitude of q_Q^{eff} can, for example, be much larger than that of q_N^{eff} for one solution (which we will call type (i)) and vice versa for another solution (type (ii)). Let us investigate this situation further. For the first kind of solution, we see from Equation 6.201 that $q_+^{\text{eff}} \approx q_Q^{\text{eff}}$ and $q_-^{\text{eff}} \approx -q_Q^{\text{eff}}$, while for the latter q_+^{eff} and q_-^{eff} are both $\approx q_N^{\text{eff}}$. This means that the effective charges of anions and cations have either different signs or the same sign for the various decay terms (different solutions κ). This situation occurs, for example, when anions and cations do not differ too much from each other apart from their signs and we have $g_{++}^{(2)}(r) \approx g_{--}^{(2)}(r)$ like for molten NaCl in Figure 6.8 and for the NaCl systems in Figure 6.12, which we investigated in Section 6.2.2.3.

From Equation 6.198, it follows that terms of type (i), where q_Q^{eff} is much larger than q_N^{eff}, contribute mainly to $H_{QQ}(r)$ and that terms of type (ii), where the reverse is true, contribute mainly to $H_{NN}(r)$. In both cases, $H_{QN}(r)$ is small because q_N^{eff} is small in one case and q_Q^{eff} is small in the other; this can also be deduced from $g_{++}^{(2)}(r) \approx g_{--}^{(2)}(r)$. Note that *all* of these correlation functions decay with the same decay lengths; only the magnitudes of the coefficients vary for the different terms with different decay lengths. Type (i) gives a large term in $H_{QQ}(r)$ and small term in $H_{NN}(r)$ with the same decay length, while the reverse is true for type (ii).

This reasoning is relevant for NaCl at 3000 K in Figure 6.12 and the current treatment based on Equation 6.201 complements the discussion of this system in Section 6.2.2.3 (see in particular the shaded text box on page 335). In this case, there are two kinds of decay parameters: $\kappa = \kappa_\Re + i\kappa_\Im$, which is complex-valued, and κ'', which is real-valued, both being solutions to Equations 6.199 and 6.200. As we have seen, the complex conjugate $\underline{\kappa} = \kappa_\Re - i\kappa_\Im$ is also a solution, and we can write the oscillatory contribution to $w_{ij}^{(2)}(r)$ as

$$\frac{q_i^{\text{eff}}(\kappa) q_j^{\text{eff}}(\kappa)}{4\pi \mathcal{E}_r^{\text{eff}}(\kappa)\varepsilon_0} \cdot \frac{e^{-(\kappa_\Re + i\kappa_\Im)r}}{r} + \frac{q_i^{\text{eff}}(\underline{\kappa}) q_j^{\text{eff}}(\underline{\kappa})}{4\pi \mathcal{E}_r^{\text{eff}}(\underline{\kappa})\varepsilon_0} \cdot \frac{e^{-(\kappa_\Re - i\kappa_\Im)r}}{r}$$

$$= 2\Re \left[\frac{q_i^{\text{eff}}(\kappa) q_j^{\text{eff}}(\kappa)}{4\pi \mathcal{E}_r^{\text{eff}}(\kappa)\varepsilon_0} e^{-i\kappa_\Im r} \right] \frac{e^{-\kappa_\Re r}}{r},$$

which can be written as in the rhs of Equation 6.175. This contribution is of type (i), so it dominates in $H_{QQ}(r)$ and gives only a small contribution to $H_{NN}(r)$. Since $q_+^{eff} \approx q_Q^{eff}$ and $q_-^{eff} \approx -q_Q^{eff}$, the pair distribution functions $g_{++}^{(2)}(r)$ and $g_{+-}^{(2)}(r)$ oscillate out of phase from each other for high densities where $1/\kappa_{\mathfrak{R}}$ is the largest decay length, that is, like in Figure 6.8. The decay parameter κ'' is of type (ii) for the present NaCl system and the contribution with decay length $1/\kappa''$ dominates in $H_{NN}(r)$ and gives only a small contribution to $H_{QQ}(r)$.

Let us now consider the long range density-density correlations near the critical point as discussed in conjunction with Figure 6.12, where we saw that the decay length $1/\kappa''$ becomes large near this point while $1/\kappa_{\mathfrak{R}}$ remains finite there. Since $1/\kappa''$ is larger than $1/\kappa_{\mathfrak{R}}$ under these conditions, the leading terms in both $H_{NN}(r)$ and $H_{QQ}(r)$ have decay length $1/\kappa''$, so both have an equally long range. The magnitude of the leading term of $H_{NN}(r)$ is proportional to $[q_N^{eff}(\kappa'')]^2/\mathcal{E}_r^{eff}(\kappa'')$, while the leading term of $H_{QQ}(r)$ is proportional to $[q_Q^{eff}(\kappa'')]^2/\mathcal{E}_r^{eff}(\kappa'')$, which is much smaller. Equation 6.200 for κ'' is

$$(\kappa'')^2 = \frac{\beta q_{\pm} q_Q^{eff}(\kappa'') n_{tot}^b}{\varepsilon_r \varepsilon_0}$$

and we see that the small value of κ'' (large decay length) is connected with the small value of $q_Q^{eff}(\kappa'')$. Thus, when the range of density-density correlations becomes longer, the magnitude of the charge-charge correlations for large r become smaller (i.e., even smaller than it is otherwise). Under conditions where the decay length diverges, the contribution to $H_{QQ}(r)$ with decay length $1/\kappa''$ vanishes and does not contribute any longer. Therefore, it is only the density-density correlations that truly become long range when the critical point is approached.[57] These remarks complete the discussion of the NaCl system of Figure 6.12.

As we have seen, $\rho_i(r)$, $n_{tot;i}(r)$, $H_{NN}(r)$, $H_{QN}(r)$ and $H_{QQ}(r)$ normally all have a common decay length and we have pointed out that there exist exceptions to this. They occur for systems like the restricted primitive model where anions and cations differ *only* by the signs of their charges, so we have $g_{++}^{(2)}(r) \equiv g_{--}^{(2)}(r)$. Hence, $\rho_N(r)$, $n_Q(r)$ and $H_{QN}(r)$ are identically zero. Compared to the cases when $g_{++}^{(2)}(r) \approx g_{--}^{(2)}(r)$, the difference is that the contributions to $\rho_Q(r)$ and $n_N(r)$ that were small in the cases with $g_{++}^{(2)}(r) \approx g_{--}^{(2)}(r)$ are now identically zero. Therefore, the decay terms of type (i) contribute *only* to $\rho_Q(r)$ and $H_{QQ}(r)$ and terms that correspond to type (ii) contribute *only* to $n_N(r)$ and $H_{NN}(r)$. This means that the charge-charge and density-density correlations have different decay lengths. The decay of the density-density correlations is decoupled from the decay of the charge-charge correlations – an exceptional behavior that is caused by the extreme symmetry of these models.

Finally, we note that the polarization response function $\chi^P(r)$ defined in Equation 6.118 is closely related to $H_{QQ}(r)$. The function $\chi^P(r_{12})$ gives the polarization charge density $\delta\rho^{pol}(\mathbf{r}_1)$ when a bulk electrolyte is exposed to a weak external electrostatic potential

---

[57] Here we are only discussing conditions near the critical point – at some distance from it – and not what happens at criticality.

$\delta \Psi^{\text{ext}}(\mathbf{r}_2)$. We have

$$
\chi^\rho(r) = -\beta \left[q_+ n_+^b \rho_+^{\text{tot}}(r) + q_- n_-^b \rho_-^{\text{tot}}(r) \right] = -\beta q_\pm n_{\text{tot}}^b \frac{\rho_+^{\text{tot}}(r) - \rho_-^{\text{tot}}(r)}{2}
$$

$$
= -\beta q_\pm n_{\text{tot}}^b \left[q_Q \delta^{(3)}(r) + \rho_Q(r) \right] = -\beta q_e^2 H_{QQ}(r), \tag{6.202}
$$

where we have used $\rho_i^{\text{tot}}(r) = q_i \delta^{(3)}(r) + \rho_i(r)$.

APPENDIX 6A: THE ORIENTATIONAL VARIABLE ω

To describe the orientation of a particle, we introduce a "particle frame" $\mathfrak{F}^P$ associated with each particle and placed with the origin at the center of mass. This frame is attached to the particle and follows its translational and rotational motions. The orientation of the particle can be specified by a comparison of the directions of the particle frame axes and axes of the laboratory frame $\mathfrak{F}^{\text{lab}}$ in the following way. We can imagine another frame $\mathfrak{F}$ centered at the particle (so its origin coincides with that of the particle frame) but oriented so that its x, y and z axes are initially parallel to the corresponding axes of $\mathfrak{F}^{\text{lab}}$. The finite rotation (or the sequence of rotations) that would bring $\mathfrak{F}$ to coincide with the particle frame $\mathfrak{F}^P$ can be used to specify the orientation of the latter and thereby the particle orientation. The rotation(s) will be specified by a variable ω and can, for example, be expressed in terms of the three Euler angles. These angles can be defined from the following set of three consecutive rotations:[58] A rotation through an angle ϕ around the z axis of $\mathfrak{F}$ is followed by a rotation through θ around the new y axis of $\mathfrak{F}$. These angles are the same as the spherical polar coordinate angles (ϕ, θ) for the z axis unit vector $\hat{z}^P$ of the particle frame and the two rotations make the rotated z axis of $\mathfrak{F}$ to coincide with the z axis of the particle frame $\mathfrak{F}^P$. A final third rotation of $\mathfrak{F}$ through the angle χ around the common z axis of $\mathfrak{F}^P$ and $\mathfrak{F}$ makes the x and y axes of $\mathfrak{F}$ to coincide with those of $\mathfrak{F}^P$. The particle frame and the rotated $\mathfrak{F}$ thereby fully coincide, so the set of angles (ϕ, θ, χ) specify the orientation of $\mathfrak{F}^P$ and thereby that of the particle. We have $0 \leq \phi < 2\pi$, $0 \leq \theta \leq \pi$ and $0 \leq \chi < 2\pi$. For a linear molecule, the last rotation through χ is redundant (the molecular axis is then selected to be the z axis of the particle frame) and for such particles the set (ϕ, θ) is sufficient to describe the orientation.

We shall, however, not select (ϕ, θ, χ) (or (ϕ, θ) for a linear molecule) to be our orientational variable ω. It is better to use $\omega' = (\phi, \cos\theta, \chi)$ to describe the orientation of the particle. We have thereby selected $\cos\theta$ rather than θ because the differential then becomes $d\omega' = d\phi\, d(\cos\theta)\, d\chi = -d\phi\, \sin\theta\, d\theta\, d\chi$, cf. the polar coordinate differential $d\phi\, \sin\theta\, d\theta$ that gives the differential solid angle. For a randomly oriented particle, there is equal probability to have its $\hat{z}^P$ axis within a certain solid angle around all different directions in space. For a linear molecule, we have $\omega' = (\phi, \cos\theta)$.

---

[58] There exist other conventions for the selection of which rotations are used to define the Euler angles. The one used here has the advantage that the first two Euler angles coincide with the angles for the z axis of $\mathfrak{F}^P$ in spherical polar coordinates.

This choice of orientational variable has the property that $\int d\omega' = \omega_{\circ}$, where $\omega_{\circ} = 8\pi^2$ except for linear molecules when we have $\omega_{\circ} = 4\pi$. It is more convenient to have a "normalized" variable ω so that $\int d\omega = 1$ in all cases. We therefore select $\omega = (\phi/2\pi, \cos\theta/2, \chi/2\pi)$ as the final choice of orientational variable, except for linear molecules when we take $\omega = (\phi/2\pi, \cos\theta/2)$. Note that ω is dimensionless.

APPENDIX 6B: VARIATIONS IN DENSITY DISTRIBUTION WHEN THE EXTERNAL POTENTIAL IS VARIED; THE FIRST YVON EQUATION

The density distribution function $n_N(\mathbf{r})$ in the canonical ensemble for a system with N particles is given by (Equation 5.42)

$$n_N(\mathbf{r}_1) = e^{-\beta v(\mathbf{r}_1)} \frac{N}{Z_N} \int d\mathbf{r}_2 \ldots d\mathbf{r}_N e^{-\beta \breve{U}_N^{\text{intr}}(\mathbf{r}^N)} e^{-\beta \sum_{\nu=2}^{N} v(\mathbf{r}_\nu)}.$$

We are going to see what happens with the density distribution when the external potential $v(\mathbf{r})$ is changed by an amount $\delta v(\mathbf{r})$ to $v(\mathbf{r}) + \delta v(\mathbf{r})$. By taking the logarithm of the equation and using $n_N(\mathbf{r}) = n^0 e^{-\beta w_N(\mathbf{r})}$, we have

$$-\beta w_N(\mathbf{r}_1) = -\beta v(\mathbf{r}_1) + \ln\left[\int d\mathbf{r}_2 \ldots d\mathbf{r}_N e^{-\beta \breve{U}_N^{\text{intr}}(\mathbf{r}^N)} e^{-\beta \sum_{\nu=2}^{N} v(\mathbf{r}_\nu)}\right]$$

$$- \ln\left[\int d\mathbf{r}_1' \ldots d\mathbf{r}_N' e^{-\beta \breve{U}_N^{\text{intr}}(\mathbf{r}'^N)} e^{-\beta \sum_{\nu=1}^{N} v(\mathbf{r}_\nu')}\right] + \text{const}$$

where we have inserted the definition of Z_N (Equation 3.10) and where const $= \ln[N/n_0]$ is a constant. The change in $w_N(\mathbf{r}_1)$ when $v(\mathbf{r})$ is altered to $v(\mathbf{r}) + \delta v(\mathbf{r})$ is equal to

$$\delta w_N(\mathbf{r}_1) = w_N(\mathbf{r}_1)|_{v(\mathbf{r})+\delta v(\mathbf{r})} - w_N(\mathbf{r}_1)|_{v(\mathbf{r})},$$

where we have indicated what external potential is used in the expression for $w_N(\mathbf{r}_1)$, and we obtain

$$-\beta\delta w_N(\mathbf{r}_1) = -\beta\delta v(\mathbf{r}_1) \tag{6.203}$$

$$+ \ln\left[\frac{\int d\mathbf{r}_2 \ldots d\mathbf{r}_N e^{-\beta \breve{U}_N^{\text{intr}}(\mathbf{r}^N)} e^{-\beta \sum_{\nu=2}^{N} v(\mathbf{r}_\nu)} e^{-\beta \sum_{\nu=2}^{N} \delta v(\mathbf{r}_\nu)}}{\int d\mathbf{r}_2 \ldots d\mathbf{r}_N e^{-\beta \breve{U}_N^{\text{intr}}(\mathbf{r}^N)} e^{-\beta \sum_{\nu=2}^{N} v(\mathbf{r}_\nu)}}\right]$$

$$- \ln\left[\frac{\int d\mathbf{r}_1' \ldots d\mathbf{r}_N' e^{-\beta \breve{U}_N^{\text{intr}}(\mathbf{r}'^N)} e^{-\beta \sum_{\nu=1}^{N} v(\mathbf{r}_\nu')} e^{-\beta \sum_{\nu=1}^{N} \delta v(\mathbf{r}_\nu')}}{\int d\mathbf{r}_1' \ldots d\mathbf{r}_N' e^{-\beta \breve{U}_N^{\text{intr}}(\mathbf{r}'^N)} e^{-\beta \sum_{\nu=1}^{N} v(\mathbf{r}_\nu')}}\right].$$

We will henceforth assume that $\delta v(\mathbf{r})$ is infinitesimally small everywhere, so we can replace $\exp[-\beta \sum_{\nu} \delta v(\mathbf{r}_\nu)]$ by $1 - \beta \sum_{\nu} \delta v(\mathbf{r}_\nu)$ in the expression. After this replacement, the square bracket (SB) in the last term, where the denominator is equal to Z_N, can be written as

$$SB_{last} = \frac{Z_N - \beta \int d\mathbf{r}'_1 \dots d\mathbf{r}'_N e^{-\beta \breve{U}_N^{intr}(\mathbf{r}'^N)} e^{-\beta \sum_{\nu=1}^N v(\mathbf{r}'_\nu)} \sum_{\nu=1}^N \delta v(\mathbf{r}'_\nu)}{Z_N}$$

$$= 1 - \beta \sum_{\nu=1}^N \int d\mathbf{r}'_1 \dots d\mathbf{r}'_N \mathcal{P}_N^{(N)}(\mathbf{r}'^N) \delta v(\mathbf{r}'_\nu),$$

where we have used the definition of $\mathcal{P}_N^{(N)}(\mathbf{r}^N)$ (Equation 5.38). All N terms in the sum in the rhs are numerically equal and each of them is therefore equal to the first term, so we can write

$$SB_{last} = 1 - \beta N \int d\mathbf{r}'_1 \left[\int d\mathbf{r}'_2 \dots d\mathbf{r}'_N \mathcal{P}_N^{(N)}(\mathbf{r}'^N) \right] \delta v(\mathbf{r}'_1)$$

$$= 1 - \beta N \int d\mathbf{r}'_1 \mathcal{P}_N^{(1)}(\mathbf{r}'_1) \delta v(\mathbf{r}'_1) = 1 - \beta \int d\mathbf{r}'_1 \, n_N(\mathbf{r}'_1) \delta v(\mathbf{r}'_1),$$

where we have used the definition (5.41) of $n_N(\mathbf{r})$.

For the square bracket (SB_{first}) in the first logarithmic term of Equation 6.203, we multiply the denominator and the numerator by $N \exp[-\beta v(\mathbf{r}_1)]/Z_N$, whereby the denominator becomes $n_N(\mathbf{r}_1)$. Then, in a similar manner as for SB_{last}, we obtain

$$SB_{first} = 1 - \beta \frac{N \sum_{\nu=2}^N \int d\mathbf{r}_2 d\mathbf{r}_3 \dots d\mathbf{r}_N \mathcal{P}_N^{(N)}(\mathbf{r}^N) \delta v(\mathbf{r}_\nu)}{n_N(\mathbf{r}_1)}$$

$$= 1 - \beta \frac{N(N-1) \int d\mathbf{r}_2 \left[\int d\mathbf{r}_3 \dots d\mathbf{r}_N \mathcal{P}_N^{(N)}(\mathbf{r}^N) \right] \delta v(\mathbf{r}_2)}{n_N(\mathbf{r}_1)}$$

$$= 1 - \beta \frac{\int d\mathbf{r}_2 \, n_N^{(2)}(\mathbf{r}_1, \mathbf{r}_2) \delta v(\mathbf{r}_2)}{n_N(\mathbf{r}_1)} = 1 - \beta \int d\mathbf{r}_2 \, n_N(\mathbf{r}_2) g_N^{(2)}(\mathbf{r}_1, \mathbf{r}_2) \delta v(\mathbf{r}_2),$$

where we have used the fact that all $N - 1$ terms in the sum are numerically equal and the definition (5.45) of $n_N^{(2)}(\mathbf{r}_1, \mathbf{r}_2)$.

We now insert these results into Equation 6.203 and since $\delta v(\mathbf{r})$ is infinitesimally small, we can replace $\ln[1 - \tau]$ by $-\tau$, where $-\tau$ is the last term in SB_{first} or SB_{last}. Furthermore, we replace the integration variable $\mathbf{r}'_1$ in SB_{last} by $\mathbf{r}_2$ and obtain

$$-\beta \delta w_N(\mathbf{r}_1) = -\beta \delta v(\mathbf{r}_1) - \beta \int d\mathbf{r}_2 \, n_N(\mathbf{r}_2) g_N^{(2)}(\mathbf{r}_1, \mathbf{r}_2) \delta v(\mathbf{r}_2)$$

$$+ \beta \int d\mathbf{r}_2 \, n_N(\mathbf{r}_2) \delta v(\mathbf{r}_2).$$

This can be written as

$$\delta w_N(\mathbf{r}_1) = \delta v(\mathbf{r}_1) + \int d\mathbf{r}_2 \, n_N(\mathbf{r}_2) h_N^{(2)}(\mathbf{r}_1, \mathbf{r}_2) \delta v(\mathbf{r}_2) \tag{6.204}$$

or, equivalently,

$$\delta n_N(\mathbf{r}_1) = -\beta n_N(\mathbf{r}_1) \left[\delta v(\mathbf{r}_1) + \int d\mathbf{r}_2 \, n_N(\mathbf{r}_2) h_N^{(2)}(\mathbf{r}_1, \mathbf{r}_2) \delta v(\mathbf{r}_2) \right] \tag{6.205}$$

since $\delta w_N(\mathbf{r}_1) = -\delta \ln n_N(\mathbf{r}_1)/\beta = -\delta n_N(\mathbf{r}_1)/[n_N(\mathbf{r}_1)\beta]$.[59] Equation 6.205 is called the **first Yvon equation**, which shows how much the density distribution $n_N(\mathbf{r}_1)$ changes when the external potential changes by a small amount $\delta v(\mathbf{r})$.

Equations 6.204 and 6.205 are, in fact, general relationships and are also valid for open systems in the grand canonical ensemble and we will from now on drop the subscript N on the functions. In the second volume of this treatise, we will encounter the *second Yvon equation*,[60] which gives the inverse relationship between $\delta n(\mathbf{r}')$ and $\delta v(\mathbf{r})$. This means that it answers the question: "What change $\delta v(\mathbf{r})$ in external potential is needed in order to produce a certain small but otherwise arbitrary change $\delta n(\mathbf{r}')$ in density distribution?" That question is well posed only for open systems where the total number of particles in the system is not fixed.

For bulk systems, Equations 6.204 and 6.205 become

$$\delta w(\mathbf{r}_1) = \delta v(\mathbf{r}_1) + n^b \int d\mathbf{r}_2 \, h^{(2)}(r_{12}) \delta v(\mathbf{r}_2) \tag{6.206}$$

and

$$\delta n(\mathbf{r}_1) = -\beta n^b \left[\delta v(\mathbf{r}_1) + n^b \int d\mathbf{r}_2 \, h^{(2)}(r_{12}) \delta v(\mathbf{r}_2) \right], \tag{6.207}$$

where $\delta n(\mathbf{r}_1)$ is the deviation from the bulk density that occurs when a bulk fluid is exposed to a *weak* external potential $\delta v(\mathbf{r})$ and where $h^{(2)}(r_{12})$ is the pair correlation function of the bulk fluid (i.e., in the absence of the external potential). Likewise, $\delta w(\mathbf{r}_1)$ is the potential of mean force that appears for a particle at location $\mathbf{r}_1$ under these condition; $w(\mathbf{r}_1)$ is zero in the bulk fluid. The second term in the rhs of Equation 6.206 tells how much the correlations between a particle at $\mathbf{r}_1$ and the surrounding particles contribute to the potential of mean force, that is, this term is the intrinsic part $\delta w^{\text{intr}}(\mathbf{r}_1)$ of the latter when external potential δv is small (cf. Equation 5.19).

All these results are readily generalized to many-component fluids, whereby one obtains

$$\delta w_i(\mathbf{r}_1) = \delta v_i(\mathbf{r}_1) + \sum_j n_j^b \int d\mathbf{r}_2 \, h_{ij}^{(2)}(r_{12}) \delta v_j(\mathbf{r}_2) \tag{6.208}$$

and

$$\delta n_i(\mathbf{r}_1) = -\beta n_i^b \left[\delta v_i(\mathbf{r}_1) + \sum_j n_j^b \int d\mathbf{r}_2 \, h_{ij}^{(2)}(r_{12}) \delta v_j(\mathbf{r}_2) \right] \tag{6.209}$$

---

[59] We have $\delta \ln n_N(\mathbf{r}) = \ln[n_N(\mathbf{r}) + \delta n_N(\mathbf{r})] - \ln n_N(\mathbf{r}) = \ln[1 + \delta n_N(\mathbf{r})/n_N(\mathbf{r})] = \delta n_N(\mathbf{r})/n_N(\mathbf{r})$ since $\delta n_N(\mathbf{r})$ is infinitesimally small.

[60] The second Yvon equation is $\delta v(\mathbf{r}_2) = -\left[\delta n(\mathbf{r}_2)/n(\mathbf{r}_2) - \int d\mathbf{r}_1 \, c^{(2)}(\mathbf{r}_1, \mathbf{r}_2) \delta n(\mathbf{r}_1) \right]/\beta$, where $c^{(2)}$ is the so-called *direct pair correlation function*.

for bulk fluids. These equations for δw_i and δn_i can be derived in an analogous manner as those for δw and δn; we will not take space to do this here. Note that a change in external potential for *any* species j induces a change in density for *all* species i; this is due to the correlations between the particles of all species. Furthermore, a change in external potential in one place changes the density at other places in the neighborhood – in principle, the density is changed everywhere, but the variation is, of course, very small far away. This is also due to the correlations.

In the present chapter, we will mainly use these results for electrolytes exposed to an external electrostatic field described by the electrostatic potential $\Psi^{\text{ext}}(\mathbf{r})$. The external potential $v_j(\mathbf{r})$ for ions of species j is then equal to $q_j\Psi^{\text{ext}}(\mathbf{r})$. In particular, we will consider what happens when a bulk electrolyte is exposed to a weak potential $\delta\Psi^{\text{ext}}(\mathbf{r})$, so we have $\delta v_j(\mathbf{r}) = q_j\delta\Psi^{\text{ext}}(\mathbf{r})$ and Equation 6.208 becomes

$$\delta w_i(\mathbf{r}_1) = q_i\delta\Psi^{\text{ext}}(\mathbf{r}_1) + \sum_j n_j^{\text{b}} \int d\mathbf{r}_2\, h_{ij}^{(2)}(r_{12}) q_j\delta\Psi^{\text{ext}}(\mathbf{r}_2)$$

$$= \int d\mathbf{r}_2 \left(q_i\delta^{(3)}(r_{12}) + \rho_i(r_{12}) \right) \delta\Psi^{\text{ext}}(\mathbf{r}_2), \tag{6.210}$$

where we have used Equation 6.25. Likewise, Equation 6.209 becomes

$$\delta n_i(\mathbf{r}_1) = -\beta n_i^{\text{b}} \int d\mathbf{r}_2 \left(q_i\delta^{(3)}(r_{12}) + \rho_i(r_{12}) \right) \delta\Psi^{\text{ext}}(\mathbf{r}_2). \tag{6.211}$$

Note that the integral in these equations is equal to the electrostatic interaction energy between $\delta\Psi^{\text{ext}}$ and the charge density $\rho_i^{\text{tot}}(r) = q_i\delta^{(3)}(r) + \rho_i(r)$ of the i-ion and its surrounding ion cloud.

APPENDIX 6C: DEFINITIONS OF THE H_{NN}, H_{QN} AND H_{QQ} CORRELATION FUNCTIONS

In this appendix, we define the density-density (NN), charge-density (QN) and charge-charge (QQ) correlation functions and derive the expressions given in Section 6.2.3 for them. We have already encountered the density-density correlation function for a one-component system in Section 5.8, where it was denoted by $H_N^{(2)}$ with subscript N (in italics), meaning the number of particles in the system. Here we we use the notation H_{NN} with subscript NN (in roman letters) meaning "density-density," so Equation 5.99 is written

$$H_{\text{NN}}(r) = n^{\text{b}} \left[n^{\text{b}} h^{(2)}(r) + \delta^{(3)}(r) \right]$$

$$= n^{\text{b}} \left[n^{\text{b}} g^{(2)}(r) - n^{\text{b}} + \delta^{(3)}(r) \right] \quad \text{(one component),} \tag{6.212}$$

which can be compared to Equation 6.187. In order to generalize this expression to a binary system, where we have N_+ cations and N_- cations, we use the **microscopic number**

density distribution

$$\check{n}^{(1)}_{\{\mathbf{r}^{N_{\text{tot}}}\}}(\mathbf{r}) \equiv \sum_{\nu=1}^{N_{\text{tot}}} \delta^{(3)}(\mathbf{r} - \mathbf{r}_\nu) \tag{6.213}$$

(introduced in Equation 5.89), where $N_{\text{tot}} = N_+ + N_-$ is the total number of particles. In multi-component systems, the species of the particles at the various locations $\mathbf{r}_\nu$ are of different kinds and we let the particles with index $1, \ldots, N_+$ be cations and those with index $N_+ + 1, \ldots, N_{\text{tot}}$ be anions. We have

$$\left\langle \check{n}^{(1)}_{\{\mathbf{r}^{N_{\text{tot}}}\}}(\mathbf{r}) \right\rangle = \left\langle \sum_{\nu=1}^{N_+} \delta^{(3)}(\mathbf{r} - \mathbf{r}_\nu) \right\rangle + \left\langle \sum_{\nu=N_++1}^{N_-} \delta^{(3)}(\mathbf{r} - \mathbf{r}_\nu) \right\rangle$$

$$= n_+(\mathbf{r}) + n_-(\mathbf{r}) = n_{\text{tot}}(\mathbf{r})$$

(cf. Equation 5.90). Since we are dealing with a bulk system, $n_{\text{tot}}(\mathbf{r}) = n^{\text{b}}_{\text{tot}}$.

We now define the **density-density correlation function** for the multi-component case, as (cf. Equation 5.93)

$$H_{\text{NN}}(\mathbf{r}, \mathbf{r}') = \left\langle \left[\check{n}^{(1)}_{\{\mathbf{r}^{N_{\text{tot}}}\}}(\mathbf{r}) - n_{\text{tot}}(\mathbf{r}) \right] \left[\check{n}^{(1)}_{\{\mathbf{r}^{N_{\text{tot}}}\}}(\mathbf{r}') - n_{\text{tot}}(\mathbf{r}') \right] \right\rangle$$

$$= \left\langle \check{n}^{(1)}_{\{\mathbf{r}^{N_{\text{tot}}}\}}(\mathbf{r}) \check{n}^{(1)}_{\{\mathbf{r}^{N_{\text{tot}}}\}}(\mathbf{r}') \right\rangle - \left(n^{\text{b}}_{\text{tot}} \right)^2, \tag{6.214}$$

where the last equality applies for a bulk phase. By proceeding analogously to the derivations in Section 5.8 and using the definition (5.91) of the microscopic pair density distribution function $\check{n}^{(2)}_{\{\mathbf{r}^N\}}(\mathbf{r}, \mathbf{r}')$ applied for $N = N_{\text{tot}}$, we have (cf. Equation 5.92)

$$\left\langle \check{n}^{(2)}_{\{\mathbf{r}^{N_{\text{tot}}}\}}(\mathbf{r}, \mathbf{r}') \right\rangle = \sum_{\nu=1}^{N_{\text{tot}}} \sum_{\substack{\nu'=1 \\ (\nu' \neq \nu)}}^{N_{\text{tot}}} \int d\mathbf{r}^{N_{\text{tot}}} \delta^{(3)}(\mathbf{r} - \mathbf{r}_\nu) \delta^{(3)}(\mathbf{r}' - \mathbf{r}_{\nu'}) \mathcal{P}^{(N_{\text{tot}})}_{N_{\text{tot}}}(\mathbf{r}^{N_{\text{tot}}}),$$

where $\mathcal{P}^{(N_{\text{tot}})}_{N_{\text{tot}}}$ is the usual probability density as defined in Equation 3.9, but with the appropriate potential energy $\check{U}^{\text{pot}}_{N_{\text{tot}}}(\mathbf{r}^{N_{\text{tot}}})$ for the multi-component system inserted. In the sum over ν, the first N_+ terms are numerically equal and the same applies to the next N_- terms; each term describes the average particle distribution of both kinds of ions around the point $\mathbf{r}_\nu$ occupied by a cation or anion, respectively. By using similar arguments as in Equation 5.92, which for the one-component case gives the result $(n^b)^2 g^{(2)}(|\mathbf{r} - \mathbf{r}'|)$, we obtain for a binary electrolyte

$$\left\langle \check{n}^{(2)}_{\{\mathbf{r}^{N_{\text{tot}}}\}}(\mathbf{r}, \mathbf{r}') \right\rangle = n^{\text{b}}_+ n_{\text{tot};+}(|\mathbf{r} - \mathbf{r}'|) + n^{\text{b}}_- n_{\text{tot};-}(|\mathbf{r} - \mathbf{r}'|) = n^{\text{b}}_{\text{tot}} n_{\text{N}}(|\mathbf{r} - \mathbf{r}'|),$$

where we have used the first line in Equation 6.186 to obtain the last equality. Using Equation 5.96, we therefore conclude that

$$\left\langle \check{n}^{(1)}_{\{\mathbf{r}^{N_{\text{tot}}}\}}(\mathbf{r})\check{n}^{(1)}_{\{\mathbf{r}^{N_{\text{tot}}}\}}(\mathbf{r}')\right\rangle = n^{\text{b}}_{\text{tot}}\left[n_{\text{N}}(|\mathbf{r}-\mathbf{r}'|) + \delta^{(3)}(\mathbf{r}'-\mathbf{r})\right]$$

and Equation 6.187 follows from the definition (6.214) of $H_{\text{NN}}(\mathbf{r},\mathbf{r}')$.

For the charge-charge correlations, we have to deal with a **microscopic charge density distribution** defined as

$$\check{\rho}_{\{\mathbf{r}^{N_{\text{tot}}}\}}(\mathbf{r}) \equiv \sum_{\nu=1}^{N_{\text{tot}}} q_{o_\nu}\delta^{(3)}(\mathbf{r}-\mathbf{r}_\nu), \tag{6.215}$$

where o_ν stands for the species ($+$ or $-$) of ion ν. The average charge density is given by the ensemble average

$$\rho(\mathbf{r}) = \left\langle\check{\rho}_{\{\mathbf{r}^{N_{\text{tot}}}\}}(\mathbf{r})\right\rangle = q_+n_+(\mathbf{r}) + q_-n_-(\mathbf{r}),$$

which is zero for a bulk electrolyte. The **charge-charge correlation function** is defined as

$$H_{\text{QQ}}(\mathbf{r},\mathbf{r}') = \frac{1}{q_{\text{e}}^2}\left\langle\left[\check{\rho}_{\{\mathbf{r}^{N_{\text{tot}}}\}}(\mathbf{r}) - \rho(\mathbf{r})\right]\left[\check{\rho}_{\{\mathbf{r}^{N_{\text{tot}}}\}}(\mathbf{r}') - \rho(\mathbf{r}')\right]\right\rangle$$

$$= \frac{1}{q_{\text{e}}^2}\left\langle\check{\rho}_{\{\mathbf{r}^{N_{\text{tot}}}\}}(\mathbf{r})\check{\rho}_{\{\mathbf{r}^{N_{\text{tot}}}\}}(\mathbf{r}')\right\rangle, \tag{6.216}$$

where the last equality applies for a bulk electrolyte, where $\rho(\mathbf{r}) = \rho(\mathbf{r}') = 0$. We have

$$\left\langle\check{\rho}_{\{\mathbf{r}^{N_{\text{tot}}}\}}(\mathbf{r})\check{\rho}_{\{\mathbf{r}^{N_{\text{tot}}}\}}(\mathbf{r}')\right\rangle = \sum_{\nu=1}^{N_{\text{tot}}} q_{o_\nu}\sum_{\substack{\nu'=1\\(\nu'\neq\nu)}}^{N_{\text{tot}}} q_{o_{\nu'}}\int d\mathbf{r}^{N_{\text{tot}}}\delta^{(3)}(\mathbf{r}-\mathbf{r}_\nu)\delta^{(3)}(\mathbf{r}'-\mathbf{r}_{\nu'})\mathcal{P}^{(N_{\text{tot}})}_{N_{\text{tot}}}(\mathbf{r}^{N_{\text{tot}}})$$

$$+ \sum_{\nu=1}^{N_{\text{tot}}} q_{o_\nu}^2\int d\mathbf{r}^{N_{\text{tot}}}\delta^{(3)}(\mathbf{r}-\mathbf{r}_\nu)\delta^{(3)}(\mathbf{r}'-\mathbf{r}_\nu)\mathcal{P}^{(N_{\text{tot}})}_{N_{\text{tot}}}(\mathbf{r}^{N_{\text{tot}}}).$$

In both sums over ν in the rhs, the first N_+ terms are numerically equal and the same applies to the next N_- terms. Apart from the factor q_{o_ν}, each term in the first sum over ν describes the average charge distribution around the point $\mathbf{r}_\nu$ occupied by a cation or anion, respectively. The entire first term in the rhs therefore equals

$$q_+n_+^{\text{b}}\rho_+(|\mathbf{r}-\mathbf{r}'|) + q_-n_-^{\text{b}}\rho_-(|\mathbf{r}-\mathbf{r}'|) = \frac{q_\pm n_{\text{tot}}^{\text{b}}}{2}\left[\rho_+(|\mathbf{r}-\mathbf{r}'|) - \rho_-(|\mathbf{r}-\mathbf{r}'|)\right],$$

where we have used Equation 6.150. In the second sum over ν, we use the same trick as we used in Equation 5.95 and write the entire term as

$$\left[q_+^2 \left\langle \sum_{\nu=1}^{N_+} \delta^{(3)}(\mathbf{r} - \mathbf{r}_\nu) \right\rangle + q_-^2 \left\langle \sum_{\nu=N_++1}^{N_-} \delta^{(3)}(\mathbf{r} - \mathbf{r}_\nu) \right\rangle \right] \delta^{(3)}(\mathbf{r}' - \mathbf{r})$$

$$= \left[q_+^2 n_+^{\mathrm{b}} + q_-^2 n_-^{\mathrm{b}} \right] \delta^{(3)}(\mathbf{r}' - \mathbf{r}) = [q_+ - q_-] \frac{q_{\pm} n_{\mathrm{tot}}^{\mathrm{b}}}{2} \delta^{(3)}(\mathbf{r}' - \mathbf{r}).$$

Equation 6.185 follows when these results are inserted into the definition (6.216) of $H_{QQ}(\mathbf{r}, \mathbf{r}')$.

The **charge-density correlation function** is defined as

$$H_{QN}(\mathbf{r}, \mathbf{r}') = \frac{1}{q_e} \left\langle \left[\breve{\rho}_{\{\mathbf{r}^{N_{\mathrm{tot}}}\}}(\mathbf{r}) - \rho(\mathbf{r}) \right] \left[\breve{n}_{\{\mathbf{r}^{N_{\mathrm{tot}}}\}}^{(1)}(\mathbf{r}') - n_{\mathrm{tot}}(\mathbf{r}') \right] \right\rangle$$

$$= \frac{1}{q_e} \left\langle \breve{\rho}_{\{\mathbf{r}^{N_{\mathrm{tot}}}\}}(\mathbf{r}) \breve{n}_{\{\mathbf{r}^{N_{\mathrm{tot}}}\}}^{(1)}(\mathbf{r}') \right\rangle, \tag{6.217}$$

where the last equality applies for a bulk electrolyte. We have

$$\left\langle \breve{\rho}_{\{\mathbf{r}^{N_{\mathrm{tot}}}\}}(\mathbf{r}) \breve{n}_{\{\mathbf{r}^{N_{\mathrm{tot}}}\}}^{(1)}(\mathbf{r}') \right\rangle = \sum_{\nu=1}^{N_{\mathrm{tot}}} q_{0\nu} \sum_{\substack{\nu'=1 \\ (\nu' \neq \nu)}}^{N_{\mathrm{tot}}} \int d\mathbf{r}^{N_{\mathrm{tot}}} \delta^{(3)}(\mathbf{r} - \mathbf{r}_\nu) \delta^{(3)}(\mathbf{r}' - \mathbf{r}_{\nu'}) \mathcal{P}_{N_{\mathrm{tot}}}^{(N_{\mathrm{tot}})}(\mathbf{r}^{N_{\mathrm{tot}}})$$

$$+ \sum_{\nu=1}^{N_{\mathrm{tot}}} q_{0\nu} \int d\mathbf{r}^{N_{\mathrm{tot}}} \delta^{(3)}(\mathbf{r} - \mathbf{r}_\nu) \delta^{(3)}(\mathbf{r}' - \mathbf{r}_\nu) \mathcal{P}_{N_{\mathrm{tot}}}^{(N_{\mathrm{tot}})}(\mathbf{r}^{N_{\mathrm{tot}}})$$

and in an analogous manner as in the previous derivations we obtain

$$\left\langle \breve{\rho}_{\{\mathbf{r}^{N_{\mathrm{tot}}}\}}(\mathbf{r}) \breve{n}_{\{\mathbf{r}^{N_{\mathrm{tot}}}\}}^{(1)}(\mathbf{r}') \right\rangle = q_+ n_+^{\mathrm{b}} n_{\mathrm{tot};+}(|\mathbf{r} - \mathbf{r}'|) + q_- n_-^{\mathrm{b}} n_{\mathrm{tot};-}(|\mathbf{r} - \mathbf{r}'|)$$

$$= \frac{q_{\pm} n_{\mathrm{tot}}^{\mathrm{b}}}{2} \left[n_{\mathrm{tot};+}(|\mathbf{r} - \mathbf{r}'|) - n_{\mathrm{tot};-}(|\mathbf{r} - \mathbf{r}'|) \right]$$

since the last sum over ν, which is proportional to $q_+ n_+^{\mathrm{b}} + q_- n_-^{\mathrm{b}}$, is zero. These results yield Equation 6.189, which is equivalent to Equation 6.191.

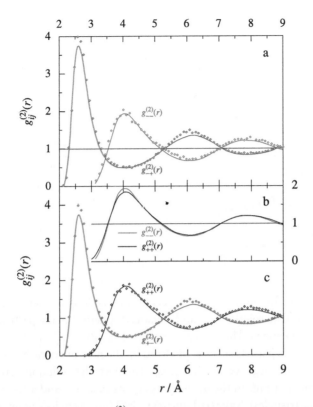

FIGURE 6.8 Pair distribution function $g_{ij}^{(2)}(r)$ for molten NaCl at 1073 K (near the freezing point). (a) Pair distributions around a Cl^- ion. Symbols are results from MD simulations and curves are results from an integral equation theory (MHNC). (b) A comparison of MHNC curves for $g_{++}^{(2)}(r)$ and $g_{--}^{(2)}(r)$. (c) Pair distributions around an Na^+ ion. Symbols show MD and curves show MHNC results.

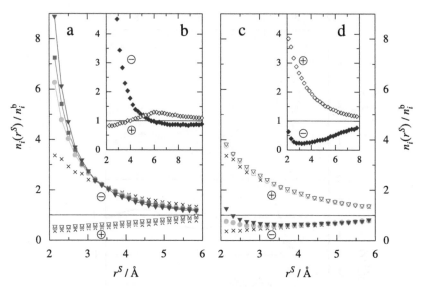

FIGURE 7.6 An example of the influence of dispersion interactions on ion distribution profiles for different aqueous electrolyte solutions around a charged spherical macroparticle with diameter 30 Å calculated by MC simulations. The profiles are plotted as $n_i(r^S)/n_i^b$, where r^S is the distance between an ion center and the particle surface. In a) and b), the macroparticle has a positive charge, $+20q_e$, and in c) and d) it has a negative charge, $-20q_e$. The electrolyte solutions have ionic strength 1.0 M and contain various 1:-1 salts in the main figures (panes a and c) and a 1:-2 salt in the insets (panes b and d): NaI (blue triangles), NaBr (red squares), NaCl (green circles), nonpolarizable monovalent ions (crosses) and Na_2SO_4 (brown diamonds). Filled symbols denote anions and open symbols denote cations. The lines are just guides for the eye. All ions are modeled as charged spheres with a diameter of 4 Å.

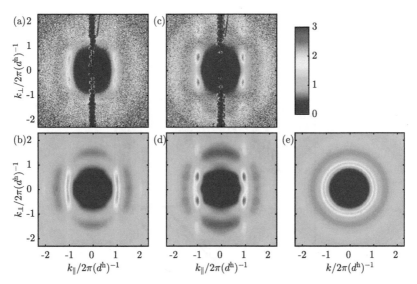

FIGURE 7.9 Experimental (top) and theoretical (bottom) anisotropic structure factors $S(\mathbf{k})$ for a dense hard-sphere fluid confined between planar walls at various separations: (a) and (b) $D^S \approx$ 2.45 d^h; (c) and (d) $D^S \approx 3.10\, d^h$; (e) in the bulk phase ($D^S = \infty$). The colors in the contour plot show the values of $S(\mathbf{k})$ according to the color legend. The fluid in the slit is in equilibrium with a bulk fluid of density $n^b = 0.75(d^h)^{-3}$. $S(\mathbf{k})$ is plotted as a function of the in-plane (lateral) component $k_\parallel$ and the out-of-plane (perpendicular) component $k_\perp$ of the scattering vector $\mathbf{k}$. The dark red feature at $k_\parallel = 0$ in the experimental data, which is due to diffraction from the array of slits with the fluid in the experiments and is given by $S_{\text{singlet}}(\mathbf{k})$, should be neglected in the comparison. The theoretical data have been obtained in the same integral equation theory as Figure 7.8.

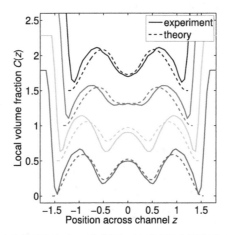

FIGURE 7.10 Experimental (solid lines) and theoretical (dashed lines) local volume fraction profiles, $C(z)$, for the same system as in Figure 7.9. The slit widths are, from top to bottom, $D^S \approx 2.45$, 2.65, 2.90, and 3.10 d^h. For clarity, consecutive sets of curves are shifted vertically by 0.5 units. The nearly vertical lines in the experimental data originate from the confining walls. The origin of z in this figure is placed in the middle of the slit, so z here corresponds to $z - D^S/2$ when the origin is located at the left surface.

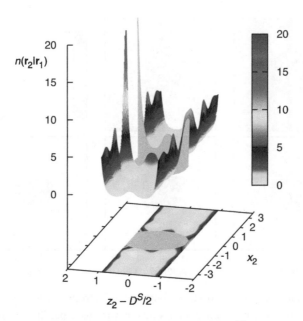

FIGURE 7.11 Plots of the density distribution $n(\mathbf{r}_2|\mathbf{r}_1) = n(z_2)g^{(2)}(\mathbf{r}_1, \mathbf{r}_2)$ at position $\mathbf{r}_2$ around a hard sphere located at $\mathbf{r}_1$ in an inhomogeneous hard sphere fluid between two hard walls at surface separation $D^S = 2.80\, d^h$ (these are the same types of plots as in Figure 5.8; see sketch of the geometry in Figure 5.7). The hard sphere at $\mathbf{r}_1$ is located in the middle of the gray circular area in the contour plot at the bottom. This area shows the region where other sphere centers cannot enter and where the density is zero. The surface plot at the top shows a different representation of the same data as the contour plot. The vertical gray area in this plot represents the discontinuity of the density function; the contact density at the sphere surface is given by the values around the top rim of this area. The system is the same as in Figure 7.8 and the calculations have been done in the same integral equation theory. The coordinates are given in units of d^h and the density in units of $(d^h)^{-3}$.

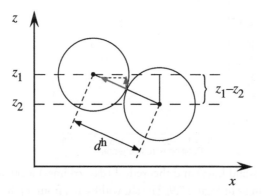

FIGURE 7.12 Sketch of a collision between two hard spheres, one with its center at z-coordinate z_1 and one with its center at z_2. The force on the former sphere due to the collision with the latter acts in the radial direction and is shown as the red arrow. The ratio between the z-component of the force (the short blue vertical arrow) and the length of the total force is equal to the ratio $(z_1 - z_2)/d^h$.

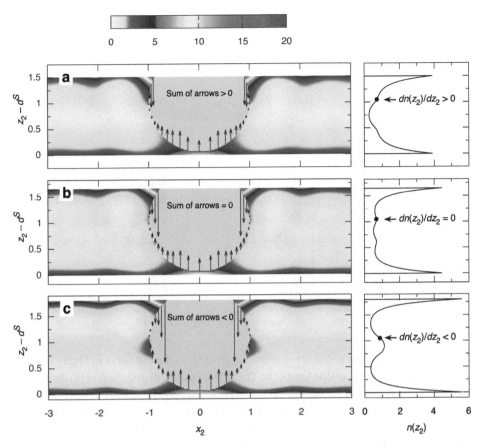

FIGURE 7.13 Contour plots of the density distribution $n(\mathbf{r}_2|\mathbf{r}_1) = n(z_2)g^{(2)}(\mathbf{r}_1,\mathbf{r}_2)$ at position $\mathbf{r}_2$ when a hard sphere is located at $\mathbf{r}_1$ in the slit between two hard walls filled with a hard sphere fluid at different surface separations: (a) $D^{\mathcal{S}} = 2.50\,d^{\mathrm{h}}$, (b) $D^{\mathcal{S}} = 2.65\,d^{\mathrm{h}}$ and (c) $D^{\mathcal{S}} = 2.80\,d^{\mathrm{h}}$. The last case is the same as in Figure 7.11. The density profiles for the respective cases are shown in the panes to the right. The dot on the profile shows the z-coordinate of the sphere, which in all cases is located at a distance of $1.55\,d^{\mathrm{h}}$ for the bottom wall (at coordinate $z_1 - d^{\mathrm{h}} = 1.05\,d^{\mathrm{h}}$). The arrows in the gray area of each contour plot depict z-components of the average collisional forces acting on the sphere. Arrows displayed at a certain z_2 coordinate represent the entire force acting on the sphere periphery at this coordinate (the sphere segment between z_2 and $z_2 + dz_2$ with a radius that lies perpendicularly to the figure plane). In pane (a), the sum of all arrows (with signs) is > 0, in (b) $= 0$ and in (c) < 0. The coordinates are given in units of d^{h} and the density in units of $(d^{\mathrm{h}})^{-3}$.

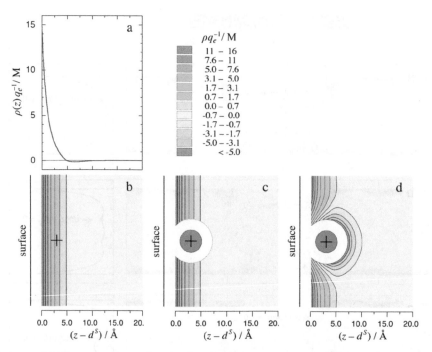

FIGURE 7.14   (a) The same charge density distribution $\rho(z)/q_e$ as the full curve in Figure 7.3b. The system is described in the caption of that figure. (b) The charge density distribution in (a) displayed as a contour plot in a cross section of the electric double-layer taken perpendicularly to the surface. The location of the surface is shown as a vertical line to the left and the bulk electrolyte lies to the right in the figure. (c) A counterion ($\oplus$) in the double layer has a cavity (excluded zone) around it, shown in white, where the centers of other ions cannot enter. The radius of the cavity is equal to the ion diameter, $d^h = 4.25$ Å. (d) The charge density distribution around a counterion in the double-layer system when its center is located at $z - d^S = 3$ Å, that is, 5.125 Å from the surface. The difference from (c) shows the polarization that the ion induces in its neighborhood, which is the charge density of the ion cloud of the ion at this position.

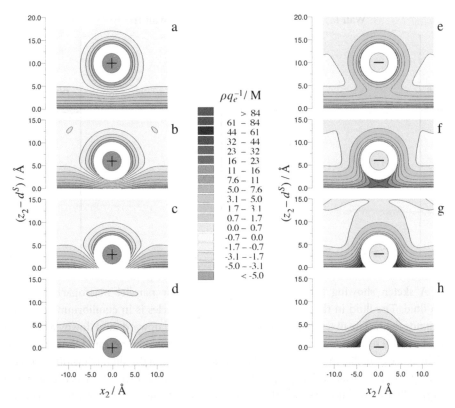

FIGURE 7.15 Panes a – d: The same charge density distribution $\rho(\mathbf{r}_2|\mathbf{r}_1;+)$ around a counterion ($\oplus$) located at $\mathbf{r}_1$ as in Figure 7.14d, but shown for various locations of the ion at different distances from the surface (pane (c) in the current figure is the same as Figure 7.14d). The ion positions are $z - d^{\mathcal{S}} = 10$ Å in (a), 6 Å in (b), 3 Å in (c) and 0 Å in (d); the latter shows a counterion in contact with the surface. Panes e – h: The corresponding charge density distribution $\rho(\mathbf{r}_2|\mathbf{r}_1;-)$ around a coion ($\ominus$) in the electric double-layer when it is located at the same distances from the surface as the counterion in the previous panes. All charge densities shown, including the density profile in Figure 7.14a, have been calculated when ion–ion correlations for all ions in the double layer are properly considered. The calculations have been done in the same integral equation theory as Figure 7.3. The density plotted in each pane has cylindrical symmetry around a vertical axis through the centre of the ion.

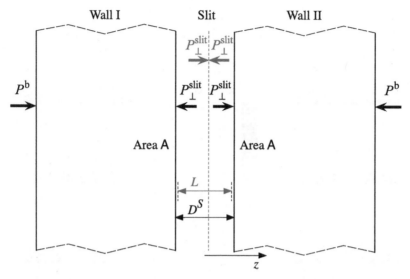

FIGURE 8.1   A sketch showing a cross section of two planar parallel macroparticles (walls) immersed in a fluid. The fluid in the slit between the macroparticles is in equilibrium with a bulk fluid that surrounds them. The macroparticles are thick and extend upwards and downwards outside the figure (not shown). The area A of each surface that faces the slit is very large and the surface separation is D^S. The space available for the centers of the fluid particles in the slit has a width L and the z axis is perpendicular to the surfaces. The perpendicular pressure component $P^{\text{slit}}_\perp$ has the same value everywhere in the slit and the bulk pressure that acts from the outside is P^b. The dashed vertical red line indicates an imagined plane parallel to the surfaces and the red arrows show the pressure acting across this plane; the right arrow shows the force (per unit area) that the part of the system to the right of the plane exerts on the part to the left of the plane and vice versa for the left arrow.

CHAPTER 7

Inhomogeneous and Confined Simple Fluids

I N THIS CHAPTER, WE will treat inhomogeneous simple fluids near or between planar walls, including hard sphere fluids, atomic fluids, and simple electrolytes. The chapter starts with a comprehensive treatment of electrolytes in contact with charged surfaces, so-called diffuse electric double-layer systems. Such systems are important in many applications and, in addition, they provide very suitable examples on how the statistical mechanical machinery can be used for inhomogeneous fluids. A treatment of planar electric double-layers in the PB approximation, also known as the Gouy-Chapman theory, is included. The concept of effective surface-charge density is introduced in a strict manner and electrostatic screening in electric double-layers is investigated in the general case by using the same general formalism as for bulk electrolytes (Section 6.2), which is applicable in exact statistical mechanics under a wide variety of conditions. Thereby, both monotonically decaying and oscillatory decaying electrostatic interactions are included. The decay of the electrostatic forces between a wall and a nonspherical macroion and between two walls is investigated in some detail.

In the PB approximation for electric double-layers, the ion-ion correlations are completely neglected. The effects of such correlations are investigated in the present chapter and several concrete examples are given. When the dielectric constants of the solvent and the material of the wall are different, there appear surface polarizations (image charge interactions). Furthermore, the polarizability of ions and walls are manifested via dispersion interactions. The effects of these kinds of interactions in electric double layers are investigated and explicit examples are given. In addition, a summary of some consequences of ion-ion dispersion interactions in electrolytes is presented.

Ion-ion correlation effects in double layers are just one example of consequences of correlations between fluid particles in inhomogeneous system. For inhomogeneous Lennard-Jones fluids, hard sphere fluids and simple electrolytes, explicit, graphical representations of the structure on the pair correlation level are presented. Concrete examples of consequences of the pair correlations for the density distribution profiles near surfaces are given, which give insight into the mechanisms behind the appearances of the density profiles.

7.1 ELECTRIC DOUBLE-LAYER SYSTEMS

An **electric double-layer** consists of a layer of positive charge and one of negative charge separated in space. In the context of electrolytes, such layers exist, for example, near the surfaces of colloidal particles or charged macroscopic walls in contact with an electrolyte solution. The surface of a particle often has a surface charge and the ion cloud outside it has a charge that is opposite to the particle – the two parts form an electric double-layer. Since the ion cloud is not localized in a single layer, one often uses the term **diffuse electric double-layer** in this case. The PB treatment of such systems in planar geometry, the **Gouy-Chapman theory**, has long have been a cornerstone in Surface and Colloid Science and other related fields of science. The theory presented in the previous Sections 6.1 and 6.2 include electric double-layers in other geometries than planar, such as colloidal particles with an internal charge distribution and spheres with a surface charge density. In the present chapter, we will treat the planar case only. After some preliminaries, we will start with the Gouy-Chapman theory and then we will proceed with more accurate approaches. For consistency with the nomenclature of the rest of the treatise, we will refer to the Gouy-Chapman theory as the PB approximation, which it indeed is.

Let us consider a smooth planar wall with a surface charge density $\sigma^{\mathcal{S}}$, where subscript $\mathcal{S}$ stands for surface, in contact with an electrolyte solution or ionic fluid. The charge density $\sigma^{\mathcal{S}}$ is assumed to be smeared out uniformly and it is located in one plane. We place the origin of the coordinate system in this plane. The z axis is perpendicular to the surface and points out into the electrolyte, while the x and y axes are parallel to the surface. The wall is assumed to be infinitely large in the lateral directions (the x and y directions). The ions are initially assumed to be charged hard spheres with the same diameters d^{h}. For one-component electrolytes, the ions can also be point-like (with diameter zero). The distance of closest approach of the ion centers to the layer of surface charge is denoted by $d^{\mathcal{S}}$. For simplicity, the dielectric constant of the wall is initially assumed to be the same as the solvent, ε_r. The case with different dielectric constants on either side of the wall surface is treated in Section 7.1.4.

The pair interaction potentials between the ions are given by Equations 6.22 and 6.23. The external potential, the ion-wall interaction potential, is given by $v_i(z) = v_i^{\mathrm{core}}(z) + v_i^{\mathrm{el}}(z)$ where

$$v_i^{\mathrm{core}}(z) = \begin{cases} \infty, & z < d^{\mathcal{S}} \\ 0, & z \geq d^{\mathcal{S}} \end{cases}$$

is a hard core–hard wall interaction potential and

$$v_i^{\mathrm{el}}(z) = -\frac{q_i \sigma^{\mathcal{S}}}{2\varepsilon_r \varepsilon_0} |z|$$

is the interaction between an ion and the uniformly smeared out surface charge density. We have here used the fact that the electrostatic potential at distance s from an infinitely large sheet of two-dimensional charge σ^S is[1] $-\sigma^S s/(2\varepsilon_r\varepsilon_0)$.

For a wall in contact with a bulk electrolyte, the ion-wall interactions make the ion densities of the various species to deviate from the bulk concentration near the wall surface. Since the surface is smooth and σ^S is uniform, the number density distribution n_j of the ions of species j depends only on the coordinate z and we have the density profile

$$n_j(z) = n_j^b g_j^{(1)}(z) = n_j^b e^{-\beta w_j(z)}.$$

Far from the surface, the electrolyte solution is in its bulk state, so we have $n_j(z) \to n_j^b$ when $z \to \infty$. The charge density $\rho(z)$ of the ion cloud near the surface is

$$\rho(z) = \sum_j q_j n_j(z) = \sum_j q_j n_j^b e^{-\beta w_j(z)}$$

and $\rho(z) \to 0$ when $z \to \infty$. The total charge density of the surface and the ion cloud is

$$\rho^{\text{tot}}(z) = \sigma^S \delta(z) + \rho(z),$$

where $\sigma^S \delta(z)$ expresses the two-dimensional surface charge density σ^S as a three-dimensional charge density by multiplication with the one-dimensional Dirac function $\delta(z)$. The mean electrostatic potential due to the surface (from the surface and its ion cloud) is

$$
\begin{aligned}
\psi(z) &= -\frac{\sigma^S |z|}{2\varepsilon_r\varepsilon_0} - \int dz' \rho(z') \frac{|z-z'|}{2\varepsilon_r\varepsilon_0} + \mathcal{K}_\infty \\
&= -\int dz' \rho^{\text{tot}}(z') \frac{|z-z'|}{2\varepsilon_r\varepsilon_0} + \mathcal{K}_\infty,
\end{aligned}
\tag{7.1}
$$

where $\mathcal{K}_\infty$ is a constant that is selected such that $\psi(z) \to 0$ when $z \to \infty$. Note that $dz' \rho(z')$ in the first integral can be interpreted as a sheet of charge of width dz' placed at z'.

7.1.1 The Poisson-Boltzmann (Gouy-Chapman) Theory

7.1.1.1 The Poisson-Boltzmann Equation for Planar Double Layers
In the PB approximation, we have $w_j(z) = q_j \psi(z)$ for $z \geq d^S$, so we obtain

$$n_j(z) = n_j^b e^{-\beta q_j \psi(z)} \quad \text{for } z \geq d^S \quad \text{(PB)} \tag{7.2}$$

---

[1] This applies to an infinitely large sheet of charge in a system where the total net charge of the entire system is equal to zero. For a sheet of surface charge density σ^S in the limit of infinite surface area A, one must sum the electrostatic potential from the sheet and the other charges system at the same time (maintaining electroneutrality); otherwise, there appears a term that goes to infinity when A $\to \infty$.

and (compare with Equation 6.56)

$$\rho(z) = \begin{cases} 0, & z < d^S \\ \sum_j q_j n_j^b e^{-\beta q_j \psi(z)}, & z \geq d^S \end{cases} \quad \text{(PB)}. \tag{7.3}$$

The PB equation is

$$-\varepsilon_r \varepsilon_0 \frac{d^2 \psi(z)}{dz^2} = \begin{cases} \sigma^S \delta(z), & z < d^S \\ \sum_j q_j n_j^b e^{-\beta q_j \psi(z)}, & z \geq d^S \end{cases} \quad \text{(PB)} \tag{7.4}$$

since $\nabla^2 \psi(z) = d^2 \psi(z)/dz^2$. The PB equation can be transformed in the following manner.

By multiplying both sides of the PB equation for $z \geq d^S$ with $2\, d\psi(z)/dz$ and by noting that

$$\frac{d}{dz}\left[\frac{d\psi(z)}{dz}\right]^2 = 2\frac{d\psi(z)}{dz}\frac{d^2\psi(z)}{dz^2} \quad \text{and}$$

$$\frac{d}{dz}\left[e^{-\beta q_j \psi(z)}\right] = -\beta q_j e^{-\beta q_j \psi(z)}\frac{d\psi(z)}{dz},$$

we can write it as

$$-\varepsilon_r \varepsilon_0 \frac{d}{dz}\left[\frac{d\psi(z)}{dz}\right]^2 = -\frac{2}{\beta}\sum_j n_j^b \frac{d}{dz}\left[e^{-\beta q_j \psi(z)}\right] \quad \text{for } z \geq d^S.$$

Since both sides contain first derivatives of the square bracket expressions, we obtain after integration with respect to z and multiplication of both sides by $-[\varepsilon_r \varepsilon_0]^{-1}$

$$\left[\frac{d\psi(z)}{dz}\right]^2 = \frac{2}{\beta\varepsilon_r \varepsilon_0}\sum_j n_j^b e^{-\beta q_j \psi(z)} + \text{const.} \quad \text{(PB)} \tag{7.5}$$

The integration constant can be determined by letting $z \to \infty$ where we have $\psi(z) \to 0$ and $d\psi(z)/dz \to 0$. This leads to $0 = 2\sum_j n_j^b/(\beta\varepsilon_r \varepsilon_0) + \text{const}$ and we finally obtain

$$\left[\frac{d\psi(z)}{dz}\right]^2 = \frac{2}{\beta\varepsilon_r \varepsilon_0}\sum_j n_j^b \left(e^{-\beta q_j \psi(z)} - 1\right). \quad \text{(PB)} \tag{7.6}$$

This relationship is often referred to as the *first integral of the PB equation*. It can also be written

$$\left[\frac{d\psi(z)}{dz}\right]^2 = \frac{2}{\beta\varepsilon_r \varepsilon_0}\sum_j \left(n_j(z) - n_j^b\right) = \frac{2k_B T}{\varepsilon_r \varepsilon_0}\left(n_{\text{tot}}(z) - n_{\text{tot}}^b\right) \quad \text{(PB)}, \tag{7.7}$$

where we have used Equation 7.2 and where $n_{\text{tot}}(z) = \sum_j n_j(z)$ is the total number density of ions at z and $n_{\text{tot}}^{\text{b}} = \sum_j n_j^{\text{b}}$ is the total bulk density of ions.

These alternative forms of the PB equation are quite useful and constitute starting points for further developments. We will start by investigating Equation 7.7 at the point of closest approach of the ion centers to the wall, $z = d^S$, which will be called the *point of contact*. The value of $d\psi(z)/dz$ at this point can be determined in the following manner. By integrating the Poisson equation $-\varepsilon_r\varepsilon_0 d^2\psi(z)/dz^2 = \rho(z)$ from $z = d^S$ to $z = \infty$, we obtain

$$\varepsilon_r\varepsilon_0 \left.\frac{d\psi(z)}{dz}\right|_{z=d^S} = \int_{d^S}^{\infty} dz\,\rho(z) = -\sigma^S, \tag{7.8}$$

where the last equality is obtained by using the fact that the charge of the ion cloud must neutralize the surface charge at the wall. If we insert this result into Equation 7.7 evaluated at $z = d^S$, we obtain

$$\frac{\left(\sigma^S\right)^2}{2\varepsilon_r\varepsilon_0} = k_BT\left(n_{\text{tot}}(d^S) - n_{\text{tot}}^{\text{b}}\right) \quad \text{(PB)}, \tag{7.9}$$

which is a relationship between the surface charge density σ^S, the total bulk density $n_{\text{tot}}^{\text{b}}$ and the ion density at the point of contact, $n_{\text{tot}}(d^S)$, that is, the *contact density*. Since $n_{\text{tot}}(z)$ has a discontinuity at $z = d^S$, the contact density should be interpreted as

$$n_{\text{tot}}(d^S) = \lim_{z\to d^{S+}} n_{\text{tot}}(z), \tag{7.10}$$

where $z \to d^{S+}$ means that the limit is taken for $z > d^S$. The consequences of Equation 7.9 will be investigated later. It can be written as

$$\frac{\beta\left(\sigma^S\right)^2}{2\varepsilon_r\varepsilon_0} = \sum_j n_j^{\text{b}}\left[e^{-\beta q_j\psi(d^S)} - 1\right] \quad \text{(PB)}, \tag{7.11}$$

which is known as **Grahame's equation**. It gives a relationship between the surface charge density, the electrostatic potential $\psi(d^S)$ at the surface and the bulk ionic densities.

Equation 7.7 can be used to obtain analytical solutions of the PB equation in some particular cases. In the next section, we specialize to a symmetric electrolyte like a 1:-1 or 2:-2 electrolyte, where such solutions exist. However, if the exponent $\beta q_j\psi(z)$ in the rhs of the PB Equation 7.4 is small, we can linearize the exponential function and obtain analytical solutions for any simple electrolyte. We then obtain from Equation 7.4 for $z > d^S$

$$\frac{d^2\psi(z)}{dz^2} = \kappa_D^2\psi(z) \quad \text{(linearized PB)}, \tag{7.12}$$

which has the solution

$$\psi(z) = \frac{\sigma^S e^{\kappa_D d^S}}{\varepsilon_r \varepsilon_0 \kappa_D} e^{-\kappa_D z} \quad \text{for } z \geq d^S \quad \text{(linearized PB)}. \tag{7.13}$$

Note that Equation 7.12 can be regarded as a special case of the Debye-Hückel Equation 6.30 in planar geometry.

{**EXERCISE 7.1**

Derive Equation 7.12 by linearizing the PB Equation 7.4. Obtain Equation 7.13 by solving Equation 7.12 for $z \geq d^S$ with the following boundary conditions: $\psi(z) \to 0$ when $z \to \infty$ and $d\psi(z)/dz$ at $z = d^S$ given by Equation 7.8.}

7.1.1.2 The Case of Symmetric Electrolytes
Let us return to the nonlinear PB case. For a symmetric electrolyte $q_+ = -q_- \equiv q$, $n_+^b = n_-^b \equiv n^b$ and $n_{tot}^b = 2n^b$. This implies that

$$\rho(z) = q_+ n_+^b e^{-\beta q_+ \psi(z)} + q_- n_-^b e^{-\beta q_- \psi(z)}$$

$$= qn^b [e^{-\beta q \psi(z)} - e^{\beta q \psi(z)}] = -2qn^b \sinh(\beta q \psi(z))$$

and

$$n_{tot}(z) = n^b [e^{-\beta q \psi(z)} + e^{\beta q \psi(z)}] = 2n^b \cosh(\beta q \psi(z)). \tag{7.14}$$

Hence, if we multiply the PB Equation 7.4 for $z \geq d^S$ by $\beta q/(\varepsilon_r \varepsilon_0)$, it can be written

$$\frac{d^2[\beta q \psi(z)]}{dz^2} = \kappa_D^2 \sinh(\beta q \psi(z)), \quad \text{(PB)} \tag{7.15}$$

where we have used the fact that $\kappa_D^2 = 2\beta q^2 n^b/(\varepsilon_r \varepsilon_0)$ for a symmetric electrolyte. Likewise, Equation 7.6 can be written [after multiplication with $(\beta q)^2$]

$$\left[\frac{d[\beta q \psi(z)]}{dz}\right]^2 = 2\kappa_D^2 \left(\cosh(\beta q \psi(z)) - 1\right). \quad \text{(PB)} \tag{7.16}$$

We can obtain an analytic solution to these equations as explained in Appendix 7A. The final result is

$$\psi(z) = \frac{2}{\beta q} \ln\left[\frac{1 + Be^{-\kappa_D(z-d^S)}}{1 - Be^{-\kappa_D(z-d^S)}}\right] \quad \text{for } z \geq d^S \quad \text{(PB)} \tag{7.17}$$

where the parameter B is a solution to the quadratic equation

$$B^2 + \frac{4\varepsilon_r \varepsilon_0 \kappa_D}{\beta q \sigma^S} B - 1 = 0, \tag{7.18}$$

whereby the solution $\mathcal{B}$ with the same sign as σ^S should be selected. As shown in the appendix, this solution can be expressed as $\mathcal{B} = \tanh\left(\beta q \psi(d^S)/4\right)$, that is, in terms of the potential $\psi(z)$ at the point of contact, $z = d^S$.

The density distribution of ions can be obtained by inserting the solution (7.17) into Equation 7.2. For the anions, we have for $z \geq d^S$

$$n_-(z) = n^b e^{\beta q \psi(z)} = n^b \left[\frac{1 + \mathcal{B}e^{-\kappa_D(z-d^S)}}{1 - \mathcal{B}e^{-\kappa_D(z-d^S)}}\right]^2 \quad (\text{PB}) \qquad (7.19)$$

and for cations

$$n_+(z) = n^b e^{-\beta q \psi(z)} = n^b \left[\frac{1 - \mathcal{B}e^{-\kappa_D(z-d^S)}}{1 + \mathcal{B}e^{-\kappa_D(z-d^S)}}\right]^2 \quad (\text{PB}). \qquad (7.20)$$

The function $\beta q \psi(z)$ is plotted in Figure 7.1a and the ionic distribution profiles $n_+(z)/n^b$ and $n_-(z)/n^b$ are plotted in Figure 7.1b,c for a positively charged surface with a quite high surface charge density ($\sigma^S = 0.266\,\text{Cm}^{-2}$) immersed in a 0.1 M aqueous 1:-1 electrolyte solution. The figures also show the corresponding results from MC simulations and we see that the PB results do not deviate much from the MC data. The PB approximation works in general quite well for 1:-1 aqueous electrolyte solutions if the ionic concentration and/or the surface charge density are not too high. However, when divalent ions are present, this approximation often has serious deficiencies as we will see later in Sections 7.1.3 and 8.5.

The electric double-layer model should, of course, be reasonably realistic and therefore we must have $d^S \neq 0$, as in Figure 7.1, so the ion centers are not allowed to approach all the way to the surface charge (like point ions would be able to do). In the solution (7.17), we have used the value of the surface charge density σ^S as the boundary condition for the PB equation at the surface. In many applications of the PB approximation for electric double-layers, it is instead customary to use the value of the electrostatic potential at the surface, usually at $z = 0$, as the boundary condition. The value $\psi(0)$ is equal to the difference in mean electrostatic potential between the surface and the bulk phase, where $\psi = 0$. In order to obtain a reasonable value for this difference, it is particularly important to use a realistic value of d^S for the following reasons and we use the PB results in Figure 7.1a as an example:

The ion densities $n_+(z)$ and $n_-(z)$ are zero for $z < d^S$ (i.e., for $z/d^h < 0.5$ in the figure), but the electrostatic potential varies a lot in the interval $0 < z/d^h < 0.5$ because the surface charge density σ^S is located at $z = 0$. As we will see shortly, $\psi(z)$ is a linear function of z in this interval so the plot of $\psi(z)$ is a straight line, which is drawn as a slanted dashed line in the figure. Therefore, we have $\psi(z) = \psi(0) + z\mathcal{K}$ in this interval, where the coefficient $\mathcal{K}$ is equal to $[\psi(d^S) - \psi(0)]/d^S$ = the slope of the linear function ($\mathcal{K}$ is negative in the present case). The dashed straight line makes a smooth continuation of the full curve in the figure because the derivative $\psi'(z) \equiv d\psi(z)/dz$ at $z = d^h$ is continuous under these conditions

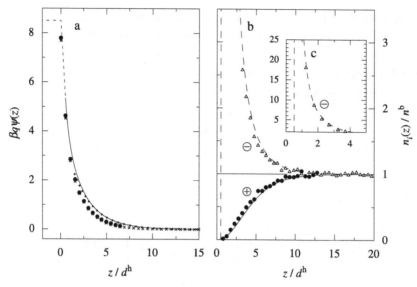

FIGURE 7.1 (a) The mean electrostatic potential plotted as $\beta q\psi(z)$ and (b, c) the ionic densities plotted as $n_+(z)/n^{\mathrm{b}}$ and $n_-(z)/n^{\mathrm{b}}$ as functions of distance z from a planar surface with uniform surface charge density $\sigma^S = 0.266$ Cm^{-2} in contact with an 0.1 M aqueous solution of a 1:-1 electrolyte at 298 K in the primitive model. The value of σ^S corresponds to one elementary charge per 60 Å^2 on the surface. The ionic diameters are $d^{\mathrm{h}} = 4.25$ Å and the distance of closest approach of the ion centers to the surface charge density is $d^S = d^{\mathrm{h}}/2$. The curves show results from the PB approximation and the symbols show results from GCMC simulations.[2] Pane c shows $n_-(z)/n^{\mathrm{b}}$ for small z on an expanded scale. In pane (a), the filled circles and crosses are for grand canonical MC (GCMC) simulations with different sizes of the simulation box and the dotted curve shows the leading asymptotic term for the PB data. The dashed lines in pane (a) display the linear parts of the potential curve for the PB case in the intervals $z < 0$ (horizontal) and $0 \le z \le d^S$ (slanted). The GCMC data are likewise linear for these z values and the potential difference $\psi(0) - \psi(d^S)$ is the same as in the PB case (Equation 7.21).

(the electric field $E_z = -d\psi(z)/dz$ is continuous). We have $\psi(0) = \psi(d^S) - \mathcal{K}d^S$, where both $\psi(d^S)$ and $\mathcal{K} = \psi'(d^S)$ are determined by the full curve. When $|\mathcal{K}|$ is large, as in the present case, the value of $\psi(0)$ is very sensitive to the magnitude of d^S, so the value of the latter matters a lot.

Let now fill in the mathematical details of this argument. In the interval $0 < z/d^{\mathrm{h}} < 0.5$, the Poisson equation is $-\varepsilon_r\varepsilon_0 d^2\psi(z)/dz^2 = 0$, which implies that $\psi(z)$ is a linear function of z there. We know from Equation 7.8 that the derivative $d\psi(z)/dz$ at $z = d^S$ is $-\sigma^S/(\varepsilon_r\varepsilon_0)$, so it follows that $\mathcal{K} = -\sigma^S/(\varepsilon_r\varepsilon_0)$ and hence we have

$$\psi(0) = \psi(d^S) + \frac{\sigma^S d^S}{\varepsilon_r\varepsilon_0}. \tag{7.21}$$

_____

[2] The GCMC data in Figure 7.1 are taken from G. M. Torrie and J. P. Valleau, *J. Chem. Phys.* **73** (1980) 5807.

Note that the derivative $d\psi(z)/dz$ is discontinuous at $z = 0$ because the surface charge density is located there; $d\psi(z)/dz$ changes by $-\sigma^S/(\varepsilon_r\varepsilon_0)$ when z passes from negative to positive z. Since the derivative of the linear part of $\psi(z)$ for $z > 0$ is equal to $\mathcal{K} = -\sigma^S/(\varepsilon_r\varepsilon_0)$, it follows that the potential must be constant, $\psi(z) = \psi(0)$ for $z < 0$.[3] This is displayed as the horizontal dashed line in Figure 7.1a. Thereby, we have determined the solution $\psi(z)$ of the PB equation for all z values. In fact, $\psi(z)$ must always be linear in the interval $0 < z/d^h < 0.5$, so the potential difference $\psi(0) - \psi(d^S)$ must always be the same as in the PB case, so Equation 7.21 has general validity.

Finally, let us consider the special case when the surface charge density σ^S is very small. Then $\psi(d^S)$ and $\mathcal{B}$ are small too and the term $\mathcal{B}^2$ in Equation 7.18 can be ignored. This means that we have $\mathcal{B} \approx \beta q \sigma^S/(4\varepsilon_r\varepsilon_0\kappa_D)$ and we obtain from Equation 7.17 in the limit $\sigma^S \to 0$

$$\psi(z) = \frac{2}{\beta q}\left[\ln\left(1 + \mathcal{B}e^{-\kappa_D(z-d^S)}\right) - \ln\left(1 - \mathcal{B}e^{-\kappa_D(z-d^S)}\right)\right]$$
$$\sim \frac{4\mathcal{B}}{\beta q}e^{-\kappa_D(z-d^S)} \sim \frac{\sigma^S}{\varepsilon_r\varepsilon_0\kappa_D}e^{-\kappa_D(z-d^S)} \quad \text{for } z \geq d^S \quad \text{(PB)},$$

(7.22)

since $\ln(1 + x) \sim x$ when $x \to 0$. We see that $\psi(z)$ is proportional to σ^S when the latter is small. The expression in the rhs is equal to the solution (7.13) of the linearized PB equation.

Let us consider the case when the ions are allowed to approach all the way to the plane of the surface charge density at $z = 0$, that is, the ions are treated like point-ions not only in their interactions between themselves but also in the interactions with the wall. Then we have $d^S = 0$ and

$$\psi(z) = \frac{\sigma^S}{\varepsilon_r\varepsilon_0\kappa_D}e^{-\kappa_D z} \quad \text{for } z \geq 0 \quad \text{(linearized PB; reference).}$$

(7.23)

This case is the simplest double-layer result in the PB approximation and it is suitable as a *reference case* when comparing with other cases, like when specifying values of effective charges. This will be done in the next section.

{**EXERCISE 7.2**

The total ion concentration $n_{tot}(d^S)$ at a positively charged planar surface has been found to be 1.171 M when the surface is in contact with a 0.100 M solution of NaCl in water at 25°C. Use the nonlinear PB approximation to calculate

a. the surface charge density σ^S of the surface,

b. the surface potential $\psi(d^S)$ and

c. the Cl$^-$ and Na$^+$ concentrations at the surface, $n_-(d^S)$ and $n_+(d^S)$, respectively.}

---

[3] The constancy of $\psi(z)$ can also be realized from the facts that the geometry is planar and the total charge for $z \geq 0$ is zero (including σ^S).

7.1.1.3 Effective Surface Charge Densities and the Decay of the Electrostatic Potential

In the linearized PB approximation, we have seen that the potential decays like

$$\psi(z) = \frac{\sigma^S e^{\kappa_D d^S}}{\varepsilon_r \varepsilon_0 \kappa_D} e^{-\kappa_D z} \quad \text{for } z \geq d^S \quad \text{(linearized PB)}$$

(Equations 7.13). This was also obtained for small σ^S in the nonlinear PB approximation (Equation 7.22), as it must. By comparing this expression with Equation 7.23 for the reference case ($d^S = 0$), we see that we can define an **effective surface charge density** $\sigma^{S\text{eff}}$ as

$$\sigma^{S\text{eff}} = \sigma^S e^{\kappa_D d^S} \quad \text{(linearized PB)}, \tag{7.24}$$

so we have

$$\psi(z) = \frac{\sigma^{S\text{eff}}}{\varepsilon_r \varepsilon_0 \kappa_D} e^{-\kappa_D z} \quad \text{for } z \geq d^S \quad \text{(linearized PB)}.$$

Then the potential away from the surface is the same as the potential at equal coordinate z in the reference system provided that the surface charge density is numerically equal to $\sigma^{S\text{eff}}$ in the latter. This is in analogy to the concept of the effective charge of a particle introduced in Section 6.1.2, where the linearized PB result for point ions was used as a reference.

Equation 7.24 says that when ion size is considered in terms of a nonzero distance of closest approach to the surface, $d^S > 0$, we have $\sigma^{S\text{eff}} > \sigma^S$. This equation is analogous to the result in Equation 6.49 for the effective charge q_i^{eff} of an ion in bulk as obtained in the linearized PB approximation (Debye-Hückel theory). In the latter case, the surrounding ions are not allowed to approach all the way to the ion center, so when $d^h > 0$ we have $q_i^{\text{eff}} > q_i$.

In the current approximation, it is rather trivial why $\sigma^{S\text{eff}} \neq \sigma^S$. As seen in the rhs of Equation 7.22, the difference from Equation 7.23 is a simple spatial translation by the distance d^S, so $e^{-\kappa_D(z-d^S)}$ appears in the former equation instead of $e^{-\kappa_D z}$. Thereby, the electrostatic potential at coordinate z differs by a factor of $e^{\kappa_D d^S}$ from the reference case, so the surface charge "appears" to have increased by the same factor. In general, however, the concept of effective surface charge is far from trivial.

Next, we proceed to the nonlinear PB approximation and examine the behavior of $\psi(z)$ for large z. We start with a *symmetric electrolyte*, whereby we can use results from the previous section. The step in Equation 7.22 that leads to the second line, that is,

$$\psi(z) \sim \frac{4\mathcal{B} e^{\kappa_D d^S}}{\beta q} e^{-\kappa_D z},$$

is valid whenever $\mathcal{B} \exp(-\kappa_D(z - d^S))$ is small. When z is large, the exponential function is small, so this expression is always valid in the limit $z \to \infty$ irrespective of the value of $\mathcal{B}$. By comparing with Equation 7.23, we see that if we define $\sigma^{S\text{eff}}$ from

$$\frac{4\mathcal{B} e^{\kappa_D d^S}}{\beta q} = \frac{\sigma^{S\text{eff}}}{\varepsilon_r \varepsilon_0 \kappa_D}$$

we obtain

$$\psi(z) \sim \frac{\sigma^{Seff}}{\varepsilon_r \varepsilon_0 \kappa_D} e^{-\kappa_D z} \quad \text{when } z \to \infty \quad \text{(PB)}, \tag{7.25}$$

which has the same form as Equation 7.23. The potential far away from the surface accordingly decays as a function of z in exactly the same fashion as in Equation 7.23, but with σ^S replaced by an effective σ^{Seff} given by

$$\sigma^{Seff} = \frac{4\mathcal{B} e^{\kappa_D d^S} \varepsilon_r \varepsilon_0 \kappa_D}{\beta q} \quad \text{(PB)}, \tag{7.26}$$

where $\mathcal{B}$ is obtained from Equation 7.18. The asymptotic decay given by the rhs of Equation 7.25 is plotted as a dotted curve in Figure 7.1a. This curve approaches the PB curve when z increases and we see that it agrees well with the PB results for $z/d^h \gtrsim 2$ in this case.

Next, we will investigate what happens in the PB approximation when the surface charge density σ^S is increased. From Equation 7.18, it follows that when σ^S is very large, the linear term is negligible and $|\mathcal{B}| \approx 1$. This means that the effective surface charge density σ^{Seff} in Equation 7.26 cannot exceed the value that corresponds to $|\mathcal{B}| = 1$. In other words, the magnitude of σ^{Seff} for a system with symmetric electrolytes cannot be larger than $\sigma^{Seff}_{Max} = 4e^{\kappa_D d^S} \varepsilon_r \varepsilon_0 \kappa_D/(\beta q)$.

It is illuminating to see what happens when σ^S is increased from zero. First, the value of σ^{Seff} increases proportionally to σ^S as expressed in Equation 7.24, but when σ^S is further increased, the value must be calculated using Equation 7.26, where $\mathcal{B}$ depends on σ^S in a nonlinear manner via Equation 7.18. The result is that the increase in σ^{Seff} deviates more and more from proportionality. The increase becomes more and more attenuated when σ^S is further increased and eventually, for large σ^S, the effective value σ^{Seff} is very close to σ^{Seff}_{Max} and stays virtually constant thereafter.

What is the physical interpretation of this behavior of the PB approximation? The attenuation in the increase in σ^{Seff} means that an amount of counterions is brought in close to the surface when the surface charge density σ^S is increased. These counterions partially neutralizes the increase in surface charge, so the net change per unit area there increases less than what σ^S does. Since σ^{Seff} cannot exceed σ^{Seff}_{Max} irrespective of the magnitude of σ^S, the amount of counterions brought to the surface will eventually compensate the entire increase in σ^S, leading to the constant value of σ^{Seff} when $\sigma^S \to \infty$. This happens even if the density of ions close to the surface thereby would exceed close-packing of spheres. This physical impossibility is a consequence of the PB approximation where the ions in the ion cloud outside the surface are treated like uncorrelated point ions in their interactions amongst themselves. Such ions can be very closely packed together, while real ions cannot. This limits the usefulness of the PB approximation to systems with low or moderate surface charge densities. This gives a restriction also for 1:-1 aqueous electrolyte solutions where this approximation otherwise is quite useful. It is, however, not only steric factors that limit the validity of the PB approximation. Whenever ion-ion correlations are important – correlations due to electrostatic interactions as well as steric factors – the PB approximation does not work. In Section 7.1.3, we will see some examples of the effects of ion-ion correlations in double layers.

The results for effective surface charge densities obtained so far apply to $z : -z$ electrolytes only. Let us now consider *any kind of electrolyte* with spherically symmetric ions in the PB approximation. The approach given in Section 6.1.2 to obtain the effective charge of a nonspherical particle P can easily be generalized to planar surfaces. Let us first recapitulate the main findings there. By subtracting the linear part of the electrostatic polarization response ρ_P^{lin} due to particle P from the total charge density ρ_P^{tot}, we defined the dressed particle charge density $\rho_P^\star$ of the particle as $\rho_P^\star = \rho_P^{tot} - \rho_P^{lin}$ (Equation 6.60). The mean electrostatic potential ψ_P due to P can then be expressed in terms of the unit screened Coulomb potential $\phi_{Coul}^\star(r)$ as (Equation 6.70)

$$\psi_P(\mathbf{r}) = \int d\mathbf{r}' \rho_P^\star(\mathbf{r}') \phi_{Coul}^\star(|\mathbf{r} - \mathbf{r}'|), \tag{7.27}$$

which has the same form as Coulomb's law. As we have seen in Section 6.2, these results hold also in the general case. Recall that in the PB approximation we have $\phi_{Coul}^\star(r) = e^{-\kappa_D r}/[4\pi \varepsilon_r \varepsilon_0 r]$ for all r (Equation 6.69) and $\rho_P^{lin} = -\varepsilon_r \varepsilon_0 \kappa_D^2 \psi_P$ (Equation 6.59). We will now use these findings in planar geometry.

Let particle P be very large and have a flat surface with a finite but huge surface area A, that is, P is a planar wall with finite lateral size. We assume that this surface has a surface charge density σ^S while the rest of P is uncharged, so we have an internal charge density of P given by $\sigma_P(\mathbf{r}) = \sigma^S \delta(z)$ for $\mathbf{r}$ inside P, including the surface of P (we have placed the origin at the plane of the surface charge as before and the z axis is directed perpendicularly out from the surface). Since Equation 7.27 is valid irrespective of the shape and internal charge distribution $\sigma_P(\mathbf{r})$ of P, it is valid also in the current case. We will obtain an infinitely large planar wall by taking the limit $A \to \infty$ while keeping σ^S constant. Consider the electrolyte outside the planar surface. In the limit $A \to \infty$, the charge density of the ion cloud depends only on z, $\rho_P(\mathbf{r}) = \rho_P(z)$, and the same is true for the mean potential $\psi_P(\mathbf{r}) = \psi(z)$. We then have $\rho_P^{tot}(\mathbf{r}) = \sigma^S \delta(z) + \rho_P(z)$ everywhere and $\rho_P^\star(\mathbf{r})$ becomes

$$\rho^\star(z) = \rho^{tot}(z) + \varepsilon_r \varepsilon_0 \kappa_D^2 \psi(z), \tag{7.28}$$

where we have dropped subscript P, and $\psi(z)$ is given by (cf. Equation 7.27)

$$\psi(z) = \int d\mathbf{r}' \rho^\star(z') \phi_{Coul}^\star(|\mathbf{r} - \mathbf{r}'|). \tag{7.29}$$

By inserting $\phi_{Coul}^\star(r)$, we obtain

$$\psi(z) = \int d\mathbf{r}' \rho^\star(z') \frac{e^{-\kappa_D|\mathbf{r}-\mathbf{r}'|}}{4\pi \varepsilon_r \varepsilon_0 |\mathbf{r}-\mathbf{r}'|}$$
$$= \frac{1}{4\pi \varepsilon_r \varepsilon_0} \int\limits_{-\infty}^{\infty} dz' \rho^\star(z') \int ds \frac{e^{-\kappa_D[s^2+(z-z')^2]^{1/2}}}{[s^2 + (z - z')^2]^{1/2}}, \tag{7.30}$$

where $\mathbf{s} = (x - x', y - y'), s = |\mathbf{s}|$ and $d\mathbf{s}$ can be replaced by $2\pi s\, ds$ when we have introduced polar coordinates in the $\mathbf{s}$ plane and performed the angle integration. We can simplify the rhs by using the identity[4]

$$\int\limits_0^\infty ds\, s\, \frac{e^{-\kappa_D[s^2+a^2]^{1/2}}}{[s^2+a^2]^{1/2}} = \frac{e^{-\kappa_D|a|}}{\kappa_D},$$

and we obtain

$$\psi(z) = \frac{1}{2\kappa_D\varepsilon_r\varepsilon_0} \int\limits_{-\infty}^\infty dz'\, \rho^\star(z') e^{-\kappa_D|z-z'|}.$$

Since $\rho^\star(z')$ decays faster than $\exp(-\kappa_D z')$ when z' increases, we can take $z' < z$ and in the limit $z \to \infty$, we obtain

$$\psi(z) \sim \frac{1}{2\kappa_D\varepsilon_r\varepsilon_0} \left[\int\limits_{-\infty}^\infty dz'\, \rho^\star(z') e^{\kappa_D z'} \right] e^{-\kappa z}.$$

The integral converges also in the lower limit $z' \to -\infty$ since $\rho^\star(z')$ is constant for negative z' because $\psi(z')$ is constant there. By defining an effective surface charge density

$$\sigma^{\mathcal{S}\text{eff}} = \tfrac{1}{2} \int\limits_{-\infty}^\infty dz'\, \rho^\star(z') e^{\kappa_D z'} \quad \text{(PB)} \tag{7.31}$$

we can write this as

$$\psi(z) \sim \frac{\sigma^{\mathcal{S}\text{eff}}}{\kappa_D\varepsilon_r\varepsilon_0} e^{-\kappa_D z} \quad \text{when } z \to \infty \quad \text{(PB)} \tag{7.32}$$

like in Equation 7.25. The difference is that the present asymptotic result for large z with $\sigma^{\mathcal{S}\text{eff}}$ defined in Equation 7.31 is valid in general for the PB approximation, while the effective charge density from Equation 7.26 is applicable for symmetric electrolytes only. There is no need to restrict the present results to equally sized ions. By doing an extension like in Equation 6.88, it follows that the result (7.32) is valid also for cases where the distances of closest approach to the surface are different for different kind of ions.

Since $n_j(z) = n_j^b e^{-\beta q_j \psi(z)}$ in the PB approximation, the deviation in density from the bulk value decays when $z \to \infty$ as

$$n_j(z) - n_j^b \sim -n_j^b \beta q_j \psi(z) \sim -\frac{\beta n_j^b q_j \sigma^{\mathcal{S}\text{eff}}}{\kappa_D\varepsilon_r\varepsilon_0} e^{-\kappa_D z} \quad \text{(PB)}, \tag{7.33}$$

where we have expanded the exponential function to linear order and used Equation 7.32.

---

[4] This identity can easily be shown by making the variable substitution $s^2 + a^2 = t^2$ in the integral.

{EXERCISE 7.3

The effective surface charge density $\sigma^{S\text{eff}}$ depends on the actual surface charge density σ^S and other parameters like the composition and concentration of the bulk electrolyte that the surface is in contact with. For a given electrolyte and a given temperature, $\sigma^{S\text{eff}}$ is a function of σ^S, so we have $\sigma^{S\text{eff}} = \sigma^{S\text{eff}}(\sigma^S)$. In general, this function depends also on the other parameters, but for a symmetric electrolyte in the PB approximation one can write this function in such a manner that it solely depends on a scaled value of σ^S.

a. Show that by defining the dimensionless scaled entities

$$\tau = \frac{\sigma^S}{\sigma_0^S}, \quad \tau^{\text{eff}} = \frac{\sigma^{S\text{eff}}}{\sigma_0^S},$$

where $\sigma_0^S = \varepsilon_r \varepsilon_0 \kappa_D / (\beta q)$, we have for the case $d^S = 0$

$$\tau^{\text{eff}} = \frac{8}{\tau} \left[\left(1 + \frac{\tau^2}{4} \right)^{1/2} - 1 \right] \quad \text{(PB, symmetric electrolytes)},$$

which is a universal expression valid for all symmetric electrolytes and all concentrations and temperatures.[5] Note that this holds only in the PB approximation and not in general.

b. In what manner can one change the formulas in (a) so that they are valid also for the case $d^S \neq 0$?}

7.1.2 Electrostatic Screening in Electric Double-Layers, General Case

In Section 6.2, we investigated electrostatic interactions and various aspects of distribution functions in bulk electrolytes, in particular the decay behavior of these functions. We found that several qualitative features found in the PB approximation are not limited to that approximation but are valid in general. In the present section, we will analyze electric double-layer systems under the same conditions for the electrolyte as in Section 6.2 (the conditions are specified in Section 6.2.1.3, see pages 315 and 317) and we will see that the corresponding findings also hold true here. However, a major difference between the PB and the general cases is that the unit screened Coulomb potential $\phi_{\text{Coul}}^\star(r)$, which equals $e^{-\kappa_D z}/[4\pi \varepsilon_r \varepsilon_0 r]$ for all r in the former (Equation 6.69), has more than one Yukawa function term as well as exponentially damped oscillatory terms in the general case, like in Equations 6.169 and 6.174. The same applies to the distribution functions and we are here considering systems where these kinds of terms are the leading ones for sufficiently large r.[6] Furthermore, the values of the decay lengths, effective charges, etc. have different values in the PB and the general cases.

---

[5] Analogous expressions exist for some unsymmetric electrolytes in the PB approximation, but the functional form is then different.

[6] For various terms in $\phi_{\text{Coul}}^\star(r)$, see comments in footnote 41 in Chapter 6.

7.1.2.1 Decay of the Electrostatic Potential Outside a Wall

To determine the decay behavior of the potential $\psi(z)$ outside a planar surface, we will, to start with, follow the same arguments as we used for the PB case leading to Equation 7.29 and, finally, to Equation 7.32. We initially have a particle P that is very large and has a flat surface with a finite but huge surface area A and a surface charge density σ^S. We place the origin at the plane of surface charge and have the z axis in the perpendicular direction out from the surface. The internal charge density is $\sigma_P(\mathbf{r}) = \sigma^S \delta(z)$ for $\mathbf{r}$ in P and the dressed particle charge density is $\rho_P^\star = \rho_P^{\text{tot}} - \rho_P^{\text{lin}} = \sigma_P + \rho_P^{\text{dress}}$ (Equations 6.129 and 6.131). The density $\rho_P^\star$ gives together with $\phi_{\text{Coul}}^\star$ the potential ψ_P via Equation 6.137 (which is the same as Equation 7.27). This equation is valid irrespective of the shape and internal charge distribution $\sigma_P(\mathbf{r})$ of P. By taking the limit $A \to \infty$ while σ^S is kept constant, we conclude in the same fashion as in the PB case that Equation 7.29 is valid in general, where $\psi(z)$ and $\rho^\star(z)$ are given by $\rho_P^\star(\mathbf{r})$ and $\psi_P(\mathbf{r})$, respectively, in the limit $A \to \infty$ like in the PB case. However, in the general case, $\rho^\star(z)$ is *not* given by Equation 7.28, which is valid only in the PB approximation. Instead, we have (Equation 6.130)

$$\rho^\star(z) = \rho^{\text{tot}}(z) - \int d\mathbf{r}' \, \psi(z') \chi^\star(|\mathbf{r} - \mathbf{r}'|),$$

where $\chi^\star(r)$ is the polarization response function of the bulk electrolyte that is in contact with the surface. As mentioned in connection to Equation 6.140, $\rho_P^\star(\mathbf{r})$ has a shorter range than $\phi_{\text{Coul}}^\star(r)$, at least if the ion density in the bulk phase is not too high and the same applies therefore to $\rho^\star(z)$ for large z. Like before, for reasons of simplicity in the theoretical treatment, it is assumed that the conditions are such that this is fulfilled.

When the electrostatic coupling is low, $\phi_{\text{Coul}}^\star(r)$ decays like a Yukawa function given by Equation 6.166. By inserting this in Equation 7.29, we obtain (cf. the first line of Equation 7.30)

$$\psi(z) \sim \int d\mathbf{r}' \rho^\star(z') \frac{e^{-\kappa|\mathbf{r}-\mathbf{r}'|}}{4\pi \mathcal{E}_r^{\text{eff}} \varepsilon_0 |\mathbf{r} - \mathbf{r}'|} \quad \text{when } z \to \infty,$$

where $\mathcal{E}_r^{\text{eff}} = \mathcal{E}_r^{\text{eff}}(\kappa)$. In exactly the same manner as we derived Equation 7.32 in the PB case, we obtain in the general case

$$\psi(z) \sim \frac{\sigma^{S\text{eff}}}{\kappa \mathcal{E}_r^{\text{eff}} \varepsilon_0} e^{-\kappa z} \quad \text{when } z \to \infty, \tag{7.34}$$

where we have defined

$$\sigma^{S\text{eff}} = \frac{1}{2} \int\limits_{-\infty}^{\infty} dz' \rho^\star(z') e^{\kappa z'} \tag{7.35}$$

like in the PB case, Equation 7.31.

When there are two real-valued solutions κ and κ' to Equation 6.163 with $\kappa < \kappa'$ and hence two Yukawa terms in $\phi_{\text{Coul}}^\star(r)$ (Equation 6.169), there are also two exponentially decaying terms in the mean potential with these decay parameters κ and κ'

$$\psi(z) \sim C e^{-\kappa z} + C' e^{-\kappa' z} \quad \text{when } z \to \infty, \tag{7.36}$$

where the constant C is equal to the coefficient in Equation 7.34 and C' is another constant that can be written in a similar manner in terms of $\mathcal{E}_r^{\text{eff}}(\kappa')$ and a kind of effective surface charge density (the explicit expression is given in Equation 7.38 below). The details are given in the following shaded text box, which can be skipped in the first reading. In the text box, we will also investigate the case of complex-valued solutions to Equation 6.163, $\kappa = \kappa_\Re \pm i\kappa_\Im$, which leads to an exponentially damped oscillatory decay of $\phi_{\text{Coul}}^\star(r)$. In this case, the mean potential decays like

$$\psi(z) \sim B e^{-\kappa_\Re z} \cos(\kappa_\Im z + \vartheta_B) \quad \text{when } z \to \infty, \tag{7.37}$$

where B and ϑ_B are real constants (Equation 7.39).

★ (The reader is referred to the text box on page 327ff for background material needed for the derivations here.) By inserting $\phi_{\text{Coul}}^\star(r)$ from Equation 6.169 into Equation 7.29, we obtain from both terms in $\phi_{\text{Coul}}^\star(r)$ in the same manner as in the derivations of Equation 7.32 and 7.34

$$\psi(z) \sim \frac{\sigma^{Seff}(\kappa)}{\kappa \mathcal{E}_r^{\text{eff}}(\kappa)\varepsilon_0} e^{-\kappa z} + \frac{\sigma^{Seff}(\kappa')}{\kappa' \mathcal{E}_r^{\text{eff}}(\kappa')\varepsilon_0} e^{-\kappa' z} \quad \text{when } z \to \infty, \tag{7.38}$$

where

$$\sigma^{Seff}(\kappa) = \frac{1}{2} \int_{-\infty}^{\infty} dz' \rho^\star(z') e^{\kappa z'}$$

is applied for both κ and κ'.

When the solutions κ and κ' merge and become complex-valued, we have $\kappa = \kappa_\Re + i\kappa_\Im$ and $\kappa' = \kappa_\Re - i\kappa_\Im$. Equation 7.38 can then be written (cf. Equations 6.173 and 6.174)

$$\psi(z) \sim 2\Re \left[\frac{|\sigma^{Seff}| e^{-i\vartheta_\sigma}}{|\kappa \mathcal{E}_r^{\text{eff}}| e^{-i(\vartheta_\mathcal{E} + \vartheta_\kappa)}\varepsilon_0} e^{-(\kappa_\Re + i\kappa_\Im)z} \right],$$

where we have written $\sigma^{Seff}(\kappa_\Re + i\kappa_\Im) = |\sigma^{Seff}| \exp(-i\vartheta_\sigma)$ and for κ in the denominator $\kappa_\Re + i\kappa_\Im = |\kappa| \exp(-i\vartheta_\kappa)$. As before, we have $\mathcal{E}_r^{\text{eff}}(\kappa_\Re + i\kappa_\Im) = |\mathcal{E}_r^{\text{eff}}| \exp(-i\vartheta_\mathcal{E})$. This yields

$$\psi(z) \sim \frac{2|\sigma^{Seff}|}{|\kappa \mathcal{E}_r^{\text{eff}}|\varepsilon_0} e^{-\kappa_\Re z} \cos(\kappa_\Im z + \vartheta_\sigma - \vartheta_\mathcal{E} - \vartheta_\kappa) \quad \text{when } z \to \infty, \tag{7.39}$$

which defines the constants B and ϑ_B in Equation 7.37.

7.1.2.2 Decay of Double-Layer Interactions: Macroion-Wall and Wall-Wall

In Section 6.2.1, we treated the decay of the interaction between two particles, for example two macroions, in the general case. There we obtained Equation 6.144 for the decay of the potential of mean force $w_{I,II}^{(2)}$ between two particles P_I and P_{II} immersed in the electrolyte. Here we will use the following part of this equation

$$w_{I,II}^{(2)}(\mathbf{r}_{I,II}, \boldsymbol{\omega}_I, \boldsymbol{\omega}_{II}) \sim \int d\mathbf{r}_1 \psi_I(\mathbf{r}_{I,1}, \boldsymbol{\omega}_I) \rho_{II}^{\star}(\mathbf{r}_{II,1}, \boldsymbol{\omega}_{II}) \quad \text{when } r_{I,II} \to \infty. \tag{7.40}$$

This formula is valid irrespective of the shape and internal charge distribution $\sigma_J(\mathbf{r}, \boldsymbol{\omega}_J)$ of P_J for $J = I$ and II. Therefore, it also holds for the case when particle I is a planar wall with internal charge density $\sigma_I(z) = \sigma_I^S \delta(z)$ and with a dressed particle charge density $\rho_I^{\star}(z)$, as for the wall in the previous section. Electrostatic interactions between charged macroparticles, including charged walls, in electrolytes are in general denoted as **electric double-layer interactions**, in particular when the internal charges of the particles and of the walls are located at the surfaces.

Let us consider a macroion P_J with orientation $\boldsymbol{\omega}_J$ placed with its center of mass at $\mathbf{r}_J$. The coordinate system is selected such that the origin is located at the plane of surface charge density of wall I and with the z axis in the perpendicular direction out from the wall. The orientation variable is, in general, given relative to the laboratory frame of reference, but with the present choice of coordinate system we can think of $\boldsymbol{\omega}_J$ as the orientation relative to wall I. We let P_J take the role of P_{II} in Equation 7.40. In the potential of mean force $w_{I,J}^{(2)}$ between P_J and wall I, only the orientation variable $\boldsymbol{\omega}_J$ will appear since the wall has a fixed orientation relative to the frame of reference.

For this case, Equation 7.40 can be written

$$w_{I,J}^{(2)}(z_J, \boldsymbol{\omega}_J) \sim \int d\mathbf{r}_1 \psi_I(z_1) \rho_J^{\star}(\mathbf{r}_1 - \mathbf{r}_J, \boldsymbol{\omega}_J) \quad \text{when } z_J \to \infty \tag{7.41}$$

since the potential from the wall depends on the z coordinate only. By inserting ψ_I from Equation 7.34, we obtain in the same limit

$$w_{I,J}^{(2)}(z_J, \boldsymbol{\omega}_J) \sim \frac{\sigma_I^{S\text{eff}}}{\kappa \mathcal{E}_r^{\text{eff}} \varepsilon_0} \int d\mathbf{r}_1 e^{-\kappa z_1} \rho_J^{\star}(\mathbf{r}_1 - \mathbf{r}_J, \boldsymbol{\omega}_J)$$

because $\rho_J^{\star}$ has a short range compared to the decay length $1/\kappa$. As before, $\mathcal{E}_r^{\text{eff}} = \mathcal{E}_r^{\text{eff}}(\kappa)$. By multiplying and dividing by $e^{-\kappa z_J}$, this equation can be written

$$w_{I,J}^{(2)}(z_J, \boldsymbol{\omega}_J) \sim \frac{\sigma_I^{S\text{eff}} e^{-\kappa z_J}}{\kappa \mathcal{E}_r^{\text{eff}} \varepsilon_0} \int d\mathbf{r}_1 e^{-\kappa(z_1 - z_J)} \rho_J^{\star}(\mathbf{r}_1 - \mathbf{r}_J, \boldsymbol{\omega}_J).$$

$$= \frac{\sigma_I^{S\text{eff}} e^{-\kappa z_J}}{\kappa \mathcal{E}_r^{\text{eff}} \varepsilon_0} \int d\mathbf{r} \, e^{-\kappa z} \rho_J^{\star}(\mathbf{r}, \boldsymbol{\omega}_J),$$

where we have used the fact that $d\mathbf{r}_1 = d(\mathbf{r}_1 - \mathbf{r}_J)$ since $\mathbf{r}_J$ is constant during the integration and where we have set $\mathbf{r} = \mathbf{r}_1 - \mathbf{r}_J$. Note that the origin of $\mathbf{r}$ is located at $\mathbf{r}_J$. Since $z = \hat{\mathbf{z}}_I \cdot \mathbf{r}$, where $\hat{\mathbf{z}}_I$ is the normal to the surface (which defines the z direction), we can write the integral as

$$\int d\mathbf{r}\, e^{-\kappa z} \rho_J^\star(\mathbf{r}, \boldsymbol{\omega}_J) = \int d\mathbf{r}\, e^{-\kappa \hat{\mathbf{z}}_I \cdot \mathbf{r}} \rho_J^\star(\mathbf{r}, \boldsymbol{\omega}_J) = Q_J^{\text{eff}}(-\hat{\mathbf{z}}_I, \boldsymbol{\omega}_J),$$

where we have used the definition of Q_J^{eff} in Equation 6.154. Thus, we have

$$w_{I,J}^{(2)}(z_J, \boldsymbol{\omega}_J) \sim \sigma_I^{\mathcal{S}\text{eff}} Q_J^{\text{eff}}(-\hat{\mathbf{z}}_I, \boldsymbol{\omega}_J) \frac{e^{-\kappa z_J}}{\kappa \mathcal{E}_r^{\text{eff}} \varepsilon_0} \quad \text{when } z_J \to \infty, \qquad (7.42)$$

which means that the magnitude of the interaction is proportional to the product of the effective interaction parameter Q_J^{eff} of particle J and the effective surface charge density $\sigma_I^{\mathcal{S}\text{eff}}$ of the wall. Note that the vector $-\hat{\mathbf{z}}_I$ points towards the wall.

Having established the decay of the interaction between the surface and particle J, for instance a macroion, we now turn to the interaction between two parallel surfaces in contact with an electrolyte. The electrolyte in the slit between the surfaces is assumed to be in equilibrium with a bulk electrolyte of a given composition. The surface separation is $D^{\mathcal{S}}$, which is taken as the distance between the planes of surface charge densities at either wall. Since Equation 7.42 is valid for a particle J of any size, shape and charge distribution placed at coordinate z_J, it is also valid for the case of a smooth planar wall with a very large but finite area A lying parallel to wall I and with a uniform surface charge density $\sigma_{II}^{\mathcal{S}}$. This will become our wall II when we let A $\to \infty$ (while keeping $\sigma_{II}^{\mathcal{S}}$ constant) in order to obtain the system with two walls. By going through this limiting procedure, the details of which can be found in the shaded text box below (which can be skipped in the first reading), we find that Equation 7.42 yields the *free energy of interaction per unit area* $\mathscr{W}_{I,II}^{\mathcal{S}}(D^{\mathcal{S}})$ between the walls I and II with infinite lateral extensions

$$\mathscr{W}_{I,II}^{\mathcal{S}}(D^{\mathcal{S}}) \sim \frac{2\sigma_I^{\mathcal{S}\text{eff}} \sigma_{II}^{\mathcal{S}\text{eff}}}{\kappa \mathcal{E}_r^{\text{eff}} \varepsilon_0} e^{-\kappa D^{\mathcal{S}}} \quad \text{when } D^{\mathcal{S}} \to \infty, \qquad (7.43)$$

where $\sigma_I^{\mathcal{S}\text{eff}}$ and $\sigma_{II}^{\mathcal{S}\text{eff}}$ are the effective surface charge densities of the two walls.

★ For the second wall, we select z_J to be the z coordinate of the surface charge density $\sigma_{II}^{\mathcal{S}}$, which means that $z_J = D^{\mathcal{S}}$. Since the orientation of the second wall is fixed, we will drop the orientation variable $\boldsymbol{\omega}_J$ from now on. For this wall, we have

$$Q_J^{\text{eff}}(-\hat{\mathbf{z}}_I) = Q_J^{\text{eff}}(\hat{\mathbf{z}}_{II}) = \int d\mathbf{r}\, e^{\kappa \hat{\mathbf{z}}_{II} \cdot \mathbf{r}} \rho_J^\star(\mathbf{r}), \qquad (7.44)$$

where $\hat{\mathbf{z}}_{II}$ is the local normal of the surface of J (the surface that faces wall I). In the local coordinate system $\mathbf{r}'$ for J, where the origin lies on the plane with surface charge density

σ_{II}^S, as it does for $\mathbf{r}$, but where the z' axis points towards wall I, we have $\mathbf{r}' \cdot \hat{\mathbf{z}}_{\text{II}} = z'$. Since σ_{II}^S is uniform, the function $\rho_j^\star(\mathbf{r})$ depends solely on coordinate z except close to the outer rim of the surface. Since we will let $A \to \infty$, we can neglect boundary effects at the rim and treat $\rho_j^\star(\mathbf{r})$ as if it only depends on z. By introducing $\rho_{\text{II}}^\star(\mathbf{r}') = \rho_j^\star(\mathbf{r})$, that is, the same function in the new coordinate system (with the new notation $\rho_{\text{II}}^\star$), we can write the integral in Equation 7.44 as

$$\int d\mathbf{r}' e^{\kappa z'} \rho_{\text{II}}^\star(\mathbf{r}') = A \int dz' \, e^{\kappa z'} \rho_{\text{II}}^\star(z') = 2A\sigma_{\text{II}}^{S\text{eff}}$$

where we have used Equation 7.35 to obtain the last equality. Using this in Equation 7.44 and inserting the result into Equation 7.42, we obtain

$$w_{\text{I},J}^{(2)}(z_J) \sim \frac{2A\sigma_{\text{I}}^{S\text{eff}}\sigma_{\text{II}}^{S\text{eff}}}{\kappa \mathcal{E}_r^{\text{eff}}\varepsilon_0} e^{-\kappa z_J}.$$

As noted earlier, we have $z_J = D^S$ and by taking the limit $A \to \infty$, we finally obtain Equation 7.43, where $\mathscr{W}_{\text{I},\text{II}}^S(D^S) = \lim_{A \to \infty} w_{\text{I},J}^{(2)}(D^S)/A$ is the free energy of interaction per unit area.

For the case of two equal surfaces, we have $\sigma_{\text{I}}^{S\text{eff}} = \sigma_{\text{II}}^{S\text{eff}}$ and the product in the prefactor of Equation 7.43 becomes a square. Then $\mathscr{W}_{\text{I},\text{II}}^S$ is repulsive provided $\mathcal{E}_r^{\text{eff}} = \mathcal{E}_r^{\text{eff}}(\kappa) > 0$, which is normally the case when κ is real. For unequal surfaces, $\mathscr{W}_{\text{I},\text{II}}^S$ can be repulsive or attractive depending on the signs of $\sigma_{\text{I}}^{S\text{eff}}$ and $\sigma_{\text{II}}^{S\text{eff}}$. Note that it is sufficient to do a calculation for a single surface in contact with an electrolyte solution in order to obtain the value of each of $\sigma_{\text{I}}^{S\text{eff}}$ and $\sigma_{\text{II}}^{S\text{eff}}$.

These results are valid also in the PB approximation provided we set $\kappa = \kappa_D$ and $\mathcal{E}_r^{\text{eff}} = \varepsilon_r$. For a symmetric electrolyte in this approximation, the values of the effective surface charge densities are explicitly given by Equation 7.26, where $\mathcal{B}$ is obtained from Equation 7.18 with σ^S equal to σ_{I}^S or σ_{II}^S, respectively.

In the general case when there are two two real-valued solutions κ and κ' to Equation 6.163, we have (cf. Equation 7.36)

$$\mathscr{W}_{\text{I},\text{II}}^S(D^S) \sim \mathscr{C}_{\text{I},\text{II}} e^{-\kappa D^S} + \mathscr{C}_{\text{I},\text{II}}' e^{-\kappa' D^S} \quad \text{when } D^S \to \infty, \tag{7.45}$$

where $\mathscr{C}_{\text{I},\text{II}}$ is the coefficient in Equation 7.43 and $\mathscr{C}_{\text{I},\text{II}}'$ is likewise defined (the full expression is given in Equation 7.47 below). In the oscillatory case where $\kappa = \kappa_{\Re} + i\kappa_{\Im}$ is complex (cf. Equation 7.37), we have

$$\mathscr{W}_{\text{I},\text{II}}^S(D^S) \sim \mathscr{B}_{\text{I},\text{II}} e^{-\kappa_{\Re} D^S} \cos(\kappa_{\Im} D^S + \vartheta_{\text{I},\text{II}}) \quad \text{when } D^S \to \infty, \tag{7.46}$$

where $\mathscr{B}_{I,II}$ and $\vartheta_{I,II}$ are constants (see Equation 7.48). The details are given in the shaded text box, which can be skipped in the first reading.

★ By inserting the second term of Equation 7.38 into Equation 7.41 and treating it in the same manner as we have treated the first term leading to Equation 7.43, we obtain

$$\mathscr{W}_{I,II}^{S}(D^{S}) \sim \frac{2\sigma_{I}^{Seff}(\kappa)\sigma_{II}^{Seff}(\kappa)}{\kappa\mathcal{E}_{r}^{eff}(\kappa)\varepsilon_{0}}e^{-\kappa D^{S}} + \frac{2\sigma_{I}^{Seff}(\kappa')\sigma_{II}^{Seff}(\kappa')}{\kappa'\mathcal{E}_{r}^{eff}(\kappa')\varepsilon_{0}}e^{-\kappa' D^{S}} \qquad (7.47)$$

in the limit $D^{S} \to \infty$. In the oscillatory case, we have $\kappa = \kappa_{\Re} + i\kappa_{\Im}$ and $\kappa' = \kappa_{\Re} - i\kappa_{\Im}$ and Equation 7.47 can then be manipulated in the same fashion as we did when we obtained Equation 7.39. The result is

$$\mathscr{W}_{I,II}^{S}(D^{S}) \sim \frac{4|\sigma_{I}^{Seff}\sigma_{II}^{Seff}|}{|\kappa\mathcal{E}_{r}^{eff}|\varepsilon_{0}}e^{-\kappa_{\Re}D^{S}}\cos(\kappa_{\Im}D^{S} + \vartheta_{I} + \vartheta_{II} - \vartheta_{\mathcal{E}} - \vartheta_{\kappa}), \qquad (7.48)$$

where $\sigma_{I}^{Seff}(\kappa_{\Re} + i\kappa_{\Im}) = |\sigma_{I}^{Seff}|\exp(-i\vartheta_{I})$ and likewise for σ_{II}^{Seff}. This expression defines the constants $\mathscr{B}_{I,II}$ and $\vartheta_{I,II}$ in Equation 7.46.

7.1.3 Ion-Ion Correlation Effects in Electric Double-Layers: Explicit Examples

In this section, we will investigate some significant effects of ion-ion correlations on the structure of electric double-layers at planar walls. In Figure 7.1, we investigated density profiles $n_{+}(z)$ and $n_{-}(z)$ for monovalent ions outside a charged surface in contact with a 0.1 M 1:-1 bulk electrolyte solution obtained from both MC simulation and from the PB approximation. This is a case where the PB approximation is quite reasonable because ion-ion correlations, which are neglected in this approximation, are not very important. On the other hand, as we noticed in Section 7.1.1.3, the PB approximation cannot work at high surface charge densities since for such cases it predicts counterion densities close to the surface that exceed close-packing of spheres. At high σ^{S}, it is therefore necessary to include ion-ion correlations in the theoretical treatment of the double layer.

As an example, the ion density distribution profiles $n_{i}(z)/n_{i}^{b}$ for a 1.0 M 1:-1 electrolyte solution in contact with a surface with a much higher σ^{S} than in Figure 7.1 is shown in Figure 7.2a. As a result of the crowding of counterions near the surface, their density profile shows a second peak at $z \approx 1.5\,d^{h}$. This means that apart from a layer of counterions in contact with the surface (at $z = 0.5\,d^{h}$), there is a second layer of counterions outside the first one. The counterions with coordinates $z \gtrsim 1.3\,d^{h}$ are attracted to the surface but cannot penetrate into the first layer, so they form a new layer. In the first layer, the density of coions is virtually zero, as seen in the figure, and this density starts to rise from around zero in the second layer, which is far less compact than the first one. Outside the second layer, the

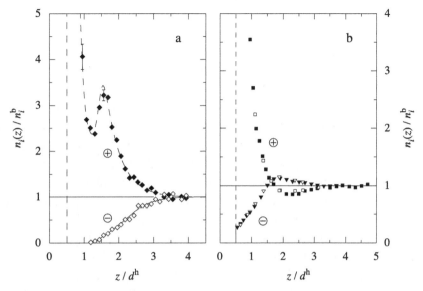

FIGURE 7.2 Ion distribution profiles $g_i^{(1)}(z) = n_i(z)/n_i^b$ in the primitive model for aqueous electrolyte solutions in contact with a planar, uniformly charged surface at room temperature ($\varepsilon_r = 78.5$). Alternatively, the systems consist of vaporized electrolytes at 23400 K ($\varepsilon_r = 1$) in contact with the surface. (a) Results for the 1.0 M 1:-1 electrolyte when the surface charge density is $\sigma^S = -0.620$ C m^{-2} (one elementary charge per 26 Å^2). Symbols show data from GCMC simulations and the curves show results of an integral equation theory:[7] filled diamonds and dashed curve show the profile for monovalent counterions and open diamonds and full curve show that for monovalent coions. (b) GCMC simulation results for the 0.5 M 2:-1 electrolyte when $\sigma^S = -0.151$ C m^{-2} (one elementary charge per 106 Å^2). Squares show the data for the divalent counterions and triangles those for the monovalent coions. The filled and open symbols denote results from simulations with different sizes of the simulation box. In all cases, the ionic diameters are $d^h = 4.25$ Å and the distance of closest approach of the ions to the surface charge density is $d^S = d^h/2$.[8]

densities of both coions and counterions decay smoothly with increasing z to the bulk value. One can say that the counterions in the first layer form a new "surface" located about 1.0 d^h outside the actual surface. The part of the system behind this "surface" has a reduced charge per unit area compared to σ^S but with the same sign as the latter. The remaining counterions and coions outside this somewhat rugged surface behave in the usual manner for a surface with a rather small surface charge density.

In this example, core-core correlations (packing effects) play a major role. The counterions are brought towards the surface not only for electrostatic reasons. They are pushed against the surface due to collisions by the surrounding ions similarly to the case of layering around a particle in Figures 5.1 and 5.2, but in the present case there is a layering close to a

---

[7] The calculations have been done in the ARHNC approximation, an integral equation theory that is introduced in Chapter 9 (in the 2nd volume).

[8] The simulation data in Figure 7.2a are taken from S. L. Carnie and G. M. Torrie, *Adv. Chem. Phys.* **56** (1984) 141 and the integral equation data are from H. Greberg and R. Kjellander, *Mol. Phys.* **83** (1994) 789. Figure 7.2b is based on data from G. M. Torrie and J. P. Valleau, *J. Chem. Phys.* **86** (1982) 3251.

planar wall. In the present example, the ions in the second layer push those in the first layer towards the surface, thereby contributing significantly to the creation of the layering there. In the same manner, ions outside the second layer contribute to its formation by their collisions. For this effect to be significant, the density must be sufficiently high, so the collision probability is large enough. In our example, the bulk density is 1.0 M, which apparently is sufficient to cause this effect near the charged surface. However, κ is real at this bulk concentration, so the oscillations do not continue after the two peaks near the surface. If oscillations induced by the conditions near the surface should continue further out and constitute more than a few peaks, they must be supported by an oscillatory decay mode in the bulk phase.

Incidentally, we may add that the primitive model of electrolyte solutions, where the solvent is a continuum, is not very realistic in this case since, in reality, the packing of solvent molecules close to the surface certainly will have a large influence on the ionic densities there. The example is nevertheless of interest since it demonstrates effects of correlations in a clear manner. However, in the case of a vaporized electrolyte (classical plasma) at high temperatures in contact with a surface (mentioned in the caption of Figure 7.2), the model is much more realistic since there is no solvent present. Then the results in Figure 7.2a are relevant, provided that the ions behave classically and have strong contact repulsions that can be modeled as a hard-core potential.

Let us now turn to a more important case where ion-ion correlations have a large influence. In aqueous electrolyte solutions at room temperature, ion-ion correlations often play a considerable role when the system contains ions with divalent or higher valencies, because the electrostatic interactions then are strong. An example is shown in Figure 7.2b, where the density distribution profiles $n_i(z)/n_i^{\mathrm{b}} \equiv g_i^{(1)}(z)$ are plotted for a 0.5 M electrolyte solution of a 2:-1 electrolyte with the divalent ions (positive) being counterions to the negatively charged surface. We see that the counterion density is high and the coion density is low close to the surface, but at $z \approx 1.6\, d^{\mathrm{h}}$ the two curves cross and in an interval thereafter the coion has a higher density than the counterions. This cannot happen in the PB approximation because the two profiles then are monotonic, so this approximation is qualitatively wrong.

The charge density outside the surface is (cf. Equation 6.148)

$$\rho(z) = \sum_i q_i n_i(z) = \sum_j q_i n_i^{\mathrm{b}} g_i^{(1)}(z) \tag{7.49}$$

$$= q_+ n_+^{\mathrm{b}} [g_+^{(1)}(z) - g_-^{(1)}(z)] = \frac{q_\pm n_{\mathrm{tot}}^{\mathrm{b}}}{2}[g_+^{(1)}(z) - g_-^{(1)}(z)]$$

where $q_\pm$ is defined in Equation 6.149. The total charge per unit area in the ion cloud outside the surface is $\int_0^\infty dz\, \rho(z) = -\sigma^S$, where the equality follows from electroneutrality. We can see from Figure 7.2b that the integral of $\rho(z)$ from the crossing point at $z \approx 1.6\, d^{\mathrm{h}}$ to infinity is negative. This follows since the coion density is larger than the counterion density up to $z \approx 3.8\, d^{\mathrm{h}}$ and since the charge density in the tail of the distribution for large z is negligible in comparison. Therefore, the charge per unit area between 0 and $z \approx 1.6\, d^{\mathrm{h}}$ must be larger than $-\sigma^S$ (the sum of the integrals on either side of $z \approx 1.6\, d^{\mathrm{h}}$ must be $-\sigma^S$), implying that a larger number of counterions is located in the immediate neighborhood of the surface than needed to neutralize the surface charge – an **overcompensation** of the surface charge. This kind of phenomenon is also called **surface charge reversal**, since the total charge of the surface together with the ions very close to the surface, in the current case up to $z \approx 1.6\, d^{h}$, is opposite in sign to the surface charge. When one talks about the charge density distribution in the double layer, it is common to describe this situation as a **charge inversion** in the profile since the coion density exceeds the counterion density in an interval outside $z \approx 1.6\, d^{h}$. In the present case, these various ways to describe the charge distribution in the double-layer system refer to different aspects of the same situation.

In the interval from $0.5\, d^{h}$ to $\approx 1.6\, d^{h}$, there is space for one layer of ions, but in contrast to the previous case, the counterion density there is substantially less than that of close-packing, so a layering of the previous type with some layers of counterions does not occur. Instead, in the second layer there is an excess of coions, brought there primarily for electrostatic reasons. Since there is a large excess of counterions in the first layer at the surface, the coions are attracted to the region outside this layer despite that the surface charge repels them. In the region $z \gtrsim 2\, d^{h}$, the profiles are like those in a system with a reversed sign of surface charge. Thus, in a way, the system behaves as if the surface charge has reversed its sign.

The large excess of counterions in the first layer, which leads to the overcompensation, is to a large extent due to the fact that each counterion near the surface repels other counterions in its neighborhood, both because of hard-core exclusion and electrostatic repulsion. This means that the surface charge of the wall is exposed near the ion and the latter can therefore be strongly attracted towards the surface with little interference from other counterions. The attraction is appreciably stronger than in the PB approximation, which does not predict this behavior. We will illustrate this in more detail later (mainly in Section 7.2.2). In general, both electrostatic forces and forces due to collisions matter in the double-layer structure, but the former are often dominating.[9]

In the 0.5 M 2:-1 bulk electrolyte solution, the decay parameter κ is complex-valued for the bulk, so the decay of the distribution functions in Figure 7.2b is exponentially damped oscillatory for large z. Therefore, there is a layering of anions and cations in the tail region, similar to what was found for the molten salt in Figure 6.8. The two primary layers for $z \lesssim 3\, d^{h}$ are, however, of a somewhat different nature. As can be seen in the figure, the amplitude of the oscillations in the tail region is very small compared to the primary layers. The existence of these layers does not depend on the presence of oscillatory decay. We will later investigate a case with monotonic decay and the same type of overcompensation of surface charge. The type of decay, monotonic or oscillatory, is a property of the bulk phase, while the presence of a few oscillations (the primary layers) near the surface depends on the influence of the surface on the electrolyte.

There is a considerable difference between systems with divalent or monovalent counterions. For a surface with the same surface charge density as in Figure 7.2b but of opposite

---

[9] There is a somewhat intricate interplay between the electrostatic and collisional forces. For example, the electrostatic attraction between anions and cations leads to a high probability of collisions between the two species. The collisions give a repulsive contribution to the potential of mean force between the ions. Therefore, there is not only an electrostatic attraction between anions and cations but also a collisional repulsion that can be strong. The two contributions counteract each other to varying degrees depending on the details of the circumstances.

sign and for the same electrolyte, the monovalent ions are counterions. In this system (not shown), the overcompensation of surface charge as seen in Figure 7.2b does not occur. Since the 0.5 M 2:-1 bulk electrolyte solution has oscillatory decay, there are oscillations in the tail region outside the surface for the ion distributions and the electrostatic potential, but the distinct features of the profiles for $z \lesssim 3\, d^{\rm h}$ that we see for divalent counterions do not occur for monovalent ones in the present case. The overcompensation can primarily happen when the electrostatic ion-ion correlations are strong enough and the ion density in bulk is not too low. The borderline between sufficiently strong and too weak ion-ion correlation effects lies approximately between divalent and monovalent counterions when we deal with electric double-layers in aqueous electrolyte solutions at room temperature and moderately high electrolyte concentrations. This is entirely fortuitous and depends on the fact that the strength of the electrostatic interactions for divalent ions in water, where $\varepsilon_r \approx 80$, happens to have the right magnitude as compared to the thermal energy $k_B T$. In solvents with appreciably smaller ε_r, the electrostatic interactions are sufficiently strong also for monovalent ions at the same temperature.

Incidentally, we can note that there exist several different mechanisms for surface charge reversal for systems with ions of various valencies. Apart from the current mechanism, it can, for example, be a specific ("chemical") adsorption of counterions at the surface – an affinity between the ions and a surface of primarily non-electrostatic origin. The "chemical" nature of such interactions can have various origins like strong chemical bonding, various types of complexation, exchange of ligands or hydrophobic bonding. In contrast to these mechanisms, the current one is due to an inherent property of the electrolyte – the ion-ion correlations in the ion cloud – and not any specific surface-ion interaction.

Figure 7.3 gives an example of charge overcompensation in a double-layer system where the ion distributions in the tail region decay in an monotonic exponential fashion. The figure shows the ion density and charge distribution profiles for a 0.35 M electrolyte solution of a 2:-2 electrolyte in contact with a charged surface. The calculations have been done in the Anisotropic Reference HNC (ARHNC) approximation, which is introduced in Chapter 9 (in the 2nd volume). In this approximation, both density profiles and pair distribution functions are self-consistently calculated for the inhomogeneous fluid. It is a very accurate approximation for this kind of system.

The qualitative features of the distribution functions close to the surface in this case are the same as in Figure 7.2b. We will analyze Figure 7.3 and the ion-ion correlation mechanism for the overcompensation of surface charge in more detail in Section 7.2.2. As a preparation for this, let us focus on the differences in Figure 7.3a between the ion density profiles calculated in the PB approximation (dotted curves) and the accurate profiles (full and dashed curves).

In the interval $0.1 \lesssim z - d^S \lesssim 3.4\,$Å, the accurately calculated counterion density exceeds that from the PB approximation. It is the large amount of counterions there that causes the overcompensation of surface charge (the differences in coion density between the accurate calculation and the PB approximation are quite small in this range). As mentioned earlier, the reason for the excess of counterions is that they are strongly attracted towards the surface, more strongly than in the PB prediction. The force on a particle is, as we have

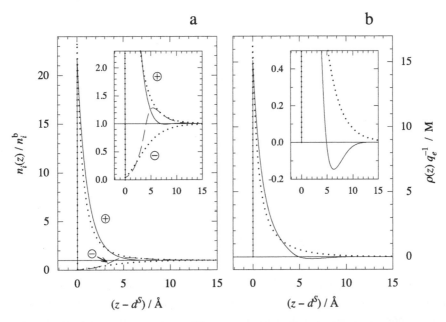

FIGURE 7.3 (a) Ion distribution profiles $g_i^{(1)}(z) = n_i(z)/n_i^b$ and (b) charge density divided by the elementary charge, $\rho(z)/q_e$, for a 0.35 M 2:-2 aqueous electrolyte solution in contact with a surface with surface charge density $\sigma^S = -0.160\,\mathrm{C\,m}^{-2}$ (one elementary charge per 100 Å²) at room temperature.[10] The ionic diameters are $d^h = 4.25$ Å and $d^S = d^h/2$. In this figure, the abscissa is equal to $z - d^S$, so zero lies at the point of closest approach of the ions to the surface. The full and dashed curves show results of integral equation theory (see text) and the dotted curves are the PB predictions for the same system. The counterion profiles are marked $\oplus$ and the coion profiles $\ominus$. The insets show the same curves on an expanded ordinate scale.

seen, related to the density profile by $\mathbf{F}(\mathbf{r}) = k_B T\,\nabla \ln n(\mathbf{r}) = k_B T\,[\nabla n(\mathbf{r})]/n(\mathbf{r})$ (Equation 5.12); in the present case, the force in the positive z direction on an ion of species i is given by $F_i(z) = k_B T\,d \ln n_i(z)/dz$. For the system in Figure 7.3, the absolute value of the attractive (negative) force $|F_+(z)|$ on counterions exceeds the PB prediction $|F_+^{PB}(z)|$ in the interval $1 \lesssim z - d^S \lesssim 6$ Å. This difference is crucial for the appearance of the overcompensation. In Section 7.2.2, we will investigate the reason for this difference and thereby the mechanism behind the overcompensation. Very close to the surface $|F_+^{PB}(z)|$ is larger than the accurate $|F_+(z)|$, which leads to a slightly larger value for the contact density $n_+(d^S)$ in the PB approximation.

The coion density is larger than the PB prediction for $z - d^S \gtrsim 2.2$ Å. Coions are repelled away from the surface, $F_-(z) > 0$, when $z - d^S \lesssim 5.4$ Å, whereby the repulsion is stronger than the PB prediction for most z values in this range. For $z - d^S \gtrsim 5.4$, the coions are instead attracted towards the surface, $F_-(z) < 0$. This is in contrast to the PB prediction which gives repulsion throughout.

[10] Figure 7.3 is based on data from R. Kjellander and H. Greberg, *J. Electroanal. Chem.* **450** (1998) 233; Errata: *ibid.* **462** (1999) 273.

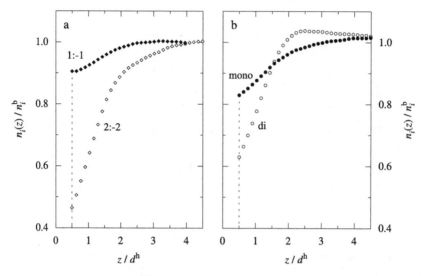

FIGURE 7.4 MC simulation results[11] for ion distribution profiles of various electrolyte solutions in contact with an uncharged surface ($\sigma^S = 0$) at room temperature: (a) for 0.5 M 1:-1 and 2:-2 electrolytes (filled and open diamonds, respectively) and (b) for a 0.5 M 2:-1 electrolyte (filled squares for monovalent ions and open squares for divalent ions). In (a), the profiles for anions and cations are identical. The ionic diameters are $d^h = 3.0$ Å and $\varepsilon_r = 80$ in all cases.

Next, we turn to another effect of ion-ion correlations, namely the appearance of ionic density distribution profiles outside an *inert uncharged* surface in contact with an electrolyte solution. The PB approximation fails to predict this since it gives $n_i(z) = n_i^b$ for all z outside an uncharged surface. Figure 7.4a shows $g_i^{(1)}(z) = n_i(z)/n_i^b$ for a 1:-1 and a 2:-2 electrolyte solution in the primitive model and we see that the ion density deviates from the bulk value near the surface, that is, $g_i^{(1)}$ deviates from 1. Since the anions and cations differ only by the sign of their charge, their density profiles are identical, $g_+^{(1)}(z) = g_-^{(1)}(z)$. This is a consequence of symmetry: if all negative and positive charges change their sign, the system remains the same because the surface is uncharged. The symmetry implies that the average charge density $\rho(z) \equiv 0$ for all z, which also follows from Equation 7.49.

Since the ion density goes down when the surface is approached from the bulk phase, the ions near the surface must experience forces that are directed away from the surface on average (to the right in the figure). To understand why, let us first consider an ion in the bulk phase where it is surrounded by a spherically symmetric ion cloud of opposite sign. The average force on the ion is then zero. For an ion close to a surface, on the other hand, the ion cloud is not spherically symmetric. Consider, for example, an ion in contact with the surface in one of the systems in the figure. The ion cloud of the ion is located solely on the side facing away from the surface (on its rhs) because no ions can be on the other side where the wall is located (on its lhs). Thus, the ion has an environment with a net charge of opposite sign on one side and no charge on the other side, and therefore it experiences

[11] Figure 7.4 is based on data from D. Henderson, D. Gillespie, T. Nagy, and D. Boda, *Mol. Phys.* **103** (2005) 2851.

an electrostatic force pulling it away from the surface. Thus, the ions are drawn away from the surface because of their correlations with the other ions; note that this is not a repulsion from the surface but instead an attraction towards the other ions. The force is stronger for a divalent ion compared to a monovalent one because the total charge of the ion cloud is twice as big for the former, which explains the difference seen in the figure. An ion a bit further away from the surface still has an unsymmetric ion cloud, but the asymmetry decreases with increasing distance from the wall and so does the average force.

In Figure 7.4b, the corresponding density distribution is shown for a 2:-1 electrolyte solution. Here $g_+^{(1)}(z) \neq g_-^{(1)}(z)$ since the ionic charges are different. Close to the surface, the ions are pulled away from it for the same reason as in the previous case and, consequently, the divalent ions are pulled away more strongly than the monovalent ones. Therefore, the charge density $\rho(z) \neq 0$ in this case. Close to the surface where the divalent ions have lower density than the monovalent ones, $\rho(z)$ must have the same sign as the latter. Due to electroneutrality, the integral of $\rho(z)$ must be zero, so there must be a region where $\rho(z)$ has the opposite sign. In the figure, this takes place for $z \gtrsim 1.5\, d^h$. This means that there is charge separation in the electrolyte phase – a kind of electric double-layer at the uncharged surface. The electrolyte solution at large z behaves like a solution near a charged surface with the same sign of charge as the monovalent ions.

Note that any asymmetry between anions and cations induces a charge separation in the electrolyte outside an uncharged surface, for example anions and cations of the same valency but of different sizes. Then one species can approach the surface closer than the other one, and in this case the sign of charge density near the surface will have the same sign at the small ions that can approach most closely.

7.1.4 Electric Double-Layers with Surface Polarizations (Image Charge Interactions)

So far, we have always assumed that the dielectric constant is the same inside the wall as it is in the electrolyte solution. In the present section, we will treat systems where it is different on either side of the surface. We will assume that the dielectric constant inside the wall is ε_r^W, while the solvent (modeled as a dielectric continuum) has dielectric constant ε_r as before. Thus, there is a **dielectric discontinuity** at the surface of the wall; the dielectric constant is ε_r^W behind and ε_r in front of the surface. This is, of course, a much better model for real systems because the material in the wall has different dielectric properties than the solvent.

When a point charge q is close to a dielectric discontinuity, it polarizes the surface, inducing a surface charge distribution there. Here we are interested in the situation when the charge is located on the solutions side of the surface, for example the charge of an ion. The electrostatic potential from the induced surface charge distribution happens to be the same as that from a point charge located behind the surface[12] [we here solely consider the potential on the solution side of the surface]. The latter charge, called an **image charge**, is located

---

[12] See, for example, Section 4.4 in J. D. Jackson, *Classical Electrodynamics* 3rd ed. (New York, Wiley, 1999).

at an equal distance from the surface as the original charge, but on the other side, and has a charge equal to $\alpha_w q$, where

$$\alpha_w = \frac{\varepsilon_r - \varepsilon_r^w}{\varepsilon_r + \varepsilon_r^w}. \tag{7.50}$$

An image charge is simply a mathematical construct used to calculate the potential from the induced surface charge distribution. The latter consists of real charges but the image charge is not a real charge.

If the charge q is located at $\mathbf{r}' = (x', y', z')$ and the image charge $\alpha_w q$ therefore is located at $\mathbf{r}'_{im} = (x', y', -z')$, the electrostatic potential due to q is consequently given by

$$\phi^{\{q\}}(\mathbf{r}; \mathbf{r}') = \frac{1}{4\pi\varepsilon_r\varepsilon_0} \left[\frac{q}{|\mathbf{r} - \mathbf{r}'|} + \frac{\alpha_w q}{|\mathbf{r} - \mathbf{r}'_{im}|} \right] \tag{7.51}$$

at position $\mathbf{r}$ on the solution side of the surface. When $\varepsilon_r^w < \varepsilon_r$, the factor α_w is positive and the presence of the dielectric continuum augments the potential from q, while the opposite is true when $\varepsilon_r^w > \varepsilon_r$. The formula for $\phi^{\{q\}}$ also applies to the case of a metal surface, in the case of which the image charge equals $-q$, so we have $\alpha_w = -1$, which corresponds to the choice $\varepsilon_r^w = \infty$.

The uniform surface charge density σ^S also has an image that gives rise to a contribution to the electrostatic potential. The image is located at $z = 0$ and has the value $\alpha_w \sigma^S$, so the total potential from σ^S and its image is equal to $-(1 + \alpha_w)\sigma^S z/(2\varepsilon_r\varepsilon_0)$ when $z > 0$.

The electrostatic part of the pair interaction potential between two ions of species i and j located at $\mathbf{r}$ and $\mathbf{r}'$, respectively, is given by

$$u_{ij}^{el}(\mathbf{r}, \mathbf{r}') = \frac{q_i q_j}{4\pi\varepsilon_r\varepsilon_0} \left[\frac{1}{|\mathbf{r} - \mathbf{r}'|} + \frac{\alpha_w}{|\mathbf{r} - \mathbf{r}'_{im}|} \right] = u_{ij}^{Coul}(|\mathbf{r} - \mathbf{r}'|) + u_{ij}^{im}(\mathbf{r}, \mathbf{r}'), \tag{7.52}$$

where $u_{ij}^{Coul}(r)$ is the usual $1/r$ Coulomb potential. The expression for the image charge potential $u_{ij}^{im}(\mathbf{r}, \mathbf{r}')$ is, in fact, symmetric in $\mathbf{r}$ and $\mathbf{r}'$ since $|\mathbf{r} - \mathbf{r}'_{im}| = [(x - x')^2 + (y - y')^2 + (z + z')^2]^{1/2}$. In addition, each ion interacts with its own image, a **self-image interaction**. This interaction is like an interaction with an external potential, but its origin is the polarization that the ion induces on the surface. We can include it in the electrostatic part of the external potential for an ion of species i, which becomes for $z > 0$

$$v_i^{el}(z) = -\frac{q_i(1 + \alpha_w)\sigma^S}{2\varepsilon_r\varepsilon_0} z + \frac{q_i^2}{4\pi\varepsilon_r\varepsilon_0} \cdot \frac{\alpha_w}{2z} \cdot \frac{1}{2}, \tag{7.53}$$

where $2z$ in the denominator of the last term is the distance between the center of the ion and its image charge. The last factor of $1/2$ in this term appears because the self-image interaction must, as we will see, be counted with only half the value compared to the interaction with an image of another ion.

An image interaction is a potential of mean force, that is, an excess chemical potential for the charge, which, in turn, is the electrostatic work required to bring the charge into its position relative to the surface (the free energy of insertion). To understand the factor of 1/2 for the self-image term, let us first consider the interaction of a charge q with the image of another charge q' placed fixed at some point $(\mathbf{r}')$. This image has the magnitude $\alpha_w q'$. To find the chemical potential for q, we can do a coupling parameter integration like we did when we treated chemical potential calculations in, for example, Section 5.7. Let us insert the charge q gradually at a fixed location $(\mathbf{r})$, so its magnitude is ξq where $0 \le \xi \le 1$. When we vary ξ by $d\xi$, the work done in the interaction with $\alpha_w q'$ is proportional[13] to $\alpha_w q' q d\xi$. The total work is hence proportional to $\int_0^1 \alpha_w q' q d\xi = \alpha_w q' q$. When we instead consider the interaction with the self-image of q, we must take into account that this image varies with ξ; its magnitude is $\alpha_w \xi q$. Therefore, we should replace $\alpha_w q'$ by $\alpha_w \xi q$, so the total work is proportional to $\int_0^1 \alpha_w \xi q q d\xi = [\int_0^1 \xi d\xi] \alpha_w q^2 = (1/2)\alpha_w q^2$. This explains the factor of 1/2 and shows the difference between the self-image interactions and the interactions with images of other charges.

An alternative manner to obtain the image interaction term in Equation 7.52 is the following. One then considers that both ions interact with the image of the other ion so one has two image terms, one with the distance $|\mathbf{r} - \mathbf{r}'_{im}|$ and the other with the distance $|\mathbf{r}' - \mathbf{r}_{im}|$. These two distances are the same so one has two identical image terms that lead to twice the value of one term, but in this case one has to do the coupling parameter integration for both ions, leading to a factor of 1/2 in front of each term. The image interaction in this case is a kind of "self-image" interaction for the *pair* of ions. Thus, the end result is the same as in Equation 7.52.

The self-image interaction term often gives rather important contributions, in particular close to the surface. When $\varepsilon_r^w < \varepsilon_r$, the self-image interaction is repulsive ($\alpha_w > 0$) and when $\varepsilon_r^w > \varepsilon_r$ it is attractive ($\alpha_w < 0$). Therefore, in the former case, the density profile close to the surface is usually lower than in the absence of a dielectric discontinuity, and in the latter case the density profile is higher. It is common to refer to the case of $\varepsilon_r^w < \varepsilon_r$ as "repulsive images" and $\varepsilon_r^w > \varepsilon_r$ as "attractive images," but this nomenclature only relates to the self-image interactions and not the interactions with the images of other ions. In particular for aqueous systems, where ε_r is large, image effects can be quite important when the medium behind the surface has low ε_r^w, which is very often the case. An interface to a gas phase is also a case of low ε_r^w; in practice, the same as for vacuum $\varepsilon_r^w = 1$. Ions are repelled from the immediate neighborhood of such an interface. The image interactions can also be quite important for the case of a metallic surface, for example an electrode. In this case, $\alpha_w = -1$, which is obtained if we set $\varepsilon_r^w = \infty$, and the self-image interactions contribute with attractive forces on the ions towards the surface.

In order to be able to treat the effects of image charges, it is necessary to use a theory that includes ion-ion correlations. Therefore, one *cannot* use the PB approximation to treat these effects – at least without making some extension of it, where effects of the correlations are included. To understand why, we will show in the shaded text box below that the effects of image charge interactions vanish in the PB approximation in planar geometry. We will also see why screening via ion-ion correlations is essential.

---

[13] The proportionality constant equals $1/(4\pi\varepsilon_r\varepsilon_0 |\mathbf{r} - \mathbf{r}'_{im}|)$, which is independent of ξ.

Let us consider the potential of mean force for ion species i in the PB theory $w_i^{PB}(z) = q_i\psi(z)$. In the absence of a dielectric discontinuity, the mean potential equals (Equation 7.1)

$$\psi(z) = -\frac{\sigma^S}{2\varepsilon_r\varepsilon_0}|z| - \int dz'\rho(z')\frac{|z-z'|}{2\varepsilon_r\varepsilon_0} + \mathcal{K}_\infty, \tag{7.54}$$

where $\mathcal{K}_\infty$ is selected such that $\psi(z) \to 0$ when $z \to \infty$. In the presence of the discontinuity, both σ^S and $\rho(z')$ have images that are located at $z' \le 0$. The image of the latter is $\alpha_w\rho(-z')$ and the mean potential becomes for $z > 0$

$$\psi(z) = -\frac{(1+\alpha_w)\sigma^S}{2\varepsilon_r\varepsilon_0}z - \int_{d^S}^{\infty} dz'\rho(z')\frac{|z-z'|}{2\varepsilon_r\varepsilon_0} - \int_{-\infty}^{-d^S} dz'\alpha_w\rho(-z')\frac{(z-z')}{2\varepsilon_r\varepsilon_0} + \mathcal{K}'_\infty, \tag{7.55}$$

where $\mathcal{K}'_\infty$ is a new constant selected such that $\psi(z) \to 0$ at infinity. Intuitively, one can realize that the images of σ^S and $\rho(z')$ can only contribute with a constant to $\psi(z)$ for $z > 0$ since they constitute planar and laterally uniform charge distributions with a total charge of zero (due to electroneutrality). The rest of the potential is the same as in Equation 7.54, which hence is still valid with a proper selection of $\mathcal{K}_\infty$. Formally, one can show this fact in the following manner.

The last integral in Equation 7.55 (without the minus sign in front) can after substitution $z' = -z''$ be written as

$$\frac{\alpha_w}{2\varepsilon_r\varepsilon_0}\left[z\int_{d^S}^{\infty} dz''\rho(z'') + \int_{d^S}^{\infty} dz''\rho(z'')z''\right] = \frac{\alpha_w}{2\varepsilon_r\varepsilon_0}\left[-z\sigma^S + \mathcal{C}'\right],$$

where we in the rhs have used electroneutrality for the first integral and where $\mathcal{C}'$ is the value of the last integral, which is independent of z. By inserting this into Equation 7.55, we obtain the same result as in Equation 7.54 since $\mathcal{K}'_\infty = \mathcal{K}_\infty + \alpha_w\mathcal{C}'/(2\varepsilon_r\varepsilon_0)$. Thus, all image interaction contributions *vanish completely* from $\psi(z)$ and therefore from $w_i^{PB}(z)$.

One might be tempted to add the self-image interaction term in Equation 7.53 to $w_i^{PB}(z)$ in order to include an effect of image interactions. Thereby, we have $n_i(z) = n_i^b\exp(-\beta[q_i\psi(z) + Bq_i^2/z])$, where $B = \alpha_w/(16\pi\varepsilon_r\varepsilon_0)$. Let us consider a symmetric electrolyte with $q_+ = -q_- = q$ and $n_+^b = n_-^b = n^b$. The average charge density is

$$\rho(z) = q_+ n_+(z) + q_- n_-(z) = qn^b \left[e^{-\beta[q\psi(z)+Bq^2/z]} - e^{-\beta[-q\psi(z)+Bq^2/z]} \right]$$

$$= qn^b \left[e^{-\beta q\psi(z)} - e^{+\beta q\psi(z)} \right] e^{-\beta Bq^2/z} = -2qn^b \sinh[\beta q\psi(z)] e^{-\beta Bq^2/z}.$$

For large z where $\psi(z)$ is small, we have $\sinh[\beta q\psi(z)] \approx \beta q\psi(z)$ and $\exp(-\beta Bq_i^2/z) \approx 1$. We obtain $\rho(z) \approx -2\beta nq^2 \psi(z)$ and the Poisson equation yields $d^2\psi(z)/dz^2 \approx \kappa_D^2 \psi(z)$, which is the same as in the linearized PB approximation. Therefore, $\psi(z)$ decays exponentially fast for large z like in the usual PB case.

For the density distribution, we obtain by expanding the exponential function

$$n_i(z) = n_i^b \left(1 - \beta \left[q_i \psi(z) + \frac{Bq_i^2}{z} \right] + \cdots \right)$$

and since $\psi(z)$ decays exponentially fast, we see that $n_i(z) - n_i^b \sim -\beta n_i^b Bq_i^2/z$ for large z. The excess number of ions per unit area in the profile (in excess to the bulk) is given by the integral $\int_{dS}^\infty [n_i(z) - n_i^b] dz$, which is infinitely large since the integral of $1/z$ diverges logarithmically at infinity. The inclusion of the self-image interaction term in this manner accordingly leads to unphysical results.

The problem is that the self-image interaction term in itself constitutes an *unscreened* Coulomb interaction potential that decays like $1/z$ when z increases. In an electrolyte, the electrostatic interactions are screened due to the correlations between the ions. The screening of the self-image potential of an ion arises from the instantaneous interaction with all other ions and their images, averaged over all configurations of the ions. In the PB mean field approximation, the images of all other ions are "smeared out" (i.e., they are given by the image of the average charge density $\rho(z)$ rather than the instantaneous distribution) and, as we saw earlier, their contributions therefore vanish completely leading to the unscreened self-image interactions.

To correctly treat the image charge interactions, one needs theories that handle pair correlations properly in inhomogeneous fluids. Some of these will be introduced in Chapter 9. Alternatively, one can do computer simulations for the double-layer system, using electrostatic interaction potentials with image interactions as given in Equations 7.52 and 7.53.

Quite typical ion distribution profiles for double layers in the presence of image charge interactions are displayed in Figure 7.5, which shows the cases with $\varepsilon_r^w = 1$ (vacuum) and $\varepsilon_r^w = \infty$ (metal) behind the surface together with the case of $\varepsilon_r^w = \varepsilon_r$ (no image charges). The corresponding values of α_w are 0.975 (vacuum), -1 (metal) and 0 (no images); cases with other ε_r^w values will have α_w between the two extremes.

For the 1:-1 electrolyte in Figure 7.5a, the effects of image interactions are quite strong close to the surface. For attractive images, $\varepsilon_r^w = \infty$, the densities of both coions and counterions there are higher than in the absence of image interactions and for repulsive images,

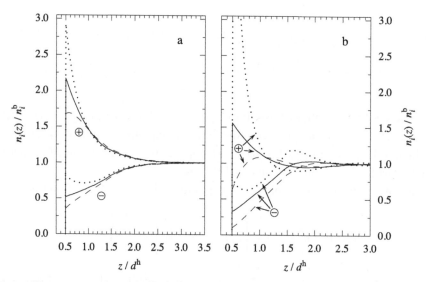

FIGURE 7.5 Illustration of image charge interaction effects on ion distribution profiles for 1.0 M aqueous solutions of (a) 1:-1 and (b) 2:-2 electrolytes in contact with a charged surface that separates regions with different dielectric constants, ε_r^w behind the surface and ε_r in the solution phase.[14] Dotted curves: $\varepsilon_r^w = \infty$ corresponding to a metallic wall; full curves: $\varepsilon_r^w = \varepsilon_r$ (no image charges); dashed curves: $\varepsilon_r^w = 1$ corresponding to vacuum behind the surface. The results have been obtained by MC simulations and we have $\sigma^S = -0.044 \, \text{C m}^{-2}$ (one elementary charge per 360 Å^2), $T = 298.15$ K, $\varepsilon_r = 78.5$, $d^h = 4.25$ Å and $d^S = d^h/2$.

$\varepsilon_r^w = 1$, their densities are lower. Further away from the surface, the influence from the images is quite low. The electrostatic potential from the images,[15] both the self-image and the images of other ions, are screened by the electrolyte – like the potential from any other charge – so they have a limited range. Since the images affect the electrostatic potential due to an ion (they increase or reduce it depending on the sign of α_w), they also influence the efficiency of the screening in the electrolyte close to the surface. Quite generally, the image charge effects are larger when the ion density is low compared to when it is high, which is a consequence of the fact that the image charge potential is more effectively screened in the latter case. Furthermore, the effects of images are quite generally larger when σ^S is low compared to when σ^S is high.

For the 2:-2 electrolyte in Figure 7.5b, the system has an overcompensation of the surface charge like in Figures 7.2b and 7.3. Compared to the 1:-1 electrolyte in Figure 7.5a, the effects of image interactions are much larger in this case, which is a consequence of the fact that the ionic charges are twice as large. The image interaction effects are strong close to the surface and they are qualitatively similar there to those in the 1:-1 case, but for the 2:-2 electrolyte the images affect the density distributions further away from the surface. This is primarily

---

[14] Figure 7.5 is based on data from A. Alawneh and D. Henderson, *Mol. Simul.* **33** (2007) 541.

[15] In reality, this is the electrostatic potential from the induced charge density at the surface, which is represented by the image potential.

due to the fact that the attractive images enhance the overcompensation of charge near the surface while repulsive images reduce it, as can be seen in the figure.

7.1.5 Electric Double-Layers with Dispersion Interactions

Ions and other particles in electrolyte systems have in reality other types of interactions in addition to the electrostatic and repulsive short-range ones. This is an important feature since it gives an element that is often crucial in applications, namely the fact that simple anions and cations differ by considerably more than their sizes and charges. Here we will consider the effects of dispersion interactions, which originate from the polarizability of the ions, other particles and any walls. In the previous section we included image charge effects, which arise from the different dielectric constants of the solvent and the walls. The dielectric constant of a medium depends on its polarizability at low frequencies. The dispersion interactions depends the polarization properties at high frequencies – primarily due to the electronic degrees of freedom. Here we will consider ion-ion and ion-wall or ion-macroparticle dispersion interactions. The strengths of these interactions depend on polarizabilities of the ions, the solvent and the walls/macroparticles.

Simple anions like halide ions have an electron cloud that is more spread-out and therefore more polarizable than the electron cloud of simple cations like alkali metal ions. For the latter, the electrons are located closer to the nucleus. Therefore, simple anions have in general larger dispersion interactions than simple cations. For example, iodide ions are more polarizable than bromide ions and bromine ions are more polarizable than chloride ions. For cations like sodium and potassium ions, the polarizability is considerably smaller and so is the difference between the cations.

By incorporating effects of polarizability of ions in the model for the electrolyte, one has included an important reason for ion specificity. Depending on the kind of anion and to a lesser extent the kind of cation, one has different strengths of interaction between the ions and various macroparticles. This influences, for example, the amount of ions of different species that become located near the particle surface. The same applies for other kinds of interfaces like the air/solution interface. The dispersion interactions between the ions of various kinds also influence the properties of bulk electrolytes significantly.

The dispersion interactions between simple ions give rise to a contribution to the pair potential that decays like $1/r^6$ for large r, like the corresponding contribution to the Lennard-Jones potential in Equation 5.6.[16] When modeling the electrolyte one can therefore use an approximate pair potential

$$u_{ij}(r) = u_{ij}^{el}(r) + u_{ij}^{sh}(r) + u_{ij}^{disp}(r),$$

where

$$u_{ij}^{disp}(r) = \frac{b_{ij}^{disp}}{r^6} \tag{7.56}$$

---

[16] The $1/r^6$ decay for dispersion interactions holds provided r is not very large, so one can neglect an effect called *retardation* which is caused by the finite speed of light (see footnote 26 in Chapter 8). Retardation causes the interaction to decay somewhat faster than $1/r^6$ with distance. In the applications considered here, this effect is entirely neglected.

and b_{ij}^{disp} is a constant that depends on the excess polarizability of the ions of species i and j (i.e., in excess of the solvent polarizability). The short-range potential $u_{ij}^{\text{sh}}(r)$ is usually set equal to a hard core potential.

Ion-wall dispersion interactions give a term in the external potential v_i between a wall and an ion of species i. For a planar wall, this term is approximately given by[17]

$$v_i^{\text{disp}}(\ell) = \frac{b_i^{\text{disp}}}{\ell^3}, \tag{7.57}$$

where ℓ is the distance from the surface and b_i^{disp} is a constant that depends on the wall medium and the excess polarizability of the ion. Note that $\ell \geq d^{\mathcal{S}}$, so the unphysical singularity at $\ell = 0$ is avoided. The distance dependence of the dispersion interaction depends on the geometry. In spherical geometry, for example, the interaction between a spherical macroparticle and a simple ion, we have the approximate expression[18]

$$v_i^{\text{disp}}(r^{\mathcal{S}}) = \frac{b_i^{\text{disp}}}{(r^{\mathcal{S}})^3 [1 + (r^{\mathcal{S}}/d_{\text{p}})]^3}, \tag{7.58}$$

where $r^{\mathcal{S}}$ is the distance between the ion center and the surface of the macroparticle and d_{p} is the sphere diameter of the latter. We here regard the interactions with the macroparticle as an external potential for the electrolyte.

Figure 7.6 shows some examples of the ionic density profiles outside the surface of a single spherical macroparticle immersed in an aqueous electrolyte solution obtained by computer simulation. In frames a and b, the macroparticle is positively charged, and in c and d it is negatively charged. The dispersion interactions are *attractive* since b_i^{disp} is negative. No image charge interactions are included.

Let us start with the cases of monovalent ions, frames a and c. The curves shown as crosses are the counterion and coion profiles in the absence of dispersion interactions, that is, for nonpolarizable ions. By comparing with the other curves (NaCl, NaBr and NaI), we can clearly see that the extra attraction to the macroparticle from the dispersion interactions has a substantial effect on the anion profiles (counterions in frame a and coions in c), while the influence on the Na$^+$ profiles is much smaller since the dispersion interactions are weaker for cations. The largest effect is found for the iodide ions (blue filled triangles), in agreement with the fact that they are more polarizable than the other ions. When the iodide ions are counterions (frame a), the dispersion interactions and the electrostatic interactions to the macroparticle charge both attract these ions towards the surface and the iodide concentration is much larger than for the case of nonpolarizable ions. For $r^{\mathcal{S}} \gtrsim 3.3$ Å, on the

---

[17] The $1/\ell^3$ decay of the dispersion interaction due to a planar wall can be obtained in the following manner, where we use a somewhat simplified argument. Consider a wall located with its surface at $z = 0$ and a particle at distance z' from the surface (at point $\mathbf{r}'$). The distance R between the particle and a point $\mathbf{r}$ inside the wall (at $z < 0$) is $R = [(z' - z)^2 + s^2]^{1/2}$, where $s = |\mathbf{s}|$ and $\mathbf{s} = (x - x', y - y')$. The total $1/r^6$ interaction between the particle and the wall is $\int_{z<0} d\mathbf{r} \, R^{-6} = \int_{-\infty}^{0} dz \int d\mathbf{s} \, [(z' - z)^2 + s^2]^{-3} = \pi (z')^{-3}/6$, which is proportional to $1/(z')^3$. By setting $z' = \ell$, we obtain the final result.

[18] This formula can be obtained in an analogous manner as the derivation in footnote 17 by doing an integration of $1/R^6$ in spherical geometry.

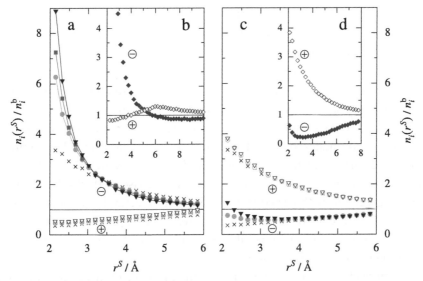

FIGURE 7.6 (**See color insert.**) An example of the influence of dispersion interactions on ion distribution profiles for different aqueous electrolyte solutions around a charged spherical macroparticle with diameter 30 Å calculated by MC simulations.[19] The profiles are plotted as $n_i(r^S)/n_i^b$, where r^S is the distance between an ion center and the particle surface. In a) and b), the macroparticle has a positive charge, $+20q_e$, and in c) and d) it has a negative charge, $-20q_e$. The electrolyte solutions have ionic strength 1.0 M and contain various 1:-1 salts in the main figures (panes a and c) and a 1:-2 salt in the insets (panes b and d): NaI (blue triangles), NaBr (red squares), NaCl (green circles), nonpolarizable monovalent ions (crosses) and Na_2SO_4 (brown diamonds). Filled symbols denote anions and open symbols denote cations. The lines are just guides for the eye. All ions are modeled as charged spheres with a diameter of 4 Å.

other hand, the iodide profile lies below the curve for the nonpolarizable case. This happens because the large amount of iodide ions close to the surface partially neutralizes the surface charge, which decreases the electrostatic interactions for ions further out from the surface: a decrease that makes the total attractive potential for iodide ions weaker there despite the presence of the dispersion attraction.

In frame c where the iodide ions are coions, they are repelled by the macroparticle charge, but the attractive dispersion interactions make the iodide profile to rise somewhat when the surface is approached. The chloride ion profiles (green filled circles) show the same behaviors in all cases, but to a lesser degree.

To conclude, we see that there is an appreciable ion specificity for the simple monovalent anions due to the dispersion interactions and little specificity for the simple monovalent cations. The effect for the anions is larger when the macroparticle is positive than when it is negative. The results in the figure have been obtained by simulations, but the PB approximation with added dispersion interactions gives very similar results[20] for these monovalent

[19] Figure 7.6 is based on data from M. Boström, F. W. Tavares, D. Bratko, and B. W. Ninham, *J. Phys. Chem. B* **109** (2005) 24489.

[20] M. Boström et al., *loc. cit.* in footnote 19.

systems. In this PB approach, the potential of mean force for the ions of species i is approximated as $w_i(r^S) \approx q_i \psi(r^S) + v_i^{\text{disp}}(r^S)$, where $\psi(r^S)$ is the mean electrostatic potential outside the macroparticle, $q_i \psi(r^S)$ is the contribution from electrostatic interactions and $v_i^{\text{disp}}(r^S)$ is the contribution from ion-macroparticle dispersion interactions.

The corresponding simulation results for a 1:-2 salt, Na_2SO_4, are shown in frames b and d of Figure 7.6. When the divalent ions (SO_4^{2-}) are counterions (frame b), the profiles show the typical charge inversion in the profile due to ion-ion correlations that we saw in Figure 7.2b. This correlation effect dominates, but the dispersion attraction adds to the large amount of counterions near the surface. In frame d where the divalent ions are coions, the effect of the attractive dispersion interactions is clearly visible in the rise of the SO_4^{2-} profile close to the surface. In agreement with the conclusions in Section 7.1.3, the importance of ion-ion correlations in the presence of divalent ions makes the PB approximation inadequate, so more advanced theories or simulations (like here) are needed to obtain correct results, at least when the divalent ions are counterions.

In addition to these effects close to the surfaces, the presence of dispersion interactions has a substantial influence on the decay of the density profiles and the electrostatic potential for large distances. This effect appears already in the PB approximation, so we will turn to that approximation as a simple illustration. We will do the analysis in *planar geometry* using the potential of mean force

$$w_i(z) = q_i \psi(z) + v_i^{\text{disp}}(z) \quad \text{(PB)} \tag{7.59}$$

with the dispersion interaction given by Equation 7.57. Furthermore, we restrict ourselves to binary $z : -z$ electrolytes, so we have $n_+^b = n_-^b \equiv n^b$, $q_+ = -q_- \equiv q$ and there are two coefficients b_+^{disp} and b_-^{disp}. The details of the analysis are given in Appendix 7B; here we will give a summary of the results. The Appendix also includes a treatment of a confined electrolyte between two surfaces. This will be of particular interest for double-layer interactions in the presence of dispersion interactions that are dealt with in Section 8.6.3.

It is shown in the Appendix that in the linearized PB approximation we have outside a single surface (see Equation 7.77)

$$\psi(z) = \psi_1(z) + \psi_2(z), \tag{7.60}$$

where $\psi_1(z) = C_1 e^{-\kappa_D z}$ and

$$\psi_2(z) \sim -\frac{1}{z^3} \cdot \frac{b_+^{\text{disp}} - b_-^{\text{disp}}}{2q} \quad \text{when } z \to \infty \quad \text{(linearized PB)}. \tag{7.61}$$

The constant C_1 in $\psi_1(z)$ can be determined from the surface charge density σ^S by using the boundary condition at the surface expressed by Equation 7.8 (see Appendix 7B). The term $\psi_1(z)$ corresponds to the electrostatic potential given by Equation 7.13 in the absence of the dispersion interactions ($b_+^{\text{disp}} = b_-^{\text{disp}} = 0$).

The second term $\psi_2(z)$ in the electrostatic potential originates from the ion-wall disper-sion interactions. It is independent of the surface charge density and other details of the surface. The appearance of this contribution shows that *the electrostatic interactions in the electrolyte are affected by the dispersion interactions*. Electrostatics cannot be treated inde-pendently from the dispersion interactions since the latter makes the electrostatic potential to decay like a power law (proportional to $1/z^3$) – a decay that dominates for large z since the exponential term $\psi_1(z)$ decays faster to zero. From the Poisson equation, it follows that the potential $\psi_2(z)$ is associated with a charge density in the electrolyte outside the surface given by

$$\rho_2(z) = -\varepsilon_r\varepsilon_0\frac{d^2\psi_2(z)}{dz^2} \sim \frac{12\varepsilon_r\varepsilon_0}{z^5} \cdot \frac{b_+^{\mathrm{disp}} - b_-^{\mathrm{disp}}}{2q} \quad \text{when } z \to \infty,$$

that also decays like a power law. The ion density profile $n_i(z)$ decays like (see Equation 7.79)

$$\frac{n_i(z) - n^{\mathrm{b}}}{n^{\mathrm{b}}} \equiv h_i^{(1)}(z) \sim -\frac{\beta}{z^3} \cdot \frac{b_+^{\mathrm{disp}} + b_-^{\mathrm{disp}}}{2} \quad \text{when } z \to \infty.$$

Note that this leading term is independent of i and is therefore exactly the same for anions and cations. The next order term, which decays like $1/z^5$, is different for the two species (shown in Equation 7.79). The charge density $\rho(z) = \sum_i q_i n_i(z)$ outside the wall decays like $\rho(z) \sim \rho_2(z)$,[21] that is, like $1/z^5$.

How can we understand the origin of $\psi_2(z)$ and $\rho_2(z)$? The dispersion interactions are not screened by the electrolyte, and therefore the ions outside the wall – including those that that are quite far from its surface – experience them. When $b_+^{\mathrm{disp}} > b_-^{\mathrm{disp}}$, the cations are more strongly repelled from (or less strongly attracted to) the wall by dispersion interactions than the anions. Even far from the surface, this gives rise to a charge separation where, on average, cations are brought somewhat further from the surface than anions: for any z' far from the surface, there is an excess of cations over anions for $z > z'$ compared to the bulk densities. In the present case, the excess is given by $\rho_2(z)$, which is positive for large z when $b_+^{\mathrm{disp}} > b_-^{\mathrm{disp}}$. This separation of charges gives rise to an electrostatic potential that is negative, which can be understood as follows. A positive charge located at z' is repelled by the excess positive charge at $z > z'$ and attracted towards a corresponding excess negative charge at $z < z'$, which is equally large (counted as total charge per unit area) because of electroneutrality. Furthermore, we can understand why the polarization caused by dispersion interactions is not affected by the details of the surface. The potential due to σ^S decays exponentially fast, as expressed in $\psi_1(z)$, and gives a negligible contribution to the potential far from the surface. The only things that matters for $\psi_2(z)$ at large z are the strength of the dispersion interactions and the ionic valency.

In the PB approach, the electrostatic interactions are treated in the mean field approxi-mation and only ion-wall dispersion interactions are considered. When ion-ion correlations

---

[21] This can be obtained from the $1/z^5$ term in Equation 7.79. The $1/z^3$ contribution to $n_i(z)$ vanishes in $\rho(z)$.

and ion-ion dispersion interactions are included, the coupling between dispersion and electrostatic interactions becomes more involved, but several of the main features seen in the PB approximation remain as explained in the shaded text box below.

Ion-ion dispersion and electrostatic interactions. The power-law decay of the electrostatic interactions is, in fact, a fundamental feature of fluids of charged particles in the presence of interparticle dispersion interactions. For small electroneutral particles, the dispersion interactions are an effect of the fluctuations in charge distribution due to, for example, the motion of electrons in atoms and molecules – basically a quantum phenomenon. This leads to the $1/r^6$ decay of their interactions.

In a quantum statistical mechanics analysis[22] of fluids that consist of particles that interact *solely* by Coulombic $1/r$ interactions,[23] it has been found that the electrostatic potential and the charge distribution around each particle *decay like power-laws* rather than in an exponential fashion as predicted by classical statistical mechanics. The same is true for decay of the pair correlations. Note that no dispersion interactions are added; they arise naturally in quantum mechanics.

More precisely, for a quantum system of charged particles in bulk, the pair correlation function $h_{ij}(r)$ and the density-density correlations decay like $1/r^6$. This $1/r^6$ decay is an effect of quantum fluctuations, essentially the same kind of mechanism that causes the dispersion interaction for electroneutral particles. Furthermore, these fluctuations make the charge distribution around a particle and the charge-density correlations decay like $1/r^8$, which implies via the Poisson equation that the electrostatic potential due to a particle decays like $1/r^6$. Finally, the charge-charge correlations decay like $1/r^{10}$ in the quantum case. Thus, the power-law decay of the electrostatic potential and the dispersion interactions are fundamentally two aspects of the same underlying physics. There are also exponentially decaying contributions, but they are dominated by the power-law contributions for large distances.

In classical statistical mechanics, the dispersion interactions are not inherent in the fundamental treatment of charge fluctuations, so one has to add these interactions as a kind of "effective" $1/r^6$ potential. As mentioned earlier, $1/r^6$ decay of the potential in spherical geometry translates into a $1/z^3$ decay for planar walls.

For electrolytes consisting of small ions that interact via both electrostatic and dispersion interactions, the latter can accordingly be included as a $1/r^6$ interaction potential in the classical treatment, so we have $u_{ij}(r) = u^{\text{sh}}(r) + q_i q_j/(4\pi\varepsilon_r\varepsilon_0 r) + b_{ij}^{\text{disp}}/r^6$, where the coefficient b_{ij}^{disp} depends on the excess polarizability of the ions of species i and j.

A classical statistical mechanical analysis of ion-ion correlations in bulk electrolytes,[24] where dispersion interactions between the ions are included, yields, in

[22] F. Cornu, *Phys. Rev. Lett.* **78** (1997) 1464; D. C. Brydges and Ph. A. Martin, *J. Phys. IV France* **10** (2000) Pr 5–53.

[23] The interaction potential used in the analysis has the divergence at $r = 0$ removed.

[24] R. Kjellander and B. Forsberg, *J. Phys. A: Math. Gen.*, **38** (2005) 5405; R. Kjellander *ibid.* **39** (2006) 4631.

fact, the same decay behavior as the quantum treatment: the pair correlation function decays like $1/r^6$, the charge-density correlations decay like $1/r^8$ and charge-charge correlations decay like $1/r^{10}$. For binary $z : -z$ electrolytes in bulk, we have the exact asymptotic results

$$h_{ij}^{(2)}(r) \sim -\frac{\beta \mathcal{K}_T^2}{r^6} b_{NN}^{disp}$$

$$\psi_i(r) \sim -\frac{\mathcal{K}_T}{r^6} \cdot \frac{\mathcal{B}_{QN}^{disp}}{q_Q} \qquad (7.62)$$

$$\rho_i(r) \sim \frac{30\varepsilon_r\varepsilon_0 \mathcal{K}_T}{r^8} \cdot \frac{\mathcal{B}_{QN}^{disp}}{q_Q}$$

where $\mathcal{K}_T = k_B T n_{tot}^b \chi_T$, χ_T is the isothermal compressibility (defined in Equation 2.131), $\mathcal{B}_{QN}^{disp} = b_{QN}^{disp} + \alpha_{QN} b_{NN}^{disp}$,

$$b_{NN}^{disp} = \frac{1}{4}\left[b_{++}^{disp} + 2b_{+-}^{disp} + b_{--}^{disp} \right]$$

$$b_{QN}^{disp} = \frac{1}{4}\left[b_{++}^{disp} - b_{--}^{disp} \right],$$

$q_Q = (q_+ + |q_-|)/2$ and α_{QN} is a state-dependent constant that contains some integrals of correlation functions. For a general binary electrolyte, we have the same results, except that

$$b_{NN}^{disp} = (x_+^b)^2 b_{++}^{disp} + 2x_+^b x_-^b b_{+-}^{disp} + (x_-^b)^2 b_{--}^{disp}$$

$$b_{QN}^{disp} = \frac{1}{2}\left[x_+^b b_{++}^{disp} + (x_-^b - x_+^b)b_{+-}^{disp} - x_-^b b_{--}^{disp} \right],$$

where $x_i^b = n_i^b / n_{tot}^b$.

In addition to the terms with power-law decay, there are exponentially decaying terms in $h_{ij}^{(2)}$, ψ_i and ρ_i like those in the absence of dispersion interactions. The exponentially decaying terms are, however, dominated by the power-law contributions for large distances.

Note that the structure of the formulas in Equation 7.62 is very similar to the structure of the equations obtained earlier for $h_i^{(1)}(z)$, $\psi_2(z)$ and $\rho_2(z)$, which are valid outside a planar wall in the linearized PB approximation. A major difference in the present case is the occurrence of $\mathcal{K}_T$, which depends on the isothermal compressibility of the fluid. This factor is not present in the PB results, where we have neglected ion-ion correlations and dispersion interactions between the ions.

The leading asymptotic decay for $h_{ij}^{(2)}(r)$ in Equation 7.62 is, in fact, the same as for a fluid of electroneutral particles that interact via $1/r^6$ interaction potentials, like Lennard-Jones fluids. These matters will be dealt with in more detail in the second volume of this treatise.

7.2 STRUCTURE OF INHOMOGENEOUS FLUIDS ON THE PAIR DISTRIBUTION LEVEL

When one talks about the *structure* of a homogeneous, isotropic fluid, one usually means the pair distribution function, that is, the average structure around each constituent particle of the fluid. More rarely, one is concerned with triplet or higher order distribution functions that describe the structure around two or more particles, although they are also aspects of the fluid structure. For inhomogeneous fluids, the primary aspect of the fluid structure is the number density distribution $n(\mathbf{r})$ – in planar geometry, the density profile $n(z)$ – while the structure on the pair distribution level is much less commonly treated. A huge amount of effort has been spent on the determination of the density profiles for various systems, while the pair distribution structure – the intrinsic fluid structure – has been much less investigated despite the fact that the latter constitutes the primary aspect of fluid structure in the homogeneous case. The main reasons for this are that several important aspects of inhomogeneous fluids primarily concern the density distribution $n(\mathbf{r})$ and that the latter is far less complicated to obtain both theoretically and experimentally than pair distributions.

In our discussion in Section 5.3 on how the equilibrium number density distribution $n(\mathbf{r})$ is set up, we saw that the force balance on each microscopic volume element of the fluid plays a fundamental role.[25] We found that the average force $\mathbf{F}(\mathbf{r})$ on each particle when located at $\mathbf{r}$ is related to the density distribution $n(\mathbf{r})$ by the relationship $\mathbf{F}(\mathbf{r}) = k_B T \nabla \ln n(\mathbf{r}) = k_B T [\nabla n(\mathbf{r})]/n(\mathbf{r})$ (Equation 5.12). This force is caused by the external potential $v(\mathbf{r})$ and the interactions with the surrounding particles, $\mathbf{F}(\mathbf{r}) = -\nabla v(\mathbf{r}) + \mathbf{F}^{\text{intr}}(\mathbf{r})$, where the intrinsic force arises from the latter interactions. We also saw that $\mathbf{F}^{\text{intr}}(\mathbf{r})$ depends on the correlations between the particles in the fluid and that it can be obtained from the pair distributions. This is a key feature of the hierarchies of distribution function dealt with in Sections 5.9 and 5.11. The force balance requirement can be dealt with in a quantitative manner via the Born-Green-Yvon equations, where the first BGY Equation 5.118 gives a relationship between $n(\mathbf{r})$ and the pair distributions via the expression (5.117) for the intrinsic force $\mathbf{F}^{\text{intr}}(\mathbf{r})$.[26] A knowledge of why $\mathbf{F}^{\text{intr}}(\mathbf{r})$ and hence $n(\mathbf{r})$ behave as they do gives important insights into the mechanisms in action in the fluid – mechanisms that determine the properties of the fluid.

Thus, we have important reasons to investigate the structure of inhomogeneous fluids on the pair distribution level. Furthermore, pair distribution functions are interesting in themselves as they constitute the intrinsic structure of the inhomogeneous fluid. The differences between the pair distributions in a bulk fluid and in the same fluid when it is exposed to an external field give fascinating insights into the basic features of the fluid.

---

[25] This is equivalent to the minimization of the free energy of the system, which is another manner to describe fundamental aspects of how the density distribution $n(\mathbf{r})$ is set up.

[26] There are several other ways to set up exact relationships between singlet density distributions and pair correlation functions of various kinds. Some of these will be derived in Part III of this treatise (in the second volume).

7.2.1 Inhomogeneous Simple Fluids

7.2.1.1 Lennard-Jones Fluids

We start with an inhomogeneous Lennard-Jones (LJ) fluid.[27] The fluid is located in the slit between two walls and is in equilibrium with a bulk fluid of density n^b. We select a coordinate system with the origin at the midplane between the walls and with the z axis perpendicular to the surfaces. In the examples we will discuss in this section, each wall consists of a solid consisting of close-packed LJ particles, but a simplified model is used for the wall-fluid inter-action where the wall structure is smoothed out in the lateral direction. The model potential, called the Steele potential,[28] therefore depends only on z and is given by

$$v^{St}(z) = \varepsilon^{LJ}\alpha_1\left[\frac{2}{5}\left(\frac{d^{LJ}}{z}\right)^{10} - \left(\frac{d^{LJ}}{z}\right)^4 - \frac{\alpha_2}{(z+\alpha_3)^3}\right],$$

where α_1, α_2 and α_3 are parameters that depend on the particle arrangement inside the wall.[29] This potential is strongly repulsive for small z, changes sign when z is increased, goes through a minimum and finally decays to zero for large z. The use of the Steele poten-tial instead of a potential from discrete particles in the wall gives only a small difference in density profile unless the slit is very narrow.

The external potential of the fluid due to the two walls is

$$v(z) = v^{St}\left(\frac{D^S}{2} + z\right) + v^{St}\left(\frac{D^S}{2} - z\right),$$

where D^S is the wall separation, which we define as the distance between the points of infi-nite wall-particle potential at either wall. Since $v^{St}(z)$ is strongly repulsive for small z, the particles cannot approach each wall closely in practice and the fluid film thickness in the slit is smaller than D^S. In the examples we will consider here, film thickness is defined as[30] $D^S - 0.69d^{LJ}$.

We have already encountered the density profile $n(z)$ of an LJ fluid in Figure 5.5b for a fluid film of thickness $6.8\,d^{LJ}$ in equilibrium with a bulk fluid with density $n^b = 0.59\,(d^{LJ})^{-3}$. The location of the large maximum in $n(z)$ next to each wall is almost entirely determined by the minimum in the interaction potential $v(z)$. The second layer near each wall arises mainly due to particle packing against the first layer; particles on either side of this layer (in the z direction) collide with the particles in the layer and keep them in place. As we will see, the layering is primarily an effect of the short-range repulsive interactions between the

---

[27] The pair interaction potential in the examples we will consider differs for practical reasons slightly from the LJ potential in Equation 5.6. It has been taken as $u(r) = u^{LJ}(r) - u^{LJ}(r_c)$ for $r < r_c$ and zero otherwise, where $u^{LJ}(r)$ is the LJ potential and $r_c = 3.5d^{LJ}$. We have $\beta u^{LJ}(r_c) = -0.002$ in the examples, so the difference from the full LJ potential is very small.

[28] W. A. Steele, *Surface Sci.*, **36** (1973) 317.

[29] The parameters are $\alpha_1 = 2\pi(d^{LJ})^2/a$, $\alpha_2 = (d^{LJ})^4/(3\Delta z)$ and $\alpha_3 = 0.61\Delta z$, where a is the area per parti-cle in each particle layer parallel to the surface inside the wall and Δz is the distance between two consecutive such layers. In the examples we will consider we have $a = (d^{LJ})^2$ and $\Delta z = d^{LJ}/\sqrt{2}$. With this choice of parameters $v^{St}(z)$ passes zero at $z = 0.84d^{LJ}$.

[30] The film thickness is defined as the sum of the particle diameter, defined as d^{LJ}, and the distance $D^S - 1.69d^{LJ}$ between the points at either wall where v^{St} is zero.

particles. The attraction to the particles in first layer plays only a minor role for the location of this layer. The corresponding is true for the next layer, but the structure becomes more fuzzy towards the middle of the slit.

In the bulk fluid, the pair distribution around each LJ particle shows a similar layer structure, but the layers have, of course, spherical shape in this case like in Figure 5.1 since the pair interaction potential $u(r)$ has this symmetry. In the inhomogeneous fluid, the distribution around a particle is determined by both the pair interaction potential $u(r)$ and the external potential $v(z)$. Therefore, the structure is a compromise between the planar symmetry of $v(z)$ and the spherical symmetry of $u(r)$. We encountered the resulting density distribution earlier in Figure 5.8 for the case of a film thickness 4.0 d^{LJ}. The figure shows the distribution function[31] $n(\mathbf{r'}|\mathbf{r}) = n(z')g^{(2)}(\mathbf{r'}, \mathbf{r})$ at point $\mathbf{r'}$ for two different locations $\mathbf{r}$ of the particle plotted in a cross-section plane made perpendicularly to the surfaces (see the illustration in Figure 5.7b).

To understand the physical meaning of a pair distribution $n(\mathbf{r'}|\mathbf{r})$ in this kind of figure, let us recapitulate the two aspects of the density distribution around a particle that we discussed at the end of Section 5.3. The distribution $n(\mathbf{r'}|\mathbf{r})$ can either be described as (i) the average density distribution when a particle is fixed at a position $\mathbf{r}$ and *all other* particles are free to move or (ii) the average density taken at those instances when a particle happens to be located at $\mathbf{r}$ while *all* particles are free to move. These two manners to describe the distribution are equivalent to each other in classical statistical mechanics.

We may illustrate the two different ways to obtain the density distribution and their equivalence as follows: For simplicity, we take a two-dimensional system where the particles move in a plane. Let us generate a huge amount of particle configurations, for instance by simulation, and make pictures of all of them with the particles represented as dots (= the particle centers). Let us go through all the pictures and pick those that have a dot at position $\mathbf{r} = \mathbf{r}_1$ (or, more practically, within a small volume element $d\mathbf{r}_1$ at $\mathbf{r}_1$). We place them in a separate pile and then record the positions $\mathbf{r'}$ of all other dots in these pictures. The average density of dots for all pictures in this pile gives $n(\mathbf{r'}|\mathbf{r}_1)$. Let us now put back all these pictures among the other ones. Next, we go through all pictures again and pick those that have a dot at another position $\mathbf{r} = \mathbf{r}_2$. The average density of dots for all pictures in this second pile gives $n(\mathbf{r'}|\mathbf{r}_2)$. When this has been repeated for a large number of points $\mathbf{r} = \mathbf{r}_\nu$ (in the limit of an infinite number of configurations), we have determined $n(\mathbf{r'}|\mathbf{r})$ for all these $\mathbf{r}$ and $\mathbf{r'}$. If we instead had generated a huge amount of particle configurations when one of the particles is kept fixed at, say, $\mathbf{r}_1$, and then calculated the average density of dots from pictures of these configurations, we would still obtain $n(\mathbf{r'}|\mathbf{r}_1)$. The average distribution is the same as the one we obtained from the first pile of pictures (in the limit of an infinite number of configurations) because the relevant probabilities depend only on the particle configurations. The two options are equivalent to each other.

Panes (a) and (b) in Figure 5.8 show the density $n(\mathbf{r'}|\mathbf{r})$ around a particle located at a point $\mathbf{r}$ in the second layer (as counted from the wall with positive z) and panes (c) and (d)

---

[31] The distribution function $n(\mathbf{r'}|\mathbf{r})$ in Figure 5.8 has been obtained by using an accurate integral equation theory for inhomogeneous fluids, whereby the pair distributions and the density profiles are calculated in a self-consistent manner. Such theories are introduced in Chapter 9 (in the 2nd volume).

show the corresponding density around a particle located between the first and the second layer. The point $\mathbf{r}$ is marked by a cross. The density distribution is strongly influenced everywhere by the presence of the particle at $\mathbf{r}$, but far from the particle, for large $|x'|$ (outside the figure), the distribution is in practice only influenced by the interactions with the walls. For such x' values, the density is given by $n(\mathbf{r}'|\mathbf{r}) \approx n(z')$ since $g^{(2)} \approx 1$ there.

The most interesting part is the density distribution $n(\mathbf{r}'|\mathbf{r})$ in the surroundings of the particle at the cross (which we will denote as "our particle") and we will give some examples of how it influences the forces that determine the density profile $n(z')$. As we can see in Figure 5.8, the density distribution in the immediate neighborhood of the particle is characterized by several strong peaks at about the same z' coordinates as the maxima in the density profile $n(z')$. They are most easily seen in panes b and d of the figure. Particles located in these peaks mainly interact repulsively with our particle; this applies to all particles located within the distance $|\mathbf{r}' - \mathbf{r}| < 1.12\, d^{\mathrm{LJ}}$ from the center of the latter since the minimum in the LJ potential occurs at the distance $2^{1/6} d^{\mathrm{LJ}} \approx 1.12\, d^{\mathrm{LJ}}$. The two peaks near the wall[32] are markedly higher than the other ones, particularly in panes c and d. This is the case because the particles in these main peaks are pushed towards our particle and the wall due to thermal motion *and* attracted by the wall.

We first consider pane (b) of Figure 5.8 where our particle is located in the middle of the second layer. Let us focus on the two peaks at the top of our particle (near the wall) and the two peaks at the bottom (near $z' = -0.4d^{\mathrm{LJ}}$). Since the particles in these peaks mainly interact repulsively with our particle, the particles at the top push our particle downwards (in the direction of negative z') and those at the bottom push it upwards. Note that only the z-components of the forces contribute – there are no net average forces parallel to the surface due to the symmetry in the particle distribution (this is illustrated in Figure 7.7, where the horizontal force components from particles to the right cancel those from particles to the left as shown in pane b of the figure). In Figure 5.8b, the forces at the top and bottom of our particle oppose each other and their action makes the particle position in the layer a relatively stable one. If, for example, the position of our particle were somewhat closer to the surface, the two peaks near the surface would be larger and the repulsion stronger, pushing our particle back into the center of the layer. The changes in the repulsive force from the bottom would enhance the effect because this force is weakened when the particle is somewhat closer to the surface. Note that the particles in the second layer (around the midsection of our particle) have little influence on this force balance because they mainly interact with forces parallel to the surface, so the net force from them is close to zero.

In contrast to the balance of forces in pane (b), the situation in pane (d) of Figure 5.8 is very different. In this case our particle is located at the bottom of the density profile minimum. As we will see, this position is unstable. The two density peaks near the surface are very strong in this case, but since the repulsions that particles there exert on our particle are directed in the radial direction (toward the cross), the z-component of the net force is

---

[32] What is seen as two peaks in the figure – one on the left side and one on the right side of our particle – are actually parts of a toroidal-shaped peak in the particle distribution around the particle. Recall that the distribution has rotational symmetry around the vertical z' axis through the cross (see the caption of Figure 5.8). We will, however, talk about "two peaks" in order to identify what is seen in the figure.

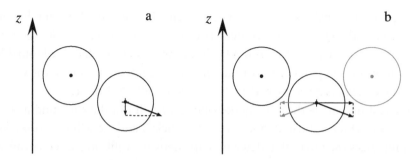

FIGURE 7.7 (a) A sketch of two particles that interact repulsively with each other when they are close together. The repulsive force that the particle to the left exerts on the particle to the right is shown as an arrow in the radial direction. The z-component of the force is also shown (small vertical arrow). (b) In this picture some parts are drawn in black and some in gray. The black parts show the same as in pane (a) except that the horizontal component of the force is shown (the component parallel to the surface). The particle to the right (drawn in gray) has equal probability to be at its position as the particle to the left due to symmetry. The force from the former on the central particle is drawn in gray. The horizontal components of the forces on the central particle cancel each other, so on average there remain only the z-components.

not so large (cf. Figure 7.7a). If, for example, the particle position were somewhat further away from the surface, the z-component of the repulsive force from these two peaks would be larger and the particle would be pushed into the second layer. This means that the probability for a particle to be located at the bottom of the profile minimum is small. Likewise, in the other minima of the density profile, the circumstances are similar and the probability for a particle to be located there is small. These are the reasons why the density is low between the layers and high in the layers – or expressed more correctly – these are the reasons why layers are formed in the first place.

Finally, we note that our particle gives rise to oscillations in the surrounding density, most notably in the lateral directions as can be seen in the layers parallel to the walls. This is most clearly visible in pane (a) of Figure 5.8 for the layer nearest to the left wall where the ridge shows an oscillatory behavior. Thus, there is a combination of oscillations in the lateral directions (layering due to our particle) and in the perpendicular direction (layering due to the walls). This is the result of the compromise between the planar symmetry of $v(z)$ and the spherical symmetry of $u(r)$ that was mentioned earlier.

7.2.1.2 Hard Sphere Fluids

Next, we will treat inhomogeneous hard sphere fluids in contact with one hard wall or between two such walls. The coordinate system is in this case selected with the origin at a hard wall surface and with the z axis perpendicular to it. The interaction potential between a hard wall and a hard sphere is

$$v(z) = \begin{cases} \infty, & z < d^{\mathcal{S}} \\ 0, & z \geq d^{\mathcal{S}}, \end{cases}$$

where $d^S = d^h/2$ is the distance of closest approach of the sphere centers to the surface. For the case of two surfaces, we have

$$v(z) = \begin{cases} \infty, & z < d^S \text{ and } z > D^S - d^S \\ 0, & \text{otherwise,} \end{cases}$$

where D^S is the distance between the surfaces. We have $D^S = L + 2d^S$, where L is the distance between points of infinite potential, that is, the width of the region that is accessible to the sphere centers in the slit.

Figure 7.8a shows the density profile outside a single wall in contact with a bulk fluid with density $n^b = 0.75\,(d^h)^{-3}$. The abscissa in the plots is selected as $z - d^S$, so it is zero at the point of closest approach. The integral equation calculations have been done in the Anisotropic Percus-Yevick (APY) approximation, which is introduced in Chapter 9. In this approximation, both density profiles and pair distribution functions are self-consistently calculated for the inhomogeneous fluid. It is an accurate approximation for this kind of system.

The formation of the first particle layer next to the wall in Figure 7.8a is due to the collision effect we have discussed earlier, namely that the spheres next to the wall are pushed towards the surface by collisions from the outside by other spheres. This leads to the high contact density at the surface. The second and the following layers are formed in essentially the same manner as we discussed in the case of LJ fluids. The fact that the profiles for LJ and hard

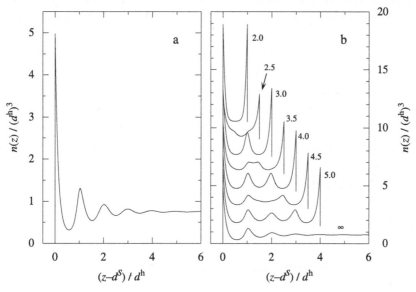

FIGURE 7.8 (a) Density profile $n(z)$ for a hard sphere fluid with density $n^b = 0.75\,(d^h)^{-3}$ in contact with a hard wall calculated by an integral equation theory (see text). (b) Density profiles for a hard sphere fluid between two hard walls in equilibrium with a bulk fluid of the same density as in (a). The surface separations of the walls are (from top to bottom) $D^S = 2.0, 2.5, 3.0, 3.5, 4.0, 4.5$ and $5.0\,d^h$. The bottom curve is for infinite surface separation and is the same as in pane (a). For clarity, consecutive curves are shifted vertically by 1.5 units.

sphere fluids are qualitatively similar (apart from the first layer) illustrates the importance of the short-range repulsive interactions in the LJ case for the formation of the liquid structure.

Figure 7.8b shows density profiles for a hard sphere fluid between two hard walls in equilibrium with a bulk fluid with the same density, $n^b = 0.75\,(d^h)^{-3}$. When the surface separation D^S/d^h is close to an integer l, there are l layers between the surfaces. For $D^S/d^h = l + 0.5$, the system is in an intermediate stage between l and $l + 1$ layers. In the case $D^S/d^h = 2.5$, there appear shoulders on the slopes on either side of the profile. When D^S/d^h is increased towards 3.0, they form, as we will see later, the layer in the middle that we have for $D^S/d^h = 3.0$. For $D^S/d^h = 3.5$, the middle layer is on the verge of splitting into the two layers that we have for $D^S/d^h = 4.0$. When $D^S/d^h = 4.5$, there is a wide minimum in the middle where a new layer in the middle will grow when D^S/d^h is increased towards 5.0.

The contact density at the surfaces is lower for the intermediate stages with $D^S/d^h = l + 0.5$ than when D^S/d^h is close to an integer and the same is true for the average density over the entire slit. The reason for this is that the spheres cannot pack efficiently for the intermediate surface separations, so there are a lot of empty spaces left between the spheres. Fully developed layers appear when they fit neatly into the slit, which occurs when D^S/d^h is close to, but not exactly equal to, an integer. This can be understood from the point of view of packing of spheres, but since the system is in a fluid state, the optimal state – with minimal free energy – is, of course, different from that in solids. We may express the situation as a frustration of the fluid structure for the intermediate separations. At, for example, $D^S/d^h \approx l + 0.5$, the slit is too wide for l fully developed layers and too narrow for $l + 1$ such layers.

For inhomogeneous hard sphere fluids, the fluid structure has been determined experimentally on both the density profile and the pair distribution levels by small-angle X-ray scattering measurements. The pair structure is then obtained in terms of the structure factor $S(\mathbf{k})$, which is a function of the scattering vector $\mathbf{k}$ (a wave vector) and gives the contribution to the scattering intensity that depends on the interparticle correlations in the fluid. We have encountered the structure factor $S(k)$ for bulk fluids in Section 5.5.2, where a concrete example was given in Figure 5.9 showing the experimental $S(k)$ for liquid argon in bulk phase. The underlying theory for $S(\mathbf{k})$ was presented in Section 5.6. In isotropic bulk fluids, S depends only on the length k of the wave vector, but for an inhomogeneous fluid it depends also on the direction of $\mathbf{k}$, so $S(\mathbf{k})$ is anisotropic (like the fluid itself). For a fluid in the slit between two planar walls, there are two distinct components of the scattering vector, parallel and perpendicular to the surface, so we have $S = S(k_\parallel, k_\perp)$; details are given in the shaded text box below, which can be skipped in the first reading.

★ The structure factor is given by (Equation 5.58)

$$S(\mathbf{k}) = 1 + \frac{1}{N} \int d\mathbf{r}_1 d\mathbf{r}_2 n(\mathbf{r}_1) n(\mathbf{r}_2) h^{(2)}(\mathbf{r}_1, \mathbf{r}_2) e^{-i\mathbf{k}\cdot(\mathbf{r}_2 - \mathbf{r}_1)}, \qquad (7.63)$$

where N is the number of particles. In planar geometry, Equation 7.63 can be written as

$$S(k_\parallel, k_\perp) = 1 + \frac{1}{N} \int d\mathbf{r}_1 d\mathbf{r}_2 n(z_1) n(z_2) h^{(2)}(\mathbf{r}_1, \mathbf{r}_2) e^{-i\mathbf{k}_\parallel\cdot(\mathbf{s}_2 - \mathbf{s}_1) - i k_\perp (z_2 - z_1)},$$

where $s_\nu = (x_\nu, y_\nu)$, $k_\parallel = |\mathbf{k}_\parallel|$, $\mathbf{k}_\parallel = (k_x, k_y)$ consists of the components of $\mathbf{k}$ in the lateral directions of the surface and $k_\perp = k_z$ is the corresponding perpendicular component.

Since the integral is taken over both $\mathbf{r}_1$ and $\mathbf{r}_2$, the structure factor measures an average pair structure for all positions of the particles. The total structure factor $S_{tot}(\mathbf{k})$ also contains the contribution $S_{singlet}(\mathbf{k})$ (defined just before Equation 5.58), which originates solely from the singlet density distribution $n(\mathbf{r}) = n(z)$ of the system. In the experiments, one obtains $S_{tot} = S_{singlet} + S_{pair}$, where $S_{pair}(\mathbf{k})$ is denoted as $S(\mathbf{k})$ in this treatise. Measurements of $S_{singlet}(\mathbf{k})$ can be used to extract information about the density profile $n(z)$.

Figure 7.9 shows the experimental $S(\mathbf{k})$ together with the corresponding theoretical result for hard spheres between hard walls. The theoretical $S(k)$ for the bulk fluid is also shown for comparison. The experimental system consists of a suspension of colloidal

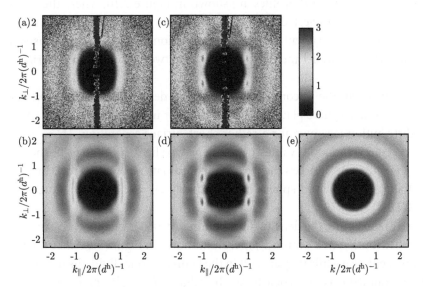

FIGURE 7.9 (**See color insert.**) Experimental (top) and theoretical (bottom) anisotropic structure factors $S(\mathbf{k})$ for a dense hard-sphere fluid confined between planar walls at various separations: (a) and (b) $D^S \approx 2.45\, d^h$; (c) and (d) $D^S \approx 3.10\, d^h$; (e) in the bulk phase ($D^S = \infty$).[33] The colors in the contour plot show the values of $S(\mathbf{k})$ according to the color legend. The fluid in the slit is in equilibrium with a bulk fluid of density $n^b = 0.75(d^h)^{-3}$. $S(\mathbf{k})$ is plotted as a function of the in-plane (lateral) component $k_\parallel$ and the out-of-plane (perpendicular) component $k_\perp$ of the scattering vector $\mathbf{k}$. The dark red feature at $k_\parallel = 0$ in the experimental data, which is due to diffraction from the array of slits with the fluid in the experiments and is given by $S_{singlet}(\mathbf{k})$, should be neglected in the comparison. The theoretical data have been obtained in the same integral equation theory as Figure 7.8.

---

[33] Figure 7.9 is reprinted with permission from K. Nygård, R. Kjellander, S. Sarman, S. Chodankar, E. Perret, J. Buitenhuis and J. F. van der Veen, *Phys. Rev. Lett.* **108**, 037802 (2012). © 2012 by the American Physical Society. The figure has been somewhat modified by K. Nygård compared to the original one.

particles in a solvent such that the particles practically behave like hard spheres. The fluid is entered in a large number of slits with the same width (an array of slits in the form of a one-dimensional grid) and the surfaces of the slits can be approximated as being hard.

The effect of confinement for the pair function can be seen by comparing panes (a)–(d) in Figure 7.9 with pane (e), which show the corresponding plot for the bulk phase. In the bulk, $S(\mathbf{k})$ is spherically symmetric (cf. Figure 5.9), while its anisotropy in the confined state is very distinct in the figure. Furthermore, in panes (c) and (d), there are rather sharp peaks since D^{S}/d^{h} is close to an integer, while the structure is less prominent in panes (a) and (b) since D^{S}/d^{h} is about half-way between integer values – in agreement with the discussion earlier.

Experimental density profiles have been extracted from an analysis of $S_{\text{singlet}}(\mathbf{k})$, which describes the diffraction from the array of slits containing the fluid. This diffraction gives the dark red feature seen around $k_{\parallel} = 0$ in Figures 7.9a and c. The density profiles are then obtained in terms of a local volume fraction of spheres, $\mathcal{C}(z)$, instead of the number density of sphere centers, $n(z)$,[34] since most of the interior of the colloidal particles scatters X-rays. The experimental and theoretical profiles are shown in Figure 7.10, where the theoretical density profiles have been converted to $\mathcal{C}(z)$ whereby the sharp features of $n(z)$ become smoothed out. Since the experimental colloidal suspension is somewhat polydisperse and the walls are not perfectly parallel and hard, deviations between theory and experiments are expected.

We will now turn to the theoretically calculated density distribution $n(\mathbf{r}_2|\mathbf{r}_1) = n(z_2)g^{(2)}(\mathbf{r}_1, \mathbf{r}_2)$ around a sphere placed at a fixed point $\mathbf{r}_1$ in a hard sphere fluid. An example of such a distribution is shown for a rather short surface separation, $D^{S} = 2.80\, d^{h}$, in Figure 7.11. The colors show the values of the density. The z coordinate in the figure is $z_2 - D^{S}/2$, so the midplane between the walls has coordinate zero in this plot. In the example shown in the figure, the fixed sphere has coordinate $z_1 = 1.55\, d^{h}$, which translates to $z_2 - D^{S}/2 = 0.15\, d^{h}$.

As we can see in the figure, the density distribution $n(\mathbf{r}_2|\mathbf{r}_1)$ is qualitatively similar to the corresponding distribution in a Lennard-Jones fluid shown in Figure 5.8. Since the particles are hard in the present case while they are soft in the Lennard-Jones fluid, the density plot has sharper features around the fixed particle here. The values of $n(z_2)g^{(2)}(\mathbf{r}_1, \mathbf{r}_2)$ around the rim of the gray area in the plot shows the *contact density* of the pair distribution, which we have encountered earlier when we treated the pressure of a bulk fluid of hard-core particles. For the bulk case, the contact value of the pair distribution is defined in Equation 5.76 as the limit of $g^{(2)}(r)$ when $r \to d^{h+}$ and the contact density equals $n^{b}g^{(2)}(d^{h+})$. In an inhomogeneous fluid, the contact density varies around the rim of the hard core region as we can see in

---

[34] For particles of diameter d, the two profiles are related by $\mathcal{C}(z) = \int dz'\, n(z - z')a(z')$, where $a(z) = \pi[(d/2)^2 - z^2]$ when $|z| \le d/2$ and zero otherwise. The function $a(z)$ is equal to the cross-sectional area of the sphere at coordinate z (counted from the sphere center) and gives the appropriate volume when multiplied by dz. If the whole sphere scatters X-rays, we have $d = d^{h}$, but if only a part of the sphere does so, say a part with diameter d', one has to set $d = d'$ and multiply $a(z)$ by the scaling factor $(d^{h}/d')^3$. This factor yields the local volume fraction for the entire sphere considering that only the part inside diameter d' scatters. The latter is the case for the experimental system we consider here (the silica colloid particles are octadecyl-grafted in order for them to behave like hard spheres in toluene, which is used as the solvent). Likewise, in the evaluation of $S(\mathbf{k})$ from the scattering intensity, a form factor given by Equation 5.67 with $r_{\mathrm{p}} = d'/2$ is used.

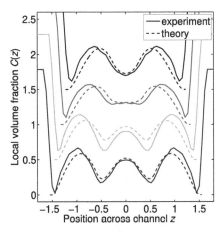

FIGURE 7.10 (**See color insert.**) Experimental (solid lines) and theoretical (dashed lines) local volume fraction profiles, $C(z)$, for the same system as in Figure 7.9.[35] The slit widths are, from top to bottom, $D^S \approx 2.45, 2.65, 2.90$, and $3.10 \, d^h$. For clarity, consecutive sets of curves are shifted vertically by 0.5 units. The nearly vertical lines in the experimental data originate from the confining walls. The origin of z in this figure is placed in the middle of the slit, so z here corresponds to $z - D^S/2$ when the origin is located at the left surface.

Figure 7.11. The reason is that both the density $n(z_2)$ and the pair distribution function $g^{(2)}(\mathbf{r}_1, \mathbf{r}_2)$ vary with $\mathbf{r}_2$ around the rim. Hard spheres interact only at contact – when they collide – and for this reason, the variation of the contact density plays a key role in the understanding of the potential of mean force in the inhomogeneous fluid. This is the subject of the rest of the current section. We will analyze in some detail how the contact density and thereby the collisional forces vary around the rim in some concrete cases. As we will see, this gives an insight in how the interparticle interactions determine the density profile.

For the surface separation in Figure 7.11, $D^S = 2.80 \, d^h$, the density profile $n(z_2)$ has a small maximum at the midplane. This maximum is not as well developed as for the case with $D^S = 3.0 \, d^h$ shown in Figure 7.8. We will shortly investigate in what manner the density distribution in the middle varies when D^S is changed between the cases $2.50 \, d^h$ (inefficient sphere packing) and $3.0 \, d^h$ (efficient sphere packing), whereby the current case is one of the steps on the way.

One can obtain an understanding of the behavior of the density distribution profile $n(z)$ and the collisional forces by using the first Born-Green-Yvon Equation 5.118, which gives the gradient of the density profile in terms of the pair distribution functions for the inhomogeneous fluid. For planar geometry, this equation is

$$
\begin{aligned}
\frac{d[\ln n(z_1)]}{dz_1} &= \beta F(z_1) \\
&= -\beta \left[\frac{dv(z_1)}{dz_1} + \int d\mathbf{r}_2 \, n(z_2) g^{(2)}(\mathbf{r}_1, \mathbf{r}_2) \frac{\partial u(\mathbf{r}_2, \mathbf{r}_1)}{\partial z_1} \right],
\end{aligned}
\tag{7.64}
$$

[35] Figure 7.10 is reprinted with permission from K. Nygård, R. Kjellander, S. Sarman, S. Chodankar, E. Perret, J. Buitenhuis and J. F. van der Veen, *Phys. Rev. Lett.* **108**, 037802 (2012). © 2012 by the American Physical Society. The theoretical data have been obtained in the same integral equation theory as Figures 7.8 and 7.9.

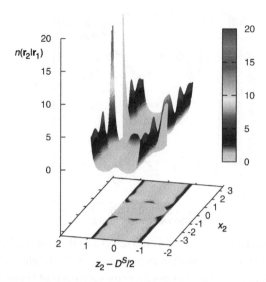

FIGURE 7.11 **(See color insert.)** Plots of the density distribution $n(\mathbf{r}_2|\mathbf{r}_1) = n(z_2)g^{(2)}(\mathbf{r}_1, \mathbf{r}_2)$ at position $\mathbf{r}_2$ around a hard sphere located at $\mathbf{r}_1$ in an inhomogeneous hard sphere fluid between two hard walls at surface separation $D^S = 2.80\, d^h$ (these are the same types of plots as in Figure 5.8; see sketch of the geometry in Figure 5.7). The hard sphere at $\mathbf{r}_1$ is located in the middle of the gray circular area in the contour plot at the bottom. This area shows the region where other sphere centers cannot enter and where the density is zero. The surface plot at the top shows a different representation of the same data as the contour plot. The vertical gray area in this plot represents the discontinuity of the density function; the contact density at the sphere surface is given by the values around the top rim of this area. The system is the same as in Figure 7.8 and the calculations have been done in the same integral equation theory. The coordinates are given in units of d^h and the density in units of $(d^h)^{-3}$.

where $F(z_1)$ is the mean force in the z direction acting on a particle at z_1. The pair distribution function $g^{(2)}(\mathbf{r}_1, \mathbf{r}_2)$ for an inhomogeneous fluid has six variables given by $\mathbf{r}_1$ and $\mathbf{r}_2$. In the current geometry, the fluid has translational and rotational symmetry in the x, y plane. Therefore, only three of these six variables are independent and we can write $g^{(2)}(\mathbf{r}_1, \mathbf{r}_2) = g^{(2)}(s_{12}, z_1, z_2)$, where $\mathbf{s}_{12} = (x_2 - x_1, y_2 - y_1)$ and $s_{12} = |\mathbf{s}_{12}|$.

For a fluid of hard spheres in contact with a hard wall, we have $v(z_1) = 0$ outside the wall, so the term $dv(z_1)/dz_1$ does not contribute inside the fluid. In this system, $F(z_1)$ is the average force on a sphere located at z_1 due to collisions by other spheres. The pair interaction potential $u(\mathbf{r}_2, \mathbf{r}_1) = u(r_{12})$ is infinitely large when $r_{12} < d^h$ and zero otherwise. To obtain an expression for $F(z_1)$, we have to deal with this behavior when evaluating the derivative of $u(r_{12})$ and we will do this analogously to the derivation of the pressure of a hard sphere bulk fluid, Equation 5.77. The derivation is done in the shaded text box below, which can be skipped in the first reading. Here we will only give the end results and interpret them physically. The expression obtained for the force on a sphere located at z_1 is

$$F(z_1) = 2\pi d^h k_B T \int_{z_1 - d^h}^{z_1 + d^h} dz_2\, n(z_2) g^{(2)\text{cont}}(z_1, z_2) \frac{z_1 - z_2}{d^h}, \qquad (7.65)$$

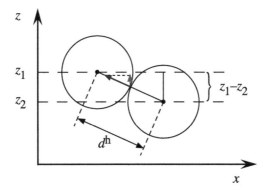

FIGURE 7.12 (**See color insert.**) Sketch of a collision between two hard spheres, one with its center at z-coordinate z_1 and one with its center at z_2. The force on the former sphere due to the collision with the latter acts in the radial direction and is shown as the red arrow. The ratio between the z-component of the force (the short blue vertical arrow) and the length of the total force is equal to the ratio $(z_1 - z_2)/d^h$.

where

$$
\begin{aligned}
g^{(2)\mathrm{cont}}(z_1, z_2) &= \Big[g^{(2)}(s_{12}, z_1, z_2)\Big]_{[s_{12}^2 + (z_2 - z_1)^2]^{1/2} = d^{h+}} \\
&= \Big[g^{(2)}(\mathbf{r}_1, \mathbf{r}_2)\Big]_{r_{12} = d^{h+}}
\end{aligned}
\tag{7.66}
$$

is the contact value of $g^{(2)}$ on the hard sphere surface for given z_1 and z_2. The particles that are in contact with each other have z coordinates z_1 and z_2, respectively.

The physical interpretation of Equation 7.65 is as follows. We consider a particle located with its center at $z = z_1$ and particles that are colliding with it (see Figure 7.12 for an example). The product $n(z_2)g^{(2)\mathrm{cont}}(z_1, z_2)$ is the average density of particles with z-coordinate z_2 in contact with the first particle. By multiplying by $k_B T$, we obtain the average force per unit area that they exert in the radial direction by collisions (momentum transfer). The factor $(z_1 - z_2)/d^h$ makes a projection of this force yielding the component in the z direction. The factor $2\pi d^h \, dz_2$ is the area of a spherical segment of radius d^h and height dz_2, so by multiplying by this factor we obtain the total force in the z direction from collisions by particles with z coordinates between z_2 and $z_2 + dz_2$. The average force components in the x and y directions cancel by symmetry (cf. illustration in Figure 7.7b). After integration, we finally obtain the total force from all particles that collide with the particle located at $z = z_1$.

★ We will derive Equation 7.65 from Equation 7.64 in a manner that is analogous to the derivation of Equations 5.75 and 5.77 for a bulk fluid, which is given in the shaded text box on page 205ff. Also in the present case, we have $g^{(2)} = y^{(2)} \exp[-\beta u]$, where $y^{(2)}$ is continuous at $r = d^h$. Since

$$
\frac{\partial u(r_{12})}{\partial z_1} = \frac{du(r_{12})}{dr_{12}} \cdot \frac{\partial r_{12}}{\partial z_1} = \frac{du(r_{12})}{dr_{12}} \cdot \frac{z_1 - z_2}{r_{12}}
$$

and

$$g^{(2)}(\mathbf{r}_1, \mathbf{r}_2)\frac{du(r_{12})}{dr_{12}} = y^{(2)}(\mathbf{r}_1, \mathbf{r}_2)e^{-\beta u(r_{12})}\frac{du(r_{12})}{dr_{12}}$$

$$= -\beta^{-1}y^{(2)}(\mathbf{r}_1, \mathbf{r}_2)\frac{de^{-\beta u(r_{12})}}{dr_{12}} = -\beta^{-1}y^{(2)}(\mathbf{r}_1, \mathbf{r}_2)\delta(r_{12} - d^h)$$

we obtain from Equation 7.64

$$\beta F(z_1) = \int d\mathbf{r}_2\, n(z_2)y^{(2)}(\mathbf{r}_1, \mathbf{r}_2)\delta(r_{12} - d^h)\frac{z_1 - z_2}{r_{12}}$$

for z_1 inside the fluid. Since we have $y^{(2)} = y^{(2)}(s_{12}, z_1, z_2)$ and $r_{12}^2 = s_{12}^2 + (z_2 - z_1)^2$, we can change the integration over $\mathbf{r}_2$ to an integration over (s_{12}, z_2); the origin for $\mathbf{s} = (x, y)$ does not matter. Since the integrand does not depend on the direction of $\mathbf{s}_{12}$, we can write $2\pi s_{12}ds_{12}$ instead of $d\mathbf{s}_{12}$ in the integral, where only the s_{12} radial integration and that over z_2 then remain. Furthermore, for constant z_1 and z_2, we can do a variable substitution $s_{12}^2 + (z_2 - z_1)^2 = (r_{12}')^2$, whereby $s_{12}ds_{12} = r_{12}'dr_{12}'$, and obtain

$$\beta F(z_1) = \int dz_2\, n(z_2)\int_{|z_1 - z_2|}^{\infty} dr_{12}'\, 2\pi r_{12}'\, y^{(2)}(s_{12}, z_1, z_2)\delta(r_{12}' - d^h)\frac{z_1 - z_2}{r_{12}'}.$$

By performing the integration over r_{12}', we obtain

$$\beta F(z_1) = \int dz_2\, n(z_2)2\pi d^h\left[y^{(2)}(s_{12}, z_1, z_2)\right]_{r_{12}'=d^h}\frac{z_1 - z_2}{d^h},$$

which is the same as Equation 7.65 since $y^{(2)}(\mathbf{r}_1, \mathbf{r}_2) = g^{(2)}(\mathbf{r}_1, \mathbf{r}_2)$ when $r_{12} \geq d^h$. The integration limits for z_2 are used in Equation 7.65 since the integrand is nonzero only when $|z_1 - z_2| \leq d^h$.

As an example of how one can use Equation 7.65 to understand the behavior of the density profiles, let us consider the case $D^S/d^h = 2.5$ in Figure 7.8b, where there appear shoulders on the slopes on either side of the profile. As mentioned earlier, when the surface separation is increased, the shoulders form a layer in the middle of the slit. By comparing with the profiles for larger surface separations in the figure, we can see that the location of the shoulder near the right surface approximately corresponds to the location of the layer next nearest to the left surface.

Figure 7.13 shows the density distribution $n(\mathbf{r}_2|\mathbf{r}_1) = n(z_2)g^{(2)}(\mathbf{r}_1, \mathbf{r}_2)$ when a hard sphere is located at $\mathbf{r}_1$ for three different surface separations. The bottom case ($D^S = 2.80\, d^h$) is the same as in Figure 7.11. To the right in Figure 7.13, the density profiles are displayed

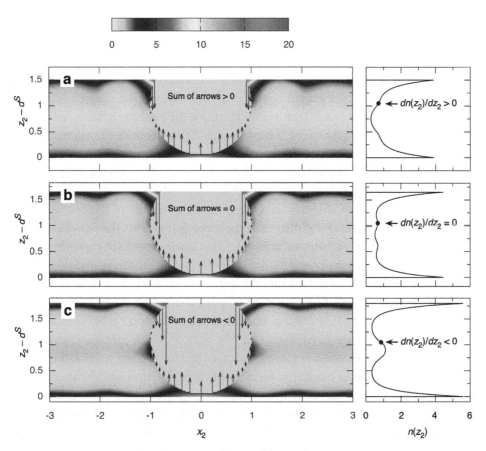

FIGURE 7.13 **(See color insert.)** Contour plots[36] of the density distribution $n(\mathbf{r}_2|\mathbf{r}_1) = n(z_2)g^{(2)}(\mathbf{r}_1, \mathbf{r}_2)$ at position $\mathbf{r}_2$ when a hard sphere is located at $\mathbf{r}_1$ in the slit between two hard walls filled with a hard sphere fluid at different surface separations: (a) $D^S = 2.50\,d^h$, (b) $D^S = 2.65\,d^h$ and (c) $D^S = 2.80\,d^h$. The last case is the same as in Figure 7.11. The density profiles for the respective cases are shown in the panes to the right. The dot on the profile shows the z-coordinate of the sphere, which in all cases is located at a distance of $1.55\,d^h$ for the bottom wall (at coordinate $z_1 - d^h = 1.05\,d^h$). The arrows in the gray area of each contour plot depict z-components of the average collisional forces acting on the sphere. Arrows displayed at a certain z_2 coordinate represent the entire force acting on the sphere periphery at this coordinate (the sphere segment between z_2 and $z_2 + dz_2$ with a radius that lies perpendicularly to the figure plane). In pane (a), the sum of all arrows (with signs) is > 0, in (b) $= 0$ and in (c) < 0. The coordinates are given in units of d^h and the density in units of $(d^h)^{-3}$.

for the three surface separations; the top one is the same as the case $D^S/d^h = 2.5$ in Figure 7.8b and the other two are for somewhat larger separations. In the middle pane to the right, the two shoulders have turned into two maxima and in the bottom pane they have merged into a peak in the middle. For the z value marked with a dot, the slope $dn(z)/dz$ is

---

[36] Figure 7.13 is reprinted from K. Nygård, S. Sarman and R. Kjellander, *J. Chem. Phys.* **141** (2014) 094501, with the permission of AIP Publishing.

positive in the top pane, equal to zero in the middle pane and negative in the bottom one. This dot marks the position of the center of the sphere at $\mathbf{r}_1$ shown in each contour plot to the left. Since $[dn(z)/dz]/n(z) = \beta F(z)$, we can deduce that the mean force on this sphere is directed upwards in the top case, is zero in the middle case and is directed downwards in the bottom one. The mean force on the hard sphere is entirely due to particle collisions at the sphere surface and the amount of such collisions determine the total force. More specifically, it is the distribution and strengths of the z component of the forces due to such collisions that matter. This is given by the rhs of Equation 7.65 and since $n(\mathbf{r}_2|\mathbf{r}_1)$ is known from the theoretical calculation, we can quantitatively evaluate the forces acting on the different parts of the sphere surface from the values of the contact density there.

The result of this evaluation is shown as arrows in the gray area of the contour plots: the arrows show the magnitude and direction of the force that acts on the periphery of the sphere. Thereby, the length of each arrow reflects (i) the density at the sphere surface at the base of the arrow (the contact density $ng^{(2)\text{cont}}$) and (ii) the fact that the force also depends on the factor $(z_1 - z_2)/d^{\text{h}}$. The latter is, for example, apparent in the region around the middle of the slit where the contact densities are not small, but the arrows are tiny. This illustrates the fact that the z component of the force is small there since the force from collisions acts in the radial direction (cf. Figure 7.7b).

Since the sum of the arrows for each sphere gives the total force F that acts on the sphere, we can see why the force varies as it does as a function of surface separation. The arrows at the bottom of the sphere are nearly the same in the three cases (the distance of the sphere from the bottom surface is the same in all cases); what matters in the current case is what happens at the top of the sphere. There, the sphere penetrates to various extents into the layer next to the top surface in the figure and it is mainly this variation that gives rise to the changes in F.

By doing the corresponding analysis for various locations of the sphere, one can understand why the profile looks like it does for each surface separation. This is the kind of analysis we did for the LJ fluid earlier in Section 7.2.1.1. The analysis in the LJ case can be made quantitative[37] by using Equation 7.64 and the numerical data for $n(\mathbf{r}_2|\mathbf{r}_1) = n(z_2)g^{(2)}(\mathbf{r}_1, \mathbf{r}_2)$ presented in Figure 5.8, but we will not pursue this here.

7.2.2 Primitive Model Electrolytes

In this section, we turn to inhomogeneous electrolytes in the primitive model and treat pair distributions (ion-ion correlations) in electric double-layers with planar geometry. Thereby, we have a charged surface with a uniform surface charge density $\sigma^{\mathcal{S}}$ in contact with an bulk electrolyte of given composition. In this section, the origin of the coordinate system is located at the plane of $\sigma^{\mathcal{S}}$ and the z axis is perpendicular to the surface.

7.2.2.1 Pair Distributions in the Electric Double-Layer
There are at least two species of ions and we consider the density distribution $n_j(z)$ of species j and the pair distribution function $g_{ij}^{(2)}(\mathbf{r}_1, \mathbf{r}_2)$ for ions of species i and j. The density of ions

---

[37] *Loc. cit.* in footnote 20 of Chapter 5, cf. Figure 5.8.

of species j at $\mathbf{r}_2$ when an ion of species i is located at $\mathbf{r}_1$ is equal to (cf. Equation 5.28)

$$n_j(\mathbf{r}_2|\mathbf{r}_1; i) = n_j(z_2)g_{ij}^{(2)}(\mathbf{r}_1, \mathbf{r}_2),$$

where the species of the ion at $\mathbf{r}_1$ is entered after the semicolon. The charge density around the i-ion is

$$\rho(\mathbf{r}_2|\mathbf{r}_1; i) = \sum_j q_j n_j(z_2)g_{ij}^{(2)}(\mathbf{r}_1, \mathbf{r}_2).$$

For bulk electrolytes, we have used the notation $\rho_i(r_{12})$ for this entity, where r_{12} is the distance from the center of the ion (cf. Equation 6.24), which means that in a homogeneous bulk phase we have $\rho_i(r_{12}) = \rho(\mathbf{r}_2|\mathbf{r}_1; i)$. For an inhomogeneous electrolyte, neither the density $n_j(\mathbf{r}_2|\mathbf{r}_1; i)$ nor the charge density $\rho(\mathbf{r}_2|\mathbf{r}_1; i)$ around a spherical ion is spherically symmetric around $\mathbf{r}_1$. They depend not only on the distance r_{12}, but on $\mathbf{r}_1$ and $\mathbf{r}_2$ separately as we saw for the pair density distributions for the LJ and hard sphere cases (Figures 5.8 and 7.11).

Let us consider the same inhomogeneous 2:-2 electrolyte solution in contact with a negatively charged surface as we dealt with in Figure 7.3. In this figure, the density and charge density profiles outside the surface are depicted for this system. Figure 7.14a shows the same charge density profile $\rho(z)$ as the full curve in Figure 7.3b. If we make a cross section perpendicularly to the surface in this system (analogous to Figure 5.7), the charge density profile can be illustrated as the contour plot in Figure 7.14b. The bulk phase is located to the right in the figure and the surface lies to the left (represented by a vertical line in the figure). The contours and the colors show values of the charge density at various places outside the surface. For $z - d^S \gtrsim 5$ Å, the charge density is negative (light blue color) because of the charge inversion associated with the overcompensation of surface charge in this system. The location $z - d^S = 0$ is the point of closest approach of the ion centers to the surface. We will consider what happens when one of the counterions is located with its center at the cross in the figure.

First of all, due to the hard core interactions between the ions, which all have the same diameter d^h, no ion can approach with its center closer than the distance d^h from the counterion. There is an exclusion zone around the ion where no ion center can be located – a cavity with radius d^h. Figure 7.14c depicts the counterion (shown as $\oplus$) and the cavity around it (shown in white). Furthermore, the counterion will repel other counterions and attract coions electrostatically. Outside the exclusion cavity, the net effect of these electrostatic and hard core interactions is that the charge distribution is changed from that in Figure 7.14c or, in other words, the ion is surrounded by an ion cloud. The resulting average charge distribution around the ion is shown in Figure 7.14d. Note that the distribution shown in this figure is the result of the interactions both with the charged surface and the counterion ($\oplus$). The polarization response of the electrolyte solely due to the surface is the charge density in Figure 7.14a, while the response due to both the surface and the counterion is the density in Figure 7.14d. The difference between these two densities is the effect of the counterion on the charge distribution outside the surface, that is, the charge distribution of the ion cloud.

The counterion shown in the figure can be any of the counterions in the system when it happens to be located with its center at the cross. Note that the charge density depicted

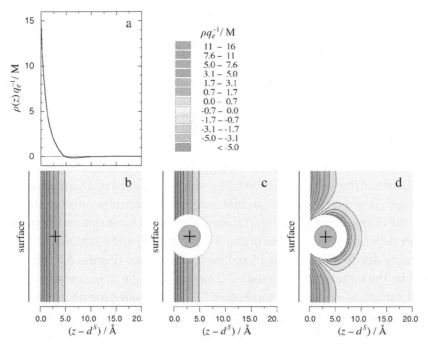

FIGURE 7.14 (**See color insert.**) (a) The same charge density distribution $\rho(z)/q_e$ as the full curve in Figure 7.3b. The system is described in the caption of that figure. (b) The charge density distribution in (a) displayed as a contour plot in a cross section of the electric double-layer taken perpendicularly to the surface. The location of the surface is shown as a vertical line to the left and the bulk electrolyte lies to the right in the figure. (c) A counterion ($\oplus$) in the double layer has a cavity (excluded zone) around it, shown in white, where the centers of other ions cannot enter. The radius of the cavity is equal to the ion diameter, $d^h = 4.25$ Å. (d) The charge density distribution around a counterion in the double-layer system when its center is located at $z - d^S = 3$ Å, that is, 5.125 Å from the surface. The difference from (c) shows the polarization that the ion induces in its neighborhood, which is the charge density of the ion cloud of the ion at this position.[38]

in Figure 7.14d is obtained when *all* ions (counterions and coions) interact and correlate with each other irrespective of where in the system they happen to be. When a counterion happens to be at some other place, the surrounding charge density distribution is different depending on the distance of the ion from the surface. Panes (a)–(d) in Figure 7.15 show some examples of such average distributions for various locations of a counterion.

Going from the top to the bottom pane, we can see how the situation changes when a counterion approaches the surface. In the bulk phase, far from the surface, the average charge density around a positive counterion is spherically symmetric and there is a dominance of negative ions in its ion cloud. When the ion is located closer to the surface, like in

---

[38] Figure 7.14 modified from R. Kjellander, *J. Phys.: Condens. Matter* **21** (2009) 424101 (http://dx.doi.org/10.1088/0953-8984/21/42/424101), with permission. Original figure: © IOP Publishing. The figure is based on data from R. Kjellander and H. Greberg, *J. Electroanal. Chem.* **450** (1998) 233; *Errata: ibid.* **462** (1999) 273. The calculations were done in the Anisotropic Reference HNC (ARHNC) approximation [introduced in Chapter 9 (in the 2nd volume)], which was also used for the data in Figure 7.3.

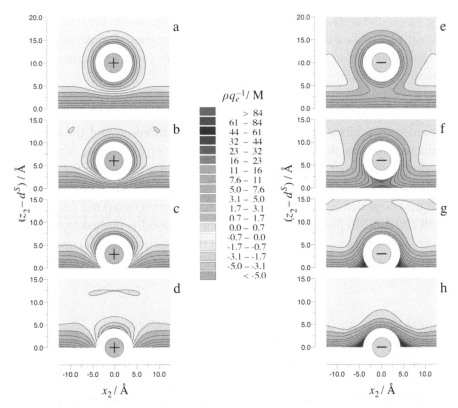

FIGURE 7.15 (**See color insert.**) Panes a – d: The same charge density distribution $\rho(\mathbf{r}_2|\mathbf{r}_1;+)$ around a counterion ($\oplus$) located at $\mathbf{r}_1$ as in Figure 7.14d, but shown for various locations of the ion at different distances from the surface (pane (c) in the current figure is the same as Figure 7.14d). The ion positions are $z - d^S = 10$ Å in (a), 6 Å in (b), 3 Å in (c) and 0 Å in (d); the latter shows a counterion in contact with the surface. Panes e – h: The corresponding charge density distribution $\rho(\mathbf{r}_2|\mathbf{r}_1;-)$ around a coion ($\ominus$) in the electric double-layer when it is located at the same distances from the surface as the counterion in the previous panes. All charge densities shown, including the density profile in Figure 7.14a, have been calculated when ion–ion correlations for all ions in the double layer are properly considered. The calculations have been done in the same integral equation theory as Figure 7.3. The density plotted in each pane has cylindrical symmetry around a vertical axis through the centre of the ion.[39]

pane (a), the density becomes somewhat distorted due to the influence from the surface. At the same time, the ion induces a distortion of the charge density close to the surface, which looks like in Figure 7.14b in the absence of a fixed ion (i.e., when all ions are free to move). When our counterion is closer to the surface, pane (b), the distortions are larger. Our positive counterion now repels other counterions quite strongly from the double-layer profile near the surface. In pane (c), this repulsion is even more pronounced and a substantial part of the other counterions cannot be near the surface because the cavity around our counterion

[39] Figure 7.15 modified from R. Kjellander, *J. Phys.: Condens. Matter* **21** (2009) 424101 (http://dx.doi.org/10.1088/0953-8984/21/42/424101), with permission. Original figure: © IOP Publishing. The figure is based on data from the same source as Figures 7.3 and 7.14, see footnote 38.

extends all the way to the surface. We also see that the negative ions in the ion cloud of our ion now are mainly located on the side away from the surface; they are repelled by the negative surface charge. Finally, in pane (d), our counterion is in contact with the surface and the distortions are maximal.

The corresponding situation when we follow one of the coions at various distances from the surface is depicted in panes (e)–(h) of Figure 7.15. Here the ion cloud around the negative coion is dominated by positive ions. In pane (e), the distortions around the ion and the surface are rather small. Note that the large green area at the top of the pane constitutes just a small positive deviation from zero (the corresponding light blue areas at the top of pane (a) and in Figure 7.14b show a small negative deviation from zero there). In pane (f), there has been built up a large density of positive ions in the region between our negative coion and the negative surface. In pane (g), this part of the ion cloud is pushed to the sides by the exclusion void around our coion, and in pane (h) the distortions are maximal. The high density of positive charge now lies in a region near the surface on the sides of our coion in the lateral directions.

7.2.2.2 Ion-Ion Correlations Forces: Influences on Density Profiles

Analogously to the case of an inhomogeneous hard sphere fluid discussed in Section 7.2.1.2, one can obtain an understanding of the density and charge distribution profiles in Figure 7.3 by using the first BGY equation to obtain the forces acting on the particles of the fluid. Thereby, one can also understand the deviations from the PB profiles shown in the same figure. Let now see how this can be achieved.

The BGY Equation 7.64 for planar geometry has the following form for a multi-component system

$$\frac{d[\ln n_i(z_1)]}{dz_1} = \beta F_i(z_1)$$

$$= -\beta \left[\frac{dv_i(z_1)}{dz_1} + \sum_j \int d\mathbf{r}_2 \, n_j(z_2) g_{ij}^{(2)}(\mathbf{r}_1, \mathbf{r}_2) \frac{\partial u_{ij}(\mathbf{r}_2, \mathbf{r}_1)}{\partial z_1} \right],$$

where the sum is over all species in the system. For $z_1 > 0$, we have $v_i(z_1) = -q_i \sigma^S z_1 / (2\varepsilon_r \varepsilon_0)$ in the current system. Since $u_{ij} = u_{ij}^{core} + u_{ij}^{el}$, where u_{ij}^{core} is the same as a hard sphere potential, the integral gives two contributions, one from hard core and one from electrostatic interactions, $F_i = F_i^{core} + F_i^{el}$, where F_i^{el} also contains the contribution from $v_i(z_1)$. The core interaction force equals

$$F_i^{core}(z_1) = -\sum_j \int d\mathbf{r}_2 \, n_j(z_2) g_{ij}^{(2)}(\mathbf{r}_1, \mathbf{r}_2) \frac{\partial u_{ij}^{core}(r_{12})}{\partial z_1}$$

$$= 2\pi d^h k_B T \int_{z_1 - d^h}^{z_1 + d^h} dz_2 \sum_j n_j(z_2) g_{ij}^{(2)\text{cont}}(z_1, z_2) \frac{z_1 - z_2}{d^h},$$

where the second line can be obtained in an analogous manner as for a hard sphere fluid, Equation 7.65,[40] and has the same physical interpretation. $F_i^{core}(z_1)$ arises from the collisions by ions of both species at the surface of an i-ion located at z_1.

The electrostatic interaction force is

$$F_i^{el}(z_1) = \frac{q_i \sigma^S}{2\varepsilon_r \varepsilon_0} - \sum_j \int d\mathbf{r}_2 \, n_j(z_2) g_{ij}^{(2)}(\mathbf{r}_1, \mathbf{r}_2) \frac{\partial}{\partial z_1}\left[\frac{q_i q_j}{4\pi \varepsilon_r \varepsilon_0 r_{12}}\right]$$

$$= \frac{q_i \sigma^S}{2\varepsilon_r \varepsilon_0} + \frac{q_i}{4\pi \varepsilon_r \varepsilon_0} \int d\mathbf{r}_2 \, \rho(\mathbf{r}_2|\mathbf{r}_1; i)\frac{z_1 - z_2}{r_{12}^3}.$$

This expression can alternatively be obtained as the force calculated via Coulomb's law from the charge distributions $\rho(\mathbf{r}_2|\mathbf{r}_1; i)$ plotted in Figure 7.15. A factor $(z_1 - z_2)/r_{12}$ makes a projection of the force on the z direction and an additional factor of $1/r_{12}^2$ originates from the force given by Coulomb's law. The x and y components of the total force vanish by symmetry.

Let us investigate F_i^{el} further. As we saw in Figure 7.14, the average charge distribution outside the surface looks very different when regarded from the perspective of the wall when all ions are free to move, pane (b), and when regarded from the perspective of an ion located close to the surface, pane (d). It is the latter perspective that is relevant when we determine $F_i^{el}(z_1)$; this is the actual environment that the ion experiences on average when it is located at z_1 and it determines the electrostatic force.

In the PB approximation, one neglects the ion-ion correlations in the double layer, which means that the force on an ion is calculated from the charge distribution in pane (b), $\rho(z)$. For an ion placed with its center at the cross in the figure, the force therefore would be

$$F_i^{PB}(z_1) = \frac{q_i \sigma^S}{2\varepsilon_r \varepsilon_0} + \frac{q_i}{4\pi \varepsilon_r \varepsilon_0} \int d\mathbf{r}_2 \, \rho(z_2)\frac{z_1 - z_2}{r_{12}^3} = -q_i \frac{d\psi(z_1)}{dz_1},$$

where $\rho(z_2)$ is inserted instead of $\rho(\mathbf{r}_2|\mathbf{r}_1; i)$ in the integral for F_i^{el}. The difference $\rho(\mathbf{r}_2|\mathbf{r}_1; i) - \rho(z_2)$ is, as mentioned earlier, the charge distribution of the ion cloud of the i-ion when $\mathbf{r}_2$ lies outside the exclusion cavity that surrounds the i-ion. Inside the cavity, where $\rho(\mathbf{r}_2|\mathbf{r}_1; i)$ is zero, this difference is equal to $-\rho(z_2)$, that is, the difference between the charge distributions in panes (c) and (d) of Figure 7.14. It is the existence of the ion cloud and the cavity around each ion that gives the difference between the PB prediction and the accurate result for all properties of the system as discussed in connection to Figure 7.3, for example the overcompensation of surface charge and the associated charge inversion in the profile.

To see why these features appear, it is useful to split F_i^{el} into two parts: (i) The force from the surface charge and the charge distribution in pane (c), which only considers the exclusion cavity. We will denote this force as $F_i^{el(Cav)}$. (ii) The remainder, $F_i^{el} - F_i^{el(Cav)}$, which

---

[40] The proof of Equation 7.65 given in the shaded text box on page 403f must be generalized to particles with $u = u^{el} + u^{core}$. This is straightforward; see the text box on page 450f, where the same technique is used to treat the collisional pressure in inhomogeneous fluids.

originates from the difference in charge distributions between panes (c) and (d), that is, the polarization charge around the i-ion induced by the interactions with it (the ion cloud). Therefore, we call this remainder $F_i^{\text{el(Pol)}}$, and we have $F_i^{\text{el}} = F_i^{\text{el(Cav)}} + F_i^{\text{el(Pol)}}$.

The charge distribution in pane (c), denoted by ρ^{Cav}, is given by $\rho^{\text{Cav}}(\mathbf{r}_2|\mathbf{r}_1; i) = 0$ when $r_{12} < d^{\text{h}}$ and $\rho^{\text{Cav}}(\mathbf{r}_2|\mathbf{r}_1; i) = \rho(z_2)$ otherwise and we have[41]

$$F_i^{\text{el(Cav)}}(z_1) = \frac{q_i \sigma^S}{2\varepsilon_r \varepsilon_0} + \frac{q_i}{4\pi\varepsilon_r\varepsilon_0} \int d\mathbf{r}_2\, \rho^{\text{Cav}}(\mathbf{r}_2|\mathbf{r}_1; i) \frac{z_1 - z_2}{r_{12}^3}. \tag{7.67}$$

Likewise, $F_i^{\text{el(Pol)}}$ is the force from the charge distribution $\rho(\mathbf{r}_2|\mathbf{r}_1; i) - \rho^{\text{Cav}}(\mathbf{r}_2|\mathbf{r}_1; i)$.

As we concluded when we discussed Figure 7.3, the large amount of counterions close to the surfaces, which causes the overcompensation of surface charge, is a consequence of a strong attractive force on the counterions, $F_+(z_1)$, which exceeds the PB prediction $F_+^{\text{PB}}(z_1)$ in the interval $1 \lesssim z_1 - d^S \lesssim 6\,\text{Å}$. This means that we have $|F_+(z_1)| > |F_+^{\text{PB}}(z_1)|$ for nearly all locations of a counterion between the positions shown in panes (b) and (d) in Figure 7.15. The question is why this is the case.

Calculations of the various force components show, in fact, that $F_+^{\text{el(Cav)}}$ is by far the dominant force contribution for all $z_1 - d^S \lesssim 5\,\text{Å}$ in the current system,[42] so the difference between $F_+(z_1)$ and $F_+^{\text{PB}}(z_1)$ mainly stems from this contribution. In the PB approximation, the charge distribution around an ion looks like in Figure 7.14b (the ion position is indicated by the cross). Other ions can be anywhere around the cross because all ions are treated as uncorrelating point ions in their interactions among themselves. Most of the ions close to the surface are counterions and those that are located between the cross and surface repel our counterion (at the cross) away from the surface. Our ion is attracted by the surface charge, but the presence of these counterions between the ion and the surface reduces the attraction. On the other hand, when ion sizes are considered, Figure 7.14b, the number of counterions between our ion ($\oplus$) and the surface is reduced because they do not fit there (as shown by the exclusion cavity). Therefore, the attraction to the surface charge is less reduced than in the PB case and the attraction is therefore appreciably larger in $F_+^{\text{el(Cav)}}$ than in the PB prediction.[43] This explains the observation.

The main reason for the large counterion attraction to the surface and hence the overcompensation is accordingly quite simple, namely that no counterions can be located between each counterion and the negative surface charge because they do not fit there. Since counterions are repelled electrostatically from each other, it would not make much difference if the hard core exclusion cavity for counterions around each counterion were smaller, that is, the core-core distance of closest approach of the counterions to each other were smaller. The electrostatic repulsions would anyway create a zone with decreased counterion density around each counterion. Therefore, counterions would still be expelled from between our

---

[41] See Equation 7.68 below for a way to write $F_i^{\text{el(Cav)}}(z_1)$ in terms of the density profile $\rho(z_2)$.

[42] The actual values of the force and its components are plotted in Figure 7.16, which can be skipped unless the reader wants to obtain quantitative information.

[43] The number of counterions outside our ion, to the right of it in Figure 7.14b, is also reduced, but the majority of the affected counterions are between the ion and the surface.

ion ($\oplus$) and the surface. In the present case, we can see in Figures 7.15c and d that the counterion density outside the cavity on either side of the counterion ($\oplus$) in the lateral directions is reduced because of this effect.

These electrostatic exclusion effects around the ion are included in $F_+^{el(Pol)}$, which also includes the electrostatic interactions with the negative coions. As regards the latter, we see in Figures 7.15c and d that when our counterion comes closer to the surface, the region of negative charge in the surrounding ion cloud is more and more restricted to the side away from the surface. When our ion is in contact with the surface, pane (d), its entire ion cloud, which has a net charge of opposite sign, is located above the ion and causes therefore an attraction away from the surface (the same kind of effect causes the depletion of ions close to an uncharged surface discussed earlier). This force counteracts the attraction towards the surface charge for our ion. It is this effect, which is completely absent from the PB approximation, that makes the contact density $n_+(d^S)$ to be smaller than the PB prediction, as observed in Figure 7.3. Otherwise, $F_+^{el(Pol)}$ has, in the present case, a quite small influence on F_+ and thereby on the density profiles for counterions.

For systems with counterions of higher valency, like trivalent ions and higher, the electrostatic interactions are much stronger and they contribute very significantly to the creation of a zone with decreased counterion density around each counterion. In such systems, the joint efforts of $F_+^{el(Pol)}$ and $F_+^{el(Cav)}$ cause the strongly attractive F_+ close to the surface when the surface charge density is sufficiently high. The mechanism is largely the same as in the present case, namely that a counterion close to the surface expels other counterions from the vicinity of the surface in the ion's neighborhood.

We have not yet mentioned the effects of the collisional force $F_+^{core}(z_1)$ on the counterions. For the system in Figure 7.15, it is attractive for $z_1 - d^S \lesssim 7$ Å, so it contributes to the attraction of counterions towards the surface in this range, but it gives a smaller contribution than $F_+^{el(Cav)}(z_1)$ for most z_1. Close to the surface, $F_+^{core}(z_1)$ must, of course, always be attractive since most ions will collide with an ion on the side away from the surface. The actual values of $F_i(z_1)$ and all its components can be found in Figure 7.16, which can be skipped unless the reader wants to obtain quantitative information. The comparison with the PB prediction is shown in Figure 7.16c.

The behavior of the force $F_-(z_1)$ on the coions and that of its components are more complicated and will not be treated in detail here. Suffice it to say that $F_-^{el(Pol)}$ and $F_-^{core}(z_1)$ have similar magnitudes but opposite signs for most z_1, so they counteract each other to a very large extent. This is an example of the competition between these two effects mentioned in footnote 9. Large values of the collisional and electrostatic forces can be inferred from Figures 7.15f–h, where there is a very large density of positive counterions on the bottom side of the coion ($\ominus$). This happens because both the surface and the coion are negatively charged and enforce each other's attraction of counterions. The net effect of $F_-^{el(Pol)}$ and $F_-^{core}(z_1)$ is, however, much smaller than each individually since they counteract each other.

The range of most interest for $F_-(z_1)$ is perhaps $z_1 - d^S \gtrsim 5$ Å, where the charge inversion of the profile takes place and the coions are attracted towards the surface, $F_-(z_1) < 0$. In this range, the net attraction arises primarily from $F_-^{el(Pol)}$, which is attractive there and

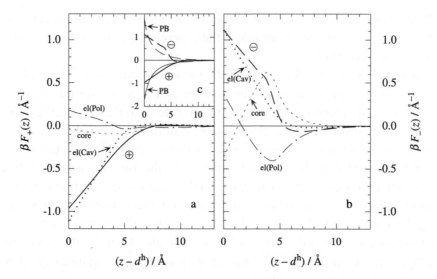

FIGURE 7.16 ★ The mean force F_i that acts on an ion in the electric double-layer system of Figures 7.3, 7.14 and 7.15.[44] The thick full line shows F_+ on a counterion (positive, $\oplus$) as a function of z and the thick dashed line shows F_- on a coion (negative, $\ominus$). The mean force F_i and its components $F_i^{el(Cav)}$ (dotted line), $F_i^{el(Pol)}$ (dash-dot) and F_i^{core} (short dashes) for a counterion ($i = +$) are shown in pane (a) and in pane (b) the same is shown for a coion ($i = -$). Pane (c) shows the mean force F_i [same as in (a) and (b)] compared with the PB prediction F_i^{PB} (thin lines) for a counterion ($\oplus$) and a coion ($\ominus$).

has a larger magnitude than $F_-^{core}(z_1)$. The electrostatic attraction towards the surface is due to the buildup of positive charge between the coion and the surface as seen in pane (f) of Figure 7.15 and to some extent in pane (e).

{EXERCISE 7.4

In Equation 7.67, we introduced $F_i^{el(Cav)}(z_1)$, which is the force acting on an ion located at z_1 inside a charge density profile $\rho(z)$ when the only ion-ion correlation effect included is the fact that the ion must be surrounded by a cavity with radius d^h where no other ions can enter, as illustrated in pane (c) of Figure 7.14. Outside this cavity, the charge density profile is assumed to be unperturbed by the presence of the ion. One can show that $F_i^{el(Cav)}$ satisfies the following relationship

$$\frac{dF_i^{el(Cav)}(z_1)}{dz_1} = \frac{q_i}{2\varepsilon_r\varepsilon_0 d^h} \int\limits_{z_1-d^h}^{z_1+d^h} dz_2\, \rho(z_2), \qquad (7.68)$$

so this derivative is proportional to the charge per unit area in a *fully* unperturbed charge density profile (without cavity) between $z_1 - d^h$ and $z_1 + d^h$. (An optional, somewhat challenging exercise is to show that Equation 7.67 implies Equation 7.68.)

---

[44] Figure 7.16 is based on data from the same source as Figures 7.3, 7.14 and 7.15, see footnotes 10 and 38.

Consider the electrostatic potential from a surface that decays for large z like $\psi(z) \sim Be^{-\kappa z}$ (the coefficient B can be written in terms of the effective surface charge density, but this is of no importance for the present task). In the PB approximation, the potential of mean force acting on an i-ion at z is $w_i^{PB}(z) = q_i\psi(z)$, while if ion-ion correlations are included, the ion acquires an effective charge q_i^{eff} and the potential of mean force decays, as we have seen, like $w_i(z) \sim q_i^{eff}\psi(z)$. Let us make a correction to the PB approximation where the cavity effect considered in Equation 7.68 is the only ion-ion correlation effect included. Show that

$$q_i^{eff} = q_i \frac{\sinh(\kappa d^h)}{\kappa d^h}$$

in this approximation, that is, if we approximate $dw_i(z)/dz = -F_i^{el(Cav)}(z)$, we have $w_i(z) \sim q_i^{eff}Be^{-\kappa z}$ for large z.}

APPENDIX 7A: SOLUTION OF PB EQUATION FOR A SURFACE IN CONTACT WITH A SYMMETRIC ELECTROLYTE

In this Appendix, we will obtain the analytic solution of the PB Equation 7.15 for a system consisting of a planar surface with surface charge density σ^S in contact with a symmetric electrolyte solution. We will thereby depart from the first integral of the PB equation in Equation 7.16. By using the identity $\sinh^2(x/2) = [\cosh(x) - 1]/2$, we can write this equation as

$$\left[\frac{d[\beta q\psi(z)]}{dz}\right]^2 = 4\kappa_D^2 \sinh^2\left(\frac{\beta q\psi(z)}{2}\right). \tag{7.69}$$

If we set $f(z) = \beta q\psi(z)/2$ and take the square root of both sides of the equation, we can write it as

$$\frac{df(z)}{dz} = -\kappa_D \sinh f(z), \tag{7.70}$$

where the minus sign follows since df/dz and f must have different signs; for instance, the derivative must be negative when $f(z)$ is positive and approaches zero when $z \to \infty$. Equation 7.70 is a separable differential equation and can be written $df/\sinh f = -\kappa_D dz$. By integration, we obtain[45]

$$\ln\left(\left|\tanh\left(\frac{f}{2}\right)\right|\right) = -\kappa_D z + \text{const.}$$

---

[45] The integral of the lhs can be obtained as follows. We have

$$\int \frac{df}{\sinh(f)} = \int \frac{df \sinh(f)}{\sinh^2(f)} = \int \frac{df \sinh(f)}{\cosh^2(f) - 1} = \int \frac{d\cosh(f)}{\cosh^2(f) - 1}$$

and since $\int dx/(x^2 - 1) = (1/2)\ln[(x-1)/(x+1)]$, we obtain our result from the fact that

$$\frac{1}{2}\ln\left(\frac{\cosh(f) - 1}{\cosh(f) + 1}\right) = \frac{1}{2}\ln\left(\frac{\sinh^2(f/2)}{\cosh^2(f/2)}\right) = \ln\left(\left|\tanh\left(\frac{f}{2}\right)\right|\right).$$

where "const" is an integration constant. By taking exp($\cdot$) of both sides, we see that this implies $\tanh(f/2) = C \exp(-\kappa_D z)$, where C is a constant that must have the same sign as $f(z)$, that is, the same sign as $\psi(z)$. It follows that

$$f(z) = 2\tanh^{-1}\left(Ce^{-\kappa_D z}\right) \equiv \ln\left[\frac{1 + Ce^{-\kappa_D z}}{1 - Ce^{-\kappa_D z}}\right] \tag{7.71}$$

since $\tanh^{-1}(x) \equiv (1/2)\ln[(1+x)/(1-x)]$. The constant C can be determined from Equation 7.8, which implies that

$$\left.\frac{df(z)}{dz}\right|_{z=d^S} = -\frac{\beta q \sigma^S}{2\varepsilon_r\varepsilon_0}.$$

From Equations 7.70 and 7.71, we see that

$$-\frac{1}{\kappa_D}\left.\frac{df(z)}{dz}\right|_{z=d^S} = \sinh f(d^S) = \sinh\ln\left[\frac{1+\mathcal{B}}{1-\mathcal{B}}\right] = \frac{2\mathcal{B}}{1-\mathcal{B}^2}.$$

where $\mathcal{B} = Ce^{-\kappa_D d^S}$. It follows that $\mathcal{B}$ satisfies the equation

$$\mathcal{B}^2 + \frac{4\varepsilon_r\varepsilon_0\kappa_D}{\beta q \sigma^S}\mathcal{B} - 1 = 0. \tag{7.72}$$

By solving this quadratic equation and selecting the solution $\mathcal{B}$ with the same sign as σ^S (and hence the same sign as ψ), we obtain $C = \mathcal{B}e^{\kappa_D d^S}$ and thus the solution $\psi(z) = 2f(z)/(\beta q)$ to the PB equation. From Equation 7.71, we have

$$\begin{aligned} \psi(z) &= \frac{2}{\beta q}\ln\left[\frac{1 + Ce^{-\kappa_D z}}{1 - Ce^{-\kappa_D z}}\right] \\ &= \frac{2}{\beta q}\ln\left[\frac{1 + \mathcal{B}e^{-\kappa_D(z-d^S)}}{1 - \mathcal{B}e^{-\kappa_D(z-d^S)}}\right] \quad \text{for } z \geq d^S \quad \text{(PB).} \end{aligned} \tag{7.73}$$

The constant $\mathcal{B}$ can also be expressed in terms of the potential $\psi(z)$ at the point of contact, $z = d^S$. From the first equality in Equation 7.71, it follows that $\tanh\left(f(d^S)/2\right) = Ce^{-\kappa_D d^S}$, which implies that $\mathcal{B} = \tanh\left(\beta q\psi(d^S)/4\right)$.

APPENDIX 7B: ELECTRIC DOUBLE-LAYERS WITH ION-WALL DISPERSION INTERACTIONS IN LINEARIZED PB APPROXIMATION

In this Appendix, we will investigate the effect of ion-wall dispersion interactions for electric double-layers. We will do this both for the electric double-layer outside a single wall in contact with a bulk electrolyte and for overlapping double-layers between two walls. Following essentially the treatment by Edwards and Williams,[46] we will thereby restrict ourselves to the linearized PB approximation for binary $z : -z$ electrolytes, so we have $n_+^b = n_-^b \equiv n^b$ and $q_+ = -q_- \equiv q$.

Let us start with a single charged surface with uniform surface charge density. The ion density of species i at distance z' from the surface is in the PB approximation equal to $n_i(z') = n_i^b e^{-\beta[q_i \psi(z') + v_i^{disp}(z')]}$ for $z' \geq d^S$, where $v_i^{disp}(z') = b_i^{disp}/(z')^3$. Provided that the exponent is small for all $z' \geq d^S$, we can linearize the exponential function and obtain

$$n_i(z') = n_i^b \left(1 - \beta \left[q_i \psi(z') + v_i^{disp}(z') \right] \right) \quad \text{(linearized PB)}. \tag{7.74}$$

The charge density is

$$\rho(z') = \sum_i q_i n_i(z') = -\beta \sum_i n_i^b q_i^2 \psi(z') - \frac{\beta}{(z')^3} \sum_i n_i^b q_i b_i^{disp}$$

$$= -\varepsilon_r \varepsilon_0 \kappa_D^2 \psi(z') - \frac{\beta n^b q [b_+^{disp} - b_-^{disp}]}{(z')^3}$$

and the Poisson equation $d^2 \psi(z')/dz'^2 = -\rho(z')/(\varepsilon_r \varepsilon_0)$ yields for $z' > d^S$

$$\frac{d^2 \psi(z')}{dz'^2} = \kappa_D^2 \left[\psi(z') + \frac{b_+^{disp} - b_-^{disp}}{2q(z')^3} \right] \quad \text{(linearized PB)}, \tag{7.75}$$

where we have used $\kappa_D^2 = 2\beta n^b q^2/(\varepsilon_r \varepsilon_0)$. In the absence of ion-wall dispersion interactions ($b_+^{disp} = b_-^{disp} = 0$), this is the usual linearized PB Equation 7.12, which has the solution given by Equation 7.13. The boundary condition we will use initially is that $\psi(z') \to 0$ when $z' \to \infty$ and that $\psi(d^S)$ has a given value. The general solution to Equation 7.75 is the sum of a particular solution to the equation and the general solution to $d^2 \psi(z')/dz'^2 = \kappa_D^2 \psi(z')$. The latter is given by $C_1 e^{-\kappa_D z'}$ with arbitrary constant C_1, where we have selected the solution that goes to zero at infinity.

Equation 7.75 has an explicit solution that can be expressed in terms of elementary functions and two nonelementary functions $\text{Shi}(x)$ and $\text{Chi}(x)$ defined as

$$\text{Shi}(x) = \int_0^x \frac{\sinh(t)}{t} dt$$

$$\text{Chi}(x) = \int_0^x \frac{\cosh(t) - 1}{t} dt + \ln(x) + C_{\text{Euler}},$$

---

[46] S. A. Edwards and D. R. M. Williams, *Phys. Rev. Lett.* **92** (2004) 248303.

where $C_{Euler} \approx 0.57722$ is the Euler constant. We will now show that there exists an expression with these functions that satisfies the equation.

We first define

$$f_1(x) = \sinh(x)\text{Chi}(x) - \cosh(x)\text{Shi}(x)$$

and it is simple to show by differentiation that

$$df_1(x)/dx = \cosh(x)\text{Chi}(x) - \sinh(x)\text{Shi}(x)$$

and

$$d^2f_1(x)/dx^2 = \sinh(x)\text{Chi}(x) - \cosh(x)\text{Shi}(x) + 1/x$$
$$= f_1(x) + 1/x.$$

This implies that the function $f_2(x)$ defined by

$$f_2(x) = f_1(x) + 1/x$$

(i.e., the rhs of the previous equation) has a second derivative given by $d^2f_2(x)/dx^2 = f_2(x) + 2/x^3$.

After these preliminaries, we will show that

$$\psi(z') = C_1 e^{-\kappa_D z'} + C_2 f_2(\kappa_D z') \tag{7.76}$$

satisfies Equation 7.75 with the given boundary conditions provided that the constants C_1 and C_2 are properly selected. Using our results for $f_2(x)$, we see that

$$\frac{d^2\psi(z')}{dz'^2} = \kappa_D^2 \left[C_1 e^{-\kappa_D z'} + C_2 \left(f_2(\kappa_D z') + \frac{2}{(\kappa_D z')^3} \right) \right]$$
$$= \kappa_D^2 \left[\psi(z') + \frac{2C_2}{(\kappa_D z')^3} \right]$$

By comparison with Equation 7.75, we see that we must select

$$C_2 = \frac{\kappa_D^3(b_+^{disp} - b_-^{disp})}{4q}$$

and by inserting $z' = d^S$ in Equation 7.76, we obtain

$$C_1 = [\psi(d^S) - C_2 f_2(\kappa_D d^S)]e^{\kappa_D d^S}.$$

If needed, one can use Equation 7.8 to express C_1 in terms of the surface charge density σ^S instead of $\psi(d^S)$. One can show that $f_2(x)$ decays asymptotically as

$$f_2(x) \sim -\left[\frac{2!}{x^3} + \frac{4!}{x^5} + \frac{6!}{x^7} + \frac{8!}{x^9} + \cdots\right] \quad \text{when } x \to \infty,$$

so $\psi(z') \to 0$ when $z' \to \infty$.

To summarize, we have for a single wall

$$\psi(z') = C_1 e^{-\kappa_D z'} + \psi_2(z'), \tag{7.77}$$

where

$$\begin{aligned}\psi_2(z') &\equiv C_2 f_2(\kappa_D z') \\ &\sim -\frac{b_+^{\text{disp}} - b_-^{\text{disp}}}{2q}\left[\frac{1}{z'^3} + \frac{12}{\kappa_D^2 z'^5} + \cdots\right]\end{aligned} \tag{7.78}$$

when $z' \to \infty$. Note that $\psi_2(z')$ is independent of $\psi(d^S)$ and hence independent of the surface charge density and other details of the surface. That information is included in C_1.

The ion density $n_i(z')$ from Equation 7.74 is given by

$$\begin{aligned}n_i(z') - n_i^b &= -\beta n_i^b\left[q_i C_1 e^{-\kappa_D z'} + q_i \psi_2(z') + v_i^{\text{disp}}(z')\right] \\ &\sim -\beta n^b\left[\frac{b_+^{\text{disp}} + b_-^{\text{disp}}}{2z'^3} - \frac{q_i\, 6(b_+^{\text{disp}} - b_-^{\text{disp}})}{q\kappa_D^2 z'^5} + \cdots\right]\end{aligned} \tag{7.79}$$

when $z' \to \infty$. Note that the first term in the square bracket is the same for anions and cations, while the second term has different signs for the two ions species (q_i/q is 1 for the cations and -1 for the anions).

For the case of *overlapping double layers* between two equal walls at distance D^S from each other, we have for $-L/2 \le z \le L/2$

$$n_i(z) = n_i^b\left(1 - \beta\left[q_i\psi(z) + v_i^{\text{disp,tot}}(z)\right]\right) \quad \text{(linearized PB)} \tag{7.80}$$

where the origin of z lies at the midplane, $L = D^S - 2d^S$ and

$$v_i^{\text{disp,tot}}(z) = v_i^{\text{disp}}\left(\frac{D^S}{2} + z\right) + v_i^{\text{disp}}\left(\frac{D^S}{2} - z\right).$$

The linearized PB equation for this case is

$$\frac{d^2\psi(z)}{dz^2} = \kappa_D^2\left[\psi(z) + \frac{b_+^{\text{disp}} - b_-^{\text{disp}}}{2q}\left(\frac{1}{((D^S/2) + z)^3} + \frac{1}{((D^S/2) - z)^3}\right)\right].$$

Since the equation is linear, its solution can be written as a sum of the general solutions of Equation 7.75 for the individual walls. Therefore, we have

$$
\psi(z) = C_3 \left[\exp\left(-\kappa_D \left(\frac{D^S}{2} + z\right)\right) + \exp\left(-\kappa_D \left(\frac{D^S}{2} - z\right)\right) \right]
$$
$$
+ C_2 \left[f_2 \left(\kappa_D \left(\frac{D^S}{2} + z\right)\right) + f_2 \left(\kappa_D \left(\frac{D^S}{2} - z\right)\right) \right],
$$

where C_2 is the same as before and C_3 is a constant that can be determined from the boundary conditions at the surfaces, $z = \pm L/2$, in an analogous manner as C_1. When $D^S \to \infty$, we have $C_3 \to C_1$ and at the midplane ($z = 0$) we have

$$
\psi(0) \sim 2 \left[C_1 \exp(-\kappa_D D^S/2) + C_2 f_2(\kappa_D D^S/2) \right]
$$
$$
\sim -\frac{b_+^{\mathrm{disp}} - b_-^{\mathrm{disp}}}{q} \left[\frac{1}{(D^S/2)^3} + \frac{12}{\kappa_D^2 (D^S/2)^5} + \cdots \right] \quad \text{when } D^S \to \infty. \tag{7.81}
$$

From Equation 7.80, it follows that the density at the midplane is given by

$$
n_i(0) - n_i^b = -\beta n_i^b \left[q_i \psi(0) + v_i^{\mathrm{disp,tot}}(0) \right]
$$
$$
\sim -\beta n^b \left[\frac{b_+^{\mathrm{disp}} + b_-^{\mathrm{disp}}}{(D^S/2)^3} - \frac{q_i 12(b_+^{\mathrm{disp}} - b_-^{\mathrm{disp}})}{q \kappa_D^2 (D^S/2)^5} + \cdots \right] \tag{7.82}
$$

when $D^S \to \infty$.

CHAPTER 8

Surface Forces

I N THIS CHAPTER, WE will treat interactions between two planar parallel walls when a fluid of spherical particles fills the slit between their surfaces. We will mainly deal with **surface forces**, that is, forces between the walls that depend on the microscopic state of the matter close to their surfaces, but we will also deal with interactions like the van der Waals interaction between the two walls themselves, which may be denoted as **body forces**. The latter are, however, affected by the state of the medium between the surfaces. Our main concern is the total interaction between the walls irrespective of its origin, so the distinction is not very important in our cases.

We will treat interactions between walls immersed in simple fluids, where the structure of the fluid in the slit gives rise to the so-called structural forces between the surfaces and interactions between charged surfaces in electrolytes, the so-called electric double-layer interactions. For the latter, we will specifically consider the effect of ion-ion correlations and the intricate coupling between electrostatic and van der Waals interactions, including ion-wall dispersion interactions. These effects cause important deviations from the predictions of the PB approximation and of the traditional theory of combined double-layer interactions and van der Waals interactions, the Derjaguin-Landau-Verwey-Overbeek (DLVO) theory, which are also treated here.

8.1 GENERAL CONSIDERATIONS

We consider systems with two walls that have a very large lateral surface area A and we will often take the limit when A becomes infinitely large. In most cases that we will deal with, the fluid in the slit between the walls is in equilibrium with a bulk phase. This kind of system is, for example, relevant when considering planar macroscopic particles (macroparticles) immersed in the fluid. The model with infinitely large surfaces is useful for such macroparticles in parallel arrangement when the distance between them is much smaller than their lateral extent. Each macroparticle is assumed to be sufficiently thick so there are negligible interactions between fluid particles on either side of it (i.e., across the macroparticle). We will start by dealing with the interactions between two such macroparticles and we can neglect

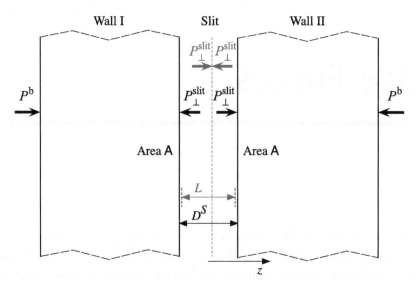

FIGURE 8.1 **(See color insert.)** A sketch showing a cross section of two planar parallel macroparticles (walls) immersed in a fluid. The fluid in the slit between the macroparticles is in equilibrium with a bulk fluid that surrounds them. The macroparticles are thick and extend upwards and downwards outside the figure (not shown). The area A of each surface that faces the slit is very large and the surface separation is D^S. The space available for the centers of the fluid particles in the slit has a width L and the z axis is perpendicular to the surfaces. The perpendicular pressure component $P_\perp^{slit}$ has the same value everywhere in the slit and the bulk pressure that acts from the outside is P^b. The dashed vertical red line indicates an imagined plane parallel to the surfaces and the red arrows show the pressure acting across this plane; the right arrow shows the force (per unit area) that the part of the system to the right of the plane exerts on the part to the left of the plane and vice versa for the left arrow.

boundary effects at the boundaries of the surfaces in the slit because the lateral dimensions are very large. Figure 8.1 shows a sketch of this kind of system.

The origin of the coordinate system is placed in the middle between the surfaces and the z axis is perpendicular to them. The number density of particles in the slit is $n(z)$ and the possible z coordinates for the centers of the particles in the slit lie in the interval $-L/2 \leq z \leq L/2$. The distance between the surfaces is D^S and the distance of closest approach of the particle centers to surface I is d_I^S and to surface II is d_{II}^S, so we have $L = D^S - d_I^S - d_{II}^S$.

8.1.1 The Disjoining Pressure and the Free Energy of Interaction

Consider a plane that is parallel to the two surfaces and located anywhere between them (at $-L/2 \leq z \leq L/2$), for instance the plane indicated by the red dashed line in Figure 8.1. The plane is a mathematical construct and does not influence the particles in any way. There is a force that acts between the two parts of the system on either side of the plane. This force is equal to the sum of the force due to momentum transfer across the plane and the interactional force between the part of the system on one side of the plane and the part on the other side of it. The average force per unit area, the pressure $P_\perp^{slit}$ in the perpendicular direction in the slit, is equal to the total force that acts across the plane divided by its area (red arrows in the figure). At equilibrium, $P_\perp^{\text{slit}}$ must be the same across any plane between the surfaces; otherwise, a slice of the liquid between two such planes would move as a whole (cf. the discussion in Section 5.3). The force per unit area due to momentum transfer across a plane located at coordinate z (the ideal part of the pressure) is $k_B T n_{\text{tot}}(z)$ and varies with z, cf. Figure 5.6. Likewise, the interactional force per unit area varies with z, but the sum of these contributions, $P_\perp^{\text{slit}}$, is constant everywhere in the slit and is equal to the average force per unit area that acts between the two walls.

The outer surface of each planar macroparticle – the surface not facing the other macroparticle – is in direct contact with the surrounding bulk phase. The fluid is inhomogeneous close to this surface but is homogeneous far from it. The pressure on this side, the outer side, is equal to the bulk pressure P^b. On each particle, there is accordingly the pressures P^b from the outside and $P_\perp^{\text{slit}}$ from inside the slit (black bold arrows in Figure 8.1). The net pressure, $P_\perp^{\text{slit}} - P^b$, is the total net force per unit area between the two particles and we introduce the pressure

$$P^S = P_\perp^{\text{slit}} - P^b, \tag{8.1}$$

called the **disjoining pressure**, to describe the interparticle force for the macroparticles. The deviation of $P_\perp^{\text{slit}}$ from P^b gives rise to the net force between the macroparticles when they are close to each other and P^S times the surface area A is the total net force. If $P_\perp^{\text{slit}}$ is larger than P^b, the surfaces are repelled from each other, $P^S > 0$, and in the reverse situation the surfaces are attracted to each other, $P^S < 0$. When the particles are far apart, the fluid in the middle of the slit is in its bulk state, so $P_\perp^{\text{slit}} = P^b$ and $P^S = 0$. The force between the macroparticles is then zero.

The **free energy of interaction per unit area** (potential of mean force per unit area), $\mathcal{W}^S(D^S)$, between the two planar surfaces at separation D^S from each other can be calculated by integrating P^S as a function of surface separation

$$\mathcal{W}^S(D^S) = \int_{D^S}^{\infty} P^S(D) dD, \tag{8.2}$$

which follows from[1] $P^S(D^S) = -d\mathcal{W}^S(D^S)/dD^S$. This requires that $P^S(D^S)$ is known for all surface separations from infinity to the D^S value in question. Note that $\mathcal{W}^S(D^S) \to 0$ when $D^S \to \infty$.

The relationship between the free energy of interaction and the pressure can be further illustrated as follows. Consider the free energy (grand potential) per unit area for the system in the slit, $\Theta^{S,\text{slit}} \equiv \Theta^{\text{slit}}/A$, at surface separation D^S and subtract the corresponding free energy per unit area of an equally wide slice of bulk fluid, that is, $-P^b D^S$ as obtained from Equation 2.122 with $V = D^S A$. This difference, which equals

$$\mathcal{W}^{\text{slit}}(D^S) \equiv \Theta^{S,\text{slit}}(D^S) + P^b D^S, \tag{8.3}$$

---

[1] For a system with surface area A, we have $P = d\mathcal{W}/dD = d(A\mathcal{W})/d(AD) = dW/dV$, where $W = A\mathcal{W}$ is the free energy and $V = AD$ is the volume.

remains finite when $D^S \to \infty$ and the free energy of interaction is given by

$$\mathscr{W}^S(D^S) = \mathscr{W}^{\text{slit}}(D^S) - \mathscr{W}^{\text{slit}}(\infty), \tag{8.4}$$

which goes to zero in the same limit. Note that this implies

$$P^S(D^S) = -\frac{d\mathscr{W}^S(D^S)}{dD^S} = -\frac{d\Theta^{S,\text{slit}}(D^S)}{dD^S} - P^b = P_\perp^{\text{slit}} - P^b$$

in accordance with Equation 8.1.

The total net force $P^S A$ is the actual force that acts between the surfaces and one can, at least in principle, measure it by techniques such as the surface-force apparatus (SFA).[2] For practical reasons, in such measurements the geometry of the surfaces is, however, not planar. One usually measures the force between two crossed cylinders. The disjoining pressure P^S between infinitely large planar surfaces and the force between surfaces with other geometries can be converted to each other by means of the so-called *Derjaguin approximation*.[3]

As an illustration of surface forces, we initially specialize to interactions between planar charged surfaces in contact with an electrolyte solution, the so-called **electric double-layer interactions**. We will return to the general case of surface forces for fluids consisting of spherical particles in Section 8.3.

8.1.2 Electric Double-Layer Interactions, Some General Matters

We consider a system with two hard, smooth walls that, in general, have different surface charge densities which are smeared out uniformly. For simplicity, the dielectric constant of the walls is initially assumed to be the same as the solvent, ε_r. To start with, we will treat cases where all ions have the same diameters and the two walls have the same surface charge density σ^S. The ions are assumed to be charged hard spheres and they interact only electrostatically with the walls apart from the hard core–hard wall interactions. The distance between the walls is D^S, as measured between the planes of surface charges at the walls. The distance of closest approach of the ion centers to each layer of surface charge is d^S, so the space available to the ions centers in the slit has the width $L = D^S - 2d^S$. The possible z coordinates for the ions span the interval $-L/2 \le z \le L/2$. The system is symmetric around the midplane between the surfaces; the total density of ions $n_{\text{tot}}(z)$ and the charge density $\rho(z)$ are equal on either side of it, $n_{\text{tot}}(-z) = n_{\text{tot}}(z)$ and $\rho(-z) = \rho(z)$. A sketch of the system is shown in Figure 8.2a.

In most cases that we will consider, the electrolyte solution in the slit between the surfaces is in equilibrium with a bulk solution of a given composition. Such a system is, for example,

---

[2] This technique is, for example, described in J. N. Israelachvili, *Intermolecular and surface forces*, 3rd edition (Academic Press, 2011).

[3] The Derjaguin approximation relates the free energy of interaction per unit area, $\mathscr{W}^S(D^S)$, between two planar surfaces at separation D^S to the force $F(D^S)$ between two particles with curved surfaces: $F(D^S) \approx 2\pi \mathcal{R}_{\text{eff}} \mathscr{W}^S(D^S)$, where $\mathcal{R}_{\text{eff}}$ is a kind of "effective" radius for the curved surfaces. For the case of two spheres with radii $\mathcal{R}_1$ and $\mathcal{R}_2$, we have $\mathcal{R}_{\text{eff}}^{-1} = \mathcal{R}_1^{-1} + \mathcal{R}_2^{-1}$ and for two orthogonally crossed cylinders with radii $\mathcal{R}_1$ and $\mathcal{R}_2$, we have $\mathcal{R}_{\text{eff}} = [\mathcal{R}_1 \mathcal{R}_2]^{1/2}$. For curved surfaces, the distance D^S is measured between the closest points on the two surfaces. The Derjaguin approximation is accurate when the radii are large compared to the surface separation.

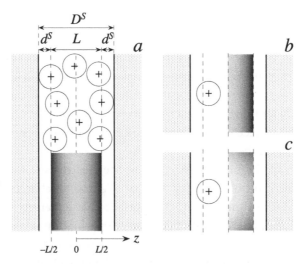

FIGURE 8.2 (a) Sketch of a slit filled with an electrolyte between two charged walls. The uniform charge density σ^S is located at the surface of each wall (thick black line) and the distance of closest approach of the ion centers to the wall is d^S; in the sketch, d^S is equal to the ion radius $d^h/2$ but this is not necessary. An instantaneous configuration of ions is shown in the top of the slit. It is assumed that σ^S is negative, so the counterions to the surfaces are positive. In the bottom, the average charge density $\rho(z)$ is drawn as a contour plot with the shade of gray indicating the charge density values. (b) In the PB approximation where ion-ion correlations in the electric double-layer are neglected, a fixed ion placed, for example, in the left half of the system does not affect the average charge distribution in the right half. (c) A sketch of how the charge density distribution may look like in the right half when ion-ion correlations are properly included for the situation in pane (b). This counterion repels other counterions and causes a depletion of them indicated by the light gray, shaded area in the right half. The figure depicts a situation where the surface distance is so short that virtually no coions remain between the surfaces.

relevant as a model for many kinds of clay particles in parallel arrangement, provided that, as before, the distance D^S between the particles is much smaller than their lateral extent. The net force per unit area between the particles is, like before, the disjoining pressure given by Equation 8.1. In actual measurements of P^S, there are several kinds of contributions to the interactions; here we are only considering the contribution to surface forces from double-layer interactions.

A slightly different case is a system with two equally charged surfaces where the slit only contains counterions to the surfaces; there is no equilibrium with some bulk electrolyte phase. The amount of counterions in the slit is then precisely what is needed to make the total system electroneutral, that is, the sum of the surface charges of the walls and the counterion charges is zero. This kind of model is relevant for planar macroscopic particles immersed in or in contact with a pure solvent. The only ions present are those between the charged surfaces. In this case, the double-layer interaction between the particles is solely due to the state of the ionic solution in the slit; the state of the solvent, modeled as a dielectric continuum, is the same everywhere and does not contribute to the pressure. Therefore, $P^b = 0$ and the pressure $P^S = P^{slit}_{\perp}$.

The pressure $P_\perp^{\text{slit}}$ in the slit can be evaluated for any plane located at any z-coordinate $z^\bullet$, where $-L/2 \leq z^\bullet \leq L/2$. As pointed out before, the value of $P_\perp^{\text{slit}}$ is the same for any such plane. The pressure consists of the ideal part due to momentum transfer across the plane at $z^\bullet$ and the part from interactions across this plane. We start by selecting the plane at $z^\bullet = L/2$, the point of closest approach of the ion centers to the right surface in Figure 8.2a.

The pressure due to momentum transfer across this plane is $k_B T n_{\text{tot}}(L/2)$, where $n_{\text{tot}}(L/2)$ is the *contact density* at the surface for ions of all species (cf. Equation 7.10). This force is the collisional force of the ions on the surface.

The interactional part of the pressure is equal to the electrostatic force per unit area between all charges to the right of the plane at $z^\bullet = L/2$ and all charges to the left of it. The only charge to the right of this plane is the surface charge density σ^S at $z = D^S/2$. The charges to the left of the plane consist of the charge distribution $\rho(z)$ of the ions in the slit and the surface charge density σ^S at $z = -D^S/2$. The force per unit area between the two surface charge densities is $(\sigma^S)^2/(2\varepsilon_r\varepsilon_0)$ and is independent of their separation D^S, as always for infinitely large uniform sheets of charge according to electrostatics; recall that electrostatic potential at distance s from σ^S is equal to $-\sigma^S s/(2\varepsilon_r\varepsilon_0)$, so the electric field given by the negative derivative is constant on either side.

The remaining part is the average force per unit area between σ^S at $z^\bullet = L/2$ and the charge distribution $\rho(z)$. We thereby start with the interaction between σ^S and the sheet of charge $\rho(z)dz$ of width dz. This sheet is also uniform in the lateral directions, so the force does not depend on where this sheet is located, that is, it is independent of z. Therefore, the force on the right surface charge σ^S from the entire charge density $\rho(z)$ is independent of how the ions are distributed as a function of z. The only thing that matters is the total charge per unit area $\int_{-L/2}^{L/2} dz\, \rho(z) = -2\sigma^S$ and the force per unit area is therefore $\sigma^S \times (-2\sigma^S)/(2\varepsilon_r\varepsilon_0)$. Together with the interaction between the two surface charge densities, we obtain a total interactional force per unit area equal to $-(\sigma^S)^2/(2\varepsilon_r\varepsilon_0)$.

The sum of the interactional and momentum transfer forces per unit area is $P_\perp^{\text{slit}}$, so we have

$$P_\perp^{\text{slit}} = k_B T n_{\text{tot}}(L/2) - \frac{(\sigma^S)^2}{2\varepsilon_r\varepsilon_0}. \tag{8.5}$$

This result is known as the **contact theorem**, a general exact relationship that can be proven formally for an electrolyte near a hard surface with a uniform surface charge density (see Section 8.5). It says that for a surface with a given σ^S in contact with a given electrolyte, the pressure $P_\perp^{\text{slit}}$ is determined by the value of the contact density.

The contact theorem can be utilized also for the outer surface of a planar particle (the side that is in contact with a bulk phase), provided this surface is smooth and hard and has a uniform surface charge density, say σ_{out}^S.[4] The charge distribution of the ion cloud outside this surface is also uniform in the lateral directions, so the analogous arguments apply: the total charge per unit area of the ion cloud is $-\sigma_{\text{out}}^S$, which gives the interactional force between the surface charge density σ_{out}^S and the ion cloud (across the outer surface) equal to $\sigma_{\text{out}}^S \times (-\sigma_{\text{out}}^S)/(2\varepsilon_r\varepsilon_0)$. The momentum transfer force per unit area is $k_B T n_{\text{tot}}^{\text{out,cont}}$,

---

[4] Remember that we assume that the particle is sufficiently thick so electrostatic interactions across the particle are negligible.

where $n_{\text{tot}}^{\text{out,cont}}$ is the contact density at the outer surface, and we have

$$P^{\text{b}} = k_B T n_{\text{tot}}^{\text{out,cont}} - \frac{(\sigma_{\text{out}}^S)^2}{2\varepsilon_r\varepsilon_0} \tag{8.6}$$

The pressure P^S is the difference between $P_\perp^{\text{slit}}$ from Equation 8.5 and this pressure. In case the surface charge densities on both sides of each particle are the same, the electrostatic terms cancel in the difference and $P^S = k_B T[n_{\text{tot}}(L/2) - n_{\text{tot}}^{\text{out,cont}}]$. Since the electrolyte in the slit and that on the outside are in different conditions, the contact pressures on either side of each particle are different. When the particle separation is changed, $n_{\text{tot}}(L/2)$ varies, while $n_{\text{tot}}^{\text{out,cont}}$ is constant. This gives the variation in double-layer interactions between two particles as a function of their separation. Even in the case where σ^S and σ_{out}^S are different, this variation is given by the changes in $n_{\text{tot}}(L/2)$ since all other terms are constant.

The decay of double-layer interactions for large surface separations has already been dealt with in Section 7.1.2.2, where the free energy of interaction per unit area $\mathscr{W}^S(D^S)$ was investigated for large D^S. The disjoining pressure between the surfaces is $P^S = -d\mathscr{W}^S(D^S)/dD^S$. For two surfaces I and II, the pressure $P_{\text{I,II}}^S$ can be obtained by taking the derivative of Equation 7.43, which yields

$$P_{\text{I,II}}^S(D^S) \sim \frac{2\sigma_{\text{I}}^{S\text{eff}}\sigma_{\text{II}}^{S\text{eff}}}{\mathcal{E}_r^{\text{eff}}\varepsilon_0}e^{-\kappa D^S} \quad \text{when } D^S \to \infty. \tag{8.7}$$

This is valid in the general case when the decay parameter κ is real, that is, the solution κ to

$$\kappa^2 = \frac{\beta}{\mathcal{E}_r^\star(\kappa)\varepsilon_0}\sum_j q_j q_j^\star n_j^{\text{b}},$$

(Equation 6.163). For the exponentially damped oscillatory case (complex κ), the decay of $P_{\text{I,II}}^S(D^S)$ can be obtained from Equation 7.46 and for the two-term plain exponential case (real κ and κ'), it can be obtained from Equation 7.45. All these results apply when the electrolyte in the slit is in equilibrium with a bulk electrolyte.

The results so far are general for the kind of electrolyte systems with charged surfaces that we discuss; we have not done any approximation for the given model. In the next section, we will treat the same systems in the PB approximation. Note that the result in Equation 8.7 is valid also in this approximation, where we have $\mathcal{E}_r^{\text{eff}} = \varepsilon_r$, $\mathcal{E}_r^\star = \varepsilon_r$, $q_j^\star = q_j$ and $\kappa = \kappa_{\text{D}}$.

8.2 POISSON-BOLTZMANN TREATMENT OF ELECTRIC DOUBLE-LAYER INTERACTIONS

In Section 7.1.1, we treated double layers near a single charged surface in contact with a bulk electrolyte in the PB approximation. Let us start with an observation regarding this approximation that follows from Equation 8.6 for the pressure on an outer surface, which is equal to the bulk pressure. Equation 7.9 can be written

$$n_{\text{tot}}^{\text{out,cont}} = \frac{\beta (\sigma_{\text{out}}^{\mathcal{S}})^2}{2\varepsilon_r \varepsilon_0} + n_{\text{tot}}^{\text{b}} \quad \text{(PB)}, \tag{8.8}$$

where $n_{\text{tot}}^{\text{out,cont}}$ is $n_{\text{tot}}(d^{\mathcal{S}})$ in the notation of Section 7.1. If we insert this into Equation 8.6, we see that $P^{\text{b}} = k_B T n_{\text{tot}}^{\text{b}}$, which is the same as the pressure of an ideal gas of density $n_{\text{tot}}^{\text{b}}$, that is, *the bulk electrolyte solution it treated as if the ions do not interact with each other!* This is a consequence of the fact that the ions in the ion cloud near the surface and therefore the ions in the bulk are treated as point ions that do not correlate with each other. This, of course, poses a severe restriction on the accuracy of the PB approximation for double-layer systems. As we have seen, this approximation works reasonably well for monovalent aqueous electrolyte solutions in many cases, but as soon as ion-ion correlations are important, it will not work well.

Incidentally, it should be noted that in our previous treatment of bulk electrolytes in the PB approximation, for example in the DH theory, we did *not* treat the solution as an ideal gas because we included correlations between a central ion and the ions in its ion cloud. The correlations between the ions in the ion cloud were ignored, but the central ion was a "representative" of the ions in the solution, so the interactional contributions to thermodynamical entities were approximately included. Also in this case, the accuracy of the PB approximation will suffer when ion-ion correlations in the ion cloud are important. In double-layer systems, the charged surface has a role in the PB approximation that corresponds to the central ion in the bulk case. All ion-ion correlations are then ignored, both in ion clouds of the surface and in the bulk phase.

8.2.1 Equally Charged Surfaces

Let us now turn to the electrolyte solution in the slit between two equally charged surfaces. As we have seen, the PB equation can be written as (Equation 7.5)

$$\left[\frac{d\psi(z)}{dz} \right]^2 = \frac{2}{\beta \varepsilon_r \varepsilon_0} \sum_j n_j^{\text{b}} e^{-\beta q_j \psi(z)} + \text{const} \quad \text{(PB)},$$

which holds in the electrolyte phase. Since $n_j(z) = n_j^{\text{b}} e^{-\beta q_j \psi(z)}$ and $n_{\text{tot}}(z) = \sum_j n_j(z)$, we can write this as

$$\left[\frac{d\psi(z)}{dz} \right]^2 = \frac{2}{\beta \varepsilon_r \varepsilon_0} n_{\text{tot}}(z) + \text{const} \quad \text{(PB)} \tag{8.9}$$

and for the present case we can determine the constant in the following manner. Since the system is symmetrical around the midplane, $d\psi(z)/dz$ is zero for $z = 0$. By setting $z = 0$ in the equation, we find that $0 = 2n_{\text{tot}}(0)/\beta \varepsilon_r \varepsilon_0 + \text{const}$, so we have

$$\left[\frac{d\psi(z)}{dz} \right]^2 = \frac{2}{\beta \varepsilon_r \varepsilon_0} [n_{\text{tot}}(z) - n_{\text{tot}}(0)] \quad \text{(PB)}. \tag{8.10}$$

An integration of the Poisson equation $-\varepsilon_r\varepsilon_0 d^2\psi(z)/dz^2 = \rho(z)$ from $z = 0$ to $z = L/2$ yields

$$-\varepsilon_r\varepsilon_0 \left.\frac{d\psi(z)}{dz}\right|_{z=L/2} = \int_0^{L/2} dz\,\rho(z) = -\sigma^S,$$

where we have used the facts that the derivative is zero at the midplane and that the charge of the ion cloud in each half of the system must neutralize the surface charge at the corresponding wall. By setting $z = L/2$ in Equation 8.10 and using this result for the derivative, we obtain

$$\frac{(\sigma^S)^2}{2\varepsilon_r\varepsilon_0} = k_B T[n_{\text{tot}}(L/2) - n_{\text{tot}}(0)]$$

and from the contact theorem (8.5), it therefore follows

$$P_\perp^{\text{slit}} = k_B T n_{\text{tot}}(L/2) - \frac{(\sigma^S)^2}{2\varepsilon_r\varepsilon_0} = k_B T n_{\text{tot}}(0) \quad \text{(PB)}. \tag{8.11}$$

Thus, the pressure in the slit can be evaluated in the PB approximation from the value of the total ion density at the midplane. Equation 8.1 and the rhs of Equation 8.11 imply that the double-layer disjoining pressure

$$P^S = k_B T(n_{\text{tot}}(0) - n_{\text{tot}}^b) \quad \text{(PB)} \tag{8.12}$$

and we see that $P^S \to 0$ when the surface separation increases since the density in the middle, $n_{\text{tot}}(0)$, goes to the bulk density in the limit of infinite separation.

Let us look a bit closer at the rhs of Equation 8.11. We have

$$P_\perp^{\text{slit}} = k_B T n_{\text{tot}}(0) \quad \text{(PB)},$$

which means that $P_\perp^{\text{slit}}$ in the PB approximation is only due to momentum transfer across the midplane; there is no contribution from an interactional force. The electrostatic interaction between the half of the system to the right of the midplane and the half to the left is zero, which can be realized as follows.

Consider the electrostatic field from the right half of the system, which encompasses the surface charge density σ^S at $z = D^S/2$ and the charge density $\rho(z)$ for $0 \le z \le L/2$. Since both charge distributions are uniform in the lateral directions, the electrostatic field for $z < 0$ does not depend on where the charges are located. Since the right half of the system has a total change of zero, the resulting field from it must be zero in the left half of the system. This means that any ion in the left half of the system does not interact electrostatically with the right half of the system, so there is no interactional force across the midplane.

Figure 8.2b shows an illustration of this situation where it, in accordance with the PB approximation, is assumed that the ion to the left does not perturb the right half of the system. The zero interactional force across the midplane is a consequence of the neglect of ion-ion correlations in the ion cloud of the surfaces in the PB approximation. In reality, the interactions between the left and the right half across the midplane contain the sum of the interactions between each ion in the right half with each ion in the left half averaged over all possible positions of the ions. An ion in the left half affects the distribution of ions in the right half as depicted in Figure 8.2c. The field that this ion feels from ions in the right half is not due to the charge distribution $\rho(z) = \sum_i q_i n_i(z)$ for $0 \leq z \leq L/2$, but rather the distribution $\sum_i q_i n_i(z) g_{ij}^{(2)}(\mathbf{r}, \mathbf{r}')$, where $\mathbf{r}'$ is the position of the ion in the left half (here of species j). Thus, the field from this charge distribution and the surface charge density σ^S at $z = D^S/2$ is *not* equal to zero. The same applies for all ions in the left half, so the electrostatic disjoining pressure across the midplane *is not zero*. It is zero in the PB approximation since we ignore ion-ion correlations and set $g_{ij}^{(2)}(\mathbf{r}, \mathbf{r}') = 1$ as a rough approximation.[5]

We have encountered a similar illustration earlier in Figure 7.14, where the electrostatic field in the PB approximation is calculated from the distribution in pane (b), while an ion at the cross would experience electrostatic field from the distribution in pane (d). Compare also with panes (a) and (b) in Figure 7.15, which shows how a more distant counterion polarizes the charge density close to the surface (at the bottom in these figures). The influence of ion-ion correlations on double-layer interactions will be treated in more detail in Section 8.5. Here we will continue with the PB approximation for the case of two surfaces with equal surface charge densities.

For very large surface separations, the density profile near one surface is in practice unaffected by the presence of the other surface. The ion density $n_{tot}(0)$ at the midplane is then equal to the bulk value and P^S is zero, as follows from Equation 8.12. When the surface separation is decreased, the density profiles from either surface start to overlap, which means that $n_{tot}(0)$ starts to deviate from n_{tot}^b and P^S becomes nonzero. For small overlaps, the deviation of $n_{tot}(0) - n_{tot}^b$ from zero is small and, as shown in the following example, one can obtain the value from a superposition of the electrostatic potentials from two single surfaces.

Example 8.1: Ion Density Near the Midplane for Large Surface Separations

For a double-layer system with two surfaces at large separation from each other, the charge distribution near each surface is only slightly affected by the presence of the other surface. Therefore, the electrostatic potential can to a good approximation be obtained by a superposition of the potentials from the two single surfaces. Each of the latter potentials will be denoted by $\psi^{single}(z')$ with z' counted from the respective surface, where ψ^{single} is equal to ψ for a single surface in contact with a bulk electrolyte treated in Section 7.1. The potential decays in an exponential manner with distance from a single surface, Equation 7.32, and as we saw earlier this leads to the exponential decay of the density, Equation 7.33. We therefore expect that the density in the middle

---

[5] For a bulk solution in the PB approximation, where we neglect the ion-ion correlations in the ion cloud of each ion, this approximation corresponds to $n(\mathbf{r}_3|\mathbf{r}_1, \mathbf{r}_2) \approx n(\mathbf{r}_3|\mathbf{r}_2) = n(\mathbf{r}_3) g^{(2)}(\mathbf{r}_3, \mathbf{r}_2)$ that we used to derive Equation 5.123 in the mean field approximation, which is the base of the PB approximation for the bulk phase. In this case, we have neglected triplet correlations, as illustrated in Figure 5.13.

of the slit between two surfaces will be exponentially decaying with increasing surface separation. We will show in this example that this indeed is the case.

Let us limit ourselves to the case where the two surfaces are equal. In the coordinate system with $z = 0$ at the midplane, the potential from the left surface (located at $z = -D^S/2$) is $\psi^{single}(D^S/2 + z)$ and that from the right surface (located at $z = D^S/2$) is $\psi^{single}(D^S/2 - z)$, so we have when D^S is large

$$\psi(z) \approx \psi^{single}(D^S/2 + z) + \psi^{single}(D^S/2 - z).$$

Since the arguments $D^S/2 \pm z$ of the potentials are large when $z \approx 0$, we can use the asymptotic formula in the rhs of Equation 7.32 and obtain

$$\psi(z) \sim \frac{\sigma^{Seff}}{\kappa_D \varepsilon_r \varepsilon_0} e^{-\kappa_D(D^S/2 + z)} + \frac{\sigma^{Seff}}{\kappa_D \varepsilon_r \varepsilon_0} e^{-\kappa_D(D^S/2 - z)}$$

$$= \frac{2\sigma^{Seff}}{\kappa_D \varepsilon_r \varepsilon_0} e^{-\kappa_D D^S/2} \cosh(\kappa_D z) \quad \text{when } z \approx 0, \ D^S \to \infty.$$

Since $n_j(z) = n_j^b e^{-\beta q_j \psi(z)}$ in the PB approximation, we have for $z \approx 0$ where the potential is small

$$n_j(z) - n_j^b \sim -n_j^b \beta q_j \psi(z) \sim -\frac{2\beta n_j^b q_j \sigma^{Seff}}{\kappa_D \varepsilon_r \varepsilon_0} e^{-\kappa_D D^S/2} \cosh(\kappa_D z)$$

when $D^S \to \infty$, where we have used the terms up to linear order (the first two terms) in the expansion $e^{-\beta q_j \psi(z)} = 1 - \beta q_j \psi(z) + [\beta q_j \psi(z)]^2/2! - \cdots$. To obtain the total ion density $n_{tot}(z) = \sum_j n_j(z)$, we need to also include the second order term of the expansion since $\sum_j q_j n_j^b = 0$ and we have when $z \approx 0$ and $D^S \to \infty$

$$n_{tot}(z) - n_{tot}^b \sim \sum_j \frac{n_j^b [\beta q_j \psi(z)]^2}{2}$$

$$\sim \frac{2\beta[\sigma^{Seff}]^2}{\varepsilon_r \varepsilon_0} e^{-\kappa_D D^S} \cosh^2(\kappa_D z) \quad \text{(PB)}, \tag{8.13}$$

where we have used the definition of κ_D, Equation 6.31. The magnitude of the deviation in n_{tot} from the bulk value in the middle between the surfaces thus decays exponentially, proportionally to $e^{-\kappa_D D^S}$, when the surface separation increases.

We can use the result (8.13) to obtain the disjoining pressure from Equation 8.12 and we obtain

$$P^S(D^S) \sim \frac{2[\sigma^{Seff}]^2}{\varepsilon_r \varepsilon_0} e^{-\kappa_D D^S} \quad \text{when } D^S \to \infty \quad \text{(PB)}. \tag{8.14}$$

This is a special case of Equation 8.7, which in the PB case becomes

$$P_{\text{I,II}}^{S}(D^{S}) \sim \frac{2\sigma_{\text{I}}^{S\text{eff}}\sigma_{\text{II}}^{S\text{eff}}}{\varepsilon_r \varepsilon_0} e^{-\kappa_D D^{S}} \quad \text{when } D^{S} \to \infty \quad \text{(PB)} \tag{8.15}$$

and holds also when the two surfaces are unequal.

{EXERCISE 8.1

The surface interactions between two cylindrical surfaces covered with mica have been measured experimentally when the surfaces were immersed in purified water. The force was recorded as a function of the distance D^{S} between the surfaces and the Derjaguin formula was used to convert the measured force to the interaction energy between two planar surfaces. It was found for large D^{S} that

$$\mathscr{W}^{S}(D^{S}) \approx 8.14 \cdot 10^{-5} e^{-\kappa D^{S}} \, \text{J m}^{-2} \tag{8.16}$$

with a decay length $1/\kappa = 145$ nm. The temperature was 20°C and the dielectric constant for water at this temperature is $\varepsilon_r = 80.2$. Because of the existence of H_3O^+ and OH^- from auto-dissociation of water as well as HCO_3^- (from some dissolved CO_2), the purified water was a very dilute 1:1 electrolyte.

The conditions for this system are such that the PB approximation is very reasonable, so in the analysis of the experimental results one can use the expressions we have obtained in the present section and all relevant PB results for single planar surfaces in Section 7.1.1. We can set $d^{S} = 0$ throughout.

a. Since the electrolyte is very dilute, $\kappa = \kappa_D$ as a very good approximation. Use this to estimate the electrolyte concentration of the water in M (mol dm^{-3}).

b. Determine the experimental effective surface charge density $\sigma^{S\text{eff}}$ of mica in water from the result in Equation 8.16.

c. Calculate the actual surface charge density σ^{S} of a single mica surface in contact with water.

d. For large surface separations, calculate the surface potential, that is, $\psi(z)$ at the point of contact of the ions at each surface. The surface potential is assumed to be independent of D^{S} for large separations.

e. Calculate the total ion concentration at each surface, $n_{\text{tot}}(z)$ at the point of contact, when the distance between the surfaces is very large. Convert this density to a concentration in M.}

When one needs the disjoining pressure for all surface separations, one must solve the PB equation for the system with two surfaces in order to obtain the electrostatic potential and thereby the density profile in the slit. We start with a special case.

Example 8.2: Only Counterions between Two Charged Walls

As a first concrete example, we will investigate a particularly simple system, namely two planar charged walls with only counterions in the slit between them. It is quite illustrative since there exists an analytic solution of the PB equation for this system. In this case, the electrolyte between the surfaces is not in equilibrium with a bulk electrolyte, so the disjoining pressure does not decay exponentially as in Equation 8.15. Instead, as we will see, the pressure decays like a power law proportionally to $1/(D^S)^2$.

Consider a system that consists of two planar hard walls, each having a uniform surface charge density σ^S, and an electrolyte solution containing spherical counterions with charge q in the slit between the two surfaces. As before, the distance between the walls is D^S, as measured between the planes of surface charges at the walls, and $L \equiv D^S - 2d^S$ is the width of the space available for the ion centers. The density distribution of the ions is given by $n(z) = n^0 \exp(-\beta w(z))$, where n^0 is the density at a point where $w = 0$. This point is arbitrary since we can always add a constant to a potential, but for simplicity we will choose it to be at the midpoint between the surfaces, $z = 0$. Thereby, we deviate from the normal convention to set $w = 0$ in the bulk, but there is no bulk phase present here so that convention cannot be followed. The charge density distribution of the ions is $\rho(z) = qn(z)$.

In the PB approximation, we have $w(z) = q\psi(z)$ for $|z| \leq L$ and the PB equation is

$$-\varepsilon_r\varepsilon_0 \frac{d^2\psi(z)}{dz^2} = \begin{cases} \sigma^S\delta(z + D^S/2), & z \leq -L/2 \\ qn^0 e^{-\beta q\psi(z)}, & |z| < L/2 \quad \text{(PB)} \\ \sigma^S\delta(z - D^S/2), & z \geq L/2 \end{cases} \quad (8.17)$$

subject to the midpoint condition $\psi(0) = 0$ and the electroneutrality condition $\int_{-L/2}^{L/2} \rho(z)dz + 2\sigma^S = 0$. Note that the latter condition implies that the number of counterions per unit area is constant independent of L.

In Appendix 8A, it is shown that the PB Equation 8.17 has an analytic solution that yields the following simple expression for the density profile:

$$n(z) = n^0 e^{-\beta q\psi(z)} = \frac{n^0}{\cos^2(K_0 z)} \quad \text{(PB)}. \quad (8.18)$$

The positive parameter K_0 is a solution to the equation

$$K_0 \tan(K_0 L/2) = \frac{\beta|q\sigma^S|}{2\varepsilon_r\varepsilon_0} \quad (8.19)$$

and it can be expressed in terms of n^0 as

$$K_0^2 = \frac{\beta n^0 q^2}{2\varepsilon_r\varepsilon_0}. \quad (8.20)$$

Note that K_0 has the dimension (length)$^{-1}$ like κ_D. It is, however, quite a different entity despite that it has a similar appearance. K_0 is defined in terms of the density n^0 at one point only and depends on L, while κ_D is a constant that is defined in terms of the bulk phase density.

The parameter K_0 can be determined for each L from Equation 8.19, which has to be solved numerically, and from K_0 we can determine n^0 using Equation 8.20. Equation 8.18 then gives the density profile for various surface separations. As shown in the appendix, one always has $|K_0 z| < \pi/2$ when $|z| \leq L/2$, so the denominator is positive. The disjoining pressure can be calculated from

$$P^S = k_B T n(0) = k_B T n^0 \quad \text{(PB)} \tag{8.21}$$

with the n^0 value just obtained.

We can easily obtain an estimate of K_0 for large L. We first write Equation 8.19 as

$$\frac{K_0 L}{2} = \arctan\left(\frac{\beta |q\sigma^S|}{2\varepsilon_r \varepsilon_0} \cdot \frac{1}{K_0}\right).$$

When L is large, n^0 must be small since the volume per counterion is large. K_0 obtained from Equation 8.20 is therefore also small, so the argument of the arctan function is large. Since $\arctan(x) \approx \pi/2$ for large x, we have $K_0 L/2 \sim \pi/2$ when $L \to \infty$, so $K_0 \sim \pi/L$. Hence, we have, using Equation 8.20,

$$n^0 \sim \frac{2\pi^2 \varepsilon_r \varepsilon_0}{\beta q^2} \cdot \frac{1}{L^2} \quad \text{when } L \to \infty \quad \text{(PB)}. \tag{8.22}$$

Note that n^0 is independent of the value of σ^S in this limit.

The result (8.22) says that the density far from the walls (near the midplane) decays like a power law $(1/L^2)$ with the surface separation. This is very different from the results we obtained for the density profile in the case of a wall in contact with a bulk solution, where the density decays exponentially fast with distance. The reason for the difference is that in the latter case there is a large amount of ions present that screen the electrostatic interactions leading to the fast decay. In the present case, there is a finite number of ions per unit area present between the walls and their density decreases when L is increased. Therefore, their influence on the electrostatic interactions is very weak for large L, leading to the very slow decay.

The double-layer interaction between the surfaces is given by Equation 8.21 and from n^0 as a function of L (obtained from the solution K_0 of Equation 8.19), we can determine P^S as a function of the surface separation. For large separations, we obtain from Equation 8.22

$$P^S \sim \frac{2\pi^2 \varepsilon_r \varepsilon_0}{\beta^2 q^2} \cdot \frac{1}{L^2} \quad \text{when } L \to \infty,$$

which we can write as

$$P^{\mathcal{S}} \sim \frac{2\pi^2 \varepsilon_r \varepsilon_0}{\beta^2 q^2} \cdot \frac{1}{(D^{\mathcal{S}})^2} \quad \text{when } D^{\mathcal{S}} \to \infty \quad \text{(PB)} \tag{8.23}$$

since $D^{\mathcal{S}} = L + 2d^{\mathcal{S}}$ and $d^{\mathcal{S}}$ is constant.

Next, we will continue with cases where the electrolyte solution in the slit is in equilibrium with a bulk solution of a given composition so the amount of anions and cations in the slit will vary with the surface separation. We will thereby treat the more general case where the surfaces can be unequal, in the case of which the system is no longer symmetric around the midplane. For very large surface separations, each surface has an ion cloud that is the same as for a single surface in contact with a bulk solution (Section 7.1) and in the middle region between the surfaces, the ionic densities are virtually equal to the bulk values. The pressure $P^{\mathcal{S}}$ is then equal to zero and there is no force between the surfaces. For smaller separations, the ion cloud of each surface starts to overlap with the cloud of the other surface, leading to a nonzero $P^{\mathcal{S}}$. As we have seen, the disjoining pressure for large surface separations is given by Equation 8.15, where each of the effective surface charge densities $\sigma_J^{\mathcal{S}\text{eff}}$ for $J = $ I and II is equal to that of a single surface in contact with a bulk electrolyte as treated earlier. We will now show how the pressure can be obtained for all surface separations.

8.2.2 Arbitrarily Charged Surfaces

We will investigate a system where the uniform surface charge densities are $\sigma_I^{\mathcal{S}}$ for wall I and $\sigma_{II}^{\mathcal{S}}$ for wall II. The PB equation for this system is[6]

$$-\varepsilon_r \varepsilon_0 \frac{d^2 \psi(z)}{dz^2} = \begin{cases} \sigma_I^{\mathcal{S}} \delta(z + D^{\mathcal{S}}/2), & z \leq -L/2 \\ \sum_i q_i n_i^b e^{-\beta q_i \psi(z)}, & |z| < L/2 \quad \text{(PB)} \\ \sigma_{II}^{\mathcal{S}} \delta(z - D^{\mathcal{S}}/2), & z \geq L/2 \end{cases}$$

subject to the condition of electroneutrality, $\int_{-L/2}^{L/2} \rho(z)dz + \sigma_I^{\mathcal{S}} + \sigma_{II}^{\mathcal{S}} = 0$. In contrast to the system treated in Example 8.2, the equation does not have any analytic solution that can be written in terms of elementary functions[7] and one has to resort to numerical solutions. Once $\psi(z)$ is determined, one can obtain the density profiles from $n_i(z) = n_i^b e^{-\beta q_i \psi(z)}$. Here we will assume that we have determined $\psi(z)$ and $n_i(z)$ for all z by solving the PB equation.

For the case with $\sigma_I^{\mathcal{S}} = \sigma_{II}^{\mathcal{S}}$ treated earlier, the double-layer disjoining pressure $P^{\mathcal{S}}$ can be determined from the midpoint density $n_{\text{tot}}(0)$, Equation 8.12, since the electrostatic

---

[6] For simplicity, we assume that all ions have the same size.

[7] In some cases, one can, however, express the solution in terms of known transcendental functions. For two surfaces with equal $\sigma^{\mathcal{S}}$ and a symmetric electrolyte, the PB equation, expressed in the form of Equation 8.10 with inserted $n_{\text{tot}}(z)$ from Equation 7.14, can be written as $df(z)/dz = \pm\kappa_D \left[\sinh^2(f(z)) - \sinh^2(f_0)\right]^{1/2}$ for $|z| \leq L/2$, where we have introduced $f(z) = \beta q \psi(z)/2$ and $f_0 = f(0)$ [this form of the PB equation corresponds to Equation 7.70 in the single-wall case]. The equation is separable and for $z \geq 0$ it leads to $\int_{f_0}^{f} [\sinh^2(f) - \sinh^2(f_0)]^{-1/2} df = \kappa_D z$, where the lhs is a so-called elliptic integral. The function $f(z)$ and hence $\psi(z)$ can thus be written in terms of the inverse function to the elliptic integral, which cannot be expressed in terms of elementary functions.

interaction across the midplane is zero in the PB approximation for this case. When $\sigma_I^S \neq \sigma_{II}^S$, the electrostatic interaction across this plane is not zero and one has to proceed differently. It is always possible to determine $P_\perp^{\text{slit}}$ and therefore P^S from the contact density at either surface via the contact theorem, Equation 8.5, with the respective surface charge density inserted, but it is worthwhile to have an alternative equation for $P_\perp^{\text{slit}}$, which we now are going to derive.

We first note that in analogy to Equation 7.8, we have[8]

$$\left.\frac{d\psi(z)}{dz}\right|_{z=-L/2} = -\frac{\sigma_I^S}{\varepsilon_r \varepsilon_0}.$$

Likewise, the derivative at $z = L/2$ is $\sigma_{II}^S/(\varepsilon_r \varepsilon_0)$. Next, we consider Equation 8.9, which can be written as

$$\left[\frac{d\psi(z)}{dz}\right]^2 - \frac{2}{\beta \varepsilon_r \varepsilon_0} n_{\text{tot}}(z) = \text{const.}$$

This means the lhs is constant independent of z in the interval $-L/2 \leq z \leq L/2$. We will use this fact for two z values: $z = L/2$, where $d\psi(z)/dz = \sigma_{II}^S/(\varepsilon_r \varepsilon_0)$, and an arbitrary point $z = z^\bullet$ in the same interval. By multiplying the equation by $\varepsilon_r \varepsilon_0/2$ and inserting these two z values, we obtain

$$\frac{\varepsilon_r \varepsilon_0}{2}\left[\frac{\sigma_{II}^S}{\varepsilon_r \varepsilon_0}\right]^2 - k_B T n_{\text{tot}}(L/2) = \frac{\varepsilon_r \varepsilon_0}{2}\left[\frac{d\psi(z)}{dz}\right]^2_{z=z^\bullet} - k_B T n_{\text{tot}}(z^\bullet).$$

From the contact theorem applied to $z = L/2$, it follows that the lhs is equal to $-P_\perp^{\text{slit}}$ and since the z component of the mean electrostatic field is $E_z = -d\psi/dz$, we have

$$P_\perp^{\text{slit}} = k_B T n_{\text{tot}}(z^\bullet) - \frac{\varepsilon_r \varepsilon_0}{2} E_z^2(z^\bullet)$$

for any $-L/2 \leq z^\bullet \leq L/2$. The first term on the rhs is the ideal pressure at the plane with coordinate $z^\bullet$, i.e., the pressure due to momentum transfer across this plane (the kinetic pressure). As shown in the shaded text box below, which can be skipped in the first reading, the last term is the electrostatic interaction pressure across this plane and we can write

$$P_\perp^{\text{slit}} = k_B T n_{\text{tot}}(z^\bullet) + P_\perp^{\text{el,mean}}(z^\bullet) \quad \text{(PB)} \tag{8.24}$$

where

$$P_\perp^{\text{el,mean}}(z^\bullet) = -\frac{\varepsilon_r \varepsilon_0}{2} E_z^2(z^\bullet) \tag{8.25}$$

---

[8] This can be realized as follows. For $z < -D^S/2$, the electrostatic potential $\psi(z)$ must be constant since the total charge for $z \geq -D^S/2$, including the surface charge density σ_I^S, is equal to zero. In the interval $-D^S/2 < z \leq -L/2$, the Poisson equation implies that $\psi(z)$ is a linear function of z. At $z = -D^S/2$, where σ_I^S is located, the derivative $d\psi(z)/dz$ changes by $-\sigma_I^S/(\varepsilon_r \varepsilon_0)$ as follows from electrostatics, so the derivative in $-D^S/2 < z \leq -L/2$ must be equal to $-\sigma_I^S/(\varepsilon_r \varepsilon_0)$.

and superscript "el, mean" indicates that it is associated with the mean electrostatic field. It is also shown in the text box that we alternatively can write this as

$$P_\perp^{\mathrm{el,mean}}(z^\bullet) = -\frac{\left(\sigma_\bullet^\mathcal{S}\right)^2}{2\varepsilon_r\varepsilon_0}, \tag{8.26}$$

where $\sigma_\bullet^\mathcal{S} = |\sigma_{\mathrm{I}\bullet}^\mathcal{S}| = |\sigma_{\bullet\mathrm{II}}^\mathcal{S}|$ is the absolute value of the total charge per unit area on either side of the plane at $z^\bullet$; the quantity $\sigma_{\mathrm{I}\bullet}^\mathcal{S}$ is the total charge to the left of $z^\bullet$ and $\sigma_{\bullet\mathrm{II}}^\mathcal{S}$ the corresponding charge to the right of $z^\bullet$

$$\sigma_{\mathrm{I}\bullet}^\mathcal{S} = \sigma_{\mathrm{I}}^\mathcal{S} + \int\limits_{-L/2}^{z^\bullet} dz\,\rho(z)$$

$$\sigma_{\bullet\mathrm{II}}^\mathcal{S} = \sigma_{\mathrm{II}}^\mathcal{S} + \int\limits_{z^\bullet}^{L/2} dz\,\rho(z). \tag{8.27}$$

Note that $\sigma_{\mathrm{I}\bullet}^\mathcal{S} = -\sigma_{\bullet\mathrm{II}}^\mathcal{S}$ due to electroneutrality and that we can write $\left(\sigma_\bullet^\mathcal{S}\right)^2 = -\sigma_{\mathrm{I}\bullet}^\mathcal{S}\sigma_{\bullet\mathrm{II}}^\mathcal{S}$. Equation 8.24 can hence be written as

$$P_\perp^{\mathrm{slit}} = k_B T n_{\mathrm{tot}}(z^\bullet) - \frac{\left(\sigma_\bullet^\mathcal{S}\right)^2}{2\varepsilon_r\varepsilon_0} \quad \text{(PB)} \tag{8.28}$$

When $z^\bullet = -L/2$ and $z^\bullet = L/2$, we have $\sigma_{\mathrm{I}\bullet}^\mathcal{S} = \sigma_{\mathrm{I}}^\mathcal{S}$ and $\sigma_{\bullet\mathrm{II}}^\mathcal{S} = \sigma_{\mathrm{II}}^\mathcal{S}$, respectively, and Equation 8.28 reduces to the contact theorem for each surface.

★ Consider the plane at coordinate $z^\bullet$ and the electrostatic interactions across this plane. The electrostatic field $E^\bullet$ from all charges to the left of the plane is given by $E^\bullet = [\int_{-L/2}^{z^\bullet} \rho(z)dz + \sigma_{\mathrm{I}}^\mathcal{S}]/(2\varepsilon_r\varepsilon_0) = \sigma_{\mathrm{I}\bullet}^\mathcal{S}/(2\varepsilon_r\varepsilon_0)$ at any point on the right-hand side of the plane. We shall consider the interactions between this field and all charges on the latter side. Since $E^\bullet$ is constant there, only the total charge per unit area to the right of the plane matters, that is, $\sigma_{\bullet\mathrm{II}}^\mathcal{S}$. Thus, the force per unit area due to the field $E^\bullet$ on the charges to the right of the plane is $\sigma_{\bullet\mathrm{II}}^\mathcal{S} E^\bullet = -\sigma_{\mathrm{I}\bullet}^\mathcal{S} E^\bullet = -2\varepsilon_r\varepsilon_0(E^\bullet)^2$.

We should, however, express this force in terms of the total field $E_z(z^\bullet)$ at the plane from *all* charges in the system. $E^\bullet$ is the field at $z^\bullet$ from all charges $(\sigma_{\mathrm{I}\bullet}^\mathcal{S})$ to the *left* of $z^\bullet$. The field at $z^\bullet$ from all charges $(\sigma_{\bullet\mathrm{II}}^\mathcal{S})$ to the *right* of $z^\bullet$ must also be equal to $E^\bullet$ because $\sigma_{\mathrm{I}\bullet}^\mathcal{S} = -\sigma_{\bullet\mathrm{II}}^\mathcal{S}$ and the direction of this field must be the same. Thus, $E_z(z^\bullet) = 2E^\bullet$, which gives the force per unit area across the plane at $z^\bullet$ equal to $-\varepsilon_r\varepsilon_0 E_z^2(z^\bullet)/2$, that is, the rhs of Equation 8.25. Alternatively, we can write this force as $\sigma_{\bullet\mathrm{II}}^\mathcal{S} E^\bullet = \sigma_{\bullet\mathrm{II}}^\mathcal{S}\sigma_{\mathrm{I}\bullet}^\mathcal{S}/(2\varepsilon_r\varepsilon_0) = -(\sigma_\bullet^\mathcal{S})^2/(2\varepsilon_r\varepsilon_0)$, which is used in Equation 8.28.

It follows from Equations 8.24 and 8.28 that $P^S = P_\perp^{slit} - k_B T n_{tot}^b$ can be evaluated from

$$P^S = k_B T[n_{tot}(z^\bullet) - n_{tot}^b] + P_\perp^{el,mean}(z^\bullet)$$

$$= k_B T[n_{tot}(z^\bullet) - n_{tot}^b] - \frac{(\sigma_\bullet^S)^2}{2\varepsilon_r\varepsilon_0} \quad \text{(PB)}. \quad (8.29)$$

This equation can be used for any $z^\bullet$ between the surfaces since the value of $P_\perp^{slit}$ is independent of $z^\bullet$. In some cases, this result can be simplified, namely when the mean electrostatic field is zero somewhere in the electrolyte. From Equation 8.25, we see that at a plane $z = z_0^\bullet$, where $E_z = -d\psi/dz = 0$ and hence $\sigma_\bullet^S = 0$, we have $P_\perp^{el,mean}(z^\bullet) = 0$ and hence $P_\perp^{slit} = k_B T n_{tot}(z_0^\bullet)$. The electrostatic interactions across such a plane are zero and we have $P^S = k_B T(n_{tot}(z_0^\bullet) - n_{tot}^b)$. Since a density is always positive, the pressure $P_\perp^{slit}$ is always repulsive in such cases. The full formula (8.29) is, however, needed if the pressure is evaluated at another plane, $z^\bullet \neq z_0^\bullet$.

In cases where no plane between the surfaces has zero E_z, the evaluation of P^S from Equation 8.29 requires the value of $E_z(z^\bullet)$ or $\sigma_\bullet^S$ irrespective of which $z^\bullet$ one uses to evaluate the pressure. This can occur when σ_I^S and σ_{II}^S have opposite signs and it is then possible for $P_\perp^{slit}$ to be either repulsive (positive) or attractive (negative).

The **free energy of interaction** (potential of mean force) per unit area, $\mathscr{W}_{I,II}^S(D^S)$, between the two surfaces I and II at separation D^S can be calculated by integrating P^S as a function of surface separation according to Equation 8.2. Alternatively, $\mathscr{W}_{I,II}^S(D^S)$ can be calculated in a manner that is analogous to the treatment of two interacting particles (macroions) in Section 6.1.4. There we calculated the free energy by first immersing the two particles in an uncharged state into the electrolyte, which was done by making holes in the electrolyte where the two uncharged particles are then inserted. Thereafter, the rest of the free energy was calculated by gradually increasing the charges of these particles from zero to the final values, whereby the electrostatic interactions between the particles and the surrounding ions increase in the corresponding manner.

In the present case, we first consider uncharged walls immersed in the electrolyte and then we do the corresponding gradual increase of electrostatic interactions between the walls and the ions in the slit by charging the surfaces. Instead of immersing the uncharged walls into the electrolyte, we have them immersed all the time but to start with, they are infinitely far apart. Then we bring them together and finish at surface separation D^S, whereby the free energy of interaction for uncharged surfaces is given by (cf. Equation 8.4)

$$\mathscr{W}_{I,II}^{S,0}(D^S) = \mathscr{W}_{I,II}^{slit,0}(D^S) - \mathscr{W}_{I,II}^{slit,0}(\infty) \quad \text{(uncharged)},$$

where superscript 0 indicates that the surfaces are uncharged. We obtain the rest of the free energy, that is, the electrostatic part defined as

$$\mathscr{W}_{I,II}^{S,el}(D^S) \equiv \mathscr{W}_{I,II}^S(D^S) - \mathscr{W}_{I,II}^{S,0}(D^S), \quad (8.30)$$

by gradually charging the surfaces from zero to the final surface charge densities, as we will see. Likewise, we define $\mathscr{W}_{I,II}^{slit,el}(D^S) \equiv \mathscr{W}_{I,II}^{slit}(D^S) - \mathscr{W}_{I,II}^{slit,0}(D^S)$ and we may note that the term $P^b D^S$ in Equation 8.3 cancels in $\mathscr{W}_{I,II}^{slit,el}$.

In the PB approximation, we only need to consider the electrostatic part, $\mathscr{W}_{I,II}^{S,el}$, of the free energy since the disjoining pressure for uncharged surfaces is equal to zero in this approximation. This is a consequence of the fact that when the surfaces are uncharged, the ion density in the slit is equal to the bulk density irrespective of separation D^S because $\psi(z)$ is zero everywhere. Therefore, $\mathscr{W}_{I,II}^{slit,0}(D^S) = \mathscr{W}_{I,II}^{slit,0}(\infty)$ and $P_\perp^{slit}(D^S) = P^b$ for uncharged surfaces. For charged surfaces, we hence have $\mathscr{W}_{I,II}^{S}(D^S) = \mathscr{W}_{I,II}^{slit,el}(D^S) - \mathscr{W}_{I,II}^{slit,el}(\infty)$ in the PB case.

In the following section (Section 8.2.3), which can be skipped in the first reading, we derive the result

$$\mathscr{W}_{I,II}^{slit,el}(D^S) = \int_0^1 d\xi \, [\upsilon_I^S \psi(-D^S/2;\zeta) + \sigma_{II}^S \psi(D^S/2;\xi)], \tag{8.31}$$

where $\psi(-D^S/2;\xi)$ and $\psi(D^S/2;\xi)$ are the values of the mean electrostatic potential ψ at the respective surface charge density when the surface charge densities are $\xi\sigma_I^S$ and $\xi\sigma_{II}^S$, respectively, for $0 \leq \xi \leq 1$. Thus, the double-layer interaction free energy can be obtained from this formula when $\psi(z;\xi)$ has been determined from the PB equation for all ξ, whereby the values of ψ at the surfaces are sufficient in order to obtain $\mathscr{W}_{I,II}^{slit,el}$. Equation 8.31 is, in fact, valid in general and not only in the PB approximation (cf. the corresponding treatment of two macroions immersed in an electrolyte in Section 6.1.4).

We can obtain the disjoining pressure from $P^S(D^S) = -d\mathscr{W}_{I,II}^{S}(D^S)/dD^S$. Note that the electrostatic part of the pressure, $-d\mathscr{W}_{I,II}^{S,el}(D^S)/dD^S$, is very different from the electrostatic interaction pressure $P_\perp^{el,mean}(z^\bullet)$ discussed earlier. The latter is the mean electrostatic interaction pressure across the plane $z = z^\bullet$ and depends on $z^\bullet$, while the electrostatic part of the pressure here is the difference in P^S between charged and uncharged surfaces.

8.2.3 Electrostatic Part of Double-Layer Interactions, General Treatment ★

In this section we will derive Equation 8.31 for the electrostatic part of the free energy of interaction per unit area, $\mathscr{W}_{I,II}^{slit,el}$, for an electrolyte between two charged walls. This equation is valid in general (cf. the corresponding treatment of two macroions in Section 6.1.4.2). The electrostatic part of the interaction free energy for the two surfaces, defined in Equation 8.30, is given by

$$\mathscr{W}_{I,II}^{S,el}(D^S) = \mathscr{W}_{I,II}^{slit,el}(D^S) - \mathscr{W}_{I,II}^{slit,el}(\infty).$$

In the general case, the mean electrostatic potential $\psi(z;\xi)$ in the system with partially charged surfaces has to be be obtained from the charge density profile $\rho(z;\xi)$ in the slit of this system as calculated either by computer simulation or in some approximate theory. To obtain the full $\mathscr{W}_{I,II}^{S}(D^S)$, one also needs to add the contribution from the interaction between uncharged walls as a function of separation and one has $\mathscr{W}_{I,II}^{slit} = \mathscr{W}_{I,II}^{slit,el} + \mathscr{W}_{I,II}^{slit,0}$. It is only in the PB approximation that the latter contribution is constant and vanishes in $\mathscr{W}_{I,II}^{S}$.

We thus assume that we initially have two uncharged walls at distance D^S from each other and we obtain $\mathscr{W}_{I,II}^{slit,el}(D^S)$ by calculating the work performed when charging the surfaces.

The electrostatic interaction interaction energy per unit area between the two surface charge densities $\sigma_{\text{I}}^{\mathcal{S}}$ and $\sigma_{\text{II}}^{\mathcal{S}}$ is

$$\mathcal{V}_{\text{I,II}}^{\text{el}}(D^{\mathcal{S}}) = -\frac{\sigma_{\text{I}}^{\mathcal{S}}\sigma_{\text{II}}^{\mathcal{S}}D^{\mathcal{S}}}{2\varepsilon_r\varepsilon_0} \tag{8.32}$$

and the interaction potentials of each of them and an ion of species j are

$$v_{\text{I},j}^{\text{el}}(z|D^{\mathcal{S}}) = -\frac{q_j\sigma_{\text{I}}^{\mathcal{S}}\left|\frac{D^{\mathcal{S}}}{2}+z\right|}{2\varepsilon_r\varepsilon_0}$$

$$v_{\text{II},j}^{\text{el}}(z|D^{\mathcal{S}}) = -\frac{q_j\sigma_{\text{II}}^{\mathcal{S}}\left|\frac{D^{\mathcal{S}}}{2}-z\right|}{2\varepsilon_r\varepsilon_0}. \tag{8.33}$$

In analogy to Equation 6.100, we have in the current case

$$\mathcal{W}_{\text{I,II}}^{\text{slit,el}}(D^{\mathcal{S}}) = \mathcal{V}_{\text{I,II}}^{\text{el}}(D^{\mathcal{S}})$$

$$+ \int_0^1 d\xi \sum_j \int_{-L/2}^{L/2} dz \left[\frac{\partial v_{\text{I},j}^{\text{el}}(z|D^{\mathcal{S}};\xi)}{\partial\xi} + \frac{\partial v_{\text{II},j}^{\text{el}}(z|D^{\mathcal{S}};\xi)}{\partial\xi}\right] n_j(z|D^{\mathcal{S}};\xi),$$

where $n_j(z|D^{\mathcal{S}};\xi)$ is the density of j-ions when the surfaces are charged to the extent ξ and where we can take $v_{\text{I},j}^{\text{el}}(z|D^{\mathcal{S}};\xi) = \xi v_{\text{I},j}^{\text{el}}(z|D^{\mathcal{S}})$ and likewise for $v_{\text{II},j}^{\text{el}}$. By going through steps analogous to those taken when we obtained Equation 6.102 and finally Equation 6.99 from Equation 6.100, we obtain Equation 8.31.

{**EXERCISE 8.2**
Fill in the details of the proof of Equation 8.31.}

8.3 SURFACE FORCES AND PAIR CORRELATIONS, GENERAL CONSIDERATIONS

In the previous section, we investigated surface forces between charged surfaces in contact with electrolytes in the Poisson-Boltzmann approximation. Here and in Section 8.4, we will give a general treatment without approximations of surface forces for fluids with spherical particles, like hard spheres or Lennard-Jones particles. In Section 8.5, we will specialize to electrolytes and present the elements of the theory of double-layer interactions beyond the PB approximation (note, however, that the results in Section 8.2.3 are valid in the general case). It is suitable at this point to reread Section 8.1 which contains material that is valid in general.

We have two planar walls, I and II, in parallel alignment at separation $D^{\mathcal{S}}$ from each other. We assume that the surfaces are smooth. The surface area A is finite but very large and there are N fluid particles of a single species in the slit between the surfaces. We can, as before, neglect boundary effects at the boundaries of the surfaces; they are negligible for the free energy and other properties we will consider since A and N are huge. We do the derivations in the canonical ensemble for simplicity, but the results are valid also in the case for systems in equilibrium with a bulk fluid provided that they are applied for a system with a given chemical potential (in principle in the limit of infinite N and A at constant ratio N/A and therefore constant chemical potential when N is very large).

Initially, the coordinate system has its origin at wall I and the z axis is perpendicular to the surfaces (note that this differs from the coordinate system used earlier in Sections 8.1 and 8.2, but the main result will be converted to the latter coordinates, see Equation 8.41). The wall-particle interaction potential $v_J(\ell)$ for $J = $ I and II only depends on the distance ℓ from the wall. When this interaction includes a hard-core potential, the distances of closest approach of the particle centers to the wall surfaces are d_I^S and d_{II}^S, respectively, and $v_J(\ell)$ is infinite when $\ell < d_J^S$. The particle centers are therefore located within $d_I^S \leq z \leq D^S - d_{II}^S$ and the width of the region available for these centers is $L = D^S - (d_I^S + d_{II}^S)$. For systems with soft walls, like the case of an LJ fluid treated earlier in Section 7.2.1.1, the wall separation D^S is defined as the distance between the points' infinite wall-particle potential at either wall. The wall-particle potential is strongly repulsive close to each surface; it normally reaches infinity gradually when ℓ approaches 0, but there can also be a gradual increase to a large value followed by a jump to infinity at $\ell = 0$. The space available for centers of the fluid particles has a width equal to D^S in these cases, but in practice the space is appreciably smaller than D^S due to the strong repulsion close to each wall.[9]

Each fluid particle interacts with the walls via the potential

$$v(z|D^S) = v_I(z) + v_{II}(D^S - z), \tag{8.34}$$

where we have explicitly shown that v depends on the wall separation. The interaction energy per unit area between the walls themselves is $\mathscr{V}_{I,II}(D^S)$, that is, the energy as it would be in the absence of the fluid. The total potential energy of the system is

$$\breve{U}_N^{pot}(\mathbf{r}^N|D^S) = \breve{U}_N^{intr}(\mathbf{r}^N) + \sum_{\nu=1}^{N}[v_I(z_\nu) + v_{II}(D^S - z_\nu)] + A\mathscr{V}_{I,II}(D^S), \tag{8.35}$$

where $\breve{U}_N^{intr}$ is the interparticle potential energy from the interactions between the N fluid particles. The density distribution $n_N(z)$ of the particles and all other singlet functions depend only on z. The volume of the system is $V = AD^S$.

The free energy is $A(T, V, N) = A(T, AD^S, N)$, but since we keep the area constant, we will instead consider the function $A = A(T, D^S, N)$ at constant A. The pressure $P_\perp^{slit}$ in the slit, i.e, the average force per unit area perpendicular to the walls, is given by

$$P_\perp^{slit} = -\frac{1}{A}\left(\frac{\partial A}{\partial D^S}\right)_{T,A,N} = \frac{k_B T}{A}\left(\frac{\partial \ln Q_N}{\partial D^S}\right)_{T,A,N}, \tag{8.36}$$

---

[9] In Section 7.2.1.1, we introduced the concept of fluid film thickness for the LJ case, see footnote 30 in that section.

where $Q_N = Z_N/(N!\Lambda^{3N})$ (Equation 3.16). By inserting the definition of the configurational partition function Z_N (Equation 3.10) and performing the derivative, one can, as shown in Appendix 8B, derive the following expression

$$P_\perp^{\text{slit}} = k_B T n(D^S - d_{\text{II}}^S)$$

$$- \int_{d_{\text{I}}^S}^{D^S - d_{\text{II}}^S} dz_1 v_{\text{II}}'(D^S - z_1)n(z_1) - \mathscr{V}_{\text{I,II}}'(D^S), \qquad (8.37)$$

where we have introduced the notation $v_{\text{II}}'(\ell) = dv_{\text{II}}(\ell)/d\ell$ and $\mathscr{V}_{\text{I,II}}'(D^S) = d\mathscr{V}_{\text{I,II}}(D^S)/dD^S$. This expression is valid in general (not only in the canonical ensemble) and can be interpreted physically as follows.

Since $n(D^S - d_{\text{II}}^S)$ is the contact density at wall II, the first term on the rhs is the pressure contribution due to collisions at its surface (the ideal term).[10] The second term is the force per unit area that the particles in the slit exerts on wall II due to the interaction potential v_{II}, whereby $-v_{\text{II}}'(D^S - z_1)$ is the force that a particle at z_1 exerts and $n_N(z_1)dz_1$ is the number of particles per unit area in a layer of width dz_1 there. The third term is the force per unit area that wall I exerts on wall II. Thus, the rhs is the pressure evaluated at coordinate $z = D^S - d_{\text{II}}^S$.

The corresponding expression holds for the pressure at $z = d_{\text{I}}^S$, which involves the interactions with wall I and the contact density $n(d_{\text{I}}^S)$ there

$$P_\perp^{\text{slit}} = k_B T n(d_{\text{I}}^S)$$

$$- \int_{d_{\text{I}}^S}^{D^S - d_{\text{II}}^S} dz_1 v_{\text{I}}'(z_1)n(z_1) - \mathscr{V}_{\text{I,II}}'(D^S), \qquad (8.38)$$

which follows by analogy. Note that Equations 8.37 and 8.38 hold *irrespective of the interparticle interactions* contained in $\breve{U}^{\text{intr}}$; it is not necessary to assume pairwise additivity.

The expressions for the pressure also hold for a single wall in contact with a bulk fluid (with $\mathscr{V}_{\text{I,II}}' = 0$), in the case of which the pressure equals the bulk pressure P^b. This can be realized by applying Equation 8.38 to a case with very large D^S. Then the fluid in the middle of the slit is in its bulk state, $\mathscr{V}_{\text{I,II}}'$ is zero and we have

$$P^b = k_B T n(d_{\text{I}}^S) - \int_{d_{\text{I}}^S}^{\infty} dz_1 v_{\text{I}}'(z_1)n(z_1) \quad \text{(single wall I)}, \qquad (8.39)$$

where we have let the integration extend to infinity since $v_{\text{I}}'(z_1)$ becomes negligible for large z_1.

---

[10] For soft walls with a potential $v_J(\ell)$ that gradually reaches infinity at this point, the contact density may be zero; nonzero contact densities appear when the potential has a jump to infinity at the contact point.

As argued in Section 8.1, the pressure at each plane in the slit, say at coordinate $z^\bullet$, must be the same at equilibrium; in the current case, this applies for $d_{\mathrm{I}}^S \leq z^\bullet \leq D^S - d_{\mathrm{II}}^S$. In order to obtain alternative expressions for $P_\perp^{\mathrm{slit}}$, we will limit ourselves to systems with pairwise additive interparticle interactions, $\breve{U}^{\mathrm{intr}} = \sum_{\nu,\nu'>\nu} u(r_{\nu\nu'})$. We can thereby use the first Born-Green-Yvon Equation 5.119, and, as shown in Appendix 8B, we can write Equation 8.37 in the following form which has a straightforward physical interpretation (see below)

$$
P_\perp^{\mathrm{slit}} = k_B T n(z^\bullet) - \int\limits_{z^\bullet}^{D^S - d_{\mathrm{II}}^S} dz_1 n(z_1) v_{\mathrm{I}}'(z_1) - \int\limits_{d_{\mathrm{I}}^S}^{z^\bullet} dz_1 n(z_1) v_{\mathrm{II}}'(D^S - z_1)
$$

$$
- \int\limits_{z^\bullet}^{D^S - d_{\mathrm{II}}^S} dz_1\, n(z_1) \int\limits_{d_{\mathrm{I}}^S}^{z^\bullet} dz_2\, n(z_2) \int d\mathbf{s}_{12}\, g^{(2)}(s_{12}, z_1, z_2)
$$

$$
\times \frac{du(r_{12})}{dr_{12}} \frac{z_1 - z_2}{r_{12}} - \mathscr{V}_{\mathrm{I,II}}'(D^S), \tag{8.40}
$$

where $\mathbf{s}_{12} = (x_2 - x_1, y_2 - y_1), s_{12} = |\mathbf{s}_{12}|$ and we have written $g^{(2)}(\mathbf{r}_1, \mathbf{r}_2) = g^{(2)}(s_{12}, z_1, z_2)$ since, as noticed in Section 7.2.1, the pair distribution function has only three independent variables in the current geometry.

Equation 8.40 can be interpreted as follows. The rhs is an expression for the pressure at a plane placed at $z^\bullet$, that is, the sum of the momentum transfer across this plane (the first term) and the average force per unit area due to the interactions between the part of the system to the left of the plane and the part to the right of it (the sum of all integrals and the last term). This can be realized as follows. The first integral is the force per unit area between wall I (in the left part) and all fluid particles to the right of $z^\bullet$ and the second integral is the force per unit area between wall II (in the right part) and all fluid particles to the left of $z^\bullet$. The third integral is the force per unit area between all fluid particles to the right of $z^\bullet$ with all fluid particles to the left of $z^\bullet$; this term will be discussed in detail below. Finally, the last term is, as before, the force per unit are a between the two walls themselves. As explained earlier, the plane at $z^\bullet$ is a theoretical construct and it does not influence the system in any way. $P_\perp^{\mathrm{slit}}$ obtained from Equation 8.41 is independent of $z^\bullet$, but the various contributions apart from $\mathscr{V}_{\mathrm{I,II}}'$ depend on $z^\bullet$; only their sum is constant.

The term with $g^{(2)}$ in Equation 8.40 warrants a further discussion. Consider a particle at z_1 to the right of the plane at $z^\bullet$; more precisely let it be located at $\mathbf{r}_1$. The density distribution of particles around it, at position $\mathbf{r}_2$, is given by $n(\mathbf{r}_2)g^{(2)}(\mathbf{r}_1, \mathbf{r}_2) = n(z_2)g^{(2)}(s_{12}, z_1, z_2)$. The interactional force between the particle at $\mathbf{r}_1$ and a particle at $\mathbf{r}_2$ is given by $-du(r_{12})/dr_{12}$ and is directed along the connecting line between them. The z component of this force is obtained by multiplication by $(z_1 - z_2)/r_{12}$. The product of these three factors appear in the term with $g^{(2)}$ in Equation 8.40. By integrating this product for all $\mathbf{r}_2$ to the left of the plane at $z^\bullet$, which is accomplished by taking $\int_{d_{\mathrm{I}}^S}^{z^\bullet} dz_2 \int d\mathbf{s}_{12}$, we obtain the z component of the mean force between the particle at $\mathbf{r}_1$ and all particles to the left of the plane. Thereby, we have taken care of all parts of the term with $g^{(2)}$, except the first part involving the integration over z_1 and the factor $n(z_1)$. This first part arises because the number of particles per unit area in a layer of width dz_1 at z_1 is given by $n(z_1)dz_1$. By taking the integral over the part of the system to the right of the plane located at $z^\bullet$, we obtain the total interactional force per unit area in the perpendicular direction across the plane, that is, between all fluid particles on the right-hand side of $z^\bullet$ with those on the left-hand side. We have thereby demonstrated the physical meaning of the term with $g^{(2)}$ in Equation 8.40.

As we will see, one can subdivide $P_\perp^{\text{slit}}$ into different physical contributions, most of which originate from various terms in the pair interaction potential $u(r)$ and the external potential $v(z)$ that will be investigated later. For future reference, these contributions are listed in the Table 8.1 together with a brief description. We have already encountered the kinetic (ideal) pressure and the mean electrostatic pressure in the PB approximation for double-layer interactions.

For future reference, it is suitable to write Equation 8.40 in a coordinate system with the origin in the middle of the slit, whereby the possible z coordinates for the ions span the interval $-L/2 \leq z \leq L/2$ and the surface separation is $D^S = L + d_I^S + d_{II}^S$. Then $z = -L/2$ corresponds to d_I^S in the old coordinate system and $z = L/2$ corresponds to $D^S - d_{II}^S$, so we obtain

$$P_\perp^{\text{slit}} = k_B T n(z^\bullet)$$

$$- \int_{z^\bullet}^{L/2} dz_1 n(z_1) v_I' \left(\tfrac{L}{2} + d_I^S + z_1\right) - \int_{-L/2}^{z^\bullet} dz_1 n(z_1) v_{II}' \left(\tfrac{L}{2} + d_{II}^S - z_1\right)$$

$$- \int_{z^\bullet}^{L/2} dz_1\, n(z_1) \int_{-L/2}^{z^\bullet} dz_2\, n(z_2) \int ds_{12}\, g^{(2)}(\mathbf{r}_1, \mathbf{r}_2)$$

$$\times \frac{du(r_{12})}{dr_{12}} \frac{z_1 - z_2}{r_{12}} - \mathcal{V}_{I,II}'(D^S), \tag{8.41}$$

TABLE 8.1　Various Physical Contributions to $P_\perp^{\text{slit}}$

$P_\perp^{\text{slit}}$ Contribution	Description	Definition
$P_\perp^{\text{kin}}$	Kinetic (ideal) pressure	$P_\perp^{\text{kin}} = k_B T n_{\text{tot}}(z^\bullet)$
$P_\perp^{\text{el,mean}}$	Mean electrostatic pressure[a]	$P_\perp^{\text{el,mean}} = -\frac{(\sigma_\bullet^S)^2}{2\varepsilon_r\varepsilon_0} = -\frac{\varepsilon_r\varepsilon_0 E_z^2(z^\bullet)}{2}$
$P_\perp^{\text{core}}$	Core-core collision pressure	Equations 8.47 and 8.58
$P_\perp^{\text{el,corr}}$	Electrostatic correlation pressure	Equations 8.50 and 8.57
$P_\perp^{\text{im}}$	Image charge interaction pressure	Equation 8.68
$P_\perp^{\text{disp}}$	Dispersion interaction pressure	Equation 8.70

[a] Equations 8.25 and 8.26.

where $-L/2 \leq z^{\bullet} \leq L/2$ and we have reintroduced $g^{(2)}(\mathbf{r}_1, \mathbf{r}_2) = g^{(2)}(s_{12}, z_1, z_2)$. Remember that $\mathbf{s}_{12} = (x_2 - x_1, y_2 - y_1)$. Equation 8.41 is, as we will see, a key result for the understanding of correlation effects in surface forces.

It is illuminating to apply Equation 8.41 to a bulk fluid where $v'_{\mathrm{I}}(z) = v'_{\mathrm{II}}(z) = 0$ and $\mathscr{V}'_{\mathrm{I,II}} = 0$. In the bulk $n(z) = n^b$, the pair distribution function is $g^{(2)b}(r_{12})$ and the pressure is P^b. We set $z^{\bullet} = 0$ (the value of $z^{\bullet}$ is irrelevant in bulk) and Equation 8.41 becomes

$$P^b = k_B T n^b - (n^b)^2 \int_0^{\infty} dz_1 \int_{-\infty}^{0} dz_2 \int d\mathbf{s}_{12}\, g^{(2)b}(r_{12}) \frac{du(r_{12})}{dr_{12}} \frac{z_1 - z_2}{r_{12}}, \tag{8.42}$$

which means that the last term, the excess pressure, is the average interactional force per unit area across a plane in the bulk fluid in the perpendicular direction. This expression is, in fact, equivalent to Equation 5.73 for the bulk pressure, that is,

$$P^b = k_B T n^b - \frac{(n^b)^2}{6} \int d\mathbf{r}\, r\, g^{(2)b}(r) \frac{du(r)}{dr} \tag{8.43}$$

as shown in the shaded text box below, which can be skipped in the first reading. Equation 8.42 gives a much clearer physical interpretation of the pressure in terms of the pair distribution than the latter equation.

The disjoining pressure for two planar particles immersed in the fluid is, as explained in Section 8.1, given by

$$P^S(D^S) = P^{\mathrm{slit}}_{\perp}(D^S) - P^b.$$

The expression (8.41) approaches (8.42) when $D^S \to \infty$ and we see that $P^S(D^S) \to 0$ as discussed before.

★ The integral in Equation 8.42 has the mathematical form

$$\int_0^{\infty} dz_1 \int_{-\infty}^{0} dz_2 \int d\mathbf{s}\, \frac{z_1 - z_2}{s^2 + (z_1 - z_2)^2} f\left(s^2 + (z_1 - z_2)^2\right) = \mathcal{I}$$

where $\mathbf{s} = (x, y)$, $s^2 + (z_1 - z_2)^2 = r^2$ and $f(r^2) = r g^{(2)}(r)[du(r)/dr]$. We shall show that $\mathcal{I} = 1/6 \int d\mathbf{r}\, f(r^2)$ where $\mathbf{r} = (\mathbf{s}, z)$. To do this, we first make a variable substitution for z_2, namely $z_1 - z_2 = z_3$ at constant z_1, so we have

$$
\mathcal{I} = \int\limits_0^\infty dz_1 \int\limits_{z_1}^\infty dz_3 \int ds \, \frac{z_3}{s^2 + z_3^2} f\left(s^2 + z_3^2\right)
$$

$$
= \left[z_1 \int\limits_{z_1}^\infty dz_3 \int ds \, \frac{z_3}{s^2 + z_3^2} f\left(s^2 + z_3^2\right) \right]_{z_1=0}^{z_1=\infty}
$$

$$
+ \int\limits_0^\infty dz_1 \, z_1 \int ds \, \frac{z_1}{s^2 + z_1^2} f\left(s^2 + z_1^2\right),
$$

where we have made a partial integration with respect to z_1 to obtain the last equality. The first term in the rhs is zero, so we have

$$
\mathcal{I} = \int\limits_0^\infty dz_1 \int ds \, \frac{z_1^2}{s^2 + z_1^2} f\left(s^2 + z_1^2\right) = \frac{1}{2} \int d\mathbf{r} \, \frac{z^2}{x^2 + y^2 + z^2} f\left(x^2 + y^2 + z^2\right),
$$

which follows since the first integral is taken over half the space and where we have set $z_1 = z$ and $s^2 = x^2 + y^2$ in the last term. Now, in the rhs, we may replace z^2 in the numerator by x^2 or y^2 and obtain the same value of the integral. Therefore, this value is $1/3$ of the value obtained if z^2 in the numerator is replaced by $x^2 + y^2 + z^2$ and the anticipated result $\mathcal{I} = 1/6 \int d\mathbf{r} f(r^2)$ follows.

{EXERCISE 8.3

Consider the term in $P_\perp^{\text{slit}}$ of Equation 8.41 that contains the pair distribution function $g^{(2)}(\mathbf{r}_1, \mathbf{r}_2)$. If we entirely neglect the pair correlations and set $g^{(2)} = 1$, this term becomes

$$
\mathcal{I}_P = - \int\limits_{z^\bullet}^{L/2} dz_1 \, n(z_1) \int\limits_{-L/2}^{z^\bullet} dz_2 \, n(z_2) \int d\mathbf{s}_{12} \frac{du(r_{12})}{dr_{12}} \frac{z_1 - z_2}{r_{12}},
$$

where $\mathbf{s}_{12} = (x_2 - x_1, y_2 - y_1)$ and $\mathbf{r}_{12}^2 = s_{12}^2 + (z_2 - z_1)^2$. Use the identity

$$
\frac{\partial u\left(\left[s_{12}^2 + (z_2 - z_1)^2\right]^{1/2}\right)}{\partial s_{12}} = \frac{s_{12}}{r_{12}} \frac{du(r_{12})}{dr_{12}}
$$

and the fact that $u(r) \to 0$ when $r \to \infty$ to show that

$$\mathcal{I}_P = 2\pi \int_{z^{\bullet}}^{L/2} dz_1 \, n(z_1) \int_{-L/2}^{z^{\bullet}} dz_2 \, n(z_2)(z_1 - z_2)u(|z_1 - z_2|), \qquad (8.44)$$

which will be of use later in Exercise 8.5.}

8.4 STRUCTURAL SURFACE FORCES

The term **structural surface forces** is commonly used for the forces that originate because the fluid between the surfaces forms a layered structure, like those for hard sphere and LJ fluids that we have encountered, and also for nonspherical particles and molecules. In this section, we will only treat the case of hard spheres between hard walls. It is particularly simple since it follows from Equations 8.37 and 8.38 that $P_{\perp}^{\text{slit}}/(k_B T)$ is equal to the contact density n^{cont} at either hard surface since the interaction terms are equal to zero. Therefore, the disjoining pressure can be obtained from

$$P^S(D^S) = k_B T \left[n^{\text{cont}}(D^S) - n^{\text{cont}}(\infty) \right] \quad \text{(hard walls)}, \qquad (8.45)$$

where $n^{\text{cont}}(\infty)$ is the contact density for a single surface in contact with a bulk fluid (infinite surface separation), so $k_B T n^{\text{cont}}(\infty) = P^b =$ the bulk pressure (cf. Equation 8.1). If we have two hard walls of finite thicknesses immersed in a bulk fluid at distance D^S from each other, P^S is the net pressure, that is, the difference between the pressure $P_{\perp}^{\text{slit}}$ in the slit and the bulk pressure P^b which acts on the outer surfaces. Equation 8.45 is, in fact, valid for any system with inert hard walls, that is, where the fluid-wall interaction *solely* contains a hard core–hard wall potential while the pair potential for the particles in the simple fluid is arbitrary.

In Figure 8.3, P^S is shown as a function of the surface separation D^S for the same hard sphere–hard wall system as in Figure 7.8. The contact densities used in the calculation of the pressure have been obtained from the same computations as the entire profiles in the latter figure. In connection to this figure, we discussed in Section 7.2.1.2 the basic reasons why the contact density varies with the surface separation D^S. The reasons for the variation in P^S are, of course, the same.

We can see in Figure 8.3 that the maxima in the disjoining pressure occur when D^S/d^h is slightly less than an integer l, at $D^S/d^h \approx l - 0.1$. This is close to – but not exactly at – the separations where the layer structure is fully developed (see below). For separations around the minima, where $P^S(D^S)$ is negative, the layer structure between the walls is frustrated because fully developed layers do not fit neatly in the slit. The width is too large for, say, l layers and too small for $l + 1$ layers, so the overall density in the slit is comparatively low because there are lots of empty spaces between the spheres. For these separations, the pressure in bulk is higher than between the surfaces; near a single surface in contact with bulk, the fluid does not experience such frustrations and $n^{\text{cont}}(\infty)$ will originate from an unconstrained structure.

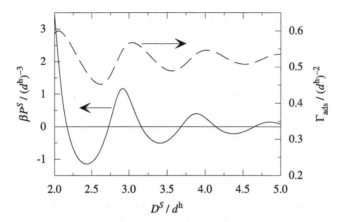

FIGURE 8.3 Disjoining pressure P^S (full curve) and adsorption excess Γ_{ads} (dashed curve) as functions of surface separation D^S for a hard sphere fluid between hard walls. The system is the same as in Figure 7.8.

Let us now consider the number of particles in the slit compared to the bulk. The **adsorption excess** in the slit, defined as $\Gamma_{\text{ads}} = \int_{-L/2}^{L/2}[n(z) - n^b]$, is the number of particles (i.e., particle centers) per unit area in excess to the number in the same volume of a bulk phase. It is plotted as a function of the surface separation D^S in Figure 8.3 and we see that it has a similar oscillatory behavior as P^S, but is slightly shifted in phase. Its maxima occur at D^S/d^h slightly larger than integer values and the minima are likewise shifted to somewhat higher separations compared to the minima in P^S. Γ_{ads} can be used as a measure of overall packing density in the slit compared to that of the bulk. It is positive, which means that the ordering influence of the walls makes the packing efficiency larger than in bulk. The maximal Γ_{ads} values occur very close to the slit widths where the layer structure is fully developed and all density peaks, apart from n^{cont} at the walls, have maximal peak heights. The extrema of Γ_{ads} give a much better indication of packing efficiency/inefficiency compared to the extrema of n^{cont} and hence of P^S.

$P_\perp^{\text{slit}}$, and hence P^S, can be calculated at any plane at $z = z^\bullet$ between the surfaces from Equation 8.41. From now on, we have the origin of the coordinate system placed at the midplane between the surfaces (like for the coordinates used in this equation). For hard spheres between hard walls, all terms in this equation, except the ideal term $k_B T n(z^\bullet)$ and the term involving the pair distribution, are equal to zero. The latter term equals

$$P_\perp^{\text{core}}(z^\bullet) = -\int_{z^\bullet}^{L/2} dz_1\, n(z_1) \int_{-L/2}^{z^\bullet} dz_2\, n(z_2)$$

$$\times \int ds_{12}\, g^{(2)}(\mathbf{r}_1, \mathbf{r}_2)\frac{du^{\text{core}}(r_{12})}{dr_{12}} \cdot \frac{z_1 - z_2}{r_{12}}, \tag{8.46}$$

where we have written $u^{\text{core}}(r)$ instead of $u(r)$ since this term appears in the pressure whenever the pair potential has a hard-core contribution. For hard spheres, the entire pair potential is $u(r) = u^{\text{core}}(r)$ and

$$P_{\perp}^{\text{slit}} = k_B T n(z^{\bullet}) + P_{\perp}^{\text{core}}(z^{\bullet}) \quad \text{(hard sphere fluid)}$$

for $-L/2 \leq z^{\bullet} \leq L/2$. When $|z^{\bullet}| = L/2$, we have $n(\pm L/2) = n^{\text{cont}}$ and $P_{\perp}^{\text{core}} = 0$, which is the case considered earlier in Equation 8.45.

As shown later in the shaded text box below on the following page, which can be skipped in the first reading, Equation 8.46 can be written as

$$P_{\perp}^{\text{core}}(z^{\bullet}) = k_B T \int_{z^{\bullet}}^{z^{\bullet}+d^h} dz_1 \, n(z_1) \int_{z_1-d^h}^{z^{\bullet}} dz_2 \, n(z_2)$$

$$\times 2\pi (z_1 - z_2) g^{(2)\text{cont}}(z_1, z_2), \tag{8.47}$$

where $g^{(2)\text{cont}}$ is the contact value of $g^{(2)}$ defined in Equation 7.66 and where the integration in practice can be restricted to $z_1 \leq L/2$ and $z_2 \geq -L/2$ since $n(z_1)$ or $n(z_2)$, respectively, is zero otherwise. This result is valid for any pair potential that contains a hard-core contribution. $P_{\perp}^{\text{core}}(z^{\bullet})$ is, as we will see, a steric contribution to the pressure due to core-core collisions of particles across the plane at $z = z^{\bullet}$ and it is always positive (repulsive). For a bulk phase, $g^{(2)\text{cont}}(z_1, z_2) = g^{(2)\text{b}}(d^{h+})$ and Equation 8.47 reduces to

$$\frac{P^{\text{core}}}{k_B T} = \frac{2\pi (d^h)^3}{3} (n^{\text{b}})^2 g^{(2)\text{b}}(d^{h+}) \quad \text{(bulk phase)}, \tag{8.48}$$

which is the hard-core term in Equation 5.75, where $g_N^{(2)} = g^{(2)\text{b}}$.

{**EXERCISE 8.4**
Show that Equation 8.47 applied to a bulk phase leads to Equation 8.48 and show thereby that the expression for $P_{\perp}^{\text{slit}}$ for a hard sphere fluid reduces to Equation 5.77 in the bulk.}

The physical interpretation of $P_{\perp}^{\text{core}}(z^{\bullet})$ from Equation 8.47 is as follows. When a particle on one side of $z^{\bullet}$ collides with a particle on the other side of $z^{\bullet}$, as depicted in Figure 8.4, there is a force across the plane at $z^{\bullet}$ due to the collision (a force due to a momentum transfer at the collision, which is different from the kind of momentum transfer considered in $P_{\perp}^{\text{kin}}(z^{\bullet})$). The pressure contribution $P_{\perp}^{\text{core}}(z^{\bullet})$ gives the total force per unit area perpendicular to the plane due to such collisions. To see this, consider a particle located at $\mathbf{r}_1$ to the right of $z^{\bullet}$. We will call it "particle 1." The pressure that acts on the surface of particle 1 due to collisions by other particles is proportional to the particle density there, that is, the contact density at the surface of the particle. This contact density is given by $n(z_2)g^{(2)}(\mathbf{r}_1, \mathbf{r}_2)$ with $r_{12} = d^{h+}$, where $\mathbf{r}_2$ is the position for a particle in contact with particle 1; one of them is shown in Figure 8.4 at the moment it collides with particle 1. The contact pressure is $k_B T n(z_2)g^{(2)}(\mathbf{r}_1, \mathbf{r}_2)$ with $r_{12} = d^{h+}$. Particle 1 must have z coordinate $z_1 \leq z^{\bullet} + d^h$ in order to be able to touch

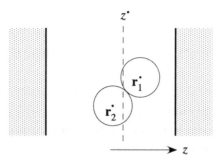

FIGURE 8.4 Collision between two particles located at either side of $z = z^\bullet$; one particle with its center at $\mathbf{r}_1$ and the other at $\mathbf{r}_2$.

particles that are located to the left of $z^\bullet$ and the latter must thereby have coordinate z_2 with $z_2 \geq z_1 - d^h$ (cf. the limits of the integrals in Equation 8.47). The force due to the collisions equals the contact pressure times the surface area at the periphery of particle 1; the part of the surface between coordinates z_2 and $z_2 + dz_2$ has the area $2\pi d^h dz_2$,[11] so the force there is $2\pi d^h dz_2 \times k_B T n(z_2) g^{(2)}(\mathbf{r}_1, \mathbf{r}_2)$ with $r_{12} = d^{h+}$. This contact force acts in the radial direction and, in order to project out the z component, one must multiply with $(z_1 - z_2)/d^h$. To obtain the total force on particle 1, we integrate with respect to z_2 and, finally, to obtain the force per unit area acting on *all* particles to the right of $z^\bullet$ (not only particle 1), we multiply by $n(z_1) dz_1$ and integrate with respect to z_1. In total, we thereby have obtained the core-core collision pressure $P_\perp^{core}(z^\bullet)$ given by Equation 8.47.

★ Here we will derive Equation 8.47 from Equation 8.46, whereby we use the same kind of arguments as in the shaded text boxes on pages 205ff and 403f but generalized to any pair potential with hard-core interactions, $u = u^{rest} + u^{core}$, where u^{rest} is the rest of the potential, for instance an electrostatic interaction. As we have seen in these text boxes, we can write $g^{(2)} = y^{(2)} \exp[-\beta u]$, where $y^{(2)}$ is continuous at $r_{12} = d^h$. Since $u = u^{rest} + u^{core}$, we can write $g^{(2)} = g_{rest}^{(2)} \exp[-\beta u^{core}]$ where $g_{rest}^{(2)} = y^{(2)} \exp[-\beta u^{rest}]$, which is also continuous at $r_{12} = d^h$. Equation 8.46 therefore contains the factor $\exp[-\beta u^{core}] \times (du^{core}/dr_{12})$, which like before equals $-\beta^{-1}\delta(r_{12} - d^h)$. Thus,

$$P_\perp^{core}(z^\bullet) = k_B T \int_{z^\bullet}^{L/2} dz_1 \, n(z_1) \int_{-L/2}^{z^\bullet} dz_2 \, n(z_2)$$

$$\times 2\pi \int_0^\infty ds_{12} \, s_{12} \, g_{rest}^{(2)}(s_{12}, z_1, z_2) \delta(r_{12} - d^h) \frac{z_1 - z_2}{r_{12}},$$

where we have written $g_{rest}^{(2)}(\mathbf{r}_1, \mathbf{r}_2) = g_{rest}^{(2)}(s_{12}, z_1, z_2)$ and inserted $2\pi s_{12} \, ds_{12}$ instead of $d\mathbf{s}_{12}$, whereby only the radial s_{12} integration remains in the last integral. By using the

---

[11] For a sphere of radius $\mathcal{R}$, the area of a spherical zone of height h equals $2\pi \mathcal{R}$h.

same argument as in the text box on page 403f and making the variable substitution $s_{12}^2 + (z_2 - z_1)^2 = (r'_{12})^2$ at constants z_1 and z_2, the last integral can be written as

$$2\pi \int_{|z_1-z_2|}^{\infty} dr'_{12}\, r'_{12}\, g_{\text{rest}}^{(2)}(s_{12}, z_1, z_2)\, \delta(r'_{12} - d^{\text{h}}) \frac{z_1 - z_2}{r'_{12}},$$

which is equal to $2\pi \left[g_{\text{rest}}^{(2)}(s_{12}, z_1, z_2) \right]_{r'_{12}=d^{\text{h}}} (z_1 - z_2)$ when $|z_1 - z_2| \leq d^{\text{h}}$ and zero otherwise (there is no contribution if $r'_{12} > d^{\text{h}}$, which is the case when $|z_1 - z_2| > d^{\text{h}}$). Since $g_{\text{rest}}^{(2)} = g^{(2)}$ when $r'_{12} \geq d^{\text{h}}$, we can write $g^{(2)}$ instead of $g_{\text{rest}}^{(2)}$ and we thus obtain the result in Equation 8.47, where the integration limits make sure that $|z_1 - z_2| \leq d^{\text{h}}$.

8.5 ELECTRIC DOUBLE-LAYER INTERACTIONS WITH ION-ION CORRELATIONS

8.5.1 Counterions between Charged Surfaces

Let us apply the results for $P_\perp^{\text{slit}}$ in Section 8.3 to double-layer interactions. To start with, we consider the case with only counterions with charge q and diameter d^{h} in the slit between the charged hard surfaces with uniform surface charge densities $\sigma_{\text{I}}^{\mathcal{S}}$ and $\sigma_{\text{II}}^{\mathcal{S}}$, respectively (compare with Example 8.2, where the PB approximation was used). The distance of closest approach of the ion centers to the planes of surface charge is $d_{\text{I}}^{\mathcal{S}}$ and $d_{\text{II}}^{\mathcal{S}}$, respectively. The surface separation is $D^{\mathcal{S}} = L + d_{\text{I}}^{\mathcal{S}} + d_{\text{II}}^{\mathcal{S}}$ and the possible z coordinates for the ion centers lie in the interval $-L/2 \leq z \leq L/2$. The electrostatic interaction potential between the surface charges is $\mathcal{V}_{\text{I,II}}^{\text{el}}(D^{\mathcal{S}}) = -\sigma_{\text{I}}^{\mathcal{S}}\sigma_{\text{II}}^{\mathcal{S}} D^{\mathcal{S}}/(2\varepsilon_r\varepsilon_0)$ (Equation 8.32) and we have the ion-surface interaction potentials $v_J^{\text{el}}(\ell) = -q\sigma_J\ell/(2\varepsilon_r\varepsilon_0)$ for $J = \text{I}$ and II, where ℓ is the distance from the surface charge. The derivatives of these potentials are

$$\mathcal{V}'_{\text{I,II}} = \mathcal{V}_{\text{I,II}}^{\text{el}\prime} = -\frac{\sigma_{\text{I}}^{\mathcal{S}}\sigma_{\text{II}}^{\mathcal{S}}}{2\varepsilon_r\varepsilon_0} \quad \text{and} \quad v'_J = v_J^{\text{el}\prime} = -\frac{q\sigma_J^{\mathcal{S}}}{2\varepsilon_r\varepsilon_0}$$

and by inserting this in Equation 8.41, we obtain

$$P_\perp^{\text{slit}} = k_B T n(z^\bullet) + \frac{\sigma_{\text{I}}^{\mathcal{S}}}{2\varepsilon_r\varepsilon_0} \int_{z^\bullet}^{L/2} dz_1 \rho(z_1) + \frac{\sigma_{\text{II}}^{\mathcal{S}}}{2\varepsilon_r\varepsilon_0} \int_{-L/2}^{z^\bullet} dz_1 \rho(z_1)$$

$$- \int_{z^\bullet}^{L/2} dz_1\, n(z_1) \int_{-L/2}^{z^\bullet} dz_2\, n(z_2) \int ds_{12}\, g^{(2)}(\mathbf{r}_1, \mathbf{r}_2)$$

$$\times \frac{du(r_{12})}{dr_{12}} \frac{z_1 - z_2}{r_{12}} + \frac{\sigma_{\text{I}}^{\mathcal{S}}\sigma_{\text{II}}^{\mathcal{S}}}{2\varepsilon_r\varepsilon_0}, \tag{8.49}$$

where $\rho(z) = qn(z)$, $u(r) = u^{\mathrm{el}}(r) + u^{\mathrm{core}}(r)$, $u^{\mathrm{el}}(r) = q^2/(4\pi\varepsilon_0\varepsilon_r r)$, $u^{\mathrm{core}}(r)$ is the hard core potential and $\mathbf{s}_{12} = (x_2 - x_1, y_2 - y_1)$. For simplicity, we start with the case of point ions, $d^{\mathrm{h}} = 0$, so we have $u(r) = u^{\mathrm{el}}(r)$. Since the ions strongly repel each other electrostatically at short separations, the results we obtain are relevant also for small ions provided d^{h} is sufficiently small; the electrostatic repulsion makes the probability for small ions to touch each other very low, so their actual size does not matter much.

In the integral with $g^{(2)}$ in Equation 8.49, we write $g^{(2)} = h^{(2)} + 1$ and we have a contribution from the integral with $h^{(2)}$ inserted and the corresponding contribution with 1 inserted. To set $g^{(2)} = 1$ would mean that one neglects ion-ion correlations in the electrolyte in the slit, which is the approximation made in the PB approximation. Therefore, the sum of all contributions to $P_\perp^{\mathrm{slit}}$ apart from that from $h^{(2)}$ will give the PB pressure given in Equations 8.24 and 8.28.[12] This means that, in addition to the PB pressure, we have in the current case an additional term

$$P_\perp^{\mathrm{el,corr}}(z^\bullet) = -\int_{z^\bullet}^{L/2} dz_1\, n(z_1) \int_{-L/2}^{z^\bullet} dz_2\, n(z_2)$$

$$\times \int d\mathbf{s}_{12}\, h^{(2)}(\mathbf{r}_1, \mathbf{r}_2) \frac{du^{\mathrm{el}}(r_{12})}{dr_{12}} \frac{z_1 - z_2}{r_{12}}, \qquad (8.50)$$

so in total we have

$$P^{\mathcal{S}} = P_\perp^{\mathrm{slit}} = k_B T n(z^\bullet) + P_\perp^{\mathrm{el,mean}}(z^\bullet) + P_\perp^{\mathrm{el,corr}}(z^\bullet)$$

$$= k_B T n(z^\bullet) - \frac{(\sigma_\bullet^{\mathcal{S}})^2}{2\varepsilon_r\varepsilon_0} + P_\perp^{\mathrm{el,corr}}(z^\bullet), \qquad (8.51)$$

where, as before, $\sigma_\bullet^{\mathcal{S}} = |\sigma_{\mathrm{I}\bullet}^{\mathcal{S}}| = |\sigma_{\bullet\mathrm{II}}^{\mathcal{S}}|$ with $\sigma_{\mathrm{I}\bullet}^{\mathcal{S}}$ being the total charge per unit area to the left of $z^\bullet$ and $\sigma_{\bullet\mathrm{II}}^{\mathcal{S}}$ the corresponding quantity to the right of $z^\bullet$ (defined in Equation 8.27). The term $P_\perp^{\mathrm{el,corr}}(z^\bullet)$ is the *pressure contribution from electrostatic ion-ion correlations* across the plane at $z = z^\bullet$, which is missing in the PB approximation. The value of $n(z^\bullet)$ is, of course, also incorrect in this approximation. The same applies to $\sigma_\bullet^{\mathcal{S}}$ when $-L/2 < z^\bullet < L/2$.

The physical interpretation of $P_\perp^{\mathrm{el,corr}}(z^\bullet)$ is the following. Consider an ion located at $\mathbf{r}_2$ to the left of the plane at $z = z^\bullet$. This ion polarizes the electrolyte on the other side of the plane by interacting electrostatically with the ions there, like in the sketch shown in Figure 8.2c where $z^\bullet = 0$. There is an ion depletion, at least close to the ion,[13] since all ions repel each other. Compared to the unperturbed average distribution (in the absence of the ion located at $\mathbf{r}_2$), this polarization of the electrolyte on the other side of the plane is effectively a charge of opposite sign, which acts with an attractive force on the ion. Summed over all ions to the left of the plane, weighted with the density $n(z_2)$, this gives $P_\perp^{\mathrm{el,corr}}(z^\bullet)$.

---

[12] This is shown explicitly in Exercise 8.5 on page 458.
[13] The pair correlation function $h^{(2)}(\mathbf{r}_1, \mathbf{r}_2)$ is negative at least for small r_{12}.

Another related interpretation is that the charge distribution on each side of the plane at $z = z^{\bullet}$ is subject to fluctuations due to the movements of the individual ions. These fluctuations induce a polarization of the electrolyte on the other side of the plane, which, for the reasons just explained, essentially has the opposite charge (in most situations). This results in an attractive force that acts across the plane. The nature of this electrostatic ion-ion correlation force is analogous to the electronic correlation interaction that results in the attractive dispersion forces in quantum systems (for example, the van der Waals interaction between noble gas atoms). The main difference is that the ions move slowly and are described by classical mechanics in the current treatment, while the electrons move quickly and must be described by quantum mechanics. Otherwise, the mechanisms for the correlation interactions are essentially the same.

If we set $z^{\bullet} = L/2$ in Equation 8.51, we have $P_{\perp}^{\mathrm{el,corr}} = 0$ and $\sigma_{\bullet}^{\mathcal{S}} = |\sigma_{\mathrm{II}}^{\mathcal{S}}|$, so this equation reduces to

$$P_{\perp}^{\mathrm{slit}} = k_B Tn(L/2) - \frac{\left(\sigma_{\mathrm{II}}^{\mathcal{S}}\right)^2}{2\varepsilon_r\varepsilon_0}. \tag{8.52}$$

This is the *contact theorem* for wall II (given in Equation 8.5), which hence holds in the general case, as noted in Section 8.1. The corresponding result for wall I is obtained when $z^{\bullet} = -L/2$. This is a very simple expression for $P_{\perp}^{\mathrm{slit}}$, but in practical calculations the two terms $k_B Tn(L/2)$ and $\left(\sigma_{\mathrm{II}}^{\mathcal{S}}\right)^2/(2\varepsilon_r\varepsilon_0)$ are often very large and nearly the same, which means that $P_{\perp}^{\mathrm{slit}}$ is the small difference between two large numbers. Therefore, one needs to calculate both of these large terms very accurately in order to obtain a reasonable accuracy for $P_{\perp}^{\mathrm{slit}}$. This limits the usefulness of Equation 8.52.

In the case of equal surfaces, $\sigma_{\mathrm{I}}^{\mathcal{S}} = \sigma_{\mathrm{II}}^{\mathcal{S}} = \sigma^{\mathcal{S}}$ and $d_{\mathrm{I}}^{\mathcal{S}} = d_{\mathrm{II}}^{\mathcal{S}} = d^{\mathcal{S}}$, the system is symmetric around the midplane, $z = 0$. If we select $z^{\bullet} = 0$, we have $\sigma_{\bullet}^{\mathcal{S}} = 0$ and therefore Equation 8.51 becomes

$$P^{\mathcal{S}} = k_B Tn(0) + P_{\perp}^{\mathrm{el,corr}}(0). \tag{8.53}$$

In the PB approximation, the last term is neglected and the pressure is proportional to the ion density at the midplane, so $P^{\mathcal{S}}$ is always repulsive for equal surfaces. The ion-ion correlation contribution is, however, always present and in the limit $L \to 0$ we have, in fact, $P_{\perp}^{\mathrm{el,corr}}(0) \to -(\sigma^{\mathcal{S}})^2/(2\varepsilon_r\varepsilon_0)$ irrespective of the valency of the ions. One can show this explicitly by investigating Equation 8.50, see Exercise 8.6 on page 458.[14]

The crucial question is: which term dominates in $P^{\mathcal{S}}$, the momentum transfer contribution $k_B Tn(0)$ or the correlation contribution $P_{\perp}^{\mathrm{el,corr}}(0)$? If the latter dominates, $P^{\mathcal{S}}$ can be attractive in contrast to the PB prediction. Since the number of counterions per unit area needed to neutralize the surface charge decreases with increasing valency, $n(0)$ is expected to be lower for counterions with high valencies compared to monovalent ions. $P_{\perp}^{\mathrm{el,corr}}(0)$ can therefore dominate at least for counterions with sufficiently high valency. It is also

---

[14] One can also understand it from the following simple argument. In Equation 8.49, the term with $g^{(2)}$ goes to zero when $L \to 0$ because of the factor $z_1 - z_2$ (all other parts of this term stay finite in this limit). When $\sigma_{\mathrm{I}}^{\mathcal{S}} = \sigma_{\mathrm{II}}^{\mathcal{S}} = \sigma^{\mathcal{S}}$, the sum of second and third terms on the rhs equals $\sigma^{\mathcal{S}} \times (-2\sigma^{\mathcal{S}})/(2\varepsilon_r\varepsilon_0)$, so the sum of all terms except the ideal one goes to $-(\sigma^{\mathcal{S}})^2/(2\varepsilon_r\varepsilon_0)$ when $L \to 0$.

of importance that many counterions are normally close to the surfaces and fewer in the middle.

It was shown in 1984 in a Monte Carlo simulation study by Guldberg, Jönsson, Wennerström and Linse[15] that the electrostatic ion-ion correlations can make $P^{\mathcal{S}}$ attractive, thereby breaking with the commonly held conception that double-layer interactions between equally charged surfaces are repulsive (as obtained in the PB approximation).[16] For aqueous electrolyte solutions at room temperature under conditions realistic for colloid systems, they found that the interactions for surfaces with monovalent counterions are repulsive (dominance by $k_B T n(0)$), while the interactions are attractive for divalent counterions at rather short surface separations, provided that the surface charge density is sufficiently large (dominance by $P_\perp^{\mathrm{el,corr}}$). The fact that the change from repulsive to attractive double-layer interactions occurs when going from monovalent to divalent ions for aqueous systems is fortuitous. When the solvent has a lower dielectric constant and/or the temperature is lower, the interactions can be attractive due to ion-ion correlations also for monovalent ions.

In order to illustrate these questions further, we generalize Equations 8.51 and 8.53 to ions with finite size, $d^h \neq 0$. Then there is a further contribution in Equation 8.49, namely a steric contribution to the pressure due to core-core collisions of ions across the plane at $z = z^\bullet$, same as $P_\perp^{\mathrm{core}}(z^\bullet)$ in the case of hard sphere fluids. It originates from u^{core} in the pair interaction $u(r) = u^{\mathrm{el}}(r) + u^{\mathrm{core}}(r)$ and is given by Equation 8.47. The total disjoining pressure therefore is

$$P^{\mathcal{S}} = P_\perp^{\mathrm{slit}} = P_\perp^{\mathrm{kin}}(z^\bullet) - \frac{\left(\sigma_\bullet^{\mathcal{S}}\right)^2}{2\varepsilon_r\varepsilon_0} + P_\perp^{\mathrm{el,corr}}(z^\bullet) + P_\perp^{\mathrm{core}}(z^\bullet), \tag{8.54}$$

where we have introduced the notation $P_\perp^{\mathrm{kin}}(z^\bullet) = k_B T n(z^\bullet)$ for the ideal term, that is, the kinetic contribution due to momentum transfer across the plane at $z = z^\bullet$ when ions pass from one side of the plane to the other.

For very small surface separations, $P_\perp^{\mathrm{kin}}(z^\bullet)$ will always dominate over the other contributions, which can be understood as follows. The average ion density in the slit must increase quite rapidly to infinity[17] for small L; it goes like $1/L$ with decreasing L since the number of ions per surface area A, $2A|\sigma^{\mathcal{S}}/q|$, is constant, while the volume LA available for the ion centers decreases. For very narrow slits, the average density and $n(z^\bullet)$ are

---

[15] L. Guldberg, B. Jönsson, H. Wennerström and P. Linse, *J. Chem. Phys.* **80** (1984) 2221.

[16] The ion-ion correlation mechanism for an attractive contribution to interactions between polyelectrolyte molecules was predicted earlier by Oosawa (in F. Oosawa, *Polyelectrolytes*, p. 123, Dekker, New York, 1971) and a similar mechanism was discussed by Hill (in T. L. Hill, *An Introduction to Statistical Thermodynamics*, p. 359–361, Addison-Wesley, Reading, 1960). Guldberg et al. showed in 1984 that the *net double-layer interaction* can be attractive between charged surfaces in actual calculations. Their findings were soon quantitively confirmed by accurate integral equation calculations (R. Kjellander and S. Marčelja, *Chem. Phys. Lett.* **112** (1984) 49).

[17] Note that the three-dimensional density n for a fluid confined to a narrow planar slit can be huge. For a fluid confined to a smooth plane, n is infinite irrespective of the number of particles per unit area since the volume available for the particle centers is zero. When the particles are confined to a very narrow slit rather than to a surface, n is finite but very large since the volume per particle is very small. For the same reason, the contact density $n(L/2)$ of a fluid in a wide slit can be huge when particles are strongly attracted to a surface and thereby restricted in motion in the perpendicular direction.

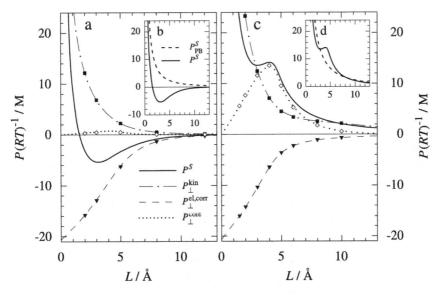

FIGURE 8.5 Pane (a) and inset (b): The disjoining pressure P^S for two equal, quite highly charged walls with divalent counterions in the slit between the surfaces as a function of the width L of the slit. The surface separation is $D^S = L + 2d^S$. The curves are calculated by an accurate integral equation theory (see text) and the symbols show the results from Monte Carlo simulations. The unit for the pressure plotted as P^S/RT is $mol\,dm^{-3}$. In inset (b), P^S is compared with the PB prediction P^S_{PB} obtained from Equation 8.21 and in pane (a) P^S is split in its components $P^S = P^{kin}_\perp(0) + P^{el,corr}_\perp(0) + P^{core}_\perp(0)$, where the kinetic (ideal) part is $P^{kin}_\perp(0) = k_B T n(0)$, the electrostatic ion-ion correlation part $P^{el,corr}_\perp(0)$ is given by Equation 8.50 with $z^\bullet = 0$ and the ionic core-core collision part $P^{core}_\perp(0)$ is given by Equation 8.47. Pane (c) and inset (d) show the corresponding plots for monovalent ions. The curve and symbol notations are the same as in (a) and (b).[18]

virtually the same. Since all terms in $P^{slit}_\perp$ apart from the ideal term are finite, it follows that $P^{slit}_\perp \sim k_B T\, 2\, |\sigma^S/(qL)|$ when $L \to 0$.

For the case of equal surfaces and $z^\bullet = 0$, Equation 8.54 becomes

$$P^S = P^{slit}_\perp = P^{kin}_\perp(0) + P^{el,corr}_\perp(0) + P^{core}_\perp(0). \tag{8.55}$$

As an example of double-layer interactions for counterions between two equally charged walls given by this equation, Figure 8.5a shows the disjoining pressure P^S and its different components for divalent counterions in aqueous solution between rather highly charged surfaces, $\sigma^S = 0.267\ C\,m^{-2}$, which corresponds to one elementary charge per 60 Å^2. The ion diameter is $d^h = 4.25$ Å and we take $d^S = d^h/2$.[19] The temperature is 298 K and ε_r is 78.8, which corresponds to water. The pressures presented in the figure are calculated with

---

[18] All plots in Figure 8.5a,b and the MC plots in c and d are based on data from R. Kjellander, T. Åkesson, B. Jönsson and S. Marčelja, *J. Chem. Phys.* **97** (1992) 1424 and the integral equation plots in Figure 8.5c and d are based on data from H. Greberg and R. Kjellander, *Mol. Phys.* **83** (1994) 789.

[19] The choice of d^S does not influence the results (when $d^S \geq 0$).

MC simulations and a very accurate integral equation theory of the same kind as for Figures 7.3 and 7.15, where both the ion density profile $n(z)$ and the anisotropic pair correlation function $g^{(2)}(\mathbf{r}_1, \mathbf{r}_2)$ are calculated explicitly and self-consistently.[20] As seen in Figure 8.5, the results of these two kinds of calculations show excellent agreement with each other.

We see in Figure 8.5a that P^S has an attractive minimum near $L = 3.2$ Å, which corresponds to a surface separation $D^S = 7.5$ Å. The attraction is due to the negative electrostatic ion-ion correlation pressure $P_\perp^{el,corr}(0)$ like in the previous case of point counterions. In the present case, the ions have nonzero size, but the contribution $P_\perp^{core}(0)$ from ion-ion collisions across the midplane is very small and can nearly be ignored. The ideal kinetic pressure $P_\perp^{kin}(0) = k_B T n(0)$ dominates for small surface separations as it must, so P^S is positive there. The PB prediction for P^S is compared in Figure 8.5b with the accurate result of the pressure shown in frame (a) of the figure and we see that the PB approximation fails completely. It cannot predict the attraction since $P_{PB}^S = k_B T n(0)$ is always positive.

When there are monovalent counterions between the surfaces instead of divalent ones, P^S is repulsive for all separations as apparent from Figure 8.5c. The comparison with the PB prediction in Figure 8.5d shows that apart from the hump in the accurate result (full curve), the two curves have similar behavior and magnitude. This is the case despite that the PB approximation neglects the ion-ion correlations that give rise to $P_\perp^{el,corr}$ and $P_\perp^{core}$; it only includes $P_\perp^{kin}(0) = k_B T n(0)$. The results in Figure 8.5c show that the terms that are missing in the PB approximation are large, so the similar magnitude of the two P^S results in frame (d) is *not* a consequence of negligible missing contributions which therefore can be neglected. Instead, there are substantial cancellations of errors in the PB approximation.

The hump in the P^S curve between $L \approx 3$ to 7 Å is due to the core-core collision pressure $P_\perp^{core}$. This contribution goes to zero when $L \to 0$ because it is a *perpendicular* pressure component; the ions in a thin slit are restricted to move in the lateral directions (directions parallel to the x, y plane) and therefore collide *laterally*. In Equation 8.47, the vanishing of $P_\perp^{core}$ for small L is brought about by the factor $(z_1 - z_2)$ in the integrand. For large surface separations, $P_\perp^{core}$, vanishes quite quickly when L increases beyond $L \approx 8$ Å, where $P_\perp^{kin}$ takes over and becomes the main contribution to P^S.

We noted earlier that $P_\perp^{el,corr}(0) \to -(\sigma^S)^2/(2\varepsilon_r\varepsilon_0)$ when $L \to 0$ irrespective of the valency of the ions and as seen in Figure 8.5a and c, $P_\perp^{el,corr}$ in the whole range shown is nearly the same for the monovalent and divalent cases. *The attractive electrostatic ion-ion correlation pressure is hence virtually equally strong for both monovalent and divalent ions!* The net repulsion for the monovalent case is instead a consequence of *larger repulsive contributions* $P_\perp^{core}$ and $P_\perp^{kin}$ to the pressure for the monovalent case compared to the divalent case. The reason for this is that the number of ions in the middle part of the slit is larger in the monovalent case compared to the divalent case. This occurs both because the number

---

[20] The integral equation data for the monovalent case were calculated in the Reference Anisotropic HNC (ARHNC) approximation, while the corresponding data for the divalent case were calculated in the slightly simpler Anisotropic HNC (AHNC) approximation. Both are introduced in Chapter 9 (in the 2nd volume). The ARHNC approximation is needed for the monovalent case since the ion density is quite large there. The AHNC approximation gives somewhat less accurate results for $P_\perp^{core}$ in this case. (Both approximations give, however, the same result for divalent ions.)

of monovalent ions needed to neutralize the surface charges is twice as high as in the former case and because there is a large attraction of divalent ions towards the surface due to strong ion-ion correlations of the same type as we discussed in connection to Figures 7.3 and 7.15 (an overcompensation of surface charge density can, however, not occur in the present system with only counterions since they balance the surface charge exactly). This attraction towards the surface regions leaves fewer ions in the middle part.

As we can see in Figure 8.5, $P_\perp^{kin}(0)$ for the monovalent case is for $L \lesssim 5$ Å indeed close to twice the value for the divalent case. For $L \gtrsim 5$ Å, we see that $P_\perp^{kin}(0)$ for the divalent case goes quickly to zero when L increases, which occurs because of the strong correlation attraction of divalent counterions towards the surfaces mentioned earlier. As regards the core-core collision pressure $P_\perp^{core}(0)$, the higher density of counterions in the monovalent case makes this contribution much more prominent than in the divalent case since the probability for collisions is larger when the density is higher. If the ions are made smaller, $P_\perp^{core}$ becomes less prominent (for point ions, $P_\perp^{core}$ is zero), but P^S is, in fact, still repulsive for monovalent ions because of the dominance of $P_\perp^{kin}(0)$ under the same conditions as in the figure.

Incidentally, we note that whenever $P_\perp^{slit}$ is attractive, the contact theorem (8.52) implies that its negative term $- \left(\sigma_{II}^S\right)^2 / (2\varepsilon_r\varepsilon_0)$, which is independent of the surface separation, must dominate over the positive contact density term $k_B T n(L/2)$. This is in stark contrast to the result in the PB approximation for equally charged surfaces, where the latter term always must dominate and give a repulsive total pressure for all separations as follows from Equation 8.11. The PB value of the contact density is hence larger than the correct value. The attractive double-layer force can hence also be explained as a consequence of a substantial *lowering* of the contact density due to ion-ion correlations compared to the PB prediction.

These findings may seem contradictory since, as we found earlier, the correlations lead to a strong attraction for the counterions towards the surface region and therefore a larger density near the surface. How can this be reconciled with a lower contact density? There is, in fact, no contradiction – the attractive force is larger than the PB prediction in most of the slit except *very* close to each surface where the force is larger in the PB case. This raises the PB value of the contact density and can be explained as follows: for a counterion in contact with the surface, there are no ions on its side towards the surface; all other counterions are located on the side away from the surface. They repel our counterion towards the surface and this force adds to the electrostatic attraction towards the surface charge, which acts in the same direction. In the PB approximation, the locations of the other counterions are unaffected by the presence of our counterion and can approach closely, which results in a strong repulsion. When correlations are considered, the repulsion towards the surface is smaller because the other counterions avoid the neighborhood of our counterion. A similar situation appears in Figure 7.3a, where the contact density for counterions is higher in the PB approximation than in the accurate theory, while the counterion density in the interval $0.1 \lesssim z - d^S \lesssim 3.4$ Å is lower in the PB case.

A final remark: It must be kept in mind that the results in Figure 8.5 are for double layers in the primitive model for electrolytes, which means that the solvent is modeled as a dielectric continuum. The molecular nature of a real solvent adds a complication to the double-layer interaction since it adds contributions from the interactions between water molecules, including electrostatic and steric interactions, and modifies the interactions between the ions. The effect of steric interactions for ions like the detailed behavior of $P_\perp^{\text{core}}$ may therefore be different in a real system. The electrostatic ion-ion correlation effects, on the other hand, have a more general nature. As we have seen, the behavior of $P_\perp^{\text{el,corr}}$ is to a large extent independent of the nature of the ions. Attractive double-layer interactions exist for real systems with multivalent ions from divalent and upwards due to essentially the same mechanism as in the primitive model.

{**EXERCISE 8.5**

Use Equation 8.44 to show that $P_\perp^{\text{slit}}$ given by Equation 8.49 with $u(r) = u^{\text{el}}(r)$ reduces to the PB result for $P_\perp^{\text{slit}}$ in Equation 8.28 when we ignore pair correlations and set $g^{(2)}(\mathbf{r}_1, \mathbf{r}_2) = 1$. Thereby, you may use Equation 8.44 with $u = u^{\text{el}}$, the definitions of $\sigma_{\mathrm{I}\bullet}^S$ and $\sigma_{\bullet\mathrm{II}}^S$ in Equation 8.27 and the fact that $\left(\sigma_\bullet^S\right)^2 = -\sigma_{\mathrm{I}\bullet}^S \sigma_{\bullet\mathrm{II}}^S.$}

{**EXERCISE 8.6**

For the case of two equal surfaces, show in the following manner that $P_\perp^{\text{el,corr}}(0)$ in Equations 8.53 and 8.55 has the limit $P_\perp^{\text{el,corr}}(0) \to -(\sigma^S)^2/(2\varepsilon_r\varepsilon_0)$ when $L \to 0$.

a. By doing the r_{12} derivative in Equation 8.50, write $P_\perp^{\text{el,corr}}(0)$ in the form

$$P_\perp^{\text{el,corr}}(0) = \frac{q^2}{2\varepsilon_r\varepsilon_0} \int_0^{L/2} dz_1\, n(z_1) \int_{-L/2}^0 dz_2\, n(z_2)$$

$$\times (z_1 - z_2) \int_0^\infty ds_{12}\, h^{(2)}(\mathbf{r}_1, \mathbf{r}_2) \frac{s_{12}}{[s_{12}^2 + (z_1 - z_2)^2]^{3/2}},$$

where $\mathbf{s}_{12} = (x_2 - x_1, y_2 - y_1)$ and $s_{12} = |\mathbf{s}_{12}|$.

b. Divide the integration interval of the last integral in two parts: $0 < s_{12} < b$ and $b \leq s_{12} < \infty$, where b is a positive number very close to 0. Show that the integral over the last interval is finite and that the corresponding contribution to $P_\perp^{\text{el,corr}}$ therefore vanishes in the limit $L \to 0$ since both z_1 and z_2 go to zero in this limit.

c. Since $g^{(2)}$ is equal to 1 for $\mathbf{r}_{12} = 0$, the function $h^{(2)}$ is equal to -1 there, so when s_{12}, z_1 and z_2 are very small, one can set $h^{(2)} = -1$. This can be done in the integral over the interval $0 < s_{12} < b$ when b and L are sufficiently small. Show that this integral goes to infinity like $-1/|z_1 - z_2|$ when $L \to 0$, so the corresponding contribution to $P_\perp^{\text{el,corr}}$ remains finite in this limit. **Hint:** Do the variable substitution $\tau = s_{12}^2 + (z_1 - z_2)^2$.

d. Show that the result in (c) leads to $P_\perp^{\text{el,corr}}(0) \to -(\sigma^S)^2/(2\varepsilon_r\varepsilon_0)$ when $L \to 0$.}

8.5.2 Equilibrium with Bulk Electrolyte

Let us generalize the theory for double-layer interactions of the previous section to the case where the electrolyte in the slit between the charged surfaces is in equilibrium with a bulk electrolyte. We then have the density distribution $n_i(z_1)$ of species i and the pair distribution function $g_{ij}^{(2)}(\mathbf{r}_1, \mathbf{r}_2)$ for ions of species i and j. In the bulk phase, the density of ions of species i is n_i^b and the pair distribution function is $g_{ij}^{(2)b}(r_{12})$.

The force in the z direction per unit area across a plane at $z^{\bullet}$ between the surfaces is, as before, given by

$$P_{\perp}^{\text{slit}} = P_{\perp}^{\text{kin}}(z^{\bullet}) - \frac{(\sigma_{\bullet}^{S})^2}{2\varepsilon_r\varepsilon_0} + P_{\perp}^{\text{el,corr}}(z^{\bullet}) + P_{\perp}^{\text{core}}(z^{\bullet}) \qquad (8.56)$$

where, in the current case, we have $P_{\perp}^{\text{kin}}(z^{\bullet}) = k_B T \sum_i n_i(z^{\bullet}) = k_B T n_{\text{tot}}(z^{\bullet})$ and where

$$P_{\perp}^{\text{el,corr}}(z^{\bullet}) = -\sum_{i,j} \int_{z^{\bullet}}^{L/2} dz_1\, n_i(z_1) \int_{-L/2}^{z^{\bullet}} dz_2\, n_j(z_2)$$
$$\times \int d\mathbf{s}_{12}\, h_{ij}^{(2)}(\mathbf{r}_1, \mathbf{r}_2) \frac{du_{ij}^{\text{el}}(r_{12})}{dr_{12}} \frac{z_1 - z_2}{r_{12}}, \qquad (8.57)$$

is the generalization of Equation 8.50 and

$$P_{\perp}^{\text{core}}(z^{\bullet}) = k_B T \sum_{i,j} \int_{z^{\bullet}}^{z^{\bullet}+d^h} dz_1\, n_i(z_1) \int_{z_1-d^h}^{z^{\bullet}} dz_2\, n_j(z_2)$$
$$\times 2\pi(z_1 - z_2) \left[g_{ij}^{(2)}(\mathbf{r}_1, \mathbf{r}_2) \right]_{r_{12}=d^{h+}}, \qquad (8.58)$$

corresponds to Equation 8.47. As before, $\mathbf{s}_{12} = (x_2 - x_1, y_2 - y_1)$. For simplicity in notation, the formulas for the pressure components are written for the case when all ions have the same size.[21] The disjoining pressure is

$$P^{S} = P_{\perp}^{\text{slit}} - P^b \qquad (8.59)$$

$$= P_{\perp}^{\text{kin}}(z^{\bullet}) - \frac{(\sigma_{\bullet}^{S})^2}{2\varepsilon_r\varepsilon_0} + P_{\perp}^{\text{el,corr}}(z^{\bullet}) + P_{\perp}^{\text{core}}(z^{\bullet}) - P^b$$

---

[21] If the different ionic species have different sizes, the integration intervals in Equation 8.57 should be $(z^{\bullet}, L_{\text{II},i}/2)$ for z_1 and $(-L_{\text{I},j}/2, z^{\bullet})$ for z_2, that is, between $z^{\bullet}$ and the z coordinate for the point of closest approach of the respective ion center to the respective wall. In Equation 8.58, d^h should be replaced by d_{ij}^h, which is the contact distance for a pair of ions, one of species i and one of species j. As regards $P_{\perp}^{\text{kin}}(z^{\bullet})$, whenever $z^{\bullet}$ is larger than $L_{\text{II},i}/2$ or smaller than $-L_{\text{I},i}/2$, the density at $n_i(z^{\bullet})$ is zero and, instead, the contact density for the respective species at the respective wall enters instead of $n_i(z^{\bullet})$ into $P_{\perp}^{\text{kin}}(z^{\bullet})$.

with the bulk pressure P^b obtained, for example, from Equations 8.39, 8.42 or 8.43 generalized in the same fashion to several species of ions. Incidentally, P^b from the latter two equations can be divided into contributions from electrostatic ion-ion correlations and core-core collisions in an analogous manner as here. The free energy of interaction (potential of mean force) per unit area, $\mathscr{W}_{I,II}^S(D^S)$, between two surfaces I and II at separation D^S is

$$\mathscr{W}_{I,II}^S(D^S) = \int_{D^S}^{\infty} P^S(D)\,dD \tag{8.60}$$

as before (Equation 8.2).

When the surface separation decreases and $L \to 0$, most ions between the surfaces leave the slit. The only ones that eventually remain are the counterions that are needed to maintain electroneutrality. Therefore, the values of $P_\perp^{slit}$ and its different components approach those of the case with only counterions between the surfaces, which was discussed in the previous section. This implies, for example, that $P_\perp^{el,corr}(0)$ for cases with equally charged surfaces will approach the universal value $-(\sigma^S)^2/(2\varepsilon_r\varepsilon_0)$ when $L \to 0$ also in this case and that $P_\perp^{slit} \sim k_B T\,2\,|\sigma^S/(q_{count}L)|$ in the same limit, where q_{count} is the charge of a counterion (assuming that there is only one kind of counterion present).

From the expression for P^S, it follows that because of the presence of the term $-P^b$, the disjoining pressure for small L will always be smaller (less repulsive or more attractive) than in the case of only counterions since the bulk pressure P^b is always positive. This means that for cases where $P_\perp^{slit}$ is attractive for small L with only counterions present, the addition of salt to the system (including the surrounding bulk phase) will make the double-layer interaction *even more attractive*. Exceptions may occur when the coions and accompanying counterions that enter the slit change the conditions there to a large extent in the opposite direction. However, for the kind of system investigated in Figure 8.5 with divalent counterions,[22] only a small number of coions enter the slit when $L \lesssim 3$ Å when the bulk concentration is increased to as high as 3 M for a 1:-2 electrolyte. These coions and the accompanying counterions make the repulsive pressure components $P_\perp^{kin}(0)$ and $P_\perp^{core}(0)$ somewhat larger than in the counterion-only case, but $P_\perp^{el,corr}(0)$ remain virtually the same for $L \lesssim 5$ Å. The end result is that the attractive minimum of P^S becomes deeper with increasing bulk concentration due to the presence of the term $-P^b$; the depth of the minimum is, for example, increased by about 20% for a 0.4 M electrolyte and doubled for a 3 M electrolyte compared with the counterion-only case. For a 2:-2 electrolyte, the corresponding increases are about 8% and 60%, respectively.

8.6 VAN DER WAALS FORCES AND IMAGE INTERACTIONS IN ELECTRIC DOUBLE-LAYER SYSTEMS

The force between two charged walls that we have considered so far is the electrostatic double-layer force due to the presence of an electrolyte in the slit between the surfaces.

---

[22] *Loc. cit.* in footnote 18.

In addition, there are other kinds of interactions between the walls, most notably *van der Waals (vdW) interactions*. The latter are forces due to the fluctuations of charges inside the walls and in the media in the slit between the surfaces, coupled to the electromagnetic fields inside and outside the bodies. When dealing with the van der Waals interactions between macroscopic bodies, one usually considers the macroscopic electromagnetic properties of matter. In the context of the primitive model of electrolyte solutions, this is congruent with the treatment of the solvent and the interior of the walls as dielectric continua characterized by their dielectric constants. For the vdW forces, the electromagnetic properties at all frequencies matter; that is, the motions/vibrations of atoms, molecules and electrons at all different time scales. The dielectric constants we have considered so far are solely due to correlations between slowly moving entities like, for example, molecules with permanent dipoles. The vdW interactions include also correlations between rapidly fluctuating charge distributions due to, for example, the motion of electrons in atoms and molecules. The latter part of the vdW interactions are denoted by *dispersion interactions*. The correlations on the slow time scale will be denoted as the *zero-frequency contributions* (static contributions), while the rest of the contributions correspond to frequencies ranging from IR to UV; in most cases, the dominant parts are from UV frequencies originating from electronic correlations. For the scope of the present chapter, an in-depth treatment of the theory of vdW forces would lead too far;[23] here we are only considering a few relevant aspects of such forces.

8.6.1 Van der Waals Interactions and Mean Field Electrostatics: The DLVO Theory

The traditional treatment of interparticle interactions for colloidal systems is the **Derjaguin-Landau-Verwey-Overbeek**[24] **(DLVO) theory**, where the total free energy of interaction is simply the sum of the double-layer interactions calculated in the Poisson-Boltzmann approximation and the vdW interaction. Consider two planar walls I and II at distance D^S from each other. They are assumed to be much thicker than the surface separation. The vdW interaction is given by

$$\mathscr{W}_{I,II}^{S,vdW} = -\frac{\mathscr{A}_{I,II}}{12\pi (D^S)^2},$$

where $\mathscr{A}_{I,II}$ is the so-called **Hamaker coefficient**, which depends on the medium behind each surface and the medium in the slit; in the DLVO theory, the latter medium is assumed to be the solvent in the absence of ions. When D^S is not very large, $\mathscr{A}_{I,II}$ is constant.

The Hamaker coefficient has contributions from the polarization properties of the media at various frequencies, including the static ones at zero frequency. The contribution to $\mathscr{A}_{I,II}$ from the electrostatic properties at zero frequency is equal to[25]

$$\mathscr{A}_{I,II}^{(0)} = \frac{1}{2} \cdot \frac{3k_B T}{2} \sum_{\nu=1}^{\infty} \frac{[\alpha_I \alpha_{II}]^{\nu}}{\nu^3}, \qquad (8.61)$$

---

[23] For a comprehensive treatise, see V. A. Parsegian, *Van der Waals Forces*, Cambridge University Press, Cambridge, 2006.

[24] The theory was established in 1941 by Boris Derjaguin and Lev Landau, two Soviet scientists, and independently in 1948 by Evert Verwey and Theodoor Overbeek, two Dutch scientists.

[25] See Equation (L2.11) in V. A. Parsegian, *loc. cit.* in footnote 23. This equation also contains the corresponding magnetic susceptibility contribution, which usually is small and can be neglected.

where

$$\alpha_J = \frac{\varepsilon_r - \varepsilon_r^J}{\varepsilon_r + \varepsilon_r^J} \qquad (8.62)$$

for $J = \text{I}$ and II, ε_r^{I} and $\varepsilon_r^{\text{II}}$ are the dielectric constants of the walls and ε_r is, as usual, the dielectric constant of the solvent of the solution in the slit. $\mathscr{A}_{\text{I,II}}^{(0)}$ is a constant, so the zero-frequency contribution to $\mathscr{W}_{\text{I,II}}^{S,\text{vdW}}$ is given by

$$\mathscr{W}_{\text{I,II}}^{S,\text{vdW}(0)} = -\frac{\mathscr{A}_{\text{I,II}}^{(0)}}{12\pi (D^S)^2} \quad \text{for all } D^S \qquad (8.63)$$

(provided the walls are infinitely thick). Expressed in terms of pressure, we have the corresponding contribution to the disjoining pressure

$$P^{S,\text{vdW}(0)} = -\frac{d\mathscr{W}_{\text{I,II}}^{S,\text{vdW}(0)}}{dD^S} = -\frac{\mathscr{A}_{\text{I,II}}^{(0)}}{6\pi (D^S)^3}. \qquad (8.64)$$

Incidentally, we note that the factor α_J is the same as α_w in Equations 7.50 and 7.51 for the image charge potential near a wall (w) with a dielectric constant different from ε_r.

The higher frequency contributions to the Hamaker coefficient $\mathscr{A}_{\text{I,II}}$ are not constant for large D^S, so the total coefficient is a function of surface separation: $\mathscr{A}_{\text{I,II}} = \mathscr{A}_{\text{I,II}}(D^S)$. The D^S dependence is such that, for intermediate surface separations, the total $\mathscr{W}_{\text{I,II}}^{S,\text{vdW}}$ decreases somewhat faster than $1/(D^S)^2$ when D^S increases;[26] for some separations, it can decrease approximately like $1/(D^S)^3$. For small and for very large D^S, the decay is, however, proportional to $1/(D^S)^2$.

For two equal walls, $\mathscr{W}_{\text{I,II}}^{S,\text{vdW}(0)}$ is attractive since $\alpha_{\text{II}} = \alpha_{\text{I}}$ and we have $(\alpha_{\text{I}})^{2v}$ in the expression for $\mathscr{A}_{\text{I,II}}^{(0)}$, which implies that this constant is positive. For unequal walls, it is, however, possible to have $\mathscr{A}_{\text{I,II}}^{(0)} < 0$ and, accordingly, a repulsive $\mathscr{W}_{\text{I,II}}^{S,\text{vdW}(0)}$. This happens when ε_r is intermediate between ε_r^{I} and $\varepsilon_r^{\text{II}}$ whereby α_{I} and α_{II} have opposite signs.

The higher frequency contributions to the entire Hamaker coefficient $\mathscr{A}_{\text{I,II}}$ can be written similarly to Equation 8.61,[27] but the rhs is then summed over a range of frequencies from IR to UV and the various ε_r and ε_r^J in the α-factors are evaluated at each of these frequencies. Furthermore, the initial factor of 1/2 is dropped. For analogous reasons as for the zero-frequency contribution, the entire vdW force is always attractive for equal walls, while it may be repulsive for unequal ones.

---

[26] The slightly faster decay of the dispersion interactions is caused by an effect called **retardation**, which is an effect of the finite velocity of light. For rapidly fluctuating charges that are distant from each other, the speed of light matters for the correlations because the charge distribution in either place may have changed during the time it takes for the field to propagate from one place to the other and back again. This diminishes the strength of the correlations and causes a more rapid decay of the interaction with distance; in the present case, it makes $\mathscr{W}_{\text{I,II}}^{S,\text{vdW}}$ decay faster with increased D^S.

[27] See Equation (L2.8) in V. A. Parsegian, *loc. cit.* in footnote 23. Like in $\mathscr{A}_{\text{I,II}}^{(0)}$, there are also magnetic susceptibility contributions, which usually are small and can be neglected.

Let now us see what the DLVO theory predicts. We have the total disjoining pressure $P^S = P^{S,PB} + P^{S,vdW}$ where $P^{S,PB}$ is the pressure in the PB approximation given by Equation 8.29 and $P^{S,vdW} = -d\mathcal{W}^{S,vdW}/dD^S$ is the contribution from the vdW interaction. Thus,

$$P^S = k_B T[n_{tot}(z^{\bullet}) - n_{tot}^b] - \frac{(\sigma_{\bullet}^S)^2}{2\varepsilon_r\varepsilon_0} + P^{S,vdW} \quad \text{(DLVO)} \qquad (8.65)$$

for any $-L/2 \leq z^{\bullet} \leq L/2$. For the special case of two equal walls with equal surface charge densities, this becomes

$$P^S = k_B T[n_{tot}(0) - n_{tot}^b] + P^{S,vdW} \quad \text{(DLVO)} \qquad (8.66)$$

when we evaluate the pressure at the midplane between the surfaces ($z^{\bullet} = 0$). Since the double-layer interaction in the PB approximation decays exponentially fast with increasing D^S, the total disjoining pressure between the walls at large D^S will always be dominated by the vdW interaction, which decays much slower.

For two equal walls, we have, using Equation 8.14 for the PB contribution from two equal surfaces,

$$P^S(D^S) \sim \frac{2[\sigma^{S\text{eff}}]^2}{\varepsilon_r\varepsilon_0}e^{-\kappa_D D^S} + P^{S,vdW} \quad \text{when } D^S \to \infty \quad \text{(DLVO)}. \qquad (8.67)$$

Since the first term is repulsive and the vdW term is attractive, it follows that the forces are always attractive for sufficiently large separations in this case. When the electrolyte concentration is low and the decay length $1/\kappa_D$ of the repulsive double-layer interaction is large, this dominance may occur first at quite large separations. At high concentrations where $1/\kappa_D$ is small, the vdW interaction dominates in a wider range of large D^S values, but the double-layer interaction dominates for smaller D^S when the surface charge density is sufficiently high. However, for very small D^S, the vdW interaction usually dominates because it goes like $-1/(D^S)^3$ for small D^S (note that the smallest possible value of D^S is nonzero when the surfaces are charged and counterions must remain in the slit, so one avoids the unphysical divergence at $D^S \to 0$). Thus, the DLVO prediction for the interaction typically varies from attractive at short surface separations, repulsive at intermediate and finally attractive at large separations.

8.6.2 The Effects of Ion-Ion Correlations on Van der Waals Interactions. Ionic Image Charge Interactions

When one adds the double layer and the vdW interactions in the DLVO theory, one implicitly assumes that they are independent of each other. In reality, this is not correct. An electrolyte screens electrostatic interactions and since the zero-frequency vdW contribution $\mathcal{W}_{I,II}^{S,vdW(0)}$ is of electrostatic origin, it is screened. This means that this term is replaced by a contribution that decays exponentially with D^S in the presence of an electrolyte in the slit. The higher frequency contributions to the full vdW interaction are, however, not screened since the ions move so slowly that they cannot correlate with the fast fluctuations in charge distribution that give rise to these contributions. The PB approximation that is used in the DLVO theory does not capture the screening of the zero-frequency vdW contribution since correlations in the electrolyte are ignored.

In most cases, the zero frequency contribution is a rather small part of the total vdW interaction, so the latter is not affected much by the presence of ions. However, for a system consisting of two walls made by hydrocarbon interacting across a slit filled with water, the zero-frequency vdW contribution can constitute more than 2/3 of the total. In this case, the presence of an electrolyte in the slit changes the vdW interaction very much.

How does a correct treatment of correlations in the electrolyte phase handle this screening, like the primitive model approach in Section 8.5.2? To see this, let us consider the ionic interactions in the system. Since the two walls have dielectric constants different from the solvent, there are image interactions like in the single surface case treated in Section 7.1.4. We therefore have to treat the interaction between the two walls in the presence of image interactions, which means that the theory in Section 8.5.2 must be generalized to this case. Here we will just give a brief outline of such a generalization. For simplicity, we will assume that the two walls are equal, so both have dielectric constant $\varepsilon_r^{\mathrm{w}} \equiv \varepsilon_r^{\mathrm{I}} = \varepsilon_r^{\mathrm{II}}$ and the surface charge density on both is σ^S. We have $\alpha_{\mathrm{I}} = \alpha_{\mathrm{II}} = \alpha_{\mathrm{w}} = (\varepsilon_r - \varepsilon_r^{\mathrm{w}})/(\varepsilon_r + \varepsilon_r^{\mathrm{w}})$, which is nonzero when $\varepsilon_r^{\mathrm{w}}$ of the walls is different from ε_r of the solvent for the electrolyte.

A significant difference between the cases of one wall and two walls as regards the interaction between charges is that for one wall each charge has one image charge, while for two walls there are an infinite number of image charges. Each image in one wall has an image in the opposite wall, which in turn has an image in the opposite wall, etc. to infinity. This

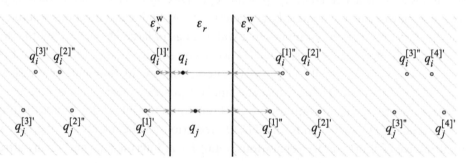

FIGURE 8.6 Sketch of two walls with equal dielectric constants $\varepsilon_r^{\mathrm{w}}$ separated by a slit filled with a fluid with dielectric constant ε_r. Two charges q_i and q_j are located in the slit (black filled circles). The image charges of these charges are also shown (gray filled circles). For q_i there are two primary images: $q_i^{[1]'} = \alpha_{\mathrm{w}} q_i$ in the left wall and $q_i^{[1]''} = \alpha_{\mathrm{w}} q_i$ in the right wall. The arrows show the distances between the charges and the surface of each wall. Each primary image has an image in the opposite wall (a secondary image): $q_i^{[2]'} = \alpha_{\mathrm{w}} q_i^{[1]'} = \alpha_{\mathrm{w}}^2 q_i$ is the image of $q_i^{[1]'}$ in the right wall (at equal distance from the right surface as the distance of $q_i^{[1]'}$ from the same surface) and $q_i^{[2]''} = \alpha_{\mathrm{w}} q_i^{[1]''} = \alpha_{\mathrm{w}}^2 q_i$ is the image of $q_i^{[1]''}$ in the left wall. Likewise, $q_i^{[3]'} = \alpha_{\mathrm{w}}^3 q_i$ is the image of $q_i^{[2]'}$ in the opposite wall and $q_i^{[3]''} = \alpha_{\mathrm{w}}^3 q_i$ is that of $q_i^{[2]''}$, etc. In all cases, the image lies at an equal distance from the surface as the charge, but on the other side of the surface. The corresponding arguments apply to q_j and its multiple images.

is illustrated in Figure 8.6, where the first few images for two charges q_i and q_j are shown. For each image of an image, there is a factor of α_w; for example, the images of order ℓ of q_i have the magnitudes $q_i^{[\ell]'} = q_i^{[\ell]''} = \alpha_w^\ell q_i$. The electrostatic potential due to q_i is the sum of the potential from the charge q_i itself and from its images of all orders. This sum is slowly converging when one adds more and more multiple images in the sum since the potential from each image goes like 1/distance.

The electrostatic pair interaction between q_i and q_j contains the direct interaction between the two charges and the interaction with the images of all orders from both charges. As discussed in connection to the image interactions from a single wall, Equation 7.52, when both ions interact with the images of both ions, each image term – including the self-image contributions – contains a factor of 1/2. The resulting pair interaction can be written (cf. Equation 7.52)

$$
\begin{aligned}
u_{ij}^{\mathrm{el}}(\mathbf{r}_1, \mathbf{r}_2; D^S) &= \frac{q_i q_j}{4\pi \varepsilon_r \varepsilon_0} \left[\frac{1}{|\mathbf{r}_1 - \mathbf{r}_2|} + \varphi^{\mathrm{im}}(\mathbf{r}_1, \mathbf{r}_2; D^S) \right] \\
&= u_{ij}^{\mathrm{Coul}}(r_{12}) + u_{ij}^{\mathrm{im}}(\mathbf{r}_1, \mathbf{r}_2; D^S),
\end{aligned}
$$

where $u_{ij}^{\mathrm{im}} = q_i q_j \varphi^{\mathrm{im}}/(4\pi \varepsilon_r \varepsilon_0)$ is the sum of the interactions of each of the two ions with the images of the other ion (described in Figure 8.6) and u_{ij}^{Coul} is the usual Coulomb potential. Likewise, the external interaction potential v_i^{el} from the walls contains a term $v_i^{\mathrm{im}}(z; D^S) = q_i^2 \varphi^{\mathrm{im}}(\mathbf{r}, \mathbf{r}; D^S)/(8\pi \varepsilon_r \varepsilon_0)$ from the self-image interactions (cf. Equation 7.53), including the multiple self-images. The functions φ^{im}, u_{ij}^{im} and v_i^{im} depend on D^S since the locations of the images depend on the surface separation.

The double-layer interaction pressure between the walls, which is given by Equation 8.59 in the absence of images, is given by the same equation with an additional term $P_\perp^{\mathrm{im}}(z^\bullet)$ that contains the explicit contributions from image interactions.[28] For the case when the disjoining pressure is evaluated at one of the surfaces ($z^\bullet = L/2$), the contribution $P_\perp^{\mathrm{im}}$ is given by the expression[29]

$$
\begin{aligned}
P_\perp^{\mathrm{im}}(L/2) = -\tfrac{1}{2} \sum_{i,j} \int_{-L/2}^{L/2} dz_1\, n_i(z_1) \int_{-L/2}^{L/2} dz_2\, n_j(z_2) \\
\times \int ds_{12}\, h_{ij}^{(2)}(\mathbf{r}_1, \mathbf{r}_2) \frac{\partial u_{ij}^{\mathrm{im}}(\mathbf{r}_1, \mathbf{r}_2; D^S)}{\partial D^S} \\
- \sum_i \int_{-L/2}^{L/2} dz_1\, n_i(z_1) \frac{\partial v_i^{\mathrm{im}}(z_1; D^S)}{\partial D^S},
\end{aligned} \tag{8.68}
$$

---

[28] The contribution $P_\perp^{\mathrm{el,corr}}$ defined in Equation 8.57 has u_{ij}^{el} set to u_{ij}^{Coul} in this case.

[29] This formula for $P_\perp^{\mathrm{im}}(L/2)$ can be derived in a similar manner as the derivation of Equation 8.37 from Equation 8.36 as done in Appendix 8B. The terms in the rhs originate from the fact that u_{ij}^{im} and v_i^{im} depend on D^S and therefore give contributions to $\partial \ln Q_N/\partial D^S$.

where $\mathbf{s}_{12} = (x_2 - x_1, y_2 - y_1)$. It can be transformed into an expression for $P_{\perp}^{\text{im}}$ evaluated by any $z^{\bullet}$ between the surfaces by use of the BGY equation, but this will not be done here. The total pressure in the slit evaluated at $z^{\bullet} = L/2$ is (cf. Equation 8.52)

$$P_{\perp}^{\text{slit}} = k_B T \sum_i n_i(L/2) - \frac{(\sigma^S)^2}{2\varepsilon_r\varepsilon_0} + P_{\perp}^{\text{im}}(L/2)$$

and the net double-layer pressure is $P_{\perp}^{\text{slit}} - P^b$.

When $P_{\perp}^{\text{slit}} - P^b$ is evaluated, it is found that the leading term decays for large surface separations like $1/(D^S)^3$ while the other contributions to $P_{\perp}^{\text{slit}} - P^b$ decay exponentially to zero (with or without oscillations). More precisely, this leading term, which is absent when there are no image interactions, makes the net double-layer pressure to decay like

$$P_{\perp}^{\text{slit}} - P^b \sim \frac{k_B T}{8\pi(D^S)^3} \sum_{\nu=1}^{\infty} \frac{\alpha_w^{2\nu}}{\nu^3} \tag{8.69}$$

when $D^S \to \infty$. This means that the double-layer interaction between the walls contains a long range repulsion caused by correlations between the ions and their images, that is, a fluctuation correlation contribution due to different dielectric properties of the wall media and the solvent for the electrolyte.

The appearance of this contribution constitutes, in fact, the manifestation of the screening of the zero frequency vdW component that we anticipated earlier. To see this, let us consider Equation 8.64 for the zero frequency component of the vdW interaction, which incidentally is also a fluctuation correlation contribution due to different dielectric properties of the media. By inserting Equation 8.61 with $\alpha_I = \alpha_{II} = \alpha_w$ into Equation 8.64, we see that the coefficient in Equation 8.69 is the same as in Equation 8.64 but with the opposite sign. Therefore, when we add the vdW and double-layer interactions, the contribution $P^{S,\text{vdW}(0)}$ in the former is canceled by the leading term in the latter. This is the manifestation of the screening of the zero frequency vdW component, so all electrostatic interactions are screened by the presence of ions. Thus, neither $P^{S,\text{vdW}(0)}$ nor the double-layer contribution in the rhs of Equation 8.69 is present in the total pressure. There remains, however, a pressure contribution from the image interactions from $P_{\perp}^{\text{im}}$, but it is decaying exponentially with D^S. Furthermore, in the total pressure, the higher frequency contributions to the vdW interactions remain, so in the total interaction between the walls,

$$P^S = P_{\perp}^{\text{slit}} - P^b + P^{S,\text{vdW}},$$

the long range attraction from the vdW force still exist, but it is somewhat weakened compared to the case without the electrolyte due to the cancellation of the zero frequency part. For systems with hydrocarbon/water/hydrocarbon interfaces, the weakening is substantial since, as we have seen, $P^{S,\text{vdW}(0)}$ is a large part of the total vdW interaction in this case. These effects are entirely missing in the DLVO theory. It is accordingly crucial to treat the ion-ion correlations in the electrolyte phase in a proper manner.

8.6.3 The Inclusion of Dispersion Interactions for the Ions

Our findings in the previous sections illustrate the interdependence of electrostatic double-layer and vdW interactions, but there is more to this story. It not only the vdW interactions that are affected by the correlations between the ions, but, inversely, the electrostatic interactions in the electrolyte are affected by the vdW interactions; more precisely, they are affected by the dispersion interactions that exist for the ions since they are polarizable objects. To include this effect, one must hence leave the primitive model for the ions and include dispersion interactions for them as we did in Section 7.1.5, where we treated electrolytes in contact with a single wall or a macroparticle. As we saw there, this is an important step since it adds a significant reason for ion specificity.

Since the dispersion interactions are not screened by the electrolyte, they give long-range effects on the distribution functions. As we saw in Section 7.1.5, the ion-wall dispersion interactions give rise to a term in the external potential $v_{Ji}^{disp}(\ell) = b_{Ji}^{disp}/\ell^3$ between wall J and an ion of species i (Equation 7.57).[30] Here ℓ is counted from the wall surface and the constant b_{Ji}^{disp} depends on the medium of wall J and the excess polarizability of the ion. We also have a dispersion interaction contribution $u_{ij}^{disp}(r) = b_{ij}^{disp}/r^6$ to the pair interactions between the ions (Equation 7.56).

From the formula (8.41) for $P_\perp^{slit}$ (generalized to more than one species), we see that the dispersion interactions in the external and pair potentials give rise to a term, which we will denote as $P_\perp^{disp}$, equal to

$$
P_\perp^{disp}(z^\bullet) = -\sum_i \left[\int_{z^\bullet}^{L/2} dz_1 n_i(z_1) v_{Ii}^{disp\prime} \left(\tfrac{L}{2} + d_I^S + z_1 \right) \right.
$$

$$
\left. + \int_{-L/2}^{z^\bullet} dz_1 n_i(z_1) v_{IIi}^{disp\prime} \left(\tfrac{L}{2} + d_{II}^S - z_1 \right) \right]
$$

$$
- \sum_{i,j} \int_{z^\bullet}^{L/2} dz_1\, n_i(z_1) \int_{-L/2}^{z^\bullet} dz_2\, n_j(z_2) \int ds_{12}\, g_{ij}^{(2)}(\mathbf{r}_1, \mathbf{r}_2)
$$

$$
\times \frac{du_{ij}^{disp}(r_{12})}{dr_{12}} \frac{z_1 - z_2}{r_{12}}, \tag{8.70}
$$

where $'$ denotes the derivative, as before. The dispersion interaction contribution to $\mathscr{V}_{I,II}'(D^S)$ is included in the vdW pressure $P^{S,vdW}$, so it does not appear in $P_\perp^{disp}$. We will split $P_\perp^{disp}$ into two parts,

$$
P_\perp^{disp}(z^\bullet) = P_\perp^{disp,v}(z^\bullet) + P_\perp^{disp,u}(z^\bullet),
$$

---

[30] Here and in what follows, we neglect the effect of retardation, cf. footnote 26.

where the first part consists of the terms with $v_{\mathrm{I}i}^{\mathrm{disp}}$ and $v_{\mathrm{II}i}^{\mathrm{disp}}$ and the second part equals the terms containing $u_{ij}^{\mathrm{disp}}(r_{12})$.

The double-layer pressure in the slit, which is given by Equation 8.56 in the absence of dispersion and image charge interactions, is given by

$$P_{\perp}^{\mathrm{slit}} = P_{\perp}^{\mathrm{kin}}(z^{\bullet}) - \frac{\left(\sigma_{\bullet}^{\mathcal{S}}\right)^2}{2\varepsilon_r\varepsilon_0} + P_{\perp}^{\mathrm{el,corr}}(z^{\bullet}) + P_{\perp}^{\mathrm{core}}(z^{\bullet}) + P_{\perp}^{\mathrm{im}}(z^{\bullet}) + P_{\perp}^{\mathrm{disp}}(z^{\bullet}),$$

where we have added $P_{\perp}^{\mathrm{im}}$ and $P_{\perp}^{\mathrm{disp}}$ (see Table 8.1 in Section 8.3 for an overview of the different contributions). As before, we have the total disjoining pressure

$$P^{\mathcal{S}} = P_{\perp}^{\mathrm{slit}} - P^{\mathrm{b}} + P^{\mathcal{S},\mathrm{vdW}},$$

where $P^{\mathcal{S},\mathrm{vdW}}$ is calculated in the absence of ions. The screening of the zero frequency contribution $P^{\mathcal{S},\mathrm{vdW}(0)}$ due to the presence of ions originates from the image charge interactions as explained earlier and the effect of ions on the dispersion interactions between the walls is included in $P_{\perp}^{\mathrm{disp}}$.

When the two walls are equal, the $b_{\mathcal{J}i}^{\mathrm{disp}}$ coefficients are the same for the interaction with either wall, $b_{\mathrm{I}i}^{\mathrm{disp}} = b_{\mathrm{II}i}^{\mathrm{disp}} \equiv b_i^{\mathrm{disp}}$, and we have $v_{\mathrm{I}i}^{\mathrm{disp}} = v_{\mathrm{II}i}^{\mathrm{disp}} \equiv v_i^{\mathrm{disp}}$ and $d_{\mathrm{I}}^{\mathcal{S}} = d_{\mathrm{II}}^{\mathcal{S}} \equiv d^{\mathcal{S}}$. Due to the symmetry around the midplane $z^{\bullet} = 0$, we have $\sigma_{\bullet}^{\mathcal{S}}(0) = 0$ and hence

$$P_{\perp}^{\mathrm{slit}} = P_{\perp}^{\mathrm{kin}}(0) + P_{\perp}^{\mathrm{el,corr}}(0) + P_{\perp}^{\mathrm{core}}(0) + P_{\perp}^{\mathrm{im}}(0) + P_{\perp}^{\mathrm{disp}}(0), \tag{8.71}$$

and we have, also due to symmetry,

$$P_{\perp}^{\mathrm{disp},v}(0) = -2\sum_i \int_0^{L/2} dz_1 n_i(z_1) v_i^{\mathrm{disp}\prime}\left(\frac{D^{\mathcal{S}}}{2} + z_1\right), \tag{8.72}$$

where we have used $D^{\mathcal{S}} = L + 2d^{\mathcal{S}}$. Note that in this expression for $P_{\perp}^{\mathrm{disp},v}(0)$, the force $-v_i^{\mathrm{disp}\prime}$ originates from the wall with surface located at $z = -D^{\mathcal{S}}/2$, while the concentration $n_i(z_1)$ is evaluated for z coordinates on the other side of the midplane, that is, for $z \geq 0$.

For large surface separations $D^{\mathcal{S}}$, the density in a large part of the middle of the slit is close to the bulk density. Then, in the limit $D^{\mathcal{S}} \to \infty$, the value of the integral in Equation 8.72 is dominated by contributions with $n_i(z_1) \approx n_i^{\mathrm{b}}$. By replacing $n_i(z_1)$ by n_i^{b} in Equation 8.72, we obtain the asymptotic result[31] when $D^S \to \infty$

$$P_\perp^{\text{disp},v}(0) \sim -2 \sum_i \int_0^{L/2} dz_1 n_i^b v_i^{\text{disp}\prime}\left(\frac{D^S}{2} + z_1\right)$$

$$\sim \frac{14 \sum_i n_i^b b_i^{\text{disp}}}{(D^S)^3}, \tag{8.73}$$

where we have used

$$\int_0^{L/2} dz_1 v_i^{\text{disp}\prime}\left(\frac{D^S}{2} + z_1\right) = v_i^{\text{disp}}\left(D^S - d^S\right) - v_i^{\text{disp}}\left(\frac{D^S}{2}\right)$$

$$\sim -\frac{7b_i^{\text{disp}}}{(D^S)^3}.$$

Thus $P_\perp^{\text{disp},v}(0)$ decays for large surface separations like the vdW interaction between two planar walls.[32] In fact, $P_\perp^{\text{disp},v}$ belongs to vdW interactions since it gives a contribution to the wall-wall dispersion interactions due to the presence of ions in the slit between the walls. It is not included in $P^{S,\text{vdW}}$ since the latter originates in the present model from the properties of the solvent in the absence of ions.

The dispersion interactions do not only give rise to the contribution $P_\perp^{\text{disp}}$, they also affect the distribution functions in a profound manner and thereby the other contributions to $P_\perp^{\text{slit}}$. As we saw in Section 7.1.5, this can be seen already in a PB treatment of electric double-layers, so as an illustration we will turn to that approximation.

In the PB approximation with included dispersion interactions, the double-layer pressure evaluated at the midplane $z = 0$ for a system with two equally charged surfaces equals

$$P^S = P_\perp^{\text{kin}}(0) + P_\perp^{\text{disp},v}(0) + P^{S,\text{vdW}} - P^b$$

$$= k_B T \sum_i \left[n_i(0) - n_i^b\right] + P_\perp^{\text{disp},v}(0) + P^{S,\text{vdW}} \quad \text{(PB)} \tag{8.74}$$

since the contributions $P_\perp^{\text{el,corr}}$, $P_\perp^{\text{core}}$ and $P_\perp^{\text{im}}$ in Equation 8.71 are zero because ion-ion correlations are neglected. Furthermore, $P_\perp^{\text{disp},u}$ is not included for the same reason so

---

[31] The deviation in n_i from n_i^b near the other wall (located at $z = D^S/2$) contributes to higher order terms in $P_\perp^{\text{disp},v}(0)$ that decay faster than $1/(D^S)^3$ with increasing D^S.

[32] In the presence of retardation, the decay is faster than $1/(D^S)^3$ (cf. footnote 26).

only $P_\perp^{\text{disp},v}$ contributes. This expression differs from Equation 8.66 in the DLVO theory by the appearance of $P_\perp^{\text{disp},v}(0)$.

The ion-wall potential of mean force $w_i(z)$ is – like in Equation 7.59 for a single wall – given by the sum of the electrostatic and dispersion contributions, but here the latter consists of a contribution v_i^{disp} from either wall. Thus, we have in the PB approximation $w_i(z) = q_j \psi(z) + v_i^{\text{disp,tot}}(z)$, where

$$
v_i^{\text{disp,tot}}(z) = v_i^{\text{disp}}\left(\frac{D^S}{2}+z\right) + v_i^{\text{disp}}\left(\frac{D^S}{2}-z\right)
$$

$$
= \frac{b_i^{\text{disp}}}{\left(\frac{D^S}{2}+z\right)^3} + \frac{b_i^{\text{disp}}}{\left(\frac{D^S}{2}-z\right)^3}. \tag{8.75}
$$

We thereby obtain the following expression for the density profiles

$$
n_i(z) = \begin{cases} 0, & z < -L/2 \\ n_i^b e^{-\beta\left[q_i\psi(z)+v_i^{\text{disp,tot}}(z)\right]}, & -L/2 \leq z \leq L/2 \quad \text{(PB)} \\ 0, & z > L/2. \end{cases} \tag{8.76}
$$

When the surface separation is large, the value of $\beta w_i(z)$ is small near the midplane, which lies far from the surfaces, so we can expand the exponential function in a Taylor series and obtain the total density at $z = 0$

$$
n_{\text{tot}}(0) = \sum_i n_i(0) = \sum_i n_i^b e^{-\beta\left[q_i\psi(0)+v_i^{\text{disp,tot}}(0)\right]}
$$

$$
\approx \sum_i n_i^b \left(1 - \beta\left[q_i\psi(0) + v_i^{\text{disp,tot}}(0)\right] + \beta^2 \left[q_i\psi(0) + v_i^{\text{disp,tot}}(0)\right]^2/2\right)
$$

$$
\approx n_{\text{tot}}^b - \sum_i n_i^b \left(\beta v_i^{\text{disp,tot}}(0) - \beta^2 \left[q_i\psi(0)\right]^2/2\right), \tag{8.77}
$$

where we have used $\sum_i n_i^b q_i = 0$ and where the last line follows since the terms with $q_i\psi(0)v_i^{\text{disp,tot}}(0)$ and $[v_i^{\text{disp,tot}}(0)]^2$ from the square term are smaller than those that are kept. In the last line, the last term decays, in fact, to zero faster than the other terms when $D^S \to \infty$ (this will be shown for a specific case later). Therefore, for large D^S, we have

$$
k_B T \left[n_{\text{tot}}(0) - n_{\text{tot}}^b\right] \approx -\sum_i n_i^b v_i^{\text{disp,tot}}(0) = -\frac{16 \sum_i n_i^b b_i^{\text{disp}}}{(D^S)^3} \quad \text{(PB)}, \tag{8.78}
$$

where we have inserted Equation 8.75. This result, together with Equation 8.73, shows that the total pressure due to the ions (the first terms in the rhs of Equation 8.74) decays like a power law, proportional to $1/(D^S)^3$. By combining the results obtained so far, we conclude that the total disjoining pressure in Equation 8.74 decays like

$$P^S \sim -\frac{2 \sum_i n_i^b b_i^{disp}}{(D^S)^3} + P^{S,vdW} \quad \text{when } D^S \to \infty \quad \text{(PB)}. \tag{8.79}$$

This equation should be compared with Equation 8.67 from the DLVO theory, that is, the same approximation in the absence of ion-wall dispersion interactions. In the latter case, the double-layer pressure decays with increasing D^S in an exponential manner, proportional to $e^{-\kappa_D D^S}$. As we will see, there is also an exponential term in the pressure for the present case, but it is dominated for large surface separations by the contributions that decay like $1/(D^S)^3$.

Let us finally investigate the electrostatic potential $\psi(z)$ and the individual ion density distribution $n_i(z)$ near the midplane, $z \approx 0$. In Example 8.1, we saw that provided D^S is sufficiently large, the ion density near the midplane is intimately linked to the decay of the potential $\psi^{single}(z')$ from single walls in contact with a bulk electrolyte, where z' is the distance from the surface. This holds also in the present case with ion-wall dispersion interactions and for large D^S, we have for two equal surfaces facing each other

$$\psi(z) \approx \psi^{single}(D^S/2 + z) + \psi^{single}(D^S/2 - z). \tag{8.80}$$

Since $\psi(z)$ and $v_i^{disp,tot}(z)$ are small near the midplane in this case, we have from Equation 8.76

$$n_i(z) - n_i^b \approx -\beta n_i^b \left[q_i \psi(z) + v_i^{disp,tot}(z) \right] \quad \text{when } z \approx 0. \tag{8.81}$$

In Section 7.1.5 and Appendix 7B, we investigated the case of a single wall in contact with a binary $z:-z$ electrolyte using the linearized PB approximation, which suffices in order to give an illustration of the principles involved. It was found that (Equations 7.60 and 7.61)

$$\psi^{single}(z') = \psi_1(z') + \psi_2(z'),$$

where $\psi_1(z') = C_1 e^{-\kappa_D z'}$ and

$$\psi_2(z') \sim -\frac{1}{(z')^3} \cdot \frac{b_+^{disp} - b_-^{disp}}{2q} \quad \text{when } z' \to \infty \quad \text{(linearized PB)}. \tag{8.82}$$

The coordinate z' is counted from the wall surface. Hence, $\psi^{single}(z')$ decays like $1/(z')^3$ for large z'.

For the case of two equal walls which face each other, it follows from Equations 8.80 and 8.82 that we have (cf. Equation 7.81)

$$\psi(0) \sim 2\psi^{single}(D^S/2) \sim -\frac{b_+^{disp} - b_-^{disp}}{(D^S/2)^3 q} \quad \text{when } D^S \to \infty.$$

From this result, we can conclude that in the rhs of Equation 8.77 the term with $[q_i \psi(0)]^2$ decays faster with increasing D^S than the other terms, as anticipated earlier.

The density at the midplane can now be obtained by inserting $\psi(0)$ and $v_i^{disp,tot}(0) = 2b_i^{disp}/(D^S/2)^3$ in Equation 8.81 for $z = 0$. Using $q_+/q = +1$ and $q_-/q = -1$, we obtain after simplification

$$n_i(0) - n_i^b \sim -\frac{\beta \, 8n^b[b_+^{disp} + b_-^{disp}]}{(D^S)^3} \quad \text{when } D^S \to \infty,$$

which is consistent with Equation 8.78 for the total ion density. This leading term is the same for anions and cations, but the next order term, which decays like $1/(D^S)^5$, is different for the two species (see Equation 7.82 in Appendix 7B, where these matters are investigated in more detail).

Due to the presence of $\psi_1(z')$ in $\psi^{single}(z')$, we see in the same fashion as in Example 8.1 that there is a term in $\psi(0)$ and hence in $n_{tot}(0) - n_{tot}^b$ that decays exponentially with D^S. This gives a contribution to P^S that decays like $e^{-\kappa_D D^S}$, but the leading term shown in Equation 8.79 dominates over this contribution for large D^S. However, for small surface separations, the exponentially decaying contributions must be included.

In general, important effects of the ion-wall dispersion interactions occur for all surface separations; the coupling to the electrostatics that gives the long-range effects we have explored here in an approximate manner is just one of the influences of such interactions. As we saw in Section 7.1.5, the amount of ions of various species near each surface can vary substantially depending on the strength of these interactions. This influences the surface forces, in particular for short surface separations. Furthermore, the ion-ion dispersion interactions have a profound influence on the decay of the distribution functions, as we saw for bulk systems in the shaded text box on page 390f. All these results illustrate how the electrostatic interactions are affected by dispersion interactions, leading to power-law decay of electrostatic potentials in electrolyte systems that occur alongside with the exponentially decaying contributions to these potentials.

APPENDIX 8A: SOLUTION OF PB EQUATION FOR COUNTERIONS BETWEEN TWO SURFACES

For a system consisting of counterions with charge q in the slit between two walls with surface charge densities σ^S, the PB equation is given by Equation 8.17. The z coordinates of the ion are located within $-L/2 \leq z \leq L/2$. By performing the same manipulations of the PB equation that lead from Equation 7.4 to Equation 7.5, we obtain in the current case

$$\left[\frac{d\psi(z)}{dz}\right]^2 = \frac{2n^0}{\beta \varepsilon_r \varepsilon_0} e^{-\beta q \psi(z)} + \text{const}$$

for $|z| \leq L/2$. The integration constant can be determined from the condition $\psi(0) = 0$ and the fact that $d\psi(z)/dz$ must be zero at $z = 0$ because the system is symmetrical around the midplane in the slit. This leads to $0 = 2n^0/\beta\varepsilon_r\varepsilon_0 + \text{const}$, so we have (compare with Equation 7.6)

$$\left[\frac{d\psi(z)}{dz}\right]^2 = \frac{2n^0}{\beta\varepsilon_r\varepsilon_0}\left[e^{-\beta q\psi(z)} - 1\right]$$

for $|z| \leq L/2$. If we set $f(z) = \beta q\psi(z)/2$, the equation can be written

$$\left[\frac{df(z)}{dz}\right]^2 = K_0^2\left[e^{-2f(z)} - 1\right]. \tag{8.83}$$

where

$$K_0^2 = \frac{\beta n^0 q^2}{2\varepsilon_r\varepsilon_0}. \tag{8.84}$$

By multiplying both sides of Equation 8.83 by $e^{2f(z)}$, it becomes

$$\left[e^{f(z)}\frac{df(z)}{dz}\right]^2 = K_0^2\left[1 - e^{2f(z)}\right], \tag{8.85}$$

where the lhs equals $[d(e^{f(z)})/dz]^2$. Therefore, we can introduce the function $\tau(z) = e^{f(z)}$ and write Equation 8.85 as

$$\left[\frac{d\tau(z)}{dz}\right]^2 = K_0^2\left[1 - \tau^2(z)\right].$$

By taking the square root of this equation, we obtain $d\tau/dz = \pm K_0\sqrt{1 - \tau^2}$, where K_0 is positive and the sign of the rhs will be determined later. This differential equation is separable and can be written $d\tau/\sqrt{1 - \tau^2} = \pm K_0 dz$. By integration, we obtain $-\arccos\tau = \pm K_0 z + \text{const}$, which implies that $\tau(z) = \cos(\mp K_0 z - \text{const})$.[33] Since $\tau(z)$ must be symmetrical around $z = 0$, we set the integration constant equal to zero and obtain $\tau(z) = \cos(K_0 z)$. As we will see later, $K_0 L/2 < \pi/2$, so we have $-\pi/2 < K_0 z < \pi/2$. Obviously, $d\tau/dz$ is positive when $z < 0$ and negative when $z > 0$, which settles the question about sign.

From these results, we obtain $f(z) = \ln\tau(z) = \ln\cos(K_0 z)$, which means that the solution to the PB equation is

$$\psi(z) = \frac{2}{\beta q}\ln\cos(K_0 z) = \frac{1}{\beta q}\ln\cos^2(K_0 z) \quad \text{(PB)} \tag{8.86}$$

---

[33] We could have selected $\arcsin\tau$ rather than $-\arccos\tau$ as a result of the integration and obtained $\tau(z) = \sin(\pm K_0 z + \text{const})$. In such a case, we select $\text{const} = \pi/2$ to fulfill the requirement of symmetry around $z = 0$ and obtain $\tau(z) = \sin(\pm K_0 z + \pi/2) = \cos(\pm K_0 z)$, which is the same result as obtained from $-\arccos\tau$.

and we have

$$n(z) = n^0 e^{-\beta q \psi(z)} = \frac{n^0}{\cos^2(K_0 z)} \quad \text{(PB)} \tag{8.87}$$

for $|z| \leq L/2$. K_0 and n^0 can be determined from the electroneutrality condition $\int_{-L/2}^{L/2} dz \, qn(z) = -2\sigma^S$, which can be written as $\int_0^{L/2} dz \, n(z) = |\sigma^S/q|$ since $n(z)$ is symmetric around $z = 0$ and since σ^S and q have opposite signs, so $\sigma^S/q < 0$. We have

$$\int\limits_0^{L/2} n(z)dz = \int\limits_0^{L/2} \frac{n^0 dz}{\cos^2(K_0 z)} = \frac{n^0}{K_0} \tan(K_0 L/2)$$

so the electroneutrality condition can be written as

$$K_0 \tan(K_0 L/2) = \frac{\beta |q\sigma^S|}{2\varepsilon_r \varepsilon_0}, \tag{8.88}$$

where we have used Equation 8.84 to eliminate n^0. This is an equation for K_0 and since $\tan(x)$ lies between 0 and ∞ when $0 < x < \pi/2$, we see that we always obtain a positive solution K_0 where $K_0 L/2 < \pi/2$. Note that the rhs of the equation contains only system parameters. When K_0 is determined, n^0 can be obtained from Equation 8.84.

APPENDIX 8B: PROOFS OF TWO EXPRESSIONS FOR $P_\perp^{\text{SLIT}}$

In this appendix, we will derive Equations 8.37 and 8.40 for $P_\perp^{\text{slit}}$. We start by deriving the former equation from Equation 8.36 and $Q_N = Z_N/(N! \Lambda^{3N})$. The configurational partition function Z_N, defined in Equation 3.10, becomes in the current case

$$Z_N = \int dx_1 dy_1 \int\limits_{d_I^S}^{D^S - d_{II}^S} dz_1 \ldots \int dx_N dy_N \int\limits_{d_I^S}^{D^S - d_{II}^S} dz_N \, e^{-\beta \breve{U}_N^{\text{pot}}(\mathbf{r}^N | D^S)}$$

and we have from Equation 8.36

$$P_\perp^{\text{slit}} = \frac{k_B T}{A} \left(\frac{\partial \ln Z_N}{\partial D^S} \right)_{T,A,N} = \frac{k_B T}{A Z_N} \left(\frac{\partial Z_N}{\partial D^S} \right)_{T,A,N}. \tag{8.89}$$

Each integral $\int_0^{D^S - d_{II}^S} dz_\nu$ gives a term in $\partial Z_N/\partial D^S$ and all these N terms have the same numerical value, so we have

$$\left(\frac{\partial \ln Z_N}{\partial D^S} \right)_{T,A,N} = \frac{N}{Z_N} \int dx_1 dy_1 \left[\int d\mathbf{r}_2 \ldots d\mathbf{r}_N e^{-\beta \breve{U}_N^{\text{pot}}} \right]_{z_1 = D^S - d_{II}^S}$$

$$- \frac{\beta}{Z_N} \int d\mathbf{r}_1 \ldots d\mathbf{r}_N \frac{\partial \breve{U}_N^{\text{pot}}}{\partial D^S} e^{-\beta \breve{U}_N^{\text{pot}}}. \tag{8.90}$$

From Equation 5.41, it follows that

$$\frac{N}{Z_N} \int d\mathbf{r}_2 \dots d\mathbf{r}_N e^{-\beta \check{U}_N^{\text{pot}}} = n_N^{(1)}(\mathbf{r}_1) = n_N(z_1),$$

which implies that the first term in Equation 8.90 equals $A\, n_N(D^S - d_{\text{II}}^S)$, where $A = \int dx_1 dy_1$ and $n_N(D^S - d_{\text{II}}^S)$ is the contact density at surface II. From Equation 8.35, it follows that

$$\frac{\partial \check{U}_N^{\text{pot}}}{\partial D^S} = \sum_{\nu=1}^{N} \frac{dv_{\text{II}}(D^S - z_\nu)}{dD^S} + A \frac{d\mathcal{V}_{\text{I,II}}(D^S)}{dD^S}$$

(all other terms in $\check{U}_N^{\text{pot}}$ are independent of D^S) and therefore the second term in the rhs of Equation 8.90 is

$$-\frac{\beta}{Z_N} \left[\sum_{\nu=1}^{N} \int d\mathbf{r}_1 \dots d\mathbf{r}_N \frac{dv_{\text{II}}(D^S - z_\nu)}{dD^S} e^{-\beta \check{U}_N^{\text{pot}}} \right.$$

$$\left. + A \frac{d\mathcal{V}_{\text{I,II}}(D^S)}{dD^S} \int d\mathbf{r}_1 \dots d\mathbf{r}_N e^{-\beta \check{U}_N^{\text{pot}}} \right].$$

All N terms in the sum over ν have the same numerical value and together they equal

$$-\frac{\beta N}{Z_N} \int d\mathbf{r}_1 \frac{dv_{\text{II}}(D^S - z_1)}{dD^S} \left(\int d\mathbf{r}_2 \dots d\mathbf{r}_N e^{-\beta \check{U}_N^{\text{pot}}} \right)$$

$$= -\beta A \int dz_1 \frac{dv_{\text{II}}(D^S - z_1)}{dD^S} n_N(z_1),$$

where we again have used Equation 5.41. By summing all contributions to Equation 8.90 and inserting the result in Equation 8.89, we obtain

$$P_\perp^{\text{slit}} = k_B T n(D^S - d_{\text{II}}^S)$$

$$- \int_{d_{\text{I}}^S}^{D^S - d_{\text{II}}^S} dz_1\, v_{\text{II}}'(D^S - z_1) n(z_1) - \mathcal{V}_{\text{I,II}}'(D^S), \tag{8.91}$$

where we have skipped subscript N on $n_N(z)$ and introduced the notation $v_{\text{II}}'(\ell) = dv_{\text{II}}(\ell)/d\ell$ and $\mathcal{V}_{\text{I,II}}'(D^S) = d\mathcal{V}_{\text{I,II}}(D^S)/dD^S$. This is Equation 8.37.

Next, we derive Equation 8.40. The first Born-Green-Yvon Equation 5.119 can be written in the current case as

$$\frac{dn(z_1)}{dz_1} = -\beta n(z_1) \left[\frac{dv(z_1|D^S)}{dz_1} + \int d\mathbf{r}_2\, n(z_2) g^{(2)}(\mathbf{r}_1, \mathbf{r}_2) \frac{\partial u(r_{12})}{\partial z_1} \right]. \tag{8.92}$$

We can write $g^{(2)}(\mathbf{r}_1, \mathbf{r}_2) = g^{(2)}(s_{12}, z_1, z_2)$, where $\mathbf{s}_{12} = (x_2 - x_1, y_2 - y_1)$ and $s_{12} = |\mathbf{s}_{12}|$, since $g^{(2)}$ has only three independent variables in the current geometry. By inserting $v(z|D^\mathcal{S}) = v_\mathrm{I}(z) + v_\mathrm{II}(D^\mathcal{S} - z)$ into Equation 8.92, we obtain

$$\frac{dn(z_1)}{dz_1} = -\beta n(z_1) \left[\frac{d[v_\mathrm{I}(z_1) + v_\mathrm{II}(D^\mathcal{S} - z_1)]}{dz_1} \right.$$

$$\left. + \int_{d_\mathrm{I}^\mathcal{S}}^{D^\mathcal{S} - d_\mathrm{II}^\mathcal{S}} dz_2 \int d\mathbf{s}_2\, n(z_2) g^{(2)}(s_{12}, z_1, z_2) \frac{\partial u(r_{12})}{\partial z_1} \right], \qquad (8.93)$$

where $\mathbf{s}_2 = (x_2, y_2)$ and we have $r_{12} = [s_{12}^2 + (z_1 - z_2)^2]^{1/2}$.

Let us select a coordinate $z^\bullet$ in the interval $d_\mathrm{I}^\mathcal{S} \leq z^\bullet \leq D^\mathcal{S} - d_\mathrm{II}^\mathcal{S}$ and integrate the BGY Equation 8.93 over z_1 from $z^\bullet$ to $D^\mathcal{S} - d_\mathrm{II}^\mathcal{S}$, whereby we obtain after multiplication by $k_B T$

$$k_B T n(D^\mathcal{S} - d_\mathrm{II}^\mathcal{S}) = k_B T n(z^\bullet) - \int_{z^\bullet}^{D^\mathcal{S} - d_\mathrm{II}^\mathcal{S}} dz_1\, n(z_1) \left[v_\mathrm{I}'(z_1) - v_\mathrm{II}'(D^\mathcal{S} - z_1) \right] \qquad (8.94)$$

$$- \int_{z^\bullet}^{D^\mathcal{S} - d_\mathrm{II}^\mathcal{S}} dz_1\, n(z_1) \int_{d_\mathrm{I}^\mathcal{S}}^{D^\mathcal{S} - d_\mathrm{II}^\mathcal{S}} dz_2\, n(z_2) \int d\mathbf{s}_{12}\, g^{(2)}(s_{12}, z_1, z_2) \frac{\partial u(r_{12})}{\partial z_1},$$

where the integration over $\mathbf{s}_2$ is replaced by one over $\mathbf{s}_{12}$ (this give the same result due to the translational symmetry of $g^{(2)}$ in the (x_2, y_2) plane). The last term of the rhs can be simplified as follows. We start by noting that

$$\frac{\partial u(r_{12})}{\partial z_1} = \frac{\partial u([s_{12}^2 + (z_2 - z_1)^2]^{1/2})}{\partial z_1} = \frac{du(r_{12})}{dr_{12}} \frac{z_1 - z_2}{r_{12}}.$$

The z_2 integration can be split into two integrals $\int_{d_\mathrm{I}^\mathcal{S}}^{D^\mathcal{S} - d_\mathrm{II}^\mathcal{S}} = \int_{d_\mathrm{I}^\mathcal{S}}^{z^\bullet} + \int_{z^\bullet}^{D^\mathcal{S} - d_\mathrm{II}^\mathcal{S}}$ and we will first investigate the contribution to Equation 8.94 from the last one:

$$- \int_{z^\bullet}^{D^\mathcal{S} - d_\mathrm{II}^\mathcal{S}} dz_1\, n(z_1) \int_{z^\bullet}^{D^\mathcal{S} - d_\mathrm{II}^\mathcal{S}} dz_2\, n(z_2) \int d\mathbf{s}_{12}\, g^{(2)}(s_{12}, z_1, z_2) \frac{du(r_{12})}{dr_{12}} \frac{z_1 - z_2}{r_{12}}.$$

All parts of the integrand and the limits are symmetric with respect to swapping of indices $1 \leftrightarrow 2$ except the factor $z_1 - z_2$. Therefore, the integrand is antisymmetric and the entire expression must therefore be equal to zero. By using this fact and inserting Equation 8.94

into Equation 8.91, we obtain after simplification

$$P_\perp^{\text{slit}} = k_B T n(z^\bullet) - \int\limits_{z^\bullet}^{D^S - d_{\text{II}}^S} dz_1\, n(z_1) v_{\text{I}}'(z_1) - \int\limits_{d_{\text{I}}^S}^{z^\bullet} dz_1\, n(z_1) v_{\text{II}}'(D^S - z_1)$$

$$- \int\limits_{z^\bullet}^{D^S - d_{\text{II}}^S} dz_1\, n(z_1) \int\limits_{d_{\text{I}}^S}^{z^\bullet} dz_2\, n(z_2) \int d\mathbf{s}_{12}\, g^{(2)}(s_{12}, z_1, z_2)$$

$$\times \frac{du(r_{12})}{dr_{12}} \frac{z_1 - z_2}{r_{12}} - \mathscr{V}_{\text{I,II}}'(D^S). \tag{8.95}$$

This is Equation 8.40.

List of Symbols

Symbol	Explanation	SI unit
A	Helmholtz energy	J
A	area	m^2
$\mathscr{A}$	Hamaker coefficient	J
A^{ex}	excess Helmholtz energy	J
c	concentration (moles per unit volume)	mol m^{-3}
c	in Chapter 4: coordination number (number of nearest neighbors)	unitless
C_P	heat capacity at constant P	J K^{-1}
C_V	heat capacity at constant V	J K^{-1}
d^{h}	diameter of hard core	m
d^{LJ}	particle size parameter in Lennard-Jones potential	m
$D^{\mathcal{S}}$	surface separation	m
$d^{\mathcal{S}}$	distance of closest approach of particle centers to a surface	m
$d\text{q}$	infinitesimally small amount of heat (added to the system)	J
$d\text{w}$	infinitesimally small amount of work (done on the system)	J
E	electrostatic field	Vm^{-1}
$\mathcal{E}_r^{\text{eff}}$	effective relative dielectric permittivity	unitless
$\mathcal{E}_r^{\star}$	dielectric factor	unitless

Symbol	Explanation	SI unit
$\mathbf{F}$	force vector, $\mathbf{F} = (F_x, F_y, F_z)$	N
F	force	N
$\mathbf{F}^{\text{intr}}$	intrinsic force = force from particles of the fluid (not from external field)	N
G	Gibbs energy	J
$g^{(2)}(r)$	pair distribution function in bulk fluid (radial distribution function)	unitless
$g^{(2)}(\mathbf{r}_1, \mathbf{r}_2)$	pair distribution function (general case)	unitless
$g^{(l)}$	l-point (l-particle) distribution function	unitless
H	enthalpy	J
h	Planck's constant ($h = 6.62607 \cdot 10^{-34}$ J s)	J s
$h^{(2)}(\mathbf{r}_1, \mathbf{r}_2)$	pair correlation function, $h^{(2)} = g^{(2)} - 1$	unitless
$H_N^{(2)}(\mathbf{r}_1, \mathbf{r}_2)$	density-density correlation function (canonical ensemble)	unitless
$H_{\text{NN}}, H_{\text{QN}}, H_{\text{QQ}}$	density-density, charge-density and charge-charge correlation functions ($H_{\text{NN}} \equiv H^{(2)}$)	unitless
i	imaginary unit	unitless
$\mathbf{k}$	wave vector, $\mathbf{k} = k\hat{\mathbf{k}}$	m^{-1}
k	wave number, $k = |\mathbf{k}| = 2\pi/\lambda$	m^{-1}
k_B	Boltzmann's constant ($k_B = 1.38065 \cdot 10^{-23}$ J K^{-1})	J K^{-1}
ℓ	in Chapter 4: number of monomers	unitless
L	width of space available for particle centers in a slit between two planar surfaces (I and II), $L = D^S - d_{\text{I}}^S - d_{\text{II}}^S$	
M	concentration unit "molar" = mol dm^{-3} (1 M $= 10^3$ mol m^{-3})	1 mol m$^{-3}=$ 10^{-3} M
m	mass	kg
N	number of particles	unitless
n	number of moles, n $= N/N_{Av}$	mol

Symbol	Explanation	SI unit
n	number density (number of particle centers per unit volume)	m^{-3}
$n(\mathbf{r}|\mathbf{r}_1,\ldots,\mathbf{r}_l)$	number density at $\mathbf{r}$ when l particles are located at positions $\mathbf{r}_1,\ldots,\mathbf{r}_l$	m^{-3}
$\check{n}^{(1)}_{\{\mathbf{r}^N\}}(\mathbf{r})$	microscopic number density distribution at $\mathbf{r}$ for particle configuration $\mathbf{r}^N$	m^{-3}
N_{Av}	Avogadro's constant ($N_{Av} = 6.02214 \cdot 10^{23}\,\mathrm{mol}^{-1}$)	mol^{-1}
n^{b}	number density in bulk phase	m^{-3}
N_i	number of particles of species i	unitless
n_i	number density of species i (number of particle centers per unit volume)	m^{-3}
$n^{(l)}$	l-point (l-particle) density distribution function ($n^{(1)} \equiv n$)	m^{-3l}
$n_N^{(N)}$	N-particle density distribution function (canonical ensemble)	m^{-3N}
n_{tot}	total number density, $n_{\mathrm{tot}} = \sum_i n_i$	m^{-3}
$n_{\mathrm{tot}}^{\mathrm{b}}$	total number density in bulk phase	m^{-3}
P	pressure = force per unit area	$\mathrm{N\,m}^{-2}$
$\mathcal{P}$	probability	unitless
$\mathscr{P}$	probability	unitless
$\mathbf{p}$	momentum vector, $\mathbf{p} = (p_x, p_y, p_z) = m\mathbf{v}$	$\mathrm{kg\,m\,s}^{-1}$ $= \mathrm{N\,s}$
p	length of momentum vector, $p = |\mathbf{p}|$	$\mathrm{kg\,m\,s}^{-1}$ $= \mathrm{N\,s}$
p	probability	unitless
P^{b}	pressure in bulk phase	$\mathrm{N\,m}^{-2}$
P_i	partial pressure of species i	$\mathrm{N\,m}^{-2}$
P_i^0	vapor pressure of pure species i	$\mathrm{N\,m}^{-2}$
$P^{\ominus}$	standard state pressure	$\mathrm{N\,m}^{-2}$
$\mathcal{P}^{(l)}$	l-point (l-particle) configurational probability density	m^{-3l}

Symbol	Explanation	SI unit
$\mathbf{p}^N$	momenta in $3N$-dimensional space, $\mathbf{p}^N \equiv \mathbf{p}_1, \mathbf{p}_2, \ldots, \mathbf{p}_N$	$\mathrm{kg\,m\,s^{-1}}$ $= \mathrm{N\,s}$
$\mathcal{P}_N^{(N)}$	N-particle configurational probability density (canonical ensemble)	$\mathrm{m^{-3N}}$
$P^{\mathcal{S}}$	disjoining pressure, $P^{\mathcal{S}} = P_\perp^{\mathrm{slit}} - P^{\mathrm{b}}$	$\mathrm{N\,m^{-2}}$
$P^{\mathcal{S},\mathrm{vdW}}$	van der Waals interaction force per unit area between two planar walls	$\mathrm{N\,m^{-2}}$
$P_\perp^{\mathrm{slit}}$	pressure perpendicular to the surfaces in a planar slit between two walls (various contributions are listed in Table 8.1 in Section 8.3)	$\mathrm{N\,m^{-2}}$
Q	canonical partition function	unitless
q	charge	C
q	single-particle partition function	unitless
q	heat (added to the system)	J
q_{e}	elementary charge (charge of a proton) ($q_{\mathrm{e}} = 1.60218 \cdot 10^{-19}\,\mathrm{C}$)	C
$Q_{\mathrm{P}}^{\mathrm{eff}}(\hat{\mathbf{r}}, \boldsymbol{\omega})$	multipolar effective charge (direction dependent) of particle P with orientation $\boldsymbol{\omega}$	C
q^{es}	q for the electronic states of a particle	unitless
q_i	charge of a particle of species i	C
q_i^{eff}	effective charge of a particle of species i	C
$q_i^\star$	dressed particle charge of species i, $q_i^\star = \int d\mathbf{r}\rho_i^\star(\mathbf{r})$	C
$q_\pm$	$q_\pm \equiv \mathsf{x}_+^{\mathrm{b}} q_+ + \mathsf{x}_-^{\mathrm{b}} |q_-|$	C
R	the universal gas constant ($R = k_B N_{Av} = 8.31446\,\mathrm{J\,K^{-1}\,mol^{-1}}$)	$\mathrm{J\,K^{-1}mol^{-1}}$
$\mathbf{r}$	cartesian coordinate vector, $\mathbf{r} = (x, y, z)$	m
r	length of vector $\mathbf{r}$, $r = |\mathbf{r}|$	m
$\hat{\mathbf{r}}$	unit vector $\hat{\mathbf{r}} \equiv \mathbf{r}/r$	unitless
$\mathbf{r}^N$	coordinates in $3N$-dimensional space, $\mathbf{r}^N \equiv \mathbf{r}_1, \mathbf{r}_2, \ldots, \mathbf{r}_N$ (particle configuration)	m
S	entropy	$\mathrm{J\,K^{-1}}$

Symbol	Explanation	SI unit
$S(\mathbf{k})$	structure factor (from pair distributions)	unitless
T	absolute temperature	K
t	time	s
U	energy	J
$\langle U \rangle, \bar{U}$	internal energy ($\langle U \rangle \equiv \bar{U}$)	J
u	energy of single particle	J
$u(r)$	pair interaction potential	J
$\mathscr{U}_i$	energy of quantum state i of a system	J
$u_{ij}(r)$	pair interaction potential for species i and j	J
u_{ij}	in Chapter 4: interaction energy between neighboring cells containing species i and j	J
$\check{U}_N^{\text{pot}}(\mathbf{r}^N)$	instantaneous potential energy for N particles	J
$\check{U}_N^{\text{tot}}(\boldsymbol{\Gamma})$	instantaneous total energy for N particles	J
V	volume	m^3
$\mathbf{v}$	velocity vector, $\mathbf{v} = (v_x, v_y, v_z)$	m s^{-1}
$v(\mathbf{r})$	external potential	J
w	potential of mean force	J
w	in Chapter 4: interaction parameter $w = u_{11} + u_{22} - 2u_{12}$	J
w	work (done on the system)	J
$w(\mathbf{r}\vert\mathbf{r}')$	potential of mean force at $\mathbf{r}$ when a particle is located at $\mathbf{r}'$	J
$w_{ij}^{(2)}$	pair potential of mean force	J
w^{intr}	intrinsic potential of mean force, that is, due to particles of the fluid (not from external potential)	J
$\mathscr{W}_{\text{I,II}}^{\mathcal{S}}$	free energy of interaction per unit area between two planar surfaces (I and II)	J m^{-2}
x_i	mole fraction of species i; for a mixture of substances 1, 2 and 3: $\mathrm{x}_i = N_i/(N_1 + N_2 + N_3) = n_i/(n_1 + n_2 + n_3)$	unitless

Symbol	Explanation	SI unit
x^b_+, x^b_-	$x^b_i = n^b_i/(n^b_+ + n^b_-)$ for $i = +, -$ in binary electrolyte bulk phase	unitless
Z	configurational partition function	m^{3N}
z_i	valency of an ion of species i, $z_i = q_i/q_e$	unitless
$Z^{(l)}_{N+l}(\mathbf{r}_1, \ldots, \mathbf{r}_l)$	configurational partition function for a system with N mobile particles when l fixed particles are located at positions $\mathbf{r}_1, \ldots, \mathbf{r}_l$	m^{3N}
β	$\beta = (k_B T)^{-1}$	J^{-1}
$\mathbf{\Gamma}$	phase space point, $\mathbf{\Gamma} = (\mathbf{r}^N, \mathbf{p}^N)$	$(m, N\,s)$
γ	activity coefficient	unitless
γ^b	activity coefficient in bulk phase	unitless
$\delta(x)$	Dirac's delta function	m^{-1}
$\delta^{(3)}(\mathbf{r})$	three-dimensional Dirac delta function, $\delta^{(3)}(\mathbf{r}) = \delta(x)\delta(y)\delta(z)$	m^{-3}
ε^{LJ}	energy parameter in Lennard-Jones potential	J
ε_r	relative dielectric permittivity (dielectric constant)	unitless
ε_0	permittivity of vacuum ($\varepsilon_0 = 8.85419 \cdot 10^{-12}\ C^2 J^{-1} m^{-1}$)	$C^2 J^{-1} m^{-1}$
ζ	activity, $\zeta = \eta e^{\beta\mu}$	m^{-3}
η	$\eta = q^{es}/\Lambda^3$	m^{-3}
Θ	grand potential	J
κ	decay parameter ($1/\kappa$ = decay length)	m^{-1}
κ_D	Debye parameter ($1/\kappa_D$ = Debye length)	m^{-1}
$\kappa_\Re, \kappa_\Im$	$\kappa_\Re = \Re(\kappa)$, $\kappa_\Im = \Im(\kappa)$ and $\kappa = \kappa_\Re + i\kappa_\Im$	m^{-1}
Λ	thermal de Broglie wavelength	m
λ	wavelength	m
μ	chemical potential	J
μ^{ex}	excess chemical potential	J

Symbol	Explanation	SI unit
$\mu^{\mathrm{ex,b}}$	excess chemical potential in bulk phase	J
$\mu^{\mathrm{ex}}(\mathbf{r}\|\mathbf{r}')$	μ^{ex} at $\mathbf{r}$ when a particle is located at $\mathbf{r}'$	J
$\mu^{(2)\mathrm{ex}}$	two-particle excess chemical potential	J
μ^{ideal}	ideal chemical potential	J
μ_i	chemical potential of species i	J
$\mu^{\ominus}$	standard state chemical potential	J
$\mu_{\pm}$	$\mu_{\pm} = \mathsf{x}^{\mathrm{b}}_{+}\mu_{+} + \mathsf{x}^{\mathrm{b}}_{-}\mu_{-}$	J
Ξ	grand canonical partition function	unitless
$\Xi^{(l)}(\mathbf{r}_1,\ldots,\mathbf{r}_l)$	grand canonical partition function for a system with fixed particles placed at $\mathbf{r}_1,\ldots,\mathbf{r}_l$	unitless
ξ	coupling parameter, $0 \leq \xi \leq 1$	unitless
ρ	charge density	$\mathrm{C\,m}^{-3}$
ρ_i	charge density around a particle of species i	$\mathrm{C\,m}^{-3}$
$\rho^{\star}_i$	dressed charge density of a particle of species i, $\rho^{\star}_i = \sigma_i + \rho^{\mathrm{dress}}_i \equiv \rho^{\mathrm{tot}}_i - \rho^{\mathrm{lin}}_i$	$\mathrm{C\,m}^{-3}$
ρ^{dress}_i	charge density of the dress of a particle of species i, $\rho^{\mathrm{dress}}_i \equiv \rho_i - \rho^{\mathrm{lin}}_i$	$\mathrm{C\,m}^{-3}$
ρ^{lin}_i	linear part of electrostatic polarization charge density due to an ion of species i	$\mathrm{C\,m}^{-3}$
ρ^{tot}_i	total charge density associated with a particle of species i, $\rho^{\mathrm{tot}}_i = \sigma_i + \rho_i$	$\mathrm{C\,m}^{-3}$
σ_i	internal charge density of a particle of species i	$\mathrm{C\,m}^{-3}$
$\sigma^{\mathcal{S}}$	surface charge density	$\mathrm{C\,m}^{-2}$
$\sigma^{\mathcal{S}}_{\bullet}$	absolute value of the total charge per unit area on either side of a plane at $z = z^{\bullet}$ between two planar surfaces	$\mathrm{C\,m}^{-2}$
$\sigma^{\mathcal{S}\mathrm{eff}}$	effective surface charge density	$\mathrm{C\,m}^{-2}$
σ_X	root mean square deviation of quantity X, $\sigma^2_X = \left\langle (X - \langle X \rangle)^2 \right\rangle$	same as X
Υ	isobaric-isothermal partition function	unitless

Symbol	Explanation	SI unit
$\Phi(U)$	number of microstates with energy $\leq U$	unitless
$\phi_{\mathrm{Coul}}(r)$	unit Coulomb potential, $\phi_{\mathrm{Coul}}(r) = (4\pi\varepsilon_r\varepsilon_0 r)^{-1}$	$\mathrm{J\,C^{-2}}$
$\phi_{\mathrm{Coul}}^{\star}(r)$	unit screened Coulomb potential	$\mathrm{J\,C^{-2}}$
φ_i	in Chapter 4: volume fraction of species i	unitless
$\chi^{\rho}(r)$	polarization response function for external electrostatic potential	$\mathrm{C^2\,J^{-1}m^{-6}}$
$\chi^{\star}(r)$	polarization response function for total electrostatic potential	$\mathrm{C^2\,J^{-1}m^{-6}}$
χ_T	isothermal compressibility	$\mathrm{m^2 N^{-1}}$
Ψ	In Chapter 1: wave function	unitless
Ψ	total electrostatic potential	$\mathrm{V = J\,C^{-1}}$
Ψ^{ext}	external electrostatic potential (from a charge distribution external to the system)	$\mathrm{V = J\,C^{-1}}$
ψ_i	mean electrostatic potential due to a particle of species i	$\mathrm{V = J\,C^{-1}}$
Ψ^{pol}	electrostatic potential from polarization charge density	$\mathrm{V = J\,C^{-1}}$
Ω	number of microstates	unitless
$\boldsymbol{\omega}$	normalized orientational variable, $\boldsymbol{\omega} = \left(\frac{\phi}{2\pi}, \frac{\cos\theta}{2}, \frac{\chi}{2\pi}\right)$ with Euler angles (ϕ, θ, χ)	
ω	density of states on energy axis	$\mathrm{J^{-1}}$
∇	gradient operator, $\nabla = \left(\frac{\partial}{\partial x}, \frac{\partial}{\partial y}, \frac{\partial}{\partial z}\right)$	$\mathrm{m^{-1}}$
∇^2	Laplace operator, $\nabla^2 = \frac{\partial^2}{\partial x^2} + \frac{\partial^2}{\partial y^2} + \frac{\partial^2}{\partial z^2}$	$\mathrm{m^{-2}}$

Index